D1698281

Colloids and Colloid Assemblies

Edited by Frank Caruso

Related Titles

P. Ajayan, L. S. Schadler, P. V. Braun

Nanocomposite Science and Technology

2003

ISBN 3-527-30359-6

G. Decher, J. B. Schlenoff (Eds.)

Multilayer Thin Films

Sequential Assembly of Nanocomposite Materials

2003

ISBN 3-527-30440-1

P. Gómez-Romero, C. Sanchez (Eds.)

Functional Hybrid Materials

2003

ISBN 3-527-30484-3

M. Komiyama, T. Takeuchi, T. Mukawa, H. Asanuma

Molecular Imprinting

From Fundamentals to Applications

2003

ISBN 3-527-30569-6

W. Krenkel (Ed.)

High Temperature Ceramic Matrix Composites

2001

ISBN 3-527-30320-0

I. Manners

Synthetic Metal-containing Polymers

2003

ISBN 3-527-29463-5

Colloids and Colloid Assemblies

Synthesis, Modification, Organization
and Utilization of Colloid Particles

Edited by Frank Caruso

WILEY-VCH

WILEY-VCH Verlag GmbH & Co. KGaA

Edited by

Prof. Frank Caruso
Department of Chemical
and Biomolecular Engineering
The University of Melbourne
Victoria 3010
Australia

■ This book was carefully produced. Nevertheless, editor, authors and publisher do not warrant the information contained therein to be free of errors. Readers are advised to keep in mind that statements, data, illustrations, procedural details or other items may inadvertently be inaccurate.

Library of Congress Card No.: applied for
A catalogue record for this book is available from the British Library.

**Bibliographic information published
by Die Deutsche Bibliothek**
Die Deutsche Bibliothek lists this publication in the Deutsche Nationalbibliografie; detailed bibliographic data is available in the Internet at <http://dnb.ddb.de>.

© 2004 WILEY-VCH Verlag GmbH & Co. KGaA, Weinheim

All rights reserved (including those of translation in other languages). No part of this book may be reproduced in any form – by photoprinting, microfilm, or any other means – nor transmitted or translated into machine language without written permission from the publishers. Registered names, trademarks, etc. used in this book, even when not specifically marked as such, are not to be considered unprotected by law.

Printed in the Federal Republic of Germany
Printed on acid-free paper

Umschlaggestaltung   SCHULZ Grafik-Design, Fußgönheim
Composition   K+V Fotosatz GmbH, Beerfelden
Printing   Druckhaus Darmstadt GmbH, Darmstadt
Bookbinding   Litges & Dopf Buchbinderei GmbH, Heppenheim

ISBN   3-527-30660-9

Foreword

Colloid Science is often described as an "enabling" discipline. The structure, theories, rules, skills or techniques of the colloid scientist allow us to think about what Gibbs called heterogeneous systems, where surface or interfacial energy considerations dominate the properties of such multiphase systems. Equally, the skills of the colloid scientist enable the industrial scientist to resolve challenging physicochemical problems in practice or in the real world. The bioscientist increasingly uses the language of colloid science to understand structure and function in the complex world of biological and medical science.

And then along comes "nanotechnology" to disturb the peaceful world of colloid science! Nanotechnology is at one level, *colloid science with pizzazz*; at another level it is an exciting way of thinking about many new worlds of science and technology.

The present book blends very cleverly the foundations of colloid science within the new perspectives given by the world of nanotechnology. Authors discussing synthesis of particles encompass particles right down to the nanometer scale. Chapters on surface modification are presented as surface engineering to achieve the desired results, again at the nanoscale. There are also significant bio-related chapters where many of the emerging successes in nanotechnology are described.

A great strength of the book is that the authors blend basic, underpinning science with new and emerging discoveries across many fields. It is perhaps the only text available to combine the science of colloids alongside the concepts that drive the assembly of such particles into two- and three-dimensional arrays and organised structures. My belief is that *Colloids and Colloid Assemblies* is an essential resource today and for many years to come to all who want to learn the principles of colloid science in action in the nanoworld.

March 2003

Thomas W. Healy
Melbourne, Australia

Preface

Colloids are ubiquitous, and for centuries they have been of scientific and technological interest. The last decade, however, has seen significant advances made in the preparation of colloids of various sizes (from several nanometers to many micrometers) with diverse and tailored composition. Colloid particles have increasingly found use as building blocks for the creation of nanoassemblies that promise to be the next generation of devices for technology-driven applications. For such "nanodevices" to be realized, it is apparent that state-of-the-art colloids with unique properties are required, and the development of flexible strategies that enable their surface modification, as well as protocols that permit their facile assembly into organized functional structures, are of paramount importance.

The topics in this book cover precisely these key aspects – the book deals with recent global developments in the synthesis, modification, organization, and utilization of colloids in the rapidly emerging fields of nanotechnology and biotechnology. The types of colloids covered include latexes, metal nanoparticles, semiconductor quantum dots, nanocapsules, and miniemulsions. Various methods for the formation of ordered and patterned particle arrays employed in advanced materials preparation are outlined. Several chapters also deal with the use of colloids in niche applications such as biolabeling, biological screening, and drug encapsulation and release. The realm of technological applications of colloids will rapidly expand, thus inspiring research in the area of colloids and colloid assemblies well into this millennium!

The collection of chapters in this book will be of interest to a multidisciplinary audience (chemists, physicists, biologists, engineers, and materials scientists) engaged in active colloid-based research. This book will also arouse the interest of scientists and engineers who wish to diversify their research. Furthermore, it serves as a reference for graduate students and the novice, providing detailed accounts of the current state of research in the various fields.

Finally, I would like to thank all of the contributors for taking valuable time from their busy schedules to write stimulating and informative chapters, and Thomas W. Healy for writing the foreword.

Melbourne, October 2003   Frank Caruso

Contents

Foreword   V

Preface   VI

1	**Latex Particles** 1	
	Klaus Tauer	
1.1	Introduction: Definitions, Historical Overview, Importance 1	
1.2	Preparation of Polymeric Latex Particles 5	
1.2.1	Secondary Polymer Dispersions – Artificial Latexes 5	
1.2.2	Primary Polymer Dispersions via Heterophase Polymerizations 8	
1.3	Conclusions, New Developments, and Perspectives 42	
1.4	Acknowledgements and Apologia 44	
1.5	References 45	
2	**Semiconductor Nanoparticles** 52	
	Andrey L. Rogach, Dmitri V. Talapin and Horst Weller	
2.1	Introduction 52	
2.2	Quantum Confinement in Semiconductor Nanoparticles 54	
2.3	Colloidal Synthesis of Semiconductor Nanoparticles 55	
2.3.1	General Remarks 55	
2.3.2	Theoretical Aspects of Nanoparticle Growth in Colloidal Solutions 56	
2.3.3	Semiconductor Nanoparticles Synthesized in Aqueous Media 62	
2.3.4	Semiconductor Nanocrystals Synthesized by the Organometallic Route 65	
2.3.5	Post-preparative Size-selective Fractionation of Nanoparticles 78	
2.4	Semiconductor Nanoparticles as Building Blocks of Superstructures 81	
2.4.1	Layer-by-layer Assembly of Water-soluble Nanoparticles 81	
2.4.2	Two-dimensional Arrays of Monodisperse Nanocrystals Formed by Self-assembly 83	
2.4.3	Colloidal Crystals of Nanocrystals 85	
2.5	Applications of Semiconductor Nanocrystals 86	
2.5.1	Nanocrystals in Light-emitting Devices 86

2.5.2	Nanocrystals in Solar Cells	87
2.5.3	Nanocrystals for Telecommunication Amplifiers	88
2.5.4	Nanocrystals as Light-emitting Species in Photonic Crystals and Microcavities	88
2.5.5	Nanocrystals for Biolabeling	89
2.6	Concluding Remarks	90
2.7	Acknowledgements	91
2.8	References	91

3 Monolayer Protected Clusters of Gold and Silver 96
Mathias Brust and Christopher J. Kiely

3.1	Introduction	96
3.2	Gold Colloids and Clusters: Historical Background	96
3.3	Synthesis	98
3.4	Structure, Reactivity and Ligand Mobility	101
3.5	Self-organization Phenomena	104
3.6	Electronic Properties	108
3.7	Current and Future Applications	112
3.8	Visions for Future Development	114
3.9	References	116

4 Sonochemical Synthesis of Inorganic and Organic Colloids 120
Franz Grieser and Muthupandian Ashokkumar

4.1	Introduction	120
4.2	Properties of Colloids	121
4.3	Bubbles (General)	122
4.3.1	Bubbles in an Acoustic Field	123
4.3.2	Bubble Temperature	125
4.4	Sonochemistry	129
4.4.1	Sonochemical Formation of Colloids	130
4.5	Conclusions	145
4.6	Acknowledgments	145
4.7	References	146

5 Colloidal Nanoreactors and Nanocontainers 150
Marc Sauer and Wolfgang Meier

5.1	Introduction	150
5.2	The Natural Approach	151
5.3	The Self-assembly Approach	152
5.4	The Template Approach	161
5.5	The Emulsion/Suspension Polymerization Approach	165
5.6	The Dendrimer Approach	168
5.7	References	172

6	**Miniemulsions for the Convenient Synthesis of Organic and Inorganic Nanoparticles and "Single Molecule" Applications in Materials Chemistry** *175*	
	Katharina Landfester and Markus Antonietti	
6.1	Introduction *175*	
6.2	Miniemulsions *177*	
6.2.1	Emulsion Stability Against Ostwald Ripening, Collisions, and Coalescence *177*	
6.2.2	Techniques of Miniemulsion Preparation and Homogenization *178*	
6.2.3	Influence of the Surfactant *180*	
6.2.4	Influence of the (Ultra)Hydrophobe *183*	
6.2.5	Inverse Miniemulsions *185*	
6.2.6	Preservation of Particle Identity Throughout Miniemulsion Polymerization *187*	
6.2.7	Checklist for the Presence of a Miniemulsion *187*	
6.2.8	Main Differences from Microemulsion and Suspension Routes and Related Processes *188*	
6.3	Materials Synthesis in Miniemulsions *190*	
6.3.1	Radical Homo- and Copolymerization of Organic Monomers *190*	
6.3.2	Non-radical Organic Polymerization Reactions in Miniemulsions *196*	
6.3.3	Hybrid Nanoparticles by Miniemulsion Technologies *198*	
6.3.4	Miniemulsions of Inorganic Droplets/Synthesis of Inorganic Nanoparticles *206*	
6.4	On the Horizon: Miniemulsion Droplets as Compartments for "Single Molecule Chemistry" *210*	
6.5	Conclusions *211*	
6.6	Acknowledgements *211*	
6.7	References *212*	
7	**Metal and Semiconductor Nanoparticle Modification via Chemical Reactions** *216*	
	Luis M. Liz Marzán	
7.1	Introduction *216*	
7.2	Why Surface Modification? *217*	
7.2.1	Chemical and Colloidal Stability *217*	
7.2.2	Tuning of Physical Properties *219*	
7.2.3	Control of Interparticle Interactions Within Assemblies *222*	
7.3	Chemical Deposition of Shells on Metals and Semiconductors *222*	
7.3.1	Metals on Metals *222*	
7.3.2	Semiconductors on Semiconductors *227*	
7.3.3	Oxides on Metals and Semiconductors *228*	
7.3.4	Polymers on Metals *235*	
7.4	Chemical Reactions at Coated Nanoparticles *236*	
7.4.1	Dissolution of Gold and Silver Cores *236*	
7.4.2	Protection of Chalcogenides against Photodegradation *238*	

7.4.3	Oxidation of Silver by Molecular Iodine and Sulfide Ions	*239*
7.5	Conclusions	*240*
7.6	Acknowledgement	*241*
7.8	References	*241*

8	**Nanoscale Particle Modification via Sequential Electrostatic Assembly**	*246*
	Frank Caruso	
8.1	Introduction *246*	
8.2	Sequential Electrostatic Assembly Applied to Particles	*248*
8.2.1	Materials for Surface Modification *251*	
8.2.2	Types of Particles *260*	
8.3	Polyelectrolyte Multilayers on Particles as Matrices for Particle Modification	*266*
8.3.1	Infiltration *266*	
8.3.2	Chemical Reactions *267*	
8.4	Potential Uses of Nanoscale Modified Particles	*270*
8.4.1	Bioscience *270*	
8.4.2	Organization and Templates for Macroporous Materials	*274*
8.5	Summary and Outlook *278*	
8.6	Acknowledgments *278*	
8.7	References *279*	

9	**Colloidal Crystals: Recent Developments and Niche Applications**	*284*
	Younan Xia, Hiroshi Fudouzi, Yu Lu, and Yadong Yin	
9.1	Introduction *284*	
9.2	Monodisperse Spherical Colloids: the Key Component	*286*
9.2.1	Polymer Latexes through Emulsion Polymerization	*287*
9.2.2	Inorganic Colloids through Controlled Precipitation	*289*
9.2.3	Core-Shell Spherical Colloids *290*	
9.3	Crystallization of Colloids under Physical Confinement	*291*
9.4	Other Methods for Fabricating Colloidal Crystals	*294*
9.4.1	Crystallization via Sedimentation in a Force Field	*295*
9.4.2	Crystallization via Repulsive Electrostatic Interactions	*296*
9.4.3	Crystallization via Attractive Capillary Forces	*298*
9.5	Control over the Spatial Orientation of Colloidal Crystals	*299*
9.6	Colloidal Crystals Embedded in Solid Matrices	*302*
9.7	Photonic Bandgap Properties of Colloidal Crystals	*303*
9.8	Concluding Remarks *307*	
9.9	Acknowledgment *311*	
9.10	References *312*	

10		**Surface-directed Colloid Patterning: Selective Deposition via Electrostatic and Secondary Interactions** *317*
		Paula T. Hammond
10.1		Introduction *317*
10.2		Observation of Selective Deposition with Simple Polyamines *320*
10.2.1		Extension of Selective Deposition to Colloids *320*
10.3		Use of Surfactants to Guide Colloidal Surface Assembly *327*
10.4		Surface Sorting with Two Different Colloid Types *330*
10.5		Surface Confinement Effects – an Additional Tool *333*
10.6		Use of Polymer Stamping as a Means to Functional Surfaces *335*
10.7		Applications, Examples and Conclusions *338*
10.8		References *339*
11		**Evolving Strategies of Nanomaterials Design** *342*
		Sean Davis
11.1		Introduction *342*
11.2		Biotemplates *343*
11.2.1		Nanohybrids *343*
11.2.2		Patterned Templates *348*
11.3		Molecular Recognition *353*
11.3.1		Programmed Assembly *353*
11.3.2		Biomolecular/Inorganic Interfaces *360*
11.4		Summary and Outlook *364*
11.5		Acknowledgments *365*
11.6		References *365*
12		**Nanoparticle Organization at the Air-water Interface and in Langmuir-Blodgett Films** *369*
		Murali Sastry
12.1		Introduction *369*
12.2		Nanoparticle Assembly at the Air-Water Interface *370*
12.2.1		Physical Methods for Generation of Nanoparticle Langmuir-Blodgett Films *371*
12.2.2		Chemical Methods for Generation of Nanoparticle Langmuir-Blodgett Films *383*
12.3		Summary *393*
12.4		Acknowledgements *393*
12.5		References *393*
13		**Layer-by-layer Self-assembly of Metal Nanoparticles on Planar Substrates: Fabrication and Properties** *398*
		Thierry P. Cassagneau
13.1		Introduction *398*
13.2		Layer-by-layer Self-assembly from Preformed Building Blocks *401*
13.2.1		Functionalization of Nanoparticles *401*

13.2.2	Layer-by-layer Self-assembly of Metal Nanoparticles 406
13.3	Alternate Polymer/Metal Particle Films Prepared by Salt Incubation into Polymer Multilayers 413
13.3.1	Polyelectrolyte/Metal Particle System 413
13.3.2	Diblock Copolymer/Metal Particle System 418
13.4	Organization of Metal Nanoparticles on Planar Surfaces 419
13.4.1	Electrostatic Self-organization 420
13.4.2	Template-assisted Organization 422
13.5	Properties of Metal Nanoparticle-containing Multilayers 423
13.5.1	Optical Properties 423
13.5.2	Electrical Properties 425
13.5.3	Magnetic Properties 431
13.6	Conclusions 432
13.7	References 433

14	**Assembly of Electrically Functional Microstructures from Colloidal Particles** 437
	Orlin D. Velev
14.1	Introduction: Electrical Forces in Colloidal Suspensions 437
14.2	Electrical Assembly of Colloidal Crystals 439
14.2.1	Assembly by Electrophoretic Particle Attraction to Surfaces 440
14.2.2	Assembly in Thin Liquid Films Between Electrode Surfaces 440
14.2.3	Annealing and Alignment of Three-dimensional Crystals by AC Fields 442
14.2.4	Dielectrophoretic Assembly of Two-dimensional Crystals Between Planar On-chip Electrodes 443
14.3	Fabrication of Electrical Circuits via Colloidal Assembly 444
14.3.1	Assembly of Electrical Circuits by Capillary Forces 444
14.3.2	Formation of Electrical Circuits by Chemical and Electrochemical Deposition 446
14.3.3	Assembly of Electrical Circuits from Nanoparticles 451
14.4	Electrically Functional Microstructures from Colloidal Particles 454
14.4.1	Chemical Sensors 454
14.4.2	Biological Sensors 455
14.4.3	Displays 457
14.4.4	Photovoltaic Cells and Light-emitting Diodes 458
14.5	Future Trends 459
14.6	Acknowledgements 459
14.7	References 460

15	**3D Ordered Macroporous Materials** 465
	Rick C. Schroden and Andreas Stein
15.1	Introduction 465
15.2	Methods of Preparation 468
15.2.1	Colloidal Crystal Templates 468

15.2.2	Sol–Gel Chemistry	468
15.2.3	Salt-precipitation and Chemical Conversion	473
15.2.4	Oxide Reduction	474
15.2.5	Nanoparticle Assembly	474
15.2.6	Solvothermal Synthesis	476
15.2.7	Emulsion Templating	476
15.2.8	Electrochemical Deposition	477
15.2.9	Electroless Deposition	478
15.2.10	Chemical Vapor Deposition	479
15.2.11	Pyrolysis and Spraying Techniques	479
15.2.12	Melt-imbibing	480
15.2.13	Organic Polymerization	480
15.2.14	Coordination Polymerization	481
15.2.15	Core-Shell Assembly and Rearrangement	481
15.3	Applications	482
15.3.1	Photonic Crystals	482
15.3.2	Surface-enhanced Raman Spectroscopy	485
15.3.3	Chemically Functional Materials	485
15.3.4	Templates for Colloid and Hollow Sphere Synthesis	488
15.4	Summary and Outlook	488
15.5	Acknowledgments	489
15.6	References	489

16	**Semiconductor Quantum Dots as Multicolor and Ultrasensitive Biological Labels**	**494**
	Warren C.W. Chan, Xiaohu Gao and Shuming Nie	
16.1	Introduction	494
16.2	Synthesis and Surface Chemistry	495
16.3	Optical Properties	497
16.4	Applications in Biology and Medicine	500
16.5	Conclusions	503
16.6	Acknowledgements	503
16.7	References	503

17	**Colloids for Encoding Chemical Libraries: Applications in Biological Screening**	**507**
	Bronwyn J. Battersby, Lisbeth Grøndahl, Gwendolyn A. Lawrie, and Matt Trau	
17.1	Introduction	507
17.1.1	Genomics	508
17.1.2	Proteomics	510
17.1.3	Drug Discovery	510
17.2	Current High-throughput Screening Technologies	511
17.2.1	Screening with Microplates	511
17.2.2	Screening with Microarrays	511

17.2.3	Screening with Colloids	513
17.3	Encoding Colloids for High-throughput Screening	517
17.3.1	Encoding Colloids for Non-combinatorial Libraries	518
17.3.2	Encoding Colloids for Combinatorial Libraries	519
17.4	Synthesis of Colloidal Silica	525
17.4.1	Synthetic and Mechanistic Aspects of Colloidal Silica Particle Formation	525
17.4.2	Determination of Particle Size	528
17.4.3	Mono-fluorescent Colloidal Silica Particles	529
17.5	Multi-fluorescent Colloidal Silica Particles	531
17.5.1	Considerations of Multiple Dyes Within Close Proximity	532
17.5.2	Selection of Dyes for Strategy I Encoding	534
17.5.3	Multi-fluorescent Reporter Colloids for Strategy I Encoding	536
17.5.4	Multi-fluorescent Layered Colloids for Strategy II Encoding	538
17.5.5	Emission and Morphology of Multi-fluorescent Layered Particles	539
17.5.6	Selection of Dyes for Strategy II Encoding	543
17.6	Optimizing Fluorescent Colloids for Use in Combinatorial Libraries	545
17.6.1	Modification of the Colloidal Silica Surface	545
17.6.2	Adhesion of Reporters to Solid Support Beads	548
17.6.3	Optical Diversity in Eleven Dimensions	549
17.7	Conclusions	554
17.8	References	555
18	**Polyelectrolyte Microcapsules as Biomimetic Models**	**561**
	Gleb B. Sukhorukov and Helmuth Möhwald	
18.1	Introduction	561
18.2	Construction and Properties of a Biomimetic Model	562
18.2.1	Polyelectrolyte Micro- and Nanocapsules	562
18.2.2	Lipid Bilayer Coupled to a Synthetic Skeleton	565
18.2.3	Coated Cells	569
18.3	Biological Processes	571
18.3.1	Protein (Enzyme) Incorporation into the Capsules	571
18.3.2	Artificial Viruses	578
18.4	Conclusions	579
18.5	References	579

Subject Index 581

List of Contributors

Prof. Markus Antonietti
Max Planck Institute of Colloids and Interfaces
14424 Potsdam
Germany
markus.antonietti@mpikg-golm.mpg.de

Dr. Muthupandian Ashokkumar
School of Chemistry
The University of Melbourne
Victoria 3010
Australia
masho@unimelb.edu.au

Dr. Bronwyn Battersby
Centre for Nanotechnology & Biomaterials
Department of Chemistry
The University of Queensland
Queensland 4072
Australia
bbattersby@uq.edu.au

Dr. Mathias Brust
The University of Liverpool
Department of Chemistry
Donnan Laboratories
Oxford Street
Liverpool, L69 7ZD
United Kingdom
m.brust@liverpool.ac.uk

Prof. Frank Caruso
Department of Chemical and Biomolecular Engineering
The University of Melbourne
Victoria 3010
Australia
fcaruso@unimelb.edu.au

Dr. Thierry P. Cassagneau
Max Planck Institute of Colloids and Interfaces
14424 Potsdam
Germany
thierry@mpikg-golm.mpg.de

Prof. Warren Chan
Institute of Biomaterials and Biomedical Engineering
4 Taddle Creek Road
Rosebrugh/Mining Building
University of Toronto
Toronto, Ontario M5S 3G9
Canada
warren.chan@utoronto.ca

Dr. Sean Davis
School of Chemistry
University of Bristol
Bristol BS8 1TS
United Kingdom
s.a.davis@bristol.ac.uk

Dr. Hiroshi Fudouzi
Materials Engineering Laboratory
National Institute of Materials Science
1-2-1, Sengen
Tsukuba 305-0047
Japan
fudouzi.hiroshi@nims.go.jp

Xiaohu Gao
Chemistry Department
Indiana University
Bloomington, IN 47405
USA
xigao@indiana.edu

Prof. Franz Grieser
School of Chemistry
The University of Melbourne
Victoria 3010
Australia
f.grieser@chemistry.unimelb.edu.au

Dr. Lisbeth Grøndahl
Centre for Nanotechnology &
Biomaterials
Department of Chemistry
The University of Queensland
Queensland 4072
Australia
l.grondahl@uk.edu.au

Prof. Paula T. Hammond
Massachusetts Institute of Technology
Department of Chemical Engineering
Cambridge, MA 02139
USA
hammond@mit.edu

Prof. Christopher J. Kiely
Department of Materials Science and
Engineering
Lehigh University
Whitaker Laboratory
5 East Packer Avenue
Bethlehem, PA 18015
USA
chris.kiely@lehigh.edu

Prof. Katharina Landfester
Organische Chemie III/
Makromolekulare Chemie
Universität Ulm
Albert-Einstein-Allee 11
89069 Ulm
Germany
katharina.landfester@chemie.uni-ulm.de

Dr. Gwendolyn A. Lawrie
Centre for Nanotechnology &
Biomaterials
Department of Chemistry
The University of Queensland
Queensland 4072
Australia
g.lawrie@uq.edu.au

Dr. Luis M. Liz Marzán
Depto. de Quimica Fisica
Facultade de Ciencias
Universidade de Vigo
36200 Vigo
Spain
lmarzan@uvigo.es

Dr. Yu Lu
Department of Materials Science and
Engineering
University of Washington
Seattle, WA 98195-2120
USA
yulu@u.washington.edu

Prof. Wolfgang Meier
Institute for Physical Chemistry
University of Basel
Klingelbergstrasse 80
4056 Basel
Switzerland
wolfgang.meier@unibas.ch

Prof. Helmuth Möhwald
Max Planck Institute of Colloids and
Interfaces
14424 Potsdam
Germany
moehwald@mpikg-golm.mpg.de

Prof. Shuming Nie
Department of Biomedical
Engineering
Emory University
Atlanta, GA 30322
USA
snie@emory.edu

Dr. Andrey L. Rogach
Department of Physics and CeNS
University of München
Amalienstraße 54
80799 München
Germany
andrey.rogach@physik.uni-muenchen.de

Dr. Murali Sastry
Materials Chemistry Division
National Chemical Laboratory
Pune 411 008
India
sastry@ems.ncl.res.in

Dr. Marc Sauer
Department of Chemistry
University of British Columbia
2036 Main Mall
Vancouver, B.C., V6T 1Z1
Canada
msauer@chem.ubc.ca

Rick C. Schroden
Department of Chemistry
University of Minnesota
207 Pleasant St. SE
Minneapolis, MN 55455
USA
schroden@chem.umn.edu

Prof. Andreas Stein
Department of Chemistry
University of Minnesota
207 Pleasant St. SE
Minneapolis, MN 55455
USA
stein@chem.umn.edu

Dr. Gleb Sukhorukov
Max Planck Institute of Colloids and
Interfaces
14424 Potsdam
Germany
gleb@mpikg-golm.mpg.de

Dr. Dimitri V. Talapin
Universität Hamburg
Institut für Physikalische Chemie
Bundesstraße 45
20146 Hamburg
Germany
talapin@chemie.uni-hamburg.de

Dr. Klaus Tauer
Max Planck Institute of Colloids and
Interfaces
14424 Potsdam
Germany
klaus.tauer@mpikg-golm.mpg.de

Prof. Matt Trau
Centre for Nanotechnology
and Biomaterials
Department of Chemistry
The University of Queensland
St. Lucia, Queensland 4072
Australia
trau@chemistry.uq.edu.au

Prof. Orlin D. Velev
Department of Chemical Engineering
Riddick Hall
North Carolina State University
Raleigh, NC 27695
USA
odvelev@unity.ncsu.edu

Prof. Horst Weller
Universität Hamburg
Institut für Physikalische Chemie
Bundesstraße 45
20146 Hamburg
Germany
weller@chemie.uni-hamburg.de

Prof. Younan Xia
Department of Chemistry
University of Washington
Seattle, WA 98195-1700
USA
xia@chem.washington.edu

Dr. Yadong Yin
Material Science Division, 11-D83
Lawrence Berkeley National Laboratory
Berkeley, CA 94720
USA
ydyin@uclink.berkeley.edu

1
Latex Particles

Klaus Tauer

1.1
Introduction: Definitions, Historical Overview, Importance

Latex particles are in a real (historical) sense natural products occurring in over 2000 plant species as secondary polymeric materials. The most important one is *cis*-1,4-polyisoprene or natural rubber. All natural rubber currently used commercially arises from a single species, that is the Brazilian rubber tree, *Hevea brasiliensis* [1]. Natural rubber has high performance properties due to its molecular structure and high molecular weight ($>10^6$ g mol^{-1}) and even today it is still impossible easily to mimic these properties by artificial products. More than 90% of the world production of natural rubber is concentrated in Asia, mainly Thailand, Indonesia, and Malaysia. More than 40 000 products based on natural rubber are in use and, hence, it is still considered to be a strategic raw material [2].

The origin of the word latex which means in Latin "liquid" or "fluid" goes back to the Greek word *"latex"* which means "droplet". It came into use in the middle of the nineteenth century by biologists to denote the milky white sap of certain trees [3]. Today the word latex has become a generic term that covers all kinds of polymer dispersions. Polymer dispersions are defined as colloidal systems where the polymer is finely distributed in a liquid dispersion medium in the form of stable individual particles. It is unimportant whether the polymer, at a given temperature, is a solid or a highly viscous fluid. The liquid forming the continuous phase can be any liquid in which the polymer is insoluble. Thus the term "poly(xyz)" latex or "poly(xyz)" dispersion denotes finely dispersed particles of polymer "xyz" in continuous phases.

It is historically proven that by 1600 BC the Mayas were able to use the sap of certain trees to make useful things such as rubber balls, glue, waterproofed clothes, medicine, and much more [4, 5]. They called this sap *"caa o-chu"* (crying wood) from which the French word *caoutchouc* was derived. From his second trip in the sixteenth century Columbus brought the first pieces of rubber to Europe, and the first batch of caoutchouc arrived in the old world in the middle of the eighteenth century. At this time a broader, scientific interest in caoutchouc arose, initiated by the report of Charles de la Condamine and Francois Fresneau on their tour through South America in 1736 and their experiences with caoutchouc and its products [6].

Colloids and Colloid Assemblies. Edited by Frank Caruso
Copyright © 2004 Wiley-VCH Verlag GmbH & Co. KGaA, Weinheim
ISBN: 3-527-30660-9

Natural rubber and natural latex played a crucial role in the development of polymer science and heterophase polymerization techniques, respectively. The main driving force in the late nineteenth and early twentieth century was given by attempts to copy Mother Nature's natural rubber (*cis*-polyisoprene) and to find a synthetic route to rubber polymers [7–9]. The birth of the first synthetic latex particles occurred in Leverkusen, Germany, at the Farbenfabriken vorm. Friedrich Bayer & Co., where in 1912 Kurt Gottlob invented a process for the preparation of synthetic rubber by polymerization of isoprene in aqueous viscous solutions [10].

Significant progress in the development of heterophase polymerization techniques was made during the following decades when natural polymers as stabilizers were replaced by oleates or alkyl aryl sulfonates and when water- as well as monomer-soluble peroxides were employed, which allowed faster and better controlled polymerization processes [11–13].

The synthesis of polymer dispersions has been developed over the decades into a large topic in its own right, with more than 90 years of history and an extensive literature including monographs, textbooks, conference proceedings, and about 1000 original papers and patent applications per year. Furthermore, over the years many completely different ways have developed to perform the preparation of latex particles. The scheme in Fig. 1.1 classifies polymer dispersions into primary and secondary dispersions. Primary dispersions are directly obtained by heterophase polymerizations and can be subdivided into latexes produced in living organisms (natural latexes) and man-made products (synthetic latexes). Besides the polyisoprenes produced in plants, poly-3-(hydroxyalkylalkanoates), produced in bacteria even on an industrial scale [14], are also natural latexes. Secondary dispersions comprise examples where polymers are dispersed only after their prepara-

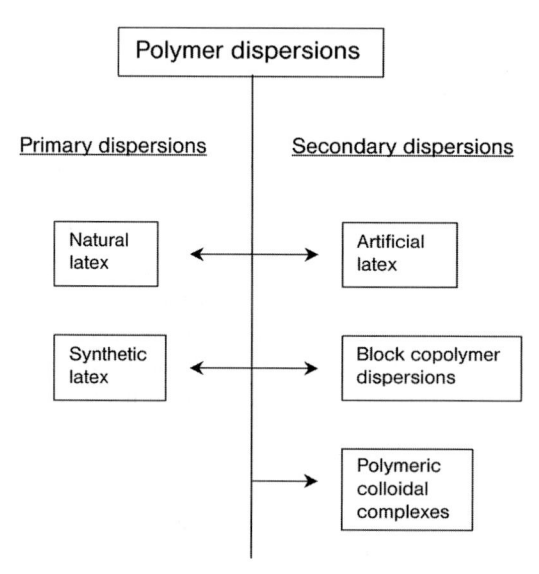

Fig. 1.1 Classification of polymer dispersions.

tion. Artificial latexes are obtained after dispersing solid polymers or emulsifying polymer solutions in a proper dispersion medium. Block copolymers in solvents selective for only one block form micelles, which are in fact sterically stabilized polymer particles and thus belong to the class of polymer dispersions [15–17]. Moreover, heterophase polymerization techniques have been developed, which directly lead to block copolymer dispersions in a very efficient way [18–21]. Polymeric colloidal complexes result from manipulating polymer solutions by inducing either hydrophobic or electrostatic interactions. Examples are the formation of hollow spheres via layer-by-layer assembly of oppositely charged polyelectrolytes onto various templates [22–25]. But also a single kind of block copolymer can form polymeric colloidal complexes via self assembly, such as poly(ethylene glycol)-b-poly(N-isopropyl acrylamide) copolymers where the poly(N-isopropyl acrylamide) block becomes insoluble in water at temperatures above 32 °C and particles are formed via hydrophobic interactions. These particles consist of a poly(N-isopropyl acrylamide) core and a stabilizing poly(ethylene glycol) shell. Such particles can easily be prepared by heterophase polymerization initiated by the poly(ethylene glycol)–ceric ion redox system [26]. Other examples of such dispersions are complexes between amphiphilic block copolymers with one polyelectrolyte block and either oppositely charged surfactants [21, 27–29] or other amphiphilic block copolymers with oppositely charged polyelectrolyte blocks [30].

Polymer dispersions are an omnipresent part of worldwide commerce as well as of scientific polymer studies. It is no exaggeration to say that they are important for our daily life, mainly because of their versatile properties and applications. The main applications of polymer dispersions are in paper making, all kinds of paints and coatings, the construction industry, adhesives, the textile and leather industries, and also in medicine and pharmaceuticals [31, 32]. These examples illustrate both the different branches of industry which make use of polymer dispersions and how the materials influence our daily life. Heterophase polymerizations are important technologies to produce high quality polymers with specifically tailored properties on a volume scale ranging from only a few milliliters up to 200 m^3.

Heterophase polymerization can be subdivided into precipitation, suspension, microsuspension, miniemulsion, emulsion, microemulsion, and dispersion polymerization (see Tab. 1.1). Precipitation polymerization is mentioned here only for the sake of completeness, since it is not one of the techniques leading to well-defined latex particles. The industrially important processes are suspension, microsuspension, precipitation (for bulk polymerizations), and emulsion polymerization for both normal and inverse processes. A clear demarcation between heterophase polymerization techniques might be possible in the following way. In the cases of suspension, microsuspension, miniemulsion, and microemulsion polymerization the monomer must be only slightly water soluble as it has to form a separate phase, mainly in the shape of spherical droplets whose size is controlled by a proper choice of the dispersing technique (stirring, ultrasonic treatment, homogenization) in combination with the stabilizing system. The droplet size decreases in the order suspension > microsuspension > miniemulsion > microemulsion po-

Tab. 1.1 Overview of various heterophase polymerization techniques

Common name	Initiator	Stabilizer	Procedures	Particle size	Ref.
Precipitation	lyophobe lyophil	none	batch	mm range	[33–36]
Suspension	lyophobe	polymeric or protective colloid	mainly batch	10–500 µm	[37–42]
Dispersion	lyophil	polymeric	batch	1–20 µm	[43–47]
Microsuspension or Minisuspension	lyophobe	polymeric plus surfactant	batch, high shear	1–10 µm	[48–50]
Emulsion	lyophobe lyophil	all kinds or none	batch, semi-batch continuous, seed	5 nm–10 µm	[51–57]
Miniemulsion	lyophil lyophobe	all kinds	batch, semibatch continuous	50–500 nm	[58–63]
Microemulsion	lyophil lyophobe	all kinds	batch, semibatch	10–100 nm	[64–67]

lymerization. The polymerization recipes are designed in such a way (for instance with monomer-soluble initiators instead of initiators that are soluble in the continuous phase) that the polymerization takes place mainly inside the preformed monomer droplets. In these techniques the stabilizers have to support the emulsification process and the stabilization of the monomer droplets, whereas in the case of an emulsion polymerization a separate free monomer phase need not necessarily be present. Moreover, the monomer can be fed continuously into the reactor either as neat monomer or as an emulsion, with the additional advantage of being able to polymerize most of the time at a conversion corresponding to the polymerization rate maximum [53]. Emulsion polymerization is commercially the most important process for effecting the preparation of polymer dispersions. Summaries of the state of the art of emulsion polymerization technology as well as with regard to kinetics and mechanism can be found in [31, 32, 51, 55, 76]. About $8 \cdot 10^6$ metric tons of polymers (dry) are sold worldwide and applied as latexes (frequently called emulsion polymers), representing a value of more than 20 billion Euro [77]. The overall amount of polymer produced by emulsion polymerization techniques is almost as twice as high, because many polymers prepared via emulsion polymerization, such as poly(vinyl chloride) and synthetic rubber, are sold and applied as bulk material. In terms of their monomer base, emulsion polymers divide between about 37% styrene–butadiene copolymers, 30% acrylic- or methacrylic-based (co)polymers, 28% vinyl acetate-based (co)polymers, and 5% others. The main application areas for emulsion polymers are paints and coatings (26%), paper/paperboard (23%), adhesives (22%), carpets (11%), and miscellaneous (18%) [32].

Of course the above demarcation between the various heterophase polymerization techniques is rather crude and hence the general term *heterophase polymeriza-*

tion as a preparation technique for latex particles is more appropriate. In any case, the detailed preparation conditions for a particular type of latex particles such as recipe, procedure, and reactor type have to be specified. Synonyms for the continuous phase are dispersion medium, homogenous phase, and serum. Any liquid can be a dispersion medium provided it is a nonsolvent for the dispersed material. In fact, although heterophase polymerizations can be carried out in any solvent, for safety and environmental reasons water is the most important continuous phase. In the following, reference to any kind of heterophase polymerization and polymer dispersion means an aqueous continuous phase, unless the term "inverse" is put in front, denoting a system with an organic continuous phase.

This chapter is an attempt to generalize the principles of preparation of latex particles rather than to treat a certain heterophase polymerization technique in detail. Thus, special emphasis is placed on primary polymer dispersions or synthetic latex particles. For detailed information on particular polymerization techniques the reader is referred to excellent monographs and reviews mentioned in Tab. 1.1, which also give typical characteristics of each procedure. A more detailed comparison between aqueous and inverse heterophase polymerizations can be found in [78]. The author hopes that the contribution will be useful to give first access as well as to indicate how to start with the synthesis of latex particles of desired properties with regard to size, size distribution, and surface properties.

1.2
Preparation of Polymeric Latex Particles

The technical term "polymer dispersion" describes a special state of matter and not a special chemical composition of the polymeric material. In other words, any kind of polymer can be obtained in the form of a polymer dispersion as either a primary or secondary dispersion (see Fig. 1.1). Many applications of latex particles require not only a specific chemical composition but also a specific particle size or particle size distribution. Particle size in the following always means the diameter of an equivalent sphere. Note that at a given solids content the particle size determines the interfacial area between polymer and continuous phase as well as the surface-to-volume ratio of the dispersed material. Thus, it is an important parameter determining both the preparation and the application of latex particles.

1.2.1
Secondary Polymer Dispersions – Artificial Latexes

Artificial latexes have until now never been of much industrial interest although several preparation methods have been developed. There are basically three driving forces for the development of artificial latexes. The first is of some economic importance because it is more advantageous to transport polymers as dry solid material instead of dispersions with about 50% by weight dispersion medium. The second has to do with the superior material properties of polymers that cannot be prepared by heterophase polymerization techniques, such as polyurethanes or chlorosulfonated polyolefins, but whose application as polymer dispersions is advantageous or desirable. The importance of composite polymer dispersions, involving combinations of different materials such as mineral substances and polymers for various potential applications, are the third reason for ongoing research on artificial latexes.

There are two principal possibilities for preparing emulsions: the disruption of a larger volume into smaller subunits (comminution) or the construction of emulsion droplets from smaller units (condensation). Both methods are of technical importance for the preparation of emulsions for polymerization processes.

Emulsification by comminution is very common and the most widely used procedure. Under the action of intense mechanical energy any combination of immiscible and mutually nonreactive liquids can be broken up into an emulsion. Comminution can be performed either by intense stirring, with homogenizers, or by ultrasonication. The most exciting question after completion of the emulsification is that concerning the structure of the emulsion, i.e. which liquid forms the dispersed phase and which the continuous one? The result of the emulsification process is influenced not only by the energy input but by the volume ratio of the liquids, the kind of emulsifying agent, and its concentration, the last effect being strongly affected by the temperature. The most important property of the emulsifying agent is its solubility or, in the case of solid, so-called "Pickering stabilizers", the wetting behavior of the liquids, expressed by the contact angle.

Bancroft summarized the results of emulsification experiments obtained until the early twentieth century in a rule of thumb, which indicated that in order to have a kinetically stable emulsion the emulsifying agent must be soluble in the continuous phase [79]. This rule finds its expression also in the empirical system of the HLB values (*h*ydrophilic–*l*ipophilic *b*alance), which was developed in the middle of the twentieth century in order to simplify the selection of effective emulsifiers for a particular emulsification task [80].

In contrast to comminution techniques the preparation of emulsions by condensation does not require mechanical energy, except sometimes gentle stirring to avoid creaming or settling. There are basically two different types of condensation methods: droplet nucleation and swelling of disperse systems. However, a combination of the two has some meaning especially for heterophase polymerizations. Note that condensation processes are mainly determined by thermodynamic principles. For a more detailed summary of the principles of preparing emulsions with regard to their application in heterophase polymerizations the reader is referred to [81].

The preparation of artificial latex particles by direct emulsification of polymer solutions has been reviewed [82]. The authors distinguish between three different methods. *First*, direct emulsification, which means that a liquid polymer or polymer solution in a volatile, water-immiscible solvent is emulsified in a surfactant solution and at the end the solvent is removed by steam stripping. *Second*, inverse emulsification, which means that as a first step the polymer or polymer solution is mixed with fatty acid and in a subsequent step is slowly fed dilute aqueous base. Thus, a water-in-polymer emulsion is formed, which, with increasing amount of base solution, is finally converted into a polymer-in-water emulsion. *Third*, self-emulsification, which is obtained with polymers containing lyophilic groups in such amounts that dispersions are formed spontaneously upon contact with the continuous phase. Prominent examples are block copolymers in selective solvents (see above). This method is based on thermodynamic principles and hence is classed as an emulsification technique by condensation. Examples of polymeric materials that have been successfully transformed into artificial latex particles by either of the first two principles above are: polystyrene, poly(vinyl acetate), epoxy resins, derivatives of cellulose, polyesters, alkyd resins, synthetic rubber, poly(vinyl butyral), silicones, polyurethanes, and many others [82]. Moreover, core–shell latex particles have been prepared by seeded emulsion polymerization using artificial latex particles as seeds. For instance, by emulsification with ultrasonication of poly(styrene-*b*-butadiene-*b*-styrene) (Kraton from Shell Chemical Company) rubbery seed particles have been prepared, which were subsequently covered with a glassy shell of poly(methyl methacrylate) [82]. Another example of potentially practical importance for the application of artificial latex particles is given in [83]. Amino-terminated telechelic polybutadiene (number average molecular weight of 3100 g mol^{-1}) dissolved in toluene was emulsified by ultrasonication into an aqueous solution of sodium dodecyl sulfate (27.4 wt.% of emulsifier relative to the amount of polybutadiene) to get stable artificial latex particles with an average diameter of about 50 nm. These particles were used as curing agents in a blend with poly(styrene-*co*-*n*-butyl acrylate-*co*-dimethyl-*m*-isopropenylbenzyl isocyanate) primary latex particles, which were prepared by emulsion polymerization.

Among the above comminution techniques ultrasonication has a special position. Ultrasonication means the application of high-frequency vibrations. In a first step larger drops are produced in such a way that instabilities of interfacial waves are enhanced, leading finally to crushing. These drops are subsequently fragmented into smaller ones by acoustic cavitation. The use of ultrasound in emulsification processes is much more efficient than the application of rotor/stator systems. Experimental results clearly show that ultrasonication leads to smaller drops, less polydisperse drop-size distribution, more stable emulsions, and less surfactant consumption for a desired drop size. Furthermore, during ultrasonication much less energy is wasted compared to all other mechanical comminution techniques [81, 82].

The preparation of artificial latex particles by condensation also requires as a first step the formation of a polymer solution. In the second step the decomposition of this solution is induced by any method that reduces the interaction between the polymer and solvent molecules. For instance temperature changes, or addition of nonsolvents, or increasing the ionic strength in polar or aqueous solutions are suitable means to induce this phase separation. Stabilization of the precipitating polymer molecules prevents the formation of a macroscopic coagulum and favors the formation of artificial latexes. The following examples should illustrate the procedure. First, solutions of sulfonated polystyrene molecules with a molecular weight of about $3 \cdot 10^5$ g mol^{-1} and a degree of sulfonation of about

5% in tetrahydrofuran ($2 \cdot 10^{-3}$ or $2 \cdot 10^{-2}$ g ml^{-1}) were mixed with an excess of water [84]. The precipitating polymer forms particles in a size range below 100 nm in diameter, which are stabilized by sulfonate groups. The particle size distribution does not depend on the kind of mixing (ultrasonication or stirring) but it does depend on both the concentration of the starting solution and the order of mixing, that is whether the solution is added to water or vice versa. Second, the formation of sterically stabilized particles of poly(ethylene glycol-b-N-isopropylacrylamide) particles by increasing either the temperature or the ionic strength [21]. Third, the formation of polystyrene particles stabilized with cetyltrimethylammonium bromide as surfactant and the utilization of these particles as seed in a subsequent emulsion polymerization [85]. Note, the first two examples belong to the above self-emulsification principle. Other examples are the formation of micelles from amphiphilic block copolymers (see above), which can be subsequently used as seed particles in emulsion polymerization [86, 87]. But also block copolymers themselves in selective solvents can form latex particles with unique morphologies. The utilization of block copolymers for various (potential) applications has recently been reviewed [88]. As condensation techniques to prepare artificial latex particles are governed by thermodynamic principles the final, stable structure should be the thermodynamically favored one. However, the formation of this equilibrium structure can take quite a long period of time, depending on the chain mobility and energetic characteristics of the particular structure (especially the depth of the minimum in the free energy in relation to thermal energy). For instance, when to a solution of poly(styrene-b-2-cinnamoylethyl methacrylate) polymer in tetrahydrofuran (good solvent for both blocks) acetonitrile (nonsolvent for both blocks) is added, precipitation takes place over several weeks, depending on the amount of acetonitrile. If the acetonitrile content is 80–90% structure formation starts with simple core–shell particles then changes via more complex morphologies such as onion-like structures until the final precipitate is formed [89].

The few examples mentioned here illustrate that there is obviously no limitation with regard to the kind of polymer that can be used to prepare well-defined artificial latex particles. The process is easier if the polymer can be dissolved but this is no limitation as it is in principle also possible to start with a solid, although the probability of degradation due to high shear and tribochemical reactions is inherent during grinding in the presence of the continuous phase.

1.2.2
Primary Polymer Dispersions via Heterophase Polymerizations

1.2.2.1 General Remarks
Heterophase polymerization techniques offer the possibility of preparing directly almost any type of latex particles that is desired. A general rule can be formulated that any polymerization recipe that works well under homogeneous conditions can be transformed into a heterogeneous polymerization system by the proper choice of a corresponding continuous phase, which should be liquid, at least at

Tab. 1.2 Heterophase polymerization chemistry

Polymerization mechanism	Example	Ref.
Enzymatic biosynthesis	cis-polyisoprene (natural rubber)	[90]
	poly-3-(hydroxyalkylalkanoates)	[14]
Condensation	polyurethanes	[91]
	silica particles (Stöber silica)	[92]
Ring opening	poly-ε-caprolactone	[93]
	poly(D,L-lactide-co-glycolide)	[94]
Ionic	cationic (methyl cyclosiloxanes)	[95]
	anionic (methyl/cyclosiloxanes styrene)	[96, 97]
Noble metal catalysts	polyethylene (bi-nuclear Ni-ylide)	[98]
	styrene (titanium complex)	[99]
Radical	Most often in water and organic continuous phases	[55, 100]

polymerization temperature. The continuous phase has to fulfill three basic requirements: it should be chemically inert, a nonsolvent for the polymer, and possess good heat transfer properties (for technical applications on larger scales). If, however, the continuous phase interferes with recipe components, the transfer from homogeneous to heterogeneous polymerization conditions requires additional effort and may require exchange of one or other recipe component. Tab. 1.2 summarizes examples of the kinds of polymerization chemistry that have been applied in heterogeneous polymerizations so far. This list is by no means complete but it confirms on the one hand the above statement with respect to the universality of heterophase polymerization and on the other the ongoing research interest over almost a century. No general statement is possible with regard to the number of components in heterophase polymerization recipes. The technique is so robust and flexible that almost any combination that fulfills the basic requirements can be tried. However, with regard to the efforts to ensure colloidal stability there is a rule that the smaller the size, for a given amount of polymer, the more effort is necessary.

Latex particles can be specifically designed with regard to various parameters as depicted in Fig. 1.2. The statement that basically all kinds of polymers can form latex particles already implies that there is no limit regarding the mechanical properties of latex particles. Tab. 1.3 [101] shows that the glass transition temperatures of various homopolymers already span almost 250 degrees and thus particles are accessible at room temperature that may be either fluids or solids. A further fine-tuning of the mechanical properties is possible by copolymerization and/or controlled cross-linking. The volume properties of particles can be modified in various ways. Solid cores with desired gradients in chemical composition and physical properties can be prepared by sophisticated methods of computer-controlled semi-batch polymerizations with predetermined monomer addition pro-

Mechanical properties
copolymer composition
glass transition
cross-linking
morphology (core shell)

Particle size
surfactant concentration
kind of surfactant
ionic strength

Interface properties
kind of stabilizer
lyophilic comonomers
kind of initiator

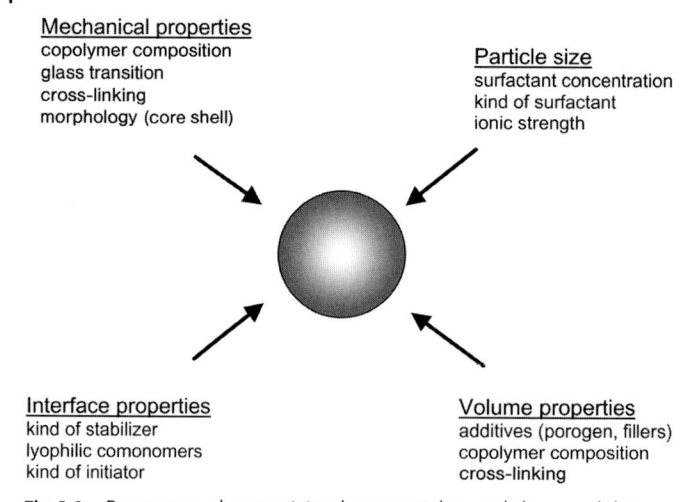

Volume properties
additives (porogen, fillers)
copolymer composition
cross-linking

Fig. 1.2 Parameters characterizing latex particles, and the possibilities for adjustment.

Tab. 1.3 Glass transition temperatures (T_g) for some homopolymers (Data from J. Snuparek, *Progr. Org. Coat.* **1996**, *29*, 225–233)

Monomer	T_g (°C)	Monomer	T_g (°C)
Acrylamide	153	n-Butyl methacrylate	20
Methacrylic acid	130	Tetradecyl acrylate	20
o-Vinyl toluene	115–125	Cyclohexyl acrylate	16
Phenyl methacrylate	110	Methyl acrylate	5–8
t-Butyl methacrylate	107	Vinyl isopropylether	–3
Acrylonitrile	96–106	n-Hexyl methacrylate	–5
Acrylic acid	106	i-Propyl acrylate	–5
Methyl methacrylate	105	Tetradecyl methacrylate	–9
p-Vinyl toluene	101	n-Octyl methacrylate	–20
Styrene	100	Ethyl acrylate	–20 to –27
Vinyl chloride	80	Vinylidene Chloride	–23
m-Vinyl toluene	72–82	Ethyl vinyl ether	–42
Ethyl methacrylate	65	n-Propyl acrylate	–52
sec-Butyl methacrylate	60	n-Butyl acrylate	–52 to –57
2-Hydroxyethyl methacrylate	55	2-Ethylhexyl acrylate	–60 to –77
t-Butyl acrylate	41	n-Dodecyl methacrylate	–65
n-Propyl methacrylate	35	n-Decyl methacrylate	–70
n-Hexadecyl acrylate	35	Ethylene	–70 to –77
Vinyl acetate	32	Butadiene	–87
2-Hydroxypropyl methacrylate	26	Octadecyl methacrylate	–100

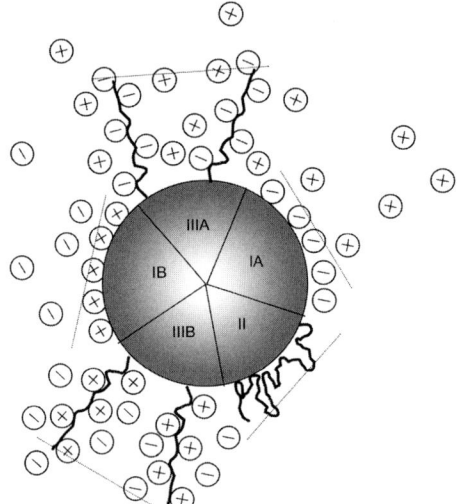

Fig. 1.3 Illustration of the latex particle-continuous medium interface and the possibilities for stabilization (not to scale).

tocols [102]. For applications where individual latex particles are required, the interfacial properties and the particle size are the more crucial parameters. The interfacial properties can be adjusted by the choice of stabilizer, initiator, and lyophilic co-monomers. For example, by changing these recipe components four different types of polystyrene latexes have been prepared and investigated in a comprehensive study [103]. The interface of an assumed model latex particle as sketched in Fig. 1.3 is characterized with regard to charge sign as well as concentration of charges and the extension of the interfacial layer into the continuous phase. The main function of the particle interface is to contribute to particle stabilization, that is to counteract the attractive van der Waals forces. Parts IA and IB of Fig. 1.3 show the most common cases of electrostatic stabilization with anionically and cationically charged stabilizers, respectively. Part II shows steric stabilization of colloidal particles without any contribution of charges, which can be perfectly realized in nonpolar, organic continuous phases with polymeric stabilizers. In contrast, parts IIIA and IIIB sketch the case of electrosteric stabilization with anionic and cationic polyelectrolytes, respectively. The dotted lines in each section of the model particle indicate the hydrodynamic layer thickness, or corona thickness, that is the extension of the interfacial layer into the continuous phase (ΔR). The particles move in the continuous phase with this size, which can be measured by dynamic light scattering. In contrast, transmission electron microscopy pictures the core size of the particles, because the stabilizer layer collapses in most cases during sample preparation onto the particle with a thickness of only about 1 nm. Each of the major cases I, II, and III contributes specifically to particle stability. For a detailed discussion of stabilization of colloidal latex particles the reader is referred to excellent contributions in [104–106]. The stabilizers or stabilizing moieties have to fulfill only one basic requirement, which is, according to

Bancroft's rule (see above), they have to be lyophilic, i.e. soluble or strongly interacting with the continuous phase. The stabilizer molecules can be either adsorbed or covalently connected with the core material, or in some special cases simply dissolved in the continuous phase [107]. But the last case, which will be not considered further, also requires a lyophilic decoration of the particle interface. Tab. 1.4 summarizes the key features of the principal kinds (I–III) of latex particle stabilization. Each kind of stabilization has at least one special advantage that makes this particular kind of stabilization useful, but for technical polymer dispersions application of a well-balanced mix of all stabilization possibilities is necessary. This is especially so if during storage and application of water-based dispersions both the ionic strength and the temperature change considerably. The most effective way to realize electrostatic stabilization is via the application of low molecular weight ionic surfactants such as alkyl sulfates (prominent example: sodium dodecylsulfate), or alkyl sulfonates, or alkyl ammonium compounds (prominent example: cetyltrimethylammonium bromide). Depending on the charge sign of the stabilizer either anionic (most prominently peroxodisulfates) or cationic initiators (such as 2,2'-azobis(2-amidinopropane)dihydrochloride) are used. But it has been shown that nonionic, water-soluble initiators such as symmetrical poly(ethylene glycol)-azo compounds might also be advantageous [108]. Moreover, the same investigations showed that much smaller particles were obtained with polymerization recipes containing ionic species than in completely nonionic polymerizations. These results clearly indicate that for effective stabilization some ions arising from the initiator, or ionic surfactants, or ionic co-monomers are needed at the particle surface. Furthermore, ionic emulsifiers are obviously more effective in stabilizing polymer particles than nonionic stabilizers. The drawback of purely electrostatic stabilization is the proneness to decrease with addition of electrolyte as the repulsive potential decreases exponentially with increasing ionic strength (see Tab. 1.4). In contrast, the use of nonionic stabilizers has a dramatic effect on the electrolyte stability of latex particles. For instance, the additional stabilization of negatively charged polystyrene particles with dodecyl hexaoxyethylene glycol monoether causes an increase in the critical coagulation concentration (determined with lanthanum nitrate) by an order of magnitude [105]. Besides increased stability against electrolytes another advantage of steric stabilization is the frequently observed reversibility of the flocculation process, which occurs if the conditions that have caused the flocculation are removed again. This reversibility requires that the stabilizers are strongly adsorbed or covalently bound to the particles. Also, an increase in the molecular weight of the steric stabilizer can kinetically retard the displacement of otherwise only weakly anchored stabilizers. Finally, as the name already indicates, electrosteric stabilization should combine features of the two other principles. However, the situation with regard to the structure of electrosteric stabilizers is ambiguous, as the charges may be either distributed statistically along the chain or arranged as a polyelectrolyte block. The first case leads to a so-called ringlet adsorption pattern (ΔR comparable with low molecular weight ionic surfactants) and the latter to so-called porcupine particles (ΔR up to more than 100 nm) [109]. Compared with poly(carboxylic acid) blocks as sta-

Tab. 1.4 Principles of stabilization of colloidal particles

Stabilization	Acting forces	Important parameters
Electrostatic (I)	Electrostatic repulsion of equally charged particles; repulsive potential (V_R) around charged particles at distance (d) decays as: $$V_R \propto f(\Psi) \exp\left(-\frac{d}{\lambda_D}\right)$$	Charge density at the interface, surface potential (Ψ), ionic strength (I_S); Debye screening length (λ_D) $$\lambda_D = \left(\frac{\varepsilon \cdot \varepsilon_0 \cdot k_B \cdot T}{\sum_i (z_i \cdot e)^2 \cdot C_{salt}}\right)^{0.5}$$
Steric (II)	Osmotic and entropic forces between overlapping stabilizer layers of approaching particles $$V_R \propto \frac{C_{S,L}^2}{v_{c,p} \cdot \rho_{S,L}^2} \cdot (\Psi_1 - \chi_{S,cp})$$ $$\cdot \left[\left(\Delta R - \frac{d}{2}\right)^2 \left(3\frac{D}{2} + 2\Delta R + \frac{d}{2}\right)\right]$$	Solution state of stabilizing polymer molecules (interaction parameter between stabilizing polymer and continuous phase); temperature, ionic strength as far as both influence the solution state of the lyophilic polymer
Electrosteric (III)	Competition between the osmotic pressure induced by counterion condensation inside the polyelectrolyte corona, which stretches the polyelectrolyte chain into the aqueous phase, and entropic polymer elasticity, which pulls the chains back to the surface, $\Delta R \propto I_S^a$; $a = -1/5$ (Pincus brush behavior)	Ionic strength, conformation and charge density of the polyelectrolyte chain, ration corona thickness to particle diameter (D); corona shrinks upon increasing ionic strength

C_{salt} is the molar bulk concentration of the ions, $k_B T$ is the thermal energy, ε and ε_0 are the permittivities in the continuous phase and in vacuum, z is the stoichiometric valency of the electrolyte, and e is the elementary charge; $C_{S,L}$ is the concentration of lyophilic polymer per unit volume inside the corona; $v_{c,p}$ is the molar volume of the continuous phase, $\rho_{S,L}$ is the density of the lyophilic polymer, Ψ_1 is an entropy parameter for mixing of the overlap region, and $\chi_{S,cp}$ is the interaction parameter between the lyophilic polymer and the dispersion medium.

bilizers, polyelectrolytes with strong acid groups (such as sulfonates) offer the advantage that their stabilizing action is practically independent of the pH. For strong polyelectrolytes tethered to spherical particles and forming a corona of thickness ΔR, Pincus derived the scaling relation given in Tab. 1.4 [110]. Note that this relation predicts extraordinary electrolyte stability compared with purely electrostatically stabilized particles, which show an exponential dependence on the ionic strength. Indeed, this was experimentally observed for various types of polyelectrolyte block copolymers such as poly(alkyl methacrylate-b-sulfonated glycidyl methacrylate) [111] but also for poly(ethyl ethylene-b-styrene sulfonate) [109]. Moreover, in the latter case a Pincus brush behavior was observed but with clear experimental evidence that a depends on the ratio of corona thickness to particle size.

Probably the most important property of polymer particles is their size and size distribution. Nowadays procedures are available to prepare latex particles in the size range between a few nanometers and a few hundreds of micrometers thus spanning five orders of magnitude. The particle size distribution can be of any shape depending mainly on both the polymerization procedure and the polymerization recipe. Note that the particle size distribution changes during the entire polymerization time owing to nucleation, coalescence or coagulation processes, and monomer consumption, which can lead to either particle growth or particle shrinkage. The latter case is especially important with regard to swollen particle size during the final stage of batch heterophase polymerizations.

Methods to adjust and to control the average size of latex particles are of special importance. Two principal methods exist to perform heterophase polymerizations, namely the presence and the absence of latex particles, or seeded and ab initio (unseeded) polymerizations, respectively. In ab initio polymerizations the nucleation of particles has to take place whereas the aim of application of seed latex is to avoid it. The size of seed latex particles can span a wide range from below 50 nanometers for industrial emulsion polymerizations up to micrometer size for preparation of monodisperse latexes for specialty applications. In the latter case the seed latexes are treated in a specific way to allow a high degree of swelling (see below). In these cases the second step polymerization will be carried out batch-wise, and proper action has to be taken to restrict the polymerization to the dispersed phase. In large-scale industrial polymerizations the application of seed particles allows control of the final particle size and size distribution by adjusting both the seed properties (number and size of seed particles) and the amount of monomers that is fed into the reactor either as neat bulk material or as emulsion (so-called semi-batch procedures). The average final particle size (D) is simply given by Eq. (1) where V_{feed} is the total volume of monomers fed during second stage polymerization, V_{seed} is the total volume of seed polymer (expressed as polymer volume), and N_{seed} is the number of seed particles.

$$D = \left[(V_{feed} + V_{seed}) \cdot \frac{6}{\pi} \cdot \frac{1}{N_{seed}}\right]^{1/3} \tag{1}$$

Equation (1) is only valid if the formation of new particles during the second-stage polymerization can be avoided. Nowadays from a commercial point of view pure batch processes play a major role for suspension, microsuspension and bulk polymerizations but only a minor role for emulsion polymerizations. The most important procedures for effecting polymer dispersions by emulsion polymerization on a technical scale are semi-batch or feed processes which are extremely flexible regarding product properties [31, 32]. Depending on the required properties with respect to particle size distribution, molecular weight distribution, chemical composition in the case of copolymerization, and particle morphology numerous feeding protocols have been developed. Almost all kinds of consecutive shell morphologies can be prepared by means of computer-controlled monomer feed streams [102].

In contrast to seeded polymerizations the control of the final particle size in ab initio polymerizations is more complicated and prone to fluctuations. The particle formation process is mainly responsible for this, because it is the least understood part of heterophase polymerizations and its investigation faces serious experimental problems. Particle nucleation occurs at an extremely low conversion or solid content. For instance, in ab initio surfactant-free styrene emulsion polymerization started with potassium peroxodisulfate, nucleation occurred at 60 °C after a prenucleation period – in which aqueous-phase polymerization takes place – of 431 s. At the moment of nucleation, which takes place within one second $1.76 \cdot 10^{13}$ particles are formed per cm^{-3} of water with an average particle size of 13 nm. The amount of polymer, or better oligomer, formed up to this moment is $2.13 \cdot 10^{-5}$ g cm^{-3} of water [112–114]. This example of an emulsion polymerization shows that an ab initio heterophase polymerization starts in the continuous phase and that it can be carried out even in the absence of both emulsifier and free monomer phase. Thus, the nucleation process can be identified as formation of the second, polymer, phase. The straightforward conclusion is that from a thermodynamic point of view nucleation requires that the reaction system, which in this situation is a solution of oligomers in the continuous phase, be brought into a thermodynamically unstable intermediate state. From a physical point of view this corresponds exactly to a situation as described by the classical nucleation theory [115, 116] or the theory of spinodal decomposition [117]. According to these theories nucleation is characterized by the necessity to surmount an energy barrier via fluctuations. Therefore a free energy of activation for such processes exists, which strongly depends on all experimental conditions such as temperature, pressure, and composition. Even minor changes in the experimental conditions can cause huge effects because the rate of the phase transition depends exponentially on the activation-free energy. This is important for carrying out heterophase polymerizations, as for instance the kind of reactor material or even the replacement of just parts of the reactor can change reaction rate and product properties. For instance, it has been shown that emulsion polymerization of methyl methacrylate is strongly influenced by the kind of reactor material (stainless steel, Teflon, glass) especially at low emulsifier concentrations (see below and [118]). Problems with regard to reproducibility have also been observed during the initial period of dispersion polymerization of methyl methacrylate in methanol initiated with 2,2′-azobisisobutyronitrile [119].

Of course, the seed particles themselves have to be prepared by ab initio polymerizations but fluctuations in their properties can easily be compensated by appropriate action [varying either V_{seed} or V_{feed}, see Eq. (1)] during the seeded polymerization.

Despite all these complications during the nucleation stage of ab initio heterophase polymerizations general rules exist about how to influence the final particle size and how to move from a given starting point in a desired direction of change. For given conditions the kind of a "rule-of-thumb graph" in Fig. 1.4 can be used as a guideline for decisions with regard to kinds of monomers, initiator,

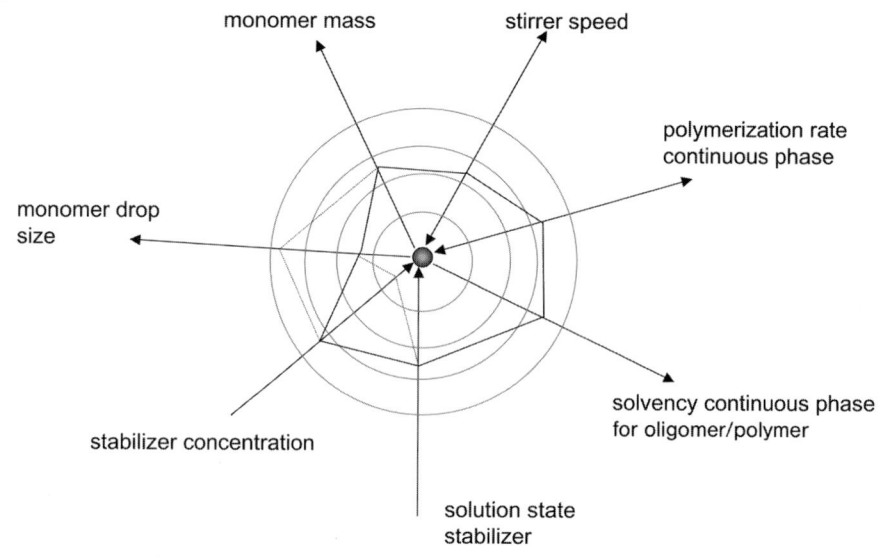

Fig. 1.4 Rule-of-thumb graph for predicting tendencies in the change of average particle size during ab initio heterophase polymerizations.

stabilizer, continuous phase, and experimental setup (mainly reactor material and shape). Provided that colloidal stability is preserved throughout all variations, that graph is to be read as follows. The average particle size increases from the center outwards. Arrowheads indicate the direction of changes in the average particle size if the corresponding parameter is "increased" or "improved". Double arrowheads mean the change in D is not unambiguous but depends on the level from which the changes were started. In other words the dependence of D on that parameter shows a minimum or maximum. For instance, increasing the surfactant concentration causes the average particle size to decrease whereas increasing the average size of monomer droplets in the case of suspension, microsuspension, miniemulsion, and microemulsion polymerization and proper initiation leads to an increase in D. This is indicated by the dotted lines in comparison with the starting situation depicted by full lines. In this sense the magnitude of the area spanned by the connecting lines between each parameter is a qualitative measure of the particle size. Note that these are general considerations, which, however, can be used to modify experiments under specific conditions (recipe, reactor) in order to shift particle sizes in a desired direction. Some practical examples of how the particle size can be controlled by various recipe components and process parameters are discussed below.

For most large-scale applications of polymer dispersions a specific *particle size distribution* is more advantageous than monodispersity. High solids content of polymer dispersions, which is desirable in order to minimize the amount of water to be transported, requires close packing of latex particles. At the same time the viscosity should not be increased too much in order to retain ease of processing.

Close packing of latex spheres depends on the potential that imparts stability to them and hence on the composition of the interface. Particles with soft potentials (that is purely electrostatic stabilization) interact over long distances and closer contact between the particles is prevented. In this sense it is not the mass fraction (solids content) but the effective volume fraction (including the hydrodynamic layer thickness) that determines viscosity. To get latexes with solids content about 70% or even higher requires polymodal particle size distributions with larger and smaller spheres (the smaller sphere have to fit in the interstices between the larger ones). This principle was successfully applied to get high solids polymer latexes up to 70% solids content and low viscosity [120]. It was found that there is a critical ratio between the larger and smaller sphere diameters, which should be above 10 in order to get both low viscosities and high solids content. The preparation of polymodal latexes requires the occurrence of multiple nucleation events in the course of the polymerization, which can be realized either by emulsifier feeding protocols [121, 122] and/or by proper choice of recipe components (co-monomers) [120].

Monodisperse latex particles are frequently required for other than commodity applications, for example for medical applications [123, 124]. Moreover, monodisperse latexes are of great value for investigations elucidating the mechanism of heterophase polymerizations (especially competitive growth experiments, see below) or proving theories of colloidal stabilization, and as secondary calibration standards in electron microscopy, light scattering, sedimentation and aerosol studies, and other fractionation techniques. But they also play an important role in investigations of a variety of colloidal phenomena. Examples are: (a) as model systems for soft-matter physics especially in understanding interactions in colloids [125–127], (b) as materials to trigger new characterization methods [128], (c) as objects to observe directly nucleation and growth of crystals [129], and (d) as components in advanced material science research [130].

1.2.2.2 Some Practical Examples – Surfactant-free Emulsion Polymerization

Stabilizer-free heterophase polymerization seems to be an attractive route to get polymeric model colloids that are free of adsorption–desorption equilibrium processes because stabilization takes place only by covalently bound stabilizing groups. The complete cleaning of latexes prepared in presence of common stabilizers by appropriate serum replacement techniques such as ultrafiltration, dialysis, or repeated centrifugation and redispersion [131] always contains the inherent danger of causing stability problems because the conditions that have led during polymerization to colloidal stability are removed. This is the reason why emulsifier-free latexes in general, and polystyrene particles specifically, have become popular model systems in colloid chemistry. It is interesting to note that the first report on the preparation of stabilizer-free emulsion polymerization appeared only in 1965 [132], more than 50 years after the first patent on heterophase polymerization had been filed. In this section we consider polymerization systems that contain only hydrophobic monomer(s) and hydrophilic initiators as active

Tab. 1.5 Surfactant-free batch ab initio emulsion polymerizations

No.	C_I (mM)	C_M (M)	T (°C)	t_{pol} (h)	$D^{a)}$ (nm)	Ref.
1	0.62 [b]	0.871	70	28.5	400	[132]
2	1.24 [b]	0.864 [c]	70	28.5	758	[133]
3	2.48 [b]	0.864 [c]	70	28.5	1374	[134]
4	2.76 [b]	0.864 [c]	70	24	536	[134]
5	2.76 [d]	0.870	70	24	431	[135]
6	2.83 [b]	1.16 [c]	66	16	3092 [e]	[136]
7	4.19 [b]	1.28 [c]	70	4.83	675 [f]	[136]
8	4.1 [d]	0.29 [c]	80	24	341 [g]	[137]
9	4.06 [h]	0.29 [c]	80	24	178	[137]
10	8.43 [d]	0.424 [i]	70	24	600 [j]	[138]

a) D determined from transmission electron microscopy
b) Potassium peroxodisulfate (KPS)
c) Styrene
d) In presence of 46 mM sodium chloride
e) In presence of 23 mM sodium chloride
f) 2,2′-azobis(2-amidinopropane)dihydrochloride (V50 from Wako)
g) 2,2′-azobis(2-(2-imidazoline-2-yl)propane)dihydrochloride (VA-044 from Wako)
h) Polydisperse particle size distribution
i) Methylstyrene (mixture 3:4 derivative 60:40)
j) Ionic strength 26 mM adjusted with sodium chloride

components in water as continuous phase, although if necessary the water may contain low molecular weight electrolytes in order to adjust the ionic strength. Ab initio surfactant-free heterophase polymerizations are especially sensitive to even slight modifications in the experimental conditions for reasons given above. Hence, the experimental conditions, including the grade or purity of all chemicals used as well as reactor materials, geometry, and pre-treatment should be described as fully as possible. The examples collected in Tab. 1.5, which are a selection of papers appearing over more than three decades, underline this statement impressively and equally illustrate the ongoing interest in emulsifier-free emulsion polymerization. For instance, the experiments in lines 1–4 of Tab. 1.5 were carried out at the same temperature and monomer concentration with only slight changes in initiator concentration, yet the polymerizations result in monodisperse polystyrene particles with surprisingly large difference in average diameter, which cannot be explained by the only slightly different polymerization times. In the first comprehensive study on surfactant-free emulsion polymerization Matsumoto and Ochi [132] observed the strong influence of the stirrer speed. For instance, the average particle size showed a tendency to increase with increasing stirrer speed while the average degree of polymerization decreased. Furthermore Matsumoto and Ochi [132] as well as Kotera et al. [133] reported an increase in D with increasing KPS concentration whereas Goodwin et al. [134] found the opposite behavior, that is a decrease in D with increasing initiator concentration. In a subsequent paper Goodwin et al. [139] summarized their comprehensive experimental data by an overall equation (Eq. 2) describing the dependence of D on the ionic strength (C_{IS} between 0.88 and 50 mM), monomer concentration (C_M between 0.58 and 0.87 M), potassium peroxodisulfate concentration (C_I between 0.29 and 2.76 mM), and temperature (T in K between 60 and 95 °C).

$$\log D = 0.238 \left[\log \frac{C_{IS} \cdot C_M^{1.723}}{C_I} + \frac{4929}{T} \right] - 0.827 \tag{2}$$

Equation (2) describes for above experimental conditions direct and indirect proportionality between D and C_{IS} and C_M on the one hand and C_I and T on the other hand, respectively. For cationic initiators (an example is depicted in line 5 of Tab. 1.5) the authors obtained a relationship analogous to Eq. (2) with different numerical values of the constants. The discrepancies between the experimental data for emulsifier-free styrene polymerization with KPS (lines 1–4 in Tab. 1.5) cannot be explained only by the minor difference in C_I. It is reasonable to assume, as had already been indicated by Matsumoto and Ochi [132] and explicitly shown almost 30 years later [136], that the hydrodynamic conditions and/or the reactor material (see below and [118]) have a strong influence on the result of surfactant-free emulsion polymerization. Tuin et al. (lines 6 and 7 in Tab. 1.5) claimed to have obtained large monodisperse polystyrene particles (average particle size above 1 μm) if the polymerizations were carried out in a flat-bottomed, all-glass reactor with an overall volume of 12 l and about 8 l reaction volume with a stainless-steel anchor stirrer at 160 revolutions per minute (rpm) and four baffles. By contrast, in a round-bottomed reactor they obtained average particle sizes below 1 μm, whether or not baffles were used. Furthermore, they observed decreasing average particle size with increasing potassium peroxodisulfate concentration as was also found by Goodwin et al. Fritz et al. (lines 8 and 9 in Tab. 1.5) prepared cationic polystyrene latexes as model drug carrier systems. They found an influence of the kind of cationic initiator as 2,2'-azobis(2-amidinopropane)dihydrochloride resulted in polydisperse particle size distribution whereas 2,2'-azobis(2-(2-imidazoline-2-yl)propane)dihydrochloride led to smaller particles with a monodisperse particle size distribution. The final example (line 10 in Tab. 1.5) describes polymerization of methylstyrene (vinyl toluene) initiated with 2,2'-azobis(2-amidinopropane)dihydrochloride. Basically these authors [138] confirmed the relations obtained from Goodwin et al. for styrene as monomer [135] regarding the dependence of the average particle size on initiator concentration and total ionic strength. Moreover, Wu et al. [138] also observed that the stirrer speed influenced the particle size distribution in such a way that at 250 and 350 rpm polydisperse and monodisperse particle size distributions were observed, respectively. It should be mentioned briefly here that surfactant-free emulsion polymerization can also be carried out in semi-batch or continuous operation mode. For instance, in a comprehensive study of semi-batch butyl acrylate polymerization initiated with sodium peroxodisulfate in the presence of sodium bicarbonate buffer it was shown that solids content up to 60% can be realized by a proper monomer feed protocol [140]. The dependence of the average particle size on the initiator concentration goes through a maximum at about 0.2% relative to monomer mass. However at lower initiator concentrations the system is not stable in a colloidal sense (there are not enough stabilizing ionic groups because of the low initiator concentration) as a huge amount of coagulum is formed. Thus, an evaluation of the maximum is problematic. A continuous emulsifier-free emulsion polymerization procedure for the synthesis of monodisperse, cross-linked particles (styrene/divinylbenzene and methyl methacrylate/ethylene glycol dimethacrylate) has been described [141]. The report indicates that the final particle size distribution is often uniform but even small changes in initiator concentration or increasing temperature (from 80 to 95 °C) caused polydisperse particle size distributions.

The above results clearly show that the synthesis of charge-stabilized, monodisperse polymeric particles as model colloids is possible by means of surfactant-free emulsion polymerization with all types of ionic initiators. Ionic groups arising only from initiator decomposition stabilize the particles. Consequently, the number of charges per latex particles (n_L) and surface charge (σ_L) density depend on both the average particle size and the average molecular weight as described by Eqs. (3) and (4), respectively [142].

$$n_L = f_T \cdot f_{CEG} \cdot d_P \cdot N_A \cdot \frac{\pi \cdot D^3}{6 \cdot M} \tag{3}$$

$$\sigma_L = \frac{F}{N_A} \cdot \frac{n_L}{\pi \cdot D^2} = f_T \cdot f_{CEG} \cdot d_P \cdot \frac{F \cdot D}{6 \cdot M} \tag{4}$$

In Eqs. (3) and (4) d_p is the polymer density, N_A is Avogadro's constant, F is the Faraday equivalent, f_T is a factor that considers whether the termination is by combination ($f_T=2$) or by disproportionation ($f_T=1$), and f_{CEG} is a factor taking into account that uncharged radicals can also start the chain growth, even in the case of ionic initiators [114, 143] ($f_{CEG} \leq 1$). The charged oligomers which are adsorbed onto the particle–water interface and are also distributed into the aqueous phase are not separately considered. Equations (3) and (4) also do not consider if there are any charged groups buried inside the particles. However, the validity of these scaling relations has been demonstrated by means of experimental data for surfactant-free styrene emulsion polymerization initiated with KPS [114, 142].

So far the only emulsifier-free latexes prepared with ionic initiators that have been considered are electrostatically stabilized particles. Under these circumstances the role of the ionic strength needs some attention as it governs the overall interaction potential between the particles (see above and Tab. 1.4). In this sense ionic initiators are acting with a double function as on the one hand they contribute to stability (covalently bound ionic groups, surface potential) and on the other hand they provide the ionic strength in the continuous phase (Debye screening length). Consequently, the influence of the concentration of ionic initiators on emulsifier-free emulsion polymerization depends on both the overall ionic strength (initiators plus non-reactive electrolytes) and especially on whether or not the experiments have been carried out at constant ionic strength (adjusted by the non-reactive electrolyte). It is straightforward that at constant overall ionic strength an increasing concentration of ionic initiators causes a decrease in average particle size because of the decreasing average molecular weight and hence an effective increase in surface charge density (see Eqs. 3 and 4), which leads to better stabilization. Several groups confirmed such behavior experimentally (see above). In contrast, if the ionic strength is not controlled an increase in the concentration of ionic initiators leads to an increase in both the average particle size and the coefficient of its variation as was comprehensively investigated in [144] provided the starting initiator concentration is high enough that results obtained at almost similar conversions can be compared. Within the rule of thumb graph (Fig. 1.4) these results represent examples of an influence on the average particle size of the solution state of the stabilizer or stabilizing groups due to changes in the ionic strength. Note that, in a stricter sense, a constant ionic strength throughout the whole polymerization requires the supply of nonreactive electrolyte in order to replace the amount of ionic initiator that is consumed during the reaction. With regard to the mechanism of emulsion polymerization the important conclusion is that the ionic strength is highest at the beginning, that is during the nucleation phase of an ab initio polymerization. The situation is the same in the presence of ionic emulsifiers, as their contribution to the ionic strength is reduced by adsorption after particle formation as well [108].

As the preparation of colloidally stable, model latex particles without any adsorbed and hence partitioned emulsifier is one of the driving forces for investigations of surfactant-free emulsion polymerization, it should be mentioned that this goal can also be achieved using ionic chain transfer agents, or ionic co-monomers, or reactive surfactants. The application of ionic chain transfer agents such as thiomalic acid has been demonstrated [114]. Ionic co-monomers, regardless of their surface activity, lead to formation of water-soluble polymeric or oligomeric materials, which also possess a certain surface activity. Here only a few examples can be mentioned. For instance, the emulsifier-free copolymerization of styrene and sodium 2-methyl-2-propensulfonate has been investigated [145]. Bimodal molecular weight distributions were encountered with the occurrence of two reaction loci, the interface and the interior of the particles, especially at early stages of the polymerization after particle nucleation. The action of another ionic co-monomer, sodium 3-allyloxy-2-hydroxyl-propanesulfonate, was investigated during the emulsifier-free copolymerization of methyl methacrylate and butyl acrylate [146]. The addition of this ionic co-monomer allows the synthesis of stable latexes with a solids content of up to 60% and almost monodisperse particles with sizes between 300 and 500 nm. The final particle size is larger the lower the concentration of both initiator (KPS) and co-monomer. The addition of ionic co-monomers during emulsifier-free emulsion polymerization is an effective way to reduce the average particle size and thus allows the synthesis of monodisperse model particles but with particle sizes down to about 120 nm [144]. A final example utilizes the application of nonionic co-monomers such as 2-hydroxyethyl methacrylate [147]. In that case copolymers with styrene contribute to steric stabilization of the particles as well as to electrosteric stabilization, as it has been shown that the hydroxyl group of the co-monomer is readily oxidized by peroxodisulfate to a carboxyl group [143, 148]. These highly mixed-charge, monodisperse particles can be assembled into robust, three-dimensionally ordered crystals. A previous study of the same monomer combination showed that the average particle size decreases with increasing concentration of nonionic co-monomer and KPS concentration but increases with increasing total ionic strength [149].

The result of emulsifier-free emulsion polymerization of hydrophobic monomers such as styrene can also be tailored by controlling the solvency of the continuous phase for the monomer and the oligomers formed during the pre-nucleation period. This possibility was successfully demonstrated by polymerizing styrene in acetone–water mixtures [150]. Okubo et al. found decreasing average particle size with both increasing acetone and KPS concentration. The optimum conditions for stable latexes ($D=160$ nm, very monodisperse) were determined to be 20 vol.% of styrene relative to the continuous phase, which is a 40:60 mixture (by volume) of acetone and water, 34 mM KPS as initiator, and 90 °C polymerization temperature.

A special case of surfactant-free emulsion polymerization is the application of symmetrical poly(ethylene glycol)–azo-initiators, so-called "PEGA-initiators", where a number placed after it denotes the molecular weight of the poly(ethylene glycol) chains. Initiators of this type can be used for the synthesis of block copolymer latex particles without any charges and additional surfactants as the poly(ethylene glycol) chains act as steric stabilizers. The final particle size depends on the molecular weight of the poly(ethylene glycol) and varies for instance from about 1 µm to 500 nm and 150 nm for PEGA200, PEGA2000, and PEGA10000 as initiator, respectively [18, 20]. This order underlines the stabilizing power of poly(ethylene glycol), which increases with increasing chain length. Thus the synthesis of nonionic model latex particles is possible.

Since the early 1990s the application of reactive surfactants in heterophase polymerization has become a huge topic in its own right. Several reviews are available [151–156] so it is only very briefly mentioned here. The basic idea is to design surfactants in such a way that they can participate in heterophase polymerizations either as co-monomers (so-called *surfmers*), or initiators (so-called *inisurfs*), or transfer agents (so-called *transurfs*), and finally be completely covalently attached to the polymer. From the application point of view the idea is to use covalent binding to avoid the migration of surfactants in the final application and prevent, for instance, the formation of hydrophilic spots with higher water-uptake in hydrophobic coatings. The polymerization properties of reactive surfactants have to be tuned carefully in order to avoid undesired effects with regard to stability and/or inhibition of the polymerization reaction.

1.2.2.3 Some Practical Examples – Monodisperse Latexes

Monodisperse latex particles are currently experiencing a renaissance, as it is believed that they might be useful for various potential applications such as photonic band gap crystals, removable templates for fabricating materials with different pore sizes, physical masks in lithographic processes, or diffracting elements in optical sensors [130]. A review covering the state of the art with regard to synthesis and application until the mid-1990s can be found in [157].

That latex particles can be prepared with very uniform sizes was first observed by chance during the course of some quantitative work with transmission electron microscopes at the end of the 1940s [158, 159]. The first monodisperse latex sample was made of polystyrene and prepared at Dow Chemical Company in 1947 [160]. This Dow latex 580-G, Lot 3584 was very popular over many years among electron microscopists [158–164]. In the following years more and more monodisperse latexes, mainly polystyrene and poly(vinyl toluene), have been prepared and their ability spontaneously to arrange themselves in close-packed crystalline-like arrays was first described in 1954 [163].

As monodisperse latex particles are in many cases the results of nucleation and growth processes, their controlled synthesis requires at least a minimum understanding of both particle nucleation and particle growth. Particle growth can occur during emulsion polymerization either by conversion of monomer into polymer (the actual polymerization reaction) or by coalescence of polymer particles. Now it is an obvious conclusion that monodispersity of latex particles results if optimum conditions exist between all three processes (nucleation, growth by monomer consumption, and growth by coalescence). Unfortunately, the prediction of these optimum conditions for a particular monomer and heterophase polymerization technique is still not possible and monodispersity of latex particles is more or less an accidental result. Nevertheless, some guidelines, even if they are crude and might not be valid in any individual case, can be formulated. The first condition is that only a single nucleation period should occur in the polymerization, in contrast to the requirement mentioned above for latexes with high solids content. A second condition is that both the number of particles generated and their growth should occur adjusted such that either $N \cdot D$ (if capture of oligomers born in the continuous phase by existing particles is determined by a diffusion mechanism) or $N \cdot D^2$ (if capture of oligomers born in the continuous phase by existing particles is determined by a collision mechanism) is high enough during the entire polymerization to avoid another nucleation step. In this sense the key is that if radical capture is high enough nucleation will cease (see [165]). The shorter the nucleation period the narrower the particle size distribution for the subsequent period of exclusive particle growth.

In this context the relative growth of particles with different diameters, so-called competitive growth, is of importance. The volume growth of a particle is generally given by Eq. (5) with K being constant and independent of particle size.

$$\frac{dv}{dt} = K \cdot D^x \qquad (5)$$

Since $v = \frac{\pi}{6} D^3$ size growth is given by Eq. (6), which leads (assuming that K and K' are independent of time, i.e. K and K' only depend on concentrations and temperature) after integration to Eq. (7). Furthermore, the size dependence of the monomer concentration is neglected. D and D_0 are the size at the beginning and the end of polymerization with duration Δt, respectively.

$$\frac{dD}{dt} = K' \cdot D^{x-2} \tag{6}$$

$$D^{3-x} - D_0^{3-x} = K'' \cdot \Delta t \tag{7}$$

A closer inspection of these equations reveals that:
- if $x<3$ smaller particles will grow faster and the particle size distribution is self-sharpening,
- if $x>3$ the distribution becomes broader as larger particles grow faster than smaller ones, and
- if $x=3$ particle growth does not depend on particle size.

There is some experimental evidence that competitive growth during seeded emulsion polymerization slightly favors smaller particles, which means that a self-sharpening of the size distribution occurs [166]. Other experimental and theoretical studies confirmed in the case of water-soluble persulfate initiator growth values of $x=2.5$ for particle sizes larger than about 150 nm which decreased towards zero when the particle size was decreased below 150 nm. However, for oil-soluble initiators $x=2.5$ for a larger size range until the particles reached a critical size needed to sustain two growing radicals [167]. This means that for larger particles particle growth does not depend on particle size and, hence, the particle size distribution evolves uniformly over time. In a series of vinyl chloride emulsion polymerizations x-values between 2 and 3, which slightly decrease with increasing peroxodisulfate concentration, have been determined [168]. Model calculations on the kinetics of competitive growth in aqueous polymerization of bidisperse seed systems with oil-soluble initiators showed that x is strongly dependent on rate of initiator decomposition, desorption rate, and the diameter ratio. If only 1% of the initiator is soluble in the continuous phase x approaches a value of 3 for low and high initiator decomposition rates. For intermediate values of initiator decomposition rate $x<3$ is obtained. On the contrary, with a completely water-insoluble initiator x is constantly equal to 3 [169]. Also more sophisticated models with regard to particle growth come to similar conclusions, that is as conversion evolves size distribution usually becomes narrower and a more monodisperse distribution can be obtained provided formation of new particles can be avoided [76].

Thus, growth of particles by monomer consumption might favor the formation of a more monodisperse size distribution after particle nucleation. In contrast, the consequence of particle growth by coalescence is more ambiguous but it strongly depends on the rate of that process. The following consideration might elucidate the situation a little. To obtain spherical latex particles requires that the result of coalescence of two spherical particles is again a spherical particle. This is a clear demarcation from coagulation or flocculation as either of these processes leads to non-spherical particles. Assuming a population of monodisperse particles that is prone to coalescence then a high coalescence rate is required if a monodisperse size distribution is to be obtained finally, for at least two reasons. Firstly, with an infinitely high rate constant the process might end, at least theoretically, with only one particle left. This undesired, but extremely monodisperse case illustrates that the higher the coalescence rate the more monodisperse the distribution at the end of that process. Secondly, the faster the coalescence process ceases the earlier volume growth by monomer consumption can lead to a self-sharpening. Note that the rate of coalescence scales with the square of the particle concentration and is thus self-reducing. There are several examples known where during heterophase polymerization a drop in the particle number occurs without, however, leading to complete coagulation. This effect is called limited flocculation, where the "system" tries to reduce its interfacial free energy, and has been observed frequently, for example in surfactant-free batch emulsion polymerization of methyl methacrylate [165], semi-batch surfactant-free emulsion polymerization of butyl acrylate [140], and also emulsion polymerization of vinyl chloride in both the presence and the absence of emulsifiers [170, 171]. In any case coalescence might favor a broad particle size distribution if N decreases so strongly that enough oligomers can no longer be captured and subsequently the conditions for secondary nucleation are fulfilled.

To obtain monodisperse particles not only should multiple nucleation events be avoided but the single nucleation period should be considered. On the one hand this should be as short as possible in order to get a particle size distribution with the optimum width but on the other hand it should be long enough for nucleation of the optimum particle number for the subsequent growth period. Obviously these conditions are best met in emulsion polymerizations with surfactant concentrations below the critical micelle concentration, as was shown in a series of papers (see [172–176]). For instance, investigations of the nucleation in styrene emulsion polymerizations gave clear evidence that the duration of the nucleation period in the presence of sodium dodecyl sulfate at concentrations above the critical micelle concentration is longer by a factor of more than 10^3 compared with emulsifier-free polymerization, where nucleation takes place within almost one second [177]. With emulsion polymerization monodisperse latex particles are easily accessible in the sub-micrometer size range. Besides the examples given in [172–174] the examples given in Tab. 1.6, which have been obtained in the present author's laboratory, are good starting points for repetition in order to get monodisperse latex particles in the given size range.

Tab. 1.6 Preparation and characterization of monodisperse polystyrene particles

MD#	m_{SDS} (g)	FG (%)	D_w (nm)	D_n (nm)	D_w/D_n
1	0.092	7.44	162.1	161.7	1.002
2	0.067	7.56	217.7	217.4	1.001
3	0.047	8.25	347.4	344.7	1.008
4	0.033	8.09	397.4	395.9	1.005

Fig. 1.5 Transmission electron microscopy pictures of monodisperse polystyrene particles prepared according to the prescriptions given in Tab. 1.6; the bar indicates 100, 200, 300, and 500 nm for MD1, MD2, MD3, and MD4, respectively.

The polymerizations were carried out in all-glass reactors of either 100 ml (MD1, MD3) or 250 ml (MD2, MD4) reaction volume. The reactors were equipped with a stirrer (stirrer speed was adjusted to 300 rpm), a reflux condenser, a nitrogen inlet and outlet, a heating jacket to control the temperature, and a valve on the bottom to remove the latex. The standard procedure was as follows: water (95 g), styrene (10 ml), and the amount of sodium dodecyl sulfate (m_{SDS}) were premixed in the reactor at reaction temperature (80 °C) for 2 min. To start the polymerization KPS (0.35 g) dissolved in water (5 g) was injected. After 4 h the polymerization was completed and after cooling-down the latex was filtered through a glass filter (pore 2) in order to remove coagulum. The latexes were characterized with regard to solids content (FG) and particle size. Number and weight average particle diameters D_n and D_w, respectively, were calculated by transmission electron micrographs (about 500 particles have been evaluated).

It might be necessary to fit especially the emulsifier concentration to the particular conditions (reactor material, reactor size and hydrodynamics, quality of chemicals, etc.) in order to reproduce the exact diameters. The size range can be expanded in either direction by changing the emulsifier concentration as described in Fig. 1.5. The size distributions are fairly monodisperse as long as the sodium dodecyl sulfate concentration is below the critical micelle concentration, which is higher the lower the ionic strength. Thus varying the initiator concentration offers additional possibilities for varying the average particle size. However, in order to get larger particles in only one step this kind of emulsion polymerization in the presence of only a single kind of surfactant is an inappropriate method. Thus, the idea was born to increase the accessible size range by reducing the stabilizing power of the ionic surfactants by adding an oppositely charged surfactant [178]. Gu and Conno described the preparation of micron-sized monodisperse polystyrene particles in the following way. The polymerization of styrene (1.1 M) was started at 70 °C with the addition of potassium peroxodisulfate (8 mM) and after 3 min cetyl trimethyl ammonium bromide (0.5 mM) was added, followed after a further 5 min (8 min after the start of the reaction) by addition of various amounts of sodium dodecyl sulfate. The final particle size depends on the amount of sodium dodecyl sulfate and goes through a maximum of about 2.5 µm at 0.425 mM. By increasing the monomer concentration Gu et al. were able to prepare monodisperse polystyrene particles with a maximum average diameter of 3.3 µm [179].

The rule-of-thumb graph in Fig. 1.4 suggests that the average particle size can be increased if the solvency power of the continuous phase for the oligomer/polymer molecules formed during the pre-nucleation period is increased. This effect was nicely demonstrated in a series of dispersion polymerizations of styrene in alcohol–water mixtures [180]. The average particle sizes (from transmission electron microscopy) obtained from experiments in which 5 ml of styrene were polymerized in the presence of 0.5 g of poly(acrylic acid) (molecular weight $2.7 \cdot 10^4$ g mol^{-1}) with 0.1 g of dibenzoyl peroxide as initiator at 78 °C in ethanol:water mixtures (ml–ml) of varying composition 50–5, 50–25, 50–27, and 50–28 were 1.5, 0.8, 0.6, and 0.5 µm, respectively. Another example of the influence of the solvency of the continuous phase can be found in [181]. These authors varied the ratio ethanol/methoxyethanol in dispersion polymerization of styrene and observed increasing particle sizes with increasing methoxyethanol content. For instance, 37.5 ml of styrene were polymerized, in the presence of 16.8 g of 25% aqueous poly(acryl amide) (molecular weight $2.5 \cdot 10^4$ g mol^{-1}) solution as stabilizer, with 1.5 g of dibenzoyl peroxide as initiator in 212.5 ml of organic phase at 68 °C. The number average particle size increases from 2.43 to 3.996, 4.95, and 5.53 µm if the composition of the organic phase (ml of ethanol + ml of methoxyethanol) changes from 212.5 + 0 to 112.5 + 100, 100 + 112.5 and 87.5 + 125, respectively, but the particle size distribution remains extremely monodisperse. There have been some attempts to quantify this behavior by means of the solubility parameters for the continuous phase. Although the precise form of the dependence of the average particle size on the polarity or the solubility parameter of the continuous phase may vary from case to case, the common trend when nonpolar monomers are polymerized in polar media is that the average particle size decreases with increasing polarity [182]. Thus, the reverse dependence can be expected for polar monomers. The solubility parameters for the continuous phase have been calculated in the following two examples and, indeed, the expected difference in the behavior for nonpolar and polar monomers has been experimentally observed. First, for copolymerizations of styrene and glycidyl methacrylate (a 12/3 weight mixture, which is nonpolar) in ethanol–water mixtures with poly(vinyl pyrrolidone) as stabilizer and 2,2'-azobisisobutyronitrile as initiator the average particle size decreases with increasing solubility parameter of the continuous phase [183]. The second example is dispersion polymerization of 2-hydroxyethyl methacrylate, a polar monomer, in mixtures of toluene with higher alcohols such as methylpropan-1-ol or butan-2-ol with cellulose acetate butyrate as stabilizer and dibenzoyl peroxide as initiator. In this case the average particle size increased with increasing solubility parameter of the continuous-phase mixture [184].

Consequently, the solubility parameter or polarity of the continuous phase alone is not a measure that allows a general statement, as one has also to consider the solubility parameter of the oligomers/polymers formed during the pre-nucleation period. Measuring the solubility parameter of a low molecular weight solvent or calculating it for solvent mixtures is no problem. In contrast, for oligomers or polymers the direct evaluation of the solubility parameter is practically impossible. But it is a matter of fact that the greatest tendency of a polymer to dissolve occurs when its solubility parameter matches that of the solvent. In typical dispersion polymerizations the monomers are soluble in the continuous phase and, hence, the situation is more complicated because the solubility parameter as well as the solvency of the continuous phase changes in the course of the polymerization owing to the decreasing monomer concentration. Thus, in order to get a uniform rule valid for nonpolar and polar monomers the solvency of the continuous phase for the oligomers/polymers is considered in Fig. 1.4 instead of only the solubility parameter of the continuous phase.

Micron-sized polymeric particles have received a lot of attention as stationary phases for chromatographic techniques and as support materials in solid-phase and combinatorial synthesis. For these purposes chloromethylstyrene particles are particularly important, because they can be modified easily by polymer-analogous reactions. Therefore the preparation of micron-sized monodisperse poly(chloromethylstyrene) particles by heterophase polymerization has been of interest for several years (for instance [185–188]). An improved dispersion polymerization method with regard to yield and monodispersity for the synthesis of poly(p-chloromethylstyrene) particles was recently published [189]. The polymerization is carried out in ethanol–methoxyethanol mixtures with poly(acrylic acid) and 2,2'-azobisisobutyronitrile as stabilizer and initiator, respectively, at 70 °C in a vibrating water bath. The variation of the methoxyethanol/ethanol volume ratio from 0.007 to 0.75 leads to an increase in the number average particle diameter from 1.74 to 3.73 μm while preserving the monodispersity. However, a still further increase in the methoxyethanol content leads to secondary particle nucleation. In that case still larger particles are formed so that consequently either $N \cdot D$ or $N \cdot D^2$ are not high enough to capture all oligomers before they reach the critical concentration for nucleation (see above).

Almong, Reich, and Levy published a benchmark paper in the development of dispersion polymerization with regard to preparation of particles with uniform sizes in 1982 [190]. The authors observed the formation of particles with a monodisperse size distribution when during dispersion polymerization of methyl methacrylate in methanol with poly(vinyl pyrrolidone) as stabilizer and 2,2'-azobisisobutyronitrile as initiator at 60 °C a cationic co-surfactant, methyl tricaprylyl ammonium chloride, was used. Instead of particles with a size distribution between 100 nm and 12 µm in the absence of the co-surfactant, monodisperse spheres of 3.9 µm diameter were formed. A similar behavior was also observed for other polymeric stabilizers such as poly(ethylene imine), poly(acrylic acid), and poly(vinyl methylether). Thus, mixed stabilizer systems are not only advantageous for aqueous emulsion polymerization but also for dispersion polymerization in polar organic media where electrostatic interactions still can contribute. This approach was also used in the present author's laboratory to prepare monodisperse polystyrene particles. During the investigations it turned out that both the molecular weight of the steric stabilizer (poly(vinyl pyrrolidone)) and the kind of initiator (2,2'-azobisisobutyronitrile or PEGA200) also strongly influence the particle size distribution. Fig. 1.6 shows transmission electron micrographs elucidating this observation. Particles were prepared according to the following recipe: ethanol (100 g), water (5 g), styrene (20 g) poly(vinyl pyrrolidone) (PVP, 0.8 g), tricapryl methyl ammonium chloride (0.228 g), and either 2,2'-azobisisobutyronitrile (AIBN; 0.2 g) or PEGA200 (0.692 g), temperature 70 °C, time 24 h, all-glass reactor, stirrer speed 80 rpm. The differences in the particles obtained are clearly visible and obviously indicate special interactions between PVP and poly(ethylene glycol) causing the strange rod-like morphology obtained for the PVP with the lower molecular weight, which seems to be the result of limited coagulation in one preferred direction.

Stöver et al. in a series of papers describe a rather unusual example of dispersion polymerization [191–194]. The authors describe a stabilizer-free dispersion polymerization of divinylbenzene homo- and copolymerization in neat acetonitrile or in acetonitrile containing porogenic co-solvents with AIBN as initiator. These polymerizations might be considered as precipitation polymerizations (see Tab. 1.1) owing to the lack of stabilizers but lead unexpectedly to well-defined, spherical particles with smooth surfaces and monodisperse size distributions. The interesting question relates to the stabilization mechanism of these particles. There are at least two possibilities. These are, firstly, charge formation due to the contact potential according to Coehn's rule [195] and, secondly, a stabilization due to polar initiator end groups such as nitrile end groups, which are compatible and hence might be interacting strongly with the continuous acetonitrile phase. Possibly also the divinylbenzene itself and its nonreactive byproducts contribute to stability and particle morphology on the one hand simply by their presence in the continuous phase (improved solvency) and on the other hand by the formation of cross-linked but also slightly swollen particles, which under these circumstances might behave comparably with lyophilic microgels. The effect of composition of the monomer mixture on particle properties was investigated for the preparation of poly(chloromethylstyrene-co-divinylbenzene) particles as described in [194]. In this study the authors differentiate between four regions with regard to appearance of the particles. At chloromethylstyrene fractions between 0.01 and 0.05

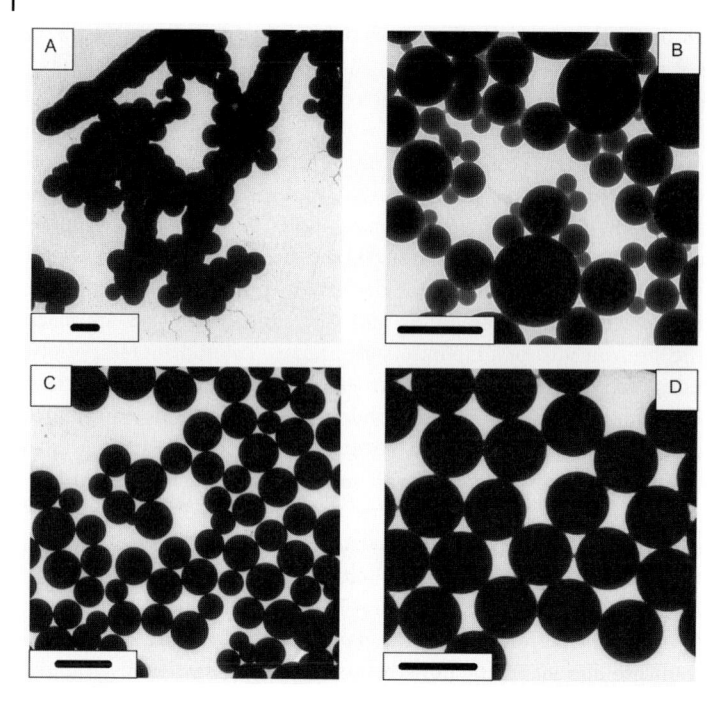

Fig. 1.6 Transmission electron micrographs of particles prepared by dispersion polymerization; A, PEGA200 and PVP with $1 \cdot 10^4$ g mol^{-1}, bar=1 μm; B, PEGA200 and PVP with $3.6 \cdot 10^5$ g mol^{-1}, bar=1 μm; C, AIBN and PVP with $1 \cdot 10^4$ g mol^{-1}, bar=5 μm; D, AIBN and PVP with $3.6 \cdot 10^5$ g mol^{-1}, bar=2 μm.

highly segmented particles and aggregates composed of tiny particles were observed. Spherical and monodisperse particles were found for divinylbenzene homopolymerization and between 0.07 and 0.6 chloromethylstyrene fractions in the monomer mixture. At chloromethylstyrene fractions between 0.7 and 0.8 polydisperse particle size distributions were obtained. At still higher chloromethylstyrene fractions up to 0.9 no precipitation took place.

A last example, worth mentioning briefly, deals with the preparation of latex particles from cyclic esters by pseudo-anionic polymerization of ε-caprolactone and lactides. These kinds of polymers are of increasing interest because they are widely used in medicine, as implantation material, bone fillers, and carriers of drugs and also because of their potential application in the packaging industry as biodegradable waste material [93, 196, 197]. Dispersion polymerizations with poly(dodecyl acrylate)-g-poly(ε-caprolactone) as stabilizer are described under anhydrous conditions in 1:9 and 1:4 volume mixtures of 1,4-dioxane and heptane in the case of ε-caprolactone at room temperature with diethylaluminium ethoxide as initiator and lactides at 95 °C with tin(II)-2-ethylhexanoate as initiator, respectively. The average particles sizes can be varied between 600 and 700 nm for poly(ε-caprolactone) particles and between 2.0 and 2.4 µm for polylactide microspheres. It has been shown that partially hydrolyzed poly(ε-caprolactone) particles can be suspended in aqueous continuous phases and loaded with drugs.

1.2.2.4 Some Practical Examples – Particles Prepared by Multi-step or Seeded Polymerizations

In addition to the above methods of producing latexes on an industrial scale, tailoring of latex particles by seeded polymerizations is also a powerful tool to prepare specialty latex particles.

The first example considered here is the so-called "shot addition" or "shot growth" technique, which has been successfully used to tailor surfaces of latex particles with lyophilic or functional monomers. The functional monomers are added during the later stages of the polymerization either all at one time (that is as a "shot") or over a longer period of time. The latter is preferred if the corresponding polymer is insoluble in the continuous phase so that secondary nucleation might occur. In this case the feed rate of the monomers has to be adjusted to fit the rate of polymerization. For instance, use of reaction calorimetry allowed the kinetically controlled synthesis of core-shell latex particles with fluid cores and cross-linked rubbery shells [198, 199].

Knowledge of the reaction rate profile allowed the start of feeding of the monomer mixture for the shell (n-butyl acrylate and 1,3-diisopropylene benzene) at about 85% conversion of the first monomer batch. Feeding was continued for 90 min in order to avoid accumulation of the monomer mixture in the continuous phase and to prevent secondary nucleation. If the cross-linking density of the shell is in the correct range the fluid core polymer [(poly(2-ethylhexyl methacrylate), molecular weight regulated with tetrabromomethane as chain transfer agent] can be extracted and thus empty shells are left.

A second example describes the preparation of cross-linked polystyrene particles with acetoxy groups at the surface by shot addition of p-acetoxystyrene almost at the end of styrene/divinylbenzene emulsion polymerization [200]. In a subsequent modification step the acetoxy groups are converted into hydroxyl groups. Finally, the poly(p-hydroxystyrene) shell of these particles is able to imbibe silver and ruthenium nanoparticles. In these composite particles the metal colloids were bound strongly to the surface and were stable in a variety of organic solvents.

Another example is the preparation of monodisperse core-shell particles, which can be used to prepare three-dimensional nanocomposites as described in [201]. In these particles core and shell can either have different properties such as glass transition temperatures and molecular weights or can contain different functionalities such as chromophores. For instance, Kalinina and Kumacheva also describe the synthesis of particles with slightly cross-linked shells and fluid cores finally forming a porous film with a regular and controlled three-dimensional arrangement of pores. Another interesting example is the attachment of chromophores to the cores leading to polymeric nanocomposites, which can be used as three-dimensional memory storage.

Another kind of composite latex particles is polystyrene particles coated with conducting polymers such as polypyrrole, polyaniline, or poly(3,4-ethylenedioxythiophen) prepared by modification of sterically stabilized polystyrene particles as described in [202].

The above examples were carried out in such a way that the monomer addition to seed particles takes place during the polymerization process with very little swelling of the first-stage particles. Two basic techniques have been developed to promote swelling of precursor particles so that polymerization finally yields monodisperse latex particles with diameters larger than 5 µm up to several tens of micrometers, which is a size range in which a high degree of monodispersity is not accessible by a single polymerization step. Highly swollen latex particles behave much more like emulsions than like suspensions. A basic understanding of this swelling process requires consideration of both the Kelvin equation (Eq. 8) and Eq. (9), which describes the swelling behavior of latex particles as derived in [203].

$$\ln \frac{p'}{p_0} = \frac{4 \cdot \sigma}{D} \cdot \frac{v''}{RT} \tag{8}$$

$$\left(\frac{4 \cdot \sigma}{D} + P_{sw}\right) \cdot \frac{v_{mon}}{R \cdot T} = -\left[\ln(1-\phi_2) + \left(1 - \frac{1}{j_2}\right) \cdot \phi_2 + \chi_{m,p} \cdot \phi_2^2 \right.$$
$$\left. + \frac{v_{mon} \cdot \rho_2}{\overline{M}_c} \left(\phi_2^{1/3} - \frac{\phi_2}{2}\right)\right] \tag{9}$$

The Kelvin equation indicates that larger objects, in coexistence with smaller ones, will grow in size at the expense of the smaller objects, which have a tendency to dissolve. This effect is known in colloid science as *Ostwald ripening*. Equation (8) shows the Kelvin equation for liquid droplets surrounded by vapor where p' is the vapor pressure outside the drop, p_0 is the bulk vapor pressure, v'' is the molar volume of the liquid, and RT is the thermal energy [204]. The smaller the drops the larger p' and hence the higher the tendency to degrade.

The term on the right hand side of Eq. (9) is the free energy of mixing between polymer and monomer where the Flory-Huggins-Rehner expression for cross-linked polymers is used. σ is the interfacial tension between the swollen particles and the continuous phase, P_{sw} is the swelling pressure, χ_{mp} is the polymer monomer interaction parameter, ϕ_2 is the polymer volume fraction in the swollen particle, RT is the thermal energy, j_2 is the average degree of polymerization of the polymer molecules, ρ_2 is the polymer density, and $\overline{M}_c$ is the average molecular weight between two cross-links in the network. If the latex particles are not cross-linked ($\overline{M}_c \Rightarrow \infty$) and if $P_{sw} \Rightarrow 0$ Eq. (9) becomes identical with the Morton-Kaizerman-Altier equation [205] as derived by Gardon [206]. Between ϕ_1, the monomer volume fraction, and ϕ_2 the relation $\phi_1 + \phi_2 = 1$ exists. The swollen (D) and unswollen particle size (D_0) are connected with ϕ_2 via $(D/D_0)^3 = 1/\phi_2$ whereas ϕ_1 is related to the monomer concentration inside the particles ($C_{M,p}$) via $\phi_1 = C_{M,p} \cdot v_{mon} = \dfrac{v_{m,p}}{v_{m,p} + v_{p,p}}$ with $v_{m,p}$ and $v_{p,p}$ being the monomer volume and the polymer volume inside the swollen particles, respectively.

The right-hand side of Eq. (9) promotes swelling whereas the terms on the left-hand side counteract swelling. The first term on the left-hand side is the interfacial free energy, which counteracts swelling due to an increase in the particle interface. The second term on the left-hand side originates from a volume work due to attractive forces between polymer chains in concentrated solutions as has been concluded from osmotic modulus measurements [207]. P_{sw} in Eq. (9) is comparable to the pressure known from the swelling of macroscopic gels, contributing together with the partial molar free interfacial energy to the equilibrium with the chemical potential of the swelling agent. ϕ_1 increases the larger the particles, the lower both the interfacial tension and the degree of polymerization of the seed polymer, and the higher the temperature, but it decreases with increasing degree of cross-linking and with increasing Flory-Huggins interaction parameter.

Recognizing the similarity between swollen latex particles and emulsion droplets John Ugelstad started a great deal of both theoretical and experimental work to develop new methods of monomer emulsion preparation (see [208]). One of the key points is the effect that the dissolution of emulsion drops can be retarded or even prevented if the drops contain a substance (compound-2), which is insoluble in the continuous phase. In this case the decrease in size increases the chemical potential of compound-2 inside the smaller drops and generates a force counteracting Ostwald ripening. Ugelstad was able to show that compound-2 has an additional effect on swelling, because the entropy term in the swelling equation (Eq. 9), which is $(1-1/j_2) \cdot \phi_2$, becomes more and more important as long as compound-2 has a low molecular weight (it could be an oligomer where j_2 is in the order of about 5). Based on these two effects, Ugelstad and co-workers developed an activated two-step swelling procedure, which enabled them to prepare and commercialize large monodisperse particles for various applications (see [123, 124]). Briefly, the procedure is as follows. In a first step an emulsion of the low molecular weight and highly water-insoluble compound-2, which may contain a solvent that is water soluble, is added to a suspension of latex seed particles. The water-soluble solvent promotes the transport of compound-2 through the aqueous phase and allows swelling, which is further facilitated if the emulsion droplets are smaller than the seed particles (action of the Kelvin equation). Then, in a second step the monomer emulsion is added. Because of the high entropy gain caused by mixing of compound-2 with the monomer inside the particles the swelling ability of the seed is activated compared to "normal" seed particles that are without compound-2. For instance, using dioctyladipate as compound-2 it was possible to swell polystyrene particles with a diameter of 1.55 µm with chlorobenzene as model compound to a diameter of about 30 µm, i.e. the seed particles imbibed more than seven-thousand times their own volume.

Another useful method to perform swelling was developed by Okubo and is called the dynamic swelling method [209]. The principle of this method is a combination of nucleation and swelling. Tiny monomer droplets, which nucleate from the supersaturated continuous phase by stepwise changing its solvency for the monomer will subsequently swell the polymer particles. The following example may clarify the principle of action where the oil phase is styrene and the seed particles consist of polystyrene. In a mixture of 6 g ethanol and 4 g water as solvent are dissolved 0.4 g styrene, 0.004 g dibenzoyl peroxide, and variable amounts of poly(vinyl alcohol) stabilizer (between 3.75 and 37.5 wt.% based on styrene). Then, the slow addition of 40 g of water decreases the solvency power for styrene more and more until the conditions for droplet nucleation are met. In the absence of seed particles the nucleated droplets will form a styrene emulsion with a fairly polydisperse size distribution. However, the situation changes completely if seed particles are present which soak up the oil as soon as it is nucleated from the aqueous phase. As an example, at a water feeding rate of 2.88 ml h^{-1} and in the presence of 3.75 wt.% poly(vinyl alcohol) together with 0.004 g polystyrene seed particles with a particle diameter of 1.8 µm, which were prepared by dispersion polymerization in an ethanol–water mixture, a monodisperse styrene emulsion is formed with a diameter of 8.5 µm. Again, as in the previously mentioned example of the activated swelling method, the presence of monodisperse polymeric seed particles leads to the formation of an emulsion with a monodisperse size distribution. Note, in the latter case no swelling promoter is necessary. The driving force for the enhanced swelling is the high Kelvin pressure inside the small styrene droplets after nucleation compared to the seed particles and the swelling proceeds via Ostwald ripening. Additionally, the freshly nucleated droplets are relatively unstable owing to the high dynamic interfacial tension of the poly(vinyl alcohol), which as a polymer only slowly adsorbs and equilibrates. Further examples of the application of the dynamic swelling methods to tailor monodisperse polymer particles including particles with various non-spherical morphologies as well as the formation of hollow particles by application of a porogen can be found in [210–219].

To summarize this section: methods are available that allow latex particles to be swollen to stable emulsion droplets. The swelling agent can be a monomer, an inert solvent, or a mixture of both. During subsequent polymerization precautions are necessary to avoid nucleation of new particles. But in any case, control of swelling enables the preparation of polymer particles with specifically tailored properties.

1.2.2.5 Some Practical Examples – Influence of Polymerization "Hardware"

A few experimental examples obtained over the last years in the author's laboratory are intended to elucidate the influence of polymerization "hardware" such as reactor material, reactor geometry, and hydrodynamic conditions on heterophase polymerizations, especially on particle and polymer properties.

Although it is a matter of fact that the reactor material has an influence on the result of an emulsion polymerization there is a lack of published results of comprehensive and systematic investigations. For instance it is known in many industrial facilities and research labs that the performance of reactors may change with lasting usage, or that virgin reactors result in different properties for the same recipe used previously, or that even the replacement of the stirrer may alter latex properties. In a systematic study the influence of different reactor materials such as glass, Teflon, and stainless steel on aqueous polymerizations of methyl metha-

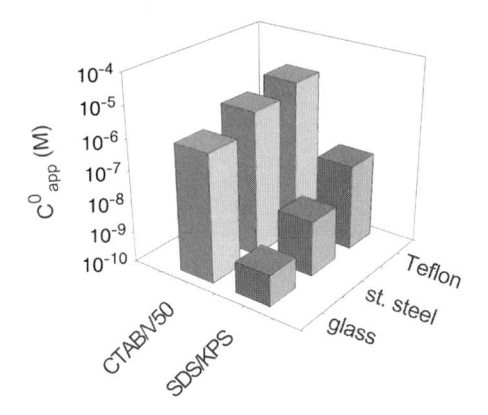

Fig. 1.7 The dependence of C_{app}^0 (M) on different reactor materials and IESs.

crylate and styrene has been investigated [118]. The influence of the reactor material on the average particle sizes has been quantified in the following way. Using the experimentally determined log–log relationship between average particle size and surfactant concentration it was possible to determine an apparent zero-surfactant concentration (C_{app}^0) with the average particle size obtained for surfactant-free polymerizations. In fact C_{app}^0 measures directly to what extent a particular reactor material in a given initiator-emulsifier system (IES) supports the formation and stabilization of particles. The C_{app}^0-values put together in Fig. 1.7 show that Teflon and a cationic IES do this much more efficiently than an anionic IES and other materials. To obtain these results the polymerizations were carried out batchwise in water (10 ml) at 60 °C in geometrically identical sealed test tubes of different materials. The starting reaction mixture was homogeneous, consisting of dissolved MMA (75 mM), corresponding amounts of emulsifier ranging from zero to concentrations well above the critical micelle concentration, and initiator. Different emulsifier-initiator combinations were investigated: sodium lauryl sulfate (SDS) and cetyl trimethyl ammonium bromide (CTAB) as emulsifiers and 2,2′-azobis(2-amidinopropane)dihydrochloride (V50) and potassium peroxodisulfate (KPS) as initiators at a concentration of 3.7 mM. The filled test tubes were placed for 2 h in a water bath whose temperature was adjusted so that the temperature inside the test tubes reached 60 °C. All runs were repeated in 10 different test tubes to investigate reproducibility and to obtain representative data. Fig. 1.7 clearly shows a large difference between anionic and cationic IES, which is obviously due to the fact that surfaces in contact with water are generally negatively charged so that interactions with cationic species are much stronger (see above, Coehn's rule). As the polymerizations were started from aqueous solutions without any stirring only the reactor material causes the differences in C_{app}^0.

The importance of hydrodynamics on heterophase polymerizations has been shown already and the easiest way to investigate its influence are polymerizations with different stirrer speeds. This is the more interesting, as the influence on the particle size is not easy to predict (see Fig. 1.4). A first example is given in Fig. 1.8

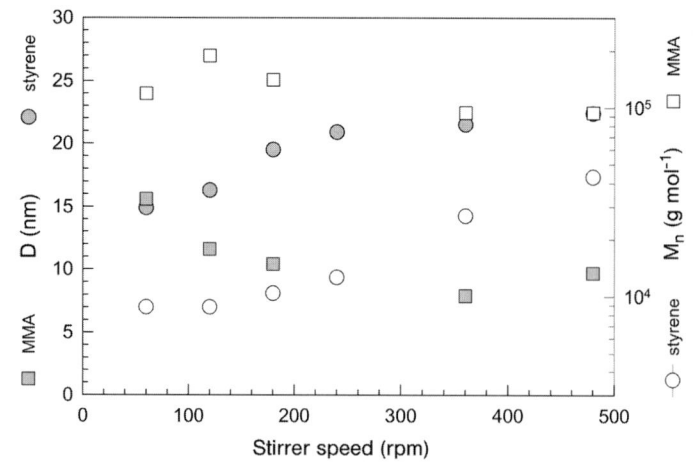

Fig. 1.8 The change of average particle size and number average molecular weight for polystyrene and poly(methyl methacrylate) latexes with change in stirrer speed during emulsion polymerization. The polymerization recipe is given in the text.

for emulsion polymerization of styrene and methyl methacrylate, respectively. The polymerizations were carried out batchwise in a reaction calorimeter (RM200S calorimeter, ChemiSens AB, Lund, Sweden), which consists of glass walls with stainless-steel bottom and stirrer. SDS (13.9 mM), styrene (115.2 mM) or methyl methylacrylate (119.8 mM) as monomer and KPS were polymerized at different stirrer speeds at 70 °C. The stirrer speed was varied between 60 and 480 rpm. At the lowest stirrer speed a bulky-free monomer phase was formed on top of the reaction mixture. The results shown in Fig. 1.8 reveal a clear difference between the two monomers, which is practically impossible to explain using prevalent opinions of the mechanism of emulsion polymerization. In the case of styrene both the average particle size and the molecular weight increase with increasing stirrer speed whereas for methyl methacrylate the opposite dependence is observed. Furthermore, reaction calorimetry reveals that the rate of styrene emulsion polymerization increases with increasing stirrer speed and that the rate for methyl methacrylate is almost unaltered. The behavior for styrene was reproduced with different hardware, that is in a completely different reactor (all-glass reactor) and also with a different operator [118]. For instance at stirrer speeds of 50 and 350 rpm the average particle size (from dynamic light scattering) and the number average molecular weight (from size exclusion chromatography) were found to be 9.1 ± 3 and 19 ± 2 nm and $3.7 \cdot 10^3 \pm 1.5 \cdot 10^3$ and $5.0 \cdot 10^4 \pm 8.7 \cdot 10^3$ g mol^{-1}, respectively. Each polymerization was repeated six times in order to get statistical reliability. This result is really remarkable as it shows that stable polystyrene particles with average diameters of about 10 nm are accessible by emulsion polymerization. Transmission electron microscopy pictures (see Fig. 1.9 A) proves the light scattering results as number and weight average diameter of 8.9 and 9.3 nm were calcu-

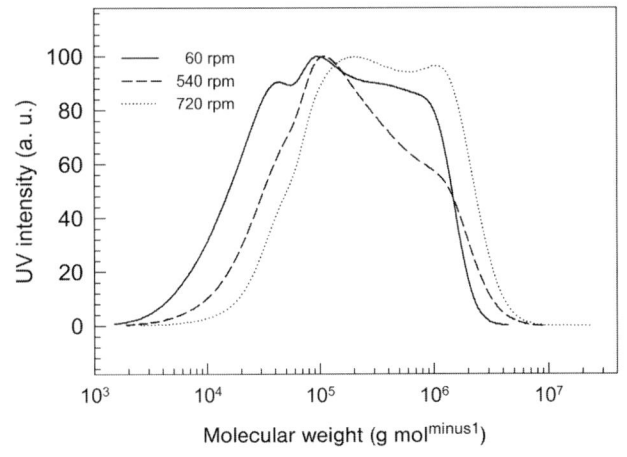

Fig. 1.12 The effect of stirrer speed on molecular weight distribution by size exclusion chromatography for polystyrene samples prepared by stabilizer-free emulsion polymerization. Polymerization recipe is given in the text.

to 360 rpm the average particle size increases but the uniformity decreases. At higher stirrer speed (540 and 720 rpm) the particles become unstable to coagulation and flocs of some 10 μm are formed, which are composed of partly fused smaller particles. This change in particle size and morphology is accompanied by changes in the molecular weight distribution as depicted in Fig. 1.12. The higher the stirrer speed the more the cumulative molecular weight distribution at the end of the polymerization is shifted towards higher values. Moreover, all these molecular weight distributions show three peaks or shoulders as characteristic features, independent of the stirrer speed.

In conclusion, the polymerization hardware has a strong influence on the particle and polymer properties and has to be carefully chosen in order to get the desired polymer latex properties. Moreover, a repetition of published prescriptions requires not only compliance with regard to recipe and purity of components but also with regard to polymerization hardware.

1.2.2.6 Some Practical Examples – Particle Design by Post-polymerization Treatment

Besides swelling, which can also be used to modify latex particles after polymerization via imbibing functional materials if necessary by means of a suitable swelling agent, controlled coagulation is another valid process for post-polymerization modification of latexes. There are two principal motivations: modification of the morphology of particles or an enlargement of their size. Both objectives are of at least potential practical interest. Such processes require close contact between latex particles and thus they always face the problem of stability control because the end product should still be a stable, colloidal polymer dispersion and not a completely coagulated system. Note, from the thermodynamic stability point of view, the process of decreasing the interfacial area is always favored.

For instance, the development of techniques for preparation of particles in the size range of about 10 μm has long been an intense research field in heterophase polymerizations. Potential areas for application of such particles are packing materials in chromatographic columns, impact modifiers, toners, pigments, or supports for biosynthesis and medical assays. Unfortunately, particles in this size range are obviously not accessible in single-stage heterophase polymerizations (see above) and, hence, controlled aggregation processes of chemically uniform latex particles became the focus of research. One of the first reports in open scientific literature was published in 1996 [220]. The authors describe a three-step synthetic route starting with emulsifier-free emulsion polymerization to produce sulfate-stabilized particles of about 500 nm in diameter. Aggregation was induced by charge neutralization with cetyl pyridinium chloride with stirring followed by addition of poly(vinyl alcohol) as steric stabilizer. In order to get particles with smooth surfaces, coalescence of the individual particles in the aggregates is induced at elevated temperature as a third step. This procedure resulted in dispersions with relatively high solids content and particles with smooth surfaces with sizes ranging between 6 and 12 μm after some "aggregated aggregates" had been easily removed by centrifugation.

In another more comprehensive study a similar procedure was investigated [221–223]. Again, the aggregation process of negatively charged latexes was induced by the addition of a cationic surfactant with stirring. If the concentration of the cationic surfactant is in the correct range, which causes only partial charge neutralization, the final aggregates are stable and still possess a negative surface potential. This indicates that the surfactant binding to the primary particles is a cooperative process and is also controlled by the hydrophobic interaction of the surfactant tails. The whole aggregation is described as a three-step process starting with formation of a gel-like network, which is broken up by shear forces and finally leads to smaller but more compact primary aggregates. These can further agglomerate into larger, secondary aggregates. Their size is mainly determined by the stirring speed and the surface chemistry of the particles (stabilization). Narrow size distribution of the secondary aggregates was only observed for latexes prepared in the presence of acrylic acid, which causes additional stabilization by the formation of an electrosteric layer.

The second of the above principal methods involves controlled aggregation between chemically different particles. Heterocoagulation is the technical term, meaning in a general sense the aggregation of oppositely charged particles that are either of the same or different sizes and are composed of either the same or different materials. Again, combined stability with ionic and nonionic stabilizers is advantageous in order to ensure stability throughout the whole process. During heterocoagulation the charge neutralization is due to oppositely charged particles and in principle various morphologies can be obtained, such as formation of a loosely aggregated network, or engulfment of smaller particles by the larger ones, or encapsulation of larger particles by the smaller ones.

Heterocoagulation means the formation of composite particles if chemically different polymeric particles are participating. A thermodynamic quantification of the heterocoagulation process is possible in a way described by Ottewill and Waters [224–226]. Assuming the system consists of water as a continuous phase (phase 2), small particles (phase 3), and oppositely charged larger particles (phase 1) then application of the principles of minimization of free energy (the reference state is characterized by both kinds of particles being separated) and introducing volume fractions leads to Eq. (10) as the criterion that defines conditions for engulfment of the smaller particles by the larger ones. The σ and ϕ terms denote the interfacial tensions between the phases and the particle volume fractions, respectively.

$$\frac{\sigma_{2,3} - \sigma_{1,3}}{\sigma_{1,2}} > \frac{1 - \phi_1^{2/3}}{\phi_3^{2/3}} \tag{10}$$

Equation (10) reveals the influence of several variables on the thermodynamics of engulfment. The process is the more favored the more hydrophobic the smaller particles, the more hydrophilic the larger particles, the less the interfacial tension between the two polymers, and the larger the volume fraction of the smaller particles that are forming the core in the composite particles. The applicability of this kind of prediction as well as of the whole process to prepare composite latex particles has been successfully demonstrated [105, 224–226].

Of course the formation of the thermodynamically favored morphology as described by Eq. (10) requires that mobility of the polymer molecules is high enough to rearrange. To achieve this chain mobility necessitates temperatures above the glass transition temperatures of the interacting polymers. Otherwise kinetically frozen morphologies result, which do not represent the free energy minimum and such particles are subjected to permanent inherent stress to change its morphology. Note, thermodynamics acts as long as the particles exist and indeed some authors describe changes in particle morphology during storage of polymer dispersions [227]. The first examples to use heterocoagulation to control particle morphology were, to the best of the present author's knowledge, published in a series of papers by Okubo [228–230] and were carried out under this kind of "frozen" conditions. The aim of these investigations was the preparation of non-spherical particle morphologies with enlarged interfaces by coagulation of smaller cationically charged particles made of styrene copolymerized with methacryloyloxyethyl trimethyl ammonium chloride as surfmer onto larger anionically charged particles made of methyl methacrylate-ethyl acrylate-methacrylic acid terpolymers at temperatures below the glass transition temperatures [230].

1.3
Conclusions, New Developments, and Perspectives

It has been a long but successful journey from the first attempts to reproduce mother nature's latex particles at the beginning of the last century to the present state of latex particle technology. Currently latex particles can be tailored to very specific needs and hence they find applications in almost all areas of our daily life as well as in all fields of applied and basic research.

Latex particles can be modified in various ways that affect their polymeric or colloidal and interfacial or volume properties. This diversity of modification possibilities accounts for the ongoing success of latex particles and the way in which new areas of application are constantly being found.

Possibly the most promising future development is in the field of combination of different materials on the various size scales accessible by latex particles. In this context, the term *different materials* does not primarily mean the modification of properties by copolymerization or layered morphologies composed of different polymers but rather combinations of polymers with other, non-polymeric materials such as all kinds of inorganic or organic compounds. The expectation is to take advantage of the combination of different material on the nanometer scale [231]. There are already several examples known in which heterophase polymerization has been carried out in the presence of preformed foreign particles. Examples are the incorporation of magnetite [232–234], calcium carbonate [235, 236], silica or metal oxides [237–239], and carbon black [240].

Recently experiments have been carried out in the present author's laboratory to prepare inorganic-polymeric composite nanoparticles by "joint-nucleation," that is the combination of precipitation and emulsion polymerization simultaneously in the same reactor. If the nucleation conditions of the inorganic and polymeric particles are matched in a proper way the in situ formation of nanocomposites is possible [241, 242]. Fig. 1.13 shows examples of the formation of calcium carbonate composite particles with poly(methyl methacrylate) and polystyrene, respectively. The composites exhibit a different morphology thus indicating the mutual influences of the nucleation processes. Poly(methyl methacrylate) interacts much more strongly with calcium carbonate as during the experiments hydrolysis to methacrylic acid and methanol takes place [243]. The methacrylic acid groups are "calciophilic" and cause a stronger interaction than in the case of polystyrene. The calcium carbonate is located both inside and at the interface of the composite particles. Thus, the calcium carbonate stabilizes the composite particles (Fig. 1.13, micrographs A1, A3), that is they retain their spherical shape in the electron beam and do not flow together as the neat poly(methyl methacrylate) particles do (micrograph A3). In contrast, tiny calcium carbonate particles are distributed inside the polystyrene composite particles (Fig. 1.13, micrograph B2).

Another possibility for changing particle morphology, which is challenging future developments as well, is the formation of anisotropic latex particles. A prominent example is the formation of nonsymmetrical latex particles especially with respect to interfacial properties such as functionality or charge distribution, or even charge

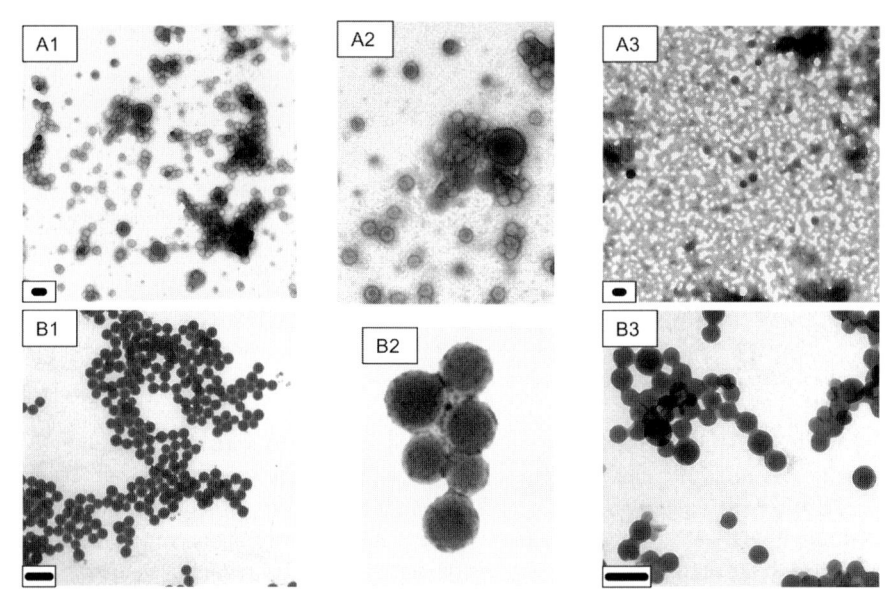

Fig. 1.13 Transmission electron micrographs of calcium nanocomosite particles with poly-(methyl methacrylate) and polystyrene (A1, A2 and B1, B2, respectively) and the corresponding homopolymer particles (A3, B3); A2 and B2 are magnifications of A1 and B1, respectively. Scale bars represent for A1, A3, 100 nm; B1, 300 nm; B3, 200 nm. For preparation conditions see [241–243].

sign. Such particles are of special interest, for instance as models for cells in living organisms. During heterophase polymerization thermodynamic effects force the particle to be spherical and to have symmetrical surface properties. However, if it were possible to prepare particles composed of two hemispheres with different properties, such as charge distribution or even sign or chemical properties, interesting applications and fascinating further experiments are imaginable. For instance, different charges should lead to particles with high dipole moments with interesting possibilities for their alignment, or different material properties might be used for chemical modifications at only one hemisphere. One approach to get such particles that has recently been published is the physical or chemical modification of homogeneous latex particles, which are adsorbed at interfaces, utilizing the Langmuir-Blodgett technique [244, 245]. Although on the one hand this is an intelligent and interesting approach, on the other hand it is a tedious procedure to modify larger amounts by this route. Another approach is described in [21], which is part of the youngest branch in the family of polymer dispersions that is via the formation of polymeric colloidal complexes (see above and Fig. 1.1). If such complexes are formed by interaction of two properly chosen double hydrophilic copolymers, in which both the ionic and the uncharged part are different in all respects (charge sign and chemistry), the resulting particles are composed of two different polymeric materials at the interface. Fig. 1.14 shows an example of this kind of unsymmetrical particles as described in [21]. These particles are made of poly(ethylene glycol)-b-

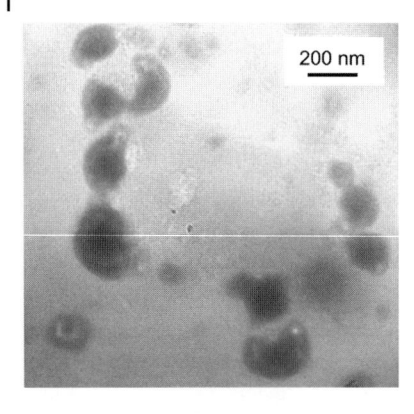

Fig. 1.14 Transmission electron micrograph of polymeric colloidal complexes made of poly-(ethylene glycol)-*b*-poly(diallyldimethyl ammonium chloride) and poly(*N*-isopropyl acrylamide)-*co*-poly(2-acrylamidopropane sulfonat), which is a statistical copolymer prepared by radical polymerization. The polymerization recipe is given in the text.

poly(diallyldimethyl ammonium chloride) and poly(N-isopropyl acryl amide)-co-poly(2-acrylamidopropane sulfonate) that is a statistical copolymer prepared by radical polymerization at 70 °C with the following recipe: 100 g of water, 11 g of N-diallyldimethyl ammonium chloride, 4.5 g of 2-acrylamidopropane sulfonate, 0.3 g of 2,2′-azobis[2-methyl-N-(2-hydroxyethyl)-propionamide] as initiator (see [21]). The formation of unsymmetrical particles by this procedure requires that the polymers at the interface do form separate phases, which is likely to occur because of the incompatibility between the polymers.

Some very first steps in another approach in the field of latex particle synthesis that, as the authors believe, might be of some future interest are described in [21, 109]. This is the direct formation of block copolymers by heterophase polymerization, as this approach allows the formation of unique latex particle morphologies.

1.4
Acknowledgements and Apologia

I would like to acknowledge deeply the co-operation of former PhD students, technical staff, and colleagues in the Max Planck Institute of Colloids and Interfaces. Particular thanks are due to Mrs Rosemarie Rafler and Mrs Sylvia Pirok for technical assistance over the last almost 25 years. I am grateful to Mrs Rona Pietschke and Dr Jürgen Hartmann for numerous electron microscopy pictures of latex particles. I wish to thank the Max Planck Society and the Max Planck Institutes of Colloids and Interfaces for allowing me to use the synthetic and analytical equipment needed as well as for providing the facilities to carry out literature searches electronically. I express my special gratitude to Mrs Dorothea Stscherbina, the librarian of the institute, for her efforts to provide copies of almost all papers requested.

The topic "latex particles" is difficult to cover in a limited space. Nevertheless, I have tried to consider the main aspects of the preparation and properties of latex particles from the very beginning to current developments and possible future developments. It is, however, a personal view of this fascinating area of colloid science and I offer my sincere apologies to all of the colleagues whose excellent contributions to the field I had to omit.

1.5 References

1 K. Cornish, D.J. Silber, O.-K. Grosjean, N. Goodman, *J. Nat. Rubb. Res.* **1993**, *8*, 275–285.
2 H. Mooibroek, K. Cornish, *Appl. Microbiol. Biotechnol.* **2000**, *53*, 355–365.
3 D.C. Blackley, *Polymer Lattices*, Vol. 1, 2nd edn, Chapman & Hall, London, **1997**, pp. 1–2.
4 D. Hosler, S.L. Burkett, M.J. Tarkanian, *Science* **1999**, *284*, 1988–1991.
5 H.-G. Elias, *Große Moleküle*, Springer, Berlin Heidelberg New York Tokyo, **1986**, pp. 5–7.
6 C. de la Condamine, F. Fresneau, *Memoires de l'Academie Royale des Sciences*, **1751**, pp. 488–510.
7 F. Hofmann, *Zeitschr. Angew. Chem. Zentralblatt Techn. Chem.* **1912**, *25*, 1462–1467.
8 K. Gottlob, *Gummi-Zeitung* **1912**, *26*, 1546–1548.
9 F.J. Pond, *J. Amer. Chem. Soc.* **1914**, *XXXVI*, 165–199.
10 German Patent, *DRP 254 672* (patented from December 12, **1912** on) to Farbenfabriken former F. Bayer & Co. in Leverkusen/Köln and in Elberfeld.
11 German Patent *DRP 558 890* (patented from January 9, **1927** on) to I.G. Farbenindustrie Akt.-Ges. in Frankfurt a.M..
12 US Patent 1 732 795 (patented October 22, **1929**; application filed September 13, 1927), R.P. Dinsmore.
13 H. Fikentscher, H. Gerrens, Schuller, *Angew. Chem.* **1960**, *72*, 856–867.
14 G.Q. Chen, G. Zhang, S.J. Park, S.Y. Lee, *Appl. Microbiol. Biotechnol.* **2001**, *57*, 50–55.
15 S. Förster, M. Antonietti, *Adv. Mater.* **1998**, *10*, 195–217.
16 K. Mortensen, *Polym. Adv. Technol.* **2001**, *12*, 2–22.
17 A. Milchev, A. Bhattacharya, K. Binder, *Macromolecules* **2001**, *34*, 1881–1893.
18 K. Tauer, *Polym. Adv. Technol.* **1995**, *6*, 435–440.
19 L. Rosengarten, K. Tauer, *Ber. Bunsenges. Phys. Chem.* **1996**, *100*, 734–737.
20 K. Tauer, M. Antonietti, L. Rosengarten, H. Müller, *Macromol. Chem. Phys.* **1998**, *199*, 897–908.
21 K. Tauer, V. Khrenov, *Macromol. Symp.* **2002**, *179*, 27–52.
22 F. Caruso, R.A. Caruso, H. Möhwald, *Science* **1998**, *282*, 1111–1114.
23 E. Donath, G.B. Sukhorukov, F. Caruso, S.A. Davis, H. Möhwald, *Angew. Chem. Int. Ed.* **1998**, *37*, 2201–2205.
24 F. Caruso, H. Möhwald, *J. Amer. Chem. Soc.* **1999**, *121*, 6039–6046.
25 F. Caruso, *Chem. A Eur. J.* **2000**, *6*, 413–419.
26 M.D.C. Topp, I.H. Leunen, P.J. Dijkstra, K. Tauer, C. Schellenberg, J. Feijen, *Macromolecules* **2000**, *33*, 4986–4988.
27 T.K. Bronich, A.V. Kabanov, V.A. Kabanov, K. Yu, A. Eisenberg, *Macromolecules* **1997**, *30*, 3519–3525.
28 E.A. Lysenko, T.K. Bronich, A. Eisenberg, V.A. Kabanov, A.V. Kabanov, *Macromolecules* **1998**, *31*, 4511–4515.
29 E.A. Lysenko, T.K. Bronich, A. Eisenberg, V.A. Kabanov, A.V. Kabanov, *Macromolecules* **1998**, *31*, 4516–4519.
30 A. Zintchenko, H. Dautzenberg, K. Tauer, V. Khrenov, *Langmuir* **2002**, *18*, 1386–1393.
31 D. Distler (ed.) *Wäßrige Polymerdispersionen: Synthese, Eigenschaften, Anwendungen*, Wiley-VCH, Weinheim, **1999**.
32 D. Urban, K. Takamura (eds.) *Polymer Dispersions and Their Industrial Applications*, Wiley-VCH, New York, **2002**.
33 G. Vidotto, A. Crosato-Arnaldi, G. Talamini, *Makromol. Chem.* **1969**, *122*, 91–104.
34 M. Carenza, G. Palma, *Eur. Polym. J.* **1985**, *21*, 41–47.
35 A. Guyot, in J.C. Salamone (ed in chief) *Polymeric Materials Encyclopedia*, Vol. 9, CRC Press, Boca Raton, **1996**, pp. 7228–7237.
36 H.D.H. Stöver, in J.C. Salamone (ed in chief), *Polymeric Materials Encyclopedia*, Vol. 9, CRC Press, Boca Raton, **1996**, pp. 7237–7238.
37 E.A. Grulke, in J.I. Kroschwitz (exc. ed.) *Encyclopedia Polymer Science and Engineering*, Vol. 16, 2nd edn, John Wiley & Sons, New York, **1989**, pp. 443–473.

38 H. Bieringer, K. Flatau, D. Reese, Angew. Makromol. Chem. **1984**, *123/124*, 307–334.
39 H. G. Yuan, G. Kalfas, W. H. Ray, *J. Macromol. Sci. – Rev. Macromol. Chem. Phys.* **1991**, *C31*, 215–299.
40 J. E. Puig, E. Mendizabal, in J. C. Salamone (ed in chief) *Polymeric Materials Encyclopedia*, Vol. 10, CRC Press, Boca Raton, **1996**, pp. 8215–8220.
41 E. Vivaldo-Lima, P. E. Wood, A. E. Hamielec, A. Penlidis, *Ind. Eng. Chem.* **1997**, *36*, 939–965.
42 P. J. Dowding, B. Vincent, *Coll. Surf. A: Physicochem. Eng. Asp.* **2000**, *161*, 259–269.
43 D. J. Walbridge, in G. C. Eastmond, A. Ledwith, S. Russo, P. Sigwalt (volume eds.) *Chain Polymerization II*, Vol. 4, Chapter 15, Pergamon Press, Oxford, **1989**, pp. 243–260.
44 H. D. H. Stöver, in J. C. Salamone (ed in chief) *Polymeric Materials Encyclopedia*, Vol. 3, CRC Press, Boca Raton, **1996**, pp. 1900–1905.
45 J. L. Cawse, in P. A. Lovell, M. S. El-Aasser (eds.) *Emulsion Polymerization and Emulsion Polymers*, John Wiley & Sons, Chichester, **1997**, pp. 743–761.
46 W.-H. Li, K. Li, H. D. H. Stöver, *J. Polym. Sci. Part A: Polym. Chem.* **1999**, *37*, 2295–2303.
47 K. E. J. Barrett (ed) *Dispersion Polymerization in Organic Media*, John Wiley & Sons, London, **1975**.
48 M. F. Cunningham, *Polym. React. Eng.* **1999**, *7*, 231–257.
49 P. C. Mork, B. Saethre, J. Ugelstad in J. C. Salamone (ed. in chief) *Polymeric Materials Encyclopedia*, Vol. 10, CRC Press, Boca Raton, **1996**, pp. 8559–8566.
50 M. J. Bunten, in J. I. Kroschwitz (exc. ed.) *Encyclopedia Polymer Science and Engineering*, Vol. 17, 2nd edn, John Wiley & Sons, New York, **1989**, pp. 329–376.
51 K. Tauer, in J. Texter (ed.) *Reactions and Synthesis in Surfactant Systems*, Marcel Dekker, New York, **2001**, pp. 429–453.
52 G. W. Poehlein, in J. I. Kroschwitz (exc. ed.) *Encyclopedia Polymer Science and Engineering*, Vol. 6, 2nd edn, John Wiley & Sons, New York, **1986**, pp. 1-51.
53 G. Markert, *Angew. Makromol. Chem.* **1984**, *123/124*, 285–306.

54 Q. Wang, S. Fu, T. Yu, *Progr. Polym. Sci.* **1994**, *19*, 703–753.
55 P. A. Lovell, M. S. El-Aasser (eds.) *Emulsion Polymerization and Emulsion Polymers*, John Wiley & Sons, Chichester, **1997**.
56 S. Sajjadi, B. W. Brooks, *Chem. Eng. Sci.* **2000**, *55*, 4757–4781.
57 H. Kawaguchi, in T. Sugimoto (ed.) *Fine Particles: Synthesis, Characterization and Mechanism of Growth*, Marcel Dekker, New York, **2000**, pp. 592–608.
58 B. Saethre, P. C. Mork, J. Ugelstad, *J. Polym. Sci. A: Polym. Chem.* **1995**, *33*, 2951–2959.
59 I. Aizpurua, J. Amalvy, M. J. Barabdiaran, J. C. de La Cal, J. M. Asua, *Macromol. Symp.* **1996**, *111*, 121–131.
60 E. D. Sudol, M. S. El-Aasser, in P. A. Lovell, M. S. El-Aasser (eds.) *Emulsion Polymerization and Emulsion Polymers*, John Wiley & Sons, Chichester, **1997**, pp. 699–722.
61 Y. Luo, J. Tsavala, J. F. Schork, *Macromolecules* **2001**, *34*, 5501–5507.
62 I. Capek, C. S. Chen, *Adv. Polym. Sci.* **2001**, *155*, 101–165.
63 M. Antonietti, K. Landfester, *Progr. Polym. Sci.* **2001**, *27*, 689–757.
64 J. M. Asua, *Progr. Polym. Sci.* **2002**, *27*, 1283–1346.
65 F. Candau, in J. I. Kroschwitz (exc. ed.) *Encyclopedia Polymer Science and Engineering*, Vol. 9, 2nd edn, John Wiley & Sons, New York, **1989**, pp. 718–724.
66 I. Capek, P. Potisk, *Eur. Polym. J.* **1995**, *31*, 1269–1277.
67 J. Santhanalakshmi, K. Anandhi, *J. Appl. Polym. Sci.* **1996**, *60*, 293–304.
68 L. M. Gan, C. H. Chew, in J. C. Salamone (ed in chief) *Polymeric Materials Encyclopedia*, Vol. 10, CRC Press, Boca Raton, **1996**, pp. 4321–4331.
69 M. Antonietti, in J. C. Salamone (ed in chief) *Polymeric Materials Encyclopedia*, Vol. 10, CRC Press, Boca Raton, **1996**, pp. 4331–4333.
70 M. Rabelero, M. Zacarias, E. Mendizabal, J. E. Puig, J. M. Dominguez, I. Katime, *Polym. Bull.* **1997**, *38*, 695–700.
71 N. Girard, Th. F. Tadros, A. I. Bailey, *Coll. Polym. Sci.* **1998**, *276*, 999–1009.
72 F. Candau, M. Pabon, J.-Y. Anquetil, *Coll. Surf. A Physicochem. Eng. Asp.* **1999**, *153*, 47–59.

73 N. Sosa, R. D. Peralta, R. G. Lopez, L. F. Ramos, I. Katime, C. Cesteros, E. Mendizabal, J. E. Puig, *Polymer* **2001**, *42*, 6923–6928.
74 C. C. Co, P. Cotts, S. Burauner, R. de Vries, E. W. Kaler, *Macromolecules* **2001**, *34*, 3245–3254.
75 F. Reynolds, K. Jun, Y. Li, *Macromolecules* **2001**, *34*, 165–170.
76 R. G. Gilbert, *Emulsion Polymerization: A Mechanistic Approach*, Academic Press, London, **1995**.
77 Freedonia, Study 1457, September **2001**.
78 D. Hunkeler, F. Candau, C. Pichot, A. E. Hamielec, T. Y. Xie, J. Barton, V. Vaskova, J. Guillot, M. V. Dimonie, K.-H. Reichert, *Adv. Polym. Sci.* **1994**, *112*, 115–133.
79 W. D. Bancroft, *J. Phys. Chem.* **1912**, *16*, 177–233.
80 W. C. Griffin, *J. Soc. Cosm. Chem.* **1949**, *1*, 311–326.
81 K. Tauer, in K. Holmberg (ed.) *Handbook of Applied Surface and Colloid Chemistry*, John Wiley & Sons, New York, **2001**, pp. 8175–8200.
82 M. S. El-Aasser, C. M. Miller, in J. M. Asua (ed.) *Polymeric Dispersions: Principles and Applications*, Kluwer Academic, Dordrecht, **1997**, pp. 109–126.
83 J. Xu, V. L. Dimonie, E. D. Sudol, M. S. El-Aasser, *J. Appl. Polym. Sci.* **1998**, *69*, 965–975.
84 M. Li, M. Jiang, C. Wu, *J. Polym. Sci. Part B: Polym. Phys.* **1997**, *35*, 1539–1599.
85 X. J. Xu, P. Y. Chow, L. M. Gan, *J. Nanosci. Nanotechnol.* **2002**, *2*, 61–65.
86 T. Rager, W. H. Meyer, G. Wegner, K. Mathauer, W. Mächtle, W. Schrof, D. Urban, *Macromol. Chem. Phys.* **1999**, *200*, 1681–1691.
87 M. Gersi, H. Schuch, D. Urban, in J. E. Glass (ed.) *Associative Polymers in Aqueous Media*, ACS Symp. Ser. 765, ACS, Washington, DC, **2000**, pp. 37–51.
88 S. Förster, T. Plantenberg, *Angew. Chem. Int. Ed.* **2002**, *41*, 688–714.
89 J. Ding, G. Liu, *Macromolecules* **1999**, *32*, 8413–8420.
90 R. A. Backhaus, *Isr. J. Botany* **1985**, *34*, 283–293.
91 F. Tiarks, K. Landfester, M. Antonietti, *J. Polym. Sci. Part A: Polym. Chem.* **2001**, *39* 2510–2524.

92 W. Stöber, A. Fink, E. Bohn, *J. Coll. Interf. Sci.* **1968**, *26*, 62–69.
93 S. Slomkowski, S. Sosnowski, M. Gadzinowski, *Coll. Surf. A: Physicochem. Eng. Asp.* **1999**, *153*, 111–118.
94 D. D. Hile, M. V. Pishko, *J. Polym. Sci. Part A: Polym. Chem.* **2001**, *39*, 562–570.
95 U. S. Patent 2 891 920, **1959**, J. F. Hyde, J. R. Wehrly.
96 D. R. Weyenberg, D. E. Findlay, J. Cekada, Jr., A. E. Bey, *J. Polym. Sci. Part C* **1969**, *27 (PC)*, 27–34.
97 M. A. Awan, V. L. Dimonie, M. S. El-Aasser, *J. Polym. Sci. Part A: Polym. Chem.* **1996**, *34*, 2633–2649.
98 Tomov, J.-P. Broyer, R. Spitz, *Macromol. Symp.* **2000**, *150*, 53–58.
99 B. Manders, L. Sciandrone, G. Hauck, M. O. Kristen, *Angew. Chem. Int. Ed.* **2001**, *40*, 4006–4007.
100 K. Landfester, H.-P. Hentze, in J. Texter (ed.) *Reactions and Synthesis in Surfactant Systems*, Chapter 23, Marcel Dekker, New York, **2001**, pp. 471–499.
101 J. Snuparek, *Progr. Org. Coat.* **1996**, *29*, 225–233.
102 D. I. Lee, *Makromol. Chem. Macromol. Suppl.* **1990**, *33*, 117–131.
103 R. Zimehl, G. Lagaly, J. Ahrens, *Coll. Polym. Sci.* **1990**, *268*, 924–933.
104 D. H. Napper, *Polymeric Stabilization of Colloidal Dispersions*, Academic Press, London, **1983**.
105 R. H. Ottewill, in P. A. Lovell, M. S. El-Aasser (eds.) *Emulsion Polymerization and Emulsion Polymers*, J. Wiley & Sons, Chichester, **1997**, pp. 59–121.
106 I. Piirma, *Polymeric Surfactants*, Surfactant Science Series 42, Marcel Dekker, New York, **1992**, pp. 1–16.
107 C. Bechinger, H.-H. von Grünberg, P. Leiderer, *Physikal. Blätter* **1999**, *55*, 53–56.
108 K. Tauer, H. Müller, *Coll. Polym. Sci.* **2003**, *281*, 52–65.
109 K. Tauer, H. Müller, L. Rosengarten, K. Riedelsberger, *Coll. Surf. A: Physicochem. Eng. Asp.* **1999**, *153*, 75–88.
110 P. Pincus, *Macromolecules* **1991**, *24*, 2912–2919.
111 L. Leemans, R. Fayt, Ph. Teyssie, N. C. de Jaeger, *Macromolecules* **1991**, *24*, 5922–5925.

112 K. Tauer, I. Kühn, *Macromolecules* **1995**, *28*, 2236–2239.
113 K. Tauer, I. Kühn, in J.M. Asua (ed.) *Polymeric Dispersions: Principles and Applications*, Kluwer Academic, Dordrecht, **1997**, pp. 49–65.
114 K. Tauer, R. Deckwer, I. Kühn, C. Schellenberg, *Coll. Polym. Sci.* **1999**, *277*, 607–626.
115 A. Laaksonen, V. Talanquer, D.W. Oxtoby, *Annu. Rev. Phys. Chem.* **1995**, *46*, 489–524.
116 D. Kashiev, *Nucleation: Basic Theory with Applications*, Butterworth-Heinemann, Oxford, **2000**.
117 K. Binder, in P. Haasen (ed.) *Material Science and Technology*, Vol. 5, VCH, Weinheim, **1991**, pp. 405–471.
118 K. Tauer, C. Schellenberg, A. Zimmermann, *Macromol. Symp.* **2000**, *150*, 1–12.
119 S. Shen, E.D. Sudol, M.S. El-Aasser, *J. Polym. Sci. Part A: Polym. Chem.* **1994**, *32*, 1097–1100.
120 R. Becker, A. Hashemzadeh, H. Zecha, *Macromol. Symp.* **2000**, *151*, 567–573.
121 F. Chu, C. Grillat, A. Guyot, *J. Appl. Polym. Sci.* **1998**, *70*, 2667–2677.
122 C. Tang, F. Chu, *J. Appl. Polym. Sci.* **2001**, *82*, 2352–2356.
123 J. Ugelstad, H.R. Mfutakamba, P.C. Mork, T. Ellingsen, A. Berge, R. Schmid, L. Holm, A. Jorgendahl, F.K. Hansen, K. Nustad, *J. Polym. Sci.: Polym. Symp.* **1985**, *72*, 225–240.
124 J. Ugelstad, P. Stenstad, L. Kilaas, W.S. Prestvik, A. Rian, K. Nustad, R. Herje, A. Berge, *Macromol. Symp.* **1996**, *101*, 491–500.
125 R. Klein, G. Nägele, *Curr. Op. Coll. Interf. Sci.* **1996**, *1*, 4–10.
126 T. Okubo, K. Kiriyama, *Ber. Bunsenges. Physikal. Chem.* **1996**, *100*, 849–856.
127 T. Palberg, *J. Phys. Cond. Matter* **1999**, *11*, R323–R360.
128 C. Sinn, R. Niehuser, E. Overbeck, T. Palberg, *Particle Syst. Charact.* **1999**, *16*, 95–101.
129 U. Gasser, E.R. Weeks, A. Schofield, P.N. Pusey, D.A. Weitz, *Science* **2001**, *292*, 258–262.
130 Y. Xia, B. Gates, Y. Yin, Y. Lu, *Adv. Mater.* **2000**, *12*, 693–713.
131 M.C. Wilkinson, J. Hearn, P.A. Steward, *Adv. Coll. Interf. Sci.* **1999**, *81*, 77–165.
132 T. Matsumoto, A. Ochi, *Kobunshi Kagaku (Tokyo)* **1965**, *22*, 481–487.
133 A. Kotera, K. Furusawa, Y. Takeda, *Kolloid-Z. u. Z. Polymere* **1970**, *239*, 677–681.
134 J.W. Goodwin, J. Hearn, C.C. Ho, R.H. Ottewill, *Br. Pol. J.* **1973**, *5*, 347–362.
135 J.W. Goodwin, R.H. Ottewill, R. Pelton, *Coll. Polym. Sci.* **1979**, *257*, 61–69.
136 G. Tuin, A.C. I.A. Peters, A. J.G. van Diemen, H.N. Stein, *J. Coll. Interf. Sci.* **1993**, *158*, 508–510.
137 H. Fritz, M. Maier, E. Bayer, *J. Coll. Interf. Sci.* **1997**, *195*, 272–288.
138 J. Xu, P. Li, C. Wu, *J. Polym. Sci. Part A: Polym. Chem.* **1999**, *37*, 2069–2074.
139 J.W. Goodwin, J. Hearn, C.C. Ho, R.H. Ottewill, *Coll. Polym. Sci.* **1974**, *252*, 464–471.
140 C.-S. Chern, C.-H. Lin, *Polym. J.* **1995**, *27*, 1094–1103.
141 J.Q. Li, R. Salovey, *J. Polym. Sci. Part A: Polym. Chem.* **2001**, *38*, 3181–3187.
142 K. Tauer, *Macromolecules* **1998**, *31*, 9390–9391.
143 K. Tauer, R. Deckwer, *Acta Polymerica* **1998**, *49*, 411–416.
144 G.T.D. Shouldice, G.A. Vandezande, A. Rudin, *Eur. Polym. J.* **1994**, *30*, 179–183.
145 H.-S. Chang, S.-A. Chen, *Makromol. Chem., Rapid Commun.* **1987**, *8*, 297–304.
146 G.-L. Tang, M.-D. Song, G.-J. Hao, T.-Y. Guo, B.-Hua Zhang, *J. Appl. Polym. Sci.* **2001**, *79*, 21–28.
147 C.E. Reese, S. Asher, *J. Coll. Interf. Sci.* **2002**, *248*, 41–46.
148 A. Martin-Rodriguez, M.A. Cabrerizo-Vilchez, R. Hildago-Álvarez, *Coll. Surf. A: Physicochem. Eng. Asp.* **1996**, *108*, 263–271.
149 S.-A. Chen, H.-S. Chang, *J. Polym. Sci. Part A: Polym. Chem.* **1990**, *28*, 2547–2561.
150 M. Okubo, A. Yamada, S. Shibo, K. Nakamae, T. Matsumoto, *J. Appl. Polym. Sci.* **1981**, *26*, 1675–1679.
151 A. Guyot, K. Tauer, in J. Texter (ed.) *Reactions and Synthesis in Surfactant Systems*, Marcel Dekker, New York, **2001**, pp. 547–575.

152 A. Guyot, *Macromol. Symp.* **2002**, *179*, 105–132.
153 A. Guyot, K. Tauer, *Adv. Polym. Sci.* **1994**, *111*, 43–65.
154 A. Guyot, K. Tauer, J. M. Asua, S. Van Es, C. Gauthier, A. C. Helgren, D. C. Sherrington, A. Montoya-Goni, M. Sjöberg, O. Sindt, F. Vidal, M. Unzué, H. A. S. Schoonbrood, E. Schipper, P. Lacroix-Desmazes, *Acta Polymerica* **1999**, *50*, 57–66.
155 K. Tauer, in J. M. Asua (ed.) *Polymeric Dispersions: Principles and Applications*, Kluwer Academic, Dordrecht, **1997**, pp 463–476.
156 K. Tauer, *Polymer News* **1995**, *20*, 342–347.
157 D. Horak, *Acta Polymerica* **1996**, *47*, 20–28.
158 R. C. Backhus, R. C. Williams, *J. Appl. Phys.* **1948**, *19*, 1186–1187.
159 R. C. Backhus, R. C. Williams, *J. Appl. Phys.* **1949**, *20*, 224–225.
160 I. M. Krieger, F. M. O'Neill, *J. Amer. Chem. Soc.* **1968**, *90*, 3114–3120.
161 G. D. Scott, *J. Appl. Phys.* **1949**, *20*, 417–418.
162 J. H. L. Watson, W. L. Grube, *J. Appl. Phys.* **1952**, *23*, 793–798.
163 T. Alfrey, Jr., E. B. Bradford, J. W. Vanderhoff, G. Oster, *J. Opt. Soc. Amer.* **1954**, *44*, 603–609.
164 E. B. Bradford, J. W. Vanderhoff, *J. Appl. Phys.* **1955**, *26*, 864–871.
165 R. M. Fitch, *Br. Polym. J.* **1973**, *5*, 467–483.
166 J. W. Vanderhoff, J. F. Vitkuske, E. B. Bradford, T. Alfrey, Jr., *J. Polym. Sci.* **1956**, *XX*, 225–234.
167 G. W. Poehlein, J. W. Vanderhoff, *J. Polym. Sci. Polym. Chem. Ed.* **1973**, *11*, 447–452.
168 J. Ugelstad, H. Flogstad, F. K. Hansen, T. Ellingsen, *J. Polym. Sci.: Symp.* **1973**, *42*, 473–485.
169 P. C. Mork, *J. Polym. Sci. Part A: Polym. Chem.* **1995**, *33*, 2305–2316.
170 K.-H. Goebel, H.-J. Schneider, W. Jaeger, G. Reinisch, *Acta Polymerica* **1982**, *33*, 49–62.
171 K. Tauer, G. Reinisch, H. Gajewski, I. Müller, *J. Macromol. Sci. Chem.* **1991**, *A28*, 431–460.
172 R. H. Ottewill, J. N. Shaw, *Kolloid-Z. u. Z. Polymere* **1967**, *215*, 161–166.
173 R. H. Ottewill, J. N. Shaw, *Kolloid-Z. u. Z. Polymere* **1967**, *218*, 34–40.
174 J. Hearn, R. H. Ottewill, J. N. Shaw, *Br. Polym. J.* **1970**, *2*, 116–120.
175 J. B. Smitham, D. V. Gibson, D. H. Napper, *J. Coll. Interf. Sci.* **1973**, *45*, 211–214.
176 D. Kumar, G. B. Butler, *Particulate Sci. Technol.* **1997**, *14*, 315–326.
177 K. Tauer, I. Kühn, H. Kaspar, *Progr. Coll. Polym. Sci.* **1996**, *101*, 30–37.
178 S. Gu, M. Konno, *J. Chem. Eng., Japan* **1997**, *30*, 742–744.
179 S. Gu, T. Mogi, M. Konno, *J. Coll. Interf. Sci.* **1998**, *207*, 113–118.
180 T. Corner, *Coll. Surf.* **1981**, *3*, 119–129.
181 K. Kobayashi, M. Senna, *J. Appl. Polym. Sci.* **1992**, *46*, 27–40.
182 C. K. Ober, K. P. Lok, *Macromolecules* **1987**, *20*, 268–273.
183 W. Yang, J. Hu, Z. Tao, L. Li, C. Wang, S. Fu, *Coll. Polym. Sci.* **1999**, *277*, 446–451.
184 D. Horak, *J. Polym. Sci. Part A: Polym. Chem.* **1999**, *37*, 3785–3792.
185 C.-H. Suen, H. Morawetz, *Macromolecules* **1984**, *17*, 1800–1803.
186 M. Okubo, K. Ikegami, Y. Yamamoto, *Coll. Polym. Sci.* **1989**, *267*, 193–200.
187 B. Verrier-Charleux, C. Graillat, Y. Chevalier, C. Pichot, A. Revillon, *Coll. Polym. Sci.* **1991**, *269*, 398–405.
188 M. D. Bale, S. J. Danielson, J. L. Daiss, K. E. Goppert, R. C. Sutton, *J. Coll. Interf. Sci.* **1989**, *132*, 176–187.
189 T. Bahar, A. Tuncel, *Polym. Eng. Sci.* **1999**, *39*, 1849–1855.
190 Y. Almong, S. Reich, M. Levy, *Br. Polym. J.* **1982**, *14*, 131–136.
191 K. Li, H. D. H. Stöver, *J. Polym. Sci. Part A: Polym. Chem.* **1993**, *31*, 2473–2470.
192 K. Li, H. D. H. Stöver, *J. Polym. Sci. Part A: Polym. Chem.* **1993**, *31*, 3257–3263.
193 W.-H. Li, H. D. H. Stöver, *J. Polym. Sci. Part A: Polym. Chem.* **1998**, *36*, 1543–1551.
194 W.-H. Li, K. Li, H. D. H. Stöver, *J. Polym. Sci. Part A: Polym. Chem.* **1999**, *37*, 2295–2303.
195 A. Coehn, *Ergebnisse der exakten Naturwissenschaften* **1922**, pp. 175–196.

196 S. Sosnowski, M. Gadzinowski, S. Slomkowski, S. Pencek, *J. Bioact. Compat. Polym.* **1994**, *9*, 345–366.

197 S. Slomkowski, S. Sosnowski, M. Gadzinowski, C. Picot, A. Elaissari, *Macromol. Symp.* **2000**, *150*, 259–268.

198 C. Schellenberg, S. Akari, M. Regenbrecht, K. Tauer, F. M. Petrat, M. Antonietti, *Langmuir* **1999**, *15*, 1283–1290.

199 C. Schellenberg, K. Tauer, M. Antonietti, *Macromol. Symp.* **2000**, *151*, 465–471.

200 M. T. Greci, S. Pathak, K. Mercado, G. K. Surya Prakash, M. E. Thomson, G. A. Olah, *J. Nanosci. Nanotechnol.* **2001**, *1*, 3–6.

201 O. Kalinina, E. Kumacheva, *Macromolecules* **1999**, *32*, 4122–4129.

202 S. F. Lascelles, S. P. Armes, *J. Mater. Chem.* **1997**, *7*, 1339–1347.

203 K. Tauer, H. Kaspar, M. Antonietti, *Coll. Polym. Sci.* **2000**, *278*, 814–820.

204 R. Defay, I. Prigogine, A. Bellemans, D. H. Everett, *Surface Tension and Adsorption*, Longmans & Green, London, **1966**, pp. 217–222.

205 M. Morton, S. Kaizerman, M. W. Altier, *J. Coll. Sci.* **1954**, *9*, 300–312.

206 L. Gardon, *J. Polym. Sci. A-1* **1968**, *6*, 2859–2879.

207 W. Burchard, *Progr. Coll. Polym. Sci.* **1988**, *78*, 63–67.

208 J. Ugelstad, P. C. Mork, K. H. Kaggerud, T. Ellingsen, A. Berge, *Adv. Coll. Interf. Sci.* **1980**, *13*, 101–140.

209 M. Okubo, M. Shiozaki, M. Tsujihiro, Y. Tsukuda, *Coll. Polym. Sci.* **1991**, *269*, 222–226.

210 M. Okubo, H. Minami, T. Yamashita, *Macromol. Symp.* **1996**, *101*, 509–516.

211 M. Okubo, H. Minami, *Coll. Polym. Sci.* **1996**, *274*, 433–438.

212 M. Okubo, T. Yamashita, T. Suzuki, T. Shimizu, *Coll. Polym. Sci.* **1997**, *275*, 288–292.

213 M. Okubo, H. Minami, *Coll. Polym. Sci.* **1997**, *275*, 992–997.

214 M. Okubo, E. Iser, T. Yamashita, *J. Appl. Polym. Sci.* **1999**, *74*, 278–285.

215 M. Okubo, H. Minami, *Macromol. Symp.* **2000**, *150*, 201–210.

216 M. Okubo, H. Minami, Y. Yamamoto, *Coll. Polym. Sci.* **2001**, *279*, 77–81.

217 M. Okubo, E. Ise, H. Yonehara, T. Yamashita, *Coll. Polym. Sci.* **2001**, *279*, 539–545.

218 M. Okubo, H. Minami, K. Morikawa, *Coll. Polym. Sci.* **2001**, *279*, 931–935.

219 M. Okubo, Z. Wang, E. Iser, H. Minami, *Coll. Polym. Sci.* **2001**, *279*, 976–982.

220 G. T. D. Shouldice, A. Rudin, A. J. Paine, *J. Polym. Sci. Part A: Polym. Chem.* **1996**, *34*, 3061–3069.

221 L. Lazar, S. A. M. Hesp, G. E. Kmiecik-Lawrynowicz, *Part. Sci. Technol.* **2000**, *18*, 103–120.

222 L. Lazar, S. A. M. Hesp, *Part. Sci. Technol.* **2000**, *18*, 121–142.

223 L. Lazar, S. A. M. Hesp, *Part. Sci. Technol.* **2000**, *18*, 143–162.

224 R. H. Ottewill, A. B. Schofield, J. A. Waters, *Coll. Polym. Sci.* **1996**, *274*, 763–771.

225 R. H. Ottewill, A. B. Schofield, J. A. Waters, N. St. J. Williams, *Coll. Polym. Sci.* **1997**, *275*, 274–283.

226 R. H. Ottewill, A. B. Schofield, J. A. Waters, *J. Disp. Sci. Technol.* **1998**, *19*, 1151–1162.

227 D. C. Sundberg, Y. G. Durant, in J. M. Asua (ed.) *Polymeric Dispersions: Principles and Applications*, Kluwer Academic, Dordrecht, **1997**, pp. 177–188.

228 M. Okubo, K. Ichikawa, M. Tsujihiro, Y. He, *Coll. Polym. Sci.* **1990**, *268*, 791–796.

229 M. Okubo, Y. He, K. Ichikawa, *Coll. Polym. Sci.* **1991**, *269*, 125–130.

230 M. Okubo, N. Miyachi, Y. Lu, *Coll. Polym. Sci.* **1994**, *272*, 270–275.

231 E. P. Giannelis, *Adv. Mater.* **1996**, *8*, 29–35.

232 M. Ochiai, M. Masui, M. Asanae, M. Tokunaga, T. Iimura, *J. Imag. Sci. Technol.* **1994**, *38*, 415–420.

233 H. K. Mahabadi, T. H. Ng, H. S. Tan, *J. Microencaps.* **1996**, *13*, 559–573.

234 K. Wormuth, *J. Coll. Interf. Sci.* **2002**, *241*, 366–377.

235 J. Yu, J. Yu, Z.-X. Guo, Y.-F. Gao, *Macromol. Rapid Commun.* **2001**, *22*, 1261–1264.

236 N. Bechthold, F. Tiarks, M. Willert, K. Landfester, M. Antonietti, *Macromol. Symp.* **2000**, *151*, 549–555.

237 Q. Wang, H. Xia, C, Zhang, *J. Appl. Polym. Sci.* **2001**, *80*, 1478–1488.

238 H. XIA, C. ZHANG, Q. WANG, *J. Appl. Polym. Sci.* **2001**, *80*, 1130–1139.
239 J. I. AMALVY, M. J. PERCY, S. P. ARMES, *Langmuir* **2001**, *17*, 4770–4778.
240 F. TIARKS, K. LANDFESTER, M. ANTONIETTI, *Macromol. Chem. Phys.* **2001**, *202*, 51–60.
241 K. PADTBERG, *On-line Verfolgung von Nukleierungsprozessen*, PhD thesis, Max Planck Institute of Colloids and Interfaces, Golm and University of Potsdam, Germany, **2001**.
242 S. NOZARI, *Joint Nucleation of Organic/Inorganic Nanoparticles*, Max Planck Institute of Colloids and Interfaces, Golm and University of Potsdam, Germany, **2002**.
243 K. TAUER, K. PADTBERG, C. DESSY, in E. S. DANIELS, E. D. SUDOL, M. S. EL-AASSER (eds.) *Polymer Colloids, Science and Technology of Latex Systems*, Chapter 8, ACS Symp. Ser., **2001**, *801*, 93–112.
244 K. FUJIMOTO, K. NAKAHAMA, M. SHIDARA, H. KAWAGUCHI, *Langmuir* **1999**, *15*, 4630–4635.
245 K. NAKAHAMA, H. KAWAGUCHI, K. FUJIMOTO, *Langmuir* **2000**, *16*, 7882–7886.

2
Semiconductor Nanoparticles

Andrey L. Rogach, Dmitri V. Talapin, and Horst Weller

2.1
Introduction

Semiconductor particles with sizes between ~1 and 10 nm have received much attention since the pioneering works of Efros [1], Ekimov [2], Henglein [3], and Brus [4] in the early 1980s. They represent a system lying in the transition regime between the bulk solid and molecules (Fig. 2.1) and appear to be a fascinating object for studying basically novel properties of matter, generally described by the term "size-quantization effect" (Section 2.2). A famous demonstration of the size-dependent properties of semiconductor nanoparticles is the drastic color change of their colloidal solutions with decreasing particle size (Fig. 2.2).

Colloidal semiconductor particles of nanometer size, which are the subject of this chapter, are referred to as nanoparticles, nanoclusters, nanocrystals, quantum dots, etc., depending mainly on the background of the researcher. We will mainly use the term *nanocrystal* throughout the text, as we are discussing the crystalline "nano-pieces" of different semiconductor materials. The term *nanocluster* will be used as a definition of extremely small, molecular-like particles of defined structure. The term *nanoparticle* will be used as a general one.

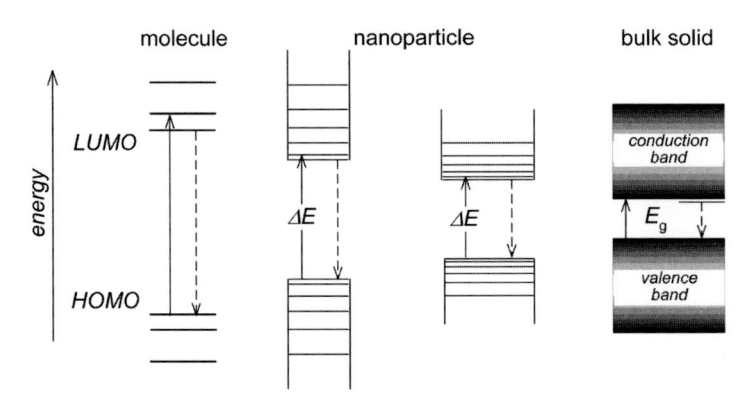

Fig. 2.1 Schematic energy diagrams illustrating the situation for a nanoparticle, in between a molecule and a bulk semiconductor.

Colloids and Colloid Assemblies. Edited by Frank Caruso
Copyright © 2004 Wiley-VCH Verlag GmbH & Co. KGaA, Weinheim
ISBN: 3-527-30660-9

Fig. 2.2 Manifestation of the size-quantization effect as a color change of aqueous colloidal solutions of CdSe nanoparticles. The particle size changes from left to right from ∼ 1.5 to ∼ 4.5 nm.

Fig. 2.3 Size-dependent luminescence of colloidal semiconductor nanoparticles. The emission color of thiol-capped CdTe nanocrystals can be adjusted by their size (2.5–4.5 nm range) to be green, yellow, orange or red; the room temperature quantum yields of the luminescence are up to 40% [6].

Colloidal synthesis of nanoparticles in a suitable solvent medium is often referred to as a "bottom-up" way to obtain nanostructured systems. In the current stage of technology, it is still very difficult to approach this size region with "top-down" lithography-based techniques. Using the chemical bottom-up approach, different high-quality semiconductor nanocrystals (high quality in this context means [5]: desired particle sizes over the largest possible range, narrow size distributions, good crystallinity, controllable surface functionalization, high luminescent quantum yields) can now be obtained on a gram scale, handled like ordinary chemical substances and used as building blocks for supramolecular structures and devices. Two general synthetic routes to semiconductor nanoparticles – aqueous and organometallic ones – are the subject of Section 2.3, together with a part discussing the theoretical aspects of the nanoparticle growth. Examples of the use of semiconductor nanoparticles as building blocks for the fabrication of different superstructures are presented in Section 2.4.

Colloidally synthesized semiconductor nanocrystals often possess a strong bandgap luminescence tunable by size as a result of the quantum confinement effect (Fig. 2.3). This makes them interesting for different applications, ranging from thin film electroluminescent devices [7–10] and optical amplifier media for telecommunication networks [11, 12] to biological fluorescent labels [13–15]. The incorporation of luminescent semiconductor nanocrystals into photonic crystals [16] and microcavities [17] has attracted considerable attention as a promising pathway to novel light sources with controllable spontaneous emission. The potential applications of semiconductor nanocrystals are briefly reviewed in Section 2.5. The reader is also referred to Chapters 7 and 16 of this book for comprehensive overviews on semiconductor nanocrystal surface modification and the use of semiconductor nanocrystals as biological label, respectively.

2.2
Quantum Confinement in Semiconductor Nanoparticles

Semiconducting properties of macrocrystalline, or "bulk", semiconductors arise from the periodic arrangement of atoms in a crystalline lattice. The spatial overlapping of the atomic orbitals leads to the formation of bands of the allowed electron and hole states separated by a forbidden gap called the band gap (Fig. 2.1). An electron excited from the valence band by the absorption of a quantum of light moves to the conduction band, leaving a positive charge (a hole). The electron and the hole experience attractive Coulomb forces and can form a so-called Wannier exciton, a state similar to a hydrogen atom and, thus, described by a two-particle Hamiltonian. The Wannier exciton is delocalized within the crystal lattice. The Bohr radius a_B of the bulk exciton depends on the effective masses of the electrons and holes m_e^* and m_h^* and on the high-frequency dielectric constant of the material ε:

$$a_B = \frac{\hbar^2 \varepsilon}{e^2} \left[\frac{1}{m_e^*} + \frac{1}{m_h^*} \right] \tag{1}$$

In a nanoparticle, the particle size is comparable to or smaller than the exciton diameter in bulk material. As a consequence, continuous energy bands split into discrete levels, the electronic excitation shifts to higher energy, and the oscillator strength is concentrated in just a few transitions. As was shown by Brus [18] using the so-called effective mass approximation, the size dependence of the energy of the first excitonic transition (or the band gap shift with respect to the bulk value E_g) upon radius R can be calculated as:

$$E(R) - E_g = \frac{\hbar^2 \pi^2}{2R^2}\left[\frac{1}{m_e^*} + \frac{1}{m_h^*}\right] - \frac{1.8e^2}{\varepsilon R} \qquad (2)$$

The first term in Eq. (2) represents the particle-in-a-box quantum localization energy and has an $1/R^2$ dependence; the second term with a $1/R$ dependence represents the Coulomb energy. Thus, the major consequence of the quantum confinement effect in semiconductors is an increase of the apparent band gap with decrease of the particle size which is experimentally observed as a blue shift in the absorption onset (Fig. 2.2). The emission spectra arising from the recombination of electron and hole from the first excited state show the same size-dependent behavior which is clearly seen in Fig. 2.3 for CdTe nanocrystals.

A very small semiconductor cluster consisting of only several atoms possesses only a few energy levels (antibonding molecular orbitals) available to excited electrons, thus being similar to molecules in its optical properties (Fig. 2.1). The use of a molecular-based nomenclature for the description of its energy structure – HOMO (highest occupied molecular orbital) and LUMO (lowest unoccupied molecular orbital) – instead of the solid-state nomenclature is even more indicative. Within the size range between nanoclusters (~1 nm) and "large" nanocrystals (~10–15 nm), effective design of the optical properties of semiconductors is possible simply through control of their size and the size deviation, which can be done either at the stage of synthesis or in post-preparative stages. These drastic changes in optical properties governed by the reduction of particle size inspired the researchers to develop advanced synthetic routes to nanoparticles of different semiconductor materials to be used in various applications.

2.3
Colloidal Synthesis of Semiconductor Nanoparticles

2.3.1
General Remarks

The availability of reliable colloidal syntheses leading to semiconductor nanoparticles being uniform in composition, size, shape, and the surface chemistry is crucial both for the study of their size-dependent properties and for their further use in different applications. At the earlier stage of nanoparticle research most work was done on II–VI compounds like CdS and CdSe. The initial concept of "keeping particles small" was the strategy of arrested precipitation, making use of water pools of inverse micelles as nanometer-sized reactors [19]. This was followed by employing stabilizing (also called capping or passivating) agents (ligands) which prevent uncontrollable growth and aggregation of nanoparticles. One successful example of both electrostatic and sterical stabilization was the use of sodium polyphosphate for the synthesis of CdS nanoparticles in water [20].

The choice of a suitable stabilizer can be considered as a key point in an advanced colloidal synthesis of semiconductor nanoparticles. Stabilizers regulate the growth rate and the size of nanoparticles, prevent them (to a lesser or greater extent) from oxidation and provide a dielectric barrier at the surface thus eliminating (partially) surface traps. Importantly, stabilizers, being organic molecules with free functional groups, make nanoparticles "soluble" in different solvents and allow their further handling (attachment to different surfaces, covalent coupling with each other and with other molecules, etc.).

The two well-established general synthetic routes we will repeatedly deal with below are an *aqueous synthesis* of II–VI semiconductor nanoparticles (CdS, CdSe, CdTe, $Cd_xHg_{1-x}Te$, and HgTe) in the presence of various short-chain thiols as stabilizing agents, and an *organometallic synthesis* of either II–VI (CdSe, CdTe) or III–V (InP, InAs) nanocrystals based on the high-temperature thermolysis of precursors in so-called coordinating solvents (long-chain alkylphosphines, alkylphosphine oxides, alkylamines) which themselves play the role of a stabilizing agent. Three examples of the reactions used in these syntheses are shown below:

$$Cd(ClO_4)_2 + H_2Te \xrightarrow{100°C, H_2O, thiol} CdTe\ nanocrystals \quad (3)$$

$$Cd(CH_3)_2 + (C_8H_{17})_3PSe \xrightarrow{300°C, HDA-TOPO-TOP} CdTe\ nanocrystals \quad (4)$$

$$InCl_3 + [(CH_3)_3Si]_3P \xrightarrow{260°C, TOP} InP\ nanocrystals \quad (5)$$

2.3.2
Theoretical Aspects of Nanoparticle Growth in Colloidal Solutions

In the case of relatively "large" nanoparticles, the crystalline core of which consists of $\sim 10^2$–10^4 atoms, a nearly continuous tunability of the particle size is possible, as an addition or a removal of a unit cell results only in a small variation of the particle's free energy. The colloidal synthesis of nanoparticles generally involves several consecutive stages: nucleation from initially homogeneous solution, growth of the pre-formed nuclei, isolation of particles reaching the desired size from the reaction mixture, post-preparative size fractionation, etc. As a rule, a temporal separation of the nucleation event from the further growth of the nuclei is required for narrow size distribution [21, 22]. The so-called hot-injection technique according to which the precursors are rapidly injected into a hot solvent with a subsequent temperature drop, satisfies this requirement [21, 23]. During further growth of the nanocrystals, several regimes can be realized depending on the system and experimental conditions, as discussed below.

2.3.2.1 Ostwald Ripening
Ostwald ripening (OR), the growth mechanism where the smaller particles dissolve and the monomer released thereby is consumed by the large particles [24, 25], takes place during most of the colloidal syntheses of both II–VI and III–V

semiconductor nanocrystals. As a result, the average nanocrystal size increases with time and the particle concentration decreases. The driving force of OR is a decrease of particle solubility with increasing size as expressed by the Gibbs-Thompson equation:

$$C(r) = C_{bulk}^0 \exp\left[\frac{2\gamma V_m}{rRT}\right] \approx C_{bulk}^0 \left(1 + \frac{2\gamma V_m}{rRT}\right) \quad (6)$$

Here, $C(r)$ and C_{bulk}^0 are the solubilities of a particle with radius r and of the bulk material, respectively, γ is the surface tension, and V_m is the molar volume of the solid. The validity of the Gibbs-Thompson equation has been proven for very small ($r \sim 1$–2 nm) colloidal particles [26, 27].

In the case of "large" ($r > \sim 25$ nm) particles the kinetics of OR can be satisfactorily described analytically in the framework of the Lifshitz-Slyozov-Wagner (LSW) theory [28, 29]. However, the LSW approach takes into account only two terms of expansion of the Gibbs-Thompson equation and fails in the description of ensembles of particles smaller than ~ 25 nm in radius due to large errors arising from non-adequate expansion of Eq. (6). The coefficient $2\gamma V_m/(RT)$ called "capillary length" is of the order of 1–3 nm for most solid–liquid interfaces [30, 31] and in the case of nanoparticles with $r=1$–5 nm the particle solubility becomes strongly nonlinear against r^{-1}. Moreover, for nano-scale particles the activation energies of the growth and dissolution processes are also size-dependent [32]. In this case a general analytical solution describing all processes occurring during the evolution of the particle ensemble cannot be obtained. Our recent study applied a Monte-Carlo simulation to describe the evolution of an ensemble of nanoparticles in a colloidal solution [32] and showed that in this case OR is characterized by some features which are not observed for large (sub-micrometer- and micrometer-sized) particles. For convenience, we will use the term "nano-OR" in our further discussion to distinguish this particular case from OR of (sub-)micrometer-sized particles adhering to the LSW theory.

OR implies that the largest particles in the ensemble have positive and the smallest ones negative growth rates. The growth/dissolution rate of a single particle of radius r is given by the following equation which seems to be valid for both nano- and (sub-)micrometer-sized particles [32]:

$$\frac{dr^*}{d\tau} = \frac{S - \exp\left[\frac{1}{r^*}\right]}{r^* + K \exp\left[\frac{a}{r^*}\right]} \quad (7)$$

where $r^* = \frac{RT}{2\gamma V_m} r$ and $\tau = \frac{R^2 T^2 D C_{bulk}^0}{4\gamma^2 V_m} t$ are the dimensionless particle radius and the time, respectively. K is the dimensionless parameter describing the impacts of diffusion and surface reaction as kinetics-limiting factors, with $K \ll 1$ corresponding to the diffusion-controlled process and $K \gg 1$ to the surface reaction-controlled one [32]. The dimensionless parameter $S = [M]/C_{bulk}^0$ describes the oversaturation

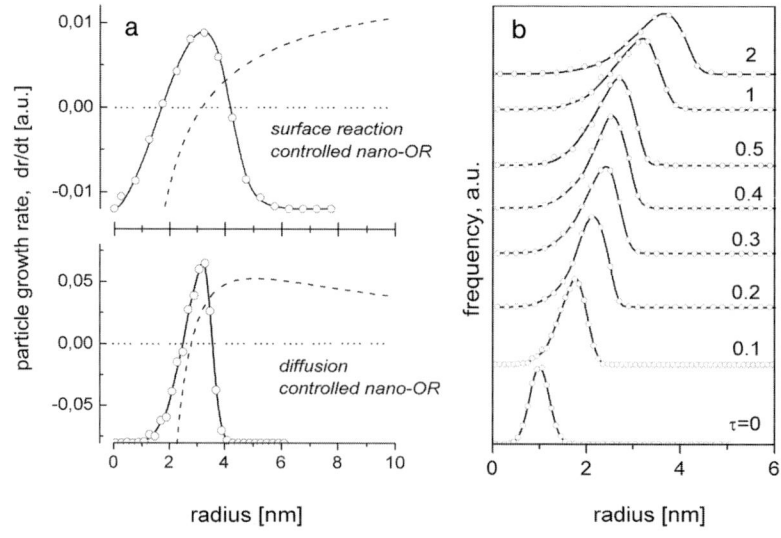

Fig. 2.4 a Dashed lines: size dependence of a single particle's growth rate during (top) surface reaction-limited and (bottom) diffusion-limited Ostwald ripening. The horizontal dotted lines correspond to zero particle growth rate (equilibrium between the nanoparticle and monomer in solution). Solid lines: corresponding "stationary" particle size distributions after prolonged growth of an ensemble of nanoparticles: the distribution shape becomes independent of time. **b** Temporal evolution of the size distribution of the nanoparticle ensemble during Ostwald ripening under diffusion control.

of the monomer in a solution with its concentration [M]. D is the diffusion coefficient for a monomer and a is the transfer coefficient of the activated complex ($0 < a < 1$).

The size-dependent particle growth/dissolution rates are shown in Fig. 2.4a for diffusion-controlled and surface reaction-controlled nano-OR. Fig. 2.4b shows the temporal evolution of the particle size distribution during diffusion-controlled nano-OR. The LSW theory predicts that upon OR the ensemble of particles tends to an asymptotic particle size distribution which is independent of the initial conditions [28]. A similar situation is realized in nano-OR: any initial particle size distribution evolves in time toward the asymmetric negatively skewed one. The asymptotic particle size distributions inherent to diffusion and surface reaction-limited nano-OR are shown in Fig. 2.4a together with the corresponding instantaneous single-particle growth rates. The resulting particle size distribution is narrower if the particle growth takes place under diffusion control.

Nano-OR proceeds considerably faster than OR in ensembles of larger particles as shown in Fig. 2.5a. The evolution of the width of particle size distribution expressed as percentage of standard deviation ($\sigma\%$) vs. average particle size is shown in Fig. 2.5b. Independently of the initial width of particle size distribution, nano-OR results in a steady-state distribution with a standard deviation indepen-

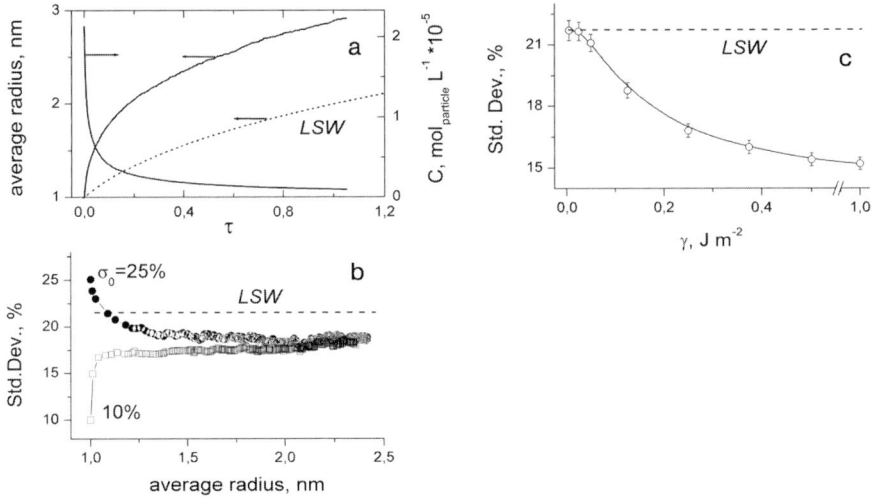

Fig. 2.5 Diffusion-controlled Ostwald ripening in the nanoparticle ensemble. **a** Evolution of the mean particle radius and the particle concentration in time. **b** Dependence of the standard deviation of the particle size distribution on the average particle size during the Ostwald ripening for different initial widths of particle size distributions. **c** Dependence of the stationary value of standard deviation of the size distribution on the surface tension. Dashed lines in all frames correspond to LSW-predicted coarsening.

dent of the average particle size. If the initial size distribution is narrower/broader than the steady-state one, the nano-OR is accompanied by a broadening/narrowing of the particle size distribution, correspondingly.

Remarkably, the steady-state particle size distribution inherent to nano-OR is narrower than that of OR in ensembles of (sub-)micrometer-sized particles. Moreover, for nanometer-sized particles the width of the steady-state size distribution depends on the surface tension at the particle–solution interface. Thus, the stationary value of $\sigma \approx 21.5\%$ is predicted by LSW for diffusion-limited OR. Fig. 2.5c shows the stationary values of standard deviation corresponding to different surface tensions at the particle–solution interface and demonstrates the decrease of the steady state value of σ with increasing γ in the case of diffusion-limited nano-OR.

The difference between nano-OR and conventional OR is caused, as expected, by the particle radius. As mentioned above, particles with $r > \sim 25$ nm fulfill the condition $2\gamma V_m/(RT) \ll r$ and higher terms of the expansion of the Gibbs-Thompson equation can be omitted, as shown in Eq. (6). However, if $2\gamma V_m/(RT) \approx r$ or even $> r$, these terms play a major role, accelerating growth and dissolution of particles. This also results in an acceleration of the coarsening rate and in the narrowing of the size distribution, as demonstrated by Fig. 2.5. If $2\gamma V_m/(RT)$ is markedly larger than r, a stationary value of σ as small as $\sim 15\%$ can be achieved (Fig. 2.5c).

Summarizing, Ostwald ripening of nanoparticles occurs faster and can result in narrower size distributions than in ensembles of (sub-)micrometer-sized particles. The narrowest size distribution can be achieved if particle growth rates are limited by diffusion of monomer from the bulk solution towards the particle surface. From the experimental point of view, OR provides a simple and precise way to achieve the desired particle size through the control of the growth duration and the proper choice of capping agents passivating the surface of the growing nanoparticles. However, it ultimately results in rather broad particle size distributions with standard deviations lying in the range of 15–21%. To prepare monodisperse nanoparticles by this method, additional post-preparative size fractionation is required. These theoretical predictions correlate well with experimental data obtained on both nanometer- and micrometer-sized particles of various materials [21, 22, 33].

2.3.2.2 Transient Regime of Particle Growth: "Focusing" and "Defocusing" of Particle Size Distribution

As mentioned in the previous section, nano-OR results in a particle size distribution above 15% which is still rather broad. Narrower size distributions can be achieved if the growth of nanoparticles occurs in a transient regime and is terminated before the equilibrium between particles and monomer is attained. If only a part of the precursors is converted to nanoparticles at the nucleation stage and a considerable amount of precursors is still present in the colloidal solution in the form of monomer, the OR regime cannot be realized. Indeed, during OR the particles are in equilibrium with the monomer in solution and the chemical potential of the monomer corresponds to that of particles lying somewhere in the middle of the size distribution. The corresponding equilibrium oversaturation of the monomer (S) can be estimated from Eq. (7) and refers to a range of ~ 1.2 to 20 for realistic experimental conditions. If the value of S exceeds this range substantially all particles in the colloidal solution will grow simultaneously at the expense of the monomer being consumed from the solution. In this case a fast increase of particle size accompanied by a strong narrowing (further referred to as "focusing") of the size distribution is observed. Fig. 2.6a shows simulations of the temporal evolution of the particle size distribution during diffusion-controlled nanoparticle growth at high initial monomer oversaturation. Very fast initial growth resulting in focusing of the size distribution from 20% standard deviation (initial value) down to 6% is followed by slow broadening (further referred to as "defocusing") of the size distribution. During the focusing stage the number of particles remains almost constant; the size distribution is nearly symmetric and fits well to normal distribution. The oversaturation drops down to some equilibrium value. Defocusing is accompanied with a transition from symmetric towards asymmetric negatively skewed size distribution keeping the mean particle size nearly constant as shown in Fig. 2.6b. The number of nanocrystals starts to decrease owing to dissolution of the smallest ones. The Ostwald ripening mechanism governs the further evolution of the nanoparticle ensemble.

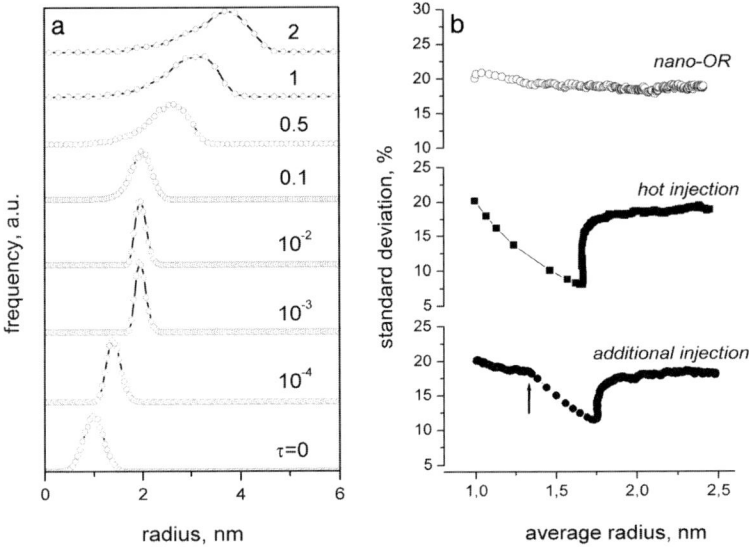

Fig. 2.6 a Temporal evolution of the size distribution of the nanoparticle ensemble at high initial oversaturation of monomer ($S_0=900$) under diffusion control. **b** Dependence of the standard deviation of particle size distribution on the average particle size for different growth regimes: Ostwald ripening ($S_0=15$, upper curve); focusing and defocusing of the size distribution at high initial oversaturation of monomer ($S_0=500$, middle curve); focusing and defocusing of the size distribution induced by additional injection of monomer ($S_0=15$; additional injection of 100% of initial amount of the precursors at the time indicated by the arrow, lower curve). All curves correspond to diffusion-limited particle growth.

Focusing of the size distribution happens if the monomer oversaturation required for nucleation differs significantly from the equilibrium values of the corresponding Ostwald ripening regime. As a result, at early stages of synthesis the nucleation rate drops much faster than oversaturation decreases and a large amount of precursors remains in the form of monomer. On the other hand, the conditions favorable to the focusing of the size distribution can be achieved by applying special experimental techniques. Thus, swift injection of cold reagents into very hot solvent results in an explosive-like nucleation which is immediately suppressed due to the fast temperature drop, thus allowing the achievement of high monomer oversaturation at early stages of nanoparticle growth (Fig. 2.6b). Similar results can be obtained by additional injections of precursors to the solution of growing nanocrystals [34]: instantaneous increase of oversaturation of monomer results in focusing of the particle size distribution (Fig. 2.6b). By applying these approaches and optimizing the reaction conditions it is possible to synthesize nanoparticles of the desired size with very narrow size distributions, as will be demonstrated below.

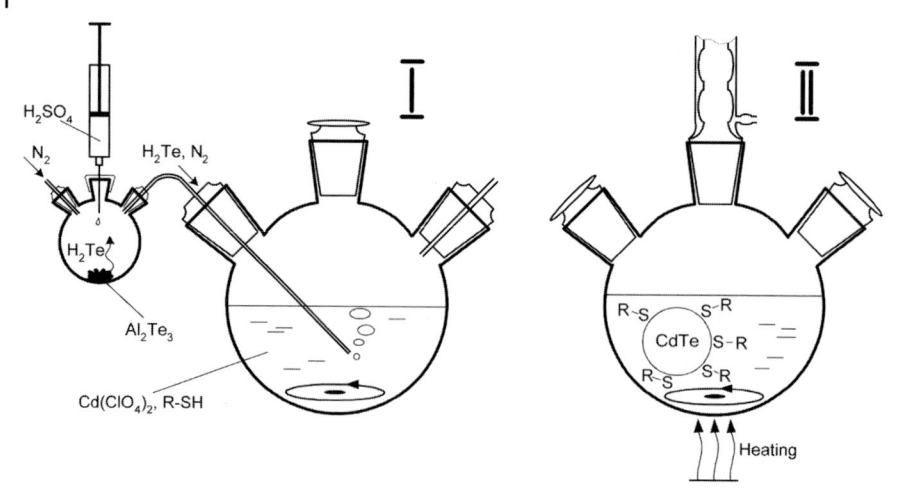

Fig. 2.7 Schematic representation of the synthesis of thiol-capped CdTe nanoparticles. First stage: formation of CdTe precursors by introducing H$_2$Te gas. Second stage: formation and growth of CdTe nanocrystals promoted by reflux (reproduced from [6], with permission from American Chemical Society).

2.3.3
Semiconductor Nanoparticles Synthesized in Aqueous Media

Various II–VI nanoparticles have been prepared in aqueous solutions using different thiols as stabilizing agents: CdS [35], CdSe [36], CdTe [33], Cd$_x$Hg$_{1-x}$Te [37] and HgTe [38]. Fig. 2.7 illustrates a typical aqueous synthesis using CdTe (see Eq. 3) as an example. In the first stage, a hydrogen chalcogenide produced by reaction between Al$_2$Te$_3$ and H$_2$SO$_4$ is introduced as gas at room temperature into the reaction vessel containing the aqueous solution of a metal salt and a suitable stabilizer at the appropriate pH. Water-soluble precursors of nanoparticles are formed in this stage, most probably in the form of some metal-chalcogen-thiol complexes. Applying heat in the second stage promotes the chemical reaction between metal and chalcogenide leading to the formation of semiconductor nuclei followed by their growth. The material added to the growing nuclei is supplied either from the existing complexes or from dissolving smaller particles following the Ostwald ripening mechanism described above. The resulting size distribution of the samples is usually relatively broad (15–20%), but can easily be improved by the so-called size-selective precipitation technique as described in Section 2.3.5.

The use of different thiols (thioethanol, thioglycerol, thioglycolic acid, dithioglycerol, mercaptoethylamine, L-cysteine, etc.) and their mixtures as stabilizing and capping ligands, as well as varying the concentrations of the reactants, the pH value of the solution, and the duration of the heat treatment, allows control over the particle size during the aqueous synthesis. As a result, the size range of 1.2–6.0 nm is generally accessible. The possibility to manipulate thiol-capped nanopar-

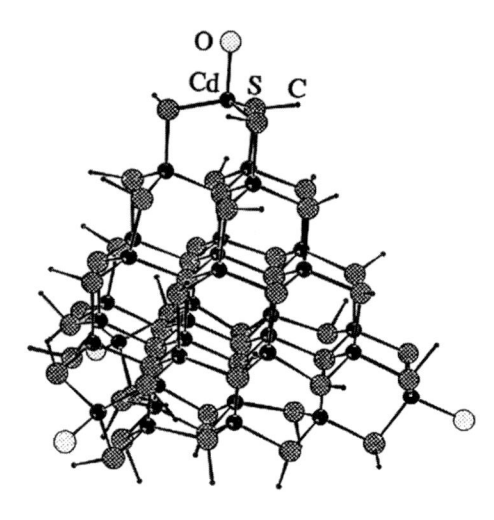

Fig. 2.8 Structure of a [Cd$_{32}$S$_{14}$(SCH$_2$CH(OH)CH$_3$)$_{36}$]·(H$_2$O)$_4$ nanocluster as derived from the single-crystal X-ray data (reproduced from [47], with permission from American Chemical Society).

ticles, e.g. by layer-by-layer assembly on planar substrates [39–41] and colloids [41, 42] or by electrophoresis [43], and to use the free functional groups of the capping thiols for conjugation [44, 45], makes them especially attractive for the fabrication of different functional materials.

Importantly, the use of thiols as capping agents allows the preparation of extremely small molecule-like nanoclusters of II–VI semiconductor compounds, e.g. [Cd$_{17}$S$_4$(SCH$_2$CH$_2$OH)$_{26}$] [46] and [Cd$_{32}$S$_{14}$(SCH$_2$CH(OH)CH$_3$)$_{36}$]·(H$_2$O)$_4$ [47]. These nanoclusters correspond to pronounced minima of the free energy vs. particle size dependence owing to their closed structural shells (the concept of so-called "clusters of magic size" in the earlier literature [48]) and are naturally "100% monodisperse". They can be crystallized in macroscopically large single crystals allowing their investigation by single-crystal X-ray analysis, including the exact determination of the atomic coordinates. Fig. 2.8 shows the inner structure of a [Cd$_{32}$S$_{14}$(SCH$_2$CH(OH)CH$_3$)$_{36}$]·(H$_2$O)$_4$ cluster representing a piece of the zinc blende lattice in the shape of a tetrahedron.

Thiol molecules can release S^{2-} in the course of the prolonged refluxing in the basic medium, which builds into the lattice of the growing nanocrystals. Thus, the positions of the powder X-ray diffraction (XRD) reflections of CdTe nanocrystals synthesized under prolonged refluxing in the presence of thioglycolic acid are intermediate between the values of the cubic CdTe and the cubic CdS phases [49]. Mixed CdTe$_x$S$_{1-x}$ nanocrystals with some gradient of sulfur distribution from inside the nanoparticles to the surface are formed under these conditions. They represent a kind of core-shell system with a naturally CdS-capped surface created by mercapto-groups covalently attached to the surface cadmium atoms. Importantly, the synthesis of such core-shell nanocrystals occurs in one step, as the sulfur originates from the stabilizing thiol molecules and is released during particle growth. At the bulk CdTe/CdS interface, the conduction band step, i.e. the offset of the absolute band position, is close to zero, whereas the valence band step is

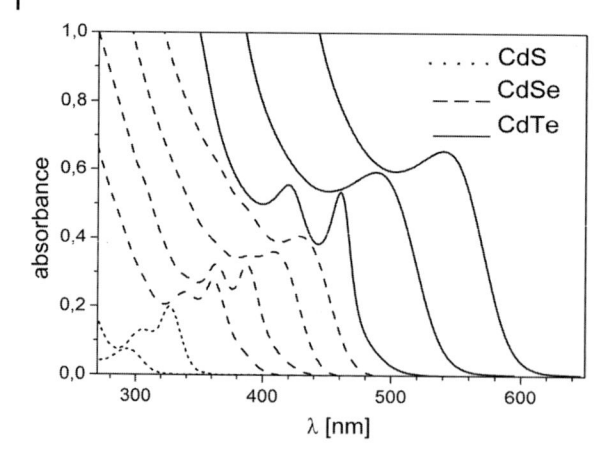

Fig. 2.9 Absorption spectra of thiol-capped CdS, CdSe, and CdTe nanoparticles.

~1 eV as given in [50]. The wave functions calculated for a CdTe/CdS system with the particle-in-a-box model [51] show a delocalization of the electron through the entire structure and the confinement of the hole to the CdTe core – the same phenomenon as reported for the organometallically synthesized core-shell CdSe/CdS nanocrystals [52], providing their photostability and electronic accessibility.

Another approach to the synthesis of mixed-phase nanoparticles using the aqueous approach can be demonstrated with $Cd_xHg_{1-x}Te$ alloy. This synthesis makes use of the chemical modification of pre-synthesized CdTe nanocrystals by Hg^{2+} ions [37]. Owing to the lower solubility of HgTe in comparison to CdTe in water, the Hg^{2+} ions substitute Cd^{2+} ions at the surface of the nanoparticles forming a $Cd_xHg_{1-x}Te$ alloy in the near-surface region, possibly with a concentration gradient decreasing towards the particle interior. The same reaction scheme combined with the growth of an outer shell of CdS as the next step has been used for the fabrication of three-layered CdS/HgS/CdS nanoparticles, so-called quantum dot quantum wells [53]. The band gap of bulk $Cd_xHg_{1-x}Te$ alloy varies approximately linearly with the composition from +1.6 eV at $x=1$ (pure CdTe) to –0.3 eV for $x=0$ (pure HgTe) [54]. This was observed experimentally for $Cd_xHg_{1-x}Te$ nanoparticles as a red shift in both absorption and luminescence spectra with increasing Hg content.

Fig. 2.9 shows a set of absorption spectra of thiol-capped CdS, CdSe, and CdTe nanoparticles of different sizes. Owing to the size-quantization effect, the absorption edge is strongly shifted towards shorter wavelengths in comparison to the corresponding bulk values (480 nm for CdS, 674 nm for CdSe, and 817 nm for CdTe), and the first excitonic transition is clearly resolved for all the samples shown. The smallest samples in each chalcogenide series are actually nanoclusters and their spectra are molecular-like.

Fig. 2.10 shows a set of photoluminescence spectra of thiol-capped CdTe, $Cd_xHg_{1-x}Te$, and HgTe nanocrystals. The entire spectral range from 500 to 2000 nm is available by varying the size and composition of the nanoparticles. In particular, the luminescence spectra of HgTe nanocrystals of different sizes cover

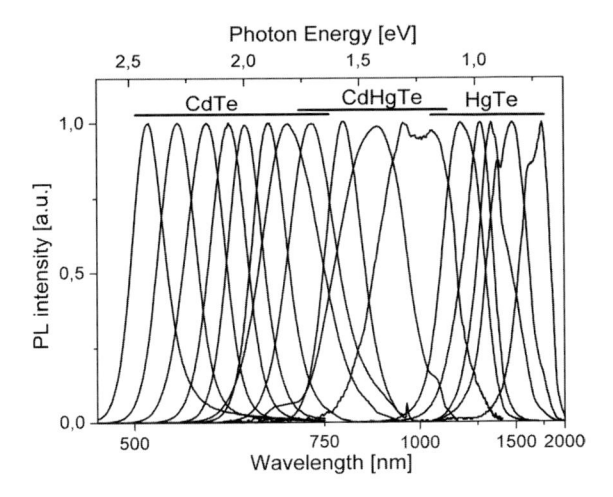

Fig. 2.10 Photoluminescence spectra of thiol-capped CdTe, $Cd_xHg_{1-x}Te$, and HgTe nanocrystals.

the spectral region between 1000 and 1800 nm making them potential candidates for application as ultra-broadband optical amplifiers in telecommunications systems [11, 12] (Section 2.5.3). The luminescence quantum yields of CdTe, Cd_x-$Hg_{1-x}Te$ and HgTe nanocrystals synthesized in water reach values of up to 40–50% [6, 38, 55].

2.3.4
Semiconductor Nanocrystals Synthesized by the Organometallic Route

Organometallic approaches to the preparation of semiconductor nanocrystals are based on the high-temperature (200–360 °C) thermolysis of organometallic precursors in the presence of stabilizing agents (so-called coordinating solvents). A narrow particle size distribution can be achieved via temporal separation of the fast nucleation event from the following slow growth of the pre-formed nuclei. Experimentally, this can be realized using the hot injection technique schematically shown in Fig. 2.11. Organometallic precursors are rapidly injected into a very hot (300–320 °C) stabilizing solvent resulting in explosive nucleation, followed by a temperature drop to ~200 °C which terminates the nucleation. The time of subsequent growth at a desired temperature and the proper choice of stabilizing agents provide control over the particle size. Additional injections of precursors may, under certain conditions, lead to the narrowing ("focusing") of the particle size distribution (Section 2.3.2).

Since a seminal paper by Murray, Norris and Bawendi in 1993 [23] where the great potential of organometallic synthesis was demonstrated for the first time, this approach has become dominant in the preparation of II–VI and III–V semiconductor nanocrystals. CdSe nanocrystals are probably the object studied most extensively among them and can be considered as a model system to illustrate the abilities of organometallic colloidal techniques.

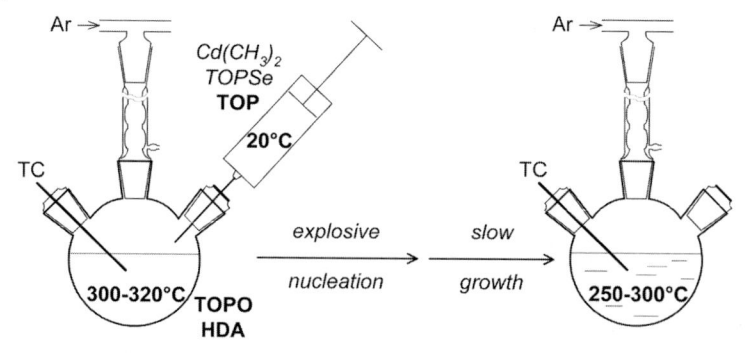

Fig. 2.11 Schematic representation of the hot injection technique employed in the organometallic synthesis of nanocrystals. The synthesis of CdSe nanocrystals in hexadecylamine-trioctylphosphine oxide-trioctylphosphine (HDA-TOPO-TOP) coordinating mixture is shown as an example. TC represents a thermocouple controlling the temperature in the reaction vessel.

2.3.4.1 CdSe Nanocrystals

The synthesis of CdSe nanocrystals by the reaction of dimethylcadmium with trioctylphosphine selenide in a trioctylphosphine oxide–trioctylphosphine (TOPO-TOP) coordinating solvent at 250–300 °C (Eq. 4) allows the preparation of colloidal nanocrystals with sizes from ~ 1.7 to 15 nm soluble in a variety of organic solvents such as toluene, hexane, chloroform, etc. [23, 56]. The solubility of the particles is provided by the shell of TOPO molecules passivating the Cd surface sites whereas the selenium surface sites are either unbonded or passivated with TOP molecules [56]. The powder diffraction patterns correspond to the hexagonal wurtzite CdSe phase (see Fig. 2.28 below). Owing to the rather high annealing temperature, all nanocrystals exhibit a high degree of crystallinity with nearly defect-free crystalline cores [21, 23, 57]. Synthesis followed by the size-selective precipitation procedure (see Section 2.3.5) allows the isolation of nearly monodisperse fractions with particle size distributions as narrow as 4–6% [23, 58]. Chemical [56, 59–61], optical [62–64], and charge transport [65] properties of TOPO-TOP-capped CdSe nanocrystals have been extensively investigated and are summarized in several reviews [21, 66].

Studies have shown that the properties of as-prepared CdSe nanocrystal colloids, especially the particle size distribution and the luminescence efficiency, crucially depend on the impurities such as alkyl phosphonic and alkyl phosphinic acids occasionally present in the technical grade TOPO used in the synthesis. Some batches of technical grade TOPO provide much better conditions for nanocrystal growth than pure TOPO [67]. These findings caused the appearance of new modifications of conventional TOPO-TOP synthesis aimed at enhanced control over the nanocrystal growth. Thus, introducing the third component hexadecylamine (HDA) to the stabilizing mixture allowed a considerable improvement of reproducibility of the growth kinetics, size distribution, and the luminescence effi-

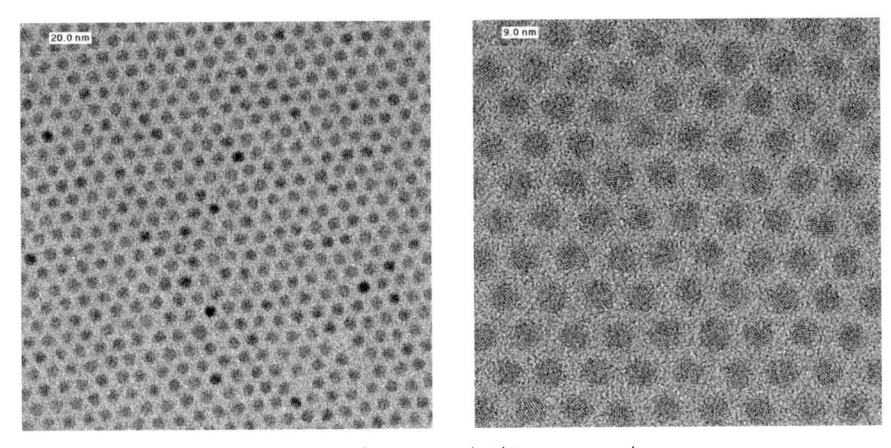

Fig. 2.12. TEM and HRTEM images of as-prepared CdSe nanocrystals grown in TDPA-HDA-TOPO-TOP mixture.

ciency of CdSe nanocrystals [68, 69]. Size distributions as narrow as 5–7% were achievable without any post-preparative size fractionation [68]. Moreover, our investigations as well as the recent data of Peng et al. [69] suggest that control over the size distribution of CdSe nanocrystals can be even further improved by introducing an acidic component, a long chain n-alkylphosphonic or carboxylic acid, to the HDA-TOPO-TOP mixture. Fig. 2.12 shows TEM and HRTEM images of CdSe nanocrystals grown in an n-tetradecyl phosphonic acid (TDPA)-HDA-TOPO-TOP stabilizing mixture, being exceptionally monodisperse as prepared.

The growth dynamics of nanocrystals can be conveniently monitored by taking aliquots from the reaction mixture at different instants of time. A set of absorption and photoluminescence spectra displaying the growth of CdSe nanocrystals in an HDA-TOPO-TOP mixture at 300 °C is shown in Fig. 2.13a. The absorption spectra possess up to five resolved electronic transitions indicating very narrow size distributions of the CdSe nanocrystals, and are comparable with the best CdSe samples obtained by the TOPO-TOP synthesis followed by several stages of size-selective precipitation [21, 23, 56]. The average size of CdSe nanocrystals at any instant of time can be estimated from the position of the first excitonic transition in the absorption spectrum using the sizing curve obtained from TEM and HRTEM data [70] (Fig. 2.13b).

The emission properties of CdSe nanocrystals can be described by four fundamental parameters: the brightness, the emission color, the color purity, and the stability of the emission [69]. The quantum size effect allows the variation of the emission color of CdSe nanocrystals from blue to red, thus covering almost the whole visible spectral region. The room temperature PL quantum yield of TOPO-TOP-capped CdSe nanocrystals lies in the range of 5 to 15% [23, 62]. The original TOPO-TOP capping ligands are not suitable for successful passivation of the surface traps of very small (< ~ 2.5 nm) CdSe nanocrystals emitting blue, whereas synthesis from the same precursors in a dodecylamine-TOP mixture at 90–150 °C

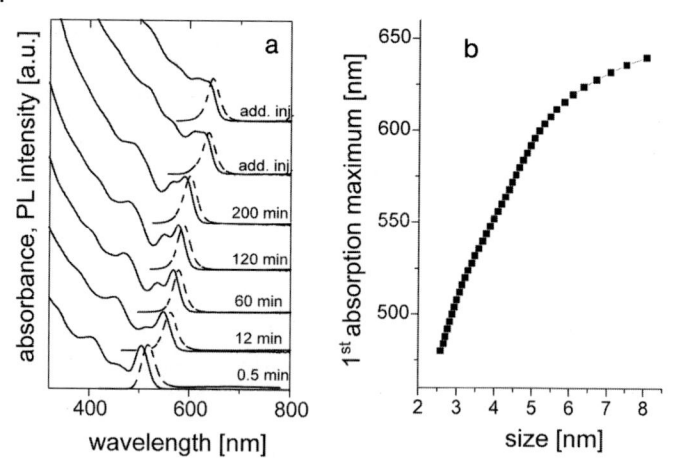

Fig. 2.13 a Room temperature absorption (solid lines) and emission (dashed lines) spectra of CdSe nanocrystals monitored during the growth at 300 °C in an HDA-TOPO-TOP mixture. "add. inj." means additional injection of monomer into the reaction mixture. **b** Experimental sizing curve for CdSe nanocrystals linking the spectral position of the first absorption maximum with the nanocrystal size (reproduced from [68] and [70], with permission from American Chemical Society).

provides CdSe nanocrystals with a bright blue emission at 460–480 nm [71]. Introducing HDA to the TOPO-TOP mixture allows considerable enhancement of the luminescence quantum efficiency of CdSe nanocrystals compared to the particles grown in TOPO-TOP [68]. The highest reported room temperature PL quantum efficiency for HDA-TOPO-TOP-capped CdSe nanocrystals is 85% [69]. In combination with the narrow, symmetric emission spectrum (full width at half maximum is 23–30 nm, see Fig. 2.13a) this makes HDA-TOPO-TOP-capped CdSe nanocrystals excellent emitters in the visible spectral region. The less sterically hindered amines provide a larger capping density which improves the surface passivation and increases the luminescence efficiency [72].

Further development of the high temperature synthetic routes to CdSe nanocrystals was explored by Peng et al., who proposed to use more environmentally benign cadmium precursors (cadmium oxide [69, 73–75], cadmium carbonate [75] and cadmium acetate [75, 76]) instead of highly toxic, expensive, and pyrophoric $Cd(CH_3)_2$. Potential interest in these new cadmium precursors is caused by so-called "green chemical principles" and is motivated by the development of "user-friendly" chemical methodologies and materials [77]. Taking into account the growing demand for high-quality semiconductor nanocrystals for various applications (Section 2.5), the alternative synthetic schemes have to be safe, simple, inexpensive and suitable for industrial up-scaling and will be of great importance in the near future.

The shape of CdSe nanocrystals prepared in TOPO-TOP or HDA-TOPO-TOP mixtures is usually rounded, from nearly spherical (Fig. 2.12) to slightly prolate

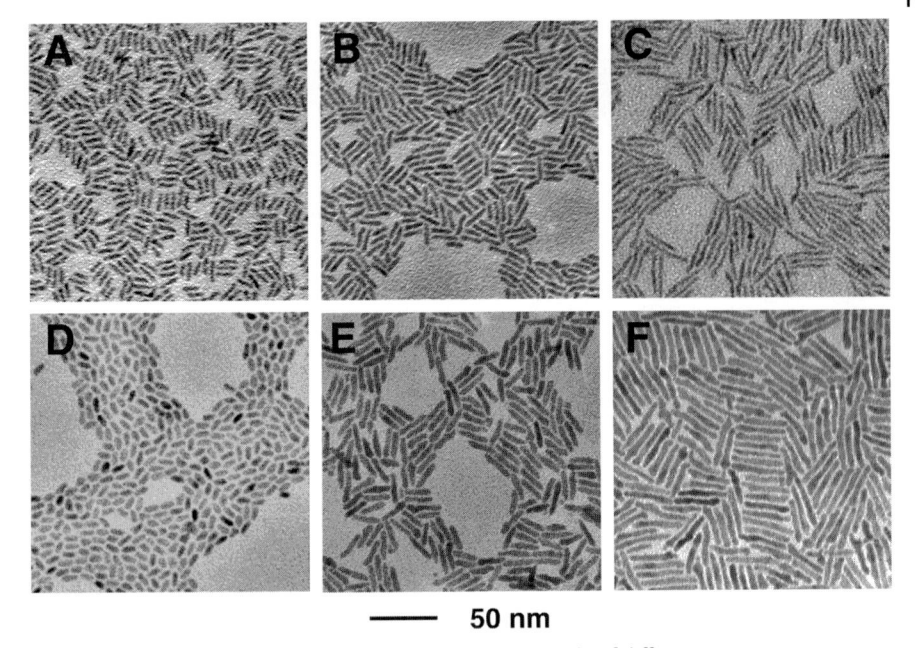

Fig. 2.14 Transmission electron micrograph of CdSe nanorods of different sizes and aspect ratios (courtesy of L. Manna and A. P. Alivisatos, reproduced from [82], with permission from American Chemical Society).

along the c-axis with the aspect ratio of $\sim$ 1.3 [21, 23]. Large (>6 nm) nanocrystals can be facetted with the overall shape assigned to the C_{3v} point group [57]. The addition of a relatively large amount (20–60 mol%) of hexylphosphonic or tetradecylphosphonic acid to the TOPO-TOP stabilizing mixture results in drastic changes of the nanocrystal shape: depending on the concentration of phosphonic acid and the heating regime, rod-, arrow-, rise-, teardrop- and tetrapod-shaped CdSe nanocrystals can be selectively synthesized (Figs. 2.14 and 2.15) [67, 74, 78]. These new objects are of great interest because they cannot be treated as purely "zero-dimensional" quantum dots [67]. Thus, CdSe nanorods (Fig. 2.14) can have an aspect ratio up to 30 and exhibit properties intermediate to those of 0D and 1D objects such as 2D quantum confinement [79, 80] and linear polarized luminescence along the c-axis [81]. They can also form liquid crystal phases [82].

Shape control of a nanocrystal becomes possible due to the difference in growth rate in different crystallographic directions. It was found that the concentration of low molecular weight cadmium compounds (monomer) in solution is a key factor in determining the shape of the resulting CdSe nanocrystals [83]. If the nanocrystal growth is kinetically overdriven by a high monomer concentration, the growth rate along the c-axis is much faster than that along any other axis, yielding rod-like particles [67, 83]. The role of the phosphonic acid lies in the formation of relatively strong soluble complexes with cadmium which increases its concentration in the solution [83]. Additionally, selective adsorption of phosphonic acid on cer-

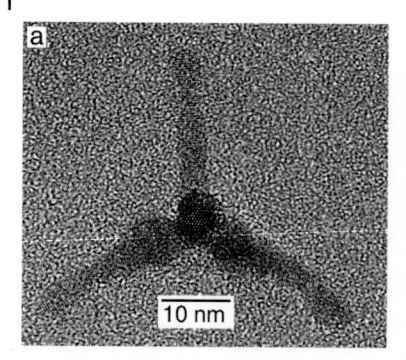

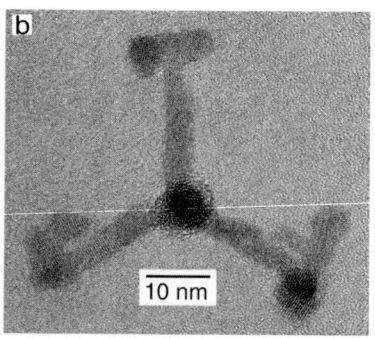

Fig. 2.15 a High-resolution transmission electron micrograph of a typical tetrapod-shaped CdSe nanocrystal, looking down the (001) direction of one arm. In image **b** a tetrapod with branches growing out of each arm is shown (Courtesy of L. Manna and A. P. Alivisatos, reproduced from [78], with permission from American Chemical Society).

tain facets blocks them for further growth [74] resulting in more complex shapes of CdSe nanocrystals.

The organometallic approaches described above result in CdSe nanocrystals capped with a shell of organic ligands (acids, amines, TOPO, and TOP). The organic molecules provide effective passivation of the surface dangling bonds and control the nanocrystal size and shape during the growth stage. However, a weakness of the organic ligands is their relatively poor stability resulting in the instability of the nanocrystal properties. HDA, TOPO, and TOP molecules in the stabilizing shell are rather labile and can be partially washed out during the post-preparative treatments. As an example, thorough washing of TOPO-TOP capped CdSe nanocrystals with methanol decreases their band edge luminescence by approximately two orders of magnitude.

A large step towards the preparation of robust highly luminescent nanocrystals was the passivation of their surface with an inorganic shell of a semiconductor with a wider band gap (usually ZnS or CdS) in analogy with the well-developed techniques for the growth of 2D quantum wells. The lattice mismatch between CdSe core and ZnS or CdS shell is small enough to allow epitaxial growth. In CdSe/ZnS and CdSe/CdS core-shell nanocrystals the large band-gap semiconductor forms a closed outer shell and the band edges of the core material lie inside the band gap of the outer material (Fig. 2.16). The outer inorganic shell provides high luminescence efficiency due to an efficient passivation of the core surface states and considerably improves the chemical stability and photostability of the nanocrystal properties. Chemical stability implies negligible changes of properties like particle size, solubility, luminescence efficiency, etc. after exposing the nanocrystals to various surroundings, e.g. to the ambient atmosphere, whereas photostability means stability of the nanoparticle properties under illumination.

The preparation of core-shell nanocrystals usually involves two stages. First, the CdSe cores are synthesized by one of the approaches described above. Then, these

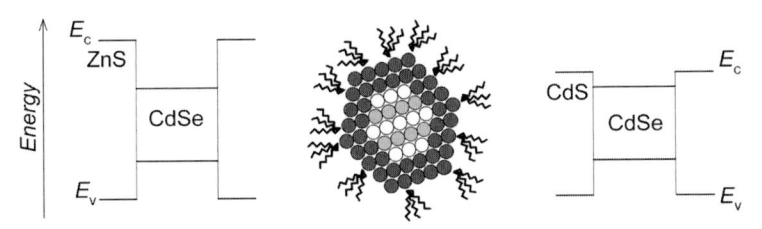

Fig. 2.16 Schematic representation of the band structure of the core shell CdSe/ZnS and CdSe/CdS nanocrystals.

cores are used as seeds for the epitaxial growth of the CdS [52], ZnS [68, 84, 85] or ZnSe [86] shell. Since CdSe nanocrystals synthesized in the HDA-TOPO-TOP mixture possess exceptionally narrow size distributions, their size fractionation and isolation from the crude solution is not required for the synthesis of core-shell particles [68]. In the case of CdSe/ZnS core-shell nanocrystals, diethyl zinc and bis-trimethylsilylsulfide are used as the ZnS precursors [68, 84]:

$$\text{CdSe-seeds} + \text{Zn}(CH_3)_2 + [(CH)_3Si]_2S \xrightarrow{\text{HDA-TOPO-TOP}, 200\,°C}$$
$$\text{CdSe/ZnS nanocrystals} \qquad (8)$$

The shell growth is carried out at relatively low temperatures (150–220 °C) to prevent Ostwald ripening of CdSe seeds. The amount of ZnS precursors necessary to obtain the desired shell thickness can be calculated from the ratio between the core and shell volumes using bulk lattice parameters of CdSe and ZnS. Fig. 2.17 compares HRTEM images of CdSe nanocrystals before and after growth of ZnS shell (∼ 1.6 monolayers of ZnS). The lattice fringes stretch through the entire nanocrystals indicating epitaxial growth. Fig. 2.17 also shows the evolution of the absorption and emission spectra of the samples upon growth of ZnS shells of different thickness. The luminescence efficiency of the CdSe/ZnS nanocrystals is considerably higher than that of the initial CdSe cores. Its maximum is usually observed for a ZnS shell thickness between 1 and 2 monolayers. For thicker shells, the strains due to the lattice mismatch between CdSe and ZnS result in dislocations, leading to a degradation of optical properties [85].

The spectral position of the emission band of CdSe nanocrystals does not shift considerably during the growth of ZnS shell showing that in the excited state both electron and hole remain confined to the CdSe core. Using CdSe nanocrystals of different sizes for growing the core-shell particles, a series of colloidal solutions emitting from blue to red with the emission band as narrow as 25–35 nm (FWHM) and room temperature PL quantum yield as high as 50–70% have been synthesized (Fig. 2.18). In fact, the ZnS epitaxial shell can also be grown at the surface of CdSe nanorods providing highly luminescent nanocrystals with polarized emission [87].

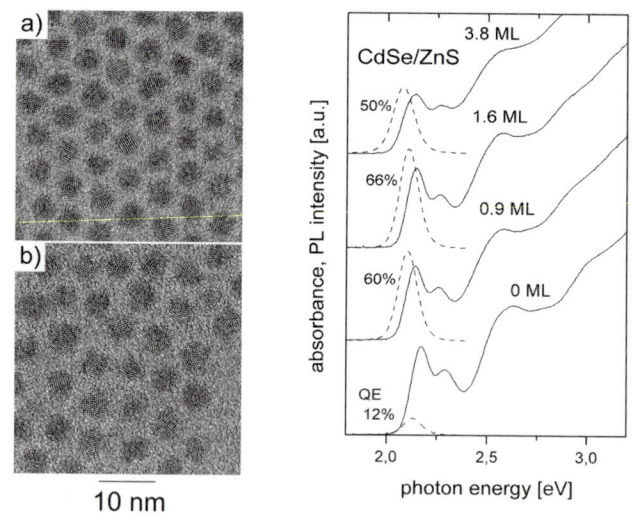

Fig. 2.17 Left: HRTEM images of **a** CdSe nanocrystals and **b** CdSe/ZnS core-shell nanoparticles grown from these CdSe cores (∼1.6 monolayers of ZnS). Right: Room-temperature absorption and emission spectra of CdSe nanocrystals before and after growth of a ZnS shell of different thickness (in monolayers, ML) (reproduced from [68], with permission from American Chemical Society).

Fig. 2.18 Size-dependent change of the emission color of colloidal solutions of CdSe/ZnS core-shell nanocrystals. The particles with the smallest (∼1.7 nm) CdSe core emit blue, the particles with the largest (∼5 nm) core emit red.

2.3.4.2 Other II–VI Nanocrystals

During the last decade plenty of organometallic routes towards different II–VI nanocrystals were proposed. However, only some of them provide nanocrystals of sufficiently high quality (in aspects described in the Introduction). Some successful organometallic synthetic approaches proposed for the synthesis of different II–VI nanocrystals are presented below.

CdS

The availability of a reliable aqueous synthesis leading to nearly monodisperse series of CdS nanocrystals [35] somewhat restricted interest in their organometallic synthesis. Wurtzite CdS nanocrystals can be prepared in a similar way to that used for CdSe, i.e. in a TOPO-TOP stabilizing mixture [23]. The precursors in this case are dimethylcadmium and bis-(trimethylsilyl)sulfide. If tetradecyl- or hexylphosphonic acid is added to the TOPO-TOP stabilizing mixture, cadmium oxide can be used as a cadmium precursor instead of $Cd(CH_3)_2$ which makes the synthesis more environmentally benign [73]. Shape control leading to rod-, bipod-, tripod-, and tetrapod-shaped CdS nanocrystals has been achieved via the thermal decomposition of $Cd(S_2CNEt_2)_2$ single molecule precursor in hexadecylamine as stabilizing solvent [88]. The synthesis of CdS/ZnS core-shell nanocrystals efficiently emitting in the blue spectral region has been proposed [89].

CdTe

CdTe nanocrystals are much less stable against degradation under ambient conditions than CdS and CdSe nanocrystals due to the oxidation of Te atoms at the nanocrystal surface. Therefore, only the proper passivation of the nanocrystal surface with organic ligands or an inorganic shell can provide processable CdTe nanocrystal colloids. Only a few reports on non-aqueous chemical routes to CdTe nanoparticles have been published [23, 71, 90–94]. Thus, the synthesis of CdTe nanocrystals from dimethylcadmium and trioctylphosphine telluride (TOPTe) in TOPO-TOP medium has been reported [23]. However, these nanocrystals exhibit inefficient luminescence, with quantum efficiencies of less than 1% [91]. Surprisingly, the CdTe nanocrystals prepared in TOPO-TOP medium have a hexagonal structure [23] whereas all other aqueous and organometallic routes yield a cubic (zinc blende) CdTe phase. The synthesis of CdTe nanocrystals from the same precursors in dodecylamine-TOP medium at 150–220 °C provides relatively monodisperse CdTe nanocrystals with sizes tunable from ~ 2.5 to 7 nm, the emission colors of which cover the visible spectral range from green to red [90]. The room temperature luminescence quantum yield of amino-capped CdTe nanocrystals reaches 65% [90]. Among other non-aqueous approaches to CdTe nanocrystals, the syntheses from cadmium oxide and TOPTe in tetradecylphosphonic acid-TOPO-TOP medium [73] and from $Cd(CH_3)_2$ and a special tellurium precursor (hexapropyl phosphorus triamide telluride) in TOPO-TOP medium [91] can be mentioned. The latter provides CdTe nanocrystals sized from ~ 4.5 to 13 nm, the luminescence of which is tunable from 580 to 770 nm; the quantum efficiency of these samples ranges from 40% to 65% [91].

ZnSe

The band-gap energy of bulk ZnSe (2.58 eV) is larger than those of CdS (2.53 eV), CdSe (1.74 eV), and CdTe (1.50 eV). Therefore, the synthesis of luminescent ZnSe nanocrystals might allow the expansion of the nanocrystal emission color to the UV spectral region. The TOPO-TOP stabilizing mixture was found to be inefficient for the stabilization of ZnSe particles [72]. On the other hand, the synthesis of ZnSe nanocrystals from diethylzinc and trioctylphosphine selenide in hexadecylamine-TOP coordination mixture at 270–310 °C provides relatively monodisperse highly luminescent ZnSe nanocrystals [72, 95]. Their band edge luminescence is tunable in size from 365 nm (3.4 eV) to 440 nm (2.8 eV), i.e. effectively covers the UV and blue spectral region. The reported values for the luminescence quantum yields are between 20% and 50% relative to Stilbene 420 [72].

PbSe

PbSe nanocrystals are of great potential interest due to the unique properties of bulk PbSe. Thus, bulk PbSe has a cubic (rock salt) crystal structure, a narrow direct band gap (0.26 eV) and a large exciton Bohr radius (46 nm, about eight times larger than that of CdSe). As a result, size-quantization effects are much more pronounced in PbSe than in cadmium chalcogenides [96], and the value of the band gap energy makes the PbSe nanocrystals interesting as emitters in the near-IR region. High-quality PbSe nanocrystals were recently synthesized from lead oleate and trioctylphosphine selenide in the diphenylether-TOP medium at 90–220°C [97]. The as-prepared nanocrystals have $\sim$ 10% size distribution which can be narrowed to 5% or even further by a size-selective precipitation procedure. The isolated monodisperse PbSe fractions sized from $\sim$ 3 nm to 15 nm exhibit a pronounced quantum size effect resulting in a shift of the first absorption maximum from $\sim$ 1200 nm for 3 nm particles to 2200 nm for 9 nm nanocrystals [97]. The best reported values for PL efficiency of PbSe nanocrystals were as high as $\sim$ 60%. The passivation of PbSe nanocrystal surface via the epitaxial growth of a PbS shell results in an improvement of both PL efficiency and nanocrystal stability [96].

2.3.4.3 III–V Nanocrystals

While II–VI semiconductor nanocrystals have been studied extensively during the last two decades, much less information is available about their III–V analogues. The reason is that III–V semiconductors are more-covalent compounds and high temperatures are usually required for their synthesis. For covalent semiconductors, the synthesis of nanocrystals by colloidal techniques becomes increasingly difficult [98]. On the other hand, III–V nanocrystals can exhibit even more pronounced quantum size effects than II–VI materials, as they have relatively covalent bonding and direct band gap structure, larger bulk exciton radii and smaller effective masses of electron and holes. Interest in III–V nanocrystals is also encouraged by "green chemical principles" [77] as some III–V compounds, e.g. InP, are more environmentally benign than the cadmium- and lead-related II–VI ones.

Papers discussing the synthesis and characterization of InP [99–105], GaP [100], and InAs [106, 107] nanocrystals have appeared during recent years. A large step toward controllable synthesis of III–V nanocrystals was made by Micic et al. who adapted the Wells' dehalosylilation reaction [108] for the synthesis of nanocrystalline InP colloids [99]. Alivisatos et al. [106] have extended this approach to InAs. The luminescent properties of as-prepared InP and InAs nanocrystals are rather poor compared to II–VI compounds, and several ways of improvement have been proposed. Thus, Micic et al. proposed to etch InP nanocrystals with HF which results in a strong enhancement of their PL efficiency [102, 103, 109]. The growth of an epitaxial shell of a higher band gap II–VI or III–V semiconductor on the core InAs nanocrystal also increases their luminescent efficiency [110, 111].

InP Nanocrystals

InP nanocrystals were synthesized via the Wells' dehalosilylation reaction [108] in presence of stabilizing agents preventing a growth of the bulk InP phase (Eq. 5). This reaction was applied for the preparation of InP nanocrystals in TOPO [101], TOPO-TOP [98, 99], and dodecylamine-TOP [71] media. In all cases the reaction pathway involves a large energetic barrier resulting in a very slow particle-nucleation stage [101]. As a consequence, no temporal separation of particle nucleation and growth via the hot-injection technique can be achieved and control over the particle size distribution in the reaction mixture is poor in comparison with that for II–VI nanocrystals. The InP particles which form after the nucleation stage are usually amorphous and annealing for 1–6 days at temperatures above 250 °C is required to improve their crystallinity [99, 101].

Synthesis of InP nanocrystals in TOPO-TOP allows InP nanocrystals sized from ~2.2 nm to 6 nm to be obtained, whereas synthesis in dodecylamine or trioctylamine results in smaller sizes [71]. Long-term (up to 3–5 days) annealing at 250–280 °C provides a high degree of particle crystallinity (Fig. 2.19). The broad size distribution (~25–30%) of as-prepared TOPO-TOP-capped InP nanocrystals does not allow the utilization of their size-dependent properties. However, the nanocrystal size distribution can be considerably improved by post-preparative size-selective precipitation. Fig. 2.19 shows a set of absorption spectra of size-selectively isolated InP nanocrystals. TEM and HRTEM investigations show that carefully performed post-preparative size fractionation allows achievement of relatively narrow (8–10%) size distributions of InP nanocrystals (Fig. 2.19). The TOPO-TOP-capped InP nanocrystals exhibit very poor luminescence, composed of a weak band edge luminescence and the luminescence from deep traps [71, 101]. The luminescence intensity increases with the oxidation of the nanocrystal surface. How-

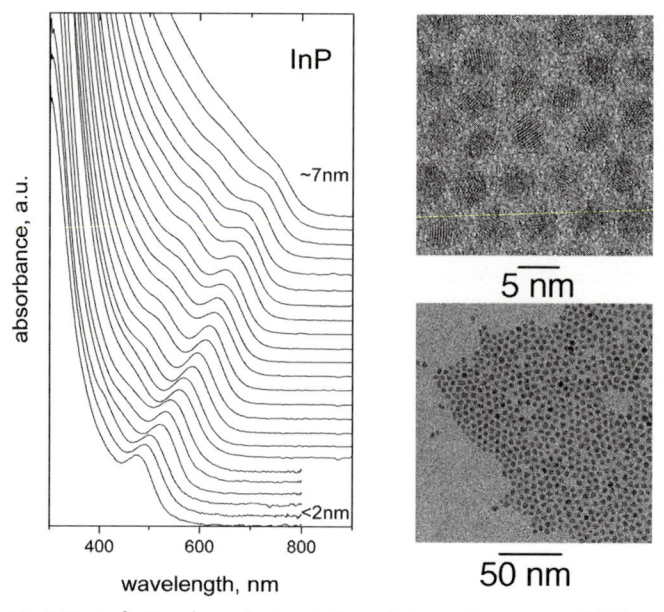

Fig. 2.19 Left: Size-dependent evolution of absorption spectra of InP nanocrystals. Right: HRTEM and TEM images of a size-selected fraction of InP nanocrystals grown in a TOPO-TOP mixture.

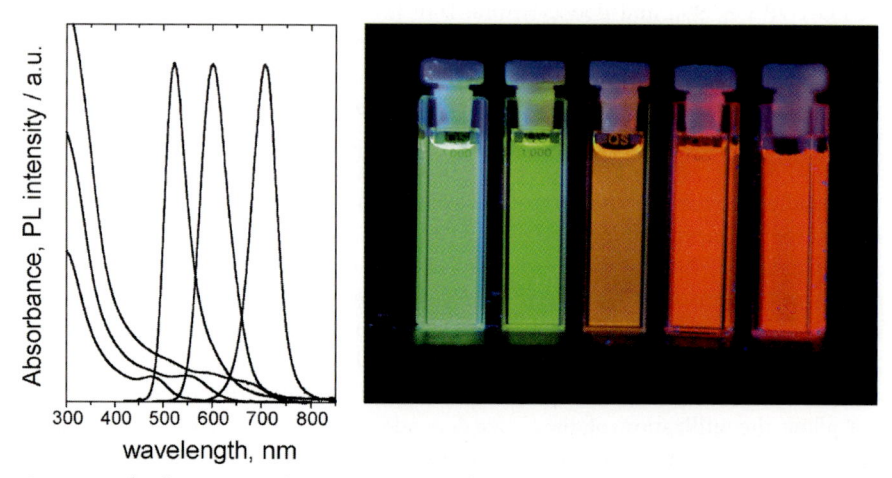

Fig. 2.20 Left: Absorption and emission spectra of HF-photoetched InP nanocrystals. Right: Size-dependent change of the photoluminescence color of HF-photoetched InP nanocrystals.

ever, the luminescence efficiency of oxidized nanocrystals does not exceed ∼ 0.01%. Fortunately, this value can be drastically improved via a treatment of the InP nanocrystal surface with certain fluorine compounds.

Low luminescence efficiency of as-prepared InP nanocrystals is usually a result of a non-adequate passivation of the nanocrystal surface. Non-passivated surface sites (often considered as so-called dangling bonds) may serve as traps for photo-generated carriers and provide non-radiative recombination pathways. A proper post-preparative passivation of the particle surface can eliminate these traps. Thus, the treatment of InP nanocrystals with HF considerably improves their band edge luminescence [109]. Our experiments have shown that the etching process has a photochemical nature, and is especially effective when performed under illumination with photons having an energy above the band gap of InP nanocrystals. The emission of photoetched InP nanocrystals is tunable from green to the near-IR region by increasing the nanocrystal size (Fig. 2.20), with a quantum efficiency between 20 and 40%.

InAs Nanocrystals

InAs nanocrystals were synthesized via the dehalosilylation reaction similar to InP nanocrystals [106]. TOP has been chosen as the reaction medium because it provides good control of the particle growth and their size distribution. The application of the hot injection technique allowed the successful separation of particle nucleation and growth, yielding size distributions of as-prepared nanocrystals of ∼ 15%. As a rule, no long-term annealing is required to obtain highly crystalline InAs particles. To isolate nearly monodisperse fractions, size-selective precipitation can be applied.

Fig. 2.21a shows the evolution of the absorption spectrum of InAs nanocrystals upon increasing the particle size from ∼ 1.8 to 4.5 nm. The band gap of bulk InAs is 0.46 eV at room temperature corresponding to the absorption onset at 2.7 μm so that the excitonic transitions of InAs nanocrystals appear in the near-IR spectral region. The absorption spectra possess several features indicating a high degree of particle monodispersity and crystallinity. A typical XRD pattern of InAs nanocrystals shows that the positions of diffraction peaks at wide angles match those of the bulk phase of cubic InAs (Fig. 2.21b). The HRTEM image shown in Fig. 2.21c demonstrates well-separated crystalline InAs particles with clearly resolved lattice fringes and a moderately narrow size distribution.

The luminescence efficiency of as-prepared InAs nanocrystals is ∼ 1% [106, 111]. It can be improved by more than one order of magnitude by growth of an epitaxial shell of either a II–VI (CdSe, ZnSe, ZnS) or a wide-band gap III–V (InP, GaAs) semiconductor at the surface of InAs nanocrystals [110, 111]. The resulting core-shell nanocrystals emit in the near-IR and exhibit a room temperature quantum efficiency of about 20% [111].

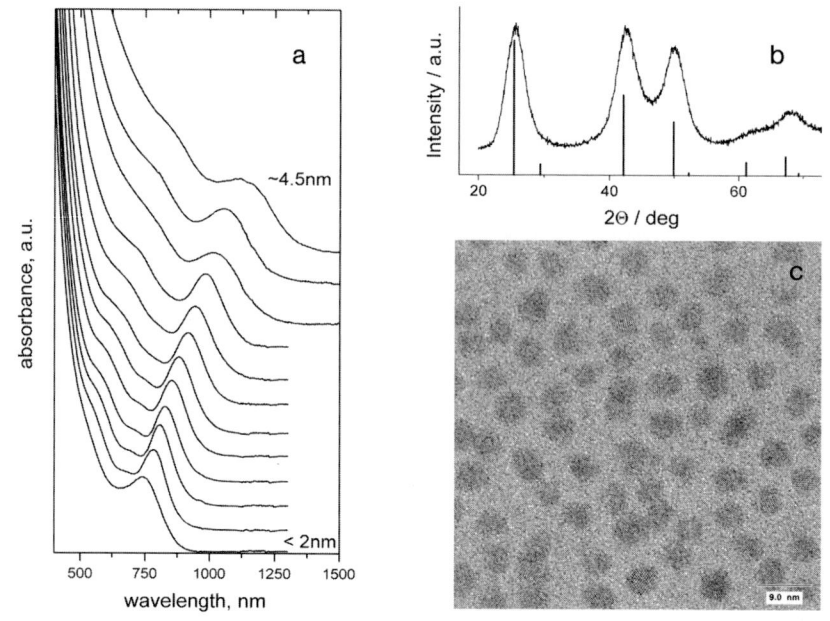

Fig. 2.21 a Size-dependent evolution of an absorption spectrum of InAs nanocrystals. **b** A typical powder X-ray diffractogram of InAs nanocrystals. The line spectrum gives the bulk InAs (zinc blende, cubic) reflections. **c** HRTEM image of a size-selected fraction of InAs nanocrystals.

GaP and GaAs Nanocrystals

The synthesis of gallium-related nanocrystals is more complex than that of indium-related ones. Very high temperatures are required to achieve satisfactory particle crystallinity. Thus, the synthesis of crystalline GaP particles from $GaCl_3$ and tris-(trimethylsilyl)phosphine in TOPO-TOP medium was possible only after long-term annealing at 370–400 °C [106]. The GaP nanocrystals prepared by this approach show interesting and unusual quantum size effects arising from the interplay between the direct and the indirect transitions [98]. However, their broad size distribution in combination with a small range of available particle sizes (from $\sim$ 2 to 3 nm) [106] did not allow detailed investigation of their optical properties. In the case of GaAs nanoparticles, although some approaches to their synthesis have been described [112, 113], their controllable preparation is still a challenge.

2.3.5
Post-preparative Size-selective Fractionation of Nanoparticles

As already mentioned above, the nanoparticle size distribution can often be improved through the post-preparative size fractionation which is based on the size-dependent variation of particle properties. The most frequently used size-selective

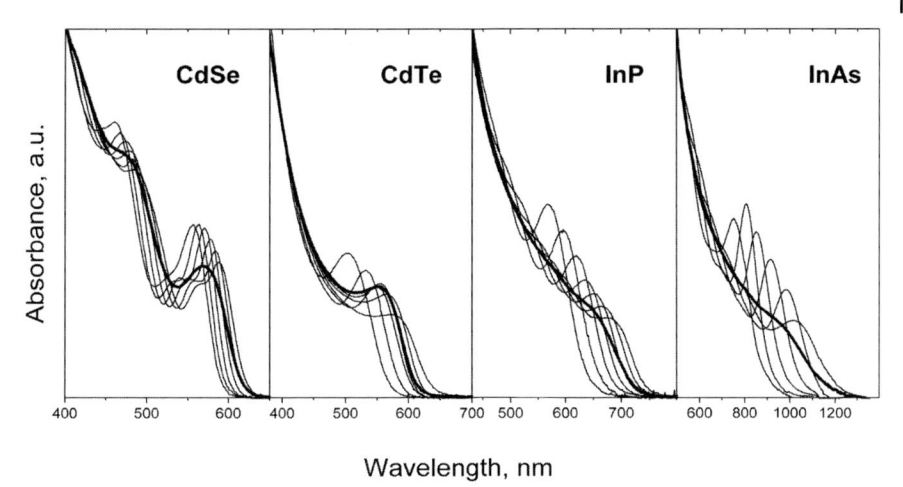

Fig. 2.22 Absorption spectra of size-selected fractions (thin lines) obtained from crude solutions (thick lines) of various II–VI and III–V nanocrystals by applying the size-selective precipitation technique. Details of preparation and solvent/nonsolvent pairs used: CdSe, TOPO-TOP synthesis [23], n-butanol/ethanol; CdTe, aqueous synthesis [33], capped by thioglycolic acid, water/isopropanol; InP, dehalosilylation reaction [102], capped with TOPO-TOP, toluene/methanol; InAs, dehalosilylation reaction [106], capped with TOP, toluene/methanol.

precipitation technique exploits the difference in solubility of smaller and larger particles [23, 114]. A typical procedure for carrying out size-selective precipitation on a nanoparticle colloid is as follows: A sample of as-prepared nanoparticles with a broad size distribution is dispersed in a solvent and a non-solvent is added dropwise under stirring until the initially optically clear solution becomes slightly turbid. The largest nanoparticles in the sample exhibit the greatest attractive van der Waals forces and tend to aggregate before the smaller particles. The aggregates consisting of the largest nanoparticles can be isolated by centrifugation or filtration and redissolved in any appropriate solvent. The next portion of non-solvent is added to the supernatant to isolate the second size-selected fraction, and so on. The procedure can be repeated several times and allows up to ~ 20 size-selected fractions to be obtained from one portion of the crude solution. Moreover, each size-selected fraction can be subjected again to size selection to further narrow the size distribution. Fig. 2.22 shows several examples of post-preparative size fractionation for different II–VI and III–V semiconductor nanocrystals. All size-selected fractions possess sharp excitonic transitions in the absorption spectra, which is direct evidence of their narrow particle size distributions. TEM and HRTEM investigations show that a carefully performed size-selective precipitation allows the achievement of size distributions as narrow as ~ 4–7%, depending on the material.

Another important feature of post-preparative size fractionation of the luminescent nanocrystals is the drastic difference in the luminescence efficiencies of size-

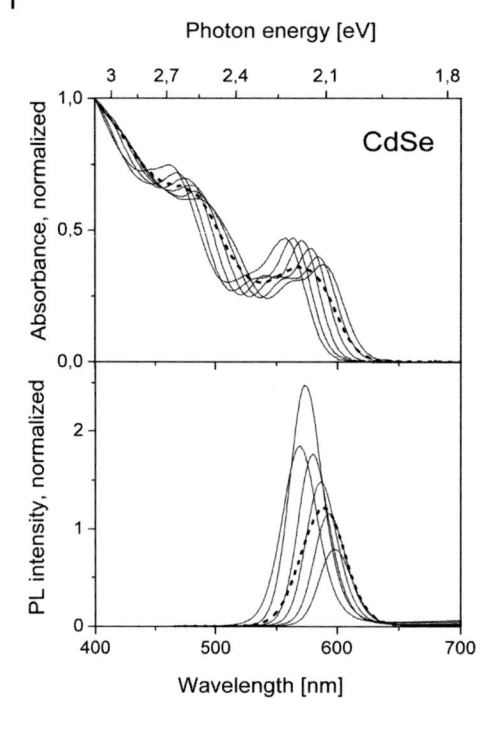

Fig. 2.23 Absorption (top) and emission (bottom) spectra of size-selected fractions of CdSe nanocrystals (solid lines). The absorption and luminescence spectra of the crude solution (dashed lines) are shown for comparison. Luminescence intensities are normalized to identical absorbance at the excitation wavelength (400 nm).

selected fractions that was observed for both organometallically prepared CdSe and InAs colloids and for CdTe nanocrystals synthesized in aqueous medium [115]. Fig. 2.23 shows, as an example, the absorption and emission spectra of six size-selected fractions obtained from a single portion of the crude solution of TOPO-TOP-capped CdSe nanocrystals. The size distribution in each fraction is, as expected, narrower than in the initial crude solution, leading to more pronounced electronic transitions in the absorption spectra and narrower luminescence bands. The absorption and emission spectra of the size-selected fractions shift to shorter wavelengths with decreasing mean particle size. It is noteworthy that the efficiency of the band-edge emission of the size-selected fractions of CdSe nanocrystals shows a clear dependence on the "fraction number". The maximum of the luminescence efficiency is always achieved for a fraction lying in the middle of the size-selected series, independent of the mean particle size of the nanocrystals in the initial crude solution. Similar variations of the luminescence intensity were observed for size-selected fractions of the crude solutions of thiol-stabilized CdTe nanocrystals and TOP-stabilized InAs nanocrystals [115].

This difference in the luminescence properties is attributed to the different averaged surface disorder of the nanocrystals originating from the Ostwald ripening growth mechanism when larger particles in the ensemble grow at the expense of dissolving smaller particles. At any stage of growth, only a fraction of particles within the ensemble has the most perfect surface and, thus, shows the most efficient photoluminescence. In an ensemble of growing nanocrystals, the fraction of particles with the highest photoluminescence corresponds to the particle size having a nearly zero average growth rate [115]. The smallest average growth rate leads to the lowest possible degree of surface disorder for any given reaction conditions.

2.4
Semiconductor Nanoparticles as Building Blocks of Superstructures

2.4.1
Layer-by-layer Assembly of Water-soluble Nanoparticles

Water-soluble semiconductor nanoparticles capped by functionalized thiols can easily be processed into high-quality thin films with thicknesses controllable on the nanometer scale using a so-called layer-by-layer (LBL) assembly approach. The LBL assembly is based on the alternating adsorption of oppositely charged species, such as positively and negatively charged polyelectrolyte pairs [116] or polyelectrolytes and nanoparticles [117]. This technique can be equally effectively applied to the coating of both macroscopically flat [39, 41] and highly curved surfaces like colloidal spheres [41, 42]. Fig. 2.24 shows a schematic illustration of the LBL assembly of polyelectrolytes and nanoparticles on planar substrates and colloidal spheres. In both cases, the consecutive electrostatic adsorption of oppositely charged species leads to the formation of multilayer composite structures.

The methodology is very general and produces large-area high-quality homogeneous films almost irrespective of the substrate with a thickness up to microns precisely controllable on the nanometer scale. It is possible to create up to 100 or more alternate "bilayers" of nanoparticles and polyelectrolytes with near nanometer precision in total film thickness. Polymer layers can be chosen to be insulating or conducting. The dense packing of nanoparticles and polyelectrolyte chains in composite films promotes regularity in the deposited layers. The equal amounts of species transferred in each deposition cycle can be seen from the se-

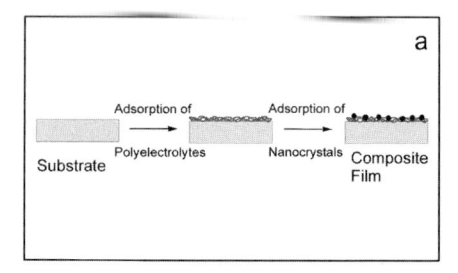

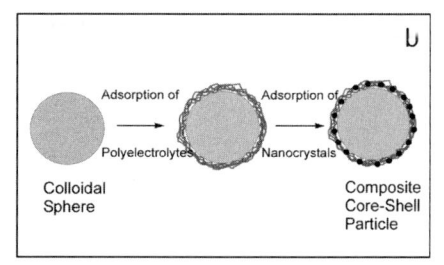

Fig. 2.24 Schematic illustration of the layer-by-layer assembly of polyelectrolytes and nanoparticles on planar substrates **a** and colloidal spheres **b** (reproduced from [41], with permission from Elsevier Science).

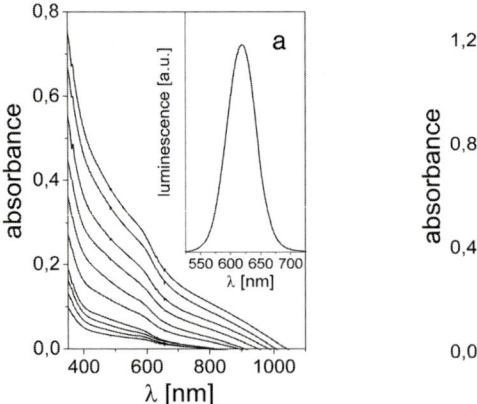

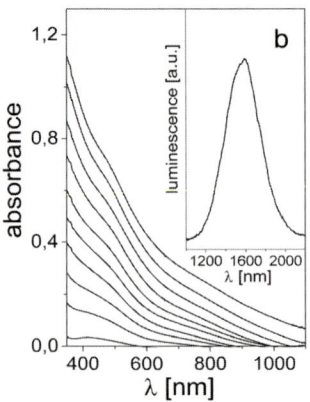

Fig. 2.25 Absorption spectra of LBL-assembled composite films produced from poly(diallyldimethylammonium chloride) and CdTe **a** or HgTe **b** nanocrystals on planar glass substrates as a function of the number of deposition cycles. The insets show the luminescence of CdTe **a** and HgTe **b** nanocrystals in the corresponding twenty-layer films (reproduced from [41], with permission from Elsevier Science).

quentially recorded absorption spectra of the multilayers of CdTe and HgTe nanocrystals deposited on planar substrates (Fig. 2.25). The optical density at a selected wavelength increases almost linearly with the number of deposited layers. Using luminescent nanoparticles it is possible to create composite films with different architectures [40], where the emission is controlled by particle size (insets in Fig. 2.25).

The LBL approach employed to cover latex microspheres represents a general technique to produce composite core-shell particles with luminescent nanocrystals being embedded in the multilayer polyelectrolyte/nanocrystal shells [41, 42]. Fig. 2.26a shows a confocal microscopy image of the four-layer polyelectrolyte/

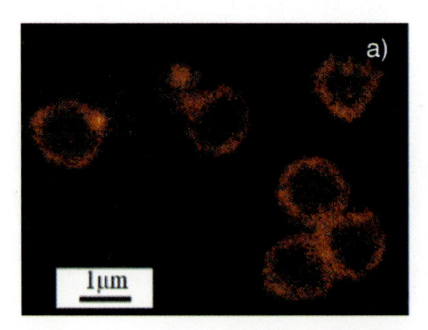

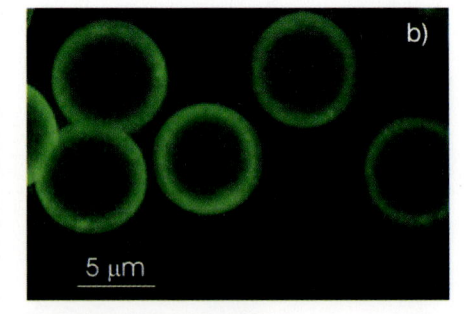

Fig. 2.26 Confocal microscopy images of the luminescent CdTe nanocrystal shells assembled by the LBL technique **a** and by the solvent-controlled precipitation technique **b** on latex spheres. (part **a** is reproduced from [42], with permission from Elsevier Science).

CdTe nanocrystal shells assembled on latex spheres of 1.5 μm diameter. The luminescence from the red-emitting nanoparticles (4 nm size) produces a ring around the latex cores. Another approach, which can effectively be used to cover colloidal particles with thicker shells of nanocrystals, is a recently developed solvent-controlled precipitation technique [118]. It is a variation of heteroaggregation when the coagulation of nanocrystals on colloidal cores is induced by the solvent-nonsolvent-pair precipitation technique. Fig. 2.26b shows a confocal microscopy image of the luminescing shells of the green-emitting CdTe nanocrystals (2.5 nm size) on latex spheres of 5 μm diameter. The coating was done by precipitation of nanocrystals from aqueous solutions by adding ethanol. Further details on the LBL coating of materials can be found in Chapters 8, 13 and 18.

2.4.2
Two-dimensional Arrays of Monodisperse Nanocrystals Formed by Self-assembly

Monodisperse nanoparticles easily organize themselves in ordered structures because of dispersive attractions caused by van der Waals forces [21, 119]. Both 2D and 3D arrays can be prepared simply by placing a drop of a colloidal solution of monodisperse nanoparticles on a suitable support and allowing the carrier solvent to evaporate slowly. Fig. 2.27 shows HRTEM images of 2D arrays of CdSe nanocrystals on carbon-coated copper TEM grids. A long-range hexagonal ordering is observed in all cases and the regular inter-particle spacing is due to the capping ligand shells around the nanocrystals. The spacing between nanocrystals can be adjusted by employing different ligands at the synthesis stage or by post-preparative cap exchange [21].

The orientation of lattice planes of nearly spherical nanocrystals forming 2D arrays is normally random (Fig. 2.27a). However, in some cases a preferred orientation is observed, e.g. for relatively large (7 nm) facetted CdSe nanocrystals elon-

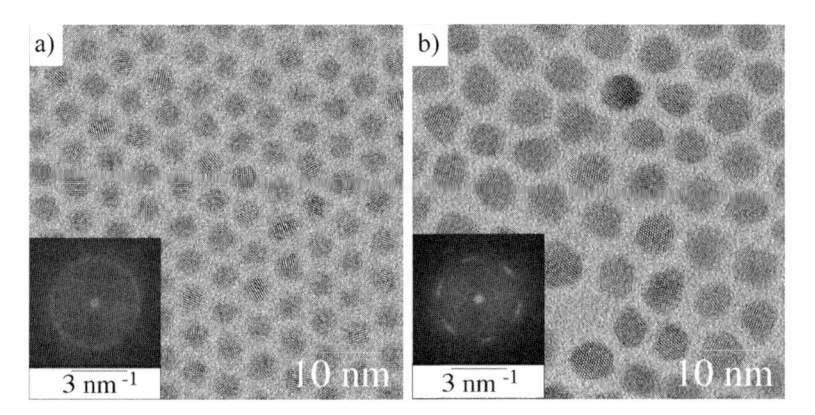

Fig. 2.27 High-resolution transmission electron microscope images of 2D arrangements of CdSe nanocrystals. **a** Fragment of an array of the 3.5 nm particles together with a correspondent FFT; **b** fragment of an array of the 7 nm particles together with a correspondent FFT (from [123]).

gated along the *c*-axis of the wurtzite structure (Fig. 2.27b). These nanocrystals are aligned on the TEM grid with the *c*-axes perpendicular to the substrate. Importantly, there is even a correlation in orientation of the lattice plane directions of single nanocrystals, which is further confirmed by six distinct reflexes in the correspondent FFT pattern being characteristic of the (100) zone (inset in Fig. 2.27b), in contrast to the isotropic FFT ring of randomly oriented CdSe particles (Fig. 2.27a).

The ordering of nanocrystals in superstructures is also reflected in small-angle XRD patterns through the appearance of Bragg diffraction peaks in the region of 2θ angles of $\sim 1°–15°$. The diffraction signal is averaged over a large number of nanocrystals in this case, whereas TEM provides information from selected areas only. A fast evaporation of a low-boiling solvent (hexane) results in the formation of a glass-like film with a local-range particle order and a liquid-like radial distribution function (Fig. 2.28a, curve 1). On the other hand, CdSe nanocrystals precipitated via slow evaporation of a higher-boiling solvent (e.g. toluene) form films with a long-range particle order exhibiting pronounced reflections in the small-angle XRD pattern (Fig. 2.28a, curve 2). Most of the reflections can be attributed to the fcc lattice of the CdSe nanocrystals.

The wide-angle part of the XRD patterns corresponds to the diffraction of X-rays on the atoms the nanocrystals consist of, and allows the estimation of the average size of the crystalline domains within each nanocrystal. The width of the diffraction peaks at wide angles is considerably broadened and increases with decreasing particle size (Fig. 2.28b). CdSe nanocrystals with sizes above ~ 4 nm exhibit XRD patterns with diffraction peaks in accord with those of hexagonal CdSe (wurtzite phase). In the case of smaller CdSe nanocrystals the XRD patterns do not permit the cubic and the hexagonal phases to be distinguished unambiguously.

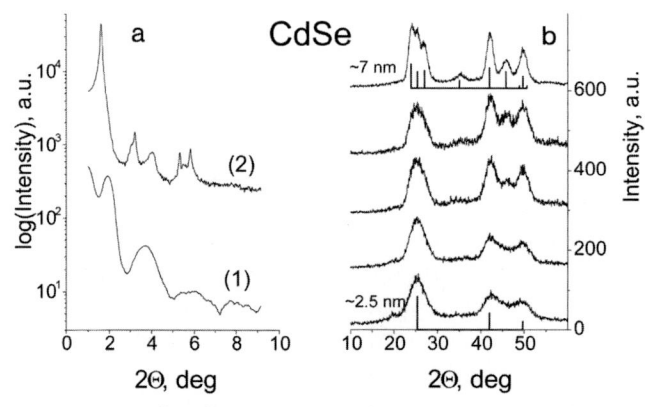

Fig. 2.28 **a** Small-angle XRD patterns of glassy (curve 1) and long-range ordered (curve 2) films of CdSe nanocrystals. **b** Wide-angle XRD patterns of a size series of CdSe nanocrystals. Vertical lines indicate bulk CdSe reflections (top: wurtzite, hexagonal; bottom: zinc blende, cubic).

2.4.3
Colloidal Crystals of Nanocrystals

Another strategy to fabricate ordered superstructures from monodisperse nanoparticles is a gentle destabilization of the colloidal dispersion. Again, dispersive attractions of nanocrystals drive their self-organization and superlattice formation. In superstructures formed by this way, individual nanoparticles playing the role of building blocks (artificial atoms in the next level of hierarchy) are aligned in a regular 3D lattice ("artificial solid"). Apart from the crystals of nanoclusters [46, 47, 120–122], there have been few reports on colloidal crystals consisting of semiconductor nanocrystals [21, 58]. Recently, a simple three-layer technique of controlled oversaturation has been used to crystallize monodisperse CdSe nanoparticles into perfectly faceted colloidal crystals with sizes of ~ 100 µm [123]. The method also works well in the case of FePt [127] and $CoPt_3$ [128] nanocrystals, indicating its wide or even general applicability.

Fig. 2.29a presents a schematic outline illustrating the concept of the crystallization procedure. A colloidal solution of nanocrystals in a solvent like toluene is placed in a vertical glass tube, and the system is slowly destabilized by diffusion of a non-solvent (e.g. methanol) into the colloid, resulting in nucleation and growth of colloidal crystals, preferentially on the walls of the tubes. The spatial distribution of local oversaturations caused by the non-solvent diffusion determines the quality of the colloidal crystals. To make the oversaturation front not as sharp as in the case of a direct solvent–non-solvent contact, a third buffer layer of propanol-2 is used between the solution of the nanocrystals and the methanol layer.

Crystalline nuclei start to form after about one to two weeks and grow slowly to colloidal crystals during one to two months. Fig. 2.29b and c show optical micrographs of colloidal crystals of CdSe nanocrystals. The images were taken with a digital camera through the objective of an optical microscope. Irregular red-colored colloidal crystals of CdSe nanoparticles formed in the absence of a buffer layer reach 80–220 µm in size (Fig. 2.29b). In the presence of a buffer layer, colloidal crystals grew in the form of perfectly faceted hexagonal orange-red colored platelets being very similar in size, about 100 µm in lateral dimension and 20 µm in height (Fig. 2.29c).

Fig. 2.30a presents a TEM image of a 3D arrangement of the 3.5 nm TOPO-capped CdSe nanocrystals at low magnification. Close packing of up to seven layers of nanocrystals into a 3D superlattice can be seen. Further investigation of this superlattice shows the two different arrangements of CdSe nanoparticles given in Fig. 2.30b and c. The FFT patterns of these arrangements (given as insets) are consistent with those expected for the (100) and (110) projection of an fcc superlattice, respectively.

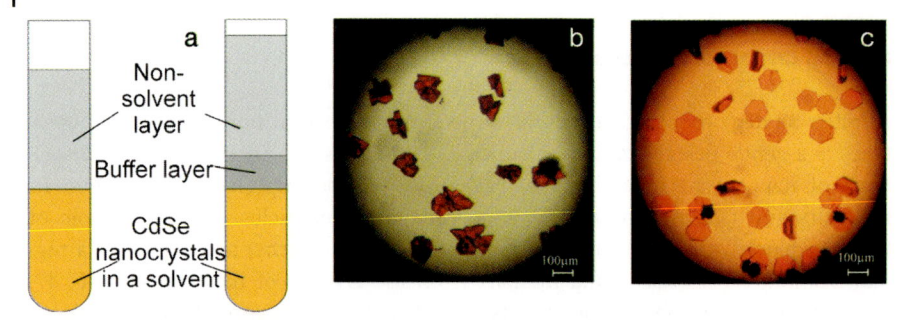

Fig. 2.29 **a** Schematic outline illustrating the concept of the oversaturation technique used for the crystallization of monodisperse nanocrystals. Left tube: the non-solvent (methanol) diffuses directly into a colloidal solution of CdSe nanocrystals in toluene. Right tube: intervening buffer layer of propanol-2 is used, allowing colloidal crystals of higher quality to be obtained. **b,c** Optical micrographs of colloidal crystals consisting of CdSe nanocrystals made by a digital camera through the objective of an optical microscope (reproduced from [123]).

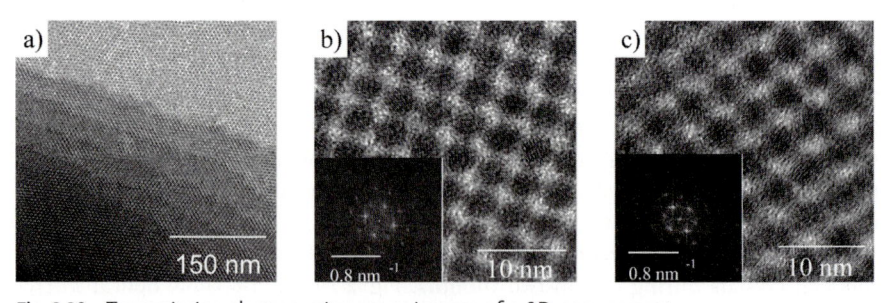

Fig. 2.30 Transmission electron microscopy images of a 3D arrangement of the 3.5 nm CdSe nanocrystals. **a** Overview picture of a fcc-like superlattice; **b** (100) projection along the superlattice with a correspondent FFT; **c** (110) projection along the superlattice with a correspondent FFT (from [123]).

2.5
Applications of Semiconductor Nanocrystals

The most up-to-date research pursuing the potential applications of semiconductor nanocrystals utilizes their strong band gap luminescence tunable by size as a result of the quantum confinement effect. Some current activities in this area are briefly outlined below.

2.5.1
Nanocrystals in Light-emitting Devices

Light-emitting devices (LEDs) based on conducting polymers have received considerable attention in recent years. This interest is motivated by a wide range of possible applications, including flat-panel displays and large-area devices [126]. Luminescent semiconductor nanocrystals were successfully integrated into the thin film polymer-based LEDs as emitting materials [7–10, 127]. Advantages of using polymer-nanocrystal composites are the processing of both polymer and nanocrystals from solution and the superior luminescent properties of the nanocrystals. Additionally, there is a possibility to tune the emission color via control of the nanocrystal size. A conjugated polymer like poly(phenylene vinylene) (PPV) usually plays the role of the hole-conducting component, facilitating the injection of holes into nanocrystals. Indeed, applying a biasing voltage (4–10 V) to the multiplayer structure with the layer sequence ITO/PPV/nanocrystals/Al results in relatively efficient electroluminescence with the narrow spectrum corresponding to the nanocrystal emission [127]. The use of CdSe/CdS core-shell nanocrystals in combination with optimized layer thickness allowed the achievement of external quantum efficiencies up to 0.22% at brightnesses of 600 cd/m^2 [7] which is close to a realistic specification for LED applications. Recently, a device emitting in the near-IR spectral region with an external efficiency of $\sim 0.5\%$ based on InAs/ZnSe core-shell nanocrystals has been reported [128]. Another strategy for the fabrication of LEDs is based on the use of aqueous thiol-capped CdTe nanocrystals assembled via the LBL technique [9] or integrated into a polyaniline or polypyrrole semiconducting matrix [10, 129].

2.5.2
Nanocrystals in Solar Cells

Photovoltaic cells based on low-cost poly- and nano-crystalline materials are of great interest for both scientific and industrial purposes. One of the directions of this research is the sensitization of wide-band gap semiconductors, mainly highly porous nanocrystalline TiO$_2$, with organic dyes which absorb visible light and inject electrons into the conduction band of TiO$_2$ (so-called Grätzel cells [130]). The problem with this approach is the limited photostability of the dye molecules in combination with their relatively low extinction coefficients. Semiconductor nanoparticles have a number of potential advantages as light-absorbing materials in Grätzel-type cells. The photo- and thermo-stability of the nanocrystals are similar to their bulk analogues and are superior to those of organic dyes. The optical absorption and band edge positions of the nanoparticles can be easily designed both by their elemental composition and by the nanocrystal size via the quantum confinement effect. Efficient charge transfer from the nanocrystals to the conduction band of wide-band gap semiconductors (TiO$_2$, ZnO, Ta$_2$O$_5$) [131–133] in combination with high extinction coefficients in the visible spectral range [13] makes them attractive for applications in Grätzel-type cells.

Another strategy for using nanoparticles in photovoltaic devices is the fabrication of blends of nanocrystals and conducting polymers [65]. The combination of electron-conducting nanocrystals and hole-conducting polymers in a single composite provides effective charge separation and transport. The same concept is used in mixtures of polymers with different electron affinities [134] and in composites of hole-conducting polymers with fullerenes [135]. Recently reported solar cells of this kind [136] utilize CdSe nanorods with high aspect ratios in order to optimize the charge transfer properties of the composites. A power energy conversion efficiency has been achieved [136].

2.5.3
Nanocrystals for Telecommunication Amplifiers

A potentially very significant application for IR-emitting nanocrystals (HgTe, InAs/CdSe) is their use as optical amplifier media for telecommunication systems based on a silica fiber which has optimal transmission windows in the 1.3 and 1.55 micron regions of the IR spectrum. One of the major limitations to the implementation of future high-capacity, ultra-broadband optical networks is the expansion of the fiber bandwidth beyond that available from the current state-of-the-art signal amplification devices, particularly erbium-doped fiber amplifiers. Erbium atoms can amplify light only corresponding to the transition energies between the fixed energy levels which do not cover the whole of the 1.3 or 1.55 micron windows. The use of colloidal semiconductor nanocrystals as "artificial atoms", where the quantum confinement effect yields size-dependent optical properties, is a radically different and alternative approach. For a given material, the spectral width, position, and profile of the luminescence band can be tailored to requirements by controlling the size and the size distribution of nanocrystals (Fig. 2.10). The challenge is to have available IR-emitting nanoparticles with high emission quantum yields which are in addition highly processable and, ideally, compatible with the existing optical fiber or planar (integrated optics) technologies. The reader is referred to the recent overview articles [11, 12] resulted from our collaboration with the British Telecom Laboratories for a more comprehensive treatment of this topic.

2.5.4
Nanocrystals as Light-emitting Species in Photonic Crystals and Microcavities

Since the pioneering works of Yablonovitch [137] and John [138], the concept of photonic crystals – dielectrics behaving with respect to electromagnetic waves like atomic crystals do with respect to electrons – has attracted much attention. The periodically varying index of refraction in photonic crystals causes a redistribution of the density of photonic states leading to appearance of photonic band gaps. Probably the most interesting basic physical phenomenon attainable in a photonic crystal is a freezing of the spontaneous decay of excited species (atoms, molecules or quantum dots) embedded therein [137]. Spontaneous emission of light is not an intrinsic property of matter but rather a result of its interaction with an electromagnetic field. If the free space around the luminescent species is replaced by a medium with the refractive index modulated on the scale comparable with the emission wavelength, the spontaneous emission rate will experience a modification and the emission will be either enhanced or inhibited. The luminophors

Fig. 2.31 Scanning electron microscopy image of a 3D colloidal crystal made from composite latex/CdTe nanocrystal spheres (from [143]).

used among other materials [16] for the studies of these phenomena are semiconductor nanoparticles with their size-dependent excitonic emission.

Artificial opals are 3D colloidal photonic crystals with stopbands in the visible and near-IR spectral ranges which are produced by chemical self-assembly (see Chapter 9). Examples of filling 3D colloidal crystals with luminescent semiconductor nanocrystals include the direct synthesis of nanoparticles in opal voids [139, 140], the infiltration of aqueous colloidal solutions of nanoparticles [141], and their chemical bath [142] or electrophoretic [43] deposition. Another strategy is the fabrication of artificial opals from composite microspheres with luminescent shells containing semiconductor nanocrystals (Fig. 2.26). Fig. 2.31 shows a scanning electron microscopy image of the 3D colloidal photonic crystals produced from latex/CdTe nanocrystal spheres by slow natural sedimentation [143].

Spherical microcavities represent another type of photonic material with modulated photonic density of states, as they are characterized by sharp well-separated cavity modes (so-called "whispering gallery" modes) [144]. Nanoparticle-doped microcavities were recently prepared using commercially available glass or latex spheres and CdSe nanocrystals [17]. The coupling of the luminescence of the nanoparticles in the microcavity with the whispering gallery modes of the latter provides a way of controlling the emission properties of nanocrystals and may lead to the realization of a whispering gallery-mode laser with an ultra-low threshold, a controllable system of optical modes, and enhanced efficiency [145].

2.5.5
Nanocrystals for Biolabeling

Recently, the possibility of employing semiconductor nanoparticles as fluorescence labeling reagents for biological imaging experiments has been demonstrated by the attachment of CdSe/CdS [13] and CdSe/ZnS [14, 146] core-shell nanocrystals to DNA molecules [13, 14] and proteins [146]. Semiconductor nanocrystals are photostable and have nearly continuous excitation spectra above the threshold of absorption, together with a strong, narrow and symmetric emission band which depends on the particle size. Therefore, many-color probes can be simultaneously

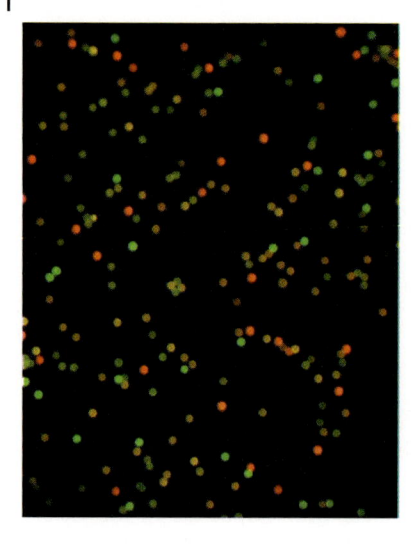

Fig. 2.32 Fluorescence micrograph of a mixture of polymer microspheres loaded with CdTe nanocrystals of different sizes. A true-color digital camera and a single light source were used to obtain the image.

excited by a single narrow-band excitation source and distinguished in a single exposure. This makes semiconductor nanocrystals superior to commonly used organic chromophores.

Multicolor optical coding for biological assays has also been proved to be possible using polymer microspheres ("reporters") loaded with specific and identifiable combinations of differently-sized CdSe/ZnS nanocrystals [15]. Water-soluble CdTe nanocrystals have been used to create the luminescence codes on polymer microcapsules [147] (Fig. 2.32). The color barcode generated on colloidal spheres can be easily decoded by fluorescence microscopy. The use of six emission colors in combination with different intensity levels could theoretically allow the encoding of a library of more than 16 million biological compounds [148]. A comprehensive treatment of this topic can be found in Chapter 16 and 17.

2.6
Concluding Remarks

During the last two decades wet chemical synthesis of semiconductor nanoparticles has developed from alchemy-like preparations to reliable chemical synthesis. Although CdSe is by far the most intensively investigated material, a large palette of other semiconductors like III–Vs, I–VIIs, and some IV–IVs are currently available in good quality. Much effort has been spent on the development of synthetic protocols allowing the preparation of highly monodisperse samples of highly crystalline nanoparticles in gram scale. Only now are we starting to understand the mechanisms determining the basics of nucleation and growth of particles on a molecular level. First theoretical treatments show, however, that the kinetics of particle growth in the nanodomain differs significantly from that of larger crystallites and that the extraordinarily large surface energy can support the genesis of extremely monodisperse samples. A complete understanding of crystal growth in the nanometer regime remains, however, a challenging field of science for the future. The present state of the art in the synthesis of nanoparticles allows for the first time the delivery of the experimental results that are needed for a fruitful development of theoretical considerations.

The second attractive field for future investigations is a better understanding and a better control of surface chemistry. Processes like surface roughening and flattening need to be explored in the same manner as surface equilibria and tight covalent binding of ligand molecules. The proper design of surface ligands determines the properties of nanoparticle composite materials almost as strongly as the particle size itself.

Self-assembly of nanoparticles, where the clusters act as artificial atoms, might open the door into a new domain of bottom-up chemistry. The number of adjustable parameters such as interparticle distance, charge, chemical composition, and Fermi level is even higher in the case of nanoparticles than in the tool-box of the periodic table. Finally one of the exciting areas in chemical nanoscience is the synthesis of conjugates formed by biological molecules and nanoparticles, a topic which has already inspired scientists, journalists and artists in the design of scifi-like nanobots healing mankind from almost any diseases. From a more serious scientific view this field requires a better understanding of the surface processes, and the development of synthetic protocols to realize the nano-bio interconnection and long studies of the interaction with living matter.

2.7
Acknowledgements

We are deeply indebted to all our colleagues and collaborators, whose names appear in the cited literature. Financial support has been provided by the Deutsche Forschungsgemeinschaft. A part of this work has appeared as a Feature Article in Advanced Functional Materials, **2002**, *12*, 653.

2.8
References

1 AL. L. EFROS, A. L. EFROS, *Sov. Phys. Semicond.* **1982**, *16*, 772.
2 A. I. EKIMOV, A. A. ONUSHCHENKO, *Sov. Phys. Semicond.* **1982**, *16*, 775.
3 A. HENGLEIN, *J. Phys. Chem.* **1982**, *86*, 2291.
4 R. ROSSETTI, S. NAKAHARA, L. E. BRUS, *J. Chem. Phys.* **1983**, *79*, 1086.
5 A. EYCHMÜLLER, *J. Phys. Chem. B* **2000**, *104*, 6514.
6 N. GAPONIK, D. V. TALAPIN, A. L. ROGACH, K. HOPPE, E. V. SHEVCHENKO, A. KORNOWSKI, A. EYCHMÜLLER, H. WELLER, *J. Phys. Chem. B* **2002**, *106*, 7177.
7 M. C. SCHLAMP, X. PENG, A. P. ALIVISATOS, *J. Appl. Phys.* **1997**, *82*, 5837.
8 H. MATTOUSSI, L. H. RADZILOWSKI, B. O. DABBOUSI, E. L. THOMAS, M. G. BAWENDI, M. F. RUBNER, *J. Appl. Phys.* **1998**, *83*, 7965.

9 M.Y. Gao, C. Lesser, S. Kirstein, H. Möhwald, A.L. Rogach, H. Weller, *J. Appl. Phys.* **2000**, *87*, 2297.
10 N.P. Gaponik, D.V. Talapin, A.L. Rogach, A. Eychmüller, *J. Mater. Chem.* **2000**, *10*, 2163.
11 M.T. Harrison, S.V. Kershaw, M.G. Burt, A.L. Rogach, A. Kornowski, A. Eychmüller, H. Weller, *Pure Appl. Chem.* **2000**, *72*, 295.
12 S.V. Kershaw, M.T. Harrison, A.L. Rogach, A. Kornowski, *IEEE J. Select. Topics Quantum Electronics* **2000**, *6*, 534.
13 M.P. Bruchez, M. Moronne, P. Gin, S. Weiss, A.P. Alivisatos, *Science* **1998**, *281*, 2013.
14 W.C.W. Chan, S. Nie, *Science* **1998**, *281*, 2016.
15 M. Han, X. Gao, J.Z. Su, S. Nie, *Nature Biotechn.* **2001**, *19*, 631.
16 S.V. Gaponenko, V.N. Bogomolov, E.P. Petrov, A.M. Kapitonov, D.A. Yarotsky, I.I. Kalosha, A. Eychmüller, A.L. Rogach, J. McGilp, U. Woggon, F. Gindele, *IEEE J. Lightwave Techn.* **1999**, *17*, 2128.
17 M.V. Artemyev, U. Woggon, R. Wannemacher, H. Jaschinski, W. Langbein, *Nano Lett.* **2001**, *1*, 309.
18 L.E. Brus, *J. Chem. Phys.* **1984**, *80*, 4403.
19 M.L. Steigerwald, A.P. Alivisatos, J.M. Gibson, T.D. Harris, R. Kortan, A.J. Muller, A.M. Thayer, T.M. Duncan, D.C. Douglass, L.E. Brus, *J. Am. Chem. Soc.* **1988**, *110*, 3046.
20 L. Spanhel, M. Haase, H. Weller, A. Henglein, *J. Am. Chem. Soc.* **1987**, *109*, 5649.
21 C.B. Murray, C.R. Kagan, M.G. Bawendi, *Annu. Rev. Mater. Sci.* **2000**, *30*, 545.
22 T. Sugimoto, *Monodisperse Particles*, Elsevier **2001**.
23 C.B. Murray, D.J. Norris, M.G. Bawendi, *J. Am. Chem. Soc.* **1993**, *115*, 8706.
24 W. Ostwald, *Z. Phys. Chem.* **1901**, *37*, 385.
25 P.W. Voorhees, *J. Stat. Phys.* **1985**, *38*, 231.
26 T. Sugimoto, F. Shiba, *J. Phys. Chem. B* **1999**, *103*, 3607.
27 V.B. Fenelonov, G.G. Kodenyov, V.G. Kostrovsky, *J. Phys. Chem. B* **2001**, *105*, 1050.
28 I.M. Lifshitz, V.V. Slyozov, *J. Phys. Chem. Solids* **1961**, *19*, 35.
29 C. Wagner, *Z. Elektrochem.* **1961**, *65*, 581.
30 A.S. Kabalnov, E.D. Shchukin, *Adv. Colloid Interfac. Sci.* **1992**, *38*, 69.
31 Y. De Smet, L. Deriemaeker, R. Finsy, *Langmuir* **1997**, *13*, 6884.
32 D.V. Talapin, A.L. Rogach, M. Haase, H. Weller, *J. Phys. Chem. B* **2001**, *105*, 12278.
33 A.L. Rogach, L. Katsikas, A. Kornowski, D. Su, A. Eychmüller, H. Weller, *Ber. Bunsen-Ges. Phys. Chem.* **1996**, *100*, 1772.
34 X. Peng, J. Wickham, A.P. Alivisatos, *J. Am. Chem. Soc.* **1998**, *120*, 5343.
35 T. Vossmeyer, L. Katsikas, M. Giersig, I.G. Popovic, K. Diesner, A. Chemseddine, A. Eychmüller, H. Weller, *J. Phys. Chem.* **1994**, *98*, 7665.
36 A.L. Rogach, A. Kornowski, M. Gao, A. Eychmüller, H. Weller, *J. Phys. Chem. B* **1999**, *103*, 3065.
37 M.T. Harrison, S.V. Kershaw, M.G. Burt, A. Eychmüller, H. Weller, A.L. Rogach, *Mater. Sci. Eng. B* **2000**, *69*, 355.
38 A.L. Rogach, S.V. Kershaw, M. Burt, M. Harrison, A. Kornowski, A. Eychmüller, H. Weller, *Adv. Mater.* **1999**, *11*, 552.
39 A.L. Rogach, D.S. Koktysh, M. Harrison, N.A. Kotov, *Chem. Mater.* **2000**, *12*, 1526.
40 A.A. Mamedov, A. Belov, M. Giersig, N.N. Mamedova, N.A. Kotov, *J. Am. Chem. Soc.* **2001**, *123*, 7738.
41 A.L. Rogach, N.A. Kotov, D.S. Koktysh, A.S. Susha, F. Caruso, *Coll. Surf. A* **2002**, *202*, 135.
42 A.S. Susha, F. Caruso, A.L. Rogach, G.B. Sukhorukov, A. Kornowski, H. Möhwald, M. Giersig, A. Eychmüller, H. Weller, *Coll. Surf. A* **2000**, *163*, 39.
43 A.L. Rogach, N.A. Kotov, D.S. Koktysh, J.W. Ostrander, G.A. Ragoisha, *Chem. Mater.* **2000**, *12*, 2721.
44 N.N. Mamedova, N.A. Kotov, A.L. Rogach, J. Studer, *Nano Lett.* **2001**, *1*, 281.
45 S. Westenhoff, N.A. Kotov, *J. Am. Chem. Soc.* **2002**, *124*, 2448.
46 T. Vossmeyer, G. Reck, L. Katsikas, E.T.K. Haupt, B. Schulz, H. Weller, *Science* **1995**, *267*, 1476.

47 T. Vossmeyer, G. Reck, B. Schulz, L. Katsikas, H. Weller, *J. Am. Chem. Soc.* **1995**, *117*, 12881.
48 A. Henglein, *Topics Current Chem.* **1988**, *143*, 113.
49 A. L. Rogach, *Mater. Sci. Eng. B* **2000**, *69*, 435.
50 A. H. Nethercot, *Phys. Rev. Lett.* **1974**, *33*, 1088.
51 D. Schooss, A. Mews, A. Eychmüller, H. Weller, *Phys. Rev. B* **1994**, *49*, 17072.
52 X. Peng, M. C. Schlamp, A. V. Kadavanich, A. P. Alivisatos, *J. Am. Chem. Soc.* **1997**, *119*, 7019.
53 A. Mews, A. Eychmüller, *Ber. Bunsen-Ges. Phys. Chem.* **1998**, *102*, 1343.
54 R. Balcerak, J. F. Gibson, W. Guiterrez, J. H. Pollard, *Opt. Eng.* **1987**, *26*, 191.
55 A. L. Rogach, M. Harrison, S. V. Kershaw, A. Kornowski, M. Burt, A. Eychmüller, H. Weller, *Phys. Stat. Sol. (b)* **2001**, *224*, 153.
56 J. E. Bowen Katari, V. L. Colvin, A. P. Alivisatos, *J. Phys. Chem.* **1994**, *98*, 4109.
57 J. J. Shiang, A. V. Kadavanich, R. K. Grubbs, A. P. Alivisatos, *J. Phys. Chem.* **1995**, *99*, 17417.
58 C. B. Murray, C. R. Kagan, M. G. Bawendi, *Science* **1995**, *270*, 1335.
59 J. Aldana, A. Wang, X. Peng, *J. Am. Chem. Soc.* **2001**, *123*, 8844.
60 M. Kuno, J. K. Lee, B. O. Dabbousi, F. V. Mikulec, M. G. Bawendi, *J. Chem. Phys.* **1997**, *106*, 9869.
61 M. Shim, P. Guyot-Sionnest, *Nature* **2000**, *407*, 981.
62 D. J. Norris, M. G. Bawendi, *Phys. Rev. B* **1996**, *53*, 16338.
63 V. I. Klimov, *J. Phys. Chem. B* **2000**, *104*, 6112.
64 V. I. Klimov, A. A. Mikhailovsky, Su Xu, A. Malko, J. A. Hollingsworth, C. A. Leatherdale, H.-J. Eisler, M. G. Bawendi, *Science* **2000**, *290*, 314.
65 N. C. Greenham, X. Peng, A. P. Alivisatos, *Phys. Rev. B* **1996**, *54*, 17628.
66 T. Trindade, P. O'Brien, N. L. Pickett, *Chem. Mater.* **2001**, *13*, 3843.
67 X. Peng, L. Manna, W. Yang, J. Wickham, E. Scher, A. Kadavanich, A. P. Alivisatos, *Nature* **2000**, *404*, 59.
68 D. V. Talapin, A. L. Rogach, A. Kornowski, M. Haase, H. Weller, *Nano Lett.* **2001**, *1*, 207.
69 L. Qu, X. Peng, *J. Am. Chem. Soc.* **2002**, *124*, 2049.
70 F. V. Mikulec, M. Kuno, M. Bennati, D. A. Hall, R. G. Griffin, M. G. Bawendi, *J. Am. Chem. Soc.* **2000**, *122*, 2532.
71 D. V. Talapin, A. L. Rogach, I. Mekis, S. Haubold, A. Kornowski, M. Haase, H. Weller, *Coll. Surf. A* **2002**, *202*, 145.
72 M. A. Hines, P. Guyot-Sionnest, *J. Phys. Chem. B* **1998**, *102*, 3655.
73 Z. A. Peng, X. Peng, *J. Am. Chem. Soc.* **2001**, *123*, 183.
74 Z. A. Peng, X. Peng, *J. Am. Chem. Soc.* **2002**, *124*, 3343.
75 L. Qu, Z. A. Peng, X. Peng, *Nano Lett.* **2001**, *1*, 333.
76 J. Aldana, Y. A. Wang, X. Peng, *J. Am. Chem. Soc.* **2001**, *123*, 8844.
77 X. Peng, *Chem. Eur. J.* **2002**, *8*, 335.
78 L. Manna, E. C. Scher, A. P. Alivisatos, *J. Am. Chem. Soc.* **2000**, *122*, 12700.
79 L.-S. Li, J. Hu, W. Yang, A. P. Alivisatos, *Nano Lett.* **2001**, *1*, 349.
80 J. Hu, L.-W. Wang, L.-S. Li, W. Yang, A. P. Alivisatos, *J. Phys. Chem. B* **2002**, *106*, 2447.
81 J. Hu, L.-S. Li, W. Yang, L. Manna, L.-W. Wang, A. P. Alivisatos, *Science* **2001**, *292*, 2060.
82 L.-S. Li, J. Walda, L. Manna, A. P. Alivisatos, *Nano Lett.* **2002**, *2*, 557.
83 Z. A. Peng, X. Peng, *J. Am. Chem. Soc.* **2001**, *123*, 1389.
84 M. A. Hines, P. Guyot-Sionnest, *J. Phys. Chem.* **1996**, *100*, 468.
85 B. O. Dabbousi, J. Rodriguez-Viejo, F. V. Mikulec, J. R. Heine, H. Mattoussi, R. Ober, K. F. Jensen, M. G. Bawendi, *J. Phys. Chem. B* **1997**, *101*, 9463.
86 M. Danek, K. F. Jensen, C. B. Murray, M. G. Bawendi, *Chem. Mater.* **1996**, *8*, 173.
87 L. Manna, E. C. Scher, L.-S. Li, A. P. Alivisatos, *J. Am. Chem. Soc.* **2002**, *124*, 7136.
88 Y.-W. Jun, S.-M. Lee, N.-J. Kang, J. Cheon, *J. Am. Chem. Soc.* **2001**, *123*, 5150.
89 J. Lee, V. C. Sundar, J. R. Heine, M. G. Bawendi, K. F. Jensen, *Adv. Mater.* **2000**, *12*, 1102.

90 D. V. Talapin, S. Haubold, A. L. Rogach, A. Kornowski, M. Haase, H. Weller, *J. Phys. Chem. B* **2001**, *105*, 2260.
91 F. V. Mikulec, M. G. Bawendi, S. Kim, Patent WO 01/07689 A2.
92 M. Pehnt, D. L. Schulz, C. J. Curtis, K. M. Jones, D. S. Ginley, *Appl. Phys. Lett.* **1995**, *67*, 2176.
93 B. Schreder, T. Schmidt, V. Platschek, U. Wikler, A. Materny, E. Umbach, M. Lerch, G. Müller, W. Kiefer, L. Spanhel, *J. Phys. Chem. B* **2000**, *104*, 1677.
94 M. Y. Shen, M. Oda, T. Goto, *Phys. Rev. Lett.* **1999**, *82*, 3915.
95 D. J. Norris, N. Yao, F. T. Charnock, T. A. Kennedy, *Nano Lett.* **2001**, *1*, 3.
96 A. Sashchiuk, L. Langof, R. Chaim, E. Lifshitz, *J. Cryst. Growth* **2002**, *240*, 431.
97 C. B. Murray, S. Sun, W. Gaschler, H. Doyle, T. A. Betley, C. R. Kagan, *IBM J. Res. & Dev.* **2001**, *45*, 47.
98 O. I. Micic, A. J. Nozik, *Handbook of Nanostructured Materials and Nanotechnology*, Nalwa H. S. (ed.), Academic Press, San Diego, **2000**, *1*, 427.
99 O. I. Micic, C. J. Curtis, K. M. Jones, J. R. Sprague, A. J. Nozik, *J. Phys. Chem.* **1994**, *98*, 4966.
100 A. A. Guzelian, J. E. B. Katari, A. V. Kadavanich, U. Banin, K. Hamad, E. Juban, A. P. Alivisatos, R. H. Wolters, C. C. Arnold, J. R. Heath, *J. Phys. Chem.* **1996**, *100*, 7212.
101 O. I. Micic, H. M. Cheong, H. Fu, A. Zunger, J. R. Sprague, A. Mascarenhas, A. J. Nozik, *J. Phys. Chem. B* **1997**, *101*, 4904.
102 O. I. Micic, K. M. Jones, A. Cahill, A. J. Nozik, *J. Phys. Chem. B* **1998**, *102*, 9791.
103 O. I. Micic, B. B. Smith, A. J. Nozik, *J. Phys. Chem. B* **2000**, *104*, 12149.
104 O. I. Micic, S. P. Ahrenkiel, A. J. Nozik, *Appl. Phys. Lett.* **2001**, *78*, 4022.
105 O. I. Micic, J. R. Sprague, C. J. Curtis, K. M. Jones, J. L. Machol, A. J. Nozik, H. Giessen, B. Fluegel, G. Mohs, N. Peyghambarian, *J. Phys. Chem.* **1995**, *99*, 7754.
106 A. A. Guzelian, U. Banin, A. V. Kadavanich, X. Peng, A. P. Alivisatos, *Appl. Phys. Lett.* **1996**, *69*, 1432.
107 U. Banin, C. J. Lee, A. A. Guzelian, A. V. Kadavanich, A. P. Alivisatos, W. Jaskolski, G. W. Bryant, Al. L. Efros, M. Rosen, *J. Chem. Phys.* **1998**, *109*, 2306.
108 R. L. Wells, C. G. Pitt, A. T. McPhail, A. P. Purdy, S. Schafieezad, R. B. Hallock, *Chem. Mater.* **1989**, *1*, 4.
109 O. I. Micic, J. Sprague, Z. Li, A. J. Nozik, *Appl. Phys. Lett.* **1996**, *68*, 3150.
110 Y. W. Cao, U. Banin, *Angew. Chem., Int. Ed.* **1999**, *38*, 3692.
111 Y. W. Cao, U. Banin, *J. Am. Chem. Soc.* **2000**, *122*, 9692.
112 M. A. Olshavsky, A. N. Goldstein, A. P. Alivisatos, *J. Am. Chem. Soc.* **1990**, *112*, 9438.
113 A. J. Nozik, H. Uchida, P. V. Kamat, C. Curtis, *Israel J. Chem.* **1993**, *33*, 15.
114 A. Chemseddine, H. Weller, *Ber. Bunsen-Ges. Phys. Chem.* **1993**, *97*, 636.
115 D. V. Talapin, A. L. Rogach, E. V. Shevchenko, A. Kornowski, M. Haase, H. Weller, *J. Am. Chem. Soc.* **2002**, *124*, 5782.
116 G. Decher, *Science* **1997**, *277*, 1232.
117 N. A. Kotov, I. Dekany, J. H. Fendler, *J. Phys. Chem.* **1995**, *99*, 13065.
118 I. L. Radtchenko, G. B. Sukhorukov, N. P. Gaponik, A. Kornowski, A. L. Rogach, H. Möhwald, *Adv. Mater.* **2001**, *13*, 1684.
119 C. P. Collier, T. Vossmeyer, J. R. Heath, *Annu. Rev. Phys. Chem.* **1998**, *49*, 371.
120 G. S. H. Lee, D. C. Craig, I. Ma, M. L. Scudder, T. D. Bailey, I. G. Dance, *J. Am. Chem. Soc.* **1988**, *110*, 4863.
121 N. Herron, J. C. Calabrese, W. E. Farneth, Y. Wang, *Science* **1993**, *259*, 1426.
122 S. Behrens, M. Bettenhausen, A. C. Deveson, A. Eichhöfer, D. Fenske, A. Lohde, U. Woggon, *Angew. Chem. Int. Ed. Engl.* **1996**, *35*, 2215.
123 D. V. Talapin, E. V. Shevchenko, A. Kornowski, N. Gaponik, M. Haase, A. L. Rogach, H. Weller, *Adv. Mater.* **2001**, *13*, 1868.
124 E. V. Shevchenko, D. V. Talapin, A. Kornowski, F. Wiekhorst, J. Kötzler, M. Haase, A. L. Rogach, H. Weller, *Adv. Mater.* **2002**, *14*, 287.
125 E. V. Shevchenko, D. V. Talapin, A. L. Rogach, A. Kornowski, M. Haase, H. Weller, *J. Am. Chem. Soc.* **2002**, *124*, 11480.

126 R. H. Friend, R. W. Gymer, A. B. Holmes, J. H. Burroughes, R. N. Marks, C. Taliani, D. D. C. Bradley, D. A. Dos Santos, J. L. Brédas, M. Lögdlung, W. R. Salaneck, *Nature* **1999**, *397*, 121.
127 V. L. Colvin, M. C. Schlamp, A. P. Alivisatos, *Nature* **1994**, *370*, 354.
128 N. Tessler, V. Medvedev, M. Kazes, S. H. Kan, U. Banin, *Science*, **2002**, *295*, 1506.
129 N. P. Gaponik, D. V. Talapin, A. L. Rogach, *Phys. Chem. Chem. Phys.* **1999**, *1*, 1787.
130 B. O'Regan, M. Grätzel, *Nature* **1991**, *353*, 737.
131 S. Hotchandani, P. V. Kamat, *J. Phys. Chem.* **1992**, *96*, 6834.
132 R. Vogel, P. Hoyer, H. Weller, *J. Phys. Chem.* **1994**, *98*, 3183.
133 R. Vogel, K. Pohl, H. Weller, *Chem. Phys. Lett.* **1990**, *174*, 241.
134 J. J. M. Halls, C. A. Walsh, N. C. Greenham, E. A. Marseglia, R. H. Friend, S. C. Moratti, A. B. Holmes, *Nature* **1995**, *376*, 498.
135 G. Yu, J. Gao, J. C. Hummelen, F. Wudl, A. J. Heeger, *Science* **1995**, *270*, 1789.
136 W. U. Huynh, J. J. Dittmer, A. P. Alivisatos, *Science* **2002**, *295*, 2425.
137 E. Yablonovitch, *Phys. Rev. Lett.* **1987**, *58*, 2059.
138 S. John, *Phys. Rev. Lett.* **1987**, *58*, 2486.
139 S. G. Romanov, A. V. Fokin, V. V. Tretijakov, V. Y. Butko, V. I. Alperovich, N. P. Johnson, C. M. Sotomayor Torres, *J. Cryst. Growth* **1996**, *159*, 857.
140 Yu. A. Vlasov, K. Luterova, I. Pelant, B. Hönerlage, V. N. Astratov, *Appl. Phys. Lett.* **1997**, *71*, 1616.
141 S. V. Gaponenko, A. M. Kapitonov, V. N. Bogomolov, A. V. Prokofiev, A. Eychmüller, A. L. Rogach, *JETP Lett.* **1998**, *68*, 142*.
142 A. Blanco, C. Lopez, R. Mayoral, F. Meseguer, A. Mifsud, J. Herrero, *Appl. Phys. Lett.* **1998**, *73*, 1781.
143 A. L. Rogach, A. S. Susha, F. Caruso, G. B. Sukhorukov, A. Kornowski, S. Kershaw, H. Möhwald, A. Eychmüller, H. Weller, *Adv. Mater.* **2000**, *12*, 333.
144 R. K. Chang, A. J. Chamillo (eds.) *Optical Processes in Microcavities*, World Scientific, Singapore, **1996**.
145 H. Cao, J. Y. Xu, W. H. Xiang, Y. Ma, S.-H. Chang, S. T. Ho, G. S. Solomon, *Appl. Phys. Lett.* **2000**, *76*, 3519.
146 H. Mattoussi, J. M. Mauro, E. R. Goldman, G. P. Anderson, V. C. Sundar, F. V. Mikulec, M. G. Bawendi, *J. Am. Chem. Soc.* **2000**, *122*, 12142.
147 N. Gaponik, I. L. Radtchenko, G. B. Sukhorukov, H. Weller, A. L. Rogach, *Adv. Mater.* **2002**, *14*, 879.
148 B. J. Battersby, D. Bryant, W. Meutermans, D. Matthews, M. L. Smythe, M. Trau, *J. Am. Chem. Soc.* **2000**, *122*, 2138.

3
Monolayer Protected Clusters of Gold and Silver

Mathias Brust and Christopher J. Kiely

3.1
Introduction

The science of metal nanoparticles is well established and represents a vast body of knowledge of both fundamental and applied aspects of the subject. Modern applications include, for example, catalysis [1], electronics [2], information storage [3], surface coatings [4], sensors [5], optical filters [6] and bio-recognition studies [7]. Particle sizes of current interest range from less than one nanometer to several hundreds of nanometers. Given the breadth of this field and the number of excellent reviews [8-10] and edited collections [11] available this chapter focuses chiefly on the more recent development of thiol-stabilized gold and silver nanoparticles [12-18]. The reason for this selection is the notion that metal nanoparticles of this type, hereafter referred to as monolayer protected clusters (MPCs), are characterized by their extraordinary stability both in solution and in the solid state, which distinguishes them from most other preparations of nanometer-sized metals. Owing to this stability MPCs have enabled scientists to carry out experiments that would have been impossible or extremely difficult to conduct using less-stable materials. The ability to handle MPCs like stable chemical compounds greatly facilitates the study of their fascinating size-dependent properties and is instrumental to the development of new applications.

3.2
Gold Colloids and Clusters: Historical Background

Colloidal gold sols have a long history and were probably first described in the literature by the Florentine glass maker and alchemist Antonio Neri in 1612 in his treatise *L'Arte Vetraria* [19]. Even earlier use of colloidal gold for decorative purposes is evidenced by the Lycurgus cup, which was made by the Romans in the fourth century AD and is now exhibited in the British Museum [20]. The first sound scientific studies of such materials were published much later by Michael Faraday in 1857 [21]. He developed new physical as well as wet-chemical preparative methods for colloidal gold and, more importantly, was the first to recognize

Colloids and Colloid Assemblies. Edited by Frank Caruso
Copyright © 2004 Wiley-VCH Verlag GmbH & Co. KGaA, Weinheim
ISBN: 3-527-30660-9

that the gold in colloidal suspensions is present in the metallic state. This conclusion was drawn from the observation that red or purple colored thin films of the dry material on a quartz surface could be compressed mechanically so that they would exhibit the well-known optical properties of continuous thin films of metallic gold (thinly beaten gold leaf), which in transmittance appear green. Since no chemical reaction was involved in this process Faraday concluded that the gold must always have been present in the metallic state. With this discovery he became the founder of the science of metal colloids, a field that continued to develop steadily, but at times of revolutionary changes in the physical sciences, appeared to be of but moderate interest.

A number of simple preparative methods for charge-stabilized gold hydrosols by reduction of a gold salt with an organic reducing agent such as formaldehyde or hydrazine were developed in the early twentieth century [22]. The most popular method to date is Turkevich's modification of Hauser and Lynn's citrate reduction route, which was first described in 1940 and leads to very uniform gold particles in the size range of ca. 10 to 40 nm [23, 24]. The preparation of virtually monodisperse gold hydrosols can be achieved by "seeded growth" techniques, which were first described in 1906 by Zsigmondy [25]. The principle of this approach is to control the reaction conditions in such a way that nucleation of new particles cannot occur and therefore reduction of metal ions only takes place by growth of deliberately added small seed colloids. Similar results are obtained when the reaction is carried out so that a very short burst of nucleation is followed by a long and slow growth period. Citrate-stabilized gold hydrosols prepared by one of the many variations of the Turkevich method are currently of interest as precursor materials for various applications including self-assembled thin films [26–28] with potentially useful spectroscopic properties (e.g. Raman enhancement) and bioconjugates [29, 30].

In parallel with the ongoing refinement of colloid chemical routes to metal nanoparticles, metal cluster chemistry was established as a branch of inorganic chemistry occupied with the synthesis and characterization of new compounds having a core of a precisely known number (≥ 3) of metal atoms surrounded usually by an organic ligand shell of well-defined structure [11]. While preparations of colloids are generally described by their size distribution, clusters are described as molecules with a precise composition and structure. Clearly, the two concepts overlap in the 1 to 5 nm range, so that small colloids with a very narrow size distribution arguably are the same as large clusters of somewhat poorly defined composition and structure. The recent literature indeed no longer distinguishes clearly between clusters, colloids and nanoparticles, and all terms are currently used according to the personal preferences of the authors. Other synonyms frequently encountered are "nanoclusters", "nanocrystals" and "quantum dots", although the latter is more commonly used to describe semiconductor nanoparticles with opto-electronic properties different from those of the bulk material due to the quantum size effect [31, 32].

One of the most important contributions of cluster chemistry to the emerging field of nanoscience has been the synthesis of the phosphine-stabilized Au_{55} cluster by Schmid and co-workers in 1981 [33]. In particular, the idea of using such clusters as components in novel nanoelectronic devices was fuelled by the pioneering work of the Schmid group and has remained to date one of the most intriguing tasks of nanotechnology [34–36]. However, a key problem of this endeavor has always been the relative lack of stability of most nanometer-sized particles. Notable exceptions are fullerenes and carbon nanotubes, which in turn are so stable that their chemical manipulation still represents a significant challenge [37]. In the case of gold colloids and clusters this problem has largely been overcome by the development of thiol-stabilization. The discovery that organic sulfur compounds spontaneously form self-assembled monolayers (SAMs) on gold surfaces was made by Nuzzo and Allara in 1983 [38] and has since become a mature science with many potential applications [39]. Ten years later Giersig and Mulvaney first employed thiols to stabilize gold colloids, from which they could assemble highly ordered hexagonal superlattices by electrophoretic deposition of the material onto amorphous carbon films [40]. Remarkably, the constituent particles of these films were fully re-dispersible, which at the time was an unusual feature for solid-state colloidal systems. A year later in 1994 a preparatively very simple two-phase wet-chemical route to such materials was developed by Schiffrin and co-workers [12]. With this new approach the gram-scale preparation of thiolate-stabilized gold nanoparticles, which could be isolated from solution, re-dispersed, purified by chromatography and chemically modified in many different ways, became routinely possible. The term MPCs mentioned above was first introduced by Murray and co-workers to clearly distinguish these nanoparticles from more conventional colloids and clusters, which are usually not of comparable stability [41]. The Murray nomenclature also highlights the fact that the new material conceptually represents a combination of cluster chemistry and surface passivation by self-assembled monolayers, in particular those of thiols on gold. In this chapter some of the most important developments of a decade of MPC chemistry will be presented along with a perspective for possible future directions this exciting area of research may take.

3.3
Synthesis

The first synthesis of MPCs reported by Schiffrin and co-workers was based on the reduction of an intensely orange Au(III) complex (tetraoctylammonium tetrabromoaurate) with sodium borohydride in the presence of dodecanethiol [12]. A two-phase liquid/liquid (water/toluene) system was chosen to carry out this reaction since it was believed that the reduction of the Au(III) ions would occur via heterogeneous electron transfer across the water/toluene interface followed by nucleation of metallic gold in the organic phase and immediate passivation of the nuclei by adsorption of thiols. The electron transfer step was believed to be assisted by the phase transfer catalyst tetraoctylammonium bromide. In reality it is now well established that the reaction does not involve heterogeneous electron transfer but rather the transfer of both the anionic gold complex and the borohydride ions to the organic phase followed by a homogeneous reduction step [13, 42, 43]. Nevertheless, most laboratories currently working with MPCs still use a variation of the original two-phase protocol although single-phase methods are available and lead to comparable results [13, 44]. MPCs prepared by any one of these methods are typically in the size range of 1 to 5 nm and give solutions which are either dark brown (1–3 nm) or deep ruby red (>3 nm) depending on the intensity of the characteristic plasmon absorption band at 525 nm, which develops with increasing cluster size. The mechanism of cluster formation in these systems is largely unexplored. Many thiols are capable of reducing Au(III) ions to Au(I) while they are themselves oxidized to disulfides. Such reactions often lead to the formation of a colorless, insoluble and probably polymeric product, which usually disappears after the addition of sodium borohydride as the gold is reduced to Au(0) [43]. On the contrary, if some aromatic thiols such as thiophenol are used as stabilizing agents, the Au(I) species can be so stable that further reduction to gold clusters by borohydride is impossible [45]. The stable Au(I) compounds formed in this case have not yet been identified.

The solubility of the MPCs in different solvents depends entirely on the choice of the stabilizing thiol ligand. However, most earlier work was restricted to the use of n-alkanethiols with a hydrocarbon chain length ranging from 5 to 18 carbon atoms. This led to the formation of either waxy or powdery dry products, which were soluble in most non-polar solvents such as toluene, hexane and chloroform, and insoluble in water, short chain alcohols and acetone. Many attempts have been made to control the size distributions of MPC preparations by varying the reaction conditions. Probably the most complete study towards this aim has been carried out by Murray and co-workers, who found that the smallest particles with the narrowest size distribution (ca. 1.5–2.5 nm) were obtained when the reducing agent (borohydride) was added fast and under vigorous stirring at low temperature [42]. Whetten and co-workers found that quantitative reduction of Au(III) to Au(I) by the thiol ligand prior to addition of borohydride was important in order to achieve small particles with a narrow size distribution [46]. In spite of these efforts, a direct synthesis, which would lead to monodisperse MPCs, comparable in size distribution, say, to freshly prepared Au_{55} clusters, has not yet been reported. Nevertheless, practically monodisperse fractions of MPCs are readily obtained in two different ways. Whetten and co-workers reported the successive fractionation of MPCs by repeated size selective precipitation, which can be achieved by the slow and gradual addition of a non-solvent, such as acetone or methanol [14, 47]. In such systems the larger particles tend to aggregate first due to the stronger van der Waals interactions between the larger spheres [48]. A more automated route to monodisperse fractions is size-exclusion chromatography, which has been applied very successfully to the separation of MPCs as well as a whole range of other nanoparticles by Wilcoxon and co-workers [49]. This group also demonstrated that the determination of particle size by careful analysis of the retention times rivals transmission electron microscopy (TEM) in accuracy, since it is possible to distinguish particles with the same metal core size but a different thickness of the organic ligand shell, which cannot be directly imaged by TEM.

Preparations of MPCs are not limited to the 1–5 nm size range. If the original two-phase synthesis is carried out in the *absence* of thiols, very stable, ruby-red solutions of gold nanoparticles in the 5–8 nm size range are obtained [50–52]. These particles are stabilized by the adsorption of bromide and bulky tetraoctylammonium ions, which, apart from balancing the charge, have a stabilizing effect similar to that of the organic ligand shell of MPCs [51]. The difference to MPCs is that this ionic capping is readily removed under a variety of conditions, for example if the dielectric constant of the medium is increased by the addition of alcohols or acetonitrile [45]. This leads to the immediate irreversible aggregation of the particles indicated by a rapid color change from red to blue followed by precipitation. An advantage of tetraoctylammonium bromide-stabilized nanoparticles is that they react readily with thiol or amine functionalities on surfaces or in solution, which facilitates the straightforward construction of self-assembled mono- and multilayer thin films of nanoparticles on various substrates as well as their conversion into stable MPCs by replacing the ionic capping agent with thiols [52–56]. Larger MPCs ranging from 10 to ca. 40 nm are accessible using classical citrate-stabilized gold hydrosols as a starting material, which can then be converted to stable MPCs by the formation of a SAM of thiols on the surface of the particles as demonstrated by Whitesides and co-workers [57]. Even larger MPCs would be desirable in view of some emerging bio-analytical applications, which rely on the excellent light-scattering properties of gold and silver nanoparticles above ca. 50 nm [58]. At present some issues of stability and uniformity in terms of size and shape distribution still have to be addressed until standard protocols for MPCs is this size range become available.

While most MPC research to date has focused on the use of gold, MPCs of silver and gold/silver alloys have also been reported since an early stage in the development of the field [17, 18, 59, 60]. The optical properties of silver may in many ways be more interesting than those of gold due to (1) the more dramatic size and shape dependence of the position and shape of the plasmon absorption band of silver, (2) the absence of d-band transitions, which in the case of gold can overlap with the free-electron contribution, and (3) the comparatively greater Raman enhancement that has been reported for silver particles [61, 62]. The synthetic protocols are comparable to those for MPCs of gold. Surprisingly, the use of silver nitrate as a starting material in the standard two-phase system leads to the formation of silver clusters in the organic phase, although no feasible phase transfer mechanism for the cationic Ag(I) exists if the cationic tetraoctylammonium phase transfer agent is used. It therefore has to be assumed that negatively charged nuclei of either metallic silver or silver bromide are transferred to the organic phase, where the formation of MPCs then proceeds to completion. This preparative aspect certainly deserves further attention in future research.

Although thiols have been the overwhelmingly preferred capping agents for MPCs of both gold and silver, other organic stabilizers have also been employed successfully. These include amines [17], thioethers [63–65] and isocyanates [66], the latter being, in fact, the preferred stabilizing agents for MPCs of platinum [66].

3.4
Structure, Reactivity and Ligand Mobility

MPCs of gold and silver consist of metal crystallites having the same fcc atomic structure as the bulk metal covered by a thiolate ligand shell of approximately twice the packing density observed on planar metal surfaces [41, 67, 68]. Depending on the size of the clusters a variety of different morphologies have been identified including decahedra, dodecahedra, icosahedra and, most commonly, truncated cubooctahedra [67]. For MPCs of gold, attempts have been made to assign precise chemical formulae to various clusters based on time-of-flight mass spectrometry in combination with a theoretical analysis of possible compositions of clusters of a given size and shape [67]. An overview of the variety of cluster shapes determined by TEM with the assistance of image simulations is presented in Fig. 3.1.

The structure of the ligand shell of MPCs of gold has been studied extensively by FTIR and NMR spectroscopy, mainly by the groups of Murray and Lennox [15, 16, 41, 42, 69–71]. The results of these investigations can broadly be summarized by the notion that the ligand shell resembles a self-assembled monolayer of thiolate molecules on a planar gold surface except for its higher packing density. For this reason MPCs have also been described as "monolayers in three dimensions" and have served as a model system for spectroscopic studies, in particular NMR, which would have been impossible using planar SAMs given the extremely small amount of material present in a monolayer on a macroscopic, planar surface [41, 69, 71]. Both ^{1}H and ^{13}C NMR spectra of MPCs exhibit very significant peak broadening, the origin of which is still debated. The signals from the nuclei closest to the gold core are those that are most broadened, which is probably due to a combination of intermolecular interactions within the ligand shell, similar to those responsible for peak broadening in solid-state materials, and a distribution of chemical shifts due to the heterogeneity of binding sites in a mixture of clusters of many different sizes and shapes [41, 69, 71].

In the solid state, the ligand shells of adjacent clusters interdigitate so that the average spacing between the metal cores is less than twice the nominal thickness of the ligand shell [16, 41]. This is evident from direct measurements of the gap between the particles by TEM and supported further by molecular dynamics simulations [68, 72].

The chemical reactivity of MPCs is determined by the functionalities present in the ligand shell and by the stability of the sulfur-gold bond at the cluster–ligand interface. The introduction of functionality is most commonly achieved via the versatile ligand place exchange reaction route developed by Murray and co-workers [73–77]. This approach is based on the observation that the thiolate ligands readily exchange with excess thiol molecules added to a solution of MPCs. Importantly, these exchange reactions do not depend on the existence of an equilibrium between ligands bound to clusters and free ligands, but on an associative mechanism in which the incoming thiol substitutes the leaving ligand via an S_N2-type reaction mechanism [41]. In this process the sulfur-bound hydrogen of the incom-

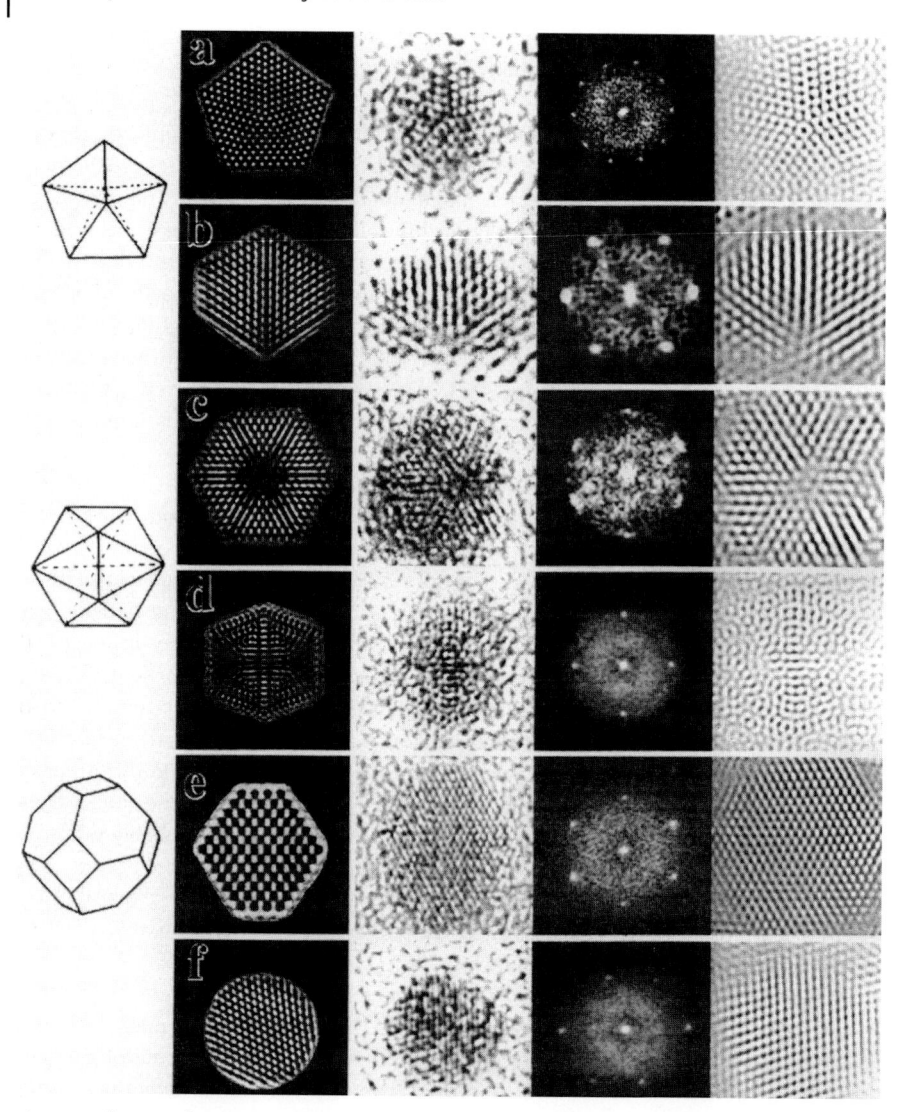

Fig. 3.1 Shape and structure analysis of MPCs of silver by TEM and image simulations. A decahedral (**a, b**), an icosahedral (**c, d**) and a truncated cubooctahedral cluster (**e, f**) are shown, each in two different orientations. For each structure, from left to right are shown: a model of the projected atomic structure; an experimentally obtained HREM image; its corresponding fast Fourier transform pattern; and a corresponding simulated HREM image (reproduced from S. A. Harfenist, Z. L. Wang, R. L. Whetton, I. Vezmar and M. Alvarez, *Adv. Mater.* **1997**, *9*, 817).

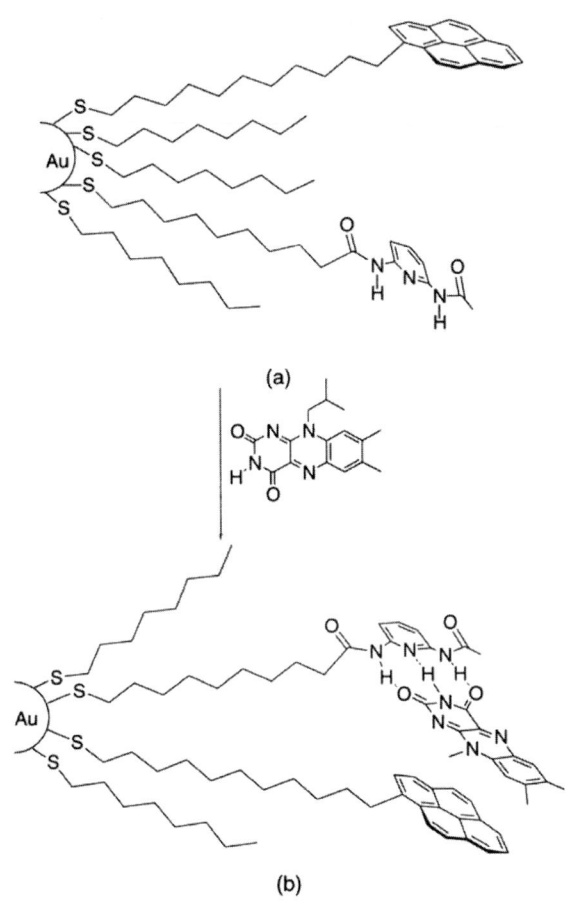

Fig. 3.2 An example of adaptive chemistry in the ligand shell of an MPC of gold. From a random distribution of pyrene and diaminopyridine ligands in the ligand shell (a), a flavine recognition site with both moieties in close proximity (b) is evolved in the presence of flavine (reproduced from [78] by kind permission of the American Chemical Society).

ing thiol is cleaved to protonate the leaving thiolate ligand. Such ligand exchange reactions can be monitored easily by NMR spectroscopy, which also confirms that no free ligand molecules are present in the solution prior to the addition of excess thiol. A large number of functionalities have been introduced into the ligand shell via ligand place exchange reactions, including not only virtually all standard functional groups [41] but also electrochemically addressable moieties [73] and fluorescent dyes [77].

A particularly intriguing aspect of the ligand shell dynamics of MPCs is the relatively high mobility of the ligand molecules on the cluster surface. This implies that MPCs with more than one type of ligand can adapt to environmental changes by reorganizing their ligand shell [78, 79]. A typical example of such an adaptive process is the common observation that MPCs containing a mixture of hydrophilic and hydrophobic ligands can initially be soluble in moderately polar solvents such as alcohols but lose all solubility after being stored for some time as dry solids [45]. This loss of solubility is probably due to a reorganization of the ligand shells that enables the clusters to form extended three-dimensional networks, which are held together in virtually any medium by either their polar or their non-polar components. Similar reorganization phenomena are also observed if such clusters are spread on the air/water interface [79]. The most striking example of adaptive chemistry in the ligand shell of MPCs is the gradual self-optimization of molecular binding sites in the presence of a specifically binding target molecule. This has been demonstrated by Rotello and co-workers, who prepared MPCs with both pyrene and diaminopyridine moieties in the ligand shell diluted by a matrix of alkane thiols [78]. Whenever the pyrene and diaminopyridine units were in close vicinity on the cluster surface as shown in Fig. 3.2, they represented an optimal recognition motif for flavine in solution due to the ability of interacting via both hydrogen bonding and π-stacking. It was found that the clusters adapted to the presence of flavine, and formed such optimized binding sites by random diffusion of the pyrene and diaminopyridine ligands on the cluster surface. This could possibly establish a generic concept for the evolutionary development of artificial receptors without the need for an often painstakingly complicated synthesis of a premeditated binding motif.

3.5
Self-organization Phenomena

Almost immediately after the first papers on the preparation of MPCs had appeared, Whetten and co-workers discovered that monodisperse fractions of the clusters self-organized into two-dimensional, hexagonal superlattices upon evaporation of the solvent as shown in Fig. 3.3 [14]. A well-known precedent for such spontaneous nanoscale crystallization processes is the formation of superlattices from monodisperse, tri-octyl phosphorous oxide (TOPO)-capped CdSe nanoparticles reported by Bawendi and co-workers in 1995 [80]. During the following decade many similar cases were described in a remarkably short period, and it is now possible to obtain highly ordered superlattices of virtually any material that can be obtained in the form of ligand-stabilized nanoparticles [51, 81–83]. The practical implications of this ability may be very significant for the field of magnetic information storage. In view of such applications, the IBM group led by Murray developed preparative methods for nanoparticle superlattices covering a range of different magnetic materials including Co and FePt alloys [82, 83].

The discovery of superlattice formation from nanoparticles was serendipitous and the driving forces for the ordering process are still subject to some debate. The formation of ordered phases of particles from concentrated colloidal solutions, such as those formed as the solvent evaporates to near dryness, requires a

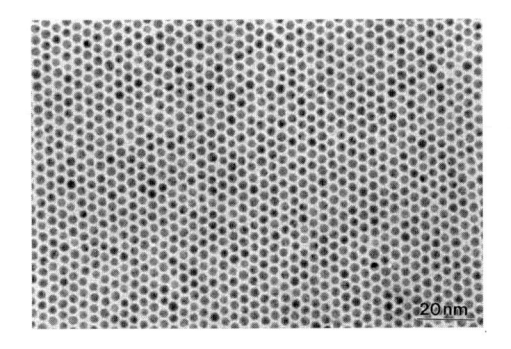

Fig. 3.3 TEM image of a typical self-assembled hexagonal superlattice of 3–4 nm MPCs of gold deposited by allowing a drop of a volatile solution of the particles to evaporate slowly on an amorphous carbon support film.

net repulsive interaction between the particles in order to prevent random disordered aggregation. However, MPCs consist of neutral metal cores with non-polar hydrocarbon ligands, which would under these conditions probably experience weak inter-particle *attraction* rather than repulsion. Indeed, it has been observed that very pure samples of MPCs do not readily form extended regular superlattices, while the deliberate addition of contaminants such as excess thiols can lead to a very significant improvement of the self-organization process [84]. It has further been pointed out by Lennox and co-workers that MPCs prepared following the standard two-phase protocol usually contain a small amount of the phase transfer agent tetraoctylammonium bromide, which is difficult to remove [16, 85].

The presence of traces of contaminants could contribute to the formation of superlattices in at least two different ways. Charged species such as bromide and/or tetraoctylammonium ions can be intimately associated with the ligand shell and thereby confer a net charge to the cluster. This would lead to repulsive inter-particle forces, which could favor the formation of ordered phases in concentrated colloidal solutions. On the other hand, the presence of neutral species such as excess thiols can promote entropically driven colloidal crystallization in very concentrated solutions by competing with the clusters for free solvent volume. The entropy increase due to the liberation of extra solvent volume following the crystallization of the MPCs overcompensates the entropy loss associated with the crystallization process. This phenomenon is well known for colloidal crystallization of micron and submicron spheres in the presence of dissolved polymer molecules or nanoparticles, which remain in solution [86].

In reality many more factors are likely to contribute to the formation of superlattices. For example, substrate effects, which have not yet been investigated systematically, almost certainly play an important role. Most preparations described in the literature have been carried out by depositing a solution of MPCs on an amorphous carbon support film and allowing the solvent to evaporate. The superlattices formed are generally two-dimensional, which implies that the crystallization process occurs when most of the solvent has evaporated and only a thin film is still wetting the substrate. It has further been observed that procedures that lead to well-ordered superlattices on amorphous carbon substrates usually give fractal or two-dimensional foam-like structures on other substrates such as oxide-

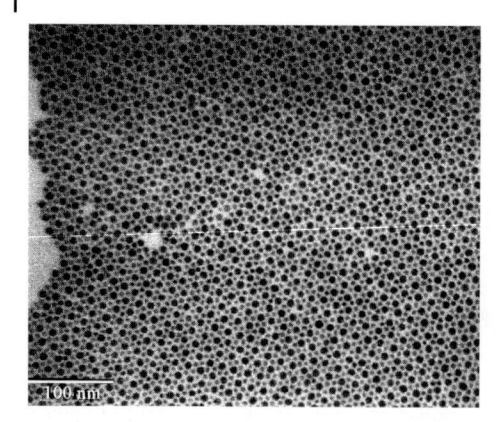

Fig. 3.4 TEM image of a bimodal superlattice spontaneously formed by evaporation of a solution containing a mixture of MPCs of different discrete sizes (adapted from [91] by kind permission of *Nature*).

covered silicon surfaces [87, 88]. Different causes for the formation of these more complex structures have been discussed including the existence of relatively narrow stability regimes, which can lead to the spontaneous formation of regular cellular structures by spinodal decomposition [87, 89]. Jaeger and co-workers demonstrated that problems represented by this complex behavior could be overcome for many different substrates if excess alkanethiol is added to the deposition solution [84]. Apart from the possible entropic effect mentioned above the thiol could form a surface monolayer, which would confer very similar chemical properties to different substrates. Depending on the amount of thiol added it could also form a viscous lubricating film, which would still allow cluster self-organization processes to occur after the solvent has evaporated. Clearly, more work is needed to understand all the factors that govern superlattice formation in the many different nanoparticle/substrate combinations that have been described.

So far, only superlattices formed from near monodisperse fractions of MPCs have been considered. These are usually hexagonally packed, although cubic packed systems have also been observed, for example in the case of highly truncated cubooctahedral MPCs of silver [90]. A number of fascinating new nanostructures with bimodal size distributions emerge when solutions of MPCs containing fractions of different discrete sizes are used instead [91]. Examples of such lattices formed from MPCs of gold are presented in Fig. 3.4. Interestingly, these structures can be rationalized by comparing the ratios of the radii of the constituent particles with the ratios of the atomic radii in ordered bimetallic alloys. For this reason bimodal nanostructures are also referred to as alloy structures [91, 92]. The AB_2-nanostructure shown in Fig. 3.4, for example, is equivalent to the atomic scale crystal structure of the intermetallic compound aluminum boride AlB_2. Likewise, some naturally occurring gem opals have equivalent colloidal crystal structures but with micrometer lattice spacings [93, 94]. A similar range of bimodal alloy structures has also been reported for ligand-stabilized semiconductor nanoparticles [92] and for mixtures of MPCs of gold and silver, in which each metal occupies a distinct lattice site according to the relative sizes of the clusters [95]. The latter example represents a case of a binary, bimodal superlattice, i.e. a true "nanoalloy".

Exploring an alternative approach to the use of solid substrates, some interesting work on nanostructure self-organization from MPCs has been carried out at the water/air interface [96–100]. Langmuir films of MPCs are readily obtained by spreading a droplet of a solution of the clusters in chloroform on the water surface and allowing the solvent to evaporate. The two-dimensional mono- and submonolayer structures formed upon compression of the films in a Langmuir trough are readily transferred either vertically or horizontally to planar supports. Apart from extended hexagonal two-dimensional superlattices [96], a surprising number of structural variants have been observed in these systems. For example, "wire-like" structures of linearly aligned MPCs of silver have been obtained by Heath and co-workers [97]. The reasons for the formation of such linear structures are not clear, but intuitively it is easy to imagine that very subtle effects (MPC shape, charge, ligand interactions, etc.) can change the type of structures obtained quite significantly due to the extremely high mobility of the clusters at the water/air interface.

An even wider range of structures is accessible if the clusters are co-deposited with surfactants, such as, for example, the phospholipid dipalmitoylphosphatidylcholine (DPPC), which forms Langmuir films with well-characterized phase transitions upon compression [98]. While the clusters in the absence of the surfactant tend to aggregate into rafts even in uncompressed films, the surfactant monolayer can to some extent function as a two-dimensional solvent and prevent the clusters from aggregation. Upon compression of the DPPC films phase transitions occur, which lead to the formation of micron-scale domains of a compact, liquid crystalline surfactant phase. As expected for a crystallization process the clusters are expelled from these areas as impurities and decorate the phase boundaries. After dense compression of the films and transfer onto freshly cleaved mica all the surfactant is present in the solid phase and the clusters form a network of lines encircling the liquid crystalline domains of the surfactant film. The evolution of these phase boundary structures upon film compression is illustrated in the AFM images in Fig. 3.5. Very similar quasi-one-dimensional structures of Au_{55} clusters embedded in a Langmuir-Blodgett film of poly(vinyl-pyrrolidone) transferred onto HOPG have at the same time been reported by Schmid and co-workers [99]. A TEM study of the lines formed in the presence of DPPC revealed at higher resolution that the constituent clusters had sintered into elongated, worm-like structures, which together formed an extended "maze" with the width of the internal "path" of molecular dimensions (Fig. 3.6) [98]. Structural instabilities and sintering of MPCs in densely compressed Langmuir films are probably due to the high lateral forces experienced by the particles under compression of the film. In the absence of surfactant, the sintering of small MPCs in Langmuir films has also been observed, but without the formation of unusual structural features [100]. Although explanations of the origin of these structures are still very speculative, they represent examples of how a surfactant or polymer film can act as a template and impose structure on both the mesoscopic and the molecular scale.

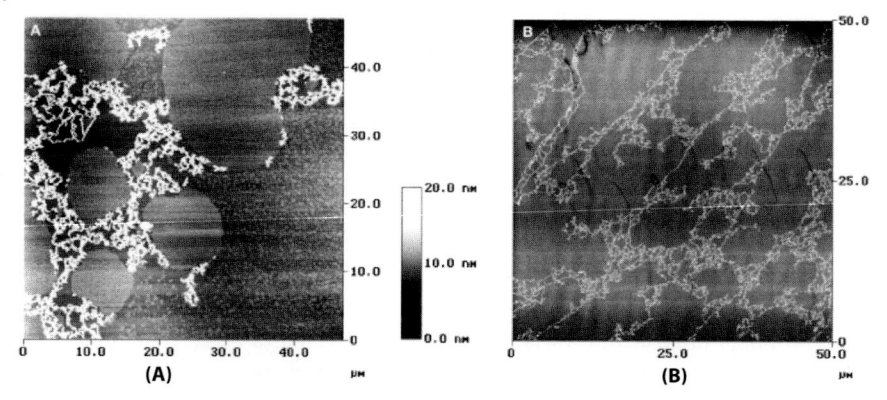

Fig. 3.5 AFM images of MPCs of gold embedded in a Langmuir-Blodgett film of dipalmitoyl-phosphatidylcholine (DPPC) transferred to freshly cleaved mica (**A**) before maximal compression of the Langmuir film was reached and (**B**) at maximal compression. In (**A**) the formation of two different phases of the phospholipid can clearly be observed as a slight difference in height. The clusters are only present in the lower-density "liquid-like" phase and finally decorate the phase boundaries in the densely compressed film (**B**) (adapted from [98] by kind permission of Adv. Mater.).

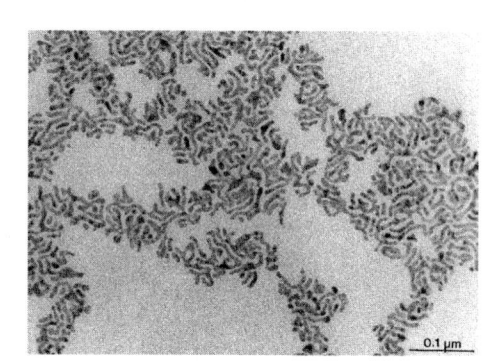

Fig. 3.6 TEM image of the "nano-maze" structure formed from MPCs of gold within a compressed Langmuir film of DPPC. The lines decorating the surfactant phase boundaries shown in Fig. 3.5 (B) consist of the structures shown here (adapted from [98] by kind permission of Adv. Mater.).

3.6
Electronic Properties

Since the first discovery of the self-organization processes described in the preceding section there has been an increasing interest in investigating the electronic properties of the various structures and materials obtained as well as those of the individual clusters. Both single clusters and ensembles of clusters exhibit interesting electronic phenomena such as distance-dependent metal–insulator transitions [101–103], electron hopping conductivity [15, 50] and size-dependent Coulomb staircase charging [104–108]. The transfer of self-assembled nanostructures to a substrate or environment that allows their electronic characterization still represents a significant challenge in many cases. Relatively large areas of well-defined hexagonal superlattices of MPCs can be obtained on a variety of substrates by transferring compressed Langmuir films from the water/air interface onto solid support materials. Heath and co-workers have employed this technique to construct a Coulomb charging device by sandwiching a two-dimensional hexagonal superlattice of monodisperse silver MPCs between planar electrodes [104]. The differential capacitance of the superlattice was shown to be a function of the bias voltage and increased in discrete steps corresponding to Coulombic charging of the clusters in the lattice by successive single-electron transfer steps. Such a behavior can only be observed with very monodisperse clusters since a size distribution would also lead to a distribution of capacitances, which would smear out the effect.

Heath and co-workers also discovered that Langmuir films of these monodisperse silver MPCs exhibit reversible, distance-dependent metal–insulator transitions, provided that the ligand molecule is short enough to allow a very close proximity of the metal cores in densely compressed films so that an extended metallic band structure can develop [101–103]. At a given surface pressure the optical properties of the films show a sharp transition from typical insulator to metal behavior. This can even be detected with the naked eye since suddenly a metallic silver mirror is formed upon compression of the films. Bard and co-workers characterized this system electrochemically in an elegant experiment using a "submarine" scanning electrochemical microscope (SECM) [109]. As a microelectrode approaches the film from underneath in the presence of an electrochemically active redox-couple, a metallic film leads to positive feedback (regeneration of the species consumed in the electrode reaction by the metallic surface film due to Nernstian equilibrium) so that the current measured by the electrode increases with decreasing distance between electrode and surface film. In contrast, an insulating film will lead to a decrease of the current as the electrode approaches since electroactive species are depleted by the electrode reaction and diffusion towards the electrode is impeded by the proximity of the surface film. While study of the optical properties revealed very sharp transitions, the behavior found in the electrochemical experiments was somewhat more gradual. This has been tentatively attributed to the possibility of a comparatively less homogeneous size distribution of the MPCs used in the electrochemical study [109].

Disordered three-dimensional materials prepared from MPCs such as self-assembled multilayer films [53, 54], compressed pellets of dithiol cross-linked particles [50] or simply drop cast films of MPCs [15] all exhibit typical electron hopping conductivity, which can be explained classically over a wide temperature range as a simple, Arrhenius type, activated transport process that is controlled chiefly by the size-dependent charging energy of the metal islands [110]. There is some evidence that highly ordered films of monodisperse clusters have significantly larger conductivities than disordered systems, possibly due to the onset of collective electronic properties comparable to those observed in compressed Langmuir films [111].

The electronic charging of individual clusters is characterized by its size-dependent Coulomb staircase behavior. This is a classical effect, which is due to the very small capacitance, C, of nanometer-sized metallic objects. The transfer of a single elementary charge, e (by addition or removal of a single electron) to a particle changes the electrostatic potential, E of the particle according to Eq. (1).

$$\Delta E = e/C. \tag{1}$$

In all systems larger than a few nanometers the value of C is large enough so that $e/C \approx 0$ and the electrostatic potential of the system can be changed continuously to any value by the transfer of an appropriate very large number of electrons. In nanosized systems, however, e/C can be of the order of a few millivolts, implying that the electrostatic potential of the system cannot be changed by an amount smaller than this value since it is impossible to transfer less than one electron. This leads to the typical Coulomb staircase of discrete single electron charging with a Coulomb blockade centred at 0 V, where no charging can occur until at least one electron can be added or removed. The potential width, ΔE_{CB}, of this range is given by Eq. (2).

$$\Delta E_{CB} = 2e/C \tag{2}$$

Coulomb staircase measurements usually have to be carried out at low temperature since the charging energy, E_c (given by Eq. 3) has to be larger than kT so that the discrete charging characteristics are not hidden by thermal noise.

$$E_c = e^2/C \tag{3}$$

The first room-temperature Coulomb blockade was reported by Reifenberger and coworkers, who prepared sufficiently small (<5 nm) gold clusters by vapor deposition onto a SAM of dithiols on a flat gold surface [105]. The experiment was carried out by positioning the tip of a scanning tunneling microscope (STM) on top of a single cluster and measuring the current/voltage characteristics of the system.

Given their stability and small size, MPCs lend themselves more than any other material to the study of single-electron charging in nanosized systems. Murray and co-workers reported the electrochemical Coulomb staircase charging of MPCs [107, 108]. A solution containing strictly monodisperse MPCs and a support electrolyte in a mixture of toluene and acetonitrile exhibited typical size-dependent electron transfer steps by cyclic voltammetry and differential pulse voltammetry. The results of this study along with the more conventional STM characterization of the charging behavior of single clusters of the same type are shown in Fig. 3.7. It was also found that the classical model of capacitance-dependent charging in equidistant steps breaks down for very small clusters (≤ 1.5 nm), probably because a transition from metallic to molecular behavior occurs in this size regime so that the position of the molecular orbitals of the clusters and not their classically calculated capacitance determines the charging characteristics [108]. The same group of researchers extended these studies to surface-immobilized MPCs and observed very similar size-dependent, discrete charging [112].

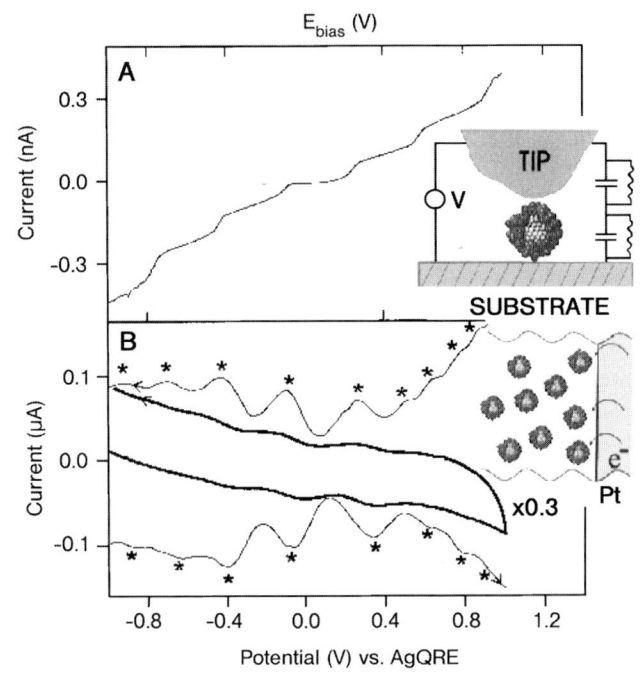

Fig. 3.7 **A** Au STM tip addressing a single cluster adsorbed on a Au-on-mica substrate (inset) and Coulomb staircase *I–V* curve at 83 K; potential is tip-substrate bias; equivalent circuit of the double tunnel junction gives capacitances C_{upper} = 0.59 aF and C_{lower} = 0.48 aF. **B** Voltammetry (CV –, 100 mV/s; DPV ▬, * are current peaks, 20 mV/s, 25 mV pulse, top and bottom are negative and positive scans, respectively) of a 0.1 mM 28 kDa cluster solution in 2:1 toluene:acetonitrile/0.05 M Hx$_4$NClO$_4$ at a 7.9×10^{-3} cm^2 Pt electrode, 298 K, Ag wire pseudoreference electrode (reproduced from [107] by kind permission of the American Chemical Society).

A somewhat more complex electrochemical system, in which gold particles were immobilized on a gold electrode such that each was covalently linked via a small number of redox-active dithiol molecules has been reported by Schiffrin and co-workers [113]. A viologen moiety served as the electrochemically addressable redox center located between the particles and the electrode, and the electron transfer characteristics of individual particles were interrogated by in-situ STM as illustrated in Fig. 3.8. Interestingly, it was found that the electronic transparency of the system depended strongly on the electrochemically controlled redox state of the central viologen groups such that the system represented an electrochemical "nano-switch". When the viologen was in the reduced radical cation state the switch was in its "on" state and current could flow between the STM tip and the gold electrode. In the original dicationic state and in the completely reduced neutral state of the viologen, on the other hand, the switch was in its "off" state and current flow was effectively blocked. At present no satisfactory models exist to ex-

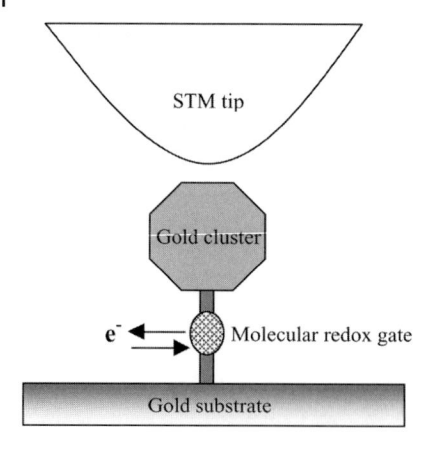

Fig. 3.8 Scheme of a self-assembled "nano-switch". Electronic transport across the device is interrogated by the tip of an STM and can be controlled electrochemically by adjusting the redox state of the gate molecules (viologen)

plain the conductance of single molecules within an environment of this complexity. In the case of the electrochemical viologen switch it is assumed that electron transfer via the cation radical state occurs by a mechanism similar to resonance tunneling, which has been observed in electrochemical in-situ STM studies of electrodes derivatized with redox active molecules such as porphyrins. These showed much higher tunneling rates through the molecules when the electrode potential corresponded to the equilibrium redox potential of the surface-bound electroactive species.

3.7
Current and Future Applications

Closing the circle all the way back to the Lycurgus cup the use of MPCs of gold, silver and alloys for decorative purposes is probably the commercially most significant application at the moment [114]. Along with Au(I) mercaptides, MPCs of gold and gold/silver alloys are used as metal paints and inks to decorate glassware, porcelain and similar items [114]. Technologically more sophisticated applications such as ink-jet printing of conductive structures are also being investigated. The development of many other commercially viable uses of these relatively new materials is currently underway.

Thin films of MPCs change their electrical conductivity rapidly and reversibly in the presence of organic vapors [5, 115, 116]. This effect is based on the swelling of the material upon gas absorption, which leads to an increase in the spacing between the metal cores. Since the typical electron hopping conductivity in these materials depends very sensitively on this distance, the absorption of organic vapor leads to a strong decrease in electrical conductivity. This behavior has been exploited technologically as a new concept for gas sensors [5, 115, 116]. Also based on the strong dependence of the conductivity on inter-particle distance, new pressure sensors and strain gauges may be developed from the same type of materials.

MPCs of gold or silver exhibit strong optical anisotropy if the particles are aligned in parallel rows [6, 117]. This is due to dipole–dipole coupling between adjacent particles in a row, which leads to a change in the plasmon absorption spectrum relative to that of the particles in solution only for the component of light that is polarized in parallel with the aligned rows of particles. Transmitted light polarized parallel to the rows will be absorbed as if it were passing through a continuous thin metal film, while light polarized perpendicular to the rows will "see" the individual particles. This effect can be exploited for polarizing filters and for displays. Two different approaches to aligning nanoparticles for this purpose have been reported [6, 117]. Foss and co-workers deposited MPCs of gold onto friction-transferred lines of PTFE on glass substrates and obtained optically anisotropic windows after a subsequent drastic heat treatment close to the melting point of the glass [117]. Dirix and co-workers achieved similar effects by the mechanical stretching of polyethylene films, into which MPCs of silver had been embedded prior to stretching [6]. The particles were found to form aligned "stretch marks" parallel to the stretch direction. The perceived color of the transmitted light was yellow if polarized perpendicular to the rows, and red if polarized in parallel. The unique optical properties of nanometer-sized gold and silver particles, which lead to a dependence of the plasmon absorption on particle size, shape, spacing and orientation, as in the above examples, are discussed in more detail in Chapter 7.

Chemical applications of MPCs are chiefly in catalysis and include asymmetric dihydroxylation reactions [118], carboxylic ester cleavage [119], electrocatalytic reductions by anthraquinone-functionalized gold particles [120] and particle-bound ring-opening metathesis polymerization [121]. These catalytic systems have in common the exploitation of the carefully designed chemical functionality of the ligand shell, rather than the potential catalytic activity of a nanostructured clean metal surface.

Probably the most widely discussed potential long-term application of nanoparticle technology is the development of new, ultimately small, electronic devices. Many promising attempts to construct devices have already been reported, based on MPCs and Au_{55} clusters [36, 122, 123]. In particular, single-electron transistor action has been demonstrated for systems that contain ideally only one particle in the gap between two contacts separated by only a few nanometers [36, 121, 123]. This central metal particle represents a Coulomb blockade and exhibits single-electron charging effects due to its extremely small capacitance. It can also act as a gate if it is independently addressable by a third terminal, for example the substrate itself, an appropriately positioned STM tip or a third metallic contact on the surface of the substrate.

Future prospects for nanoelectronics notwithstanding, it is very likely that the most important medium- and short-term applications of MPCs will be in the field of biological sciences. These, however, require the further development of water-soluble MPCs, which should be as stable and chemically versatile as their hydrophobic analogues, to which to date the vast majority of reports has been dedicated. Recently reported ligands for the preparation of water-soluble MPCs include glutathione [124], thiopronine [125], thiolated sugars [126], carboxylic acids [55] and poly(ethylene glycol) derivatives [127–129] as well as dimethylaminopyridine (DMAP) [130].

Based on thiol derivatization of 13-nm classical citrate-stabilized gold nanoparticles, Mirkin and co-workers obtained stable, water-soluble particles, which were surrounded by a ligand shell of thiol-modified DNA oligonucleotides [131–134]. These particles aggregated and changed color from red to blue prior to precipitation in the presence of linker DNA, which was complementary to the DNA on the surface of the particles. This effect has been exploited to develop extremely sensitive colorimetric methods of DNA analysis capable of detecting trace amounts of a particular oligonucleotide sequence. With this approach it is even possible to distinguish between perfectly complementary DNA sequences and those that exhibit different degrees of base pair mismatches [134]. The relations between structure and optical properties in these systems have been studied quantitatively by Lazarides and Schatz, who carried out electrodynamics calculations involving aggregates of a large number of particles [135]. Alivisatos and co-workers reported a similar DNA-controlled linkage of particles to each other using smaller (ca. 3 nm) phosphine-stabilized gold clusters [136–139]. In addition, this group devised methods of linking discrete numbers of particles to each other by DNA and separating the aggregates according to the number of connected particles by gel electrophoresis [138, 139]. The Mirkin system, on the other hand, has been exploited further for the "scannometric" detection of DNA [140]. Here the particles are attached to surface-immobilized DNA via hybridization with a target sequence and detected by an ordinary flat-bed scanner.

By far the most sensitive detection of gold and silver nanoparticles is achieved by measuring the light scattered by the particles [141–143]. Single-particle detection by dark field microscopy is routinely possible provided that the particles are not too small to be effective Mie scatterers. An optimum particle size around 60 nm has been reported by Yguerabide and co-workers, who pioneered the use of light-scattering particles for bio-analytical applications [141, 142]. These light-scattering probes can be employed as an alternative to fluorescent dyes in binding assays, bio-labeling experiments and gene chips. They have the advantages of a comparatively higher sensitivity and the absence of photobleaching, which represents a severe limitation to the use of fluorescent dyes in many applications.

A further potentially important biological application of MPCs is gene delivery. Rotello and co-workers have demonstrated that DNA fragments can be bound reversibly to amine-functionalized MPCs via pH-dependent electrostatic interactions and that such bioconjugates are capable of transfecting cells efficiently [144].

3.8
Visions for Future Development

It was stated in the preceding section that biological applications of MPCs represent a very exciting area for future developments. As well as those outlined above, more adventurous, and maybe to some extent futuristic, possibilities can be envisaged. For example, MPCs can, in principle, be prepared to cover the entire size range of intracellular components from small proteins to cell nuclei. This makes them ideal, not only as passive, optically easily detectable, intracellular probes, but potentially also as active nano-machines that could be employed to manipulate processes in vivo on the intracellular level. Their stability and chemical versatility are ideal properties that would allow for complex, multifunctional, bioactive objects of sub-cellular dimensions to be synthesized. Biomolecules such as DNA, enzymes, antibodies and polysaccharides could be arranged around a core-particle in such a way that they would form units with complex biological functions, which could include combinations of recognition, activation, inhibition, synthesis and signaling. Such artificial organelles would be effective in manipulating intracellular target processes with high specificity. Likewise, outside the biological context, MPCs with a number of synergistically cooperating ligands could be devised to act as artificial enzymes, which could present a new and sophisticated approach to enantioselective catalysis and to other challenges of modern preparative chemistry. A new impulse for the chemical synthesis of nanostructures may result from the possibility of manipulating DNA/nanoparticle conjugates with DNA-processing enzymes (restriction endonucleases, ligases), which provides opportunities for the systematic multi-step synthesis of nanostructures [145–147].

In the area of nanoelectronics MPCs may play a role as stable components of novel electronic devices. However, it remains to be seen if so-called bottom-up construction, i.e. self-assembly, of miniaturized devices will establish itself as a viable technology that could complement conventional silicon technology. The potential for a new generation of electronics based on single-electron devices, which are assembled from chemically prepared components including MPCs has, in principle, been demonstrated. The remaining problems of long-term stability, integration and addressability, which would have to be overcome before electronic applications of such a new technology could realistically be envisaged, are nevertheless still very significant. Fortunately, the development of new nanoelectronic concepts and devices is currently a very active area of research, which will continue to yield spectacular new insight into the fundamental physics of such systems and will certainly also lead to important future applications, which may or may not include the use of MPCs.

Finally, there should be a significant potential for the future development of nanostructured multicomponent systems. These are materials or structures that contain more than one type of nanoparticle, for example, carbon nanotubes and MPCs, which in combination may lead to unique structural and/or functional properties. To date, very little work has been carried out on multicomponent systems, probably because well-defined spatial arrangements of such nanocomponents are still difficult to achieve. It is quite clear, however, that major breakthroughs in the development of new functional materials in such diverse areas as energy conversion, optical switching, information storage, signal processing, cell adhesion, sensors, actuators and displays can be expected from integrating different nanostructured components into a device or material.

3.9 References

1. M. Haruta, N. Yamada, T. Kobayashi, S. Iijima, *J. Catal.* **1989**, *115*, 301.
2. D. L. Feldheim, C. D. Keating, *Chem. Soc. Rev.* **1998**, *27*, 1.
3. H. J. Richter, *J. Phys. D* **1999**, *32*, R147.
4. J. H. Fendler, *Nanoparticles and Nanostructured Films: Preparation, Characterization and Applications*, Wiley-VCH, Weinheim, **1998**.
5. S. F. Cheng, L. K. Chau, *Anal. Chem.* **2003**, *75*, 16.
6. Y. Dirix, C. Bastiaansen, W. Caseri, P. Smith, *Adv. Mater.* **1999**, *11*, 223.
7. D. J. Maxwell, J. R. Taylor, S. Nie, *J. Am. Chem. Soc.* **2002**, 9606.
8. J. Schulz, A. Roucoux, H. Patin, *Chem. Rev.* **2002**, *102*, 3757.
9. M. P. Pileni, *J. Phys. Chem. B* **2001**, *105*, 3358.
10. G. Schmid, L. F. Chi, *Adv. Mater.* **1998**, *10*, 515.
11. *Clusters and Colloids* (ed. G. Schmid), VCH, Weinheim, **1994**.
12. M. Brust, M. Walker, D. Bethell, D. J. Schiffrin, R. Whyman, *J. Chem. Soc., Chem. Commun.* **1994**, 801.
13. M. Brust, J. Fink, D. Bethell, D. J. Schiffrin, C. J. Kiely, *J. Chem. Soc., Chem. Commun.* **1995**, 1655.
14. R. L. Whetten, J. T. Khoury, M. Alvarez, S. Murthy, I. Vezmar, Z. L. Wang, P. W. Stephens, C. L. Cleveland, W. D. Luedtke, U. Landman, *Adv. Mater.* **1996**, *8*, 428.
15. R. H. Terrill, T. A. Postlethwaite, C.-H. Chen, C.-D. Poon, A. Terzis, A. Chen, J. E. Hutchison, M. R. Clark, G. Wignall, J. D. Londono, R. Superfine, M. Falvo, C. S. Johnson Jr., E. T. Samulski, R. W. Murray, *J. Am. Chem. Soc.* **1995**, *117*, 12537.
16. A. Badia, S. Singh, L. Demers, L. Cuccia, G. B. Brown, R. B. Lennox, *Chem. Eur. J.* **1996**, *2*, 359.
17. J. R. Heath, C. M. Knobler, D. V. Leff, *J. Phys. Chem. B* **1997**, *101*, 189.
18. B. A. Korgel, S. Fullam, S. Connolly, D. Fitzmaurice, *J. Phys. Chem. B* **1998**, *102*, 8379.
19. A. Neri, *L'Arte Vetraria*, Vol. 7, Ch. 129, Florence, **1612**.
20. P. Mulvaney, *MRS Bull.* **2001**, *26*, 1009.
21. M. Faraday, *Philos. Trans. R. Soc. London* **1857**, *147*, 145.
22. R. Zsigmondy, *Kolloidchemie*, Verlag von Otto Spamer, Leipzig, **1918**.
23. E. A. Hauser, J. E. Lynn, *Experiments in Colloid Chemistry*, McGraw Hill, New York, **1940**.
24. J. Turkevich, P. C. Stevenson, J. H. Hillier, *Discuss. Faraday Soc.* **1951**, *11*, 55.
25. R. Zsigmondy, *Z. phys. Chemie* **1906**, *56*, 65.
26. R. G. Freeman, K. C. Grabar, K. J. Allison, R. M. Bright, J. A. Davis, A. P. Guthrie, M. B. Hommer, M. A. Jackson, P. C. Smith, D. G. Walker, M. J. Natan, *Science* **1995**, *267*, 1629.
27. G. Chumanov, K. Sokolov, T. M. Cotton, *J. Phys. Chem.* **1996**, *100*, 5166.
28. A. Doron, E. Katz, I. Willner, *Langmuir* **1995**, *11*, 1313.
29. A. Csaki, R. Möller, W. Straube, J. M. Köhler, W. Fritzsche, *Nucleic Acids Res.* **2001**, *29*, art. no. 81.
30. C. Niemeyer, *Angew. Chem. Int. Ed.* **2002**, 4129.
31. L. E. Brus, *J. Phys. Chem.* **1994**, *98*, 3575.
32. H. Weller, *Angew. Chem. Int. Ed. Engl.* **1993**, *32*, 41.
33. G. Schmid, R. Pfeil, R. Boese, F. Bandermann, S. Meyer, G. H. M. Calis, J. W. A. van der Felden, *Chem. Ber.* **1981**, *114*, 3634.
34. L. F. Chi, M. Hartig, T. Drechsler, T. Schwaack, S. Seidel, H. Fuchs, G. Schmid, *J. Appl. Phys. A* **1998**, *66*, 187.
35. U. Simon, G. Schön, G. Schmid, *Angew. Chem. Int. Ed. Engl.* **1993**, *32*, 250.
36. M. Schumann, Y. Liu, T. Raschke, C. Radehaus, G. Schmid, *Nano Lett.* **2001**, *1*, 405.
37. M. S. Dresselhaus, G. Dresselhaus, P. C. Eklund, *Science of Fullerenes and Carbon Nanotubes*, Academic Press, New York, **1995**.
38. R. G. Nuzzo, D. L. Allara, *J. Am. Chem. Soc.* **1983**, *105*, 4481.

39 H. O. Finklea, in *Electroanalytical Chemistry* 1996, *19*, 109 (A. J. Bard, I. Rubinstein eds.).
40 M. Giersig, P. Mulvaney, *Langmuir* 1993, *9*, 3408.
41 A. C. Templeton, W. P. Wuelfing, R. W. Murray, *Acc. Chem. Res.* 2000, *33*, 27.
42 M. J. Hostetler, J. E. Wingate, C.-Z. Zhong, J. E. Harris, R. W. Vachet, M. R. Clark, J. D. Londono, S. J. Green, J. J. Stokes, G. D. Wignall, G. L. Glish, M. D. Porter, N. D. Evans, R. W. Murray, *Langmuir* 1998, *14*, 17.
43 R. L. Whetten, M. N. Shafigullin, J. T. Khoury, T. G. Schaaff, I. Vezmar, M. M. Alvarez, A. Wilkinson, *Acc. Chem. Res.* 1999, *32*, 397.
44 Y. Kang, K. Kim, *Langmuir* 1998, *14*, 226.
45 M. Brust et al., unpublished observations.
46 R. L. Whetten, M. N. Shafigullin, J. T. Khoury, T. G. Schaaff, I. Vezmar, M. M. Alvarez, A. Wilkinson, *Acc. Chem. Res.* 1999, *32*, 397.
47 E. Gutierrez, R. D. Powell, F. R. Furuya, J. F. Hainfeld, T. G. Schaaff, M. N. Shafigullin, P. W. Stephens, R. L. Whetten, *Eur. Phys. J.* 1999, *D9*, 647.
48 P. C. Ohara, D. V. Leff, J. R. Heath, W. M. Gelbart, *Phys. Rev. Lett.* 1995, *75*, 3466.
49 J. P. Wilcoxon, J. E. Martin, P. Provencio, *Langmuir* 2000, *16*, 9912.
50 M. Brust, D. Bethell, D. J. Schiffrin, C. J. Kiely, *Adv. Mater.* 1995, *7*, 795.
51 J. Fink, C. J. Kiely, D. Bethell, D. J. Schiffrin, *Chem. Mater.* 1998, *10*, 922.
52 C. Demaille, M. Brust, M. Tsionsky, A. J. Bard, *Anal. Chem.* 1997, *69*, 2323.
53 D. Bethell, M. Brust, D. J. Schiffrin, C. J. Kiely, *J. Electroanal. Chem.* 1996, *409*, 137.
54 M. Brust, D. Bethell, C. J. Kiely, D. J. Schiffrin, *Langmuir* 1998, *14*, 5425.
55 D. I. Gittins, F. Caruso, *Chem. Phys. Chem.* 2002, *3*, 110.
56 D. I. Gittins, F. Caruso, *Adv. Mater.* 2000, *12*, 1947.
57 C. S. Weisbecker, M. V. Merritt, G. W. Whitesides, *Langmuir* 1996, *12*, 3763.
58 J. Yguerabide, E. E. Yguerabide, *J. Cell Biochem.* 2001, *37*, 71.
59 M. J. Hostetler, C. J. Zhong, B. K. H. Yen, J. Anderegg, S. M. Gross, N. D. Evans, M. Porter, R. W. Murray, *J. Am. Chem. Soc.* 1998, *120*, 9396.
60 S. W. Han, Y. Kim, K. Kim, *J. Coll. Int. Sci.* 1998, *208*, 272.
61 *Surface Enhanced Raman Scattering* (eds. R. K. Chang, T. E. Furtak), Plenum, New York, 1982.
62 B. Vlčková, D. Tsai, X. Gu, M. Moskovits, *J. Phys. Chem.* 1996, *100*, 3169.
63 X.-M. Li, M. R. de Jong, K. Inoue, S. Shinkai, J. Huskens, D. N. Reinhoudt, *J. Mater. Chem.* 2001, *11*, 1919.
64 E. J. Shelley, D. Ryan, S. R. Johnson, M. Couillard, D. Fitzmaurice, P. D. Nellist, Y. Chen, R. E. Palmer, J. A. Preece, *Langmuir* 2002, *18*, 1791.
65 M. Hasan, D. Bethell, M. Brust, *J. Am. Chem. Soc.* 2002, *124*, 1132.
66 S. L. Horswell, I. A. O'Neil, D. J. Schiffrin, *J. Phys. Chem. B* 2001, *105*, 941.
67 C. L. Cleveland, U. Landman, M. N. Shafigullin, P. M. Stephens, R. L. Whetten, *Z. Phys. D* 1997, *40*, 503.
68 W. D. Luedtke, U. Landman, *J. Phys. Chem.* 1996, *100*, 13323.
69 A. Badia, W. Gao, S. Singh, L. Demers, L. Cuccia, L. Reven, *Langmuir* 1996, *12*, 1262.
70 A. Badia, L. Cuccia, L. Demers, F. Morin, R. B. Lennox, *J. Am. Chem. Soc.* 1997, *119*, 2682.
71 A. Badia, L. Demers, L. Dickinson, F. G. Morin, R. B. Lennox, L. Reven, *J. Am. Chem. Soc.* 1997, *119*, 11104.
72 Z. L. Wang, *J. Phys. Chem. B* 2000, 1153.
73 M. J. Hostetler, S. J. Green, J. J. Stokes, R. W. Murray, *J. Am. Chem. Soc.* 1996, *118*, 4212.
74 R. S. Ingram, M. J. Hostetler, R. W. Murray, *J. Am. Chem. Soc.* 1997, *119*, 9175.
75 A. C. Templeton, M. J. Hostetler, E. K. Warmoth, S. Chen, C. M. Hartshorn, V. M. Krishnamurthy, M. D. E. Forbes, R. W. Murray, *J. Am. Chem. Soc.* 1998, *120*, 4845.
76 M. J. Hostetler, A. C. Templeton, R. W. Murray, *Langmuir* 1999, *15*, 3782.
77 A. C. Templeton, D. E. Cliffel, R. W. Murray, *J. Am. Chem. Soc.* 1999, *121*, 7081.

78 A. K. Boal, V. M. Rotello, *J. Am. Chem. Soc.* **2000**, *122*, 734.
79 K. Nørgaard, M. J. Weygand, K. Kjaer, M. Brust, T. Bjørnholm, *Discuss. Faraday Soc.* **2003**, submitted for publication.
80 C. B. Murray, C. R. Kagan, M. G. Bawendi, *Science* **1995**, *270*, 1335.
81 M. P. Pileni, *Langmuir* **1997**, *13*, 3266.
82 S. Sun, C. B. Murray, *J. Appl. Phys.* **1999**, *85*, 4325.
83 S. Sun, C. B. Murray, D. Weller, L. Folks, A. Moser, *Science* **2000**, *287*, 1989.
84 X. M. Lin, H. M. Jaeger, C. M. Sorensen, K. J. Klabunde, *J. Phys. Chem. B* **2001**, *105*, 3353.
85 C. Waters, A. J. Mills, K. A. Johnson, D. J. Schiffrin, *Chem. Commun.* **2003**, 540.
86 A. D. Dinsmore, A. G. Yodh, D. J. Pine, *Nature* **1996**, *383*, 239.
87 P. Moriarty, M. D. R. Taylor, M. Brust, *Phys. Rev. Lett.* **2002**, *89*, art. no. 248303.
88 J. N. O'Shea, M. A. Phillips, M. D. R. Taylor, P. Moriarty, M. Brust, V. R. Dhanak, *Appl. Phys. Lett.* **2002**, *81*, 5039.
89 G. Ge, L. Brus, *J. Phys. Chem. B* **2000**, *104*, 9573.
90 J. Fink, C. J. Kiely, M. Brust, D. Bethell, D. J. Schiffrin, unpublished results.
91 C. J. Kiely, J. Fink, M. Brust, D. Bethell, D. J. Schiffrin, *Nature* **1998**, *396*, 444.
92 A. L. Rogach, D. V. Talapin, E. V. Shevchenko, A. Kornowski, M. Haase, H. Weller, *Adv. Funct. Mater.* **2002**, *12*, 653.
93 J. V. Sanders, M. J. Murray, *Nature* **1978**, *275*, 201.
94 J. V. Sanders, M. J. Murray, *Phil. Mag. A* **1980**, *42*, 705.
95 C. J. Kiely, J. Fink, J. G. Zheng, M. Brust, D. Bethell, D. J. Schiffrin, *Adv. Mater.* **2000**, *12*, 640.
96 C. P. Collier, T. Vossmeyer, J. R. Heath, *Annu. Rev. Phys. Chem.* **1998**, *49*, 371.
97 S. W. Chung, G. Markovich, J. R. Heath, *J. Phys. Chem. B* **1998**, *102*, 6685.
98 T. Hassenkamp, K. Norgaard, L. Iverson, C. J. Kiely, M. Brust, T. Bjornholm, *Adv. Mater.* **2002**, *14*, 1126.
99 T. Reuter, O. Vidoni, V. Torma, G. Schmid, L. Nan, M. Gleiche, L. Chi, H. Fuchs, *Nano Lett.* **2002**, *2*, 709.
100 S. Chen, *Langmuir* **2001**, *17*, 2878.
101 C. P. Collier, R. J. Saykally, J. J. Shiang, S. E. Henrichs, J. R. Heath, *Science* **1997**, *277*, 1978.
102 J. J. Shiang, J. R. Heath, C. P. Collier, R. J. Saykally, *J. Phys. Chem. B* **1998**, *102*, 3425.
103 G. Markovich, C. P. Collier, S. E. Henrichs, F. Remacle, R. D. Levine, J. R. Heath, *Acc. Chem. Res.* **1999**, *32*, 415.
104 G. Markovich, D. V. Leff, S. W. Chung, H. M. Soyez, B. Dunn, J. R. Heath, *Appl. Phys. Lett.* **1997**, *10*, 3107.
105 M. Dorogi, J. Gomez, R. Osifchin, R. P. Andres, R. Reifenberger, *Phys. Rev. B* **1995**, *52*, 9071.
106 R. P. Andres, T. Bein, M. Dorogi, S. Feng, J. I. Henderson, C. B. Kubiak, W. Mahoney, R. G. Osifchin, R. Reifenberger, *Science* **1996**, *272*, 1323.
107 R. S. Ingram, M. J. Hostetler, R. W. Murray, T. G. Schaaff, J. T. Khoury, R. L. Whetten, T. P. Bigioni, D. K. Guthrie, P. N. First, *J. Am. Chem. Soc.* **1997**, *119*, 9279.
108 S. Chen, R. S. Ingram, M. J. Hostetler, J. J. Pietron, R. W. Murray, T. G. Schaaff, J. T. Khoury, M. M. Alvarez, R. L. Whetten, *Science* **1998**, *280*, 2098.
109 B. M. Quinn, I. Prieto, S. K. Haram, A. J. Bard, *J. Phys. Chem. B* **2001**, *105*, 7474.
110 C. A. Neugebauer, M. B. Webb, *J. Appl. Phys.* **1962**, *33*, 74.
111 R. C. Doty, H. B. Yu, C. K. Shih, B. A. Korgel, *J. Phys. Chem. B* **2001**, *105*, 8291.
112 F. P. Zamborini, J. F. Hicks, R. W. Murray, *J. Am. Chem. Soc.* **2000**, *122*, 4514.
113 D. L. Gittins, D. Bethell, D. J. Schiffrin, R. J. Nichols, *Nature* **2000**, *408*, 67.
114 P. T. Bishop, *Gold Bull.* **2002**, *35*, 89.
115 H. Wohltjen, A. W. Snow, *Anal. Chem.* **1998**, *70*, 2856.
116 S. D. Evans, S. R. Johnson, Y. L. Cheng, T. Shen, *J. Mater. Chem.* **2000**, *10*, 183.
117 A. H. Lu, G. H. Lu, A. M. Kessinger, C. A. Foss Jr., *J. Phys. Chem. B* **1997**, *101*, 9139.
118 H. Li, Y.-Y. Luk, M. Mrksich, *Langmuir* **1999**, *15*, 4957.
119 L. Pasquato, F. Rancan, P. Scrimin, F. Mancin, C. Frigeri, *Chem. Commun.* **2000**, 2253.

120 J. J. Pietron, R. W. Murray, *J. Phys. Chem. B* **1999**, *103*, 4440.
121 M. Bartz, J. Küther, R. Seshadri, W. Tremel, *Angew. Chem. Int. Ed.* **1998**, *37*, 2466.
122 T. Sato, H. Ahmed, D. Brown, B. F. G. Johnson, *J. Appl. Phys.* **1997**, *82*, 1007.
123 S. H. M. Persson, L. Olofsson, L. Hedberg, *Appl. Phys. Lett.* **1999**, *74*, 2546.
124 T. G. Schaaff, G. Knight, M. N. Shafigullin, R. F. Borkman, R. L. Whetten, *Phys. Chem. B* **1998**, *102*, 10643.
125 A. C. Templeton, S. W. Chen, S. M. Gross, R. W. Murray, *Langmuir* **1999**, *15*, 66.
126 J. M. de la Fuente, A. G. Barrientos, T. C. Rojas, J. Rojo, J. Cananda, A. Fernandez, S. Penades, *Angew. Chem. Int. Ed.* **2000**, *40*, 2257.
127 A. G. Kanaras, F. S. Kamounah, K. Schaumburg, C. J. Kiely, M. Brust, *Chem. Commun.* **2002**, *20*, 2294.
128 M. Bartz, J. Kuther, G. Nelles, N. Weber, R. Seshadri, W. Tremel, *J. Mater. Chem.* **1999**, *9*, 1121.
129 W. P. Wuelfing, S. M. Gross, D. T. Miles, R. W. Murray, *J. Am. Chem. Soc.* **1998**, *120*, 12696.
130 D. I. Gittins, F. Caruso, *Angew. Chem. Int. Ed.* **2001**, *40*, 3001.
131 C. A. Mirkin, R. L. Letsinger, R. C. Mucic, J. J. Storhoff, *Nature* **1996**, *382*, 607.
132 J. J. Storhoff, R. Elghanian, R. C. Mucic, C. A. Mirkin, R. L. Letsinger, *J. Am. Chem. Soc.* **1998**, *120*, 1959.
133 R. C. Mucic, J. J. Storhoff, C. A. Mirkin, R. L. Letsinger, *J. Am. Chem. Soc.* **1998**, *120*, 12674.

134 J. J. Storhoff, C. A. Mirkin, *Chem. Rev.* **1999**, *99*, 1849.
135 A. A. Lazarides, G. C. Schatz, *J. Phys. Chem. B* **2000**, *104*, 460.
136 A. P. Alivisatos, K. P. Johnsson, X. Peng, T. E. Wilson, C. J. Loweth, M. P. Bruchez Jr., P. G. Schultz, *Nature* **1996**, *382*, 609.
137 C. J. Loweth, W. B. Caldwell, X. G. Peng, A. P. Alivisatos, P. G. Schultz, *Angew. Chem. Int. Ed.* **1999**, *38*, 1808.
138 D. Zanchet, C. M. Micheel, W. J. Parak, D. Gerion, A. P. Alivisatos, *Nano Lett.* **2001**, *1*, 32.
139 D. Zanchet, C. M. Micheel, W. J. Parak, D. Gerion, S. C. Williams, A. P. Alivisatos, *J. Phys. Chem. B* **2002**, *106*, 11758.
140 T. A. Taton, C. A. Mirkin, R. L. Letsinger, *Science* **2000**, *289*, 1757.
141 J. Yguerabide, E. E. Yguerabide, *Anal. Biochem.* **1998**, *262*, 137.
142 J. Yguerabide, E. E. Yguerabide, *Anal. Biochem.* **1998**, *262*, 157.
143 S. Schultz, D. R. Smith, J. J. Mock, D. A. Schultz, *Proc. Natl. Acad. Sci. USA* **2000**, *97*, 996.
144 K. K. Sandhu, C. M. McIntosh, J. M. Simard, S. W. Smith, V. M Rotello, *Bioconjugate Chem.* **2002**, *13*, 3.
145 C. S. Yun, G. A. Khitrov, D. E. Vergona, N. O. Reich, G. F. Strouse, *J. Am. Chem. Soc.* **2002**, *124*, 7644.
146 J. M. Perez, T. O'Loughin, F. J. Simeone, R. Weissleder, L. Josephson, *J. Am. Chem. Soc.* **2002**, 2856.
147 A. G. Kanaras, Z. Wang, A. D. Bates, R. Cosstick, M. Brust, *Angew. Chem. Int. Ed.* **2003**, *42*, 191.

4
Sonochemical Synthesis of Inorganic and Organic Colloids

Franz Grieser and Muthupandian Ashokkumar

4.1
Introduction

Standard text books refer to colloids as being particles, not necessarily solid, with dimensions in the range of micrometers to nanometers. The wide use of colloidal particles in our society, ranging from personal products, such as facial creams, to industrial catalysts, such as Ni, Fe, etc., has meant that the synthesis of colloidal particles has been the subject of extensive investigation [1–7].

Colloid science is concerned with systems which are heterogeneous in nature. The overall nature of a colloidal system will depend on a number of factors of varying degrees of importance. These factors include particle size and shape, surface properties, particle–solvent interactions, particle–particle interactions, and the optical properties of the particle–solvent system. In more recent years, there has been a considerable interest shown in studying colloidal particles a few nanometers in size. These particles exhibit quantum size effects and have attracted attention because of their unique material properties and the potential to exploit them in what has been termed, somewhat loosely, "nanotechnology".

A brief outline of some of the properties of colloids is given in Section 4.2, with the main emphasis being on solid particles in aqueous systems. The bulk of this chapter deals with the use of ultrasound in the preparation of different types of colloids in aqueous solutions. The sonochemical synthesis of colloidal particles is a relatively new technology although the use of ultrasound in colloid chemistry dates back to the early 1920s [8]. Sonochemical processes are by their nature very complex, and always involve micro bubbles. Section 4.3 deals with some of the properties of bubbles, particularly when exposed to an ultrasound field. Section 4.4 introduces the basics of sonochemistry and how it can be used to form colloids. The chapter concludes with examples of colloids that have been produced in both aqueous and nonaqueous systems. Also a brief outline on some of the mechanisms that are believed to be involved in the relevant chemical reactions initiated by ultrasound are given.

4.2
Properties of Colloids

From a surface chemistry perspective, a proper characterization of the particle–solution interface is vital to an understanding of the properties of a colloidal system. Most colloids will acquire a surface electric charge when dispersed in a polar medium such as water. The electric charge can be brought about through the ionization of surface functional groups on the colloid, ion adsorption, or by the partial and unequal ion dissolution of the colloidal material.

An important consequence of the charge being on the particle surface is that an electrical double layer around the colloid is produced in polar solutions. The double layer consists of a surface charge and a diffuse ion region in which ions are distributed in response to the electrical forces operating in the system and random thermal motion.

There are several theoretical models that exist to describe and account for the features of the electrical double layer. The Gouy-Chapman model is the simplest quantitative treatment of the electrical double layer. Fig. 4.1 shows a schematic of the electrical double layer and the corresponding variation of the electrostatic potential away from the charged surface.

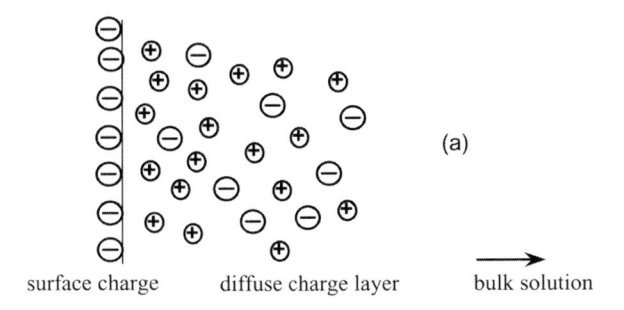

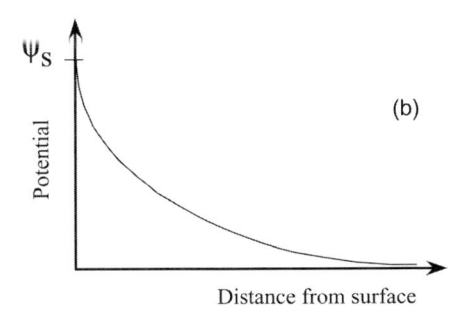

Fig. 4.1 **a** Schematic representation of the electric double layer of a charged surface in a polar environment; **b** the electrical potential extending from a Gouy-Chapman-like interface.

More sophisticated quantitative models of the charged surface incorporate the finite size of ions and the nonspecific adsorption of the ions and other molecules onto the surface. These types of models have been extended to include diffuse adsorbed charges such as those encountered when a polyelectrolyte adsorbs onto to a colloidal particle [9–11].

With respect to particle–particle interactions, which determine colloid stability, the most widely accepted quantitative model is that described by the Derjaguin-Landau-Verwey-Overbeek (DLVO) theory. This theory treats the interaction of overlapping electrical double layers and forces of attraction between particles. The forces of attraction are London–van der Waals interactions between particles across a fluid with different optical/dielectric properties than those of the particles.

Under solution conditions of low ionic strength ($<10^{-3}$ M) and moderate surface potentials (>25 mV) the electrostatic repulsion between particles is normally sufficient to prevent the attractive forces from causing the particles to aggregate.

DLVO theory is usually an excellent predictor of colloid stability although there are well-known cases where it fails. This failure occurs in systems where "structural forces" such as depletion forces operate. Depletion forces come about when certain types of solutes, e.g. polymers, micelles, etc., are present between the interacting surfaces. Depletion forces are in essence driven by the osmotic pressure difference between the gap of the interacting surfaces and the bulk solution. The physical exclusion of the solute from between the interacting particles leads to solvent flow from between the gap in order to dilute the bulk medium and this induces particle flocculation.

There are many excellent textbooks available dealing with colloidal systems, which clearly reveal the significant maturity of this branch of chemistry. The synthesis of colloidal particles with specific desired properties, however, still remains a challenge.

4.3
Bubbles (General)

Gas bubbles are inherently present in liquids. However, a free gas bubble in a liquid does not remain stable. It either dissolves or floats to the surface of the liquid, depending on the size of the bubble. The dissolution of the bubble in the liquid is controlled by the Laplace pressure (P_L), viz., the difference between the internal pressure of the bubble and the pressure in the surrounding fluid. The Laplace pressure can be calculated by the relation, $P_L = 2\sigma/R$, where σ is the surface tension of the liquid and R is the radius of the bubble. For example, if $R=1$ μm and $\sigma=72$ mN/m, then $P_L>1$ atm. This pressure, if significant in magnitude, will result in the diffusion of gas/vapor from the bubble into the liquid eventually leading to the complete dissolution of the bubble. However, for larger bubbles, e.g. $R=1$ cm, $P_L<0.001$ atm. Such large bubbles will rise to the surface due to buoyancy before they have time to dissolve.

4.3.1
Bubbles in an Acoustic Field

In the presence of an applied acoustic field, (cavitation) bubbles that are within a certain size range are forced to oscillate and grow in size. Acoustic cavitation is the ultrasound-driven formation, growth and collapse of microbubbles. Bubble "formation" does not have to be the creation of a cavity by overcoming the cohesive forces of a liquid. For example, if a "void" is to be created in water by moving water molecules apart by their van der Waals distance, a negative pressure of about 1500 atm is needed. This is equivalent to an acoustic power of about 6.5×10^9 W m^{-2}. However, in practice, acoustic pressures of the order of a few atmospheres are sufficient to achieve cavitation. This is due to the pre-existence of bubbles in a liquid.

4.3.1.1 Bubble Growth

Bubbles that are inherently present as small nuclei grow towards a critical size (controlled by several parameters, such as ultrasound frequency, acoustic pressure, viscosity of the medium, etc.) under the influence of an applied ultrasonic field. This ultrasound-driven growth of the bubbles is due to "rectified diffusion" [12], which can be described as "the slow growth of a pulsating gas bubble due to an average flow of mass into the bubble as a function of time". Crum [12] explained this "rectification of mass" in terms of two effects, viz., an "area effect" and a "shell effect", both of which occur as a direct consequence of the applied sound field. The scheme shown in Fig. 4.2 provides a pictorial view of these two different effects. An acoustic field delivers positive and negative pressures in the liquid medium. A gas bubble, "trapped" in the liquid, is forced to expand when the surrounding liquid experiences the negative pressure (rarefaction cycle) of the sound wave. At this stage, the internal pressure of the bubble is very low, resulting in the evaporation of solvent molecules into the bubble as well as the diffusion of dissolved gas into the bubble from the surrounding liquid. Thus the rarefaction cycle leads to the "intake" of gas and vapor molecules. The same bubble is compressed when the surrounding liquid experiences the positive pressure (compression cycle) of the sound wave. At this stage, the internal pressure of the bubble is high, which leads to the expulsion of the gas/vapor molecules from the bubble into the surrounding liquid. Thus the compression cycle leads to the "loss" of bubble mass. However, as indicated in the scheme, the surface area of the bubble is at a minimum in the compressed state compared to that of the expanded state, where the bubble has a maximum surface area. Since mass (transport) diffusion across the bubble wall is proportional to the surface area, there is more "intake" of the material during the expansion cycle than is "expelled" during the compression cycle, resulting in a "net growth" of the bubble.

A change in the surface area of the bubble alone, however, is not sufficient to explain rectified diffusion. During bubble oscillations, the concentration of the gas in the surrounding liquid is also subject to variation, leading to a dynamic change

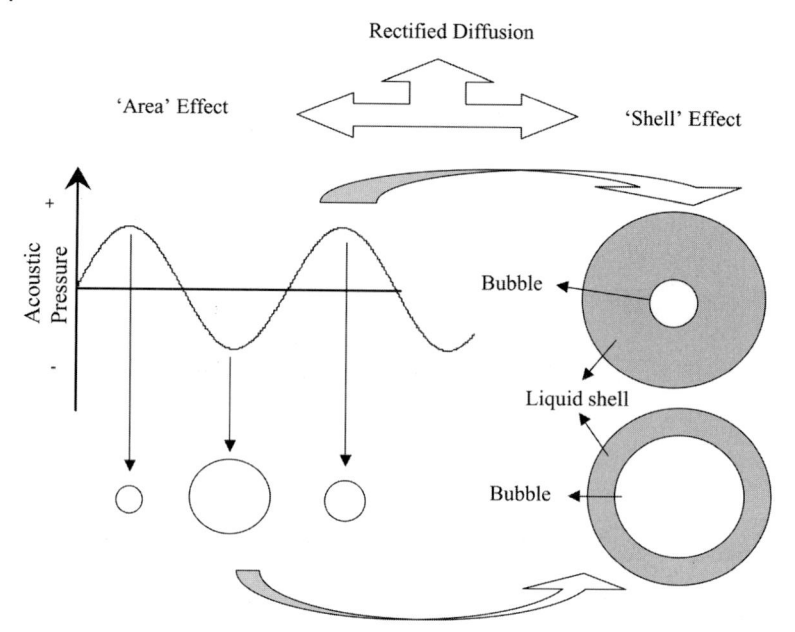

Fig. 4.2 Schematic description of the growth of a bubble in an acoustic field by "area" and "shell" effects [12].

in the gas concentration in the surrounding liquid. The result is a concentration gradient of the gas in the surrounding liquid.

In fact, the concentration of dissolved gas in the surrounding liquid "shell" can be expected to play a major role in rectified diffusion. In order to describe this effect, it can be assumed that the bubble oscillates within a liquid shell as shown in Fig. 4.2. When the bubble is compressed, the concentration of the gas inside the bubble is relatively high compared to that in the liquid shell. This results in the diffusion of material from the bubble to the liquid. However, since the liquid shell is larger than the bubble, a much longer time is needed to establish the equilibrium of the gas concentration between the bubble and the "liquid shell". In addition to this, as the gas diffuses out of the bubble, a concentration gradient is formed between the bubble wall and the edge of the liquid shell. The diffusion rate of gas in a liquid is proportional to the gradient of the concentration of the dissolved gas. This concentration gradient also slows down the diffusion of the material from the bubble into the surrounding liquid. When the bubble is in its expanded state, the liquid shell becomes thinner (relative to the size of the bubble) with a relatively higher gas concentration. Since the gas concentration inside the bubble is lower, material diffuses into the bubble from the surrounding liquid shell. An overall net gas accumulation inside the bubble results, owing to the changes in the thickness of the liquid shell accompanied by changes in the gas concentration in the surrounding liquid.

The major difference between the "area effect" and the "shell effect" is that the area effect considers only the changes in the surface area of the bubble during an acoustic cycle, whereas the shell effect considers the changes in both the surface area of the bubble and the concentration of dissolved gas in the surrounding liquid shell. Crum [12] notes that both effects have to be considered to theoretically model the rectified diffusion process. The kinetics of the bubble growth and collapse is also a crucial factor that can be expected to control rectified diffusion. Single bubble dynamics experiments [13] show that bubble growth is a slower process than bubble collapse. For example, at 20 kHz, the growth cycle takes about 10 µs, whereas the collapse happens in about 1 µs. The slower growth of the bubble will also lead to the diffusion of more material into the bubble, compared with the amount of material expelled in the short time of the collapse phase, resulting in a net growth of the bubble.

4.3.1.2 Bubble Collapse

As mentioned earlier, rectified diffusion leads to the growth of the bubble to a critical size, at which the bubble oscillation frequency matches that of the driving ultrasound frequency. A simple relation between the frequency of the ultrasound and the radius of the bubble (Eq. 1) can be used to calculate the critical size of the bubble for a given ultrasound frequency. At the critical size,

$$F \times R \approx 3 \tag{1}$$

(where F=frequency in Hz, R=radius of the bubble in meters)

For example, a bubble can reach a maximum size of about 150 µm at 20 kHz. However, bubbles never reach the theoretical maximum size as calculated from Eq. (1). As mentioned earlier, there are several other factors that also control the size of the bubble [13–15]. Once the critical size is reached, the bubble implodes (collapses). Bubble implosion, from a thermodynamic consideration, is important because a large change in bubble volume occurs. For example, single bubble dynamics experiments at 20 kHz have shown that the bubble radius changes from ~ 60 µm (R_{max}) to ~ 0.6 µm (R_{min}), when the collapse occurs. Since the bubble collapse happens in a very short time domain (~ 1 µs), the "work done" (PdV) leads to a "near" adiabatic heating of the contents of the bubble, which results in the generation of very high temperatures (>5000 K) and pressures (>100 atm) within the bubble.

4.3.2
Bubble Temperature

Based on a simple thermodynamic model for bubble collapse and assuming adiabatic compression takes place, the maximum theoretical temperature within the bubble (T_{max}) can be calculated using Eq. (2) [13]:

$$T_{max} = T_0 \left\{ \frac{P_m(\gamma - 1)}{P_v} \right\} \quad (2)$$

T_0 is the ambient solution temperature, P_m is the pressure in the liquid (a sum of the hydrostatic and acoustic pressures), $\gamma = \frac{C_p}{C_v}$ is the specific heat capacity ratio of the gas/vapor mixture within the bubble and P_v is the pressure in the bubble at its maximum size. For simplicity, P_v is usually assumed to be equal to the vapor pressure of the liquid, although it is known that gas molecules originally dissolved in the fluid will also be present. For example, if we assume that $P_m = 101\,325$ Pa, $P_v = 307$ Pa, the ambient solution temperature = 298 K, and $\gamma = 1.4$, then the theoretical estimate of T_{max} is approximately 39 000 K.

It should be noted that Eq. (2) overestimates T_{max}, because it does not take into account the heat leaking from the bubble into the surrounding fluid or the thermal conductivity of the gases or the energy consumed in the decomposition of the vapor/gas within the bubble. More elaborate models of bubble collapse exist that consider the "missing" components in Eq. (2). However, these models also make a number of assumptions that have yet to be experimentally verified.

Determination of the temperature within the cavitation bubble by different experimental techniques has been made by a number of research groups. By fitting experimentally recorded single bubble sonoluminescence spectra to a blackbody emitter, temperatures in the range of 5000 K to > 50 000 K [13–16] have been estimated. However, experimental estimates of the temperature within the collapsing bubbles based on multibubble sonochemistry and emissions from excited species, are reported to be between 750 K and 6000 K. The reason for this range is in part due to the different methods employed to determine the temperature and in part due to the different experimental conditions/systems used.

Sonolysis of methane in argon-saturated water has been used by Henglein [17] to estimate the bubble core temperature. The initial reaction step in the sonolysis of methane is the formation of $CH_3^{\bullet}$ and $H^{\bullet}$ radicals (Eq. 3). Also, it is well known that the sonolysis of water leads to the homolytic cleavage of water molecules to form $H^{\bullet}$ and $OH^{\bullet}$ radicals (Eq. 4; known as primary radicals), which can react with methane and produce $CH_3^{\bullet}$ radicals (Eqs. 5 and 6).

$$CH_4 \rightarrow CH_3^{\bullet} + H^{\bullet} \quad (3)$$

$$H_2O \rightarrow H^{\bullet} + OH^{\bullet} \quad (4)$$

$$H^{\bullet} + CH_4 \rightarrow CH_3^{\bullet} + H_2 \quad (5)$$

$$OH^{\bullet} + CH_4 \rightarrow CH_3^{\bullet} + H_2O \quad (6)$$

Further reactions of $CH_3^{\bullet}$ radicals in the gas phase to form ethane (Eq. 7) and ethylene (Eq. 8) are known from flame and combustion studies.

$$CH_3^{\bullet} + CH_3^{\bullet} \rightarrow C_2H_6 \quad (7)$$

$$CH_3^{\bullet} + CH_3^{\bullet} \rightarrow C_2H_4 + H_2 \quad (8)$$

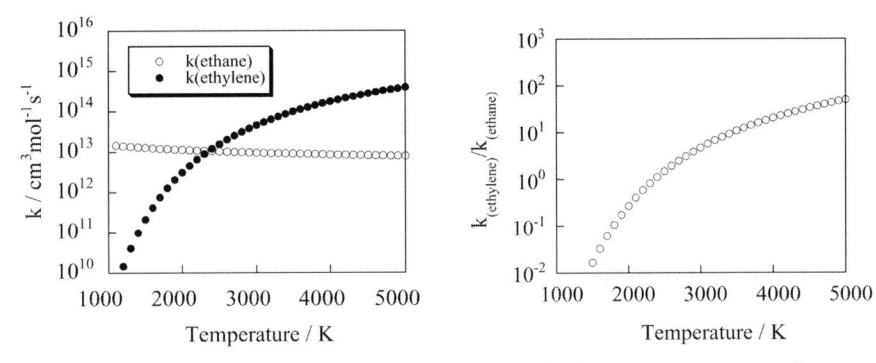

Fig. 4.3 a Temperature dependence of the rate constants for the reactions represented by Eqs. (5) and (6); **b** temperature dependence of the ratio, $k_{ethylene}/k_{ethane}$ (adapted from Figs. 11 and 12 of [17]).

The temperature dependence of the rate constants for the reactions represented by Eqs. (7) and (8) are given by the relations [17] $k_{ethane} = 2.4 \times 10^{14}\ T^{-0.4}\ cm^3\ mol^{-1}\ s^{-1}$ and $k_{ethylene} = 1.0 \times 10^{16}\ \exp(-E/RT)\ cm^3\ mol^{-1}\ s^{-1}$ (where $E = 134 \times 10^3\ J\ mol^{-1}$), respectively. The temperature dependence of these rate constants are shown in Fig. 4.3 a. As shown in this figure, $k_{ethylene}$ increases significantly with an increase in temperature, whereas k_{ethane} has only a weak temperature dependence. Further analysis of the temperature dependence of these two rate constants shows that the ratio $k_{ethylene}/k_{ethane}$ increases with an increase in temperature (Fig. 4.3 b). With the initial concentration of $CH_3^{\bullet}$ being constant, the product ratio yield$_{(ethylene)}$/yield$_{(ethane)}$ ($= k_{ethylene}/k_{ethane}$) can be used to estimate the maximum bubble temperature using Fig. 4.3 b. Depending upon the percentage of methane and argon present in water, the temperature was estimated to be in the range of 1930 K to 2720 K (the temperature decreasing with an increase in the percentage of the methane used).

Tauber et al. [18] used a similar technique in their study of the sonolysis of t-butanol (Eq. 9) in water, and estimated the bubble temperature to be in the range 2300 K to 3600 K. They suggested that the sonolysis of t-butanol generates $CH_3^{\bullet}$ as shown in Eq. (9).

$$(CH_3)_3COH\ \rightarrow\ CH_3^{\bullet} + H^{\bullet} + CO \tag{9}$$

The lower temperature was obtained for solutions with the highest concentration of t-butanol (0.5 M). Both these studies [17, 18] show that the core temperature of the bubble is strongly dependent upon the level of organic solute in the system. Recent investigations [19–26] on the effect of volatile solutes on single bubble and multibubble sonoluminescence confirm that the organic solutes consume the thermal energy produced on bubble collapse for bond breakage, leading to a decrease in bubble temperature. Misík et al. [27] have followed the kinetic isotope effect in an EPR spin-trapping study of the sonolysis of H_2O/D_2O mixtures to estimate the bubble temperature. Sonication of a 50/50 mixture of H_2O/D_2O led to the generation of H• and D• atoms (radicals) by the homolytic cleavage of H_2O and D_2O (Eqs. 10 and 11).

$$H\text{-O-H} \xrightarrow{k_H} H^\bullet + OH^\bullet \qquad (10)$$

$$D\text{-O-D} \xrightarrow{k_D} D^\bullet + OD^\bullet \qquad (11)$$

Spin traps (ST) such as 5,5-dimethyl-1-pyrroline-N-oxide were used to trap these radicals to quantitatively determine the amounts of H• and D• radicals generated (Eqs. 12 and 13).

$$ST + H^\bullet \rightarrow (ST\text{-H})^\bullet \qquad (12)$$
$$ST + D^\bullet \rightarrow (ST\text{-D})^\bullet \qquad (13)$$

Under the assumption that all H• and D• radicals generated were trapped by the spin traps, $[H^\bullet]=[(ST\text{-H})^\bullet]$ and $[D^\bullet]=[(ST\text{-D})^\bullet]$. The temperature dependence of the ratio of the rate constants, k_H/k_D, for the homolytic cleavage of O-H and O-D bonds of water molecules, as obtained by a semiclassical treatment (Eq. 14) was used to estimate the bubble temperature

$$\frac{k_H}{k_D} = \exp\left\{\frac{B}{RT}\right\} \qquad (14)$$

where $B=5.2\times10^3$ J mol^{-1}. Since $k_H/k_D=[(ST\text{-H})^\bullet]/[(ST\text{-D})^\bullet]$ under the experimental conditions used, the bubble temperature could be estimated using the amounts of H• and D• formed and Eq. (14). The cavitation temperatures thus determined were found to be dependent on the specific spin trap used and fell in the range of 1000–4600 K. They attributed the differences in the values obtained as being due to the sampling of different regions of the cavitation "hot spot" by the different spin traps used.

Misík and Riesz [28] also made use of the kinetic isotope effect in the ultrasound-induced production of radicals in organic liquids to estimate the temperatures during cavitation. The temperature region where hydrogen radical abstraction occurred in n-dodecane was estimated to be about 750 K, whereas the region where benzyl radical formation occurred in toluene was about 6000 K. An interesting aspect of this work, as in the H_2O/D_2O study described above, was the determination of a mean temperature in different regions of a "hot spot" depending on which radicals were detected. Using comparative rate thermometry in alkane solutions, Suslick et al. [29] postulated that there are two regions of sonochemical reactivity: a gas phase zone within the collapsing cavity with an estimated temperature and pressure of 5200 ± 650 K and 500 atm, respectively, and a thin liquid layer of 200-nm thickness surrounding the collapsing cavity with an estimated temperature of ~1900 K. It is also of interest that characteristic emission bands have been observed from a number of electronically excited transition metal atoms formed within the bubbles by the sonolysis of volatile metal carbonyls in organic liquids [30]. The relative intensities and line broadening of the emission bands observed have also been used to calculate the temperature (5000 K) and pressure (500 atm) generated within a collapsing bubble.

4.4 Sonochemistry

Ultrasound induced cavitation events are accompanied by several effects in the liquid [31–40]. These effects can be grouped into two categories, viz., physical and chemical, as shown below in Fig. 4.4.

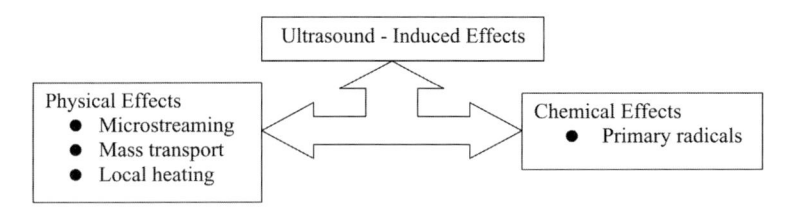

Fig. 4.4 Ultrasound-induced physical and chemical effects.

The ultrasound-driven growth and collapse of microbubbles is accompanied by the generation of microstreaming of the liquid, which facilitates mass transport within the liquid medium. The extreme temperatures created inside collapsing bubbles also extend to the surrounding liquid shells, causing local heating. These ultrasound-induced physical effects have been found to be useful in several organic and inorganic synthetic reactions [31–40].

In addition to these physical effects, chemical reactions are also induced by the cavitation phenomenon both inside the collapsing bubbles and in the liquid medium. As has been described earlier, extreme temperatures and pressures are generated within the cavitation bubbles. Solvent and solute molecules present within the bubbles are decomposed under these extreme conditions and generate several highly reactive radicals. For example, if the sonicated medium is water, H$^\bullet$ and OH$^\bullet$ (primary) radicals are generated [31–41] as represented by Eq. (4). It can be theoretically expected that sonication of water generates equal amounts of H$^\bullet$ and OH$^\bullet$ radicals. However, it has been experimentally shown in air-saturated aqueous solutions that there is always a higher amount of OH$^\bullet$ radicals generated compared to that of H$^\bullet$. This can be explained by Eqs. (15) and (16) involving H$^\bullet$ and O$_2$ leading to the formation of H$_2$O$_2$.

$$H^\bullet + O_2 \rightarrow HO_2^\bullet \tag{15}$$

$$HO_2^\bullet + HO_2^\bullet \rightarrow H_2O_2 + O_2 \tag{16}$$

It should also be noted that even in the absence of oxygen, the reaction between H and water molecules (Eq. 17) can also lead to a higher OH• yield.

$$H^\bullet + H_2O \rightarrow 2OH^\bullet \tag{17}$$

If solutes like alcohols are present in the sonication medium, secondary radicals are likely to be generated in the medium by the reaction of the primary radicals (Eq. 18) that escape into the liquid during the collapse of the bubbles [40–43]. These secondary radicals have also been found to be useful for achieving specific chemical reactions.

$$RHOH + H^\bullet(OH^\bullet) \rightarrow ROH^\bullet + H_2(H_2O) \tag{18}$$

Henglein's review [41] on various aspects of cavitation chemistry provides details on the detection of different free radicals during sonication of aqueous and non-aqueous solutions. Both the physical and chemical effects of the ultrasound induced cavitation phenomenon have been fruitfully exploited to achieve useful products [31–57] including the synthesis of various organic and inorganic colloidal materials [58–127].

4.4.1
Sonochemical Formation of Colloids

As described in the previous section, sonication of a liquid medium leads to the generation of oxidizing and reducing radicals. These radicals have been used in the synthesis of colloidal particles, examples of which are summarized in Fig. 4.5. In aqueous systems, both primary radicals, generated through sonication within the bubbles, and secondary radicals, generated by the reaction of the primary radicals with other solution solutes, have been shown to be involved in the reduction of metal ions in the bulk solution. In the case of non-aqueous media where volatile complexes of the metals have been used, reaction occurs inside the cavitation bubbles.

4.4.1.1 Aqueous Media
Synthesis of a number of metals, nonmetals, metal compounds and polymers in colloidal form has been achieved by the sonochemical technique. In the preparation of metal colloids, a common practise is to sonicate a solution containing selected metal ions. The metal ions are reduced during sonication to form metal atoms in the solution phase, which then aggregate to form metal colloids. Among the primary radicals (H• and OH•) generated during the sonication of water, only H atoms can be expected to be involved in the reduction of metal ions to form metal colloids (Eqs. 19 and 20).

$$M^+ + H^\bullet \rightarrow M + H^+ \tag{19}$$

$$nM \rightarrow (M)_n \qquad (20)$$

The initial work on the ultrasonic formation of Ag reported by Henglein and co-workers [60] can be considered as an example of this reaction mechanism. Solutions containing silver ions were sonicated at 1 MHz under various mixed argon-hydrogen atmospheres. However, the mechanism of the reduction of silver ions to form silver atoms was not given. Since the primary aspect of this study was to use silver ions as hydrogen atom scavengers, the reactions occurring to produce the colloid are likely to be

$$Ag^+ + H^\bullet \rightarrow Ag + H^+ \qquad (21)$$

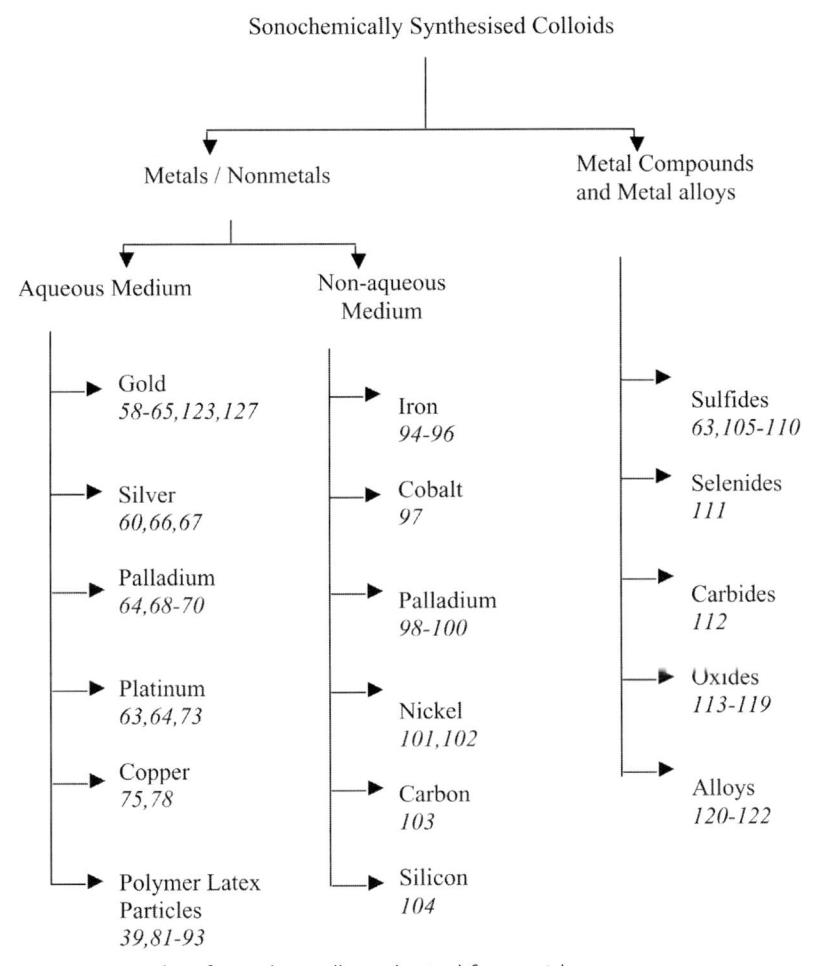

Fig. 4.5 Examples of sonochemically synthesized fine particles (reference numbers are given in italics (adapted from [126])).

Sonochemically Synthesised Colloids

- Metals / Nonmetals
  - Aqueous Medium
    - Gold *58-65,123,127*
    - Silver *60,66,67*
    - Palladium *64,68-70*
    - Platinum *63,64,73*
    - Copper *75,78*
    - Polymer Latex Particles *39,81-93*
  - Non-aqueous Medium
    - Iron *94-96*
    - Cobalt *97*
    - Palladium *98-100*
    - Nickel *101,102*
    - Carbon *103*
    - Silicon *104*
- Metal Compounds and Metal alloys
  - Sulfides *63,105-110*
  - Selenides *111*
  - Carbides *112*
  - Oxides *113-119*
  - Alloys *120-122*

$$n\text{Ag} \rightarrow (\text{Ag})_n \tag{22}$$

However, ionic silver clusters [$(\text{Ag})_2^+$ and $(\text{Ag})_x^{n+}$], formed by the reactions represented in Eqs. (23) and (24), will also compete with H atoms in addition to free Ag^+ ions.

$$\text{Ag} + \text{Ag}^+ \rightarrow (\text{Ag})_2^+ \tag{23}$$

$$(\text{Ag})_m + n\text{Ag}^+ \rightarrow (\text{Ag})_x^{n+} \tag{24}$$

Similarly, the formation of colloidal gold by the sonication of an aqueous solution containing AuCl_4^- ions is also suggested to be due to the reduction of AuCl_4^- ions by H atoms (Eq. 25) [127].

$$\text{AuCl}_4^- \xrightarrow{\text{H}^\bullet} \text{Au} + \text{products} \tag{25}$$

Baigent and Mueller [59] were the first to use an ultrasonic method to prepare colloidal Au particles with a diameter of <10 nm. Henglein and coworkers [60] also reported the production of colloidal gold following the sonication of aqueous solutions containing AuCl_4^- ions. The formation of gold colloids was initially identified by the visual observation of a dark precipitate, as the colloids were unstable in the sonicated solution and coagulated to form a precipitate. However, these gold nanoparticles were stabilized when poly(vinylpyrrolidone) (PVP) was present in the medium and the characteristic absorption spectrum of colloidal gold could be recorded. It was also reported that the efficiency of gold colloid formation was highly dependent on the nature of the dissolved gas [63, 126]. It is known that only about 20% of the primary radicals that are generated within the collapsing bubbles are responsible for the secondary chemical reactions in the solution phase. It has been estimated that about 80% of the primary radicals recombine within the bubble. However, in the presence of surface-active solutes such as alcohols and surfactants more of these primary radicals can be trapped [63] resulting in the formation of secondary radicals (Eq. 18). One interesting aspect here is that the reaction of these surface-active solutes with both the primary radicals result in the formation of reducing alcohol radicals. This is particularly an advantage in the synthesis of metal colloids by the reduction of metal ions. In the absence of the reaction of Eq. (18), the only reducing radical is $\text{H}^\bullet$ (Eq. 19). The surface-active solutes react with the oxidizing $\text{OH}^\bullet$ radicals and generate reducing radicals such as $\text{ROH}^\bullet$ (Eq. 18). This leads to an overall higher sonochemical reduction of the metal ions as a consequence of the reaction:

$$\text{ROH}^\bullet + \text{M}^+ \rightarrow \text{M} + \text{RO} + \text{H}^+ \tag{26}$$

For example, Fig. 4.6 shows that the presence of ethanol enhances the amount of gold produced. This particular advantage has been used in the synthesis of several metals and metal compounds by Grieser and coworkers, who have carried out an extensive investigation of the sonochemical formation of gold sols [61–63, 127]. In

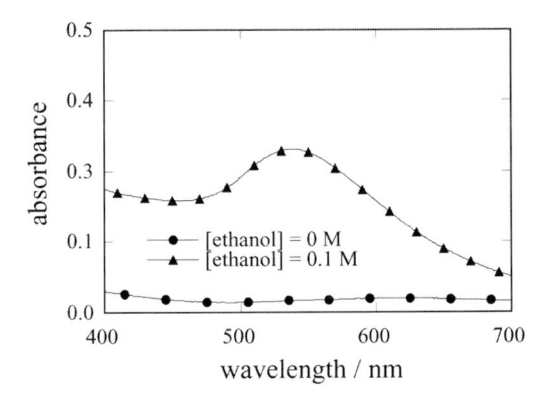

Fig. 4.6 Absorption spectra of argon-saturated aqueous $AuCl_4^-$ solutions (0.2 mM) after 5 min sonication (at 500 kHz) in the presence and in the absence of 0.1 M ethanol. The absorption band centered at ~ 530 nm is due to Au colloids.

their investigations, dilute aqueous solutions of $AuCl_4^-$ were sonicated under different experimental conditions, and most notably it was found that the presence of aliphatic alcohols as co-solutes enhanced the rate of formation of colloidal gold.

It has been suggested that the added alcohols scavenge the primary radicals at the bubble/solution interface thereby preventing the recombination of the primary radicals within the bubble. The so-formed secondary radicals, after diffusing away from the bubble/solution interface, reduce the $AuCl_4^-$ (Eq. 27) in bulk solution to form gold atoms, which then agglomerate to form colloidal gold. The exact mechanism involved in Eq. (27) is still not fully understood [127] but it is unlikely that three separate secondary radicals (ROH$^\bullet$) reduce a single Au(III) to produce a gold atom.

$$AuCl_4^- \xrightarrow{ROH^\bullet} Au + products \quad (27)$$

It has also been found in the study carried out by Caruso [63] that the higher the alcohol concentration (or the more hydrophobic the alcohol) the smaller the particle size of the colloid produced (Fig. 4.7). The general trends revealed in that study were that the particle size decreases with an increase in alcohol concentration for any particular alcohol, and the particle size, on average, decreases with increasing chain length of the alcohol. In a separate series of experiments it was found that aliphatic alcohols adsorb onto a gold surface and the more hydrophobic the alcohol the more readily it adsorbs [63].

Further investigation [63] on the sonochemical formation of colloidal gold showed that there were several factors that affected the efficiency of formation of gold colloids. One of these factors was the amount of surface-active solute that was present at the bubble/water interface, which played an important role in controlling the sonochemical efficiency. For a given concentration of aliphatic alcohol, a higher rate of gold colloid formation was observed for the alcohol having the longer alkyl chain (Fig. 4.8).

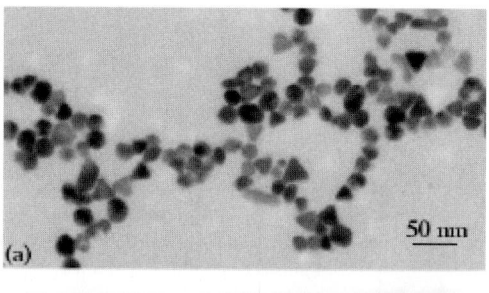

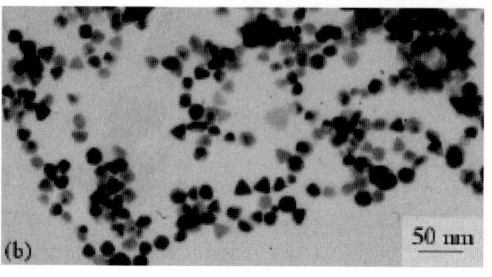

Fig. 4.7 Electron micrographs showing the gold particles obtained following the sonication of **a** 2×10^{-4} M $AuCl_4^-$ and 0.2 M ethanol and **b** 2×10^{-4} M $AuCl_4^-$ and 1.0 M ethanol, for 10 min at 20 kHz (reprinted from [127] with permission from the American Chemical Society, Copyright 2002).

Fig. 4.8 shows that at lower alcohol concentrations, the amount of $AuCl_4^-$ reduced is greater for butan-1-ol than for ethanol. However, the same amount of reduction is reached at higher ethanol concentrations. In order to understand the importance of the surface concentration of these solutes, the Gibbs surface excesses of these alcohols [128] have to be considered. The Gibbs surface excess provides the two-dimensional concentration of a surface-active solute at the air/water interface, which can be determined using experimental surface tension data. Since longer-chain-length alcohols are more surface active (more hydrophobic), a given surface excess value is reached at a lower concentration of butan-1-ol than of ethanol. The data shown in Fig. 4.8 can now be clearly understood based on this argument. For a lower alcohol concentration, say 10 mM, it can be expected

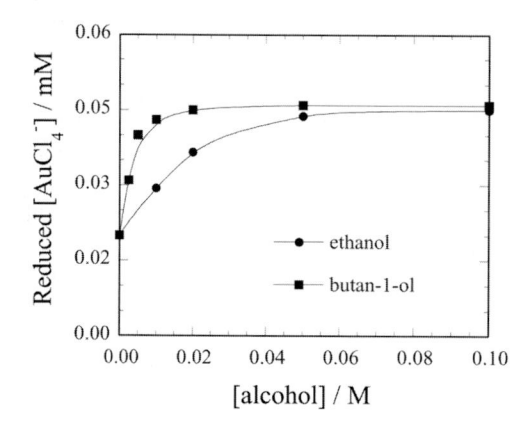

Fig. 4.8 Amount of $AuCl_4^-$ reduced as a function of the bulk alcohol concentration following 7.5 min sonication at 515 kHz in argon-saturated solutions (adapted from Fig. 4.2 of [21]).

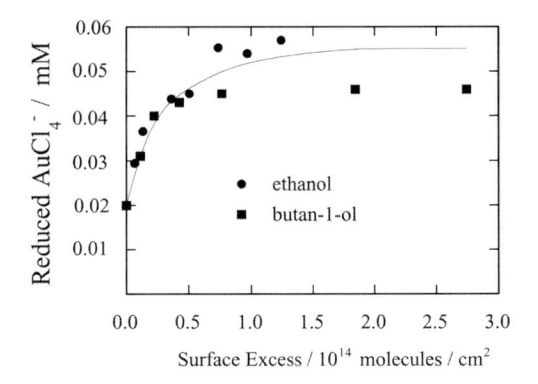

Fig. 4.9 Amount of gold chloride reduced as a function of the surface excess of the alcohols (adapted from Fig. 4.10 of [21]).

that the number of butan-1-ol molecules at the bubble/solution interface would be greater, leading to more primary radical scavenging and hence to a higher reduction rate. If the data shown in Fig. 4.8 is replotted in terms of surface excess of the alcohols, a direct correlation between the amount of $AuCl_4^-$ reduced and the Gibbs surface excess of the alcohols can be established (Fig. 4.9). This investigation [63] yielded a very significant insight into ultrasound-initiated reactions, viz., interfacial chemistry plays a major role in sonochemistry.

The importance of having solutes adsorbed at the bubble/solution interface in enhancing the yield of colloidal gold was further illustrated by using a range of surfactants instead of alcohols in the sonochemical production of colloidal gold and other metal colloids [63]. The maximum yield of colloid was found to occur at the critical micelle concentration of the surfactant system being used, which correlates directly with the maximum adsorption of the solute at the air/water interface.

Nagata et al. [66] have extended the investigation of the use of surface-active solutes for the sonochemical formation of colloidal silver. Ultrasonic irradiation at 200 kHz of an aqueous solution of $AgClO_4$ or $AgNO_3$ containing surface-active agents such as alcohols and sodium dodecylsulfate resulted in the production of 10–20-nm diameter silver particles. The surface-active molecules act as H/OH radical scavengers as well as stabilizing agents. It has been suggested that isopropyl alcohol radicals that are produced according to Eq. (18) are involved in the silver ion reduction (Eq. 28).

$$(CH_3)_2COH^\bullet + Ag^+ \rightarrow Ag + (CH_3)_2CO + H^+ \tag{28}$$

This reaction is followed by the coagulation of the Ag atoms to produce the colloids. This mechanism is most certainly incorrect considering the redox levels involved in the reaction, and all the other known reactions that occur in an Ag^+/Ag system [129]. Gedanken and coworkers [67] have similarly prepared amorphous 20-nm-sized silver particles by sonicating an aqueous silver nitrate solution and postulated that the reduction of Ag^+ to form Ag occurs by hydrogen radicals as suggested by Henglein [60].

The use of ultrasound for the preparation of nanosized palladium particles has been found to be a successful method for the preparation of Pd with a high catalytic activity. The sonochemical preparation of palladium nanoparticles has been reported by Nagata and coworkers [64], using a procedure similar to that for the sonochemical preparation of gold colloids. In a typical experiment, aqueous solutions containing palladium salts were sonicated in the presence of radical scavengers such as isopropyl alcohol and stabilizers such as PVP. The exact mechanism of the reduction of palladium ions to palladium metal has not been elucidated in detail. However, it is reasonable to suggest that secondary reducing radicals such as isopropyl alcohol radicals, which are produced by the reaction of isopropanol with the OH• radicals, are involved in the reduction of palladium ions to produce palladium colloids (Eq. 29).

$$Pd(II) + reducing\ radicals \rightarrow Pd(0) \tag{29}$$

Pd particles in the size range of 5–100 nm were also ultrasonically prepared from $PdCl_2$ in the presence of stabilizers, such as sodium dodecylsulfate (SDS), poly(oxyethylene sorbitan monolaurate), poly(vinylpyrrolidone), etc. [68]. The size of the Pd particles could be controlled by adjusting the nature and concentration of the stabilizing agents. In a continuation of this work, Okitsu et al. have reported on the sonochemical preparation of size-controlled palladium nanoparticles on an alumina surface [69]. The above method of preparation was used in the presence of α-Al_2O_3 particles, with an average diameter of 0.5 μm, and resulted in the formation of highly dispersed Pd nanoparticles immobilized on the alumina colloids. The TEM photographs presented by Okitsu et al. show that the Pd particles, of an average size of 5 nm, are uniformly distributed on the alumina surface. Bianchi et al. have used 20-kHz ultrasound to prepare palladium supported on carbon catalysts using a similar procedure to that just described [70].

Colloidal platinum has a number of uses, one of which is as a catalyst for a number of redox processes [71]. The reactivity and catalytic activity of colloidal platinum are different depending on the method of production [72]. Sonochemical formation of colloidal platinum has been investigated in some detail by Grieser and coworkers [63, 73]. For the production of Pt colloids using 20-kHz ultrasound, a standard sonifier with a high gain stepped horn (19-mm diameter tip) was used to deliver the ultrasound to solutions of $PtCl_6^{2-}$ ions in the presence and absence of alcohols. In $PtCl_6^{2-}$ solutions in the absence of alcohol the only reducing radical available is H•, and this in only very small amounts in bulk solution. However, the presence of alcohol as co-solute enhances the sonochemical reduction of the Pt(IV) by as much as a factor of four, and there is a clear trend in efficacy with increasing alkyl chain length, and hence hydrophobicity, of the alcohol. This behavior is very similar to that observed for the sonochemical reduction of gold(III) chloride. These results strongly suggest that a radical pathway is the dominant process for the formation of colloidal Pt in aqueous solution. Unfortunately, identifying the elementary steps leading to colloidal Pt is not yet possible. It is known that there is ligand exchange of the chloride ion in platinum chloride complexes, for H_2O in aqueous solutions with time [130]. However, the complete scheme for ligand exchange between $PtCl_4^{2-}$ and $PtCl_6^{2-}$, which is initiated by light or redox catalysts, has not been derived [131]. Although it is known that $PtCl_4^{2-}$ can be produced from the reduction of $PtCl_6^{2-}$ the reaction scheme for $PtCl_4^{2-} \rightarrow Pt(0)$ is not known [131, 132].

Reisse and coworkers have made use of a sonoelectrochemical method to synthesize copper colloids [75]. An overview of the sonoelectrochemical reactions and mechanism is provided in [76]. The experimental technique simply involves the immersion of a sonic horn into a conventional electrochemical cell with the sonic horn acting as one of the electrodes. In the experiments of Reisse and coworkers a titanium horn (working electrode) was attached to a 20-kHz ultrasound generator. In sonoelectrochemical reactions the use of ultrasound increases the mass transfer of ionic species from the bulk solution to the electrode surface. Combining ultrasonic vibration of the horn during electrodeposition of metals onto its surface was found to produce ultrafine metal particles [75, 77]. It was also reported that the catalytic activity of copper particles prepared by the sonoelectrochemical method was very high compared with those commercially available.

Dhas et al. [78] have reported on the sonochemical synthesis, characterization and properties of metallic copper nanoparticles. Sonication of an aqueous solution of copper(II) hydrazine carboxylate results in the formation of a mixture of metallic copper and copper oxide particles. The formation of Cu_2O along with the copper nanoparticles in the sonochemical process has been attributed to the partial oxidation of copper by the in situ generated H_2O_2 during the cavitation process. However, Dhas and coworkers found that the formation of Cu_2O can be eliminated by purging the solution with a mixture of argon and hydrogen gases (95% argon, 5% hydrogen). It was suggested that this procedure decreases the hydrogen peroxide produced because of the scavenging action of hydrogen towards OH radicals. The sonochemically prepared copper powders consisted of porous aggregates (50–70 nm), which contained an irregular network of copper nanoparticles. The catalytic activity towards the Ullmann reaction (condensation of aryl halides) of these copper nanoparticles was tested, and the efficiency of the reaction was found to be >80%.

The photochemical and photocatalytic properties of sulfide semiconductor particles are highly dependent upon the particle size and method of preparation. Several investigations have been carried out using ultrasound to synthesize sulfide semiconductor particles. Grieser and coworkers [63, 105–107] have synthesized quantum-sized CdS particles by the irradiation of an aqueous solution containing Cd^{2+}/thiol/sodium hexametaphosphate with 20-kHz ultrasound. Ultrasonic irradiation of an aqueous solution of Cd^{2+}/thiol/stabilizer resulted in the formation of CdS semiconductor colloids of <5 nm [105]. Fig. 4.10 shows the spectral changes with increasing sonication time of an aqueous solution containing 2-mercaptopropionic acid and Cd^{2+}(aq). The absorption onset initially is at 340 nm, and with increasing sonication time red-shifts, indicating that the particles increase in size with increasing sonication time.

From the band-edge energy, these particles are estimated to be about 2 nm in diameter after sonication for 60 min. The mechanism responsible for particle for-

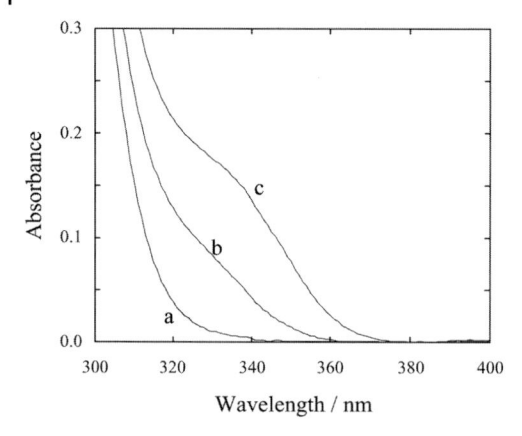

Fig. 4.10 Absorbance spectra of a Cd^{2+} (2×10^{-4} M)/2-mercaptopropionic acid (0.1 M)/HMP (1.5×10^{-4} M) solution sonicated for different times: **a** 0 min, **b** 30 min and **c** 60 min. Bulk solution temperature and pH were $22\pm4\,°C$ and 5.2, respectively (adapted from Fig. 4.4 of [126]).

mation starts with the primary radicals (H$^\bullet$) generated within the bubble being scavenged by the thiol to produce H_2S, which then reacts with aqueous Cd^{2+} to form CdS colloids. Hexametaphosphate (HMP) was used to stabilize the colloids. The low local concentration of sonochemically generated H_2S was attributed to the formation of CdS nanoparticles showing Q-state properties. The following reactions were suggested:

$$H^\bullet + CH_3CH(SH)COOH \rightarrow H_2S + CH_3CHCOOH \tag{30}$$

$$Cd^{2+} + H_2S \rightarrow CdS + 2H^+ \tag{31}$$

$$n CdS \xrightarrow{HMP} (CdS)_n \cdot xHMP \tag{32}$$

This group has also used similar procedures to sonochemically produce CuS and PbS [63], and indicated that the process was suitable as a general method for producing metal sulfide colloids.

Since polymerization reactions are induced by radicals, it is obvious that one could use the primary radicals formed during the cavitation events in polymer synthesis. As noted by Price and coworkers [39, 81] ultrasound had been used in polymer chemistry as early as 1933 [82, 83]. However, a lack of a clear understanding of cavitation events has limited the application of ultrasonic techniques to polymer and other synthetic reactions [39, 81, 84–93].

Biggs and Grieser [85] studied the sonochemical emulsion polymerization of styrene-in-water emulsions in the absence of any added chemical initiators. Latex particles in the size range of 50 nm were produced. Cooper et al. [87] later synthesized butyl acrylate/vinyl acetate copolymer latex nanoparticles using ultrasound as a polymerization initiation source. Comparing the thermal and sonochemical polymerization of the above monomers, and to account for the observed narrow size distribution of the latex particles under sonication, it was suggested [87] that the site of radical initiation and reaction is important in determining the polymer produced. Under ultrasonic cavitation conditions monomer droplets are the most likely sites of initiation of polymerization. Since cavitation causes violent shearing of the emulsion, small monomer droplets of narrow size distribution are suggested to be formed which then control the size of the polymer particles produced. Ooi and Biggs [92] later extended this investigation to synthesize polystyrene latex particles from oil-in-water emulsions of styrene.

Bradley and Grieser [93] have examined in some detail the mechanism involved in the synthesis of poly(methyl methacrylate) (poly(MMA)) and poly(butyl acrylate) latex particles in the size range 40–150 nm. In a typical ultrasonically initiated latex synthesis experiment, a three-component solution was prepared containing monomer, surfactant/emulsifier (dodecyltrimethylammonium chloride) and water. The authors have suggested that the initiation of polymerization occurs through the generation of monomeric radical species in an o/w emulsion. The individual steps that take place are summarized diagrammatically in Fig. 4.11. The H/OH radicals generated by the homolysis of water are intercepted before they reach the aqueous phase by the solutes in the boundary layer of the bubble/solution interface. In an emulsion these solutes are expected to be monomer and surfactant molecules. The resulting H/OH radicals can undergo addition to the monomer adsorbed at the bubble–solvent interface. Thereafter, the outcome is the formation of monomeric radicals that have three alternative pathways to initiate polymerization. They can add to a monomer molecule and undergo polymerization in the bulk phase, enter into a droplet or enter a micelle. Based on their results and

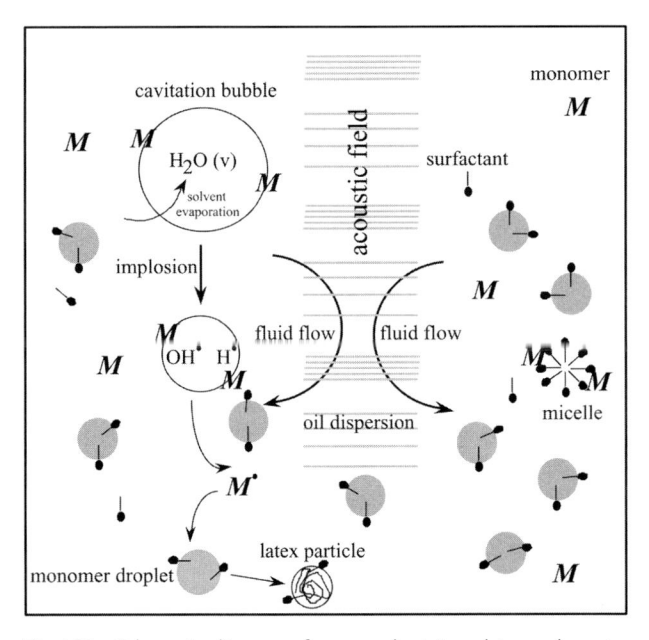

Fig. 4.11 Schematic diagram of proposed miniemulsion polymerization pathway induced by ultrasound (adapted from Fig. 4.5 of [93]).

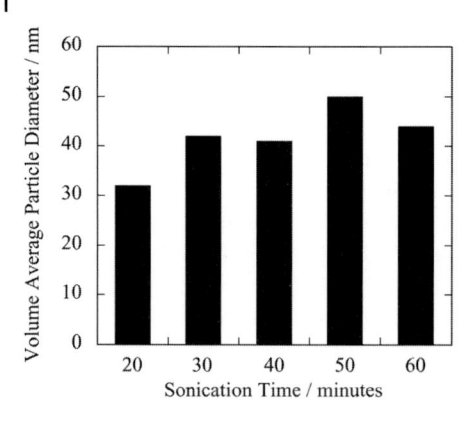

Fig. 4.12 Volume average particle diameter as a function of sonication time for poly(MMA) latex (adapted from Fig. 4.3 of [93]).

analysis, the authors have suggested that the polymerization process involves a miniemulsion system, in which continuous nucleation of particles takes place throughout the monomer-to-polymer conversion reaction.

One of the major findings in this study is that the average particle diameter of the latex particles remained in a narrow size distribution irrespective of the length of the percentage monomer conversion up to a sonication time of 60 min (see Fig. 4.12).

Investigations by Chou and Stoffer [90, 91] on the ultrasound-initiated emulsion polymerization of methyl methacrylate demonstrated that the rate of polymerization and polymer yield were significantly enhanced under sonication conditions compared to conventional high-temperature polymerization methods. The effects of various factors, such as acoustic intensity, surfactants, etc., on the ultrasound-initiated polymerization reactions were also investigated.

4.4.1.2 Non-aqueous Media

In most of the aqueous systems studied to date, sonochemical synthesis of metal colloids was achieved by the reactions of solute with primary and secondary radicals in the bulk solution phase. Suslick et al. [94–97] followed a different technique in order to synthesize metal nanoparticles in non-aqueous solvents. The general procedure followed by Suslick's group and others who have used the method is schematically shown in Fig. 4.13. Ultrasonic irradiation of organic liquids containing dissolved organometallic complexes led to the formation of metal colloids. The volatile complexes evaporate into the bubble during the expansion cycle. The high-temperature conditions of the bubble during the collapse result in the decomposition of the metal complexes to produce metal atoms.

An example for the process shown in Fig. 4.13 is the sonochemical formation of iron from iron carbonyl [95, 96]. Volatile $Fe(CO)_5$, dissolved in octanol, decane and other organic solvents, was sonochemically decomposed within the core of the bubble. Fine Fe colloids were produced and stabilized using PVP or oleic acid. The mechanism of Fe formation from the metal carbonyl was given as

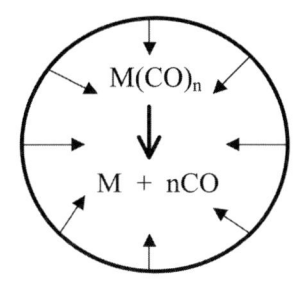

Fig. 4.13 Decomposition of gaseous metal carbonyl within the bubble to form metal atoms owing to the high temperature generated during the collapse.

$$Fe(CO)_5 \rightarrow Fe(CO)_n + nCO \tag{33}$$

$$Fe(CO)_3 + 2\,Fe(CO)_5 \rightarrow Fe_3(CO)_{12} + CO \tag{34}$$

$$Fe(CO)_n \rightarrow Fe_{(s)} + nCO \tag{35}$$

It was observed that the sonochemically produced Fe particles were amorphous and this was attributed to the fast cooling rates of the hot spots. This fast cooling rate causes solidification of the atom cluster before crystallization can take place. It has been estimated that acoustic cavitation can provide cooling rates greater than 10^9 K/s. Another interesting and important advantage of this method of synthesis of fine particles is that the particles thus prepared show higher catalytic activity towards selected reactions. For example, it was observed [95] that the sonochemically synthesized amorphous iron powders were found to be active catalysts for Fischer-Tropsch hydrogenation of carbon monoxide and for hydrogenolysis and dehydrogenation of saturated hydrocarbons. As shown in Fig. 4.14, sonochemically synthesized amorphous iron is a better catalyst than crystalline iron powder for the hydrogenolysis and dehydrogenation of cyclohexane.

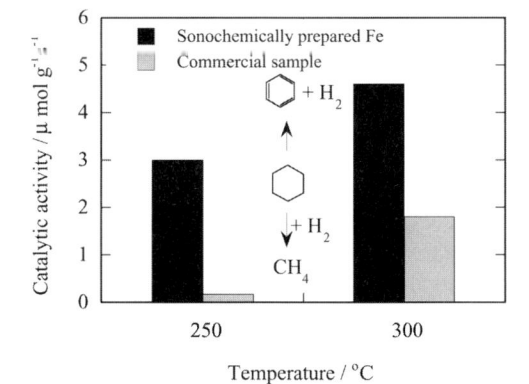

Fig. 4.14 The catalytic activity of amorphous and crystalline iron powder as a function of temperature for the cyclohexane dehydrogenation and hydrogenolysis reactions. Sonochemically prepared Fe is amorphous iron powder, prepared sonochemically from $Fe(CO)_5$; commercial sample is crystalline iron powder (5-μm diameter, Aldrich Chemicals) (adapted from Fig. 4.4 of [95]).

Based on the general procedure for particle formation in non-aqueous solvents as developed by Suslick and coworkers, Diodati et al. [98] have prepared Pd particles by sonicating a solution of 4% (by weight) of palladium acetylacetonate in toluene. In spite of the authors' extensive report on the characterization of the particles produced, a reaction pathway for the sonochemical formation of Pd was not suggested.

Dhas and Gedanken [99] have used a 1:2 molar ratio mixture of palladium acetate and a quaternary ammonium bromide salt in THF or methanol as a sonication medium to produce palladium metallic clusters. The Pd nanoparticles thus prepared showed a higher catalytic activity for the hydrogenation of cyclohexene in diethyl ether than that achieved with a standard commercial Pd catalyst.

The same group has also reported [100] on the sonochemical preparation of carbon-activated palladium metallic clusters using an organometallic precursor, tris-μ-[dibenzylideneacetonate]dipalladium in mesitylene. It was observed, from a detailed study of these particles by different analytical techniques such as XRD, TEM, etc., that each particle has a metallic core, covered by a carbonic shell that plays an important role in the stability of the nanoparticles.

Gedanken and coworkers [101] prepared nickel particles by sonicating nickel tetracarbonyl in decane. Sonochemical deposition of nickel nanoclusters on silica microspheres [102] has also been achieved by the sonication of a suspension containing nickel tetracarbonyl and silica microspheres in decalin with high-intensity ultrasound.

Nonmetals such as carbon and silicon have also been sonochemically fabricated. Ultrasound irradiation of liquid chlorobenzene in the absence and in the presence of solids such as Ni, Zn, etc. has been found to produce carbon nanotubes in the form of a black carbon polymer along with disordered carbon [103]. It has been suggested that the polymer and the disordered carbon are formed during the cavitational collapse process and annealed by the turbulent flow and shock waves from the cavitating bubbles. The mechanism responsible for the formation is essentially unknown.

Luminescent silicon nanoparticles were produced [104] by sonicating a colloidal solution of sodium and tetraethyl orthosilicate. The highly aggregated 2.5-nm particles that were produced exhibited a luminescence similar to that of porous silicon.

Sonochemistry has also been used to synthesize metal compounds. Dhas and Gedanken [108] followed a slightly modified procedure to that of Grieser's group to synthesize CdS nanoparticles. A slurry of cadmium sulfate, thiourea and silica was sonicated. This method led to the production of luminescent CdS-capped SiO_2 particles, which were characterized by a number of analytical techniques.

Suslick's group has reported on the sonochemical synthesis of nanostructured MoS_2 in a non-aqueous medium [109]. Ultrasonic irradiation of a slurry of molybdenum hexacarbonyl and sulfur in tetramethylbenzene under an argon atmosphere produced MoS_2 particles, which were characterized by SEM and TEM techniques. This sonochemically prepared MoS_2 was found to exhibit excellent catalytic activity for thiophene hydrosulfurization.

Fullerene-like MoS_2 nanoparticles were sonoelectrochemically synthesized by Gedanken's group [110]. The experimental unit that this group used was similar to that of Reisse and coworkers [75]. A Ti horn was used as both cathode and the ultrasound emitter. The electrolyte used was an aqueous solution of 50 mM ammonium tetrathiomolybdate and 1 M sodium sulfate at pH 6. The MoS_2 particles produced by the sonoelectrolysis of this electrolyte had fullerene-like morphology as determined by TEM. It has been suggested that amorphous MoS_2 was initially formed by electrodeposition of the material onto the sonic probe cathode. The subsequent sonic shock, and accompanying events, then removed this deposition, simultaneously changing the morphology of the particles.

Li et al. have used 18-kHz ultrasound to synthesize silver, copper and lead selenides [111]. Ultrasonic irradiation of a mixture of $AgNO_3$ or CuI or $PbCl_2$ and selenium in ethylenediamine was carried out for 10 h to produce the respective metal selenides. These materials were characterized by XRD and TEM. The authors made no comment on the possible mechanisms or the efficiency of the sonochemical reactions.

Molybdenum carbide, having a similar heterogeneous catalytic activity to that of platinum, was synthesized by sonicating a slurry of molybdenum hexacarbonyl at 20 kHz [112]. The catalytic and selectivity of the sonochemically produced molybdenum carbide were examined for the dehydrogenation versus hydrogenolysis of cyclohexane. It was found that the sonochemical decomposition of molybdenum hexacarbonyl produced nanometer-sized Mo_2C powder consisting of highly porous aggregates of 2-nm particles. The activity and the selectivity of these particles as a dehydrogenation catalyst were found to be comparable to those of Pt.

A wide variety of metal oxides have been sonochemically synthesized [113–119]. Enomoto et al. [113] could enhance the production of Fe_3O_4 from Fe(II) hydroxide by ultrasound. The authors have suggested that the increase in the efficiency of this reaction in the presence of ultrasound is due to the rapid ion transport and extreme local temperatures produced by the cavitation process.

Gedanken's group has reported the sonochemical synthesis of Mo_2O_5 [114], $CoFe_2O_4$ [115], Eu_2O_3 and Tb_2O_3 on SiO_2 and Al_2O_3 [116], Fe_2O_3 on Al_2O_3 [117] and Co_3O_4 and CoO [118]. In most cases, a non-aqueous solvent containing the metal complex of choice was sonicated in the presence of air. Despite the fact that a detailed analytical investigation on the materials synthesized was undertaken, very little attention was given to the possible sonochemical mechanisms responsible for particle formation. Gedanken's group has recently reported [119] on the sonochemical synthesis of fullerene-like Tl_2O and Tl_2OCl_2 from an aqueous solution of $TlCl_3$.

The sonochemical preparation of Fe–Co alloys has been achieved by sonicating non-aqueous solutions containing $Fe(CO)_5$ and $Co(CO)_3(NO)$ complexes [120]. The composition of the Fe–Co alloy could be controlled by changing the concentration ratio of the precursors in the solution under sonication. The catalytic properties of the sonochemically prepared Fe, Co and Fe–Co were compared with respect to the cyclohexane reaction, i.e. hydrogenolysis of cyclohexane to form methane or dehydrogenation of cyclohexane to form benzene. It was found that the catalytic activ-

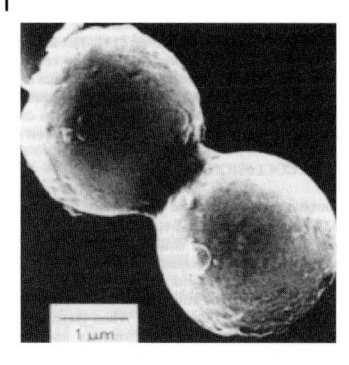

Fig. 4.15 Scanning electron micrograph of 5-μm diameter Zn powder. Neck formation from localized melting is caused by high-velocity collisions (reprinted from [122] with permission from the American Association for the Advancement of Science, Copyright 1990).

ity of the alloy lies in between the catalytic activities of the pure metal powders. However, the Fe–Co alloy seems to possess the highest selectivity towards the formation of benzene. The reason for this selectivity was suggested to be the production of small amounts of carbon on the surface of the catalyst during sonication. Gedanken's group [121] followed a similar procedure to the above to produce a Co–Ni alloy by sonicating decalin containing $Co(CO)_3(NO)$ and $Ni(CO)_4$.

A novel method for the sonochemical formation of alloys was reported by Suslick and coworkers [122]. Interparticle collisions between metal particles resulted in the formation of a "neck", when colloidal metal particle suspensions were irradiated by ultrasound (Fig. 4.15). This is one of the examples that shows how mechanical effects generated by acoustic cavitation can be used to cause a process that normally requires extreme conditions.

Hydrocarbon liquids containing metal particles such as zinc, tin, nickel, chromium, etc. were irradiated for 30 min at 15 °C in freshly distilled decane under argon with high-intensity ultrasound (direct immersion ultrasonic horn; 20 kHz, 50 W/cm^2). The particles, driven by the turbulent flow induced by cavitation-generated shock waves, collided with each other at about half the speed of sound.

In Fig. 4.15, "neck" formation between two individual Zn particles was due to the localized melting caused by the high-velocity collisions, followed by rapid cooling (cooling time is estimated as 1 ns) of the molten collision zone. When Sn and Fe particles were ultrasonically irradiated, fusion between Sn and Fe particles occurred. Analysis of the neck region showed alloy formation between Sn and Fe, mainly consisting of Sn. The maximum temperatures during interparticle collisions were estimated to be between 2600 and 3400 °C based on the interparticle collision experiments using different metal particles. An upper limit of 3400 °C was suggested since no neck formation could be observed between W particles (melting point 3410 °C) during ultrasound irradiation.

4.5
Conclusions

Sonochemical synthesis of colloidal materials is a novel and successful synthetic technique in both aqueous and non-aqueous systems. The extreme conditions that are generated during the cavitation process have been exploited to achieve reactions/products of interest. For colloid formation occurring in an aqueous medium, it is known that the primary radicals H$^{\bullet}$ and OH$^{\bullet}$ are involved either directly or indirectly in the reactions leading to the formation of the colloids. However, mechanistic details of the sonochemical formation of most colloids produced to date remains sketchy. The use of surface-active agents in aqueous solutions enhances colloid formation in many systems. They effectively scavenge the primary radicals produced by cavitation and form secondary radicals, which can reduce metal ions in bulk solution. The sonochemically synthesized polymer latex particles have a narrow size distribution range. In a non-aqueous medium colloid formation occurs by quite a different pathway to that in aqueous solutions. Volatile metal complexes, which are soluble in the non-aqueous solvent diffuse into the bubble during the cavitation process, wherein the high local temperatures generated during bubble collapse decompose the complexes to produce fine metal particles. In both aqueous and non-aqueous systems, the cavitation-induced physical effects play a significant role in determining the properties of the particles produced. The narrow size distribution of polymer particles and the formation of alloys in the "neck" are some of the examples of the role of physical effects. With the significant development that has been achieved in the field of sonochemical synthesis of colloidal materials in the past few decades, as evidenced from the increasing number of publications in this area, it is evident that this technology could be adopted and expanded for industrial purposes.

4.6
Acknowledgments

Financial support from the Australian Research Council and the collaborative support of the EC (COST Chemistry D10 program) are gratefully acknowledged.

4.7 References

1 BELLONI, J.; MOSTAFAVI, M.; REMITA, H.; MARIGNIER, J.L.; DELCOURT, M.O. *New. J. Chem.* **1998**, *22*, 1239–1255.
2 CARDENASTRIVINO, G.; ALVIAL, M.; KLABUNDE, K.J.; PANTOJA, O.; SOTO, H. *Coll. Polym. Sci.* **1994**, *272*, 310–316.
3 TSUJI, T.; IRYO, K.; NISHIMURA, Y.; TSUJI, M. *J. Photochem. Photobiol. A Chemistry* **2001**, *145*, 201–207.
4 NAKAMURA, K.; MORI, Y. *J. Chem. Eng. Jpn.* **2001**, *34*, 1538–1544.
5 CHEN, C.W.; SERIZAWA, T.; AKASHI, M. *Chem. Mater.* **2002**, *14*, 2232–2239.
6 MATIJEVIC, E.; PARTCH, R.E. in *Fine Particles: Synthesis, Characterization, and Mechanisms of Growth*, Surfactant Science Series, SUGIMOTO, T. (ed.) Marcel Dekker NY, **2000**, pp. 97–113.
7 ZIMMERMANN, C.J.; PARTCH, R.E.; MATIJEVIC, E. *J. Aerosol Sci.* **1991**, *22*, 881–886.
8 PRAKASH, S.; GHOSH, A.K. *Ultrasonics and Colloids*, Asia Publishing House, London, **1962**.
9 OSHIMA, H. *Adv. Coll. Interface Sci.* **1995**, *62*, 189–235.
10 OSHIMA, H. *Coll. Polym. Sci.* **1997**, *275*, 480–485.
11 OSHIMA, H.; NAKAMURA, M.; KONDO, T. *Coll. Polym. Sci.* **1992**, *270*, 873–877.
12 CRUM, L.A. *Ultrasonics* **1984**, *22*, 215–223.
13 LEIGHTON, T. *The Acoustic Bubble*, Academic Press, London, **1994**.
14 BRENNEN, C.E. *Cavitation and Bubble Dynamics*, Oxford University Press, Oxford, **1995**.
15 YOUNG, F.R. *Cavitation*, Imperial College Press, London **1999**.
16 CRUM, L.A.; MASON, T.J.; REISSE, J.L.; SUSLICK, K.S. (eds.) *Sonochemistry and Sonoluminescence*, NATO ASI Series C, Vol. 524, Kluwer Publishers, London, **1999**.
17 HART, E.J.; FISCHER, C.-H.; HENGLEIN, A. *Radiat. Phys. Chem.* **1990**, *36*, 511–516.
18 TAUBER, A.; MARK, G.; SCHUCHMANN, H.-P.; VON SONNTAG, C., *J. Chem. Soc., Perkin Trans. II* **1999**, pp. 1129–1135.
19 ASHOKKUMAR, M.; HALL, R.; MULVANEY, P.; GRIESER, F. *J. Phys. Chem. B* **1997**, *101*, 10845–10850.
20 ASHOKKUMAR, M.; MULVANEY, P.; GRIESER, F. *J. Am. Chem. Soc.* **1999**, *121*, 7355–7359.
21 BARBOUR, K.; ASHOKKUMAR, M.; CARUSO, R.A.; GRIESER, F. *J. Phys. Chem. B* **1999**, *103*, 9231–9236.
22 ASHOKKUMAR, M.; VINODGOPAL, K.; GRIESER, F. *J. Phys. Chem. B* **2000**, *104*, 6447–6451.
23 ASHOKKUMAR, M.; CRUM, L.A.; FRENSLEY, C.A.; GRIESER, F.; MATULA, T.J.; MCNAMARA III, W.B.; SUSLICK, K.S. *J. Phys. Chem. A* **2000**, *104*, 8462–8465.
24 GRIESER, F.; ASHOKKUMAR, M. *Adv. Coll. Interface Sci.* **2001**, *89/90*, 423–438.
25 ASHOKKUMAR, M.; GUAN, J.; TRONSON, R.; MATULA, T.J.; NUSKE, J.W.; GRIESER, F. *Phys. Rev. E* **2002**, *65*, 463101–463104.
26 TRONSON, R.; ASHOKKUMAR, M.; GRIESER, F. *J. Phys. Chem. B* **2002**, *106*, 11064–11068.
27 MISÍK, V.; MIYOSHI, N.; RIESZ, P. *J. Phys. Chem.* **1995**, *99*, 3605–3611.
28 MISÍK, V.; RIESZ, P. *Ultrasonics Sonochem.* **1996**, *3*, 25–37.
29 SUSLICK, K.S.; HAMMERTON, D.A.; CLINE. R.E. Jr. *J. Am. Chem. Soc.* **1986**, *108*, 5641–5642.
30 SUSLICK, K.S.; MCNAMARA III, W.B.; DIDENKO, Y.T. in *Sonochemistry and Sonoluminescence*, NATO ASI Series C, Vol. 524, (CRUM, L.A.; MASON, T.J.; REISSE, J.L.; SUSLICK, K.S., eds.) Kluwer Publishers, London, **1999**, pp. 191–204.
31 ASHOKKUMAR, M.; GRIESER, F. *Chem. Eng. Rev.* **1999**, *15*, 41–83 and references therein.
32 MASON, T.J.; LORIMER, J.P. *Sonochemistry: Theory, Applications and Uses of Ultrasound in Chemistry*, Ellis Horwood, Chichester, **1988**.
33 MASON, T.J. (ed.) *Sonochemistry: The Uses of Ultrasound in Chemistry*, Royal Society of Chemistry, Cambridge, **1990**.
34 MASON. T.J. *Practical Sonochemistry: User's Guide to Applications in Chemistry and Chemical Engineering*, Ellis Horwood, Chichester, **1991**.

35 MASON, T. J. (ed.) *Advances in Sonochemistry*, Vols. 1–5, JAI Press. Connecticut, **1999**.
36 SUSLICK, K. S. (ed.) *Ultrasound: Its Chemical, Physical and Biological Effects*, VCH Publishers, New York, **1988**.
37 SUSLICK, K. S.; CRUM, L. A. in *Handbook of Acoustics* (CROCKER, M. J. ed.) John Wiley & Sons, New York, **1998**, pp. 243–253.
38 SUSLICK, K. S.; PRICE, G. J. *Ann. Rev. Mater. Sci.* **1999**, *29*, 295–326.
39 PRICE, G. J. (ed.) *Current Trends in Sonochemistry*, Royal Society of Chemistry, Cambridge, **1992**.
40 GRIESER, F. in *Studies in Surface Science and Catalysis* (KAMAT, P. V.; MEISEL, D. eds.) **1996**, *103*, 57–77.
41 HENGLEIN, A. *Advances in Sonochemistry*, **1993**, *3*, 17–83.
42 MAKINO, K.; MOSSOBA, M. M.; RIESZ, P. *J. Phys. Chem.* **1983**, *87*, 1369–1377.
43 KRISHNA, C. M. ; KONDO, T.; RIESZ, P. *J. Phys. Chem.* **1989**, *93*, 5166–5172.
44 SAKAKIBARA, M.; WANG, D.; IKEDA, K.; SUZUKI, K. *Ultrasonics Sonochem.* **1994**, *1*, S107–S110.
45 SWAMY, K. M.; SUKLA, L. B.; NARAYANA, K. L.; KAR, R. N.; PANCHANADIKAR, V. V. *Ultrasonics Sonochem.* **1995**, *2*, S5–S9.
46 BOLDYREV, V. V. *Ultrasonics Sonochem.* **1995**, *2*, S143–S145.
47 LICKISS, P. D.; MCGRATH, V. *Chem. Britain*, March **1996**, 47–50.
48 GIBBS, W. W. *Sci. Am.*, June **1996**, 22–24.
49 TSUJINO, J.; UEOKA, T.; HASEGAWA, K.; FUJITA, Y.; SHIRAKI, T.; OKADA, T.; TAMURA, T. *New Ultrasonics* **1996**, *34*, 177–185.
50 KAMAT, P. V.; VINODGOPAL, K. *Langmuir* **1996**, *12*, 5739.
51 UMEMURA, S.; KAWABATA, K.; SASAKI, K.; YUMITA, N.; UMEMURA, K.; NISHIGAKI, R. *Ultrasonics Sonochem.* **1996**, *3*, S187–S191.
52 TIEHM, A.; NEIS, U. (eds.) *TUHH Reports on Sanitary Engineering*, **1999**.
53 POOL, R. *Science* **1994**, *266*, 1804.
54 MOSS, W. C.; CLARKE, D. B.; WHITE, J. W.; YOUNG, D. A. *Phys. Lett. A* **1996**, *211*, 69–74.
55 CRUM, L. A.; MATULA, T. J. *Science* **1997**, *276*, 1348–1349.
56 LUCHE, J.-L. (ed.) *Synthetic Organic Sonochemistry*, Plenum Press, NY, **1998**.

57 HENGLEIN, A.; FISCHER, C. H. *Ber. Bunsen-Ges. Phys. Chem.* **1984**, *88*, 1196–1199.
58 GRIESER, F.; ASHOKKUMAR, M.; SOSTARIC, J. Z. in *Sonochemistry and Sonoluminescence* (CRUM, L. A.; MASON, T. J.; REISSE, J. L.; SUSLICK, K. S., eds.) NATO ASI series C, **1997**, *524*, 345–362.
59 BAIGENT, C. L.; MULLER, G. A. *Experientia* **1980**, *36*, 472–473.
60 GUTIERREZ, M.; HENGLEIN, A.; DOHRMANN, J. K. *J. Phys. Chem.* **1987**, *91*, 6687–6690.
61 YEUNG, S. A.; HOBSON, R.; BIGGS, S.; GRIESER, F. *J. Chem. Soc., Chem. Commun.* **1993**, 378–379.
62 GRIESER, F.; HOBSON, R.; SOSTARIC, J. Z. MULVANEY, P. *Ultrasonics* **1996**, *34*, 547–550.
63 CARUSO, R. A. *Colloidal Particle Formation using Sonochemistry*, Ph.D. Thesis, University of Melbourne, **1997**.
64 OKITSU, K.; MIZUKOSHI, Y.; BANDOW, H.; MAEDA, Y.; YAMAMOTO, T.; NAGATA, Y. *Ultrasonics Sonochem.* **1996**, *3*, S249–S251.
65 NAGATA, Y.; MIZUKOSHI, Y.; OKITSU, K.; MAEDA, Y. *Rad. Res.* **1996**, *146*, 333–338.
66 NAGATA, Y.; WATANABE, Y.; FUJITA, S.-I.; DOHMARU, T.; TANIGUCHI, S. *J. Chem. Soc., Chem. Commun.* **1992**, 1620–1622.
67 SALKAR, R. A.; JEEVANDAM, P.; ARUNA, S. T.; KOLTYPIN, Y.; GEDANKEN, Y. *J. Mater. Chem.* **1999**, *9*, 1333–1335.
68 OKITSU, K.; BANDOW, H.; MAEDA, Y. *Chem. Mater.* **1996**, *8*, 315–317.
69 OKITSU, K.; NAGAOKA, S.; TANABE, S.; MATSUMOTO, H.; MIZUKOSHI, Y.; NAGATA, Y. *Chem. Lett.* **1999**, 271–272.
70 BIANCHI, C. L.; GOTTI, E.; TOSCANO, L.; RAGAINI, V. *Ultrasonics Sonochem.* **1997**, *4*, 317–320.
71 HENGLEIN, A.; LINDIG, B.; WESTERHAUSEN, J. *Radiat. Phys. Chem.* **1984**, *23*, 199–205.
72 HENGLEIN, A.; ERSHOV, B. G.; MALOW, M. *J. Phys. Chem.* **1995**, *99*, 14129–14136.
73 CARUSO, R. A.; ASHOKKUMAR, M.; GRIESER, F. *Colloids Surfaces A* **2000**, *169*, 219–225.
74 MIZUKOSHI, Y.; OSHIMA, R.; MAEDA, Y.; NAGATA, Y. *Langmuir* **1999**, *15*, 2733–2737.
75 REISSE, J.; FRANCOIS, H.; VANDERCAMMEN, J.; FABRE, O.; DE MESMAEKER, A. K.; MAERSCHALK, C.; DELPLANCKE, J. L. *Electrochim. Acta* **1994**, *39*, 37–39.

76 COMPTON, R.G.; EKLUND, J.C.; MARKEN, F.; WALLER, D.N. *Electrochim. Acta* **1996**, *41*, 315–320.
77 DURANT, A.; DELPLANCKE, J.L.; WINAND, R.; REISSE, J. *Tetrahedron Lett.* **1995**, *36*, 4257–4260.
78 DHAS, N.A.; RAJ, C.P.; GEDANKEN, N.A. *Chem. Mater.* **1998**, *10*, 1446–1452.
79 SUSLICK, K.S.; GRINSTAFF, M.W. *J. Am. Chem. Soc.* **1990**, *112*, 7807–7809.
80 GRINSTAFF, M.W.; SUSLICK, K.S. *Proc. Natl. Acad. Sci. USA* **1991**, *88*, 7708–7710.
81 PRICE, G.J.; WEST, P.J.; SMITH, P.F. *Ultrasonics Sonochem.* **1994**, *1*, S51–S57.
82 FLOSDORF, E.W.; CHAMBERS, L.A. *J. Am. Chem. Soc.* **1933**, *55*, 3051–3052.
83 GYORGYI, A.S. *Nature*, **1933**, *131*, 278.
84 PORTENLANGER, G.; HEUSINGER, H. *Ultrasonics Sonochem.* **1994**, *1*, S125–S129.
85 BIGGS, S.; GRIESER, F. *Macromolecules* **1995**, *28*, 4877–4882.
86 PRICE, G.J. *Ultrasonics Sonochem.* **1996**, *3*, S229–S238.
87 COOPER, G.; GRIESER, F.; BIGGS, S. *J. Colloid Interface Sci.* **1996**, *184*, 52–63.
88 KODA, S.; AMANO, T.; NOMURA, H. *Ultrasonics Sonochem.* **1996**, *3*, S91–S95.
89 KATOH, R.; YOKOI, H.; USUBA, S.; KAKUDATE, Y.; FUJIWARA, S. *Ultrasonics Sonochem.* **1998**, *5*, 69–72.
90 CHOU, H.C.J.; STOFFER, J.O. *J. Appl. Polym. Sci.* **1999**, *72*, 797–825.
91 CHOU, H.C.J.; STOFFER, J.O. *J. Appl. Polym. Sci.* **1999**, *72*, 827–834.
92 OOI, S.K.; BIGGS, S. *Ultrasonics Sonochem.* **2000**, *7*, 125–133.
93 BRADLEY, M.; GRIESER, F. *J. Coll. Interface Sci.* **2002**, *251*, 78–84.
94 SUSLICK, K.S.; GAWIENOWSKI, J.J.; SCHUBERT, P.F.; WANG, H.H. *Ultrasonics* **1984**, *22*, 33–36.
95 SUSLICK, K.S.; CHOE, S.-B.; CICHOWLAS, A.A.; GRINSTAFF, M.W. *Nature* **1991**, *353*, 414–416.
96 SUSLICK, K.S.; FANG, M.; HYEON, T. *J. Am. Chem. Soc.* **1996**, *118*, 11960–11961.
97 SUSLICK, K.S.; HYEON, T.; FANG, M.; CICHOWLAS, A.A. in *Adv. Catal. Nanostruct. Mater* (WILLIAM, W.R., ed.) Academic Publishers, San Diego, **1996**, pp. 197–212.
98 DIODATI, P.; GIANNINI, G.; MIRRI, L.; PETRILLO, C.; SACCHETTI, F. *Ultrasonics Sonochem.* **1997**, *4*, 45–48.
99 DHAS, N.A.; GEDANKEN, A. *J. Mater. Chem.* **1998**, *8*, 445–450.
100 DHAS, N.A.; COHEN, H.; GEDANKEN, A. *J. Phys. Chem. B* **1997**, *101*, 6834–6838.
101 KOLTYPIN, Y.; KATABI, G.; CAO, X.; PROZOROV, R.; GEDANKEN, A. *J. Non-Cryst. Solids* **1996**, *201*, 159–162.
102 RAMESH, S.; KOLTYPIN, Y.; PROZOROV, R.; GEDANKEN, A. *Chem. Mater.* **1997**, *9*, 546–551.
103 KATOH, R.; TASAKA, Y.; YUMURA, M.; IKAZAKI, F.; KAKUDATE, Y.; FUJIWARA, S. *Ultrasonics Sonochem.* **1999**, *6*, 185–187.
104 DHAS, N.A.; RAJ, C.P.; GEDANKEN, A., *Chem. Mater.* **1998**, *10*, 3278–3279.
105 HOBSON, R.A.; MULVANEY, P.; GRIESER, F. *J. Chem. Soc., Chem. Commun.* **1994**, pp. 823–824.
106 SOSTARIC, J.Z.; HOBSON, R.A.; MULVANEY, P.; GRIESER, F. *J. Chem. Soc., Faraday Trans.* **1997**, *93*, 1791–1795.
107 SOSTARIC, J.Z. *Interfacial Effects on Aqueous Sonochemistry and Sonoluminescence*, Ph.D. Thesis, University of Melbourne, **1999**.
108 DHAS, N.A.; GEDANKEN, A. *Appl. Phys. Lett.* **1998**, *72*, 2514–2516.
109 MDLELENI, M.M.; HEYON, T.; SUSLICK, K.S. *J. Am. Chem. Soc.* **1998**, *120*, 6189–6190.
110 MASTAI, Y.; HOMYONFER, M.; GEDANKEN, A.; HODES, G. *Adv. Mater.* **1999**, *11*, 1010–1013.
111 LI, B.; XIE, Y.; HUANG, J.; QIAN, Y. *Ultrasonics Sonochem.* **1999**, *6*, 217–220.
112 HYEON, T.; FANG, M.; SUSLICK, K.S. *J. Am. Chem. Soc.* **1996**, *118*, 5492–5493.
113 ENOMOTO, N.; AKAGI, J.-I.; NAKAGAWA, Z. *Ultrasonics Sonochem.* **1996**, *3*, S97–S103.
114 DHAS, N.A.; GEDANKEN, A. *J. Phys. Chem. B* **1997**, *101*, 9495–9503.
115 SHAFI, K.V.P.M.; GEDANKEN, A.; PROZOROV, R.; BALOGH, J. *Chem. Mater.* **1998**, *10*, 3445–3450.
116 PATRA, A.; SOMINSKA, E.; RAMESH, S.; KOLTYPIN, Y.; ZHONG, Z.; MINTI, H.; REISFELD, R.; GEDANKEN, A. *J. Phys. Chem. B* **1999**, *103*, 3361–3365.
117 ZHONG, Z.Y.; PROZOROV, T.; FELNER, I.; GEDANKEN, A. *J. Phys. Chem. B* **2000**, *103*, 947–956.
118 JEEVANANDAM, P.; KOLTYPIN, Y.; GEDANKEN, A.; MASTAI, Y. *J. Mater. Chem.* **2000**, *10*, 511–514.

119 Avivi, S.; Mastai, Y.; Gedanken, A. *J. Am. Chem. Soc.* **2000**, *122*, 4331–4334.

120 Suslick, K. S.; Fang, M. M.; Hyeon, T.; Mdleleni, M. M. in *Sonochemistry and Sonoluminescence* (L. A. Crum, T. J. Mason, J. L. Reisse, K. S. Suslick eds.), Kluwer Publishers, London, NATO ASI Series C, **1999**, *524*, 291–320.

121 Shai, K. V. P. M.; Gedanken, A.; Prozorov, R. *J. Mater. Chem.* **1998**, *8*, 769–773.

122 Doktycz, S. J.; Suslick, K. S. *Science* **1990**, *247*, 1067–1069.

123 Okitsu, K.; Yue, A.; Tanabe, S.; Matsumoto, H.; Yobiko, Y. *Langmuir* **2001**, *17*, 7717–7720.

124 Kumar, R. V.; Palchik, O.; Koltypin, Y.; Diamant, Y.; Gedanken, A. *Ultrason. Sonochem.* **2002**, *9*, 65–70.

125 Xia, X.; Shen, L. L.; Guo, Z. P.; Liu, H. K.; Walter, G. *J. Nanosci. Technol.* **2002**, *2*, 45–46.

126 Ashokkumar, M.; Grieser, F. in *Encyclopedia of Surface and Colloid Science* (Hubbard, A. T. ed.), Marcel Dekker, NY, **2002**, pp. 4760–4774.

127 Caruso, R. A.; Ashokkumar, M.; Grieser, F. *Langmuir* **2002**, *18*, 7831–7836.

128 Adamson, A. W. *Physical Chemistry of Surfaces*, 5th edn, Wiley Interscience, New York, **1990**, pp. 71–86 (The Gibbs surface excess, Γ_s, can be calculated from the Gibbs-Duhem Equation which, at constant pressure and temperature and for a neutral solute, has the form $\Gamma_s = -1/kT \, (d\gamma/d \ln C_s)$, where γ is the air/water surface tension (N/m) at the bulk solute concentration, C_s (M)).

129 Belloni, J.; Mostafavi, M.; Marignier, J. L.; Amblard, J. *J. Imag. Sci.* **1991**, *35*, 68–74.

130 Coley, R. F.; Martin, D. S. Jr. *Inorg. Chim. Acta* **1973**, *7*, 573–577.

131 Janata, E.; Henglein, A.; Ershov, B. *J. Phys. Chem.* **1996**, *100*, 1989–1992.

132 Canterford, J. H.; Colton, R. *Halides of the 2nd and 3rd Row Transition Metals*, Wiley Interscience, London, **1968**.

5
Colloidal Nanoreactors and Nanocontainers

Marc Sauer and Wolfgang Meier

5.1
Introduction

In recent years materials with well-defined structures in the submicrometer regions have attracted increasing interest. The idea in this context is to tailor the composition, structure and function of materials with control at the nanometer level, which may lead to new properties for well-known standard materials and hence to new applications.

In this chapter we will focus on nanometer-sized hollow spheres, so-called nanocontainers. Such particles are of increasing interest due to their potential for encapsulation of larger quantities of guest molecules or large-sized guests within their empty core domain. They can be useful in applications in areas such as biological chemistry, synthesis and catalysis.

Similar and very effective nanocontainers such as micelles, vesicles and viruses are used by Nature in biological systems, for example for transport and delivery of sensitive compounds such as hormones or DNA [1]. However, these highly functionalized assemblies of proteins, nucleic acids and other (macro)molecules perform complicated tasks that are still too complex to be completely understood. Although the particles seems to be of perfect design after millions of years of evolution, their direct use in many technical applications is often not feasible. In many cases this is due to the non-covalent interactions responsible for their formation and hence their limited stability. This leads, for example, to a rapid clearance of conventional lipid vesicles from the blood after their intravasal administration [2]. Many applications, however, require more stable particles.

Therefore a great effort has been devoted in recent years to prepare size- and shape-persistent nanocapsules, and so far now a huge variety of different techniques has been described, each of them having its special advantages and, of course, disadvantages.

This chapter is divided into sections each dealing with a different approach commonly used to attain the desired particle morphology. This division is somewhat arbitrary, especially since the different methods overlap to some extent. However, Section 5.2 gives a short introduction to natural nanocontainers, which are of major importance for the understanding of natural processes in spite of their

Colloids and Colloid Assemblies. Edited by Frank Caruso
Copyright © 2004 Wiley-VCH Verlag GmbH & Co. KGaA, Weinheim
ISBN: 3-527-30660-9

limited technical applicability. The following three sections describe approaches in which exclusively non-covalent interactions (that is van der Waals, electrostatic or hydrophobic interactions) are used to imprint the characteristic shape on the resulting particles. The final section deals with single polymers that have by design intrinsic hollow sphere morphology in the nanometer range.

5.2
The Natural Approach

Although many naturally available nanocontainers are not suitable for technical applications, much interesting research work has been published that describes the use of various natural systems within nanomaterial science [3]. Especially in the field of biomineralization, significant progress has been made with so-called bio-nanoreactors [4–7]. Among the different types of possible mineralization vessels ferritin is probably the most intensively studied and best understood [8]. Nature uses ferritin as a storage device for iron, where it is encapsulated as hydrated iron(III)oxide within a multisubunit protein shell. After the iron has been removed from inside the protein shell the remaining cavity has been successfully used to generate various nanometer-sized inorganic particles, for example manganese oxide or uranyl oxyhydroxide crystals [9].

Virus particles, such as the tobacco mosaic virus (TMV) or the cowpea chlorotic mottle virus (CCMV), are another type of biological structure that can be applied as a template for biomineralization processes. As indicated in Fig. 5.1 several routes were explored for the synthesis of nanotubes composites that use TMV templates [10].

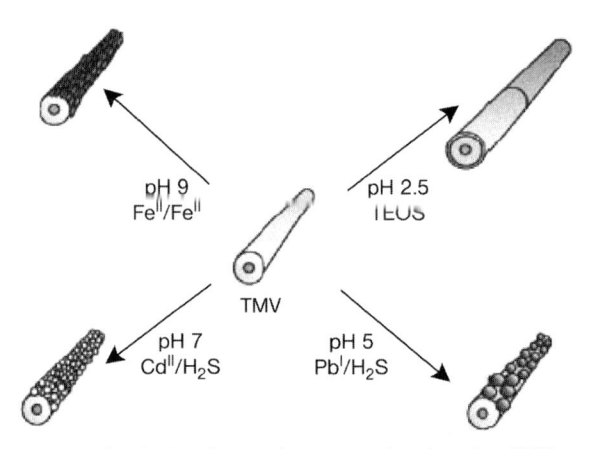

Fig. 5.1 Synthesis of nanotubes composites by using TMV templates. (TEOS = tetraethoxysilane) (reproduced from W. Shenton, T. Douglas, M. Young, G. Stubbs, S. Mann, *Adv. Mater.* **1999**, *11*, 253 with permission).

pH < 6.5 ⬅——————➡ pH > 6.5

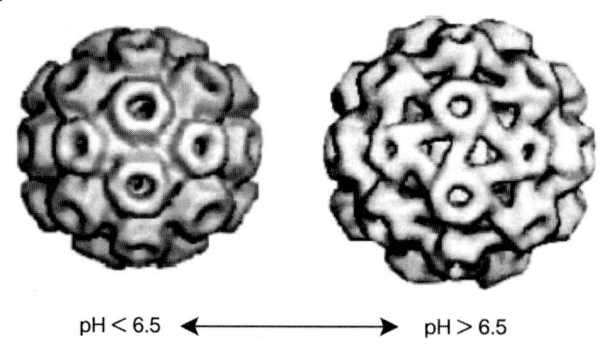

Fig. 5.2 Cryoelectron microscopy and image reconstitution of the Cowpea chlorotic mottle virus (CCMV). The structure in an unswollen condition at low pH values (left) and in a swollen condition at high pH values (right). Swelling results in the formation of 60 pores with a diameter of 2 nm (reproduced from T. Douglas, M. Young, *Nature* **1998**, *393*, 152 with permission).

The spherical protein cage of the CCMV shows a completely reversible pH-induced swelling transition of approximately 10%, if it is incubated in media of a pH value above 6.5 (Fig. 5.2) [11]. During this swelling, gated pores are opened along the viral shell and allow a free molecular exchange between the viruses interior and the surrounding bulk medium. In contrast, no exchange of larger molecules occurs at pH values lower than 6.5. This property has been used for controlled host-guest encapsulation and crystal growth of inorganic and organic materials within such viruses [12].

These examples demonstrate clearly that bio-nanocontainers may be suitable for the formation of new materials. However, in addition to their low mechanical stability, isolating and handling larger quantities of such bioreactors present major difficulties.

5.3
The Self-assembly Approach

Owing to their amphiphilic nature and molecular geometry, lipid molecules can aggregate in dilute aqueous solution into spherically closed bilayer structures, so-called vesicles or liposomes [2]. It is quite reasonable that the hollow-sphere morphology of these aggregates should render them suitable as precursors for the preparation of size- and shape-persistent nanocontainers. This can be realized using different concepts. Fig. 5.3 gives an overview of various methods. For example, lipids that are functionalized with polymerizable groups can be polymerized within such vesicular structures. As a result of the polymerization reaction the individual lipid molecules are interconnected via covalent bonds, which stabilize the shell-forming membrane considerably [13, 14]. In some sense the polymerized

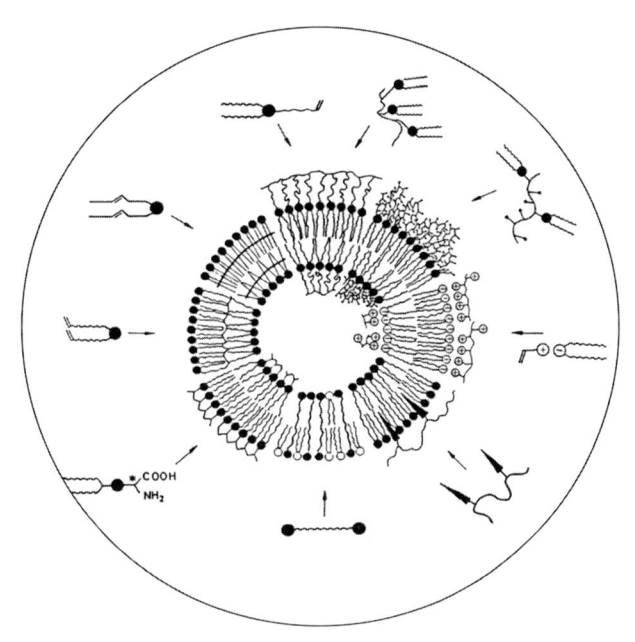

Fig. 5.3 Schematic representation of the different possibilities for stabilizing lipid vesicles (reproduced from H. Ringsdorf, B. Schlarb, J. Venzmer, *Angew. Chem.* **1988**, *100*, 117 with permission).

lipids can be regarded as mimicking the role of the cytoskeleton or of the murein network (that is the polymer structures which Nature uses to stabilize biological cell membranes) [15]. In contrast to these natural polymer scaffolds that are simply attached to the lipid membrane via hydrophobic interactions, the polymerized lipids are, however, all covalently attached to the polymer chain within the membrane. It is obvious that this considerably obstructs their lateral mobility within the membrane.

Interestingly, the polymerization of reactive lipids in bilayers formed from mixtures of different lipids may induce phase-separation phenomena within the membranes. This can be exploited to produce labile domains in a controlled manner in the shell of the polymerized vesicles. These domains can be plugged or unplugged using external stimuli, which renders such particles suitable for applications in the area of controlled or triggered release [16–19]. Since the pioneering studies on polymerized vesicles in the late 1970s and early 1980s this area has developed into a broad and active field of research. A detailed discussion of all its different aspects would be inappropriate here, but several recommendable reviews and books have already appeared on this topic to which the interested reader may refer [13, 14].

In an analogous fashion to lipids, amphiphilic block copolymers can also aggregate in aqueous solution to form vesicular structures [20–25]. Block copolymer vesicles may be significantly more stable than those formed from conventional lipids, owing to the larger size and the lower dynamics of the underlying polymer molecules [26]. Nevertheless, similarly to conventional lipids, they are held together solely by non-covalent interactions. Hence they can disintegrate under certain conditions (for example dilution or presence of detergent) and dissolve as individual block copolymer molecules. It is obvious that, analogous to the reactive lipids, block copolymer molecules could also be modified with polymerizable groups. Subsequent polymerization of the resulting "macromonomers" interconnects them via covalent bonds, which stabilize the whole particle. Such block copolymer-based nanocapsules possess great potential in a broad area of applications, such as controlled-release systems, confined nanometer-sized reaction vessels or templates for the design of new materials. In this context it is particularly interesting that the physical properties of the polymer shells can be controlled to a large extent by the block length ratio or the chemical constitution of the underlying polymer molecules [27].

Up to now, however, only a few papers dealing with such particles have appeared. In one of them the formation of vesicles from a poly(isoprene)-β-poly(2-cinnamoyl methacrylate) (PI-PCEMA) diblock copolymer in hexane–tetrahydrofuran mixtures has been exploited as a starting point [28]. Converting these vesicles into stable, water-soluble polymer nanocapsules requires, however, a rather costly procedure. The PCEMA blocks were photocross-linked and then in a second step the PI blocks had to be hydroxylated to make these hollow nanospheres water-soluble. The radii of the nanocapsules were about 50 to 60 nm and changed only very slightly during these conversions. After loading the nanocontainers with Rhodamine B the dye could be quantitatively released in water and water–ethanol mixtures in such a way that the release rate could be tuned by the composition of the water–ethanol mixtures. Ethanol seems to be a good solvent for the underlying block copolymers. Hence higher ethanol content in the solvent mixture increases the solvent quality and the cross-linked shell of the cross-linked particles swells increasingly. As a result shell permeability increases and release of the dye becomes faster. This feature makes such systems attractive for applications, since prefabricated particles can be loaded effectively in an organic solvent and the encapsulated material subsequently slowly released in water.

In a similar approach a rather simple one-step procedure has been used to prepare vesicular structures from poly(2-methyloxazoline)-β-poly(dimethylsiloxane)-β-poly(2-methyloxazoline) (PMOXA-PDMS-PMOXA) triblock copolymers directly in aqueous solution. The size of the resulting vesicles could be controlled from 50 nm to 100 µm [20]. The underlying triblock copolymers were modified with polymerizable methacrylate end-groups without changing their aggregation behavior in water. These "macromonomers" were polymerized within the vesicles using a UV-induced free-radical polymerization, which did not lead to any measurable changes in size, size distribution or even molecular weight of the particles [20]. Obviously the polymer chain reaction occurs mainly intravesicularly. Intervesicle reactions such as exchange of individual triblock copolymer molecules or a chain propagation reaction involving more than one vesicular aggregate play only a minor role on the timescale of the experiment.

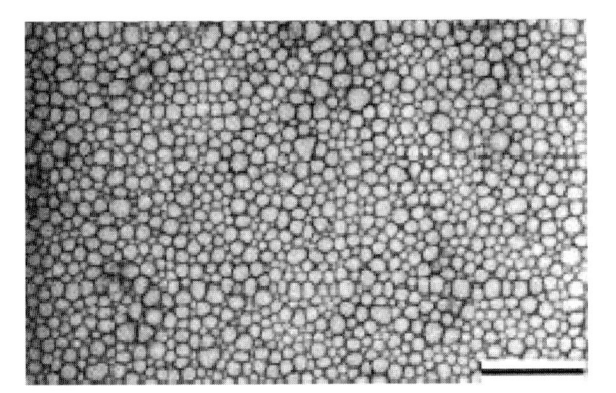

Fig. 5.4 Transmission electron micrograph of polymerized ABA-triblock copolymer vesicles. The length of the bar corresponds to 1 μm (reproduced from C. Nardin, T. Hirt, J. Leukel, W. Meier, *Langmuir* **2000**, *16*, 1035 with permission).

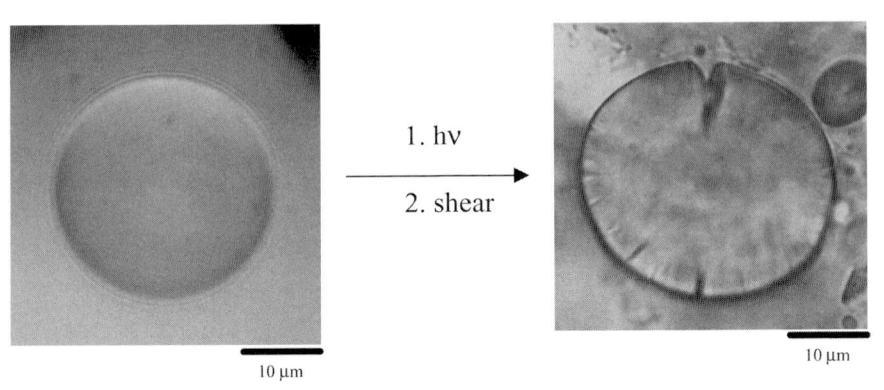

Fig. 5.5 Reflection intensity contrast micrograph of a giant, polymerized PMOXA-PDMS-PMOXA triblock copolymer vesicle (left) before and (right) after shear-induced rupture.

The covalently cross-linked polymer network structures result in particles that possess solid-state properties like shape persistence. Therefore they are able to preserve their hollow sphere morphology even after their isolation from the aqueous solution and redispersion in an organic solvent, like chloroform, tetrahydrofuran or ethanol. This is illustrated in Fig. 5.4, which shows a transmission electron micrograph of polymerized triblock copolymer vesicles isolated from water and redispersed in ethanol. Ethanol is a good solvent for the whole underlying block copolymer molecule. Hence, non-polymerized vesicles would immediately disintegrate under such conditions into singly dissolved polymer molecules. In contrast to their fluid-like, non-polymerized precursors a shear deformation of the polymerized vesicles may lead to cracks characteristic of solid materials [29]. This is directly demonstrated in the reflection intensity contrast micrographs of

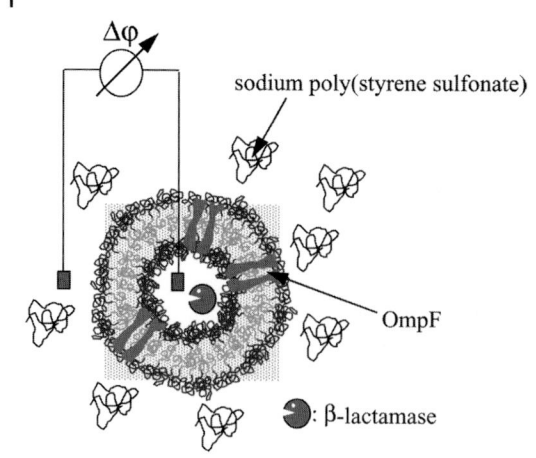

Fig. 5.6 Schematic view of a PMOXA-PDMS-PMOXA nanoreactor with encapsulated β-lactamase and of the Donnan potential induced by polyelectrolyte present in the external solution.

Fig. 5.5, which show a giant polymerized triblock copolymer vesicle before and after shear-induced rupture. This shape-persistence allows the use of such vesicles as a delivery system since they can be loaded with guest molecules in an organic solvent. After isolation of the loaded polymer shells and redispersion in aqueous media the encapsulated material can be subsequently released.

A more specific uptake and release of substances could be realized by reconstitution of membrane proteins in the completely artificial surroundings of a block copolymer membrane [30–32]. It has been shown that the functionality of the incorporated proteins is fully preserved even after polymerization of the vesicular membrane. With regard to the wide range of membrane proteins that Nature provides, such copolymer–protein hybrid materials possess great potential in many areas of application, for example gene therapy [33] or biomineralization [34].

By encapsulating an enzyme (β-lactamase) in the aqueous core domain of the nanocapsule and reconstitution of a membrane protein (OMPF) in the polymer shell a novel kind of nanoreactor could be designed [31]. Furthermore, the channels of this protein close reversibly above a critical transmembrane potential.

By simply adding a large enough polyelectrolyte, which cannot pass the protein channels, a Donnan-potential is created across the vesicular membrane. If this potential exceeds the critical voltage necessary for closure of the protein the substrates can no longer enter the vesicular interior and the reactor is deactivated.

Decreasing the potential below the critical voltage (for example by dilution) reactivates the reactor. The possibility of triggering the activation (or deactivation) of such systems is highly interesting for applications, since it allows local and temporal control of the uptake and release of substrate.

Recently the use of polypeptide-based diblock copolymers, poly(butadiene)-β-poly(L-glutamate) (PB-PGA), for the preparation of vesicular structures has been described [35, 36] (Fig. 5.7). It is particularly interesting here that the solvating chains are polypeptides and form vesicles, or so-called "peptosomes", composed of modified protein units. A subsequent polymerization of the polybutadiene block

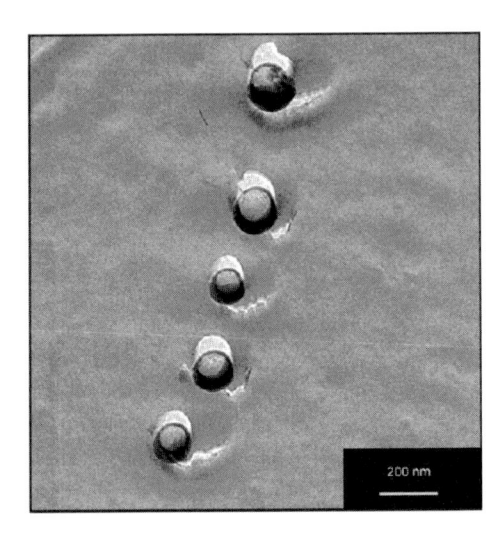

Fig. 5.7 Freeze-fracture electron micrograph of a vesicular solution of poly(butadiene)$_{85}$-β-poly(glutamate)$_{55}$ in pure water (pH ~ 6.0) (reproduced from H. Kukula, H. Schlaad, M. Antonietti, S. Förster, *J. Am. Chem. Soc.* **2002**, *124*, 1658 with permission).

leads to shape-persistent nanocontainers that, because of their polypeptide block, possess a certain pH- and ionic strength-sensitivity [36]. By changing this parameter the particle's dimension changes reversibly as a result of the polypeptide block's transition from a compact helical secondary structure at low pH values to a more extended random-coil conformation at pH values above 7.

A similar concept uses just the geometry of the vesicular aggregates as a template [37–44]. In this case it is not the amphiphilic molecules themselves that are polymerized to freeze in the whole superstructure of the supramolecular aggregate. They simply provide a geometrically restricted environment for dissolving and polymerizing conventional monomers.

It is well known that vesicles or liposomes are able to solubilize hydrophobic substances to a certain degree. Such compounds are usually dissolved in the hydrophobic part of the lipid bilayer. If such substances also carry polymerizable groups, their subsequent polymerization should lead to the formation of polymer chains entrapped in the interior of the membrane. In contrast to polymerizable lipids, the polymer chains are now simply dissolved within the alkane part of the bilayer-forming lipids. Hence, they are of minor influence on the overall physical properties of the membranes.

One special feature of vesicular polymerization of conventional hydrophilic or hydrophobic monomers is that the different compartments provided by the self-assembly of the lipid molecules generally serve only as a template, which determines the size and shape of the resulting polymers. Thus, it is possible to use nearly every natural or synthetic lipid, without any modification. Although in most reported studies on vesicular polymerization of conventional monomers, synthetic lipids like dioctadecyl diammonium bromide (DODAB), chloride (DODAC), sodium di-2-ethylhexyl phosphate (SEHP) or even spontaneously formed vesicles prepared from mixtures of cationic and anionic surfactants have usually been used, any natural lipid would provide a suitable matrix. Moreover, a combination of polymerizable lipids and conventional monomers could be incorporated in templating vesicles to yield hybrids with interesting new polymer structures.

Of course the incorporation of hydrophobic substances into lipid bilayers should not exceed certain saturation concentration [45]. Above this concentration the monomer is no longer homogeneously distributed within the bilayers. This has been shown for toluene in phospholipid vesicles at concentrations above the saturation level. Moreover, exceeding this saturation concentration may also disrupt the whole bilayer structure, thus converting the system into a conventional emulsion or even leading to the formation of a separate monomer phase in the presence of intact vesicles.

It is obvious that the overall thickness of the lipid membranes has to increase upon the solubilization of hydrophobic substances. The maximum swelling of the membrane leads typically to an increase from about 3 to 5 nm [46]. This is, however, a negligible effect compared to the overall diameter of typical small unilamellar vesicles, which is about 100 nm. Therefore no dimensional changes of the underlying vesicles can usually be detected upon swelling of the lipid bilayers of the vesicles as long the monomer concentration stays below the saturation value in the membranes.

It has been shown that the hydrophobic portion of lipid bilayers can be selectively swollen by a variety of hydrophobic monomers, such as styrene [37, 38, 40, 41], alkyl(meth)acrylates [39, 42, 43] or even lipofullerenes carrying polymerizable octadecadiine side chains [44]. A cross-linking polymerization of such monomers leads to the formation of two-dimensional polymer networks in the interior of the membranes. Such networks can act as a polymeric scaffold that increases the mechanical stability of lipid membranes considerably without impeding the mobility of the lipid molecules within the aggregates.

The free radical polymerization of the hydrophobic monomers incorporated into the lipid membranes of vesicles can be initiated, similarly to the polymerization of reactive lipids, by UV irradiation or by thermal or redox chemical radical generation.

The polymerization process itself seems, however, to be rather sensitive towards the composition of the system, that is, the chemical constitution of the monomer and the lipid and the concentration of the monomers in the bilayer, which, of course, limits its possible applications. While polymerization of alkyl (meth)acrylates in dioctadecyl-ammonium chloride vesicles [42, 43] or the reaction of lipofullerenes in dipalmitoyl-phosphatidylcholine vesicles [44] clearly leads to the formation of polymer hollow spheres, in the case of styrene/divinylbenzene in dioctadecylammonium bromide vesicles an intravesicular phase separation occurs during polymerization, thus leading to the formation of so-called parachute-like structures [41, 47].

An interesting aspect of the formation of cross-linked polymer particles in vesicular dispersions arises from the fact that in contrast to linear polymers they should be able to retain their structure even after their isolation from the lipid matrix. Although the particles contract considerably after their isolation from the lipid membrane they preserve their spherical shape. Their dimensions always remain, however, directly proportional to those of the underlying vesicles. This is not too surprising, since the polymer chains are expected to be forced into a nearly two-dimensional conformation in the interior of the lipid membrane. After their liberation from the membrane the polymer chains can gain entropy by adopting a three-dimensional conformation. To do this, such spherically closed polymer shells have to shrink and the thickness of their shells increases. Up to now it has, however, not been fully clarified why the polymers retain their spherical shape (without collapsing) even in the dry state.

The extent of the observed contraction of the particles depends sensitively on the cross-linking density of the polymer network structure. The contraction increases with increasing cross-linking density, thereby showing the same scaling behavior as branched polymers upon variation of their number of branches. For the highest cross-linking densities the particles contract to about 1/10 of the original size of the templating vesicles.

In the context of possible applications it would be of great interest to have detailed information about the permeability of these polymer hollow spheres. It has to be expected that besides the chemical constitution of the polymer backbone, the mesh size, that is, the cross-linking density of the polymer network structure, also plays an important role. Only molecules that are smaller than this mesh size should be able to diffuse across the polymer shell. Molecules that are larger cannot pass the polymer membrane for geometrical reasons.

Whereas the templating vesicles directly determine the size and shape of the resulting polymer particles, the polymer scaffold can be modified rather easily using conventional chemical reactions. This allows, for example, the conversion of poly(*tert*-butyl acrylate) hollow spheres into poly(acrylic acid) hollow spheres. The resulting polyelectrolyte nanocapsules can swell as a response to changes in the pH of their environment [48]. This pH-dependent transition exerts a considerable influence on the permeability of such polyelectrolyte shells and can be used to selectively entrap and release specific substances.

Although such a vesicular polymerization represents a rather elegant approach to producing polymeric nanocontainers its possible technical application is expected to be rather limited. The reason for this lies in the low economic efficiency of this method. Apart from the often energy-consuming vesicle preparation procedure, the synthesis of one gram of pure polymer nanocapsules also requires approximately 1.5–2 times their weight of lipid, a reaction volume of about 300–400 mL of water and additionally several liters of organic solvents (for purification).

Not only vesicular structures but also micelles can be used for the controlled formation of nanocapsules. It is, for example, well known that block copolymers may assemble to polymeric micelles with diameters in the 10 to 100 nm range. An important issue in making these self-assembled systems useful for specific applications is again their capability to respond to external stimuli, which allows them to interact with their surrounding environment. Very recently the formation of three-layer micelles in water from a polystyrene-β-poly(2-vinylpyridine)-β-poly(ethylene oxide) (PS-β-P2VP-β-PEO) triblock copolymers was reported [49]. It was demonstrated that the pH-sensitivity of the P2VP shell can be used to tune the micelle size from a hydrodynamic diameter (D_h) of 75 nm at pH >5 to 135 nm at pH <5. This effect is based on the electrostatic repulsion between the charged P2VP blocks and is not completely reversible because of the formation of salt with each pH cycle. Although this feature makes the system useful for encapsulation and release of active species, and has already been used for the synthesis of well-defined gold nanoparticles, its long-term stability is still questionable.

One major aspect of block copolymers is that they can be modified so that either the interior or the exterior blocks within the micelles contain polymerizable groups. For example poly(isoprene)-β-poly(acrylic acid) (PI-PAA) diblock copolymers form micelles in aqueous solution with a PI core and a PAA shell. It has been shown that the PAA shell can be cross-linked with α,ω-diamino-poly(ethylene glycol) [50]. Similarly, a poly(isoprene)-β-poly(2-cinnamoylethyl methacrylate)-β-poly(tert-butyl acrylate) (PI-PCEMA-PTBA) triblock copolymer forms micelles with a PTBA corona, PCEMA shell and a PI core in THF-methanol mixtures [51]. In this case the micellar structure could be locked in by UV-cross-linking of the PCEMA within the micelles. Subsequently the PI cores of both the cross-linked PI-PAA and the PI-PCEMA-PTBA micelles could be degraded by ozonolysis into small fragments that could diffuse into solution and leave nanospheres with a central cavity. A schematic representation of the whole process is given in Fig. 5.8.

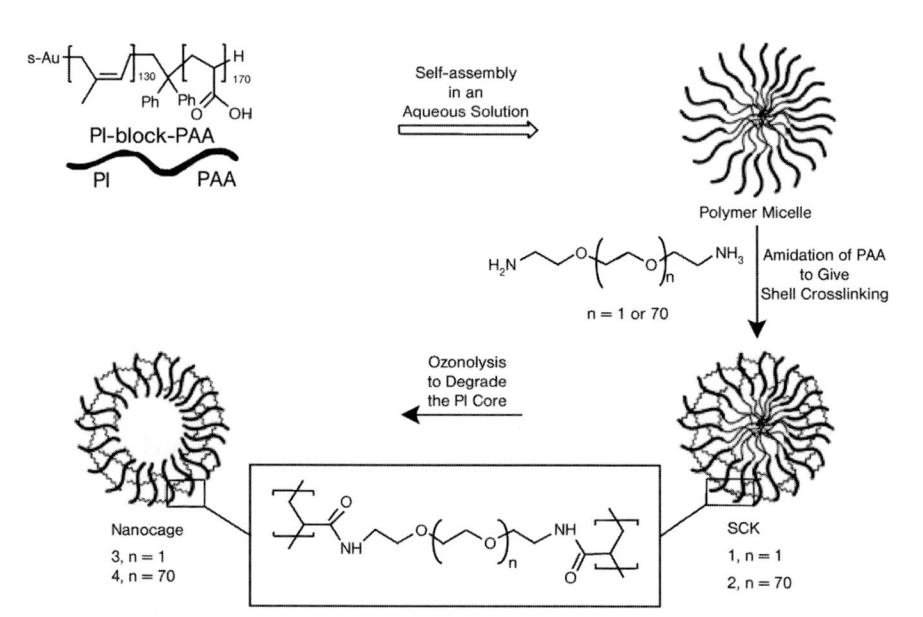

Fig. 5.8 Procedure for the preparation of nanocages from amphiphilic diblock copolymers. The shell of the final nanocages consists of cross-linked poly(acrylamide) (reproduced from H. Huang, E. E. Remsen, T. Kowalewski, K. L. Wooley, J. Am. Chem. Soc. **1999**, 121, 3805 with permission).

The potential of such systems for encapsulation of smaller molecules has been demonstrated by loading the cross-linked PCEMA-PTBA capsules with Rhodamine B [51]. The incorporation of the dye into the central cavity of the particles could readily be visualized by TEM.

The degradation of the shell cross-linked PI-PAA micelles leads to water-soluble cross-linked poly(acrylamide) hollow spheres which considerably increase the D_h of their shells after removal of the core [50]. The increase of D_h from 27 to 133 nm has been explained by the fact that the cross-linked poly(acrylamide) shells can be regarded as a hydrogel that swells when the core domain fills with water after the PI core has been removed. The diameter of the hollow-sphere products depends sensitively on both the degree of polymerization of the block copolymers originally used to form the micelle and the nature of the cross-linking diamine used to prepare the shell cross-linked micelles.

5.4
The Template Approach

Another possibility for generating polymer nanocapsules is to form a polymer shell around a preformed template particle that can subsequently be removed, thus leaving an empty polymer shell. There are several methods of realizing such a template synthesis of hollow polymer particles.

A convenient way is to exploit the well-known polyelectrolyte self-assembly at charged surfaces. This chemistry uses a series of layer-by-layer (LbL) deposition steps of oppositely charged polyelectrolytes [52]. The driving force behind the LbL method at each step of the assembly is the electrostatic attraction between the added polymer and the surface. One starts with colloidal particles carrying surface charges, for example a negative surface charge. Polyelectrolyte molecules having the opposite charge (for example polycations) are readily adsorbed due to the previously mentioned electrostatic interactions with the surface. Not all of the ionic groups of the adsorbed polyelectrolyte are consumed by the electrostatic interactions with the surface. As a result the original surface charge is usually overcompensated by the adsorbed polymer. Hence, the surface charge of the coated particle changes its sign and is now available for the adsorption of a polyelectrolyte of again opposite charge (that is a polyanion). As sketched in Fig. 5.9 such sequential deposition produces ordered polyelectrolyte multilayers.

The size and shape of the resulting core-shell particle is determined by the template colloidal particle, where the formation of particles with diameters ranging from 0.2 to 10 µm has been reported. The thickness of the layered shell is determined by the number of polyelectrolyte layers, and can be adjusted in the nanometer range. Up to now a variety of substances, such as synthetic polyelectrolytes, biopolymers, lipids, and inorganic particles have been incorporated as layer constituents to build the multilayer shells on colloidal particles [53–55].

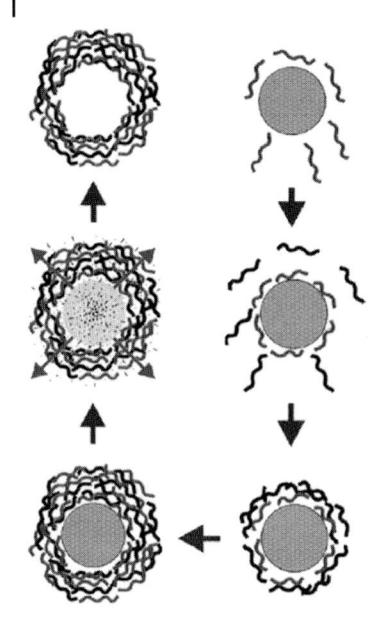

Fig. 5.9 Illustration of the procedure for preparing hollow spheres using layer-by-layer deposition of oppositely charged polyelectrolytes on colloidal particles (reproduced from E. Donath, G. B. Sukhorukov, F. Caruso, S. A. Davis, H. Möhwald, *Angew. Chem. Int. Ed.* **1998**, *37*, 2201 with permission).

However, to avoid a polyelectrolyte-induced particle flocculation one has to work at rather low particle concentrations and excess polyelectrolyte not adsorbed to the surface has to be removed carefully after each step.

After completed deposition of a predefined number of layers the colloidal core can be dissolved and removed. If human erythrocytes are used as colloidal templates they can be oxidatively decomposed in an aqueous solution of sodium hypochlorite. In the case of weakly cross-linked melamine-formaldehyde particles the core dissolves when exposed to an acidic solution of pH < 1.6. Products of both decomposition methods are expelled through the shell wall and removed by several centrifugation and washing cycles. Although the core removal takes place under harsh conditions neither the acid treatment nor the oxidative decomposition affects the polyelectrolyte layer shell as can be seen in Fig. 5.10 [52].

They clearly preserve their hollow sphere morphology and possess shape persistence. It has been shown that small dye molecules can readily permeate such layered polyelectrolyte shells, while larger sized polymers with molecular weights larger than 4000 Da obviously do not [56–58].

Nevertheless, it has to be expected that their long-term stability depends sensitively on the surrounding environment of the particles. Especially in biological fluids (like for example blood plasma), or in media of high ionic strength, which may screen the ionic interactions responsible for maintaining their integrity, the long-term stability of such polyelectrolyte shells may be rather limited.

However, this sensitivity towards the physico-chemical conditions of the surrounding medium was successfully used to get control over the permeability of multilayer shells consisting of poly(styrene sulfonate) and poly(allylamine) hydrochloride. By this control it is possible to release prior encapsulated biomacromole-

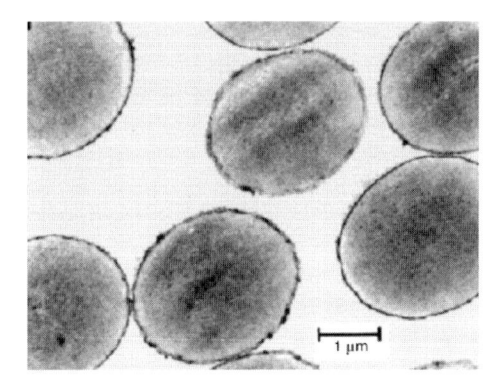

Fig. 5.10 Transmission electron micrograph of polyelectrolyte hollow spheres. The shell of the particle consists of nine layers (poly(styrene sulfonate)poly(allylamine hydrochloride))$_4$/poly(styrene sulfonate) (reproduced from E. Donath, G. B. Sukhorukov, F. Caruso, S. A. Davis, H. Möhwald, *Angew. Chem. Int. Ed.* **1998**, *37*, 2201 with permission).

cules or fluorescent probes from these capsules in a controlled manner or to use them as templates for the selective crystallization of various dyes [59–61].

Another interesting point is that the capsules could be filled with oils by sequential exchange of the solvent. These oil-filled capsules could be dispersed in water due to their amphiphilic nature, thus leading to a stable, surfactant-free oil-in-water emulsion [62, 63].

Functionalized polystyrene latex particles carrying surface charges are also suitable substrates for the polyelectrolyte self-assembly technique. In one case inorganic particles were incorporated into the adsorbed shells by a sequential adsorption of nanometer-sized SiO_2 particles with negative surface charge and cationic poly(diallyldimethylammonium chloride) (PDAD-MAC) [64]. Layers with a thickness ranging from tens to hundreds of nanometers could be prepared by this procedure. Removing the polystyrene core leaves SiO_2/PDADMAC nanocomposite shells and, after calcination, even pure SiO_2 hollow spheres. Both the composite and the purely inorganic capsules can be expected to show interesting physical properties such as enhanced mechanical stability or exceptional permeability behavior.

Similarly poly(N-vinylpyrrolidone)-stabilized polystyrene latex particles were coated with thin overlayers of poly(aniline) to produce electrically conductive core-shell particles [65]. In contrast to the LbL deposition method, where preformed polymers are adsorbed to a surface, here the polymer is formed in situ by oxidative coupling of monomers at the particle surface. The poly(aniline) layers had, however, a rather non-uniform and inhomogeneous morphology. Therefore extraction of the polystyrene core left poly(aniline) shells that displayed a so-called "broken egg-shell" morphology, that is, they were largely disrupted.

Gold nanoparticles have also successfully been used as templates for nucleation and growth of surrounding poly(pyrrole) and poly(N-methylpyrrole) shells [66]. Etching the gold leaves structurally intact hollow polymer particles with a shell thickness governed by the polymerization time. This formation of polymer-coated colloids involves trapping and aligning the particles in the pores of membranes by vacuum filtration and a subsequently performed polymerization of a conducting polymer inside the pores (Fig. 5.11).

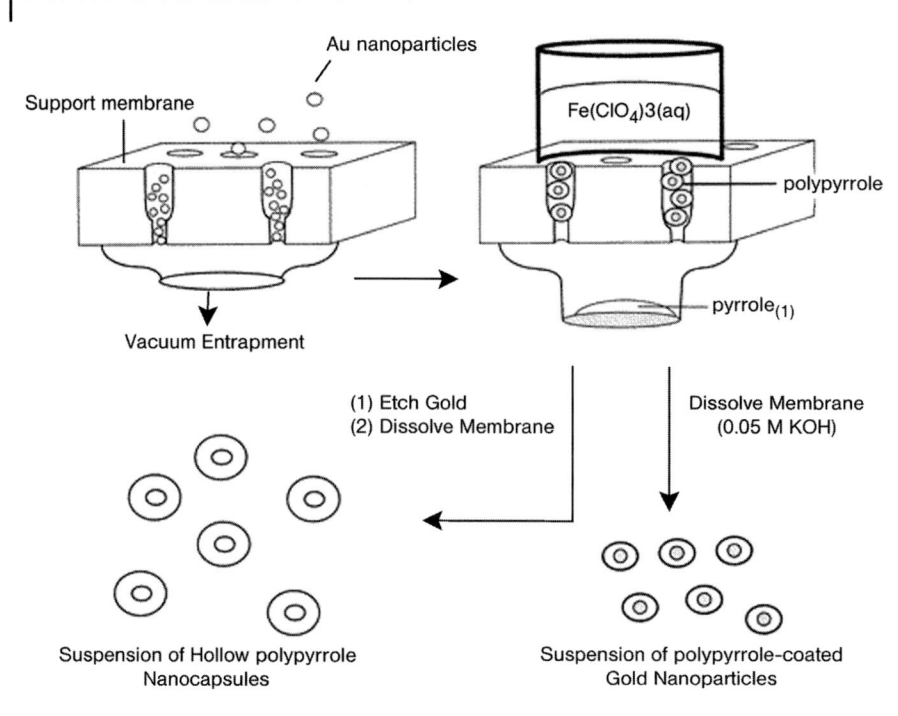

Fig. 5.11 Schematic diagram of the membrane-based method for synthesizing gold-core/polymer-shell nanoparticles. The particles are first trapped and aligned in the membrane pores by vacuum filtration and subsequently coated with poly(pyrrole), which occurs via polymerization of the monomer vapor when it diffuses into the membrane and interacts with the initiator $(Fe(ClO_4)_3)$. The membrane is then dissolved, leaving behind nanoparticle composites. The gold can also be etched first and the membrane then dissolved, resulting in hollow poly(pyrrole) nanocapsules (reproduced from M. Marinakos, J. P. Novak, L. C. Brouseau III, A. B. House, E. M. Edeki, J. C. Feldhaus, D. L. Feldheim, *J. Am. Chem. Soc.* **1999**, *121*, 8518 with permission).

In a first step the gold nanoparticles are filtered into a porous Al_2O_3 support membrane with a pore size of 200 nm. After an initiator ($FeClO_4$) has been poured into the top of the membrane, monomer (pyrrole or *N*-methylpyrrole) is placed underneath the membrane. Upon diffusion of the monomer vapor into the membrane it contacts the initiator to form polymer, preferentially occurring on the surface of the gold particles. The resulting polymer-encapsulated nanoparticles can be isolated by dissolution of the membrane in basic solution. The shell thickness is dependent on the polymerization time and can be controlled in a range from 5 to 100 nm. In order to obtain hollow particles the gold was etched, for example with a $KCN-K_3[Fe(CN)_6]$ solution, prior to dissolution of the membrane material. In this context it was shown that the gold particles are not only useful as a template material. For example, ligands attached to the gold surface prior to polymerization remained trapped inside the hollow particles after gold etching, which demonstrates their potential as protective shells for sensitive compounds like enzymes.

Obviously the ions of the gold etchant were able to diffuse across the polymer shell, the permeability of which could even be tuned by the oxidation state of the polymer. However, even rather small organic guest molecules like Rhodamin B were not released again from the interior of the particles, even over a period of three weeks. That means that there is only very restricted access to substances enclosed in the interior of the shells, and in addition the particle type appears to be limited to those that fit in membrane supports used, which both limit their applications. Nevertheless, the approach shows promise for coating of various templates like metals and biomacromolecules.

5.5
The Emulsion/Suspension Polymerization Approach

This section describes the formation of nanocontainers by several suspension and emulsion polymerization techniques. Although in most cases these methods have been shown to lead to particles with diameters of several micrometers, recent developments demonstrate that nanometer-sized polymer hollow spheres are also accessible.

One example was described by Okubo et al. [67, 68] where the penetration and release behavior of various solvents respectively into or from the interior of monodisperse cross-linked poly(styrene)/poly(divinylbenzene) composite particles could be investigated. The hollow structure of the particles is in this case achieved by seeded polymerization utilizing the so-called dynamic swelling method. For example the polymerization of divinylbenzene in toluene/divinylbenzene swollen polystyrene latex particles or in poly(styrene) containing toluene droplets leads to the formation of hollow poly(divinylbenzene) (PDVB) particles. This is a result of a microphase separation due to the limited compatibility of the chemical different polymers in solution, which leads to the formation of a PDVB shell around a toluene-polystyrene core. After evaporation of the toluene a cavity remains in the centers of such particles.

Another rather convenient method leading to hollow polymer particles proceeds via emulsion polymerization. Usually a two-stage process via seeded latex particles with physical or time separation between the two steps is applied. In a first step the core particles are synthesized by a conventional emulsion polymerization. Then in a second step a different monomer or monomer mixture is added and a cross-linked shell is polymerized around the previously formed core particle. Although the preparation of such core-shell particles seems to be quite simple in theory, it is rather difficult in practice. This holds particularly if one is interested in well-defined and homogeneous particle morphologies, which are a basic requirement for the preparation of hollow polymer particles. It has been demonstrated that in this context both thermodynamic and kinetic factors are of crucial importance [69, 70]. The direct influence of many process parameters controlling

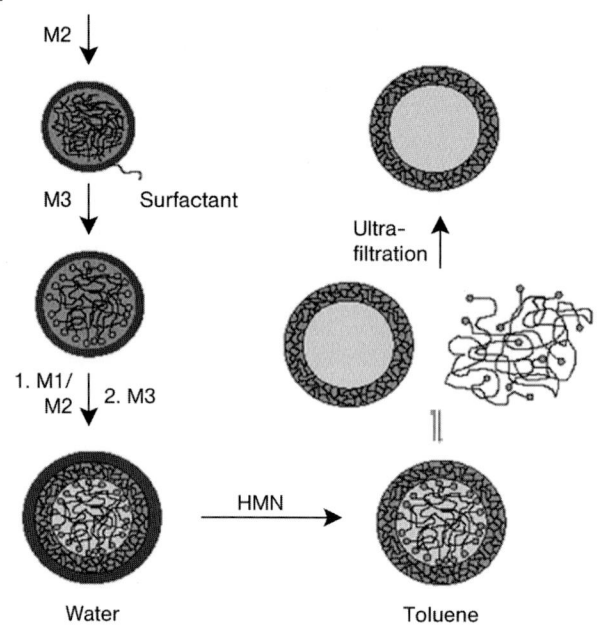

Fig. 5.12 Preparation of organosilicon nanocapsules. M1: MeSi(OMe)$_3$, M2: Me$_2$Si(OMe)$_2$, M3: Me$_3$SiOMe, HMN: hexamethyldisilazane (reproduced from O. Emmerich, N. Hugenberg, M. Schmidt, S. S. Sheikov, F. Baumann, B. Deubzer, J. Weiss, J. Ebenhoch, *Adv. Mater.* **1999**, *11*, 1299 with permission).

the particle morphology, such as surface polarity, mode of monomer addition, role of surfactant and chain transfer agent, effect of polymer cross-linking and initiator, has already been demonstrated [71–77]. Additionally, to end up with a hollow polymer sphere one has finally to remove the core-forming material. Since core and shell are, however, frequently chemically rather similar this is another critical step of the preparation procedure. Usually rather aggressive reaction conditions such as prolonged alkali and acid treatment at high temperatures are required to degrade the particle core [78, 79]. Although using these methods forms clearly hollow particles, the question remains to what extent the polymer shells survive intact under these conditions.

A rather elegant approach to core removal under very mild conditions has recently been demonstrated. The authors report the synthesis and characterization of nanometer-sized hollow organosilicon particles [80]. The synthesis of these particles followed a two-step procedure similar to that described above. The core of the particles was formed by a rather low molecular weight polydimethylsiloxane (PDMS) around which a cross-linked organosilicon shell was formed in a second step. The PDMS core-forming material from the interior of the particles could be removed quantitatively by ultrafiltration. The remaining organosilicon nanoboxes possess typical diameters of 50 nm and a shell thickness of about 6 nm (Fig. 5.12).

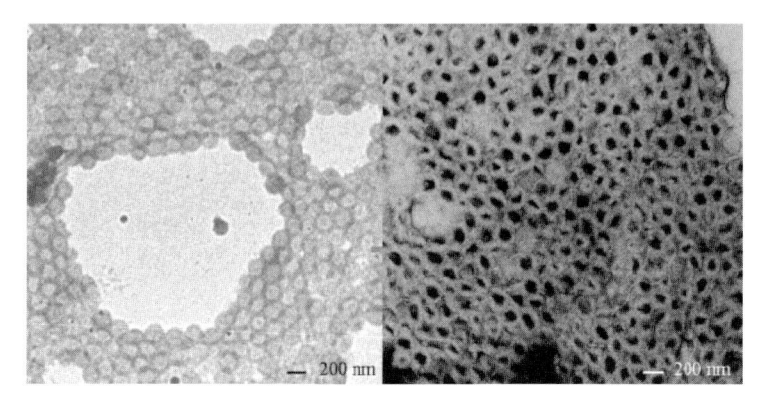

Fig. 5.13 Transmission electron micrograph of nanocapsules (left) before and (right) after loading with pyrene (reproduced from M. Sauer, D. Streich, W. Meier, *Adv. Mater.* **2001**, *13*, 1649 with permission).

Interestingly, they could be refilled with PDMS chains with a molecular weight of about 6000 Da, that is rather large molecules, which reflects an obviously rather high porosity of the polymer shells. Hence, typical low molecular weight substances are expected to be released very fast from such particles. Nevertheless, these organosilicon capsules represent a very promising system for applications in various areas. The formation of nanocontainers with a stimuli-sensitive permeability could be achieved by a slight variation of this approach. Here, a poly(ethyl-hexylmethacrylate) (PEtHMA) core was covered by a cross-linked poly(*tert*-butylacrylate) shell. Because of their low molecular weight, that is, below 10000 Da, the single PEtHMA chains are able to diffuse across the particles shell and can be removed by ultrafiltration. The isolated poly(*tert*-butylacrylate) spheres could successfully be used for encapsulation of dye molecules (Fig. 5.13) and converted into stimuli-responsive polyelectrolyte nanocapsules by selective saponification of the *tert*-butyl ester groups [81].

The preparation of nanometer-sized hollow polymer particles as a one-step emulsion process was recently described by applying a miniemulsion technique [82]. Such miniemulsions are typically formed by subjecting an oil/water/surfactant/co-surfactant system to a high shear field created by devices such as an ultrasonifier and a microfluidizer. The droplets thus formed generally range in size from 50 to 500 nm (in contrast to microemulsions where the droplet sizes vary usually from 10 to 100 nm). When monomer is used as the oil-phase, free-radical polymerization can be carried out subsequently after the phase has been broken by high shear into droplets. If one uses for the synthesis an oil and a monomer in such a way that the two components are miscible in the monomeric state, but immiscible as soon as polymerization takes place, phase separation occurs and results in an oily core surrounded by a polymer shell. Fig. 5.14 shows an image of nanocontainers that encapsulated a liquid during their formation [83].

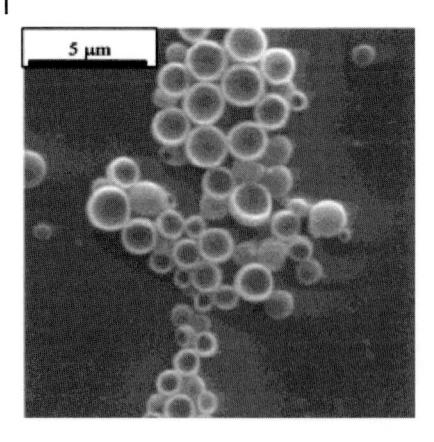

Fig. 5.14 Encapsulation of liquid materials by the miniemulsion process to form nanocapsules (reproduced from K. Landfester, *Adv. Mater.* **2001**, *13*, 765 with permission).

As already mentioned earlier, one crucial step in emulsion polymerization is to obtain reproducibility of the particles' structure and homogeneity. In conventional emulsion polymerization this is mainly controlled by the concentration of surfactant and initiator. Although it was expected that in miniemulsions particle formation would be independent of these parameters, this has not proven to be the case [84]. However, the use of miniemulsions is of increasing interest and shows a great potential for the design of new materials, since it is not restricted to a single (radical) polymerization procedure and allows a lower-cost reaction process (for example, by comparison to microemulsions less surfactant has to be employed).

5.6
The Dendrimer Approach

Dendrimers are highly branched cascade molecules that emanate from a central core through a stepwise repetitive reaction sequence [85–88]. The question of the existence of a central cavity within dendrimers, which would make them nanocapsules, has also been debated frequently. Indeed, molecules have been encapsulated in noncovalent fashion within dendrimers, but this does not mean that dendrimers have a permanent and rigid cavity.

However, by design such a molecule consists of three topologically different regions: a small initiator core of low density and multiple branching units, the density of which increases with increasing separation from the core, thus eventually leading to a rather densely packed shell. Hence, at some stage in the synthesis of such a dendrimer the space available for construction of the next generation is not sufficient to accommodate all of the atoms required for complete conversion. That means, dendrimers that have internal cavities with a dense outer shell may be synthesized by controlling the chemistry of the last step. This has been demonstrated by the preparation of the fifth-generation poly(propylene imine dendrimer) [89] shown in Fig. 5.15.

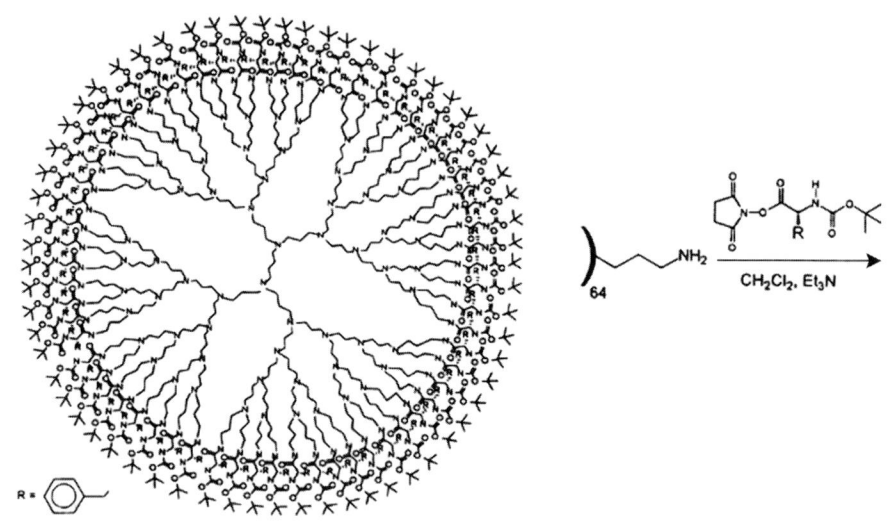

Fig. 5.15 A dendritic box capable of encapsulating small guest molecules during construction (reproduced from J. F. G. A. Jansen, D. A. F. J. van Boxtel, E. M. M. de Brabander-van den Berg, E. W. Meijer, *J. Am. Chem. Soc.* **1995**, *117*, 4417 with permission).

Because of their dense outer shell these molecules could be regarded as dendritic boxes, which were capable of retaining guest molecules trapped during synthesis. Subsequent guest diffusion out of the box was slow, since the dendrimer shell is close-packed owing to the bulky H-bonded surface groups. Guest molecules could diffuse out of the boxes if the *tert*-butyl groups were removed, but only if they were sufficiently small. Thus Rose Bengal remained in the containers while *p*-nitrobenzoic acid leaked out [89].

Closely related to such dendritic boxes are amphiphilic dendrimers or hyperbranched molecules consisting of a hydrophobic (hydrophilic) core and a hydrophilic (hydrophobic) shell, so called unimolecular micelles [90–95]. In contrast to classical micelles, which are thermodynamic aggregates of amphiphilic molecules and therefore dynamic assemblies of small molecules, these unimolecular micelles are static and retain their cohesion regardless of concentration. Owing to their amphiphilic nature these systems are able to solubilize selectively guest molecules within their core domain. For example, dendritic molecules with a hydrophobic interior and an oligoethylene glycol periphery have been used to entrap hydrophobic drugs, such as indomethacin (Fig. 5.16) [96]. For an overview of dendrimer application in diagnostics and therapy see [97].

A different type of encapsulation, involving the formation of metal nanoparticles within dendrimers, has been used to prepare inorganic-organic composite structures that are useful in catalytic applications. The formation of Cu-nanoparticles could be realized by the use of poly(amidoamine) starburst dendrimers. Owing to the permeability of the outer shell to small molecules and ions, Cu^{2+} ions

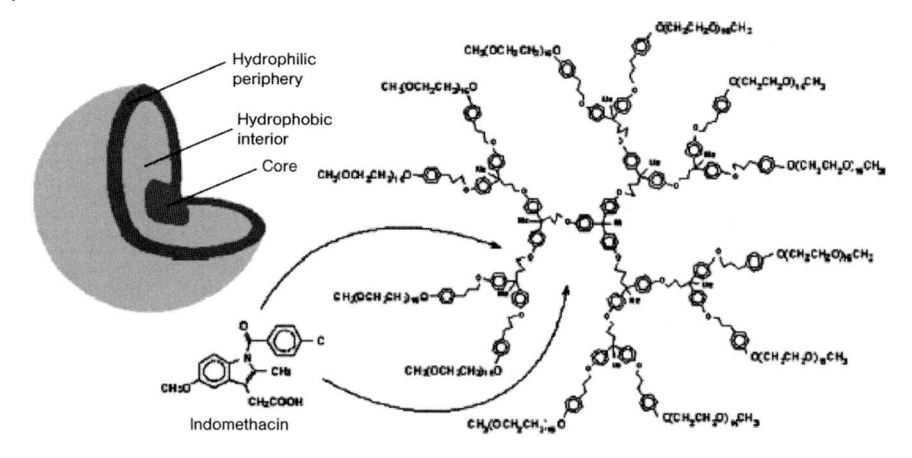

Fig. 5.16 Dendritic "unimolecular micelle" used for the slow release of indomethacin (reproduced from J. M. Fréchet, *Proceedings of the National Academic of Sciences USA* **2002**, *99*, 4782 with permission).

could diffuse into the interior of the dendritic boxes, where they were converted into Cu-nanoparticles upon reduction. Because, compared to the single ions, these had rather large dimensions with diameters of ca. 2 nm, the Cu-nanoparticles were too bulky to leak out of the dendritic cavity [98].

A different dendritic nanoreactor consists of a nonpolar aliphatic periphery and a polar inner functionality that can be used to catalyze the E_1 elimination of a solution of tertiary alkyl halides in a nonpolar solvent. The alkyl halide has some polarity and becomes concentrated within the polar dendrimer, where it is converted into an alkene. As a result of its low polarity and the existence of a gradient of polarity between the dendrimer interior and its exterior, the alkene product is readily expelled from the dendrimer back into the nonpolar solvent [99].

Although dendrimers show interesting and unique properties they are, however, generally not really hollow particles, because their core covalently links the dendritic wedges of the molecule. It is obvious that this core is of crucial importance for the integrity of the whole molecule. Hence, removing the core requires another connection between the outer zones of the dendrimer. Indeed, applying similar concepts as shown in the approaches of Wooley [50] and Liu [51], it is also possible to produce truly hollow structures from dendrimers. This has been demonstrated using a polyether dendrimer with a trimesic acid ester core. This polymer contains three cleavable ester bonds at its core and robust ether bonds throughout the rest of the molecule. As shown schematically in Fig. 5.17 [100], the hollow particles were formed by selective cross-linking of homoallyl ether groups at their periphery and subsequent degradation of the core region by hydrolysis. An interesting possibility offered by this method is that the remaining functional groups in the interior of the container system could serve as eno-receptors available for molecular recognition. This approach allows a high control over the

1a : X = H,H; Y = O
1b : X = O; Y = H,H

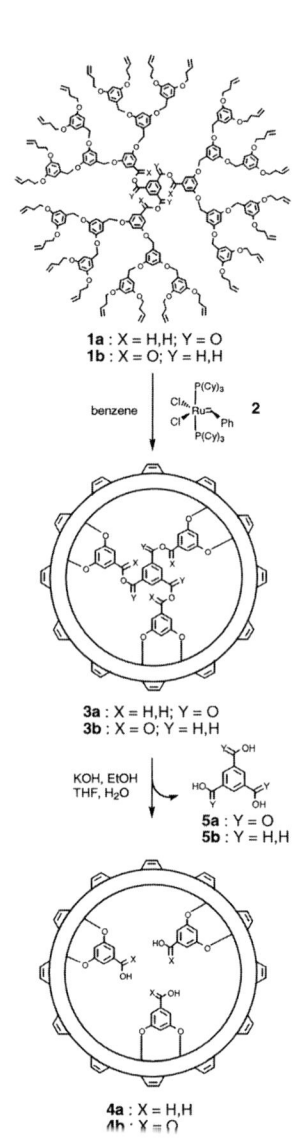

benzene | Cl⁄⁄⁄Ru=Ph with P(Cy)₃ ligands **2**

3a : X = H,H; Y = O
3b : X = O; Y = H,H

KOH, EtOH
THF, H₂O

5a : Y = O
5b : Y = H,H

Fig. 5.17 Preparation of a cored dendrimer (reproduced from M. S. Wendland, S. C. Zimmermann, *J. Am. Chem. Soc.* **1999**, *121*, 1389 with permission).

4a : X = H,H
4b : X = O

size and geometry of the formed nanocapsules. However, the preparation of the particles requires a rather costly and tedious procedure, which clearly presents a limiting factor for possible applications. Nevertheless, the quest for practical applications for dendrimers is becoming increasingly intense, but still limited to highly specific areas.

5.7 References

1 M. Rosoff, *Vesicles*, Marcel Dekker, New York, **1996**.
2 D. D. Lasic, *Liposomes from Physics to Applications*, Elsevier, Amsterdam, **1993**.
3 C. Niemeyer, *Angew. Chem. Int. Ed.* **2001**, *40*, 4128.
4 S. Mann, *Biomimetic Materials Chemistry*, VCH, New York, **1996**.
5 S. Hyde, *The Language of Shape*, Elsevier, Amsterdam, **1997**.
6 S. Mann, J. Webb, R. J. P. Williams, *Biomineralization: Chemical and Biological Perspectives*, VCH, Weinheim, **1989**.
7 S. Mann, *Angew. Chemie* **2000**, *112*, 3532.
8 N. D. Chasteen, P. M. Harrison, *J. Struct. Biol.* **1999**, *126*, 182.
9 F. C. Meldrum, V. J. Wade, D. L. Nimmo, B. R. Heywood, S. Mann, *Nature* **1991**, *349*, 684.
10 W. Shenton, T. Douglas, M. Young, G. Stubbs, S. Mann, *Adv. Mater.* **1999**, *11*, 253.
11 T. Douglas, M. Young, *Nature* **1998**, *393*, 152.
12 T. Douglas, M. Young, *Adv. Mater.* **1999**, *11*, 679.
13 H. Ringsdorf, B. Schlarb, J. Venzmer, *Angew. Chemie* **1988**, *100*, 117.
14 D. F. O'Brien, B. Armitage, A. Benedicto, D. E. Bennett, H. G. Lamparski, Y. S. Lee, W. Srisri, T. H. Sisson, *Acc. Chem. Res.* **1998**, *31*, 861.
15 B. Alberts, D. Bray, J. Lewis, M. Raff, K. Roberts, J. D. Watson, *Molecular Biology of the Cell*, Garland, New York, *1983*.
16 M. K. Pratten, J. B. Lloyd, G. Hörpel, H. Ringsdorf, *Makromol. Chem.* **1985**, *186*, 725.
17 L. W. Seymour, K. Kataoka, A. V. Kabanov, *Self-Assembling Complexes for Gene Delivery: From Laboratory to Clinical Trial*, Wiley, Chichester, **1998**.
18 P. Lemieux, S. V. Vinogradov, S. L. Gebhard, N. Guerin, G. Paradis, H. K. Nguyen, B. Ochietti, Y. G. Suzdaltseva, E. V. Bartakova, T. K. Bronich, Y. St-Pierre, V. Y. Alakhov, A. V. Kabanov, *J. Drug Target* **2000**, *8*, 91.
19 A. V. Kabanov, V. A. Kabanov, *Adv. Drug Delivery Rev.* **1998**, *30*, 49.
20 C. Nardin, T. Hirt, J. Leukel, W. Meier, *Langmuir* **2000**, *16*, 1035.
21 L. Zhang, A. Eisenberg, *Science* **1995**, *268*, 1728.
22 K. Yu, A. Eisenberg, *Macromolecules* **1998**, *31*, 3509.
23 M. Regenbrecht, S. Akari, S. Förster, H. Möhwald, *J. Phys. Chem. B* **1999**, *103*, 6669.
24 H. Shen, A. Eisenberg, *J. Phys. Chem. B* **1999**, *103*, 9473.
25 M. Maskos, J. R. Harris, *Macromol. Rapid Commun* **2001**, *22*, 271.
26 B. M. Discher, Y. Y. Won, D. S, Ege, J. C. M. Lee, F. S. Bates, D. E. Discher, D. A. Hammer, *Science* **1999**, *284*, 1143.
27 N. Hadjichristidis, *Block Copolymers: Synthetic Strategies, Physical Properties, Applications*, Wiley, New York, **2002**.
28 J. Ding, G. Liu, *J. Phys. Chem. B* **1998**, *102*, 6107.
29 C. Nardin, W. Meier, *Chimica* **2001**, *55*, 142.
30 W. Meier, C. Nardin, M. Winterhalter, *Angew. Chem. Int. Ed.* **2000**, *39*, 4599.
31 C. Nardin, J. Widmer, M. Winterhalter, W. Meier, *Eur. Phys. J. E* **2001**, *4*, 403.
32 C. Nardin, S. Thoeni, J. Widmer, M. Winterhalter, W. Meier, *Chem. Commun.* **2000**, 1433.
33 A. Graff, M. Sauer, P. van Gelder, W. Meier, *PNAS* **2002**, *99*, 5064.
34 M. Sauer, T. Haefele, A. Graff, C. Nardin, W. Meier, *Chem. Commun.* 2001, 2452.
35 H. Kukula, H. Schlaad, M. Antonietti, S. Förster, *J. Am. Chem. Soc.* **2002**, *124*, 1658.
36 F. Checot, S. Lecommandoux, Y. Gnanou, H. A. Klok, *Angew. Chem. Int. Ed.* **2002**, *41*, 1339.
37 J. Murtagh, J. K. Thomas, *Faraday Discuss. Chem. Soc.* **1986**, *81*, 127.
38 J. Kurja, R. J. M. Nolte, I. A. Maxwell, A. L. German, *Polymer* **1993**, *34*, 2045.
39 N. Poulain, E. Nakache, A. Pina, G. Levesque, *J. Polym. Sci. A: Polymer Chem.* **1996**, *34*, 729.

40 J.D. MORGAN, C.A. JOHNSON, E.W. KALER, *Langmuir* **1997**, *13*, 6447.
41 M. JUNG, D. HUBERT, P.H.H. BOMANS, P.M. FREDERIK, J. MEULDIJK, A. VAN HERK, H. FISCHER, A.L. GERMAN, *Langmuir* **1997**, *13*, 6877.
42 J. HOTZ, W. MEIER, *Langmuir* **1998**, *14*, 1031.
43 J. HOTZ, W. MEIER, *Adv. Mater.* **1998**, *10*, 1387.
44 M. HETZER, H. CLAUSEN-SCHAUMANN, S. BAYERL, T.M. BAYERL, X. CAMPS, O. VOSTROWSKY, A. HIRSCH, *Angew. Chem. Int. Ed.* **1999**, *38*, 1962.
45 E. BRÜCKNER, H. REHAGE, *Prog. Colloid Polymer Sci.* **1988**, *109*, 21.
46 T.J. MCINTOSH, S.A. SIMON, R.C. MACDONALD, *Biochim. Biophys. Acta* **1980**, *597*, 445.
47 M. JUNG, D.H.W. HUBERT, P.H.H. BOMANS, P.M. FREDERIK, A. VAN HERK, A.L. GERMAN, *Adv. Mater.*, **2000**, *12*, 210.
48 M. SAUER, W. MEIER, *Chem. Commun.* **2001**, 55.
49 J.F. GOHY, N. WILLET, S. VARSHNEY, J.X. ZHANG, R. JEROME, *Angew. Chem. Int. Ed.* **2001**, *40*, 3214.
50 H. HUANG, E.E. REMSEN, T. KOWALEWSKI, K.L. WOOLEY, *J. Am. Chem. Soc.* **1999**, *121*, 3805.
51 S. STEWART, G.J. LIU, *Chem. Mater.* **1999**, *11*, 1048.
52 E. DONATH, G.B. SUKHORUKOV, F. CARUSO, S.A. DAVIS, H. MÖHWALD, *Angew. Chem. Int. Ed.* **1998**, *37*, 2201.
53 G. DECHER, *Science* **1997**, *227*, 1232.
54 P. BERTRAND, A. JONAS, A. LASCHEWSKY, R. LEGRAS, *Macromol. Rapid Commun.* **2000**, *21*, 319.
55 F. CARUSO, *Adv. Mater.* **2001**, *13*, 11.
56 P. RILLING, T. WALTER, R. POMMERSHEIM, W. VOGT, *J. Membr. Sci.* **1997**, *129*, 283.
57 G.B. SHUKHORUKOV, M. BRUMEN, E. DONATH, H. MÖHWALD, *J. Phys. Chem.* **1999**, *103*, 6434.
58 G.B. SUKHORUKOV, E. DONATH, S. MOYA, A.S. SUSHA, A. VOIGT, J. HARTMANN, H. MÖHWALD, *J. Microencapsulation* **2000**, *17*, 177.
59 F. CARUSO, D. TRAU, H. MÖHWALD, R. RENNEBERG, *Langmuir* **2000**, *16*, 1485.
60 G.B. SUKHORUKOV, A.A. ANTIPOV, A. VOIGT, E. DONATH, H. MÖHWALD, *Macromol. Rapid Commun.* **2001**, *22*, 44.
61 G.B. SUKHORUKOV, L. DÄHNE, J. HARTMANN, E. DONATH, H. MÖHWALD, *Adv. Mater.* **2000**, *12*, 112.
62 E. DONATH, G.B. SUKHORUKOV, H. MÖHWALD, *Nachr. Chem. Tech. Lab.* **1999**, *47*, 400.
63 S. MOYA, G.B. SUKHORUKOV, M. AUCH, E. DONATH, H. MÖHWALD, *J. Coll. Int. Sci.* **1999**, *216*, 297.
64 F. CARUSO, R.A. CARUSO, H. MÖHWALD, *Science* **1998**, *282*, 1111.
65 C. BARTHELET, S.P. ARMES, S.F. LASCELLE, S.Y. LUK, H.M.E. STANLEY, *Langmuir* **1988**, *14*, 2032.
66 M. MARINAKOS, J.P. NOVAK, L.C. BROUSEAU III, A.B. HOUSE, E.M. EDEKI, J.C. FELDHAUS, D.L. FELDHEIM, *J. Am. Chem. Soc.* **1999**, *121*, 8518.
67 M. OKUBO, Y. KONISHI, H. MINAMI, *Colloid Polymer Sci.* **1988**, *276*, 638.
68 M. OKUBO, H. MOINAMI, *Colloid Polymer Sci.* **1996**, *274*, 433.
69 D. SUNDBERG, A.P. CASASSA, J. PANTAZOPOULOS, M.R. MUSCATO, *J. Appl. Polymer Sci.* **1990**, *41*, 1429.
70 Y.C. CHEN, V.N. DIMONIE, M.S. EL-AASER, *Macromolecules* **1991**, *24*, 3779.
71 V.L. DIMONIE, M.S. EL-AASSER, J.W. VANDERHOFF, *Polymer Mater. Sci. Eng.* **1988**, *58*, 821.
72 I. CHO, K.W.J. LEE, *J. Appl. Polymer Sci.* **1985**, *30*, 1903.
73 M. OKUBO, A. YAMADA, T. MATSUMOTO, *J. Polymer Sci., Polymer Chem.* **1980**, *18*, 3219.
74 S. LEE, A. RUDIN, *Polymer Latexes*, ACS Symposium Series 492, American Chemical Society, Washington DC, **1992**.
75 M.P. MERKEL, V.I. DIMONIE, M.S. EL-AASSER, J.W. VANDERHOFF, *J. Polymer Sci., Polymer Chem.* **1987**, *25*, 1755.
76 J.E. JÖNSSON, H. HASSANDER, L.H. JANSSON, B. TÖRNEL, *Macromolecules* **1994**, *27*, 1932.
77 D.C. SUNDBERG, A.P. CASSASA, J. PANTAZOPOULOS, M.R. MUSCATO, B. KRONBERG, J. BERG, *J. Appl. Polymer Sci.* **1990**, *41*, 1425.

78 X.Z. Kong, C.Y. Kan, H.H. Li, D.Q. Yu, Q. Juan, *Polymer Adv. Technol.* **1997**, *8*, 627.

79 T. Dobashi, F. Yeh, Q. Ying, K. Ichikawa, B. Chu, *Langmuir* **1995**, *11*, 4278.

80 O. Emmerich, N. Hugenberg, M. Schmidt, S.S. Sheikov, F. Baumann, B. Deubzer, J. Weiss, J. Ebenhoch, *Adv. Mater.* **1999**, *11*, 1299.

81 M. Sauer, D. Streich, W. Meier, *Adv. Mater.* **2001**, *13*, 1649.

82 F. Tiarks, K. Landfester, M. Antonietti, *Langmuir* **2001**, *17*, 908.

83 K. Landfester, *Adv. Mater.* **2001**, *13*, 765.

84 P.A. Lovell, M.S. El-Aasser, *Emulsion Polymerization and Emulsion Polymers*, Wiley, New York, **1997**.

85 D.A. Tomalia, H. Baker, J. Dewald, J.M. Hall, G. Kallos, R. Martin, J. Ryder, *Polymer J.* **1985**, *17*, 117.

86 G.R. Newkome, Z. Yao, G.R. Baker, V.K. Gupta, *J. Org. Chem.* **1985**, *50*, 2003.

87 G.R. Newkome, C.N. Moorefield, F. Vögtle, *Dendritic Molecules: Concepts, Synthesis, Perspectives*, VCH, Weinheim, **1986**.

88 J.M. Fréchet, D.A. Tomalia, *Dendrimers and other Dendritic Molecules*, Wiley, Winchester, **2001**.

89 J.F.G.A. Jansen, D.A.F.J. van Boxtel, E.M.M. de Brabander-van den Berg, E.W. Meijer, *J. Am. Chem. Soc.* **1995**, *117*, 4417.

90 A. Sunder, M. Krämer, R. Hasselmann, R. Mülhaupt, H. Frey, *Angew. Chemie* **1999**, *111*, 3758.

91 C.J. Hawker, K.L. Wooley, J.M.J. Fréchet, *J. Chem. Soc. Perkin. Trans.* **1993**, *1*, 1287.

92 G.R. Newkome, N. Moorefield, G.R. Baker, M.J. Saunders, S.H. Grossmann, *Angew. Chem. Int. Ed. Engl.* **1991**, *30*, 1178.

93 S. Mattei, P. Seiler, F. Diederich, V. Gramlich, *Helv. Chim. Acta* **1995**, *78*, 1904.

94 S. Stevelmanns, J.C.M. van Hest, J.F.G.A. Jansen, D.A.F.J. van Boxtel, E.M.M. de Brabander-van den Berg, E.W. Meijer, *J. Am. Chem. Soc.* **1996**, *118*, 7398.

95 M. Liu, K. Kono, J.M.J. Fréchet, *J. Controlled Release* **2000**, *65*, 121.

96 J.M. Fréchet, *Proc. Natl. Acad. Sci. USA* **2002**, *99*, 4782.

97 S.E. Striba, H. Frey, R. Haag, *Angew Chem. Int. Ed.* **2002**, *41*, 1329.

98 M. Zhao, L. Sun, R.M. Crooks, *J. Am. Chem. Soc.* **1998**, *120*, 4877.

99 M.E. Piotti, F. Rivera, R. Bond, C.J. Baker, J.M.J. Fréchet, *J. Am. Chem. Soc.* **1999**, *121*, 9471.

100 M.S. Wendland, S.C. Zimmermann, *J. Am. Chem. Soc.* **1999**, *121*, 1389.

6
Miniemulsions for the Convenient Synthesis of Organic and Inorganic Nanoparticles and "Single Molecule" Applications in Materials Chemistry

Katharina Landfester and Markus Antonietti

6.1
Introduction

The synthesis and application of both polymer and inorganic nanoparticles, dispersed in a non-solvent, has experienced unparalleled growth in materials science. This is because such dispersions, usually in water, allow a secure and environmentally benign application of the materials in the consumer's environment. Equally, nanoparticles offer the additional advantages of their surface and their size, bringing a new scale into materials construction, accompanied in many cases by new and profitable additional properties, such as hardness or low sintering temperatures, or an additional nanostructure implemented by the particles themselves. Thus materials can be generated employing rational structure design not only on the molecular scale, but also on the mesoscale [1].

The most widely applied nanoparticles in current technology are polymer latexes, which are usually made by emulsion polymerization [2]. It is not our intention to dismiss emulsion polymerization, since it has created incredible wealth and many applications, but in the present context one has to be aware that a very restricted set of reactions can be performed in this way. Strictly speaking, emulsion polymerization is suitable for radical homopolymerization of a small set of barely water-soluble monomers, but it is of limited value in radical copolymerization, let alone in all other polymer reactions or the synthesis of inorganic nanoparticles. The reason for this is the polymerization mechanism, because the polymer particles are the product of a kinetically controlled growth and all the monomer has to be transported by diffusion through the water phase. The kinetics dictate, even for radical copolymerization, serious disadvantages such as lack of homogeneity, and restrictions in the accessible composition range have to be accepted.

There is a variety of other techniques to generate polymer and inorganic nanoparticle dispersions, such as reactions from microemulsions or inverse micelles, which will be discussed in more detail below. All of them are, however, subject to serious disadvantages such as excessive use of surfactant, insufficient colloidal stability, or costly procedures so that although they find niche applications, these cannot really be generalized.

Colloids and Colloid Assemblies. Edited by Frank Caruso
Copyright © 2004 Wiley-VCH Verlag GmbH & Co. KGaA, Weinheim
ISBN: 3-527-30660-9

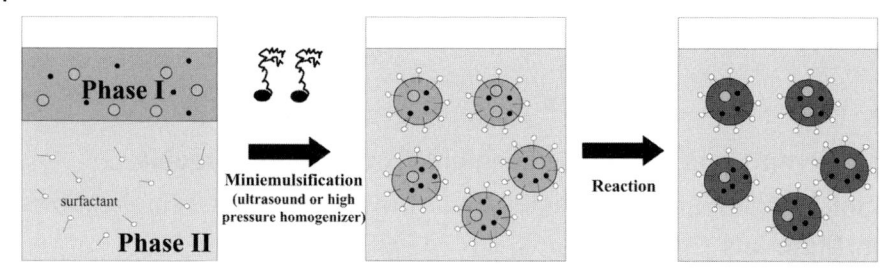

Fig. 6.1 The principle of miniemulsion synthesis of nanoparticles.

It is therefore a long-standing vision or concept to generate small, homogeneous, and stable droplets of monomer(s) or other precursors which are then transferred by (as many as possible) reactions to the final nanoparticle, keeping their particular identity without serious exchange kinetics involved. Here, reactions should proceed in a highly parallel fashion, i.e. the synthesis is performed in 10^{18}–10^{20} nanocompartments per liter as in a hypothetical bulk state, where the continuous phase can still serve for transport of reactants, side products, and heat. This is the so-called "nanoreactor" concept [3], since every droplet behaves as an independent reaction vessel without being seriously disturbed by all the other events. Within the concept of "nanoreactors", both thermodynamic aspects and shear history, but not mass transport, control the design and inner structure of the nanoparticles.

This situation is realized to a great extent in miniemulsion polymerization, as schematically illustrated in Fig. 6.1.

Although there are already a few overviews of the role of miniemulsion in radical polymerization [4–6], it is the intention of this chapter to set out the rapidly developing understanding and experimental progress that have occurred, so as to allow an actual assessment of the status and the possibilities of the field. Miniemulsions indeed are not restricted to radical polymerization in water, but open the way to new systems via a simple, cheap and reliable liquid/liquid technology both in direct (aqueous solvent) and inverse (organic or hydrocarbon solvent) situations. The main focus is laid on a detailed description of the working principles of miniemulsions, and a summary of the current approaches to the synthesis of both polymer and metal and ceramic nanoparticles. We also will show that the principle is highly favorable for the mutual encapsulation of nanoparticles, thus generating organic/inorganic nanocomposite particles with high stability and processibility.

6.2
Miniemulsions

6.2.1
Emulsion Stability Against Ostwald Ripening, Collisions, and Coalescence

A stable emulsion of very small droplets is for historical reasons called a miniemulsion (as proposed by Chou et al. [7]). To create such a state, the droplets must be stabilized both against molecular diffusion degradation (Ostwald ripening – a monomolecular process or τ_1 mechanism) and against coalescence by collisions (a bimolecular process or τ_2 mechanism).

When an oil-in-water emulsion is created by mechanical agitation of a heterogeneous fluid containing surfactants, a distribution of droplet sizes results. Even when the surfactant provides sufficient colloidal stability of droplets against collision, the fate of the droplet distribution is determined by their different droplet or Laplace pressure, which increases with decreasing droplet size, resulting in a net mass flux by molecular diffusion of the dispersed phase through the continuous phase. If the droplets are not stabilized against diffusional degradation, small ones will disappear, increasing the average droplet size. The idea that unstable droplets of aerosols or fog can be stabilized by the presence of a non-volatile third component was first formulated by Köhler in 1922 [8]. A thermodynamic description of this phenomenon was later given by La Mer et al. in 1952 [9], still for the aerosol case. These old papers are usually forgotten, and it is to the credit of Reiss and Koper that they reviewed this original literature in a valuable article describing different thermodynamic scenarios for the evolution of nanodroplets in equilibrium with a gas phase [10].

In 1962, it was proposed that unstable emulsions (the liquid/liquid situation) may also be stabilized with respect to Ostwald ripening by the addition of small amounts of a third component which should be exclusively located in the dispersed phase [11]. This stabilization effect has been theoretically described by Webster and Cates [12]. Emulsion evolution is driven by the competition between the osmotic pressure of the trapped species and the Laplace pressure within the droplets. Davis et al. [13] suggested that the added material reduces the total vapor pressure as defined by Raoult's law. In order for the system to reach equilibrium, the constituting oil will leave the small droplets and pass to larger droplets. This loss will cause an increase in the mole fraction of the ultrahydrophobe in the small droplets and a decrease in the large droplets. Thus the small droplets will have an osmotically reduced vapor pressure with respect to the larger droplets, which will continue until pressure equilibrium is obtained. In no case, this mechanism prevents the disappearance of particles.

Modification with an osmotically active agent is the key to the production of stable emulsions and miniemulsions. The increased stability is technologically used in other fields such as anesthetic/analgesic emulsions, the stability of which is provided by a white mineral or other ripening inhibitor [14]. Since the rate of Ostwald ripening depends on the size, the polydispersity, and the solubility of the dispersed phase in the continuous phase, a hydrophobic oil dispersed as small droplets with a low polydispersity already shows slow net mass exchange. But by adding an "ultrahydrophobe", the stability can still be significantly increased by the counteracting osmotic pressure. This was shown for fluorocarbon emulsions which were based on perfluorodecaline droplets stabilized by lecithin. By adding to the fluorocarbon a still less water-soluble species, e.g. perfluorodimorphinopropane, the droplets' stability was increased and could be introduced as stable blood substitutes [15, 16].

Besides the molecular diffusion of the dispersed phase, destabilization of emulsions can also occur by collision and coalescence processes. The handling of this problem is the standard question in colloid science and is usually solved by addition of appropriate surfactants, which provide either electrostatic or steric stabilization to the droplets.

6.2.2
Techniques of Miniemulsion Preparation and Homogenization

Mechanical emulsification starts with a premix of the fluid phases containing surface active agents and further additives. The emulsification includes two mechanistic steps: First, deformation and disruption of droplets, which increase the specific surface area of the emulsion, and secondly, the stabilization of these newly formed interfaces by surfactants.

The energy transferred by mechanical homogenization or by the application of an ultra-turrax is usually not sufficient to obtain small and homogeneously distributed droplets [17]. A much higher energy is required, significantly higher than the difference in surface energy $\gamma \Delta A$ (where γ is the surface or interfacial tension and ΔA the area of the newly formed interface), since the viscous resistance during agitation absorbs most of the energy and creates heat [18, 19]. Present day high-force dispersion uses ultrasonication, especially for the homogenization of small quantities, although rotor-stator dispersers with special rotor geometries or high-pressure homogenizers are preferred for the emulsification of larger quantities. Different machines are commercially available for emulsification.

Using a high-pressure homogenizer with an orifice valve [20] it was shown that the time droplets spend in the laminar flow is long enough for a large number of disruption steps to take place. During the deformation and break-up of a single droplet hardly any surfactant molecules adsorb at the newly forming interface, because the adsorption time is longer than the disruption step. This is why a special mechanical design is necessary to ensure either highly turbulent flow after disruption or sufficient residual time in the elongational flow to allow surfactant adsorption at the newly formed droplets. Then, the total disruption process can be facilitated by surfactant.

Power ultrasound emulsification was first reported in 1927 [21]. There are several possible mechanisms of droplet formation and disruption under the influence of longitudinal density waves, but cavitation is the mechanism generally regarded as crucial under practical conditions [22, 23]. Imploding cavitation bubbles

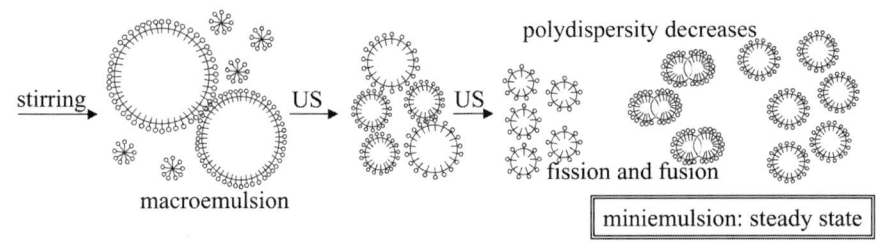

Fig. 6.2 Scheme for the formation of a steady-state miniemulsion by ultrasound.

cause intense shock waves in the surrounding liquid and the formation of liquid jets of high velocity with enormous elongational fields [24]. At constant energy density, the droplet size decreases when stabilizers are added, whereas the viscosity of the oil in w/o emulsions has no effect [25].

In precursor miniemulsions for nanoparticle synthesis, the monomer droplets change quite rapidly in size throughout sonication and reach a pseudo-steady state. Once this state is reached, the size of the monomer droplet is no longer a function of the amount of applied mechanical energy, assuming a required minimum is used. In the early stages of homogenization, the polydispersity of the droplets is still quite high, but by constant fusion and fission processes, the polydispersity decreases (see Fig. 6.2) [26].

The process of homogenization can be followed by different methods, e.g. by turbidity and by surface tension measurements. With increasing sonication time the droplet size decreases and therefore the entire oil/water interface increases. Since a constant amount of surfactant has now to be distributed at a larger interface, the interfacial tension as well as the surface tension at the air/dispersion interface increases. In the example given in Fig. 6.3, the surface tension reaches a value close to 60 mN m^{-1} indicating that the coverage of the droplets is indeed very low (calculated as 10% of a dense surfactant layer). This limiting value, however, depends on the size of the droplets [26]. The surface tension measurement is sensitive to the total oil/water interface, but it cannot distinguish between poly-

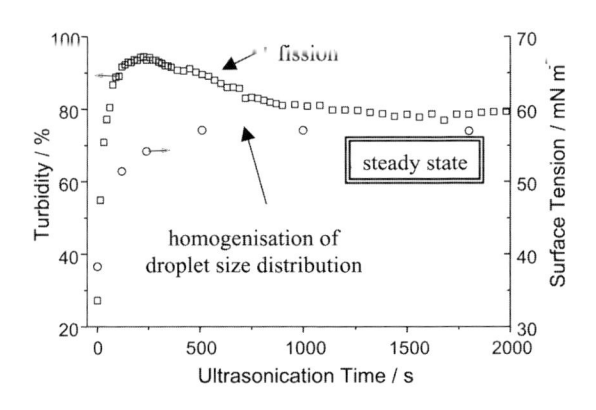

Fig. 6.3 Homogenization process followed by surface tension and turbidity measurements up to the steady state.
(□ turbidity, ○ surface tension)

disperse and monodisperse systems as long as the interfacial area is the same. Supplementing turbidity measurements (included in Fig. 6.3), which are sensitive to the size and size distribution of the droplets, equilibrate later than the surface tension measurements, indicating the complexity of the underlying equilibration process.

All experimental details really indicate that the droplet size and size distribution are controlled by a Fokker-Planck-type dynamic rate equilibrium of droplet fusion and fission processes, i.e. primary droplets directly after fission are much smaller but colloidally unstable, whereas larger droplets have a higher probability of being broken up. This also means that miniemulsions come to the minimal particle size possible under the applied conditions (surfactant load, volume fraction, temperature, salinity, ...). The resulting nanodroplets are at the critical borderline between stability and instability, which is why miniemulsions have been called "critically stabilized" [26, 27]. Practically speaking, this means that miniemulsions potentially make use of the surfactant in the most efficient way possible.

Another consequence of this dynamic rate equilibrium is that the final droplet size depends on the amount of the dispersed phase. Typically, the particles become smaller with decreasing amount of the dispersed phase when droplet collision is rate determining. For monomer miniemulsions with solid contents between 5 and 25% and constant monomer/surfactant ratio (1.7 rel.% sodium dodecylsulfate SDS), a variation from 67 nm (5 wt.% solid content) to 88 nm (25 wt.% solid content) has been reported [28].

6.2.3
Influence of the Surfactant

Colloidal stability is usually controlled by the type and amount of surfactant employed. This is why in miniemulsions, the fusion–fission rate equilibrium during sonication, and therefore the size of the droplets directly after primary equilibration, depends on the amount of surfactant. For the model surfactant sodium dodecylsulfate (SDS) and styrene at 20 wt.% dispersed phase, the minimum droplet size covers a range from 180 nm (0.3 rel.% SDS relative to styrene) down to 32 nm (50 rel.% SDS) (Fig. 6.4). Here, it is anticipated that rapidly polymerized latexes are typical of the parental miniemulsion. As compared to emulsion and microemulsion polymerization, those particles are – with respect to the amount of surfactant used – very small, comparable to the best emulsion polymerization recipes. A latex with a particle size of 32 nm is already translucent and very close to the size obtained in a microemulsion polymerization process with no hydrophobe, but a four-fold amount of a SDS/alcohol mixture [29].

At the same time, the surface tension of the resulting miniemulsions also systematically depends on particle size, as shown in Fig. 6.4b. The miniemulsion-based polystyrene latexes exceeding 100 nm size have a surface tension close to that of pure water. This is because the bare particle surface is so large that adsorption equilibrium ensures a very low concentration of free surfactants in the continuous phase. Smaller droplets with their higher surface coverage have a higher

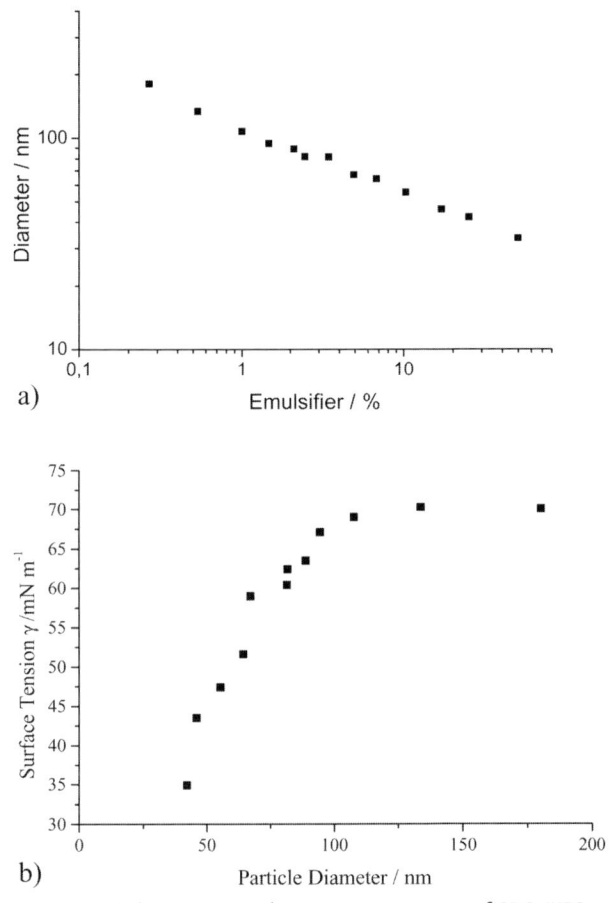

Fig. 6.4 **a** Polystyrene particle size versus amount of SDS (KPS as initiator); **b** surface tension of the latexes in dependence of size.

equilibrium concentration of free surfactant, but the concentration usually stays well below the cmc (critical micelle concentration) value. In other words: in miniemulsions, there are no free micelles present. This is very important for the chemical reactivity and the mechanisms of particle synthesis in such systems.

The surface tension is also coupled with the surfactant coverage of the particle surface. From the stoichiometry of the dispersion and the particle size, one can calculate the average stabilized oil-water surface area per surfactant molecule, A_{surf}. In the presented data set [26], it is found that the whole range from a dense surfactant monolayer (A_{surf} about 0.4 nm^2) for small particles to very incompletely covered latex particles (A_{surf} about 5 nm^2) is obtained. This reflects the fact that, at comparable volume fractions, smaller particles have a higher particle number density, a shorter average surface-to-surface distance, a higher relative mobility, and lower potential barriers, and therefore rely on denser surfactant layers to be-

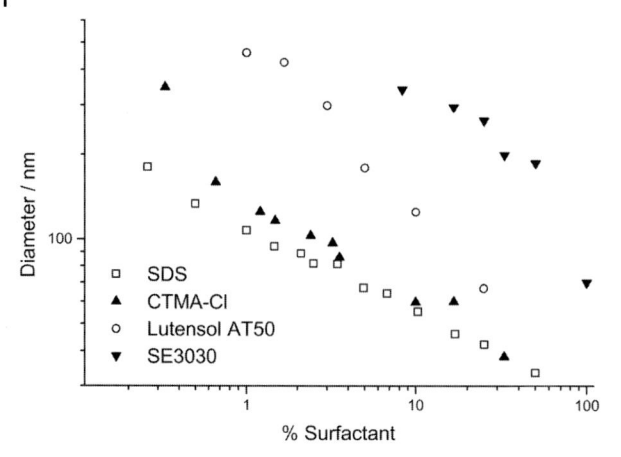

Fig. 6.5 Variation of the particle size by changing the relative amount and type of surfactant in a styrene miniemulsion at a volume fraction of the dispersed phase of 0.2.

come colloidally stable. It has to be stated that this is also true for the final nanoparticle dispersions after the polymerization step.

Beside the anionic SDS, cationic surfactants such as octadecyl pyridinium bromide or cetyl trimethylammonium chloride are also used for the preparation of miniemulsions [7, 30–32], essentially following the behavior described for SDS. A new class of cationic surfactants with sulfonium headgroups was also effectively employed for the synthesis of miniemulsion polymers [33]. Nonionic surfactants or poly(ethylene oxide) derivatives rely on larger amounts of surfactants, but also result in larger, very well-defined latexes [32, 34]. Particle sizes between 135 and 280 nm were realized.

Besides standard surfactants, polymeric emulsion stabilizers can also be employed (an emulsion stabilizer does not necessarily show pronounced surface activity and might not dissolve in either the oil or the water phase). Lim and Chen used poly(methyl methacrylate-β-(diethylamino)ethyl methacrylate) diblock copolymer as surfactant where particles with sizes between about 150 and 400 nm were produced [35]. Stable vinyl acetate miniemulsions were made employing nonionic polyvinyl alcohol (PVOH) as surfactant [36]. The favorable use of an amphiphilic statistical copolymer composed of octadecyl acrylate and acrylic or methacrylic acid groups has been demonstrated by Baskar et al. [37]. In this case, the polymer acts as a surfactant and a hydrophobe at the same time, which is technically important.

In all these cases, the nanodroplet size can be varied over a wide range by variation of the relative amount of surfactant to monomer [38]. This is summarized in Fig. 6.5, which also shows that different size ranges are accessible depending on the type of surfactant. Latexes synthesized with sodium dodecyl sulfate and cetyl trimethylammonium bromide show about the same size-concentration curve, i.e.

the efficiency of the surfactant and the size-dependent surface coverage are very similar, independent of the sign of charge.

The efficiency of the two nonionic surfactants (where SE3030 is a polymeric stabilizer) is lower, and the whole size-concentration curve is shifted to larger sizes. This is attributed to the lower efficiency of steric stabilization as compared to electrostatic stabilization and the fact that steric stabilization relies on a more dense surfactant packing to become efficient. As can be derived from surface tension measurements of the latexes and surfactant titrations, the nonionic particles are nevertheless also incompletely covered by surfactant molecules and show surface tensions well above the values of saturated surfactant layers.

6.2.4
Influence of the (Ultra)Hydrophobe

It was stated above that Ostwald ripening can efficiently be counterbalanced by the addition of a hydrophobic agent to the dispersed phase, which counteracts the Laplace pressure of the droplet. It is important to choose an agent which does not readily diffuse from one droplet to another and is therefore trapped in a single droplet. The effectiveness of the hydrophobe increases with decreasing water solubility in the continuous phase, and there is a low but final hydrophobe content required in order to become operative. Using an adequate amount of hydrophobe and employing an efficient homogenization process, a steady-state miniemulsification is reached. After stopping sonication, a rather rapid and minor equilibration process has to occur where the effective chemical potential in each droplet (which can be expressed as an effective net pressure) is equilibrating. Since the particle number is fixed, it is not the average size that is influenced by this process, but the particle size distribution. It is assumed that this process is very fast and is usually not seen in the experiment. It can be calculated that the Laplace pressure within the resulting nanodroplets is still larger than the osmotic pressure created by the hydrophobe [28]. Therefore, the droplet size after steady-state miniemulsification is not given by an effective zero droplet pressure, i.e. $p_{Laplace} - \Pi_{osm} = 0$ (which indeed would represent a real thermodynamic equilibrium), but rather characterized by a state of equal pressure in all droplets. Minor statistical differences of the pressure directly after sonication are presumably rapidly equilibrated, since changing the particle size leads to adaptation of the Laplace pressure with R^{-1}, whereas the osmotic pressure varies with R^{-3}, which means that a locally stable situation is always reached.

After that initial period, the equality of droplet pressures makes such systems insensitive to net mass exchange by diffusion processes. However, the net positive character of the pressure makes the whole system sensitive to all other changes of droplet size. It is an experimental observation [26] that steady-state-homogenized miniemulsions which are critically stabilized undergo very slow droplet growth on the timescale of hundreds of hours, presumably by collisions or hydrophobe exchange. The droplets seem to grow until a zero effective pressure is reached. It is, however, possible to obtain immediate long-term colloidal stability of miniemulsions by addition of an appropriate second dose of surfactant after the dispersion step. This dose is not used to increase the particle number, but goes to the bare interface of the preformed miniemulsion droplets in order to decrease the interface tension between the oil and the water phases and thus the coupled Laplace pressure. Such post-stabilized miniemulsions do not change their droplet size on the timescale of months [26]. This leads to the conclusion that most miniemulsions described in the literature are only thermodynamically metastable, i.e. with respect to conservation of particle number they are in a local minimum of the chemical potential. This minimum, however, is deep enough to allow most chemical reactions without significant change of the particle size and structure.

Another implication of this mechanism is that the minidroplet size depends only weakly, if at all, on the amount of the hydrophobe. It is not that doubling the amount of hydrophobe decrease the radius by a factor of 2 (as expected from a zero effective pressure), it is just that the effective pressure (pressure difference) has to be the same in every droplet, a mechanism which in principle does not depend on the amount of hydrophobe. This insensitivity of the particle size to the amount of the hydrophobe is well documented in the literature [39–41]. It was found that a minimum molar ratio of the hydrophobe to the monomer of about 1:250 is required in order to build up a sufficient osmotic pressure in the droplets to exceed the influence of the initial structures made in a potential synthesis of polymer chains. This also explains the fact that a small amount of high molecular weight polymer in the droplets, e.g. polystyrene, stabilizes the miniemulsions, but can barely act as an osmotic stabilizing agent throughout a particle polymerization reaction [42, 43].

The choice of potential ultrahydrophobes is widespread: beside hexadecane and polymers as model compounds [42, 44–47], long chain alcohols [48–53], transfer agents [54, 55], hydrophobic dyes [56], water-insoluble comonomers [57–59], initiators or oxidation agents [60, 61], plasticizers, or other additives can be used. Silanes, siloxanes, isocyanates, polyester, fluorinated alkanes, and many others have been found to be very efficient in suppressing Ostwald ripening. It is evident that the less water-soluble the hydrophobe is, the more effective it is as an osmotic pressure agent. The stability of miniemulsion systems has been studied using shelf-life stability [35, 55, 62, 63] and ultracentrifugation stability studies [64] and by measuring levels of free surfactant.

Measuring the primary droplet size in miniemulsions is an important issue in the literature. The size of the *polymer particles* is easily determined by light scattering or microscopic methods since the dispersions can be diluted without changing the particles. Few attempts have been made at measuring *droplet sizes* in emulsions directly. Ugelstad et al. [44] and Azad et al. [30] stained miniemulsions with OsO_4 and used transmission electron microscopy. However, the treatment can alter the sizes. Goetz and El-Aasser [65] made some attempts to determine droplet sizes using light scattering and transmission electron microscopy. Miller et al. [66] observed by CHDF that for cetyl alcohol systems there is a period of rapid increase of average droplet size increase followed by a region in which the droplet size increases slowly. This behavior was not found in the case of hexadecane. Nevertheless, for those measurements the emulsions had to be diluted, which seriously changes the system. Even if the emulsion is diluted with monomer-saturated water [67], the size of the droplets will change slightly owing to different solubility effects. The high shear during CHDF measurements can also lead to coalescence of the droplets.

Some indirect methods for measuring the droplet size have also been used. The interfacial area and therefore the droplet size was determined by measuring the critical micelle concentration of styrene miniemulsions prepared with DMA as hydrophobe and SDS as surfactant and was compared to styrene emulsions prepared without a hydrophobe [68]. Erdem et al. determined droplet sizes on concentrated miniemulsions via soap titration which could be confirmed by CHDF measurements [69]. Droplet sizes without diluting the system can be estimated by SANS measurements [70].

6.2.5
Inverse Miniemulsions

The concept of miniemulsion stabilization is not restricted to direct miniemulsions, but can also be extended to inverse miniemulsions where the osmotic pressure is built up by an agent insoluble in the continuous oily phase, a so-called "lipophobe". Ionic compounds, simple salts, or sugars show a low solubility in organic solvents and can be used as lipophobes in water-in-oil miniemulsions [71]. Another adaptation of the process is that for the dispersion of polar monomers in non-polar dispersion media, surfactants with low HLB values are required. A number of surfactants was screened, including standard systems such as $C_{18}E_{10}$, sodium-bis(2-ethylhexyl)-sulfosuccinate (AOT), sorbitan monooleate (Span80), and the nonionic block copolymer stabilizer poly(ethylene-co-butylene)-b-poly(ethylene oxide) (PE/B-b-PEO, see Fig. 6.6). PE/B-b-PEO turned out to be most efficient because of its polymeric and sterically demanding nature, providing maximal steric stabilization which is the predominant stabilization mechanism in inverse emulsions.

Here, the extraordinarily high droplet stability against exchange can be demonstrated in a very illustrative way by the formation (or suppression) of Prussian Blue nanoparticles from inverse miniemulsions. One miniemulsion with droplets

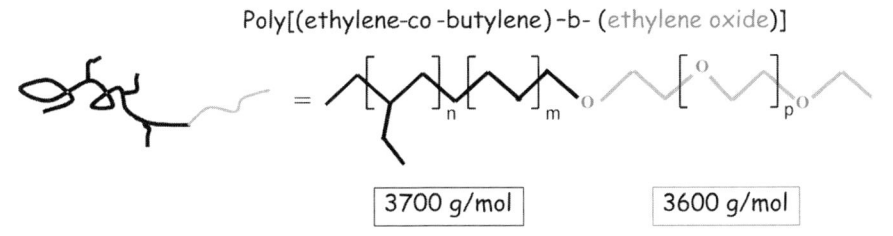

Fig. 6.6 Schematic presentation of the amphiphilic blockcopolymer efficient for the stabilization of inverse miniemulsions and salt melts.

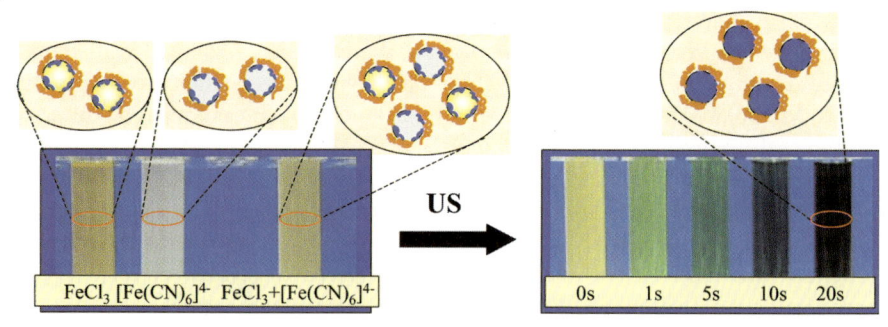

Fig. 6.7 a Mixing of 2 inverse miniemulsions, one containing FeCl$_3$ solution, the other [Fe(CN)$_6$]$^{4-}$; **b** controlled fusion and fission of the droplets during ultrasonication.

containing a FeCl$_3$ solution, and one miniemulsion containing a [Fe(CN)$_6$]$^{4-}$ solution are mixed. Fig. 6.7a shows that the mixed miniemulsion stays colorless for weeks, which indicates that the droplets with FeCl$_3$ and the droplets with [Fe(CN)$_6$]$^{4-}$ stay separated as colloidal entities on the timescale of most chemical reactions. Repetition of the same experiment with two microemulsion or micellar solutions would lead to an immediate reaction because of unblocked droplet exchange. In miniemulsions, the exchange can be stimulated by mechanical energy, such as ultrasonication used to prepare the original miniemulsions. In this case, fusion and fission processes are induced, and with increasing ultrasonication the miniemulsion indeed turns blue (see Fig. 6.7b).

Also in the inverse case, the droplet size throughout the miniemulsification process runs into an equilibrium state (steady-state miniemulsion) which is characterized by a dynamic rate equilibrium between fusion and fission of the droplets. This can be determined by turbidity measurements. As expected and in accordance with a droplet fission/droplet fusion picture, the approach to the steady state of turbidity is very similar for the systems with salt or without salt as an osmotically active agent [71]. High stability of the droplets after the high shear treatment, however, depends only on the osmotic agent. The type of lipophobe has no influence on the stability of the inverse miniemulsion. The droplet size depends, contrary to regular miniemulsions, on the amount of osmotic agent [72]. It seems that in inverse miniemulsions, the droplets experience a real zero effective pressure (the osmotic pressure counterbalances the Laplace pressure), which makes them very stable. This was speculatively attributed to the different stabilization mechanism and mutual particle potentials, which makes a pressure equilibration directly after the ultrasonication possible. This is why we believe that inverse miniemulsions are not only critically stabilized, but are genuinely stable systems.

The fusion/fission mechanism of minidroplet formation also results in the typical triangular relation between the amount of surfactant, the resulting particle size, and the surface coverage. With increasing amounts of surfactant, the particle size decreases. The smaller the particles are, the higher is the coverage of the particles by surfactant. For inverse miniemulsions, these relations depend in addition on the amount of hydrophobe. Nevertheless, for inverse miniemulsions the surfactant is also used in a very efficient way, at least as compared to inverse microemulsion [73, 74] or inverse suspension polymerization [75]. Again, the surface coverage of the inverse minidroplets with surfactant is incomplete. Interface tension values of an inverse latex in cyclohexane to water [72] translate into a surface coverage of the inverse droplets of only 30%, and prove the absence of empty inverse micelles. Again, this is important for the mechanisms of nanoparticle synthesis.

6.2.6
Preservation of Particle Identity Throughout Miniemulsion Polymerization

The idea of reactions in miniemulsions, as already stated, is to perform the particle build-up in each of the small stabilized droplets, without major secondary nucleation from the continuous phase or mass transport processes involved. Preservation of particle number and particle identity is therefore a key issue. Ugelstad et al. [51] first published results where droplets with sizes of less than 0.7 μm were nucleated leading to polystyrene latexes, and even for the rather imperfect recipes of the early days, a situation very close to a 1:1 copying of the monomer droplets to polymer particles was obtained. From the current point of view, it is possible either to polymerize a freshly prepared, steady-state miniemulsion with minimal particle size, freezing the critically stabilized state by rapid polymerization, or to make the system fully stable by adding an adequate second dose of surfactant (calculated from surface tension measurements) to saturate the particle surface and avoid the presence of free micelles. The growth of the minidroplets is then effectively suppressed, and polymerization avoids any "racing" situation of growing particles.

The preservation of particle character was tested with small angle neutron scattering (SANS) experiments [70], which were applied to characterize the droplet or particle sizes before and after the polymerization without diluting the system. It was found that the part of the scattering function describing the outer particle form does not change, whereas inner droplet structure and interparticle potentials do change. SANS was also used to check that a small amount of added polystyrene does not have any effect on the droplet size and the droplet size distribution. Neutron scattering experiments were also used to determine that the hexadecane is located homogeneously within each droplet and does not have the character of a co-surfactant.

6.2.7
Checklist for the Presence of a Miniemulsion

In some crucial cases, it might be not obvious whether the system represents a miniemulsion or not, so a short checklist summarizing the characteristics of miniemulsions is provided:

1. Steady-state dispersed miniemulsions are stable against diffusional degradation, but critically stabilized with respect to colloidal stability.

2. The interface energy between the oil and water phases in a miniemulsion is significantly larger than zero. The surface coverage of the miniemulsion droplets by surfactant molecules is incomplete.

3. The formation of a miniemulsion requires high mechanical agitation to reach a steady state given by a rate equilibrium of droplet fission and fusion.

4. The stability of miniemulsion droplets against diffusional degradation results from an osmotic pressure in the droplets which controls the solvent or monomer evaporation. The osmotic pressure is created by the addition of a substance which has an extremely low solubility in the continuous phase. This crucial prerequisite is usually not present in microemulsions, but can also be added there to increase the stability.

5. Such miniemulsions can still undergo structural changes by changing their average droplet number to end up in a situation of zero effective pressure, however only on very long timescales. This secondary growth can be suppressed by an appropriate second dose of surfactant added after homogenization.

6. Nanoparticle formation in miniemulsions occurs only by droplet nucleation.

7. During particle formation, the growth of droplets in miniemulsions can be suppressed. In miniemulsions the redistribution of starting product is balanced by a high osmotic background of the hydrophobe which makes the influence of the initial structures less important.

8. The amount of surfactant or inherent surface-stabilizing groups required to form a miniemulsion is comparatively small, e.g. for SDS between 0.25 and 25% relative to the monomer phase, which is well below the surfactant amounts required for microemulsions or reactions in lyotropic phases.

6.2.8
Main Differences from Microemulsion and Suspension Routes and Related Processes

Reactions in microemulsions starts from a thermodynamically stable, spontaneously formed state, which relies on high amounts of special surfactants or mixtures which possess an interfacial tension at the oil/water interface of close to zero. By contrast, miniemulsions are critically stabilized, require high shear to reach a steady state, and have an interfacial tension that is much larger than zero. The high amount of surfactant which is required for microemulsion preparation also leads to complete surfactant coverage of the particles. Since a reaction usually does not start in all nanodroplets simultaneously, the first structures form only in some droplets. The osmotic and elastic influence of these chains usually destabilize the fragile microemulsions and lead to an increase of the particle size, the formation of empty micelles, and secondary nucleation. The final product are often very small latexes, 5 to 50 nm in size, coexisting with a majority of empty micelles. For more details about organic polymerization microemulsions, the reader is referred to the review literature [73, 76, 77].

For the build-up of inorganic nanoparticles from microemulsions of micellar phases [78, 79], another problem occurs. Those surfactant assemblies dissolve and rebuild themselves on rather short timescales (in the millisecond region). Any chemical reaction slower than this exchange time, and therefore practically all chemical reactions, does not experience the size-quantized structure of the medium, but a quasi-continuous environment with some special diffusional restrictions. Therefore, the so-synthesized inorganic nanostructures are usually significantly bigger than the structural elements of the original complex fluid.

Suspension polymerization involves much larger monomer droplets (1 µm–1 mm) dispersed in the continuous phase. Nucleation occurs predominantly in the droplets, and each polymerizing droplet behaves as an isolated batch polymerization reactor. This principle of microreactors is similar to the idea of miniemulsions. Indeed, the main difference is the size and stability of the particles formed. Owing to the very large particle sizes involved, the coupled droplet pressures are rather low, and Ostwald ripening occurs at a much lower rate. This is also why addition of hydrophobes is not common in suspension polymerization.

Monodisperse emulsions of intermediate-sized droplets with a high stability can be obtained by shearing a crude, polydisperse emulsion at high concentrations with a well-defined low shear rate, such as is available in a Couette apparatus [80, 81]. The viscosity of the premixed emulsion and the related dissipation of mechanical energy must be sufficient if a monodisperse emulsion is to be obtained, which is also called a Bibette emulsion. It was found that the thinner the gap in the Couette apparatus is, the more monodisperse is the final emulsion. The final drop size depends on the applied shear rate, the viscoelasticity of the premixed emulsion, and the interfacial tension between the dispersed and the dispersing phase, but is usually around 1 to 10 µm. In principle, this process allows the establishment of monodisperse emulsions with larger droplet sizes, and reactions in Bibette emulsions can be performed according to the general principles predominantly leading to droplet nucleation. The use of mixing tools such as the ultra-turrax, or simple stirring, do not result in a narrow distribution since the homogeneity of shear is not sufficient.

Emulsions with a narrow droplet size distribution can also be obtained by membrane emulsification. The droplets can possess a high stability when the droplets are uniform and a hydrophobe is added; again, hexadecane has been employed for this purpose [82]. The droplet size can be varied by pressing the crude preemulsions through membranes with different pore sizes, but the membrane material and process parameters such as the membrane pressure and the viscosity of the continuous phase also influence the droplet sizes. The polymerization of such emulsions can lead to monodisperse polymer particles, as reported by Omi et al. using a particular microporous glass membrane [83, 84]. Several pore sizes from 0.5 to 18 µm were employed for the preparation of stable emulsions. The reaction mechanism is similar to that of an ordinary suspension polymerization, with the advantage that an initial narrow size distribution of the droplets is employed.

Relatively uniform biodegradable polylactide microspheres were prepared by employing the same membrane emulsification technique [85]. In this case, polylactide was dissolved in dichloromethane, and a hydrophobe was also added. This mixture was dispersed in an aqueous phase by SDS and poly(vinyl alcohol), and the solvent was evaporated.

6.3
Materials Synthesis in Miniemulsions

The use of miniemulsions in principle allows that all kind of monomers can be used for the formation of particles that are not miscible with the continuous phase. In the case of prevailing droplet nucleation, or at the start of the polymer reaction in the droplet phase, each miniemulsion droplet can indeed be treated as a small nanoreactor. This makes possible a whole variety of polymerization reactions which lead to polymer nanoparticles (on a much wider scale than classical heterophase polymerization).

6.3.1
Radical Homo- and Copolymerization of Organic Monomers

Radical polymerization of a miniemulsion is, as already stated, very similar to suspension polymerization. The monomer is miniemulsified, and a radical initiator is added. This initiator can be soluble in either oil or water.

Usually, a water-soluble initiator is used to start the polymerization from the water phase. The start from the continuous phase is similar to conventional emulsion polymerization where mainly water-soluble initiators are used. Free primary radicals are formed in the water phase. Varying the initiator concentration leads to surprisingly similar reaction profiles, as revealed by calorimetry [86], but the particle nucleation interval is longer when less initiator is present. Once all the droplets are nucleated, the monomer is steadily consumed. The average radical number per particle, $\bar{n}=0.5$, during this interval is independent of the amount of initiator, and an increase of the initiator concentration cannot result in an acceleration of the polymerization process in this interval. It was also found that the chain length of the resulting polymer is inversely proportional to the square root of the initiator concentration [86], underlining that the reaction in a miniemulsion is very direct and close to an ideal radical polymerization, i.e. the proposed nanoreactor concept is fulfilled here quite ideally.

This is typical for hydrophobic monomers, but not necessarily true in general, since an increased initiator concentration also results in an increased probability for homogeneous nucleation. In the case of more water-soluble monomers, for example MMA and vinyl chloride [87], secondary particle formation was observed.

Here, the number of new particles increases with the concentration of the initiator.

In the case of an oil-soluble initiator, the initiator is dissolved in the monomeric phase prior to miniemulsification. This is the preferred choice for monomers with either high water solubility or with an extremely low water solubility [e.g. lauryl methacrylate (LMA)] where the monomer concentration in the water phase is not high enough to create oligoradicals that can enter the droplets. However, because of the finite size of the droplets, direct radical recombination is a problem that must be faced. The ability of initiators with different water solubilities, namely lauroyl peroxide, benzoyl peroxide and azobis(isobutyronitrile) (AIBN), to stabilize monomer droplets against degradation by molecular diffusion, and their efficiency for polymerization, was investigated in [60]. Upon heating, the initiator decomposes, and it was found that sufficiently long polymer chains are formed only when radicals appear in the monomer droplets one at a time as opposed to the normal case of pair generation in which, owing to initiator decomposition, two radicals appear in the monomer droplets at the same time. Single radicals can be formed by desorption of one of the radicals formed by initiator decomposition or by entry of a radical from the aqueous phase. This makes oil-soluble initiators effective only when one or both of the formed radicals are sufficiently hydrophilic to undergo desorption. Comparing different oil-soluble initiators, the probability of nucleation is much larger for AIBN than in the cases of lauroyl and benzoyl peroxide.

As a model monomer for the radical homopolymerization of oily monomers, styrene is described in many papers. The polymerization of acrylates and methacrylates is also well known. A miniemulsion also easily allows the polymerization of the ultrahydrophobic monomer LMA [28] without any carrier materials as would be necessary in emulsion polymerization [88].

The polymerization of more hydrophilic monomers is also possible, as shown for MMA and vinyl acetate [36, 41, 89]. In the case of monomers with pronounced water solubility, nucleation in water should be efficiently suppressed in order to avoid secondary nucleation in the water phase. This can be achieved, for example, by using an oil-soluble initiator or the addition of a termination agent to the continuous phase.

Fig. 6.8 illustrates the ability to control particle size by the surfactant concentration (here SDS is employed) for different monomers in direct miniemulsions. For MMA, BA and styrene, similar size-relative surfactant concentration curves are found, whereas the hydrophobic monomer LMA leads to larger particles. This is because the larger interfacial tension between the monomer and the water phase has to be compensated by a higher coverage by surfactant molecules.

For simultaneous better control of the polymer molecular weight and structure (in addition to the controlled particle properties), living free-radical polymerization represents a promising technique. Different possible systems known from bulk polymerizations have been used in miniemulsions. The living free-radical polymerization of, for example, styrene via the miniemulsion approach allows elimination of the drawback of the bulk system where an increase in polydispersity was

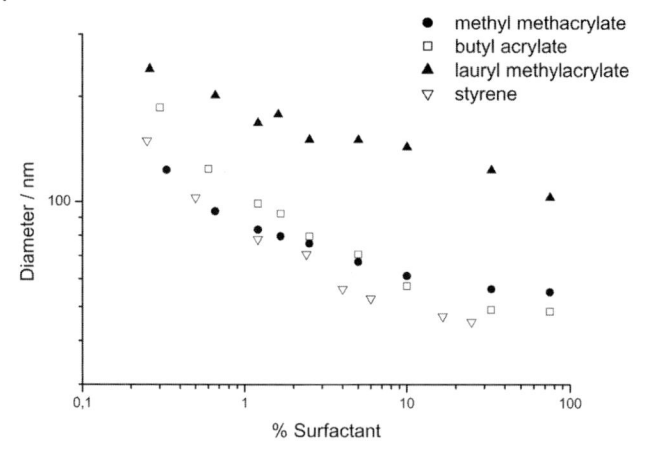

Fig. 6.8 The use of different monomers in miniemulsions: the dependence of the surfactant concentration on the particle size.

found at high conversions owing to the very high viscosity of the reaction medium [90].

Four different approaches for controlled radical polymerization have been adapted to the miniemulsion polymerization process:

1. The controlled free-radical miniemulsion polymerization of styrene was performed in aqueous dispersions using a *degenerative transfer process* with iodine exchange [91, 92]. An efficiency of 100% was reached. It has also been demonstrated that the synthesis of block copolymers consisting of polystyrene and poly(butyl acrylate) can be easily performed [93]. This allows the synthesis of well-defined polymers with predictable molar mass, narrow molar mass distribution, and complex architecture.

2. In a *stable free-radical polymerization* (SFRP) the initiated polymer chains are reversible capped by a stable radical, for example the 2,2,6,6-tetramethyl pyridin-1-oxyl radical (TEMPO). Stable polystyrene dispersions were prepared by miniemulsion polymerization with an optimized ratio and amount of surfactant, hydrophobe, nitroxide, and KPS as initiator at 135 °C [94]. TEMPO in combination with BPO (benzoyl peroxide) was used by Prodpran et al. at 125 °C [95]. At a TEMPO to BPO 3:1 molar ratio, polymers with the lowest polydispersity (1.3) were achieved. In order to decrease the reaction temperature below 100 °C, an acrylic β-phosphonylated nitroxide in combination with the $KPS/Na_2S_2O_5$ redox initiator system was used. It was demonstrated that water-soluble alkoxyamines were initially formed during the induction period and were progressively transferred into the organic phase [96].

3. *Living radical polymerizations* in miniemulsions have also be conducted using reversible addition-fragmentation chain transfer (RAFT) [97]. Miniemulsion stability was found to be a key issue, and the authors' system was only stable

with nonionic surfactants. The polydispersity was usually below 1.2. The living character is further exemplified by the formation of block copolymers. The increased polymerization rate of the compartmentalized miniemulsion system leads to an improved block copolymer purity compared to that of homogeneous systems.

4. *Reverse atom transfer radical polymerization* (ATRP) of butyl methacrylate was successfully conducted in miniemulsions using the water-soluble initiator V50 and the hydrophobic ligand 4,4'-di(5-nonyl)-4,4'-bipyridine (dNbpy) to complex the copper ions. Although the forming radical mediator Cu(II) complex had a large water-partitioning coefficient, the rapid transfer of Cu(II) between the organic and aqueous phases assured an adequate concentration of the deactivator in the organic phase. As a result, controlled polymerization was achieved [98, 99].

With respect to radical polymerization of partly crystalline polymers, polyacrylonitrile is a very interesting polymer for many engineering applications, such as fiber spinning or for housing and packaging application. Even though it is produced in a radical polymerization process as a mainly atactic polymer, it has the tendency to form crystallites: the crystallinity degree is reported to be between 28 and 34% [100]. A peculiarity of polyacrylonitrile is that it is insoluble in its monomer. This makes it very difficult to homopolymerize acrylonitrile in an emulsion polymerization process since nucleated polymer particles cannot grow further by monomer swelling.

Polymerization in a miniemulsion is a very suitable technique for avoiding this problem since each droplet acts as a nanoreactor. As a result, it is possible to obtain pure polyacrylonitrile (PAN) nanoparticles in the size range between 100 nm $< d < 180$ nm depending on the amount of surfactant [101]. Compared to a standard styrene miniemulsion, the solubility of acrylonitrile in water is rather high (7.35%). For a miniemulsion with 20 wt.% acrylonitrile, just about 70% of the monomer is located inside the droplets, whereas 30% is dissolved in the water phase. This is no restriction for a miniemulsion polymerization process, as the use of a hydrophobic initiator (V59) allows preservation of the droplets as the reaction sites (droplet nucleation). During the reaction the entire monomer in the water phase is consumed, as shown by NMR measurements, and the droplets are expected to grow by the amount of monomer consumed from the water phase.

Owing to the insolubility of the polymer in the monomer, the formed polymer precipitates and crystallizes during the polymerization within the droplets. Pure PAN latexes have a crumpled appearance where the single polymer nanocrystals remain in the final structure and are easily identified by their sharp edges and flat surfaces.

In the case of inverse systems, hydrophilic monomers such as hydroxyethyl methacrylate, acrylamide, and acrylic acid were miniemulsified in a non-polar medium, e.g. cyclohexane or hexadecane [71, 72]. Rather small and narrow distributed latexes in a size range between 50 nm $< d < 200$ nm were made with the nonionic amphiphilic block copolymers as shown in Fig. 6.6. Depending on the sys-

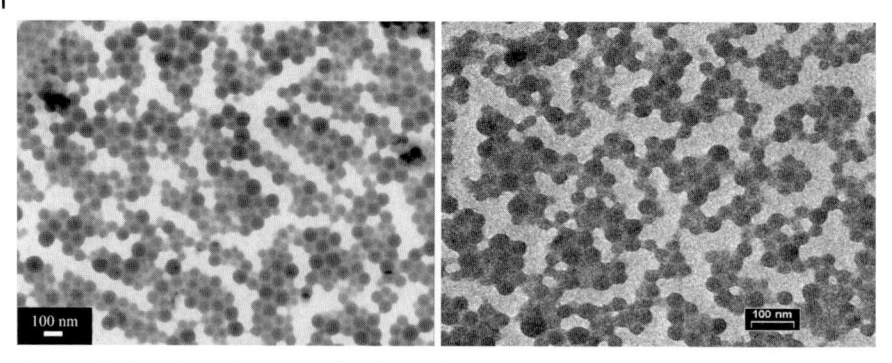

Fig. 6.9 Hydrophilic polymer particles obtained in inverse miniemulsion polymerization in Isopar: **a** polyacrylamide; **b** poly(acrylic acid).

tem, the surfactant loads can be as low as 1.5 wt.% per monomer, which is very low for an inverse heterophase polymerization reaction and clearly underlines the advantages of the miniemulsion technique.

For the moderately hydrophilic hydroxyethyl methacrylate, cyclohexane and hexadecane were chosen as the continuous phase. As initiator, PEGA200 which is soluble in the monomer phase but not in cyclohexane, was found to be applicable. Small amounts of water act as an osmotic agent. Rather small inverse latex particles in the size range between 80 nm and 160 nm and narrow size distributions are obtained. The systems are stable down to 1.6 wt.% surfactant with respect to monomer; at lower amounts the systems tend to coagulate. It is remarkable that the final dispersions are stable for a long time even at low surface coverages.

For the synthesis of acrylamide in a miniemulsion polymerization, the solid crystalline monomer had to be dissolved in water, and therefore a higher amount of water was used in the synthesis. Cyclohexane was chosen as the continuous phase. After sonication the miniemulsions show only a low stability (less than 1 h) without the addition of a strong lipophobe (1 M NaCl), which increases the stability of the miniemulsions to the timescale of several days to months. Polymerization with AIBN from the continuous phase resulted in stable polymer dispersions, as shown in Fig. 6.9. The high homogeneity and rather well-defined character of those latexes is clearly evident. Again, even surfactant loads as low as 1.8 rel.% with respect to the dispersed phase result in stable latexes.

Acrylic acid was polymerized in inverse miniemulsions together with 4 wt.% of the crosslinking agent diethylene glycol diacrylate in order to obtain homogeneous polyelectrolyte microgels, which can also be redispersed and characterized in water. In this case, again, the solubility of acrylic acid in cyclohexane is not negligible. For formulation of a stable inverse acrylic acid-containing miniemulsion, a 5 M NaOH solution was used, because the partitioning of acrylic acid was shifted to the water phase, and only 7% of the acrylic acid was found in the cyclohexane phase. The sodium salt of acrylic acid dissolved in water could also be easily be used for the preparation of miniemulsions. Starting from a critical surfactant amount of 2.5 rel.% to prevent the formed polymer particles from aggregation, stable latexes of about 100-nm diameter were produced. Increasing the surfactant amount leads to smaller particles. The surfactant efficiency or stabilized area per surfactant molecule, A_{surf}, is very high. The quality and the low dispersity of the particles prepared from the sodium salt of acrylic acid is illustrated in Fig. 6.9b. Those particles have an average size of 50 nm and a polydispersity of less than 10%.

Compared to classical inverse heterophase polymerization techniques, such as polymerization in inverse microemulsions [73] or dispersion polymerization [102, 103], polymerization of inverse miniemulsions is favored by the very efficient use of surfactant and the copying process from the droplets to the particles. This leads to a homogeneous structure and composition of the resulting particles (no kinetic effects are involved). The latter feature is especially important for homogeneous crosslinking or copolymerization in inverse heterophase polymerization.

As well as allowing radical homopolymerization, it is a strength of miniemulsions that they also allow the copolymerization of monomers with very different water solubilities, something which is impossible with classical emulsion polymerization. This is especially important for more sophisticated applications in polymer science, e.g. the controlled synthesis of nanofine latexes with controlled surface functionality or the use of macromonomers in heterophase polymerization.

Miniemulsion copolymerization of 50:50 styrene/MMA monomer mixture was carried out using hexadecane as hydrophobe [104]. Copolymerization of styrene and butyl acrylate was successfully performed out using the redox initiator system $(NH_4)_2S_2O_8/NaHSO_3$ at lower temperature [105]. Inaba et al. prepared a series of model styrene/n-butyl acrylate copolymer latexes with glass transition temperatures at room temperature. The functional monomer dimethyl m-isopropenyl benzyl isocyanate (TMI) was used as monomer/crosslinking agent for further film formation. A small amount of methacrylic acid was introduced in some formulations in order to enhance the crosslinking reaction. A redox initiation system was used to reduce premature crosslinking during the polymerization [106].

The copolymerization of monomers where one monomer acts as the hydrophobe has been reported. MMA was copolymerized with p-methyl styrene, vinyl hexanoate, or vinyl 2-ethylhexanoate. The resulting copolymer composition tended to follow the predictions of the reactivity ratios, i.e. the reaction progresses much as a bulk reaction [57]. Wu and Schork used monomer combinations with large differences in reactivity ratios and water solubility: vinyl acetate/butyl acrylate, vinyl acetate/dioctyl maleate, and vinyl acetate/n-methylol acrylamide. The miniemulsion system follows the integrated Mayo-Lewis equations more closely than the comparable emulsion system [89].

Miniemulsion polymerization allowed the synthesis of particles in which PBuA and PMMA macromonomer were copolymerized [107, 108]. The macromonomer acts as compatibilizing agent for the preparation of core/shell PBuA/PMMA particles. The degree of phase separation between the two polymers in the composite particles is affected by the amount of macromonomer used.

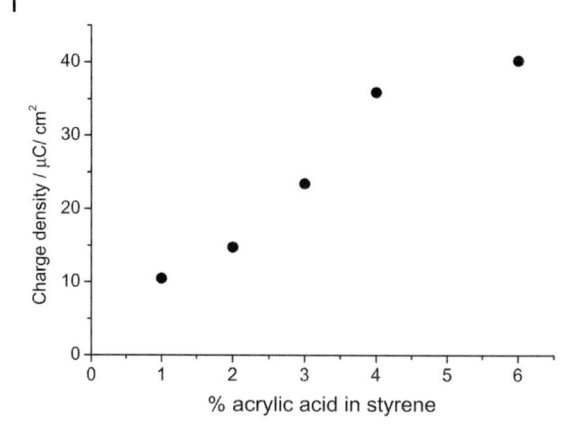

Fig. 6.10 The copolymerization of styrene with acrylic acid leads to charged particles.

Acrylic acid can be copolymerized with styrene in order to obtain functionalized particles that can be used, for example, by binding antibodies on them. In order to avoid a reaction of the acrylic acid in the aqueous phase, the oil-soluble initiator ADVN was used. Owing to the hydrophilic character of the acrylic acid, its tendency to be on the particle surface is very high. As shown in Fig. 6.10, the charge density on the surface of a dialyzed sample increases with increasing amounts of acrylic acid.

Polymerization of two monomers with different polarities but in similar ratios is a difficult task owing to solubility problems. However, using the miniemulsion process it is possible to start from very different spatial monomer distributions, resulting in very different amphiphilic copolymers in dispersion [109]. The monomer, which is insoluble in the continuous phase, is miniemulsified in order to form stable and small droplets with a low amount of surfactant. The monomer with the opposite hydrophilicity dissolves in the continuous phase (and not in the droplets). As examples, the formation of acrylamide/methyl methacrylate (AAm/MMA) and acrylamide/styrene (AAm/Sty) copolymers using the miniemulsion process has been demonstrated [109].

6.3.2
Non-radical Organic Polymerization Reactions in Miniemulsions

6.3.2.1 Polyaddition Reactions

As already indicated, the existence of stable liquid "nanoreactors" enables the miniemulsion process to be extended towards other chemical reactions. For instance, monomeric components can be mixed together, and polyaddition and polycondensation reactions performed *after* miniemulsification in the miniemulsified state.

The successful transfer of the principle of miniemulsion polymerization to polyadditions of epoxy-resins was shown in [110] where mixtures of different epoxides with varying diamines, dithiols or diols were heated to 60 °C to form the respective

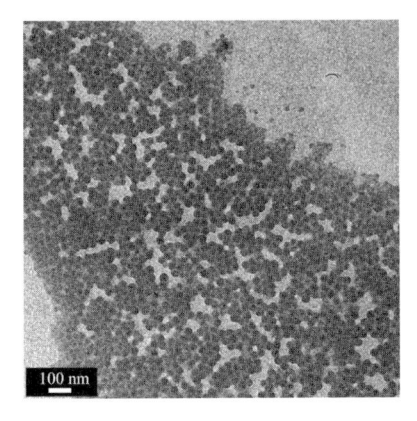

Fig. 6.11 Typical latex obtained in a polyaddition process in miniemulsion (Epikote E828 (bisphenol-A-diglycidylether) and 4,4'-diaminobibenzyl).

polymers. The requirement for the formulation of stable miniemulsions is that both components of the polyaddition reaction show a relatively low water solubility. The diepoxide Epikote E828, the triepoxide Decanol Ex-314 and the tetraepoxide Ex-411 were successfully used as epoxy components. As amino components Jeffamin D2000 [an $-NH_2$ terminated poly(propylene oxide) with $M_w = 2032$ g mol^{-1}], 4,4'-diaminobibenzyl, 1,12-diaminododecane, and 4,4'-diaminodicyclohexylmethane have been used. Jeffamine D400 with its lower molecular weight was too water soluble and could not be used. As other addition components besides amines, 1,6-hexanedithiol and Bisphenol A were used. The hydrophobic components that are required for the formulation of stable miniemulsions are usually the applied epoxides themselves, as they have very poor water solubility and provide sufficient droplet stabilization. In case of the dithiol and the bisphenol products, GPC-elution to judge the quality of the polyaddition reaction was possible, and the final polymers reveal molecular weights of about 20 000 g mol^{-1}, with a dispersity of close to 2. This means that unexpectedly ideal reaction conditions are preserved during the reaction in miniemulsion, and that the proximity of the interface to water does not really disturb the reaction. Depending on the amount of surfactant, particle sizes between ca. 800 nm (0.85% SDS) down to 36 nm (25% SDS) were made.

In general, it appears that the more hydrophobic the reaction partners the smaller the latexes. Small latex particles with a diameter down to 30 nm can be synthesized from a number of the most hydrophobic combinations. This is also about the minimum size limit which can be made by radical processes, i.e. we reach a limit given by the fundamental laws of colloidal stability. Fig. 6.11 shows as a typical example the reaction product of Epikote E828 (bisphenol-A-diglycidylether) and 4,4'-diaminobibenzyl; small particles with a relatively narrow size distribution are obtained.

Using non-stoichiometric ratios of the diepoxide and diamine components (e.g. Epikote E828/Jeffamin D2000 1:1.22) allows highly functionalized particles to be obtained, with residual primary and secondary amine groups. Such particles with very small dimensions are interesting from a technical point of view, since they can act as a polyfunctional network component (highly precondensed, amine-rich coupling component) for a number of applications, e.g. in glues or dental fillings.

Also polyurethanes latexes can be made by direct miniemulsification of a monomer mixture of diisocyanate and diol in an aqueous surfactant solution followed by heating [110]. This is somewhat special since one might expect a suppression of polymerization by side reactions between the very reactive diisocyanates and the aqueous continuous phase. However, polymer dispersions are obtained when the reactants have a low water solubility. The reaction between diisocyanate and diol is slower than the time needed for the miniemulsification step, and the side reaction of the diisocyanate with water in the dispersed state is slower than the reaction with the diol.

For the third requirement, one has to consider that the ratio of diisocyanate molecules located at the droplet/water interface to the molecules inside the droplets is comparatively small. It was speculated that after a potential reaction of the interfacial isocyanate molecules with water, the resulting more hydrophilic molecule can form a passivating layer, which slows down further reactions of the isocyanate molecules inside the droplet with the water.

6.3.2.2 Anionic and Metal-catalyzed Polymerization Reactions

For the anionic polymerization of phenyl glycidyl ether (PGE) in miniemulsion, didodecyldimethylammonium hydroxide was used as an "inisurf" which acts as a surfactant and an anionic initiator by means of its hydroxy counterion. As revealed by ^{1}H NMR and FTIR spectroscopy, genuine α,ω-dihydroxylated polyether chains were produced. The average molecular weight could be increased by varying the initiator concentration and the type and concentration of surfactants, or by adding an alcohol as a co-stabilizer. With increasing conversion, the polymer chains increased but remained small, with a critical polymerization degree of $DP_{max} = 8$.

Ethylene can be polymerized in an aqueous miniemulsion with the aid of an organo-transition metal catalyst at ethylene pressures of 10–30 bar and temperatures of 45–80 °C resulting in large particles of about 600 nm [111]. A maximum productivity of 2520 kg PE per g atom active metal was achieved, which represents about 60% of the productivity of the same catalyst when used in ethylene suspension polymerization in an organic phase.

6.3.3
Hybrid Nanoparticles by Miniemulsion Technologies

As a type of a "colloidal" generalization of the copolymerization principle, miniemulsification also enables the convenient and direct incorporation of major amounts of water-insoluble materials, such as resins or other nanoparticles, by dissolution or dispersion in the organic phase. In a following step, the physical mixture is jointly dispersed and reacted. That way, so-called hybrid or nanocomposite particles are formed. The principle of hybrid nanoparticle formation is shown in Fig. 6.12.

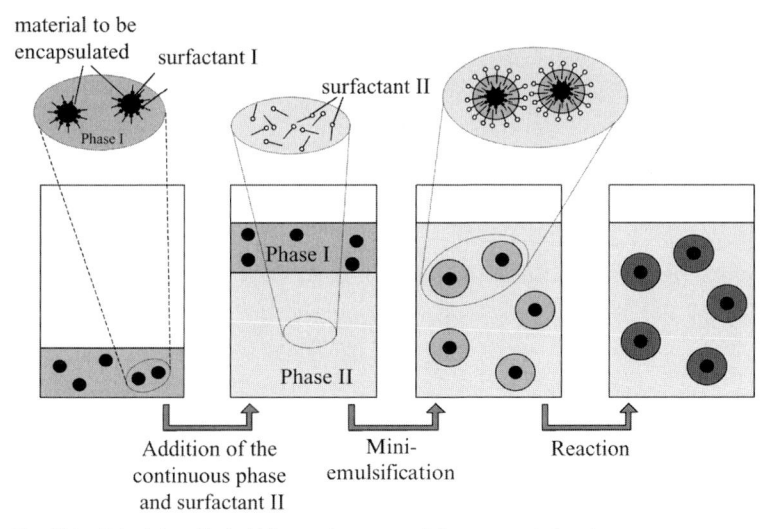

Fig. 6.12 Principle of hybrid formation or particle encapsulation by miniemulsion polymerization.

Depending on the chemistry and aggregate state of both dispersed components, the technically relevant systems can be differentiated into polymer–polymer hybrids, encapsulated nanoparticles, and nanocapsules filled with an organic liquid or solid.

6.3.3.1 Polymer–Polymer Hybrid Nanoparticles

This class of hybrids is easiest to make, since one only has to dissolve one component as a pre-polymer or a resin into the monomer of the second, which is subsequently reacted. This was done, for example, with acrylic monomers (MMA, butyl acrylate, and acrylic acid) in the presence of alkyd resin in order to produce stable polymer–polymer hybrid latex particles incorporating an alkyd resin into acrylic coating polymers [67]. Throughout the reaction, the resin simultaneously acts both as a hydrophobe and to allow the stabilization of the miniemulsion. The predominant form was poly(acrylate-g-alkyd resin) which also confirms the monomer droplet nucleation mechanism [112]. Tsavalas et al. [113] showed in a similar system that despite a high degree of crosslinking (>70%), residual double bonds were present in the polymer–polymer hybrid latex for curing reactions during film formation. The presence of unsaturated resin was found to favor chain transfer reactions at higher conversions leading to both inhibition of the reaction and grafting [113]. The in situ grafting between a polyester resin and an acrylic polymer in a nearly zero-VOC environment offers a novel, water-based, crosslinkable latex coating, incorporating properties from both water-based and solvent-based systems. Polymerizing acrylic monomers in the presence of oil-modified polyurethane leads also to a grafting onto the polyacrylics, resulting in dispersions suitable for stable water-borne latexes with good adhesion properties and fair hardness properties [63].

Oil–acrylate hybrid emulsions were formed using a fatty acid hydroperoxide as the interfacial initiator system for the miniemulsion polymerization of acrylates. Initiation at the droplet interface results in the formation of polyacrylates modified with triglycerides. These molecules act as compatibilizers between oil and the PMMA phase, resulting in more homogeneous particles. Only a small number of the available sites for crosslinking of the oils are used for initiation of the miniemulsion polymerization and therefore the auto-oxidative crosslinking properties of the oils remain unchanged [114].

6.3.3.2 Polymer–Particle Hybrids

For the encapsulation of pigments by miniemulsification, two different approaches can be used. In both cases, the pigment/polymer interface as well as the polymer/water interface have to be carefully chemically adjusted in order to obtain encapsulation as a thermodynamically favored state. The design of the interfaces is mainly dictated by the use of two surfactant systems which govern the interfacial tensions, as well as by employment of appropriate functional comonomers, initiators, or termination agents. The sum of all interface energies has to be minimized. In order to achieve the incorporation of a pigment into the latex, the size of the monomer droplets has also to be adjusted to incorporate the pigment according to its lateral dimensions.

Therefore, two steps have to be controlled for successful preparation of a particle-containing miniemulsion (see Fig. 6.12). First, the already hydrophobic or hydrophobized particulate pigment with a size of up to 100 nm has to be dispersed in the monomer phase. Hydrophilic pigments require a hydrophobic surface to be dispersed in the hydrophobic monomer phase, which is usually promoted by a surfactant system I with low HLB value. Then, this common mixture is miniemulsified in the water phase, employing a surfactant system II with high HLB which has a higher tendency to stabilize the outer interface towards water.

Erdem et al. described the encapsulation of TiO_2 particles via miniemulsions. First, TiO_2 was dispersed in the monomer using OLOA 370 (polybutene-succinimide) as stabilizer [115, 116]. The presence of TiO_2 particles within the droplets limited the droplet size, and complete encapsulation was not achieved (encapsulation extents were reported to be between 73 and 83%). Also the amount of encapsulated material was very low: a TiO_2 to styrene weight ratio of 3:97 could not be exceeded [117].

Nanoparticulate hydrophilic $CaCO_3$ was coated with a layer of stearic acid prior to dispersing the pigments into the oil phase [38]. The –COOH groups act as good linker groups to the $CaCO_3$, and the tendency of the stearic acid to go to the second polymer/water interface was found to be low. 5 wt.% of $CaCO_3$ could be completely encapsulated into polystyrene particles. It was shown that the weight limit was given by the fact that at this concentration, each polymer particle already contained one $CaCO_3$ colloid which was encapsulated in the middle of the latex (Fig. 6.13a).

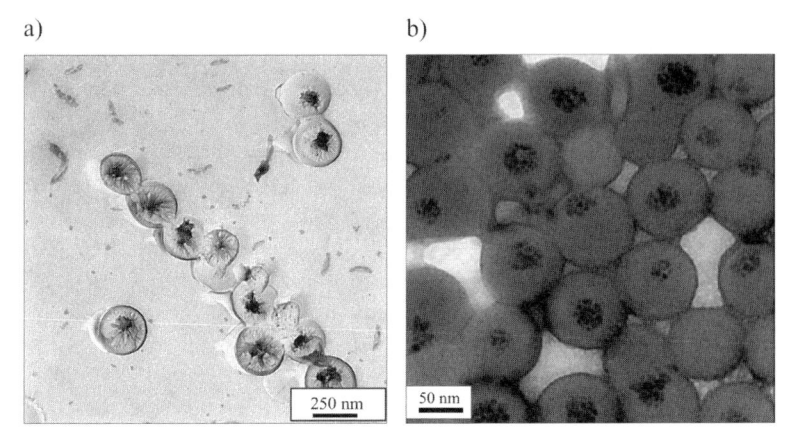

Fig. 6.13 **a** Encapsulation of one colloid per polymer particle: $CaCO_3$ nanoparticles in PS latexes; **b** encapsulation of many colloids per particle: Fe_3O_4 in PS latexes.

The encapsulation of magnetite particles into polystyrene particles was efficiently achieved by a miniemulsion process using oleoyl sarcosine acid [118] or oleic acid as the first surfactant system to handle the interface magnetite/styrene, and sodium dodecylsulfate to stabilize the interface styrene/water, thus creating a polymer-coated ferrofluid (Fig. 6.13b). Since the magnetite particles were very small (ca. 10 nm), each polymer particle was able to incorporate many inorganic nanoparticles.

Carbon black is a rather hydrophobic pigment, and encapsulation of various carbon blacks in polymer shells can be achieved by direct dispersion of the pigment powder in the monomer phase prior to emulsification [38]. Here, full encapsulation of non-agglomerated carbon particles can be provided by appropriate choice of the hydrophobe. In this case the hydrophobe not only acts as the stabilizing agent for the miniemulsion process, but also mediates to the monomer phase by partial adsorption. The main drawback of the direct dispersion of carbon black in the monomer is that the carbon is still highly agglomerated in the monomer, and a maximum of 8 wt.% carbon black has been found to be incorporated. At higher amounts, the viscosity of the carbon-in-monomer dispersion was too high and a miniemulsion could not be obtained.

To increase the amount of encapsulated carbon to up to 80 wt.%, another approach was developed [119]. Here, both monomer and carbon black were independently dispersed in water using SDS as a surfactant and later mixed in any ratio between the monomer and carbon. Then, this mixture was co-sonicated, and the controlled fission/fusion process characteristic of miniemulsification destroyed all aggregates and liquid droplets, and only hybrid particles composed of carbon black and monomer remained, presumably because this species shows the highest stability. This controlled droplet fission and heteroaggregation process can be realized by high-energy ultrasound or high-pressure homogenization, as discussed earlier.

TEM and ultracentrifuge results showed that this process results almost completely in effective encapsulation of the carbon: only rather small hybrid particles, but no free carbon or empty polymer particles, were found. Hybrid particles with high carbon contents are not spherical, but adopt the typical fractal structure of carbon clusters, coated with a thin, but homogeneous, polymer film. The thickness of the monomer film depends on the amount of monomer, and the exchange of monomer between different surface layers is – as in miniemulsion polymerization – suppressed by the presence of the added osmotic agent. Therefore, the process is best described as a polymerization in an adsorbed monomer layer created and stabilized as a miniemulsion ("ad-miniemulsion polymerization"). As the ultrahydrophobe for this process, which provided both osmotic stability of the monomer ad-layer and efficient binding to the monomer, 3 wt.% of a hydrophobic polyurethane was found to be a suitable choice.

The completeness of the surface coverage of the encapsulated products can be analyzed by BET measurements on the dialyzed and dried products. For the pure carbon black used in the experiment, the BET surface area was determined to be 250 m^2 g^{-1}. A pure PS latex with a diameter of 100 nm shows a BET surface area of about 50 m^2 g^{-1}. For the dialyzed and dried carbon/styrene hybrid made by the ad-miniemulsion polymerization process, a BET surface area of 58 m^2 g^{-1} was determined, which speaks for the close-to complete encapsulation of the carbon, blocking the inner micropores for adsorption [119]. Those systems are very interesting for high-resolution electronic printing applications.

6.3.3.3 Encapsulation of a Liquid – Direct Formation of Polymer Nanocapsules by Miniemulsion Routes

Polymerization in miniemulsion can also be performed in the presence of an oil which is inert to the polymerization process. During polymerization, oil and polymer can demix, and a variety of structures such as an oil droplet, encapsulated by a polymer shell, sponge-like architectures or "dotted" oil droplets can be formed. The synthesis of hollow polymer nanocapsules as a convenient one-step process using miniemulsion polymerization has the advantage of being thermodynamically controlled [120] and is based on early thermodynamic considerations on the behavior of two immiscible oils in water published by Torza and Mason [121]. They showed that the resulting equilibrium configuration of two immiscible liquid droplets, designated as phases 1 and 3, suspended in a third immiscible liquid, phase 2, is readily predicted from the interfacial tensions σ_{ij} and the resulting spreading coefficients $S_i = \sigma_{jk} - (\sigma_{ij} + \sigma_{ik})$. Chemical control of the expected particle morphology for an encapsulation process is therefore a complex parameter field. Recognizing the dramatic effect that common emulsifiers have on the interfacial tension between water and organic liquids or solids, it is not surprising to find that the preferred particle morphology reacts sensitively to the chemical natures of the emulsifier, the polymer, and the oil, as well as to additives such as an employed additional hydrophobe, the initiator, or possible functional comonomers.

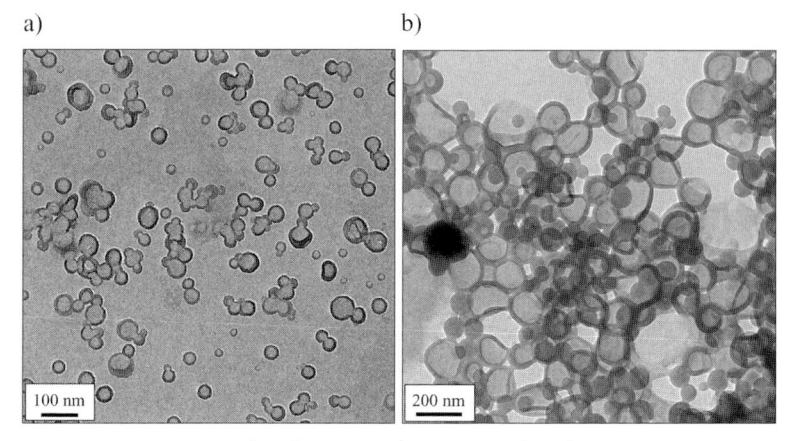

Fig. 6.14 TEM micrographs of nanocapsules: **a** PMMA/hexadecane/SDS; **b** PS/hexadecane/SE3030/KPS and 1 wt.% acrylic acid as a comonomer.

In the case of PMMA and hexadecane as a model oil to be encapsulated, the pronounced differences in hydrophilicity are suitable for direct nanocapsule formation. PMMA is regarded as rather polar (but is not water soluble), whereas hexadecane is very nonpolar so that the spreading coefficients are of the right order to stabilize a structure in which a hexadecane droplet core is encapsulated by a PMMA shell surrounded by water [120]. Miniemulsions were obtained by mixing the monomer MMA and hexadecane together with the hydrophobic, oil-soluble initiator AIBN and miniemulsifying the mixture in an aqueous solution of SDS. The polymerization leads to polymer capsules with 160-nm diameter and a narrow particle size distribution of less than 15% gaussian width, as shown in Fig. 6.14a. In the miniemulsion state the monomer and hexadecane are miscible, but phase separation occurs during the polymerization process owing to the immiscibility of hexadecane and PMMA. Nanocapsules with a higher shell stability can be obtained by using up to 10 wt.% of EGDMA (ethylene glycol dimethacrylate) as crosslinking agent.

In the case of styrene as a monomer and hexadecane as model oil, the cohesion energy density of the polymer phase is closer to that of the oil and therefore the structure of the final particles depends more sensitively on the parameters which critically influence the interfacial tensions. A variety of different morphologies in the styrene/hexadecane system can be obtained by changing monomer concentration and type and amount of surfactant, as well as initiator and functional comonomer and thus the corresponding spreading coefficients. The most perfect directly liquid-filled capsules of this examination were obtained by using only 1 wt.% of the very hydrophilic acrylic acid, which is enough to saturate the capsule surface with carboxylic groups and to reveal hollow shell structures with constant capsule thickness (Fig. 6.14b).

For some applications, the gas permeation or chemical sensitivity of polymer capsules is still too high for efficient encapsulation. Here, employment of crystal-

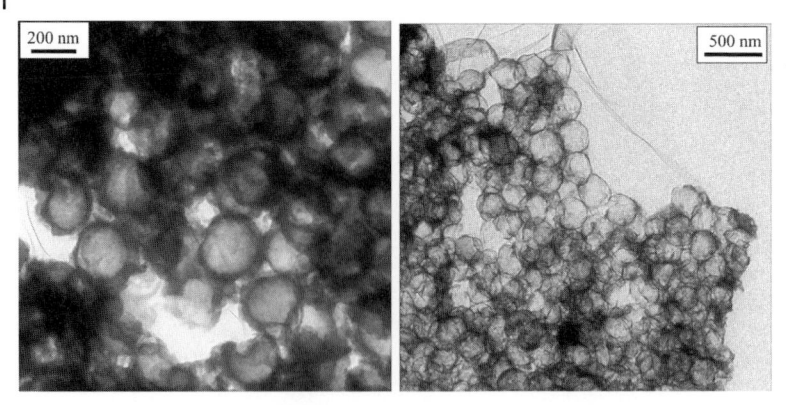

Fig. 6.15 TEM of armored latexes: **a** sealed clay on PBA; **b** sealed clay on PMMA, after removal of PMMA.

line inorganic materials, such as clay sheets with 1.5-nm thickness, can be recommended. Since those clay sheets are fixed like scales onto the soft, liquid miniemulsion droplet, the resulting objects have been called "armored latexes" [122]. Since clays carry a negative surface charge, miniemulsions stabilized with cationic sulfonium surfactants represented a convenient way to generate those "armored" minidroplets or crystalline nanocapsules. Owing to their high stability against changes in the chemical environment, it has been possible to use miniemulsion droplets themselves (as well as already polymerized latex particles) as templates for the plating procedure. Fig. 6.15a shows the outcome of such a plating process.

Here, a synthetic monodisperse model clay with small lateral extensions was employed. As a result, the liquid droplets or the polymer particles are completely covered with clay plates, which is also macroscopically observable by the suppression of film formation or coalescence.

The silica nanotiles can be connected and sealed by a subsequent condensation reaction with silicic acid which reacts with itself, and also with residual surface OH-groups on the clay plates. To prove the successful construction of the hollow shell structure, the polymer template was removed and the free-standing capsule analyzed. This was done using a PMMA miniemulsion latex as a template, which depolymerizes by increased illumination with the electron beam in TEM. Fig. 6.15b shows the emptied crystalline hulls obtained by electron degradation.

6.3.3.4 Miniemulsions Stabilized with Inorganic Nanoparticles

Addition of water-dispersible inorganic nanoparticles, e.g. colloidal silica, throughout miniemulsification leads to a potential structural complexity which covers the whole range from embedded particles (as in the case of calcium carbonate and carbon blacks) to surface bound inorganic layers (as in the case of clays).

The structure created by the ternary system oil/water/nanoparticles follows the laws of spreading thermodynamics, as they hold for ternary immiscible emulsions

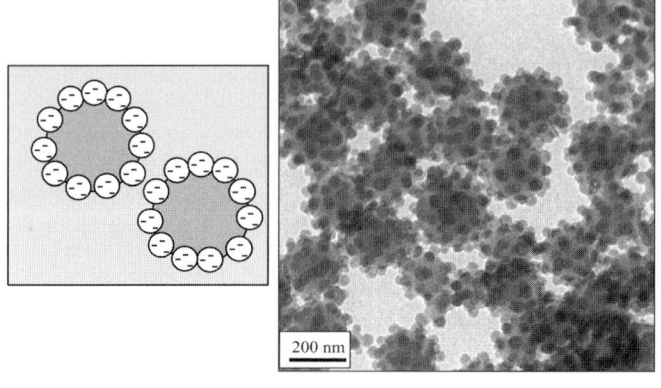

Fig. 6.16 TEM pictures for latexes using silica particles as stabilizers for monomer droplets (pickering stabilization with a monomer to silica ratio of 1:0.32).

(oil 1/oil 2/water). The only difference is that the interfacial area and the curvature of the solid nanoparticles stay constant, i.e. an additional boundary condition is added.

When, in addition to charges, the inorganic nanoparticles also possess a certain hydrophobic character, they become enriched at the oil–water interface, which is the physical basis of the stabilizing power of special inorganic macromolecules to act as so-called Pickering stabilizers [123–125]. In other words: The surface energy of the system oil/nanoparticles/water has to be lower than the sum of the binary combinations oil/water and water/nanoparticles to enable miniemulsion formation to occur, even without a surfactant. Since all three terms can be adjusted by the choice of the monomer and the potential addition of surfactants, this spans the composition diagram and a variety of morphologies can occur.

Silica nanoparticles are ideal as model nanoparticles for the systematic examination of compositional phase behavior since they are easy to obtain and to control with respect to their surface structure and interacting forces. The latter is done either by variation of pH, which changes the surface charge density, or by adsorption of cationic organic components, changing the polarity of the objects.

It has been shown that silica nanoparticles in the absence of any surfactant could act as Pickering stabilizers for a miniemulsion process [126]. The high quality and small overall particle size obtained only under alkaline conditions (pH = 10) and in presence of the basic comonomer 4-vinylpyridine is shown in Fig. 6.16.

The compulsory use of an aminic coupler was introduced by Armes [127, 128], who made similar-looking particles by precipitation polymerization. Purely hydrophobic droplets cannot be stabilized using colloidal silicas. Using miniemulsion polymerization as an easy-to-analyze model, it can be shown that starting from 6 wt.% of a basic monomer the interfacial energy between the monomer phase and silica particles becomes favorable, and only a few free silica particles are observed [126].

The particle size depends on the amount of silica in the expected way: The higher the silica content, the smaller the resulting stable hybrid structures. Comparatively small compound particles, between 120-nm and 220-nm diameter, with rather narrow size distributions can be obtained, speaking for the high stabilization power of the silica particles as Pickering stabilizers. Since these systems are free from low molecular weight surfactants, the measured surface tension was as high as 71.4 mN m^{-1}, which is practically the value of pure water. The ζ-potential of the nanocomposite dispersion versus pH corresponded to the potential of the pure silica sol.

The addition of a surfactant to the same system resulted in a more complex zoo of structures [126]. Nonionic surfactants are preferentially bound to the silica nanoparticles owing to a preferential interaction between the silica and the ethylene oxide chains [129] which screens any interaction with the monomer mixture. Also addition of anionic SDS leads to electrostatic repulsion and competition between the surfactant and silica nanoparticles. Only at pH = 3 (where the silica particles are practically uncharged) does coupling between the nanoparticles and the oil phase occur, and very small hybrid particles with a diameter of 68 nm are obtained, significantly smaller than those made with a similar recipe without silica [26]. It must be emphasized that silica particles alone at pH = 3 do not stabilize the emulsion. This means that in this case SDS and the weakly charged silica act in a synergistic fashion, i.e. as surfactant and co-surfactant.

The most pronounced structure changes, however, were observed with the cationic surfactant cetyltrimethylammonium chloride (CTMA-Cl). Charge coupling as well as induced dipole interaction cause this surfactant to bind strongly to silica over the whole pH range. For standard amounts of cationic surfactant in the percent region, there is not enough CTMA in the recipe to counterbalance the negative charges of the system at pH = 10, and a hedgehog morphology is found with small overall diameters of 90 nm. Measurements of the ζ-potential reveal that those compound particles are still negatively charged. The CTMA molecules are presumably arranged in patches between the polymer spheres and the silica particles. A dense binding of CTMA to the silica results in a hydrophobization of this spot and provides binding to the monomer droplet.

Starting from a calculated surface coverage of 75% with CTMA, the silica particles become incorporated into the droplet. The hybrids now have a raspberry morphology, although it is rather heterogeneous with respect to loading with silica. Using ultracentrifugation experiments in a density gradient, a homogeneous distribution of particles within a density range from 1.05 (bare polymer) to 1.20 g cm^{-3} (stable hybrids) are found, whereas particles with higher densities (pure silica) are not detected.

6.3.4
Miniemulsions of Inorganic Droplets/Synthesis of Inorganic Nanoparticles

Miniemulsification is not restricted to organic monomers, but can also be applied to low melting salts and metals to obtain salt or metal colloids of high homogeneity with diameters between 150 nm and 400 nm [130]. We regard this as a significant generalization, as it allows a cheap and reliable synthesis and handling of inorganic or metallic powders (and their incorporation into coatings, inks or nanocomposites) by simple and convenient polymer technologies.

The extension of miniemulsions from water or polar monomers as a dispersed phase in oils or hydrocarbons as a continuous phase to salt melts or concentrated salt solutions is nevertheless demanding, since those liquids show higher cohesion energies, surface tensions, and mutual attractions than the corresponding organic components. For that, a well-chosen steric stabilizer has to be employed, the polar part of which has to be miscible with salt melts, whereas the nonpolar part has to be sufficiently long and tightly packed to provide adequate steric stabilization. Again, it has been found that amphiphilic block copolymers [131] with a poly(ethylene oxide) block are most suitable. For the salt, one can choose from a wide variety of salts or metals which melt below the boiling or chemical decomposition temperature of the continuous phase (which can be selected to be as high as 250–300 °C for short periods). It is also possible to decrease the melting point by adding ternary components to the salts.

Examples from [130] include $Fe(III)Cl_3 \cdot H_2O$ (37 °C), $ZrOCl_2 \cdot 8H_2O$ ($-6H_2O$ 150 °C), gallium (39 °C), and Wood's metal (70 °C). All these systems can be heated above their melting points and can be miniemulsified in organic solvents. Cooling below the melting temperature results in recrystallization of the particles, while keeping the particulate character or the integrity of the inorganic miniemulsion droplets. Some examples of such inorganic miniemulsions are shown in Fig. 6.17.

A $3:1$ $ZrOCl_2$:water mixture melts at about 70 °C. The molten salt was added to Isopar M at 75 °C. A stable miniemulsion was obtained using 10 wt.% of the amphiphilic block copolymer PE/B-β-PEO, which transforms throughout cooling into a dispersion of single $ZrOCl_2$ nanocrystals. TEM pictures show (see Fig. 6.17a) that the particles are of uniform polyhedral crystalline shape.

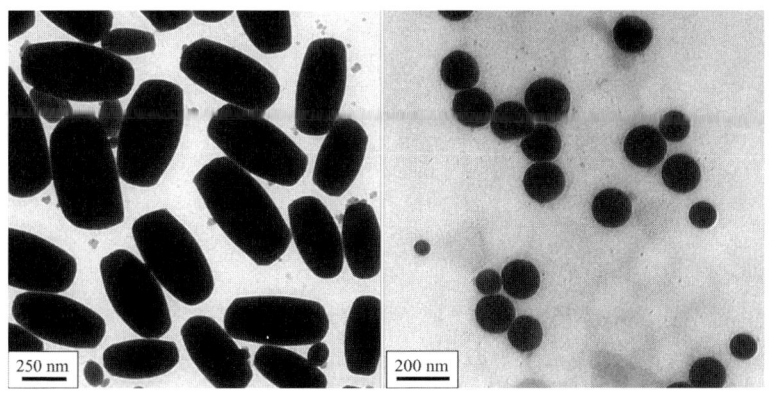

Fig. 6.17 Particles obtained in an inverse miniemulsion process consisting of **a** $ZrOCl_2 \cdot 8H_2O$, and **b** Woods metal.

For the preparation of nanosized metal dispersions the same procedure was used. Molten metals have very strong cohesion forces, which make them very difficult to disperse in an organic phase by conventional techniques. Miniemulsification, however, was applied to disperse low-melting alloys, like Wood's metal (composition: Bi50, Pb25, Cd12.5, Sn12.5) (mp=70°C) or Rose's metal (composition: Bi50, Pb28, Sn22) (mp=110°C). Because of the very high density difference between the metal [ρ(Wood's metal) = 9.67 g cm^{-3}] and the continuous phase (ρ(Isopar M)=0.87 g cm^{-3}), the weight content of the metal was increased to 50 wt.% (with respect to the overall dispersion) to obtain significant volume fractions (for TEM see Fig. 6.17b). The resulting dispersions are still stable, and their averaged overall density was determined to be 1.325 g cm^{-3}, which is consistent with the number calculated from the stoichiometry. The application of these metal dispersions to paper as a conducting ink results in homogeneous films with metallic gloss and rather high conductivity. For such experiments, a lower surfactant concentration of 10 wt.% with respect to metal seems to be the most suitable, since the lower layer thickness of the potentially insulating polymer surfactant leads to better particle contacts and higher conductivities.

As another example, iron(III)-chloride hexahydrate was melted by heating above 37°C and miniemulsified in the continuous phase (Isopar, cyclohexane, etc.) to a stable miniemulsion using at least 5 wt.% (with respect to salt) of the block copolymer stabilizer (see Fig. 6.18). The process of miniemulsification was followed by turbidity measurements, and a steady miniemulsion state was reached. Decreasing the temperature leads to nanoscopic salt crystals dispersed in a continuous oil phase. The average size of these particles is about 350 nm, a typical number for inverse dispersions. Due to the high density of the particles [ρ(FeCl$_3$) = 1.82 g cm^{-3}), sedimentation occurs throughout days, but can be reversed to the single entities by stirring.

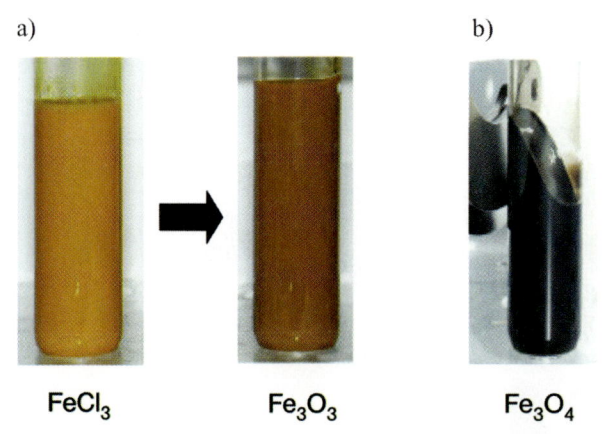

Fig. 6.18 a Formation of Fe$_2$O$_3$ particles from miniemulsified FeCl$_3 \cdot$H$_2$O droplets in Isopar by adding a base; **b** A dispersion of magnetite particles in Isopar obtained from FeCl$_2$/FeCl$_3$ droplets shows ferrofluidic behavior.

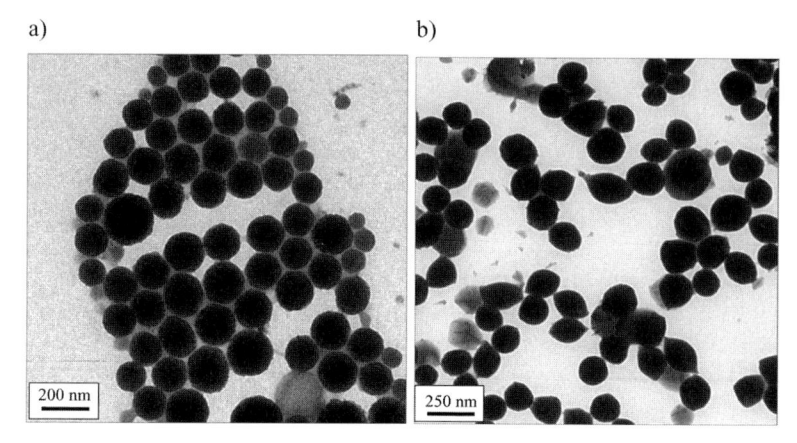

Fig. 6.19 TEM pictures of the products already shown in Fig. 6.18 ("inorganic polymerizations"): **a** Fe_2O_3 particles obtained from $FeCl_3$ droplets; **b** Fe_3O_4 particles obtained from $FeCl_2/FeCl_3$ droplets.

The formation of high-melting iron oxides can be achieved by a further reaction where the low-melting iron salt is used as a precursor, e.g. from iron(III)-chloride hexahydrate to iron(III)-oxide by addition of a base (pyridine or methoxyethylamine) which can mix with the continuous phase. It has been shown that in the dispersed state heterophase reactions such as precipitations or oxidations can be performed, which essentially occur with preservation of the colloidal entities as single nanoreactors, very similar to polymerization reactions of organic monomers. In the case presented, the crystal water enters into the reaction, while pyridine from the continuous phase neutralizes the eliminated HCl. Obviously, the high interface area of the miniemulsion is sufficient to promote this reaction.

Formation of Fe_2O_3 is accompanied with an increase of particle density ($\rho(Fe_2O_3) = 5.24$ g cm^{-3}). However, light scattering values and TEM pictures show that the droplets do not shrink as a whole, but instead show a hollow aggregate structure with interstitial cavities between primary particles (Fig. 6.19a). The aggregate size could be varied between 150 nm and 390 nm.

The confinement of two species in stoichiometric amounts within the nanodroplets also allows the syntheses of mixed species. A mixture of Fe^{2+} and Fe^{3+} salts leads to the formation of magnetite, Fe_3O_4. The final dispersion with a particle size of 200 nm is black and is ferrofluidic, as is illustrated in Fig. 6.18b. As is seen in the TEM pictures (Fig. 6.19b), the superstructure is anisotropic (lemon shaped), and the lemons are composed of needle-shaped nanocrystals arranged as bundles along the main axis.

6.4
On the Horizon: Miniemulsion Droplets as Compartments for "Single Molecule Chemistry"

The obvious fact that miniemulsions allow a very convenient and effective separation of objects in compartments of the size of 30- to 300-nm diameter also opens some general new perspectives for the chemistry of large and complex units. In miniemulsion droplets, it is in principle possible to isolate complex polymers or colloids strictly from each other and to react each single molecule for itself with other components, still working with significant amounts of matter and technically relevant mass fluxes. This has been called "single (polymer) molecule chemistry" [132].

In this mode of operation, single molecule chemistry usually takes place in a highly parallel fashion, since 3D space is compartmentalized in small "nanoreactors" in each of which the same reaction takes place, each on a single molecule. Although this is hardly used in classical chemistry, it is the regular case in biochemical reactions since practically all reactions take place in different compartmentalized areas of the cell [133]. The approach is not restricted to organic synthesis, but includes more complex physicochemical processes, such as protein folding, which mainly takes place as a single molecular event in the nanocompartments [134]. Mimicking those processes in polymer chemistry would open a door to gaining better control of the outcome of a demanding complex process or chemical reaction.

The potential advantage of nanoreactors is that a chemical reaction, confined to a nanosized environment, can have a different outcome as compared to the same reaction or process in free 3D space:

- The size of the structure to be made is limited by the size of the confinement and the number of precursor molecules in each nanoreactor. This has already been shown to be interesting for polymerizations, precipitations, and crosslinking reactions.

- The relation between inter- and intramolecular reactions can be shifted (cycling reactions and internal microgels). An intramolecular reaction is exclusively obtained as long as there is only a single polymer in each nanoreactor. For small confinements, this is guaranteed even at rather high concentrations. In a droplet of 30 nm, a 10 wt.% solution of a polymer or protein with $M_w = 850\,000$ g mol^{-1} is essentially single molecular. The principle of compartmentalization therefore makes high dilutions for single molecular reactions unnecessary.

- Protein folding, polymer crystallization, or more complex processes such as DNA hybridization or polymer–polymer complexation and templating occur with the restriction of having a single polymer chain. This simplifies the topology of the resulting structure.

A conceptual comparison of the miniemulsion system to classical nanoreactors such as reverse micelles, microemulsions, block copolymer micelles, and microgels is given in the literature [132]. It may be summarized by stating that miniemulsions are suitable for the design of "real" single molecule experiments, since the lifetime of a single droplet can be on the scale of days to months, whereas efficient reactor-filling factors are still of the order of 20–40%, and isolation is as convenient as breaking the miniemulsion.

6.5
Conclusions

The main aim of this chapter was to provide a critical review of the literature in the fascinating field of miniemulsions. We have shown that the use of high shear, appropriate surfactants, and the addition of a hydrophobe to suppress the influence of Ostwald ripening are key factors for the formation of the small and stable nanodroplets with low polydispersity.

We have also shown that the advantage of miniemulsions is the formation of particulate nanostructures which are hardly accessible by other synthetic approaches. Non-radical polymerizations and the formation of hybrid materials by the encapsulation of resins, inorganic materials, or liquids are some examples of the wide applicability of miniemulsions, which can also answer some technologically relevant questions. With the miniemulsification of molten inorganic materials and a subsequent reaction, miniemulsions cross the border between the classical disciplines and open new possibilities for the fabrication of solid particles for material science.

In our opinion, the miniemulsion concept is still advancing in polymer and materials chemistry since there are numerous additional possibilities for both fundamental research and application. As a vision one may think of single polymer molecules trapped, handled, and folded in each small droplet which enables new types of physicochemical experiments and handling of complex matter.

6.6
Acknowledgements

The authors would like to thank the Max Planck Society and the Fonds der chemischen Industrie for continued support.

6.7 References

1 M. Antonietti, C. Goltner, *Angew. Chem. Int. Ed. Eng.* **1997**, *36*, 910–928.
2 D. C. Blackley, *Polymer Latices*, 2nd edn; Chapman & Hall, London, **1997**.
3 M. Antonietti, E. Wenz, L. Bronstein, M. Seregina, *Adv. Mater.* **1995**, *7*, 1000–1005.
4 P. L. Tang, E. D. Sudol, M. E. Adams, C. A. Silibi, M. S. El-Aasser in *Polymer Latexes* (E. S. Daniels, E. D. Sudol, M. S. El-Aasser, eds.), American Chemical Society, Washington, **1992**, pp. 72–98.
5 E. D. Sudol, M. S. El-Aasser in *Emulsion Polymerization and Emulsion Polymers* (P. A. Lovell, M. S. El-Aasser, eds.), Johm Wiley & Sons Ltd, Chichester, **1997**, pp. 699–722.
6 F. J. Schork, G. W. Poehlein, S. Wang, J. Reimers, J. Rodrigues, C. Samer, *Colloids Surf. A: Physicochem. Eng. Asp.* **1999**, *153*, 39–45.
7 Y. J. Chou, M. S. El-Aasser, J. W. Vanderhoff, *J. Dispers. Sci. Technol.* **1980**, *1*, 129–150.
8 H. Köhler, *Geofyiske publikesjoner* **1922**, *2*, 3–15.
9 V. K. LaMer, R. Gruen, *Trans. Faraday Soc.* **1952**, *48*, 410–415.
10 H. Reiss, G. J. M. Koper, *J. Phys. Chem.* **1995**, *99*, 7837–7844.
11 W. I. Higuchi, J. Misra, *J. Pharmaceutical Sci.* **1962**, pp. 459–466.
12 A. J. Webster, M. E. Cates, *Langmuir* **1998**, *14*, 2068–2079.
13 S. S. Davis, H. P. Round, T. S. Purewal, *J. Coll. Interf. Sci.* **1981**, *80*, 508–511.
14 K. Welin-Berger, B. Bergenstahl, *International Journal of Pharmaceutics* **2000**, *200*, 249–260.
15 M. Postel, J. G. Riess, J. G. Weers, *Art. Cells, Blood Subs., and Immob. Biotech.* **1994**, *22*, 991–1005.
16 K. C. Lowe, *Art. Cells, Blood Subs., and Immob. Biotech.* **2000**, *28*, 25–38.
17 B. Abismail, J. P. Canselier, A. M. Wilhelm, H. Delmas, C. Gourdon, *Ultrasonics Sonochemistry* **1999**, *6*, 75–83.
18 P. Walstra, *Chem. Eng. Sci.* **1993**, *48*, 333–349.
19 S. E. Friberg, S. Jones in *Kirk-Othmer Encyclopedia of Chemical Technology*, 4 (J. I. Kroschwith, ed.), Wiley: New York, **1994**, *9*, pp. 393–413.
20 S. Brösel, H. Schubert, *Chem. Eng. Process.* **1999**, *38*, 533–540.
21 R. W. Wood, A. L. Loomis, *Phil. Mag.* **1927**, *4*, 417.
22 C. Bondy, K. Söllner, *Trans. Faraday Soc.* **1935**, *31*, 835–842.
23 T. J. Mason, *Ultrasonics Sonochemistry* **1992**, *30*, 192–196.
24 W. Lauterborn, *Ultrasonics Sonochemistry* **1997**, *4*, 65–75.
25 O. Behrend, K. Ax, H. Schubert, *Ultrasonics Sonochemistry* **2000**, *7*, 77–85.
26 K. Landfester, N. Bechthold, F. Tiarks, M. Antonietti, *Macromolecules* **1999**, *32*, 5222–5228.
27 K. Landfester, *Macromol. Symp.* **2000**, *150*, 171–178.
28 K. Landfester, *Macromol. Rapid Comm.* **2001**, *22*, 896–936.
29 J. S. Guo, M. El-Aasser, J. Vanderhoff, *J. Polym. Sci., Polym. Chem. Ed.* **1989**, *24*, 861–874.
30 A. R. M. Azad, J. Ugelstad, R. M. Fitch, F. K. Hansen in *Emulsion Polymerization* (I. Piirma, Gardon, J. L., ed.), ACS: Washington, DC, **1976**, *24*, pp. 1–23.
31 Y. J. Chou, M. S. El-Aasser in *Polymer Colloids II* (R. M. Fitch, ed.), Plenum Press, New York London, **1980**, pp. 599–618.
32 K. Landfester, N. Bechthold, F. Tiarks, M. Antonietti, *Macromolecules* **1999**, *32*, 2679–2683.
33 B. Z. Putlitz, H.-P. Hentze, K. Landfester, M. Antonietti, *Langmuir* **2000**, *16*, 3214–3220.
34 C. S. Chern, Y.-C. Liou, *Polymer* **1999**, *40*, 3763–3772.
35 M.-S. Lim, H. Chen, *J. Polym. Sci., Polym. Chem. Ed.* **2000**, *38*, 1818–1827.
36 S. Wang, F. J. Schork, *J. Appl. Polym. Sci.* **1994**, *54*, 2157–2164.
37 G. Baskar, K. Landfester, M. Antonietti, *Macromolecules* **2000**, *33*, 9228–9232.

38 N. Bechthold, F. Tiarks, M. Willert, K. Landfester, M. Antonietti, *Macromol. Symp.* **2000**, *151*, 549–555.
39 J. Delgado, M.S. El-Aasser, J.W. Vanderhoff, *J. Polym. Sci., Polym. Chem. Ed.* **1986**, *24*, 861–874.
40 C.S. Chern, T.J. Chen, *Colloid Polym. Sci.* **1997**, *275*, 546–554.
41 J. Delgado, M.S. EL-Aasser, C.A. Silibi, J.W. Vanderhoff, *J. Polym. Sci., Polym. Chem. Ed.* **1990**, *28*, 777–794.
42 J.L. Reimers, F.J. Schork, *J. Appl. Polym. Sci.* **1996**, *60*, 251–262.
43 J. Reimers, F.J. Schork, *J. Appl. Polym. Sci.* **1996**, *59*, 1833–1841.
44 J. Ugelstad, P.C. Mork, K.H. Kaggerud, T. Ellingsen, A. Berge, *Adv. Colloid Interf. Sci.* **1980**, *13*, 101–140.
45 K. Fontenot, F.J. Schork, *J. Appl. Polym. Sci.* **1993**, *49*, 633–655.
46 K.J.J. Fontenot, F.J. Schork in *4th International Workshop on Polymer Reaction Engineering* (K.H. Reichert, Moritz, H. (ed.), Dechema Monographs: Weinheim, Germany, **1992**, *127*, 429–439.
47 J.L. Reimers, A.H.P. Skelland, F.J. Schork, *Polym. Reaction Eng.* **1995**, *3*, 235–260.
48 J. Ugelstad, F.K. Hansen, S. Lange, *Makromol. Chem.* **1974**, *175*, 507–521.
49 C.D. Lack, M.S. El-Aasser, J.W. Vanderhoff, F.M. Fowkes, *ASC Symp. Ser.* **1985**, *272*, 345–356.
50 C.D. Lack, M.S. El-Aasser, C.A. Silebi, J.W. Vanderhoff, F.M. Fowkes, *Langmuir* **1987**, *3*, 1155–1160.
51 J. Ugelstad, M.S. El-Aasser, J.W. Vanderhoff, *J. Polym. Sci., Polym. Lett. Ed.* **1973**, *11*, 503–513.
52 Y.T. Choi, Ph.D. Thesis, Lehigh University, **1981**.
53 M.S. El-Aasser, C.D. Lack, Y.T. Choi, T.I. Min, J.W. Vanderhoff, F.M. Fowkes, *Colloids and Surfaces* **1984**, *12*, 79–97.
54 D. Mouran, J. Reimers, J. Schork, *J. Polym. Sci., Polym. Chem. Ed.* **1996**, *34*, 1073–1081.
55 S. Wang, G.W. Poehlein, F.J. Schork, *J. Polym. Sci., Polym. Chem. Ed.* **1997**, *35*, 595–603.
56 C.S. Chern, T.J. Chen, Y.C. Liou, *Polymer* **1998**, *39*, 3767–3777.
57 J. Reimers, F.J. Schork, *Polym. Reaction Eng.* **1996**, *4*, 135–152.
58 C.S. Chern, T.J. Chen, *Colloids Surf. A: Physicochem. Eng. Asp.* **1998**, *138*, 65–74.
59 C.-S. Chern, Y.-C. Liou, *Macromol. Chem. Phys.* **1998**, *199*, 2051–2061.
60 J.A. Alduncin, J. Forcada, J.M. Asua, *Macromolecules* **1994**, *27*, 2256–2261.
61 J.L. Reimers, F.J. Schork, *Industrial Eng. Chem. Res.* **1997**, *36*, 1085–1087.
62 I. Aizpurua, J.I. Amalvy, M.J. Barandiaran, *Colloid Surf. A: Physicochem. Eng. Asp.* **2000**, *166*, 59–66.
63 J.W. Gooch, H. Dong, F.J. Schork, *Journal of Applied Polymer Science* **2000**, *76*, 105–114.
64 M.S. El-Aasser, C.D. Lack, Y.T. Choi, J.W. Vanderhoff, F.M. Fowkes, *Colloid Surf.* **1984**, *12*, 79–97.
65 R.J. Goetz, M.S. El-Aasser, *Langmuir* **1990**, *6*, 132–136.
66 C.M. Miller, J. Venkatesan, C.A. Silibi, E.D. Sudol, M.S. El-Aasser, *J. Colloid Interf. Sci.* **1994**, *162*, 11–18.
67 S.T. Wang, F.J. Schork, G.W. Poehlein, J.W. Gooch, *J. Appl. Polym. Sci.* **1996**, *60*, 2069–2076.
68 H.-C. Chang, Y.-Y. Lin, C.-S. Chern, S.-Y. Lin, *Langmuir* **1998**, *14*, 6632–6638.
69 B. Erdem, Y. Sully, E.D. Sudol, V.L. Dimonie, M.S. El-Aasser, *Langmuir* **2000**, *16*, 4890–4895.
70 K. Landfester, N. Bechthold, S. Förster, M. Antonietti, *Macromol. Rapid Commun.* **1999**, *20*, 81–84.
71 K. Landfester, M. Willert, M. Antonietti, *Macromolecules* **2000**, *33*, 2370–2376.
72 M. Willert, Ph.D. thesis, Universität Potsdam, **2001**.
73 F. Candau, In *Polymerization in Organized Media* (E.C. Paleos, ed.), Gordon and Breach Science Publisher: Philadelphia, **1992**, pp. 215–282.
74 F. Candau, *Macromol. Symp.* **1995**, *92*, 169–178.
75 D.J. Hunkeler, J. Hernandez-Barajas, *Polymeric Materials Encyclopedia* (J.C. Salamone, ed.), CRC Press, New York, **1996**, *9*, 3322–3333.
76 M. Antonietti, R. Basten, S. Lohmann, *Macromol. Chem. Phys.* **1995**, *196*, 441–466.

77 F. Candau, M. Pabon, J.-Y. Anquetil, Colloids Surf. A: Physicochem. Eng. Asp. **1999**, *153*, 47–59.

78 M. P. Pileni, Langmuir **1997**, *13*, 3266–3276.

79 C. Petit, M. P. Pileni, Journal of Physical Chemistry **1988**, *92*, 2282–2286.

80 T. G. Mason, J. Bibette, Langmuir **1997**, *13*, 4600–4613.

81 C. Mabille, V. Schmitt, P. Gorria, F. L. Caldeon, V. Faye, B. Deminiere, J. Bibette, Langmuir **2000**, *16*, 422–429.

82 H. Yuyama, T. Watanabe, G.-H. Ma, M. Nagai, S. Omi, Colloids Surf. A: Physicochem. Eng. Asp. **2000**, *168*, 159–174.

83 S. Omi, K. Katami, A. Yamamoto, M. Iso, J. Appl. Polym. Sci. **1994**, *51*, 1–11.

84 S. Omi, K. Katami, T. Taguchi, K. Kaneko, M. Iso, Macromol. Symp. **1995**, *92*, 309–320.

85 G. H. Ma, M. Nagai, S. Omi, Colloid Surf. A: Phys. Chem. Eng. Asp. **1999**, *153*, 383–394.

86 N. Bechthold, K. Landfester, Macromolecules **2000**, *33*, 4682–4689.

87 B. Saethre, P. C. Moerk, J. Ugelstad, J. Polym. Sci., Polym. Chem. Ed. **1995**, *33*, 2951–2959.

88 S. Rimmer, Macromol. Chem. Phys. **2000**, *150*, 149–154.

89 X. Q. Wu, F. J. Schork, Industrial & Engineering Chemistry Research **2000**, *39*, 2855–2865.

90 K. Matyjaszewski, S. G. Gaynor in Applied Polymer Science 21th Century (C. D. Craver, C. E. Carraher, eds.), Elsevier Science Ltd., Oxford, **2000**, pp. 929–978.

91 M. Lansalot, C. Farcet, B. Charleux, J.-P. Vairon, Macromolecules **1999**, *32*, 7354–7360.

92 A. Butte, G. Storti, M. Morbidelli, Macromolecules **2000**, *33*, 3485–3487.

93 C. Farcet, M. Lansalot, R. Pirri, J. P. Vairon, B. Charleux, Macromol. Rapid Commun. **2000**, *21*, 921–926.

94 P. J. MacLeod, R. Barber, P. G. Odell, B. Keoshkerian, M. K. Georges, Macromolecular Symposia **2000**, *155*, 31–38.

95 T. Prodpran, V. L. Dimonie, E. D. Sudol, M. S. El-Aasser, Macromol. Symp. **2000**, *155*, 1–14.

96 C. Farcet, M. Lansalot, B. Charleux, R. Pirri, J. P. Vairon, Macromolecules **2000**, *33*, 8559–8570.

97 H. de Brouwer, J. G. Tsavalas, F. J. Schork, M. J. Monteiro, Macromolecules **2000**, *33*, 9239–9246.

98 K. Matyaszewski, J. Qiu, N. V. Tsarevsky, B. Charleux, J. Polym. Sci., Polym. Chem. Ed. **2000**, *38*, 4724–4734.

99 J. Qiu, T. Pintauer, S. G. Gaynor, K. Matyjaszewski, B. Charleux, J. P. Vairon, Macromolecules **2000**, *33*, 7310–7320.

100 G. Hinrichsen, H. Orth, Kolloid Z. Z. Polym. **1971**, *247*, 844.

101 K. Landfester, M. Antonietti, Macromol. Rapid Comm. **2000**, *21*, 820–824.

102 K.-H. Reichert, W. Baade, Angew. Makromol. Chem **1984**, *123/124*, 381–386.

103 W. Baade, K.-H. Reichert, Makromol. Chem., Rapid Commun. **1986**, *7*, 235–241.

104 V. S. Rodriguez, M. S. El-Aasser, J. M. Asua, C. A. Silibi, J. Polym. Sci., Polym. Chem. Ed. **1989**, *27*, 3659–3671.

105 H. Huang, H. Zhang, J. Li, S. Cheng, F. Hu, B. Tan, J. Appl. Polym. Sci. **1998**, *68*, 2029–2039.

106 Y. Inaba, E. S. Daniels, M. S. El-Aasser, J. Coatings Technol. **1994**, *66*, 63–74.

107 P. Rajatapiti, V. L. Dimonie, M. S. El-Aasser, J.M.S.-Pure Appl. Chem. **1995**, *A32(8/9)*, 1445–1460.

108 P. Rajatapiti, V. L. Dimonie, M. S. El-Aasser, M. S. Vratsanos, J. Appl. Polym. Sci. **1997**, *63*, 205–219.

109 M. Willert, K. Landfester, Macromol. Chem. Phys. **2002**, *203*, 825–836.

110 K. Landfester, F. Tiarks, H.-P. Hentze, M. Antonietti, Macromol. Chem. Phys. **2000**, *201*, 1–5.

111 A. Tomov, J. P. Broyer, R. Spitz, Macromolecular Symposia **2000**, *150*, 53–58.

112 X. Q. Wu, F. J. Schork, J. W. Gooch, J. Polym. Sci., Polym. Chem. Ed. **1999**, *37*, 4159–4168.

113 J. G. Tsavalas, J. W. Gooch, F. J. Schork, Journal of Applied Polymer Science **2000**, *75*, 916–927.

114 E. M. S. van Hamersveld, J. van Es, F. P. Cuperus, Colloid Surf. A: Physicochem. Eng. Asp. **1999**, *153*, 285–296.

6.7 References

115 B. ERDEM, E.D. SUDOL, V.L. DIMONIE, M.S. EL-AASSER, *J. Poly. Sci., Poly. Chem.* **2000**, *38*, 4419–4430.
116 B. ERDEM, E.D. SUDOL, V.L. DIMONIE, M.S. EL-AASSER, *J Polym. Sci., Polym. Chem. Ed.* **2000**, *38*, 4431–4440.
117 B. ERDEM, E.D. SUDOL, V.L. DIMONIE, M.S. EL-AASSER, *J Polym. Sci., Polym. Chem. Ed.* **2000**, *38*, 4441–4450.
118 D. HOFFMANN, K. LANDFESTER, M. ANTONIETTI, *Magnetohydrodynamics* **2001**, *37*, 217–221.
119 F. TIARKS, K. LANDFESTER, M. ANTONIETTI, *Macromol. Chem. Phys.* **2001**, *202*, 51–60.
120 F. TIARKS, K. LANDFESTER, M. ANTONIETTI, *Langmuir* **2001**, *17*, 908–917.
121 S. TORZA, S.G. MASON, *J. Coll. Interf. Sci.* **1970**, *33*, 6783.
122 B.Z. PUTLITZ, K. LANDFESTER, H. FISCHER, M. ANTONIETTI, *Adv. Mater.* **2001**, *13*, 500–503.
123 W. RAMSDEN, *Proc. Roy. Soc. London* **1903**, *72*, 156–164.
124 S.U. PICKERING, *J. Chem. Soc. Commun.* **1907**, *91*, 2001.
125 T.R. BRIGGS, *Ind. Eng. Chem. Prod. res. Dev.* **1921**, *13*, 1008.
126 F. TIARKS, K. LANDFESTER, M. ANTONIETTI, *Langmuir* **2001**, *17*, 5775–5780.
127 C. BARTHET, A.J. HICKEY, D.B. CAIRNS, S.P. ARMES, *Adv. Mater.* **1999**, *11*, 408–410.
128 M.J. PERCY, C. BARTHET, J.C. LOBB, M.A. KHAN, S.F. LASCELLES, M. VAMVAKAKI, S.P. ARMES, *Langmuir* **2000**, *16*, 6913–6920.
129 G.J.D.A. SOLER-ILLIA, C. SANCHEZ, *New Journal of Chemistry* **2000**, *24*, 493.
130 M. WILLERT, R. ROTHE, K. LANDFESTER, M. ANTONIETTI, *Chem. Mater.* **2001**, *13*, 4681–4685.
131 S. FÖRSTER, M. ANTONIETTI, *Advanced Materials* **1998**, *10*, 195–217.
132 M. ANTONIETTI, K. LANDFESTER, *ChemPhysChem.* **2001**, *2*, 207–210.
133 K. STÄHLER, J. SELB, F. CANDAU, *Langmuir* **1999**, *15*, 7565.
134 F.U. HARTL, *Nature* **1996**, *381*, 571.

7
Metal and Semiconductor Nanoparticle Modification via Chemical Reactions

Luis M. Liz Marzán

7.1
Introduction

Metal and semiconductor nanoparticles are of great importance owing to their special optical and electronic properties. As discussed in Chapters 2 (semiconductors) and 3 (metals), these special properties mainly arise from either the size-dependent variation of the band gap in the case of semiconductors (providing strong and tunable luminescence) or the confinement of conduction electrons in the case of metals (giving rise to plasmon resonances in the visible spectrum). Given their small size, it is clear that the surface area of any kind of nanoparticles is extremely large (1 g of spherical bulk gold has a surface area of about 0.66 cm^2, but if it is divided into *nanospheres* with a diameter of 10 nm, the surface area is increased to 31 m^2!), and therefore their properties will be to a large extent determined by surface properties. Thus, modification of the nanoparticle surface can significantly alter the properties and behavior of metal and semiconductor nanoparticles. Traditionally, surface modification has been performed through physisorption of suitable molecules (typically surfactants or polymers), or through true chemical modification, so that a new material is actually bound to the surface, which often results in the formation of an outer, continuous shell. In this chapter we focus on the latter approach, so that the use of chemical reactions for the surface modification of nanoparticles is extensively reviewed and several examples are discussed in more detail.

In Section 7.2, the principal reasons for surface modification are outlined, as a foundation upon which the following sections are based. Section 7.3 reviews the deposition of various types of coating shells on metal and semiconductor nanoparticles, while Section 7.4 describes several examples of chemical reactions that can affect metal and semiconductor nanoparticles, and how the presence of a porous shell affects such reactions. Finally, general conclusions are summarized.

Colloids and Colloid Assemblies. Edited by Frank Caruso
Copyright © 2004 Wiley-VCH Verlag GmbH & Co. KGaA, Weinheim
ISBN: 3-527-30660-9

7.2
Why Surface Modification?

The surface modification of colloid particles of various nature and size has become a very common practice during the last few decades for a number of reasons. This section describes the most relevant objectives that lead to the design of procedures aiming to a controlled surface modification.

7.2.1
Chemical and Colloidal Stability

Major problems to be overcome in nanoparticle dispersions are the degradation of the material through chemical etching and the agglomeration caused by strong van der Waals attractive forces.

Chemical etching is not an important limitation in the case of noble metal nanoparticles, since these particles are stable in most environments, but it will be shown below that even they can be dissolved if not properly protected. The problem is much worse in the case of other transition metals, such as iron or nickel, since they are readily oxidized, so that their magnetic properties dramatically change. Similarly, many semiconductor nanoparticles, such as metal chalcogenides, are air sensitive, so that they can be completely dissolved in the presence of oxygen and light. These processes will be discussed below, as well as possible means to prevent them.

On the other hand, colloidal stability is always a relevant issue when inorganic (lyophobic) nanoparticles are synthesized in liquid solvents. The main forces affecting colloidal stability are (attractive) van der Waals forces and (repulsive) double layer and steric interactions. Such forces have been extensively discussed in the literature. In the case of aqueous dispersions (and, in general, in polar solvents), double-layer interactions will be primarily involved in the separation of colloid particles from each other, while for dispersions in non-polar solvents, steric effects due to adsorbed organic chains will be of major relevance.

Particle size has significant effects upon colloid stability. As a consequence, compared to micron-sized particles, nanosized particles exhibit different flocculation and coagulation behavior. Hence, they require different mechanisms to achieve colloid stability. Deryagin and Landau [1] and Verwey and Overbeek [?] independently developed a theory (known as DLVO theory) to calculate the energy (V_T) of interaction between charged spherical particles as a function of interparticle separation and other parameters. V_T is considered to be the sum of the double-layer repulsion V_R and van der Waals attraction V_A. For small, nanosized particles (radius a, separation D) this gives:

$$V_T = 2\pi\varepsilon_0\varepsilon_r a\psi_0^2 \exp[-\kappa D] - \frac{Aa}{12D} \tag{1}$$

where ε_0 is the permittivity of vacuum, ε_r is the relative permittivity of the medium in the diffuse layer, ψ_0 the surface potential, and κ^{-1} the double layer thickness.

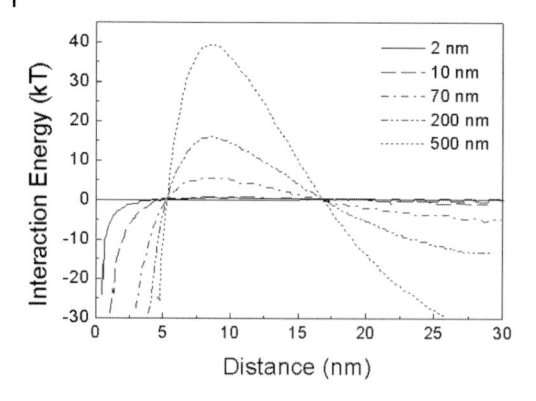

Fig. 7.1 DLVO total interaction energy vs interparticle separation curves as the particle size is varied from the nanosized to micronsized domains. The profiles are calculated from Eq. (1), based on the following parameters: $A=2\times10^9$ J, $\psi_0=-35$ mV, $\kappa^{-1}=10$ nm, $\varepsilon=78.5$.

According to Eq. (1) (valid only for $\kappa a \ll 1$ and for small double layer interactions, i.e. dilute dispersions), both the repulsive and attractive interaction energies increase with particle size as the particles approach each other.

However, it is the balance between these energies that determines particle stability against coagulation. The total interaction energy of two particles with varying particle size is plotted in Fig. 7.1. In colloid chemistry, a barrier of 15–20 kT is usually sufficient to ensure colloid stability [3].

From the figure, we can see that the secondary minimum is deeper for micronsized particles, which can lead them to weak flocculation. Conversely, for nanosized particles, the secondary minimum is shallower, and hence repeptization will occur more readily. From the same plot, it can also be seen that the barrier height diminishes with decreasing particle size. This means that nanoparticles are more susceptible to coalescence or "salting out". Additionally [4], the critical coagulation concentration of electrolyte decreases rapidly with decreasing particle size. Thus, smaller particles require a larger zeta-potential to obtain the same repulsive interaction energy. On the other hand, larger particles are more stable against coalescence because they possess a larger barrier height. However, conversely again, the primary minimum associated with coagulation is smaller for nanoparticles; thus stabilization by adsorbed polymers, surfactants, or chemisorbed complexing agents, such as thiols or small carboxylic acids, is much more effective than for larger particles. Small molecules should then be sufficient to prevent coalescence of nanoparticles. This means that a steric barrier provided by chemisorbed molecules for particles smaller than 10 nm in diameter is predicted to be sufficient to offset the van der Waals interactions leading to coagulation at the primary minimum. However, if two particles of this diameter coalesce in solution, they will not be able to separate again since their thermal energy will be insufficient to allow them out of the primary minimum.

Thus, surface modification to increase the surface charge, and in turn the zeta-potential, is a powerful technique to enhance colloidal stability of nanoparticles in polar solvents, where the adsorption of organic (macro)molecules is not very efficient.

7.2.2
Tuning of Physical Properties

The physical properties of a composite material are the result of an interplay between the individual properties of its consituent materials. A clear example is found in the optical properties of metal nanoparticles. Since they depend (among other parameters) upon the dielectric properties of their environment, deposition of a different material on the metal particle surface can strongly affect the optical response, especially if the deposited material is another metal, so that electrons can easily be transferred from one to the other. It is also desirable in many cases that a certain material displays a certain property, which can be accomplished by tailored deposition of a selected coating material. To understand the effects of coating layers on the optical behavior of metal nanoparticles, some background is required on the origin of such a behavior and the main parameters that influence it.

The mechanism for the absorption of light by small metal particles is based on the assumption that the plasma oscillation of the collective conduction electrons of the particles is caused by the restoring force resulting from the induced polarization on the surface of the particle during the interaction between the alternating electromagnetic field and the particle. Such absorption of light in the UV-visible region by metal particles is sensitive to particle size [5–9], shape [10–17], dispersion medium [6–9, 18, 19], particle material [14], encapsulating layers [20–27], electron density on the particles [28–30], and temperature [6, 31–33]. Restricting ourselves to colloid spheres in dilute solutions, the basic equations describing the optical properties are contained in what is frequently termed the "Mie theory for light absorption and scattering by spheres" [34]. Combination of Mie theory with the classical Drude model for the dielectric properties of metals leads to a fairly simple calculation of the optical response of metal nanoparticles in solution. The extinction cross-section of a single particle, C_{ext}, results from both scattering and absorption of light by the particle. Although both processes occur simultaneously, for particles which are very small compared with the wavelength of light, only the absorption is significant. We only consider here particles in the nanosize regime in which scattering is negligible.

Mie [34] calculated the extinction cross-section, C_{ext}, of a solution of dilute small spherical metal particles embedded in an isotropic, non-absorbing medium of dielectric constant $\varepsilon_m = n_m^2$ to be [35–37]:

$$C_{ext} = \frac{2\pi}{m^2} \sum_{n=1}^{\infty} (2n+1)\operatorname{Re}(a_n + b_n) \qquad (2)$$

where $m = 2\pi/\lambda$ is the wavenumber, and a_n and b_n are the scattering coefficients which are functions of the particle radius R and the incident wavelength λ in terms of Ricatti-Bessel functions. Based on the dipole approximation, for spherical particles very small compared with the wavelength of light ($2\pi R/\lambda \ll 1$) whereby scattering is insignificant, only the first, electric-dipole term in Eq. (2) is important, and C_{ext} is reduced to

$$C_{ext} = \frac{24\pi^2 R^3 \varepsilon_m^{3/2}}{\lambda} \frac{\varepsilon''}{(\varepsilon' + 2\varepsilon_m)^2 + \varepsilon''^2} \quad (3)$$

where ε' and ε'' are the real and imaginary parts of the complex dielectric function of the metal particle. Other assumptions made, for Eq. (2) to be valid, are that the particles are monodisperse and isolated. That is, the separation between the particles must be larger than some 10 particle diameters; thus in the case of 10-nm particles, this corresponds to a maximum concentration of ca. 10^{15} particles/L. Equation (3) determines the shape of the absorption band of the particles, with the width and height of the resonance governed by ε'' and the peak position by ε' at the resonance. Given that ε'' is small or does not change greatly around the absorption band, the maximum absorption occurs when $\varepsilon' = -2\varepsilon_m$.

The real and imaginary dielectric components in $\varepsilon = \varepsilon' + i\varepsilon''$ are functions of the frequency of the incoming electromagnetic radiation. According to the Drude model [38], the real and imaginary parts of the complex dielectric function may be related to the light angular frequency $\omega = 2\pi\nu$ as

$$\varepsilon' = \varepsilon^\infty - \frac{\omega_p^2}{\omega^2 + \omega_d^2}, \quad \varepsilon'' = \frac{\omega_p^2 \omega_d}{\omega(\omega^2 + \omega_d^2)} \quad (4)$$

where ε^∞ is the high frequency dielectric constant and ω_p is the plasma frequency given by

$$\omega_p^2 = \frac{Ne^2}{m\varepsilon_0} \quad (5)$$

in terms of the electron concentration N, electron charge e, electron mass m and permittivity of vacuum ε_0. ω_d is the scattering frequency, which for the bulk metal is given by

$$\omega_d = \frac{v_F}{R_{bulk}} \quad (6)$$

where v_F is the velocity of conduction electrons at the Fermi level, and R_{bulk} is the mean free path for conduction electrons in the bulk metal. The scattering of free electrons is directly related to the metal conductivity in the Drude formalism by

$$\sigma = \frac{Ne^2 R_{bulk}}{mv_F} = \frac{Ne^2}{m\omega_d} \quad (7)$$

For particles with sizes between 3 and 20 nm, the absorption spectra do not depend strongly on particle size, since the quadrupole and higher-order terms in the Mie summation become significant only when particles are larger than 20 nm [35]. However, below this size range the spectra of some metal colloids do show a size dependence because the collision of conduction electrons with the particle surface becomes more appreciable as the particle radius, R, becomes smaller than the electron mean free path in the bulk metal, $1/R_{bulk}$. Consequently, the effective electron mean free path, R_{eff}, becomes size dependent as

$$\frac{1}{R_{eff}} = \frac{1}{R} + \frac{1}{R_{bulk}} \tag{8}$$

which means

$$\sigma(R) = \frac{Ne^2 R}{mv_F} \tag{9}$$

This was verified by Kreibig for gold and silver particles [5–8]. Kreibig demonstrated that decreasing R increases ω_d which consequently both broadens the surface plasmon band and decreases its intensity but causes only a very small shift in the peak position. This effect is most marked for metals with relatively long electron mean free paths such as the alkali metals, copper, or silver, but is small for most other metals [39].

Using Eq. (3), the optical properties of spherical metal particles can be readily calculated from sets of the optical constants, the real and imaginary parts of the complex dielectric function, $\varepsilon = \varepsilon' + i\varepsilon''$. Equation (3) can then be extended to calculating the spectra of semiconductor nanocrystals when ε' and ε'' are known. Unlike metal particles, semiconductor particles possess lower conduction electron concentrations (i.e. small ω_p), and consequently, the surface plasmon absorption occurs in the IR regime rather than in the visible part of the spectrum and there is only a subtle change in color as the particle size decreases below the wavelength of visible light. For semiconductor particles below 5 nm in radius, quantum size effects dominate and there are color changes due to changes in allowed energy levels such that the absorption is due to the electronic transition across the band gap which can be strongly sensitive to the variation in particle size. Thus, Mie theory will not correctly predict the spectra for semiconductor quantum dots.

Finally, we briefly refer to the modifications of Mie theory to account for the optical properties of particles coated with a different material such as small molecules, surfactant molecules, polymers, dielectrics (silica, tin dioxide, titanium dioxide), and metals. In such cases, Eq. (3) is modified to [37]:

$$C_{ext} = 4\pi R^2 m$$
$$\times \operatorname{Im}\left\{\frac{(\varepsilon_{shell} - \varepsilon_m)(\varepsilon_{core} - 2\varepsilon_{shell}) + (1-g)(\varepsilon_{core} - \varepsilon_{shell})(\varepsilon_m + 2\varepsilon_{shell})}{(\varepsilon_{shell} + 2\varepsilon_m)(\varepsilon_{core} + 2\varepsilon_{shell}) + (1-g)(\varepsilon_{shell} - 2\varepsilon_m)(\varepsilon_{core} - \varepsilon_{shell})}\right\}$$
$$\tag{10}$$

with R being the radius of the coated particle, ε_{shell} the dielectric function and g the volume fraction of the coating material, ε_{core} the dielectric function of the core particles, and ε_m the dielectric constant of the surrounding medium. One can see that when $g=0$, Eq. (10) reduces to Eq. (3) for an uncoated sphere, and for $g=1$, Eq. (10) yields the extinction cross-section for a sphere of the shell material in the medium with dielectric constant ε_m.

7.2.3
Control of Interparticle Interactions Within Assemblies

The collective properties of nanoparticle assemblies are influenced to a large extent by the separation between the basic units, i.e. the nanoparticles. Thus, control of the interparticle separation leads directly to control over the properties of the assembly. Several strategies have been devised to establish careful control of the position of colloid particles on a substrate, such as lithography or microcontact printing. However, probably the simplest way to achieve it comprises coating the particles with a shell of an inert material, so that the thickness of the shell directly determines the interparticle separation in close-packed assemblies. Several uses have been made of this concept, principally with silica as a coating material, because of its optical inertness, insulating character, and capability of sintering, so that more stable structures can be achieved. Since the formation of such nanostructures is not within the scope of this chapter, the reader is referred to another recent review by the author [40].

7.3
Chemical Deposition of Shells on Metals and Semiconductors

7.3.1
Metals on Metals

As explained above, the main interest in metal nanoparticles is due to their optical and electronic properties, which arise from the collective oscillation of conduction electrons, as a response to their interaction with electromagnetic radiation of the appropriate frequency. This is mainly a surface phenomenon, and as such is greatly influenced by the nature of the medium that composes the immediate surroundings of the nanoparticle. Additionally, different metals have different dielectric functions, and therefore the plasmon resonance will occur at different frequencies. Therefore, a great deal of study has focussed on the optical effects that arise when one metal is deposited on the surface of nanoparticles made of a different metal. In what follows, we shall use the symbol @ to link the core with the shell material (core@shell).

The first studies on core-shell bimetallic nanoparticles were carried out by Morriss and Collins [41], who synthesized colloidal Au@Ag spheres with a constant gold core diameter of 5.9 nm, and silver shells grown epitaxially and varying in thickness from 0.5 to 23.15 nm. The plasmon absorption band of the core, initially at 518 nm, blue-shifted to 496 nm and broadened when the shell was 0.5 nm thick. For a 1-nm silver shell, the band comprised two peaks, at 390 and just below 500 nm, which then merged to form a single broad peak around 400 nm when the shell thickness increased to 1.75 nm. Increasing the shell to 8.75 nm led the band to become sharper but it remained at the same position. However, as the shell thickness increased to 23.15 nm, it became broader and red-shifted to

425 nm. Furthermore, Morriss and Collins found that their experimental spectra were in reasonable agreement with the theory, but their spectra were generally broader and lower in intensity. The discrepancy was due to not considering the damping in the silver shell caused by the smaller mean free path of the conducting electrons (see above). Much later, Kreibig and coworkers, carried out more refined calculations on the same Au–Ag system [42]. All their calculations agreed with respect to the major features of the spectra: a pronounced dip at 325 nm, a blue shift of the gold plasmon band at small silver coverage, and the formation of a broad band between about 350 and 500 nm, which initially had shoulders at both ends, but then developed into a single strong band moving from shorter wavelengths to 390 nm with increasing silver thickness.

One of the most refined and tunable procedures for the synthesis of metal nanoparticles in aqueous solution comprises the reduction of metal salts by organic free radicals generated radiolytically under an inert atmosphere, in the presence of a suitable stabilizer. This process, which was studied in detail by Henglein and co-workers, also allowed them to perform a systematic study on the controlled deposition of metal layers on preformed nanoparticles. Examples of the resulting encapsulated particles include Au@Cd [43], Au@Pb [43, 44], Au@Sn [45], Au@Tl [43], Ag@Pb [21, 46], Ag@Cd [47], and Ag@In [21]. They observed a variety of optical and chemical properties of the particles in dilute aqueous solutions. As each of the less noble metals (Cd, Pb, Sn, Tl, and In) was deposited as shells onto either of the noble core particles (Au and Ag), the absorption band of the core particles blue-shifted towards that of a pure colloid of the shell material. The blue shift was ascribed to the donation of electron density from adatoms to the core metal particle [21, 47]. They consistently found that only a few monolayers of shell material were required to alter the absorption band of the respective cores significantly. The band of the core particles continued to blue-shift until it was masked completely by the shell, with the exception of Tl in Au@Tl whose spectrum still possessed some residual band features of the Au core. For example, as few as two to three layers of Cd were required to mask the band of Ag and Au in Ag@Cd and Au@Cd, and three layers of Pb to mask completely the band of Au in Au@Pb. Another interesting feature they discovered was the chemical behavior of the shell, in particular for the innermost layer. The innermost layers were reoxidized by air, Ag^+, MV^{2+} and H_2O_2 more slowly than if the shell metal were a pure colloid [21, 43, 48]. The slow reoxidation of the shell was ascribed to the fact that the mantle metal becomes more "noble" than if it were a monometallic colloid due to the process of underpotential deposition [43, 46, 47]. Moreover, unexpectedly, the innermost shell could not be completely reoxidized by any of the above oxidants as the adatoms bonded very strongly to the core atoms. While the core and shell materials of the majority of the coated particles retained their separate crystalline phases, some alloying took place in Au@Sn particles, especially when a large amount of Sn was deposited. Conversely, in the case of Ag@Au [20], a dramatic red shift towards the absorption band of pure Au particles was observed when the more noble Au was deposited on Ag cores, as was also observed by Sinzig et al. [42].

Au@Ag particles [43] were generated chemically rather than radiolytically when Ag$^+$ reoxidized the shell of Au@Pb and Au@Tl, thereby being reduced to Ag atoms, which were subsequently deposited on the same cores in the place of the less noble metals. However, Ag$^+$ could not completely reoxidize the Pb shell of Au@Pb as the adatoms in the innermost shell were strongly bonded to the Au atoms. In turn, the first Ag shell of the newly formed Au@Ag particles could not be reoxidized by strongly oxidizing H_2O_2, which normally readily dissolves pure Ag particles. Trimetallic Au@Pb@Cd particles [43] were constructed through the deposition of Cd onto Au@Pb particles. The absorption band was seen to be located between those of monometallic Pb and Cd particles after deposition of three layers of Cd, which was normally sufficient to mask completely the band of the core particles in Ag@Cd and Au@Cd particles. Obviously, one can see that this method can produce encapsulated particles of any number of layers. A major problem using Cd and Tl as shells is that particles coated with these metals agglomerate readily.

One of the more difficult noble metals to prepare in colloidal form is mercury. Seeking to observe the optical properties of Hg colloids, Henglein et al. [23] used silver particles as a template on which they attempted to deposit mercury by γ-radiolysis. They observed that the silver plasmon band blue-shifted when a couple of layers of mercury were deposited on the colloidal silver particles. Further deposition only led to the undesirable formation of mercury precipitates. They suggested that the first couple of layers of mercury remained intact on the silver surface because the two metals formed an amalgam. A similar experiment was later performed in the presence of gold nanoparticles [49], and the observations (absorption spectra and TEM micrographs) were interpreted as a limited penetration of mercury and formation of a labile mercury layer on the surface.

Henglein has also demonstrated the generation of Au@Pt and Pt@Au nanoparticles [50]. Although a gold layer on platinum clusters could be deposited radiolytically, the deposition of platinum on gold had to be performed by hydrogen reduction of aged $PtCl_4^-$, and the resulting core-shell structures seem to be formed by Pt islands that join together if sufficient metal is deposited. Successive radiolytic reduction has also led to the formation of trimetallic Pd@Au@Ag nanoparticles with interesting optical features [51].

Ag@Au and Au@Ag core-shell nanoparticles were also radiolytically generated, and it was demonstrated that laser-induced heating, mediated by electron–phonon coupling, leads to melting and alloying, so that the core-shell morphology gradually disappears as more laser pulses are absorbed, which means that intermediate stages with a graded interlayer can also be designed [52]. Fig. 7.2 shows an example of gold nanoparticles coated with silver. The UV-visible spectra show the influence of the shell on the optical properties, while in the TEM image the silver shell is observed as a lighter area surrounding the darker core.

The radiolytic reduction of metals was also studied by Belloni and coworkers [53, 54]. Most of these systems were reported to be alloys [53, 55], though they also discussed how to tune the experimental conditions for the design of core-shell structures, based on the redox potentials of the different metals [54].

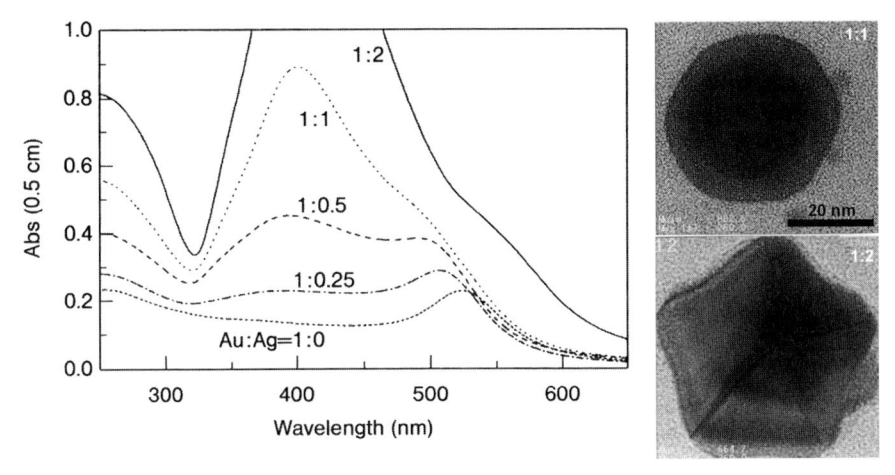

7.3 Chemical Deposition of Shells on Metals and Semiconductors

Fig. 7.2 a Absorption spectra of Au@Ag particles of different molar compositions. The gold concentration was 1.7×10^{-4} M, and silver was deposited in different amounts; b high-resolution electron micrographs of Au@Ag particles. Upper: Au:Ag=1:1; lower: Au:Ag=1:2 (reproduced with permission from [52], copyright (2000) American Chemical Society).

As a general rule, it is very difficult to deposit a noble metal onto a less noble metal colloid particle. For example, it was observed that a Au^{3+} solution oxidized 10-nm colloidal silver, forming colloidal gold [24]. To avoid the problem, Mulvaney et al. [20] employed $Au(CN)_2^-$ ($E^0 = -0.61$ V) which is less noble than $Ag(CN)_2^-$ ($E^0 = -0.31$ V) for the radiolytic deposition of gold onto silver seeds. During the gold deposition, the silver absorption plasmon band red-shifted towards that of pure gold particles. Based on the same approach, Michaelis et al. [25] used $Ag(CN)_2^-$ to deposit silver on palladium core particles. Formaldehyde was employed as a reductant in this case, which, unlike radiolysis, avoided the unnecessary formation of free silver particles. Moreover, Michaelis et al. showed that the palladium colloids acted as a catalyst or "microelectrode" for silver deposition, since silver was immediately deposited in the presence of palladium particles, but not when the core particles were absent. At least four layers of silver were necessary to mask the band of the palladium core. Wang and Toshima [56], relying on a different approach, not only generated the easily formed Pt@Pd particles but also the more difficult Pd@Pt particles, this time using molecular hydrogen as sacrificial protection for core Pd particles during the deposition of Pt.

In 1995 Lee et al. [57], deliberately deposited a very thin skin of palladium on 5-nm gold colloids (Au@Pd). They believed that having the palladium skin as thin as possible would allow the bimetallic particles to be most effective as a catalyst. The palladium shell was structurally different from its bulk counterpart, owing to electronic perturbation of the shell by the gold core. Subsequently, they observed a very large increase in the catalytic performance of the bimetallic colloids in the cyclization of acetylene to form various isomers of butane, n-hexane, and benzene, even at room temperature. When the Au@Pd particles were annealed at increasing temperatures in the range 300–573 K, the progressive alloying of Au and Pd was accompanied by a significant acceleration of the acetylene cyclization, with the production of benzene being strongly favored over the other products. Moreover, particle growth became the principal process in the temperature interval 573–873 K. The H_2-reductive condensation of a carbon-supported molecular cluster precursor, $PtRu_5C(CO)_{16}$ into a bimetallic Pt–Ru particle, led to a different particle morphology, depending on the reaction temperature [58]. At 473 K, Pt@Ru particles were produced, whereas at 673 K, Ru@Pt particles were formed instead. Pd coating on Au was also achieved using sonochemistry [59], with great enhancement of the catalytic activity for hydrogenation reactions.

Bright et al. [60] observed a dramatic improvement in SERS of films consisting of Au@Ag particles. Gold colloids were selected as cores because they provided uniform SERS effects throughout the entire film, while silver was chosen as a shell because it provided the material with the best SERS signal amplitude. Similarly, Freeman et al. [61] observed that their Au@Ag particles at a composition of 95% Au and 5% Ag gave an increase in SERS intensity but further silver deposition led to complete loss of signal.

Synthesis of Au@Ag and Ag@Au was also carried out by various chemical methods in solution. Rivas et al. [62] achieved both structures by citrate reduction, and based the characterization on TEM, UV-visible spectroscopy and SERS. Silver shells were also deposited on citrate-stabilized 12-nm gold nanoparticles in aqueous solution, using ascorbic acid as a reductant and cetyltrimethylammonium chloride as additional stabilizer during the growth [63]. The resulting core-shell particles were quite monodisperse, and the thickness of the silver shell was controlled through the concentration of seeds used. Coating of silver with gold was achieved by Srnová-Sloufová et al. [64], using hydroxylamine as reductant, in a similar approach as that developed by Brown and Natan for the seeded growth of Au nanoparticles [65]. Mirkin et al. [66] have shown that, in the presence of preformed silver nanoparticles passivated with bis(p-sulfonatophenyl) phenylphosphine, simple reduction of gold salts by sodium borohydride leads to the formation of Ag@Au nanoparticles. Such core-shell particles are expected to be useful for DNA diagnostics. An interesting report on the Au/Ag system was made by Xia et al. [67], who observed that addition of Au^{3+} ions onto colloid dispersions of Ag nanoparticles leads to oxidation of the silver cores and reduction of gold on the surface, in such a way that hollow gold particles with the shape of the original silver cores are obtained. The process can be applied to silver nanorods, thus promoting the formation of gold nanotubes. In a more complicated, multistep process, Schierhorn and Liz-Marzán [68] achieved the synthesis of bimetallic colloids with tailored intermetallic separation. This was achieved by depositing an insulating silica shell on uniform gold cores, and then applying an electroless plating procedure [69] to deposit silver on the outer silica surface, so that the silica shell served as an effective spacer between the two noble metals, and their spectral features could be combined at will.

A different approach to metal@metal colloids was suggested by González-Penedo et al. [70], based on the successive reduction of Fe^{2+} or Co^{2+} and Ag^+ within the aqueous microdroplets of inverse microemulsions. A similar approach was later used by Seip and O'Connor [71] for the synthesis of Fe@Au using cationic surfactants. This is a method that continues to be explored by other authors because of the apparent advantages that can be gained by the use of a confined space for the reaction.

7.3.2
Semiconductors on Semiconductors

In a similar fashion to metals, semiconductor nanoparticles are of great importance for technological applications owing to their special optical and electronic properties. The observation of quantum size effects in these systems, due to the size-dependent separation between the valence and conduction bands, has led to the establishment of the term *quantum dots*, which is generally accepted nowadays. The preparation of core-shell structures for semiconductor nanoparticles is the primary means for tailoring their optical properties, as well as for 2D/3D organization [72–76]. Coating of II–VI nanoparticles with a different semiconducting material was shown to have a profound impact on the photophysics of the nanocrystalline core [77–85]. A broad range of intensities and spectral emission characteristics can be obtained by varying the thickness and the band gap of the overlayer. Deposition of a semiconductor with a large band gap relative to the core typically results in the enhancement of luminescence from the nanoparticles owing to the suppression of radiationless recombination mediated by surface states [77–86], while the degree of charge-carrier confinement does not change. The red shift observed in the position of absorption bands, even for quite thick layers of ZnS on CdSe [77], CdS on CdSe [79] and ZnSe on CdSe [78] is very small, as expected for a marginal perturbation of the confinement regime. Conversely, a layer from a small band-gap semiconductor provides an additional area of delocalization for both the electron and the hole [81–88]. The relaxation of the confinement regime results in a red shift of spectral features. These studies have been carried out with PbS and HgS layers deposited on CdS and CdSe nanoparticles. Recently, CdS nanoparticles with a monomolecular coating of MoS_4 have been investigated, which demonstrated both enhancement of the excitonic emission and a noticeable red shift of absorption features [89].

Various methods had been suggested for the synthesis of core-shell II–VI nanoparticles. Hoener et al. [90] used bis-2-ethylhexyl sodium sulphosuccinate (AOT) microemulsions to prepare CdSe nanoparticles, and subsequently grow ZnSe. The structure was confirmed by XPS and Auger spectroscopy. The microemulsion method was used again by Han et al. [87] for the coating of CdS with Ag_2S, which resulted in a product with large nonlinear absorption. Tian et al. [88] prepared CdS@CdSe and CdSe@CdS by sequential addition of H_2S and H_2Se to aqueous $Cd(ClO_4)_2$ solutions containing $(NaPO_3)_6$. The structural investigation was based on spectroscopy, electrochemical pulse radiolysis, and electron microscopy. Hao et al. [91] also used the microemulsion method to prepare CdSe and CdSe@CdS nanoparticles, observing a notable enhancement of the excitonic luminescence upon coating.

In 1993, Murray et al. reported an organometallic method of nanoparticle preparation based on TOP/TOPO mixtures [92]. Several techniques of making core-shell nanoclusters modifying this technology have subsequently appeared [77–79]. Briefly, to a dispersion of CdSe in this solvent, preformed CdS particles dispersed in pyridine are added at 100 °C, which leads to a core-shell CdSe@CdS structure accompanied by a large increase in the photoluminescence quantum yield. The quality of the product so prepared seems to be related to the crystallinity of the semiconductors, which is enhanced at the high temperature used. A similar procedure was used for the synthesis of CdSe@ZnS with very high luminescence quantum yield [77]. O'Brien and co-workers [93] devised a novel method for the preparation of highly monodisperse CdSe@CdS involving the successive thermolysis of [Cd{Se$_2$CNMe(Hex)}$_2$] and [Cd{S$_2$CNMe(Hex)}$_2$] in TOPO. The same group extended the single-precursor approach to other core-shell combinations, such as CdSe@ZnS, CdSe@ZnSe, using bis(hexyl(methyl)dithio-/-diselenocarbamato) cadmium(II)/zinc(II) compounds [94]. CdS@CdSe nanowires were also recently prepared by Xie et al. [95] by treatment of CdS nanowires with Se in tributylphosphine.

Recently, Spanhel and co-workers [96] devised a procedure based on chalcogen atom exchange (Te for S) when 3 to 4-nm tributylphosphine-stabilized CdS nanoparticles are reacted with bis(trimethylsilyl)-telluride in the presence of excess CdCl$_2$. The resulting core-shell structure displays a much higher luminescence intensity, high colloid stability, and decreased air sensitivity, compared to the original CdS nanoparticles.

Apart from the II–VI semiconductors, core-shell quantum dots were also made with InAs cores, with band gaps in the near IR. Banin's team [97, 98] explored the coating of these quantum dots with InP and CdSe using a high-temperature procedure, finding a drop in luminescence quantum yield for InP, but an enhancement using CdSe, which was explained in terms of the quality of the outer surface.

7.3.3
Oxides on Metals and Semiconductors

The reasons for the deposition of metal oxide shells on the surface of metal and semiconductor nanoparticles can be various, as was shown in the Introduction. In this section we present a survey of several procedures that have allowed the encapsulation of metal and semiconductor nanoparticles with metal oxides, as well as the main advantages that have been acquired thereby.

7.3.3.1 Coating of Metal Particles with Silica
The use of silica as a coating material has a long tradition in colloid science. One of the major reasons is the anomalously high stability of silica colloids, specially in aqueous media, but other reasons are the easy control of the deposition process (and therefore on the shell thickness), its processibility, chemical inertness, con-

trollable porosity, and optical transparency. All these properties make silica an ideal and low-cost material for tailoring surface properties, while basically maintaining the physical properties of the underlying cores.

One of the great advantages of a core-shell geometry for colloid particles is that the stability of the colloid will be largely determined by the nature of the shell material. In order to explain the reasons why the encapsulation of various nanoparticles with silica enhances their colloidal stability, we briefly describe here the basic colloid stability of silica sols. The only effect the core exerts upon two approaching spheres is the van der Waals attraction, which can only occur when the thickness of the silica shell is below a few nanometers, within which van der Waals forces are predominant. Contrary to DLVO theory, it has been observed that silica sols in the size range of 10–100 nm possess a remarkable stability at very high salt concentration (e.g. 0.15 M NaCl) even at their isoelectric point (pH_{iep}). Furthermore they sometimes coagulate in alkaline solutions where their electrical surface charge is high. The main factors favoring the remarkable stability of silica sols are that their Hamaker constant is much lower than those of metals, latex, and other oxide particles, and, more importantly, at the silica–water interface at $pH \geq 10.5$ a polymeric silicate layer is present on which highly hydrated cations, in particular Li^+ and Na^+, are very tightly bound. The silicate layer confers both steric and electrostatic protection to silica particles in the pH regime above pH_{iep} and below pH 10.5, while at pH_{iep} only the steric barrier is present [99–107].

Even though the core properties of silica-coated nanoparticles are unaffected should they become coagulated, for many purposes such as formation of 2D and 3D crystals, it is still desirable for them to remain dispersed. Such dispersibility can be achieved by using silica as an encapsulating shell. Thus, silica coating has been employed for nearly half a century [108] for the enhancement of the colloid stability of different particles. However, only recently has it been possible to adapt these procedures to the coating of nanoparticles [109, 110]. The coating of very small particles is complicated because of the extremely high curvature. Additionally, in the case of coating noble metal nanoparticles with an oxide, one faces the difficulty of the chemical near inertness encountered at the nanoparticle surface. This means that a procedure must be designed to overcome the very low tendency of the core and shell materials to bind to each other. Several routes have been followed with better or worse results. Ohmori and Matijevic [111] prepared Fe@SiO_2 particles by an indirect path comprising the initial coating of hematite spindles by hydrolysis of tetraethoxysilane (TEOS) in ethanol [112], followed by reduction of the core within a furnace at 450 °C, while passing a stream of H_2 gas. The high temperature promoted sealing of the pores of the silica shell, thus avoiding further oxidation of the metallic core. Patil and coworkers [113] synthesized Au@SiO_2 and Ag@SiO_2 particles using a gas aggregation source. Though the technique can be employed to synthesize a whole range of particles, it suffers from many important disadvantages: (1) The gas chamber, operated under inert atmosphere, requires extremely high operating temperatures, up to 1500 °C; (2) the operating setup is complex and expensive; (3) it seems that silica-encapsulated metal particles may be difficult to handle (e.g. to transfer into liquid solutions) as they are prepared on solid substrates; (4) particles cannot be produced on a large scale, which may limit some investigations requiring large amounts of samples. A much simpler procedure was used by Liz-Marzán and Philipse [114] for the synthesis of Au@SiO$_2$ particles. The procedure was based on the formation of nanosized gold particles in the presence of small silica spheres, which led to heterocoagulation. The composite particles were then diluted with ethanol, where extensive growth was achieved by means of the Stöber method [115], so that Au was ultimately embedded as a core. This synthetic route resulted in a mixture of unlabeled and labeled silica particles, with a low proportion of the labeled particles, but by using an index matching technique, these colloids could be applied to the study of the dynamics of concentrated silica colloids.

A method that has provided substantially better results was later designed [26, 27]. The method was based on the use of silane coupling agents [116] as surface primers to provide the nanoparticle surface with silanol groups, and therefore to render the surface of noble metals compatible with silicate moieties. For metal nanoparticles, (3-aminopropyl) trimethoxysilane (APS) was used as a coupling agent, because of the large complexation constant for noble metal amines. Once the surface had been modified, slow deposition of a thin silica shell was effected from a solution of sodium silicate (as described by Iler [108]). Subsequent transfer into ethanol led to sudden condensation of unreacted silicate [117], which can promote the formation of multiple-core composites, but can also be adjusted to yield nicely concentric Au@SiO$_2$ core-shells. The shells could then be grown by careful addition of ammonia and tetraethoxysilane, resulting in monodisperse colloids as shown in Fig. 7.3. This system proved ideal for a systematic study of optical properties [27, 31]. Having hydrophobized the silica surface of their Au@SiO$_2$ colloids with trimethoxysilane propylmethacrylate (TPM), Liz-Marzán et al. were also able to stabilize the particles in pure ethanol and mixtures of toluene and ethanol in various proportions without aggregation [27], which would not otherwise have been possible for uncoated particles. The same method was later extended to Ag@SiO$_2$ [118], though new difficulties arose because of the dissolution of the Ag cores when concentrated ammonia was added to increase the thickness of the silica shell. Optimization of the procedure for silver cores included dialysis of the metallic seed colloid and control of pH during the several stages of the process.

The same team [24] demonstrated that dissolution by ammonia is due to the possibility of chemical reactions taking place on the cores because of the porosity of the shells, which will be discussed in Section 7.4. The enhanced colloidal stability provided by the silica shell allowed a number of further studies, such as the temperature effect on the optical properties [31], the incorporation of preformed metal nanoparticles within silica gels [119] for the synthesis of materials with large optical nonlinearities [120], or the assembly of metal nanoparticles on larger colloids [121]. Another advantage of this method is that the first deposition in water is slow enough to preserve the shape of the original core in the initial stages. Advantage of this fact was taken by Chang et al. [122], who were able to apply the same procedure to coat gold rods prepared by electrochemical reduction. Similar work was performed by Murphy et al. on chemically synthesized nano-

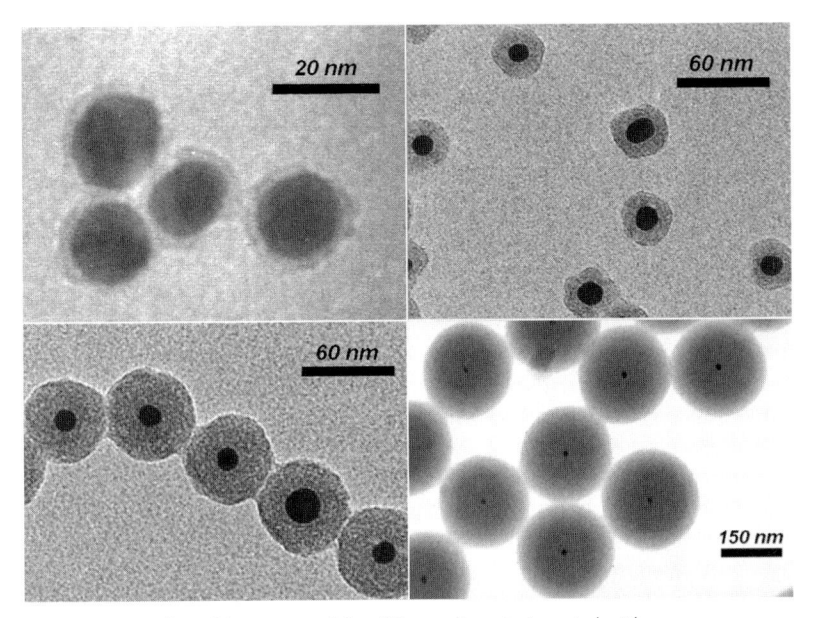

Fig. 7.3 Examples of Au nanoparticles (15-nm diameter) coated with silica shells of various thickness.

rods [123]. Yau et al. [124] studied single-electron tunneling through the silica shell of Au@SiO$_2$ with STM. They observed Coulomb staircases due to nonlinearity in the current steps.

Recently, the method was modified by Mann et al. [125] to incorporate chemical functionalities in the shell, by means of the copolymerization of silane-coupling agents with TEOS during the final growth step. Hardikar and Matijevic [126] demonstrated that for larger silver particles (60±5 nm) stabilized with a complicated compound (Daxad 19), no coupling agent is necessary for silica coating through TEOS hydrolysis in 2-propanol. A similar procedure was also recently applied by Xia and coworkers to coat polyvinylpyrrolidone-stabilized silver nanorods with silica [127]. Again, the problem of silica dissolution by ammonia was reported by these authors, who reported that coated particles (rather than hollow silica shells) could only be obtained if they were separated from the ammonia solution using centrifugation for some 20 min after addition of the base. The same authors reported that this procedure can also be used to coat commercial gold nanoparticles, but the nature of the initial surface coating of these particles was not specified [128]. Mine et al. have also recently found [129] that standard, citrate-stabilized Au nanoparticles can be homogeneously coated with relatively thick shells without the use of a silane-coupling agent, through careful tailoring of the concentrations of ammonia and TEOS, as well as by choosing the proper addition sequence. This leads to a cleaner system (no chemical alteration of the metal surface), but almost invariably implies the formation of some core-free silica particles in addition, which is undesirable for some applications.

Makarova et al. [130] trapped fluorescent dye probes inside silica shells by adsorbing the dye onto the surface of gold cores prior to silica coating by the previously described method and then dissolving off the cores, leaving the dye molecules locked inside the SiO_2 shell. They observed a significant difference in the fluorescence of the dye molecules in bulk solution, attached to the gold cores in Au@SiO_2, and trapped inside the hollow SiO_2 nanospheres. A further advance by the same group was the use of surfactants to create mesoporous silica shells on Au nanoparticles coated with a thin silica layer [131]. Ostafin and co-workers explicitly mention the need for an initial coating in order to stabilize the metal colloids during the growth of the mesoporous shell.

Other methods have been later designed for the preparation of Ag@SiO_2. The group of Adair [132] reported the synthesis of homogeneous Ag@SiO_2 particles within microemulsions. This method comprised the formation of Ag nanoparticles within the droplets of W/O microemulsions, followed by polymerization of TEOS at the same droplets. The problem with this technique is that the required removal of surfactants is expensive, tedious and time consuming. In addition, the presence of surfactants, which cannot be separated completely from Ag@SiO_2 particles, can complicate the systems under investigation. A similar method was used by Martino et al. [133] to prepare silica gels loaded with Au nanoparticles. Pastoriza-Santos and Liz-Marzán prepared Ag@SiO_2 particles through reduction of Ag^+ by N,N-dimethylformamide (DMF) [134] and DMF/ethanol mixtures [135], in the presence of APS as a stabilizer. High temperatures were needed to achieve a fast reduction of the silver salt, prior to condensation of the organo-silica shell. The obtained degree of monodispersity of the silver cores was high, and the particles could be easily extracted into water or ethanol. The silica shells also contained amine groups, which may be useful for some applications.

In a similar spirit of growing ceramic materials on the surface of metal particles, Lee et al. [136] reported the growth of $CaCO_3$ on Au nanoparticles protected by 4-mercaptobenzoic acid. The hydrophilic acid groups act as nucleation sites for crystal growth, in a biomineralization process. Using different ratios of Ca^{2+} and Mg^{2+} ions, these authors showed that different gold-containing nanostructures are formed.

7.3.3.2 Coating of Semiconductor Nanoparticles with Silica

Relatively few accounts have been published related to the coating of semiconductor nanoparticles with silica. Such a coating can serve not only to eliminate surface defects, but also to greatly enhance the stability, specially in aqueous solvents. A first report on the preparation of CdS@SiO_2 was published by Chang et al. [137, 138], using the microemulsion method. These authors showed several possibilities for the preparation of composite nanoparticles with different morphologies. Liz-Marzán and coworkers [139] extended the method previously developed for metals [26], based on the use of silane-coupling agents, to link silica to the surface of CdS. The basic difference comprises the use of a mercaptosilane instead of an aminosilane, based on a higher affinity for the specific surface to be coated.

They showed [139] that the presence of the silica shell rendered the CdS cores stable against photodegradation. A variation of this method was later used by Alivisatos et al. [140] to coat their highly luminescent CdSe@CdS, which can then be used as fluorescent biological labels. Authors within this group performed a detailed study of an elaborate multistep coating process [141], showing that there is plenty of colloid chemistry behind it, and that it can be applied to quantum dots of different sizes without a significant damping of the luminescence intensity, so that the labeling efficiency can be very high.

The same method with slight modifications was used by Rogach et al. [142] for the coating of CdTe and CdSe@CdS nanocrystals, which resulted in the inclusion of multiple cores within every silica sphere, what the authors call "raisin bun"-type nanoparticles. Unfortunately, the luminescence intensity dramatically drops in this case upon silica deposition. Further studies on this system [143] have allowed the deposition of homogeneous coatings, and furthermore indicated that upon illumination of the colloid with daylight the luminescence rises again, thus allowing for the formation of nanostructured fluorescent systems, either using layer-by-layer assembly or a sol-gel procedure.

7.3.3.3 Coating of Metal Nanoparticles with Semiconductors

We finally include in this subsection metals surrounded by semiconductor shells, even though some of these shells are not composed of oxides, but rather of chalcogenides. In 1997 Kamat and Shanghavi [144] reported on the preparation of Au@CdS nanoparticles by surface modification of Au colloids with mercaptonicotinic acid followed by addition of Cd^{2+} and exposure to H_2S. The result was the formation of very small CdS particles attached to the surface of the Au cores, but also free in solution. Nayral et al. prepared Sn@SnO_2 particles to be used as potential gas sensors [145]. The core was allowed to undergo complete oxidation to form pure tin dioxide particles. With a view to conferring novel optical properties on particles due to both local field effects of the metal and further enhancing the nonlinear optics of quantum dots, Nayak et al. [146] generated Au@CdSe particles in TOP/TOPO organic solvents, though a large proportion of particles were a mixture of individual Au and CdSe particles, as detected by TEM. Tada et al. [147] loaded TiO_2 particles with $AgNO_3$ and then irradiated the sol with a Hg lamp to reduce the adsorbed Ag^+ ions. Unfortunately, they showed no evidence of the actual structure of the resulting particles. Hirano et al. [148] encapsulated nanosized gold with insulating boron nitride at extremely high temperatures, and proposed that these particles would have potential applications for single-electron transistors.

Pastoriza-Santos et al. [149] synthesized Ag@TiO_2 particles by a modification of the procedure previously used for the synthesis of Ag@SiO_2 in DMF [134]. In this case, the reduction of $AgNO_3$ took place in DMF/ethanol mixtures in the presence of titanium tetrabutoxide, which condensed on the surface of the silver cores. Layer-by-layer assembly of these coated nanoparticles followed by dissolution of the silver cores with ammonia, subsequently led to the development of ion-selective and biocompatible films [150], which were found useful for monitoring the

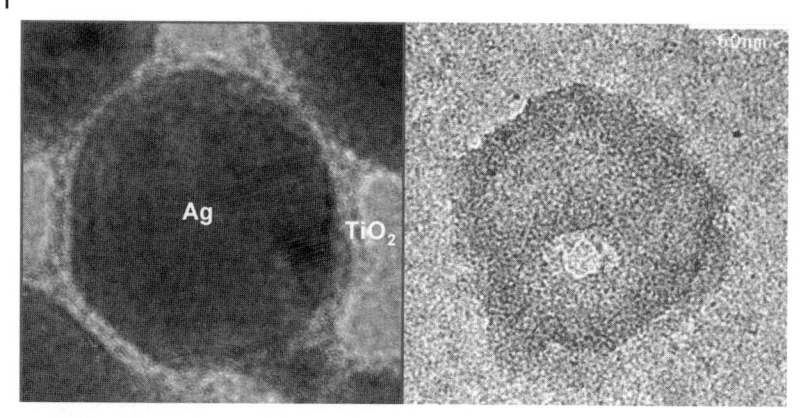

Fig. 7.4 HRTEM images of Ag@TiO$_2$ nanoparticles (left) before and (right) after dissolution with ammonia.

diffusion of dopamine, an important substance in neurochemical processes. Fig. 7.4 shows TEM micrographs of one Ag@TiO$_2$ nanoparticle and one TiO$_2$ nanoshell obtained through ammonia dissolution of the core.

Coating of gold nanoparticles with titania was also achieved by Mayya et al. [151], using a different TiO$_2$ precursor (titanium(IV) bis(ammonium lactato) dihydroxide), which is negatively charged and readily complexes with a positively charged polyelectrolyte, poly(dimethyldiallylammonium chloride), previously assembled on the nanoparticles' surface. Upon complexation on the surface, controlled hydrolysis can be carried out, thus providing good control of the morphology of the shell. The same authors applied this procedure to coat Ni nanorods with titania [152], and upon dissolution of the Ni cores they were able to obtain uniform TiO$_2$ nanotubes.

Another interesting application of metal nanoparticles surrounded by semiconductor shells is the fabrication of composite nanoparticles with a large electronic capacitance. The idea is that there is a large difference between the Fermi level of the core and the conduction band energy of the shell, so that electrons diffusing through the shell can be trapped in the core for a long time. Mulvaney et al. [153] explored this possibility with Au nanoparticles encapsulated in a polycrystalline SnO$_2$ shell. These authors demonstrated that charge can be injected by cathodic polarization using γ-radiolysis, which resulted in a blue shift of the plasmon band (as predicted by theory, see above), which remained blue-shifted in the absence of oxygen, but red-shifted back when open to air because of oxygen reduction by the electrons stored in the core. The number of electrons stored can be as high as 1500–2000, as calculated from the blue shift of the surface plasmon band and from the discharge using zwitterionic viologen as an electron acceptor. Fig. 7.5 shows a high-resolution TEM image of one Au@SnO$_2$ nanoparticle, as well as the effect of oxygen addition upon cathodic polarization of the nanoparticles.

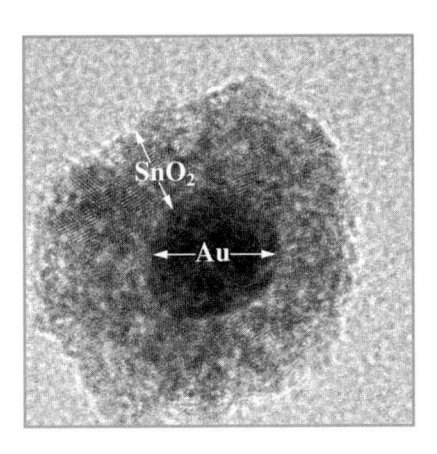

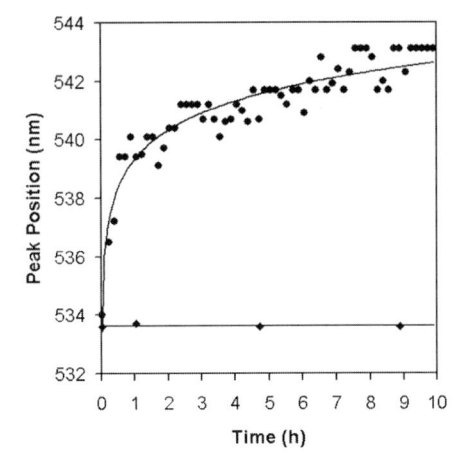

Fig. 7.5 Left: HRTEM image of a 15-nm Au nanoparticle coated with polycrystalline SnO_2 (av. thickness 10 nm). Right: Effect of oxygen (circles) on the position of the plasmon band of Au@SnO_2 nanocapacitors, which have been cathodically polarized by radiolytically generated CH_3CHOH radicals. As a reference, the plasmon band of the same sample stored under nitrogen (diamonds) is plotted over the same period of time.

7.3.4
Polymers on Metals

Organic polymeric shells have also been successfully applied to encapsulating metal nanoparticles. These procedures can in principle be used for the stabilization and nanostructuration of the nanoparticles, just as in the case of metal oxides, but suffer from some limitations, such as dissolution in some organic solvents, or much higher sensitivity to high-temperature processing.

Marinakos and coworkers [154–156] coated gold particles by the conducting polymers poly(N-methylpyrrole) and poly(pyrrole), and formed 1D chains, each containing 2–3 polymer-coated gold particles. They also created conducting polymer nanobubbles by elimination of the cores. However, the technique of Marinakos and coworkers suffers from its inability to produce particles on a much larger scale. Selvan et al. [157, 158] used block copolymers to encapsulate gold cores with pyrrole, obtaining a rather heterogeneous result. They observed that annealing leads to larger, single cores, which, however, are not well centered in the composite particles. Two other papers are those by Menzel et al. [159] reporting the deposition of colloidal gold on an MPS-modified substrate, followed by self-assembly of diacetylene and photopolymerization, and finally Quaroni and Chumanov [160], who coated Ag particles by polymerization of styrene and/or methacrylic acid in emulsions of oleic acid. Mandal et al. [161] covered 2-bromoisobutyryl bromide-modified gold nanoparticles with poly(methylmethacrylate) using living radical polymerization, which results in a noticeable red shift of the surface plasmon band.

Polystyrene coating of Au nanorods was performed by Murphy et al. [123] by polymerization of styrene using ammonium persulfate as initiator. Further dissolution of the Au nanorod cores led to formation of polymeric nanotubes.

7.4
Chemical Reactions at Coated Nanoparticles

Apart from using chemical reactions for the modification of metal and semiconductor nanoparticles, it is of much interest how the presence of a shell material affects chemical reactions being performed on the core. Additionally, such in situ chemical reactions performed on a specific composite material can lead to the formation of new composite materials which are hard to achieve otherwise. In this section we explore the effect of silica shells on several selected chemical reactions that can take place in metal and/or semiconductor nanoparticles.

In the specific case of silica shells, the main factor allowing or preventing access to the cores by reactants in the surrounding solution is the porosity of the shell. Thus control of shell porosity would lead to control of which reactions are possible and which are not. Here, we restrict ourselves to reactions that occur on nanoparticles coated with silica, either formed from condensation of sodium silicate solution, or from TEOS hydrolysis and condensation in ethanol. The shells formed from sodium silicate possess larger pores than is the case for TEOS-derived shells, and thus the reactants can more readily diffuse through the pores and reach the core. Extensive studies [162, 163] on the properties of silica particles grown using TEOS have clearly demonstrated that the particles are significantly less dense than bulk silica, with micropores of the order of several nanometers.

7.4.1
Dissolution of Gold and Silver Cores

One of the few chemical reactions that metallic gold can undergo is oxidation by cyanide ions in the presence of oxygen, and this reaction is widely used in the mining industry for ore extraction. The result of the oxidation is a gold-cyanide complex, which is non-absorbing in the visible. The reaction involved in basic conditions is the following:

$$4\,M + 8\,CN^- + O_2 + 2H_2O \quad \rightarrow \quad 4\,M(CN)_2^- + 4\,OH^- \tag{11}$$

where M stands for Au or Ag. Note that dissolution of the core requires the transport through the silica shell of both molecular oxygen and cyanide ions, and for complete dissolution outward diffusion of metal cyanide anions.

Two features make this process of special relevance: (1) dissolution of the metal cores leads to the formation of hollow, rigid shells; and (2) since the products do not absorb in the visible, the reaction can be monitored through UV-visible spectroscopy by following the gradual disappearance of the plasmon band.

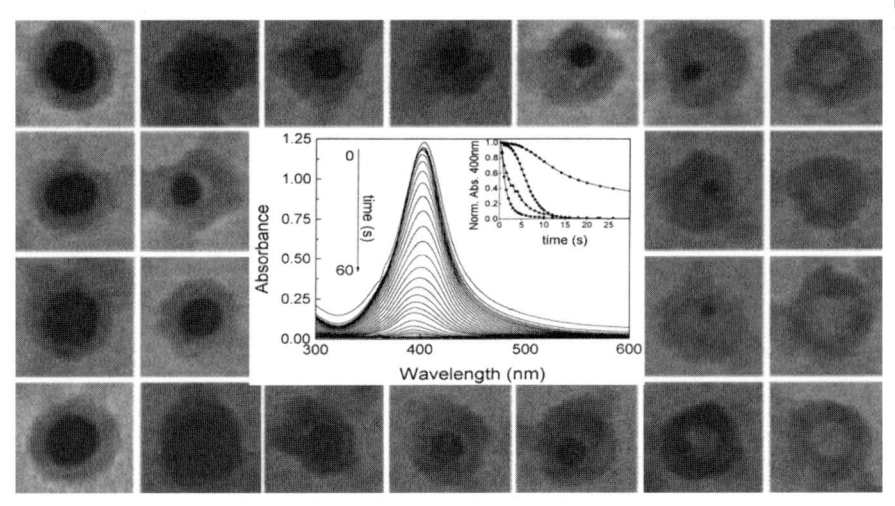

Fig. 7.6 Gallery of TEM images showing the gradual dissolution of Au cores from Au@SiO$_2$, due to sequential exposure to a KCN solution. The inset shows a typical spectral evolution during silver dissolution, as well as kinetic traces for Ag@SiO$_2$ with various shell thicknesses (inset reproduced with permission from [24], copyright (1998) American Chemical Society).

Transmission electron microscopy has undoubtedly shown [118] that the metal cores are indeed oxidized by cyanide ions present in solution. The process was monitored by sequential dipping of a labeled carbon-coated grid in a cyanide solution, drying, and observation of "the same" particles. Fig. 7.6 shows a series of electron micrographs of various Au@SiO$_2$ particles after exposure to 1 mM cyanide ion. After 10 min, the gold cores are completely dissolved, and hollow nano-sized silica shells remained.

A systematic study on the effect of the thickness of the silica shell on the kinetics of oxidation of Au [164] and Ag [24] nanoparticles was performed using time-resolved UV-visible spectroscopy, which showed that citrate-stabilized sols are oxidized very quickly by cyanide ion. Silica shells grown by silicate deposition in water promote a thickness-dependent retardation of the reaction (see inset of Fig. 7.6), while the effect of those grown by TEOS hydrolysis (not shown) is much more dramatic. This can easily be related to the pore size of the different silicas. Sodium silicate forms oligomers in solution, which condense to form silica, yielding relatively large pores, whilst TEOS silica is formed from the condensation of monomers arising from the slow hydrolysis of TEOS. For similar silica shells, the thickness dependency could be modeled based on diffusion of molecules through spherical shells, observing a qualitative agreement with the experimental observations.

It has additionally been observed that boiling of the M@SiO$_2$ sol also leads to a slowing down of the reaction, presumably due to compression of the silica shell through pore shrinkage.

A similar example of metal core dissolution, which was mentioned above, is found when ammonia is added to a Ag@SiO$_2$ colloid [24, 150]. Ammonia catalyzes the oxidation of silver by air, though the reaction:

$$4\,\text{Ag} + 8\,\text{NH}_3 + \text{O}_2 + 2\,\text{H}_2\text{O} \rightarrow 4\,\text{Ag(NH}_3)_2^+ + 4\,\text{OH}^- \tag{12}$$

is noticeably slower than that for cyanide oxidation even when a much higher concentration (0.1 M) is used. As was indicated above, this reaction represents a serious inconvenience for the growth of thick silica shells on silver particles, since the Stöber method [115] requires using high ammonia concentrations.

7.4.2
Protection of Chalcogenides against Photodegradation

In contrast to the previous example, in which a chemical reaction was employed for the transformation of the core material, we present here the inhibition of an undesired reaction, which has practical applications with respect to the stabilization of CdS semiconductor particles by means of silica coating [139].

CdS can be degraded under the influence of light in the presence of dissolved oxygen. The action of oxygen consists of an oxidation of sulfide radicals arising through the formation of hole–anion pairs at the particle surface [165]:

$$\text{S}_\text{S}^{\bullet-} + \text{O}_2 \rightarrow \text{S}_\text{S} + \text{O}_2^{\bullet-} \tag{13}$$

where the subscript S indicates a surface atom. This process can be inhibited through the addition of excess sulfide ions, though an alternative and cleaner way is to carry out the reaction under nitrogen atmosphere. When the synthesis is performed in the absence of oxygen, the colloid is stable for weeks, allowing for modifications to be performed on it without further precautions.

The influence of a silica coating on the stability of CdS colloids with respect to photodegradation was monitored through the temporal decay of the UV-visible spectrum. Silica-coated sols proved to be far more stable against photodegradation than the uncoated ones, so that samples could be stored in light for several months with negligible variation of their absorption spectra. This can be related to the very small pore size of silica prepared by means of Stöber synthesis, which implies growth through monomer addition. This makes it very difficult for O$_2$ molecules to reach the particle surface, thus severely restricting the oxidation process. Such protection has been of great importance in the preparation of nanostructured materials with uniformly distributed semiconductor cores [166], as well as for luminescent quantum dots used as biomarkers [141].

7.4.3
Oxidation of Silver by Molecular Iodine and Sulfide Ions

Giersig et al. [118] have examined the effect of adding molecular iodine, a strong oxidant, to an aqueous dispersion of Ag@SiO$_2$ particles. Molecular iodine oxidizes silver thus:

$$2\,Ag + I_2 \rightarrow 2\,AgI \tag{14}$$

After the addition of molecular I$_2$ to a dispersion of Ag@SiO$_2$ particles with a shell of thickness 10 nm and initial Ag cores of 9-nm diameter, it was observed [118] that the surface plasmon band of the silver core located at 390 nm disappears, and after 30 min the spectrum reveals the typical 420 nm peak of colloidal β-AgI [167], thereby confirming that the observed absorption changes are due to metallic silver oxidation and formation of colloidal AgI, and demonstrating that molecular iodine can diffuse through the 10-nm silica shell and completely oxidize the silver cores. This reaction has the fundamental mechanistic problem that the molar volume of the hexagonal β-AgI phase is 41 cm^3 mol^{-1}, whilst that of face-centered cubic silver is just 10.26 cm^3 mol^{-1}. This means that the core volume must quadruple in order to accommodate the nascent AgI nucleus. The dynamics of this nanoscale physical transformation has also been monitored directly by TEM [118, 168], showing that, after contact with I$_2$, a small nucleus erupts onto the surface of the silica shell and then expands. At the same time, the silver core is seen to shrink, and eventually to disappear altogether. By the completion of the reaction, a single particle can be seen adsorbed to the exterior of a silica nanoshell. Spectroscopic monitoring of the reaction in solution at 390 nm showed that the reaction lasted 20–25 min, slightly longer than the time found by direct TEM analysis. Conversely, the reaction of uncoated silver particles with iodine was completed within several seconds. The final product particle exhibits an increased lattice spacing of 3.4 Å, and a change in crystal symmetry from cubic to hexagonal, which can be unambiguously assigned to the [110] plane of hexagonal AgI. From the TEM images, it is clear that large volume changes occur during the reaction. The adsorbed AgI particle has an apparent diameter of 17.5 nm, as compared to a silver core of just 9.0 nm diameter. If we assume that the AgI is roughly hemispherical, then its volume is indeed about four times that of the core, as predicted from the respective molar volumes.

Whilst surface nucleation of AgI clearly occurs, the mechanism itself is less obvious. A pore transport model was suggested [118], based on the known semiconducting properties of AgI. In the first stage, molecular iodine diffuses through the pores to the silver surface where it begins to corrode the metal surface. Since the volume of the corroded core is larger than the untarnished Ag, the pore rapidly begins to fill with AgI. The expansion exerts strong pressure on the pore, causing it to dilate, or even to rupture, which in turn facilitates the access of molecular iodine to the silver particle surface. The initial AgI nucleation must occur at the silver particle surface. Once AgI has formed, the system becomes a three-phase one, with iodine on the outer surface and a reservoir of silver metal on the inner surface. Though AgI is primarily an ionic conductor [169], there is a measurable contribution from electronic charge carriers. Thus electrons can be transported through the AgI layer, and the AgI will not act as a passivation layer on the silver surface. In fact, it is well known that silver metal deposited onto silver halide crystals spontaneously diffuses into the crystal [170]. The silver particle can thus act as an electron source for the growing AgI nucleus. As AgI fills the small access pores, both electrons and silver ions (or silver atoms) diffuse through the AgI lattice to the AgI/SiO_2 pore interface thereby facilitating reaction. As soon as one pore fills up completely, continued corrosion on the external silica surface is possible and subsequent reaction then occurs preferentially at this point. Therefore surface AgI formation is inevitable if silver metal can migrate through the nascent AgI forming in the pores and access the external solution. According to this simple model, the AgI actually forms at the surface and not in the core at all. The AgI phase that emerges onto the external silica surface grows by spherical diffusion in the bulk solution, where the I_2 concentration is much higher, and diffusion is unimpeded.

A similar process takes place when sodium sulfide solution is added to a Ag@SiO_2 colloid [24]. Sulfide ion is particularly aggressive towards silver metal, and oxidation of the naked sols is instantaneous. The final spectrum resembles that of colloidal Ag_2S, which has been reported by Henglein et al. [171]. Silver sulfide is a difficult colloid to prepare in solution. It has a small band gap and readily coalesces. The permeable silica shell not only allows the semiconductor to be synthesized directly as a core but simultaneously acts to control the particle size. It may be possible to synthesize a number of semiconducting materials as quantized particles in aqueous solution by admission of chalcogenide gases to silica-coated metal or metal oxide precursors via this route.

7.5
Conclusions

This review of the modification of metal and semiconductor nanoparticles via chemical reactions intended to show that surface modification through deposition of shells on preformed nanoparticles is a very useful technique for a large number of applications. Although we have basically concentrated here on the description of optical properties, other applications have also been mentioned where the presence of an outer shell clearly leads to an improvement of the efficiency, as compared to naked particles. We have also shown that new materials can be synthesized by starting with a core-shell geometry and selectively performing chemical reactions, either on the core or on the shell.

7.6
Acknowledgement

The author of this chapter is indebted to many collaborators and students, who have either performed part of the work described or held numerous discussions related to the synthesis and processing of surface-modified nanoparticles.

7.8
References

1. B. V. DERYAGIN, L. LANDAU, *Acta Physicochim. URSS* **1941**, *14*, 633–662.
2. E. J. W. VERWEY, J. TH. G. OVERBEEK, *Theory of the Stability of Lyophobic Colloids*, Elsevier, New York, **1948**.
3. P. HIEMENZ, *Principles of Colloid and Surface Chemistry*, Marcel Dekker, New York, **1986**.
4. G. R. WIESE, T. W. HEALY, *Trans. Faraday Soc.* **1970**, *66*, 490–499.
5. U. KREIBIG, U. GENZEL, *Surf. Sci.* **1985**, *156*, 678–700.
6. U. KREIBIG, *J. Phys. F: Met. Phys.* **1974**, *4*, 999–1014.
7. U. KREIBIG, *Physik B* **1978**, *31*, 39–47.
8. U. KREIBIG, *J. Phys. (Paris)* **1977**, *C2 38*, 97–103.
9. P. MULVANEY, *Langmuir* **1996**, *12*, 788–800.
10. I. LISIECKI, F. BILLOUDET, M. P. PILENI, *J. Phys. Chem.* **1996**, *100*, 4160–4166.
11. Y. Y. YU, S. S. CHANG, C. L. LEE, C. R. WANG, *J. Phys. Chem. B* **1997**, *101*, 6661–6664.
12. R. C. JIN, Y. W. CAO, C. A. MIRKIN, K. L. KELLY, G. C. SCHATZ, J. G. ZHENG, *Science* **2001**, *204*, 1901–1903.
13. I. PASTORIZA-SANTOS, L. M. LIZ MARZÁN, *Nano Lett.* **2002**, *2*, 903–905.
14. J. A. CREIGHTON, D. G. EADON, *J. Chem. Soc., Faraday Trans.* **1991**, *87*, 3881–3891.
15. B. M. I. VAN DER ZANDE, M. R. BÖMER, L. G. J. FOKKINK, C. SCHÖNEBERGER, *J. Phys. Chem. B* **1997**, *101*, 852–854.
16. C. A. FOSS, JR., G. L. HORNYAK, J. A. STOCKERT, C. R. MARTIN, *J. Phys. Chem.* **1994**, *98*, 2963–2971.
17. S. LINK, M. B. MOHAMED, M. A. EL-SAYED, *J. Phys. Chem. B* **1999**, *103*, 3073–3077.
18. S. UNDERWOOD, P. MULVANEY, *Langmuir* **1994**, *10*, 3427–3430.
19. L. M. LIZ MARZÁN, I. LADO TOURIÑO, *Langmuir* **1996**, *12*, 3585–3589.
20. P. MULVANEY, M. GIERSIG, A. HENGLEIN, *J. Phys. Chem.* **1993**, *97*, 7061–7064.
21. A. HENGLEIN, P. MULVANEY, A. HOLZWARTH, T. E. SOSEBEE, A. FOJTIK, *Ber. Bunsenges. Phys. Chem.* **1992**, *96*, 754–759.
22. N. TOSHIMA, *Macromol. Symp.* **1996**, *105*, 111–118.
23. L. KATSIKAS, M. GUTIÉRREZ, A. HENGLEIN, *J. Phys. Chem.* **1996**, *100*, 11203–11206.
24. T. UNG, L. M. LIZ MARZÁN, P. MULVANEY, *Langmuir* **1998**, *14*, 3740–3748.
25. M. MICHAELIS, A. HENGLEIN, P. MULVANEY, *J. Phys. Chem.* **1994**, *98*, 6212–6215.
26. L. M. LIZ MARZÁN, M. GIERSIG, P. MULVANEY, *J. Chem. Soc., Chem. Commun.* **1996**, 731–732.
27. L. M. LIZ MARZÁN, M. GIERSIG, P. MULVANEY, *Langmuir* **1996**, *12*, 4329–4335.
28. T. UNG, M. GIERSIG, D. DUNSTAN, P. MULVANEY, *Langmuir* **1997**, *13*, 1773–1782.
29. T. UNG, L. M. LIZ MARZÁN, P. MULVANEY, *J. Phys. Chem. B* **1999**, *103*, 6770–6773.
30. P. MULVANEY, T. LINNERT, A. HENGLEIN, *J. Phys. Chem.* **1991**, *95*, 7843–7846.
31. L. M. LIZ MARZÁN, P. MULVANEY, *New J. Chem.* **1998**, 1285–1288.
32. R. H. DOREMUS, *J. Chem. Phys.* **1964**, *40*, 2389–2396.
33. S. LINK, M. A. EL-SAYED, *J. Phys. Chem. B* **1999**, *103*, 4212–4217.
34. G. MIE, *Ann. Phys.* **1908**, *25*, 337–445.
35. M. KERKER, *The Scattering of Light and Other Electromagnetic Radiation*, Academic Press, New York, **1969**.
36. H. C. VAN DER HULST, *Light Scattering by Small Particles*, Wiley, New York, **1957**.
37. C. F. BOHREN, D. R. HUFFMAN, *Absorption and Scattering of Light by Small Particles*, Wiley, New York, **1983**.

38 C. Kittel, *Introduction to Solid State Physics*, Wiley, New York, **1956**.
39 N. W. Ashcroft, N. D. Mermin, *Solid State Physics*, Holt, Rinehardt & Winston, Philadelphia, **1976**.
40 L. M. Liz Marzán, in *Nanoscale Materials* (L. M. Liz Marzán, P. V. Kamat, eds.), Kluwer Academic, Boston, 227–246.
41 R. H. Morriss, L. F. Collins, *J. Chem. Phys.* **1964**, *41*, 3357–3363.
42 J. Sinzig, U. Radtke, M. Quinten, U. Kreibig, *Z. Phys. D* **1993**, *26*, 242–245.
43 F. Henglein, A. Henglein, P. Mulvaney, *Ber. Bunsenges. Phys. Chem.* **1994**, *98*, 180–189.
44 P. Mulvaney, M. Giersig, A. Henglein, *J. Phys. Chem.* **1993**, *97*, 7061–7064.
45 A. Henglein, M. Giersig, *J. Phys. Chem.* **1994**, *98*, 6931–6935.
46 A. Henglein, A. Holzwarth, P. Mulvaney, *J. Phys. Chem.* **1992**, *96*, 8700–8702.
47 A. Henglein, P. Mulvaney, A. Holzwarth, *J. Phys. Chem.* **1992**, *96*, 2411–2414.
48 P. Mulvaney, M. Giersig, A. Henglein, *J. Phys. Chem.* **1992**, *96*, 10419–10424.
49 A. Henglein, M. Giersig, *J. Phys. Chem. B* **2000**, *104*, 5056–5060.
50 A. Henglein, *J. Phys. Chem. B* **2000**, *104*, 2201–2203.
51 A. Henglein, *J. Phys. Chem. B* **2000**, *104*, 6683–6685.
52 J. H. Hodak, A. Henglein, M. Giersig, G. V. Hartland, *J. Phys. Chem. B* **2000**, *104*, 11708–11718.
53 J. L. Marignier, J. Belloni, M. O. Delcourt, J. P. Chevalier, *Nature* **1985**, *317*, 344–345.
54 J. Belloni, M. Mostafavi, H. Remita, J. L. Marignier, M. O. Delcourt, *New J. Chem.* **1998**, 1239–1255.
55 M. Treguer, C. de Cointet, H. Remita, J. Khatouri, J. Amblard, J. Belloni, R. de Keyzer, *J. Phys. Chem. B* **1998**, *102*, 4310–4321.
56 Y. Wang, N. Toshima, *J. Phys. Chem. B* **1997**, *101*, 5301–5306.
57 A. F. Lee, C. J. Baddeley, C. Hardacre, R. M. Ormerod, R. M. Lambert, G. Schmid, H. West, *J. Phys. Chem.* **1995**, *99*, 6096–6102.
58 M. S. Nashner, A. I. Frenkel, D. Somerville, C. W. Hills, J. R. Shapley, R. G. Nuzzo, *J. Am. Chem. Soc.* **1998**, *120*, 8093–8101.
59 Y. Mizukoshi, T. Fujimoto, Y. Nagata, R. Oshima, Y. Maeda, *J. Phys. Chem. B* **2000**, *104*, 6028–6032.
60 R. M. Bright, D. G. Walter, M. D. Musick, M. A. Jackson, K. J. Allison, M. J. Natan, *Langmuir* **1996**, *12*, 810–817.
61 R. G. Freeman, M. B. Hommer, K. C. Grabar, M. A. Jackson, M. J. Natan, *J. Phys. Chem.* **1996**, *100*, 718–724.
62 L. Rivas, S. Sánchez-Cortes, J. V. García-Ramos, G. Morcillo, *Langmuir* **2000**, *16*, 9722–9728.
63 L. Lu, H. Wang, Y. Zhou, S. Xi, H. Zhang, J. Hu, B. Zhao, *Chem. Commun.* **2002**, 144–145.
64 I. Srnová-Sloufová, F. Lednicky, A. Gemperle, J. Gemperlová, *Langmuir* **2000**, *16*, 9928–9935.
65 K. R. Brown, M. J. Natan, *Langmuir* **1998**, *14*, 726–728.
66 Y. W. Cao, R. Jin, C. A. Mirkin, *J. Am. Chem. Soc.* **2001**, *123*, 7961–7962.
67 Y. Sun, B. T. Mayers, Y. Xia, *Nano Lett.* **2002**, *2*, 481–485.
68 M. Schierhorn, L. M. Liz Marzán, *Nano Lett.* **2002**, *2*, 13–16.
69 Y. Kobayashi, V. Salgueiriño-Maceira, L. M. Liz Marzán, *Chem. Mater.* **2001**, *13*, 1630–1633.
70 A. González-Penedo, I. Lado Touriño, M. A. López-Quintela, J. Quibén Solla, J. Rivas Rey, J. M. Greneche, *High Temp. Chem. Processes* **1994**, *3*, 507–515.
71 C. T. Seip, C. J. O'Connor, *Nanost. Mater.* **1999**, *12*, 183–186.
72 A. P. Alivisatos, *Science* **1996**, *271*, 933–937.
73 L. L. Beechroft, C. K. Ober, *Chem. Mater.* **1997**, *9*, 1302–1317.
74 N. Herron, D. L. Thorn, *Adv. Mater.* **1998**, *10*, 1173–1184.
75 U. Woggon, *Optical Properties of Semiconductor Quantum Dots*, Springer-Verlag, Berlin, **1997**.
76 T. Vossmeyer, S. Jia, E. Delonno, M. R. Diehl, X. Peng, A. P. Alivisatos, J. R. Heath, *J. Appl. Phys.* **1998**, *84*, 3664–3670.
77 B. O. Dabbousi, J. Rodriguez-Viejo, F. V. Mikulec, J. R. Heine, H. Mattoussi, R. Ober, K. F. Jensen, M. G. Bawendi, *J. Phys. Chem. B* **1997**, *101*, 9463–9475.

78 M. Danek, K.F. Jensen, C.B. Murray, M.G. Bawendi, *Chem. Mater.* **1996**, *8*, 173–180.
79 X. Peng, M.C. Schlamp, A.V. Kadavanich, A.P. Alivisatos, *J. Am. Chem. Soc.* **1997**, *119*, 7019–7029.
80 M.A. Hines, P. Guyot-Sionnest, *J. Phys. Chem.* **1996**, *100*, 468–471.
81 A. Eychmüller, A. Hasselbarth, H. Weller, *J. Luminescence* **1992**, *53*, 113–115.
82 A. Hasselbarth, A. Eychmüller, M. Eichberger, M. Giersig, A. Mews, H. Weller, *J. Phys. Chem.* **1993**, *97*, 5333–5340.
83 V.F. Kamalov, R. Little, S.L. Logunov, M.A. El-Sayed, *J. Phys. Chem.* **1996**, *100*, 6381–6384.
84 H.S. Zhou, H. Sasahara, I. Homma, H. Komiyama, J. Haus, *Chem. Mater.* **1994**, *6*, 1534–1541.
85 A.P. Alivisatos, *J. Phys. Chem.* **1996**, *100*, 13226–13239.
86 G. Counio, S. Esnouf, T. Gacoin, J.P. Boilot, *J. Phys. Chem.* **1996**, *100*, 20021–20026.
87 M.Y. Han, W. Huang, C.H. Chew, L.M. Gan, X.J. Zhang, W. Ji, *J. Phys. Chem. B* **1998**, *102*, 1884–1887.
88 Y. Tian, T. Newton, N.A. Kotov, D.M. Guldi, J.H. Fendler, *J. Phys. Chem.* **1996**, *100*, 8927–8939.
89 D. Diaz, J. Robles, T. Ni, S.-E. Castillo-Blum, D. Nagesha, O.-J.A. Fregoso, N.A. Kotov, *J. Phys. Chem. B* **1999**, *103*, 9854–9858.
90 C.F. Hoener, K.A. Allan, A.J. Bard, A. Campion, A.M. Fox, T.E. Mallouk, S.E. Webber, J.M. White, *J. Phys. Chem.* **1992**, *96*, 3812–3817.
91 E. Hao, H. Sun, Z. Zhou, J. Liu, B. Yang, J. Shen, *Chem. Mater.* **1999**, *11*, 3096–3102.
92 C.B. Murray, D.J. Norris, M.G. Bawendi, *J. Am. Chem. Soc.* **1993**, *115*, 8706–8715.
93 N. Revaprasadu, M.A. Malik, P. O'Brien, G. Wakefield, *Chem. Commun.* **1999**, 1573–1574.
94 M.A. Malik, P. O'Brien, N. Revaprasadu, *Chem. Mater.* **2002**, *14*, 2004–2010.
95 Y. Xie, P. Yan, J. Lu, Y. Qian, S. Zhang, *Chem. Commun.* **1999**, 1969–1970.
96 B. Schreder, T. Schmidt, V. Patschek, U. Winkler, A. Materny, E. Umbach, M. Lerch, G. Müller, W. Kiefer, L. Spanhel, *J. Phys. Chem. B* **2000**, *104*, 1677–1685.
97 Y.-W. Cao, U. Banin, *Angew. Chem. Int. Ed.* **1999**, *38*, 3692–3694.
98 Y.-W. Cao, U. Banin, *J. Am. Chem. Soc.* **2000**, *122*, 9692–9702.
99 R.K. Iler, *The Chemistry of Silica*, Wiley, New York, **1979**.
100 L.H. Allen, E. Matijevic, *J. Colloid Interface Sci.* **1969**, *31*, 287–296.
101 J. Depasse, A. Watillon, *J. Colloid Interface Sci.* **1970**, *33*, 431–439.
102 G. Vigil, Z. Xu, S. Steinberg, J. Israelachvili, *J. Colloid Interface Sci.* **1994**, *165*, 367–385.
103 F. Küspert, *Ber.* **1919**, *35*, 2815.
104 D.N. Furlong, P.A. Freeman, A.C.M. Lau, *J. Colloid Interface Sci.* **1981**, *80*, 20–31.
105 H.E. Bergna, in *The Colloid Chemistry of Silica* (H.E. Bergna, ed.), American Chemical Society, Washington DC, **1994**, pp. 1–47.
106 T.W. Healy, A.M. Homola, R.O. James, R.J. Hunter, *Faraday Discuss. Chem. Soc.* **1978**, *65*, 156–163.
107 T.W. Healy, in *The Colloid Chemistry of Silica* (H.E. Bergna, ed.), American Chemical Society, Washington DC, **1994**, pp. 147–159.
108 R.K. Iler, US Patent No. 2, 885, 366, **1959**.
109 A.P. Philipse, M.P. van Bruggen, C. Pathmamanoharan, *Langmuir* **1994**, *10*, 92–99.
110 A.P. Philipse, A.M. Nechifor, C. Pathmamanoharan, *Langmuir* **1994**, *10*, 4451–4458.
111 M. Ohmori, E. Matijevic, *J. Colloid Interface Sci.* **1993**, *160*, 288–292.
112 M. Ohmori, E. Matijevic, *J. Colloid Interface Sci.* **1992**, *150*, 594–598.
113 A.N. Patil, R.P. Andres, N. Otsuka, *J. Phys. Chem.* **1994**, *98*, 9247–9251.
114 L.M. Liz Marzán, A.P. Philipse, *J. Colloid Interface Sci.* **1995**, *176*, 459–466.
115 W. Stöber, A. Fink, E. Bohn, *J. Colloid Interface Sci.* **1968**, *26*, 62–69.
116 E.P. Plueddermann, *Silane Coupling Agents*, Plenum Press, New York, **1991**.
117 P.A. Buining, L.M. Liz Marzán, A.P. Philipse, *J. Colloid Interface Sci.* **1996**, *179*, 318–321.

118 M. Giersig, T. Ung, L. M. Liz Marzán, P. Mulvaney, *Adv. Mater.* **1997**, *9*, 570–575.

119 Y. Kobayashi, M. A. Correa-Duarte, L. M. Liz Marzán, *Langmuir* **2001**, *17*, 6375–6379.

120 S. T. Selvan, T. Hayakawa, M. Nogami, Y. Kobayashi, L. M. Liz Marzán, Y. Hamanaka, A. Nakamura, *J. Phys. Chem. B*, **2002**, *107*, 10157–10162.

121 F. Caruso, M. Spasova, V. Salgueiriño-Maceira, L. M. Liz Marzán, *Adv. Mater.* **2001**, *13*, 1090–1094.

122 S. S. Chang, C. W. Shih, C. D. Chen, W. C. Lai, C. R. Wang, *Langmuir* **1999**, *15*, 701–709.

123 S. O. Obare, N. R. Jana, C. J. Murphy, *Nano Lett.* **2001**, *1*, 601–603.

124 S.-T. Yau, P. Mulvaney, W. Xu, G. M. Spinks, *Phys. Rev. B* **1998**, *57*, R15124–R15127.

125 S. R. Hall, S. A. Davis, S. Mann, *Langmuir* **2000**, *16*, 1454–1456.

126 V. V. Hardikar, E. Matijevic, *J. Colloid Interface Sci.* **2000**, *221*, 133–136.

127 Y. Yin, Y. Lu, Y. Sun, Y. Xia, *Nano Lett.* **2002**, *2*, 427–430.

128 Y. Lu, Y. Yin, Z.-Y. Li, Y. Xia, *Nano Lett.* **2002**, *2*, 785–788.

129 E. Mine, A. Yamada, Y. Kobayashi, M. Konno, L. M. Liz Marzán, *J. Colloid Interface Sci.* **2003**, *264*, 385–390.

130 O. V. Makarova, A. E. Ostafin, H. Miyoshi, J. R. Norris, Jr, D. Meisel, *J. Phys. Chem. B* **1999**, *103*, 9080–9084.

131 R. I. Nooney, T. Dhanasekaran, Y. Chen, R. Josephs, A. E. Ostafin, *Adv. Mater.* **2002**, *14*, 529–532.

132 T. Li, J. Moon, A. A. Morrone, J. J. Mecholsky, D. R. Talhman, J. H. Adair, *Langmuir* **1999**, *15*, 4328–4334.

133 A. Martino, S. A. Yamanaka, J. S. Kawola, D. A. Loy, *Chem. Mater.* **1997**, *9*, 423–429.

134 I. Pastoriza-Santos, L. M. Liz Marzán, *Langmuir* **1999**, *15*, 948–951.

135 I. Pastoriza-Santos, L. M. Liz Marzán, *Pure Appl. Chem.* **2000**, *72*, 83–90.

136 I. Lee, S. W. Han, H. J. Choi, K. Kim, *Adv. Mater.* **2001**, *13*, 1617–1620.

137 S. Chang, L. Liu, S. A. Asher, *J. Am. Chem. Soc.* **1994**, *116*, 6745–6747.

138 S. Chang, L. Liu, S. A. Asher, *J. Am. Chem. Soc.* **1994**, *116*, 6739–6744.

139 M. A. Correa-Duarte, M. Giersig, L. M. Liz Marzán, *Chem. Phys. Lett.* **1998**, *286*, 497–501.

140 M. Bruchez, Jr., M. Moronne, P. Gin, S. Weiss, A. P. Alivisatos, *Science* **1998**, *281*, 2013–2016.

141 D. Gerion, F. Pinaud, S. C. Williams, W. J. Parak, D. Zanchet, S. Weiss, A. P. Alivisatos, *J. Phys. Chem. B* **2001**, *105*, 8861–8871.

142 A. L. Rogach, D. K. Nagesha, J. W. Ostrander, M. Giersig, N. A. Kotov, *Chem. Mater.* **2000**, *12*, 2676–2685.

143 Y. Wang, Z. Tang, M. A. Correa-Duarte, L. M. Liz Marzan, N. A. Kotov, *J. Am. Chem. Soc.* **2003**, *125*, 2830–2831.

144 P. V. Kamat, B. Shanghavi, *J. Phys. Chem. B* **1997**, *101*, 7675–7679.

145 C. Nayral, T. Ould-Ely, A. Maisonnat, B. Chaudret, P. Fau, L. Lescouzères, A. Peyre-Lavigne, *Adv. Mater.* **1999**, *11*, 61–63.

146 R. Nayak, J. Galsworthy, P. Dobson, J. Hutchinson, *J. Mater. Res.* **1998**, *13*, 905–908.

147 H. Tada, K. Teranishi, S. Ito, *Langmuir* **1999**, *15*, 7084–7087.

148 T. Hirano, T. Oku, K. Suganuma, *J. Mater. Chem.* **1999**, *9*, 855–857.

149 I. Pastoriza-Santos, D. Koktysch, A. Mamedov, N. A. Kotov, L. M. Liz Marzán, *Langmuir* **2000**, *16*, 2731–2735.

150 D. S. Koktysch, X. Liang, B.-G. Yun, I. Pastoriza-Santos, R. L. Matts, M. Giersig, C. Serra-Rodríguez, L. M. Liz Marzán, N. A. Kotov, *Adv. Funct. Mater.* **2002**, *12*, 255–265.

151 K. S. Mayya, D. I. Gittins, F. Caruso, *Chem. Mater.* **2001**, *13*, 3833–3836.

152 K. S. Mayya, D. I. Gittins, A. M. Dibaj, F. Caruso, *Nano Lett.* **2001**, *1*, 727–730.

153 G. Oldfield, T. Ung, P. Mulvaney, *Adv. Mater.* **2000**, *12*, 1519–1522.

154 S. M. Marinakos, L. C. Brousseau, A. Jones, D. L. Feldheim, *Chem. Mater.* **1998**, *10*, 1214–1219.

155 S. M. Marinakos, J. P. Novak, L. C. Brousseau, A. B. House, E. M. Edeki, D. L. Feldheim, *J. Am. Chem. Soc.* **1999**, *121*, 8518–8522.

156 S. M. Marinakos, D. A. Shultz, D. L. Feldheim, *Adv. Mater.* **1999**, *11*, 34–37.
157 S. T. Selvan, J. P. Spatz, H. A. Klok, M. Möller, *Adv. Mater.* **1998**, *10*, 132–134.
158 S. T. Selvan, *Chem. Commun.* **1998**, 351–352.
159 H. Menzel, M. D. Mowery, M. Cai, C. E. Evans, *Adv. Mater.* **1999**, *11*, 131–134.
160 L. Quaroni, G. Chumanov, *J. Am. Chem. Soc.* **1999**, *121*, 10642–10643.
161 T. K. Mandal, M. S. Fleming, D. R. Walt, *Nano Lett.* **2002**, *2*, 3–7.
162 A. van Blaaderen, A. P. M. Kentgens, *J. Non-Cryst. Solids* **1992**, *149*, 161–178.
163 A. van Blaaderen, A. Vrij, *J. Colloid Interface Sci.* **1993**, *156*, 1–18.
164 L. M. Liz Marzán, P. Mulvaney, *Rec. Res. Devel. Phys. Chem.* **1998**, *2*, 1–32.
165 A. Henglein, *Ber. Bunsenges. Phys. Chem.* **1982**, *86*, 301–305.
166 M. Alejandro-Arellano, A. Blanco, T. Ung, P. Mulvaney, L. M. Liz Marzán, *Pure Appl. Chem.* **2000**, *72*, 257–267.
167 P. Mulvaney, *Colloids Surf. A* **1993**, *81*, 231–238.
168 M. Giersig, L. M. Liz Marzán, T. Ung, D. Su, P. Mulvaney, *Ber. Bunsenges. Phys. Chem.* **1997**, *101*, 1617–1620.
169 H. Hoshino, M. Shimoji, *J. Phys. Chem. Solids* **1974**, *35*, 321–326.
170 A. L. Laskar, *Mater. Sci. Forum* **1984**, *1*, 59–74.
171 L. Spanhel, H. Weller, A. Fojtik, A. Henglein, *Ber. Bunsenges. Phys. Chem.* **1987**, *91*, 88–94.

8
Nanoscale Particle Modification via Sequential Electrostatic Assembly

Frank Caruso

8.1
Introduction

Colloid particles have been of interest for centuries, from both scientific and technological perspectives. Over the years they have aroused the interest of scientists, which has led to an enhanced understanding of the principal factors governing colloidal interactions and stabilization [1–5]. Such studies have impacted the disciplines of chemistry, physics, biology, pharmacy, chemical engineering, and materials science. Technologically, they have been, and continue to be, routinely utilized in diverse areas, including catalysis, photography, separations, diagnostics, agriculture, and mining. They are also widely used as additives and pigments (e.g. in foods, cosmetics, inks, and paints) [1–5]. Recently, it has become apparent that for significant advances to be made in high-technology applications in the rapidly emerging fields of nanotechnology and biotechnology, tailor-made building blocks with designed and functional properties are required. Colloid particles are as such useful building blocks, and although the existing suite of available particles (e.g. latexes, silica, gold, semiconductors) are increasingly finding use as components in the construction of advanced materials and devices for technology-driven applications [6–18], considerable interest has focused on strategies that allow the preparation of colloids with unique and tailored functional properties (e.g. optical, mechanical, thermal, electrical, magnetic, catalytic, biological) [17, 18].

A promising route to prepare colloids with novel physical and chemical properties involves taking core particles and modifying their surface by depositing various materials [17, 18]. This process can be referred to as particle coating, or nanoscale coating when the deposited materials form coatings of thickness in the nanometer range. In coating particles, control and optimization of the surface characteristics of the particles is of utmost importance, particularly with respect to colloidal stability. Unwanted levels of particle aggregation need to be avoided, as stable colloidal dispersions are required for subsequent exploitation of the modified particles in applications. Not only do coated (i.e. surface-modified) colloids often display improved properties over the single-component particles, but they can also be prepared with enhanced colloidal stability. This makes them attractive candidates for use in numerous areas, as exemplified by their application in encapsu-

Colloids and Colloid Assemblies. Edited by Frank Caruso
Copyright © 2004 Wiley-VCH Verlag GmbH & Co. KGaA, Weinheim
ISBN: 3-527-30660-9

lation, catalysis, drug delivery, electronics, separations, diagnostics, and in the synthesis of combinatorial libraries [17–24]. Controlling the surface properties of colloidal materials has now become a major research area encompassing studies in the pharmaceutical, mining, semiconductor, biological, and medical fields.

Both organic and inorganic coatings have been formed on particles by a variety of techniques [17, 18]. Polymerization-based methods are the most commonly used to form polymer coatings: these include polymerization of monomers adsorbed onto the particle surface [25, 26], heterocoagulation-polymerization [27], and emulsion polymerization [18, 28]. Inorganic coatings are most often formed by chemical reactions, viz. the precipitation and surface reactions of inorganic precursors with the particle surface [29–31]. Chapter 7 deals with the formation of coatings on particles via chemical reactions.

A significant advance in the surface modification of particles originated from the application of the now well-known "layer-by-layer" (LbL) method to colloid particles [17]. This approach, adapted from that used to modify planar supports [32–35], entails the nanoscale sequential coating of colloid particles with multiple layers of oppositely charged materials, utilizing electrostatic interactions for layer buildup. It represents a facile, non-covalent approach that is suitable for modifying particles of different size, shape, and composition. In addition, it provides stepwise nanoscale control over the thickness, composition, and structure of the deposited layers. Control over these parameters provides the basis for the preparation of surface-modified colloid particles with tailored chemical and physical properties.

This chapter provides an overview of the utilization of the LbL technique to modify the surface of colloid particles. Section 8.8.2 will deal with the principles of the LbL assembly technique, specifically focusing on its application to colloids. In the following sub-sections, emphasis is placed on the variety of layer constituents and core particles that can be used in the process. A number of systems are presented, showing the applicability of polyelectrolytes, nanoparticles, proteins and/or low molecular weight compounds as layer deposition materials. The control that the LbL method permits over the composition and wall thickness of the layers deposited is also detailed. Examples involving the use of core particles of different size, shape, and composition are given. In Section 8.8.3, the use of polyelectrolyte multilayers as matrices for confining materials at or near the particle surface via adsorption and chemical reactions is discussed. Finally, Section 8.8.4 focuses on promising research directions for application of LbL surface-modified particles, for example, biocatalysis, immunodiagnostics, drug delivery, photonics, and porous materials.

8.2
Sequential Electrostatic Assembly Applied to Particles

The origins of LbL electrostatic assembly of charged species onto planar surfaces can be traced back to the work of Iler in the mid-1960s [32]. In that work it was shown that particles could be sequentially deposited onto solid substrates. In the early 1990s, the pioneering work of Decher and Hong showed that polycations and polyanions (charged polymers) could also be LbL assembled onto solid supports [33, 34]. Following that work, there was a rapid expansion in the area as it became apparent that a broad range of charged materials could be assembled by the LbL technique [35]. The approach was applied to numerous other charged species, including nanoparticles, biomolecules, and dyes. For example, polyelectrolyte-composite multilayer films were formed by the sequential adsorption of a charged material and an oppositely charged polyelectrolyte. The presence of polyelectrolyte interlayers, which mediate film growth, was found to be important in the regular formation and stability of composite films where nanoparticles and dyes were used. Several recent reviews have been published on the preparation, characterization, and utilization of LbL-assembled thin films on *planar* supports (e.g. glass, gold, and graphite) [35–37]. Further details concerning the LbL assembly of nanoparticle-polyelectrolyte multilayer films can be found in Chapters 2 and 13.

Application of the LbL method to *particles* involves the sequential adsorption of charged materials onto particles [17]. Surface functionality and optical, catalytic, and magnetic properties can be imparted to particles through the LbL-deposited materials. Several groups have utilized the LbL technique to modify colloid particles. In 1995, Keller et al. reported the construction of alternating composite multilayers of exfoliated zirconium phosphate sheets and charged redox polymers on silane-modified silica particles [38]. Several years later, Chen and Somasunduran deposited nanosized alumina particles in alternation with poly(acrylic acid), which acts as the bridging polymer, on submicrometer-sized alumina core particles [39]. Dokoutchaev et al. alternately assembled metal nanosized particles (Au, Pd and Pt) and oppositely charged polyelectrolyte onto polystyrene (PS) microspheres [40]. Other related studies, which date back to 1994, have involved biomaterials. It was shown that sequentially deposited polycation/polyanion multilayer coatings on preformed alginate-based microcapsules are effective in controlling their permeability [41], and that enzymes could be retained within alginate microcapsules upon coating with polyelectrolyte multilayers [41, 42]. It was also demonstrated that the number of alginate and chitosan layers deposited onto the microparticles determines the diffusion of enzymes through the layers [43].

In a series of studies beginning in 1998, it has been shown that the LbL method allows the deposition of multiple layers of various polyelectrolytes on colloids in a controlled fashion [45–51]. By substituting one of the charged polyelectrolyte components with preformed inorganic nanoparticles [52–63], proteins [64–68], or low molecular weight substances [69, 70], it was demonstrated that a wide range of nanocomposite layers could be prepared on submicrometer-sized particles via this approach. Additionally, the use of molecular precursors that can be LbL as-

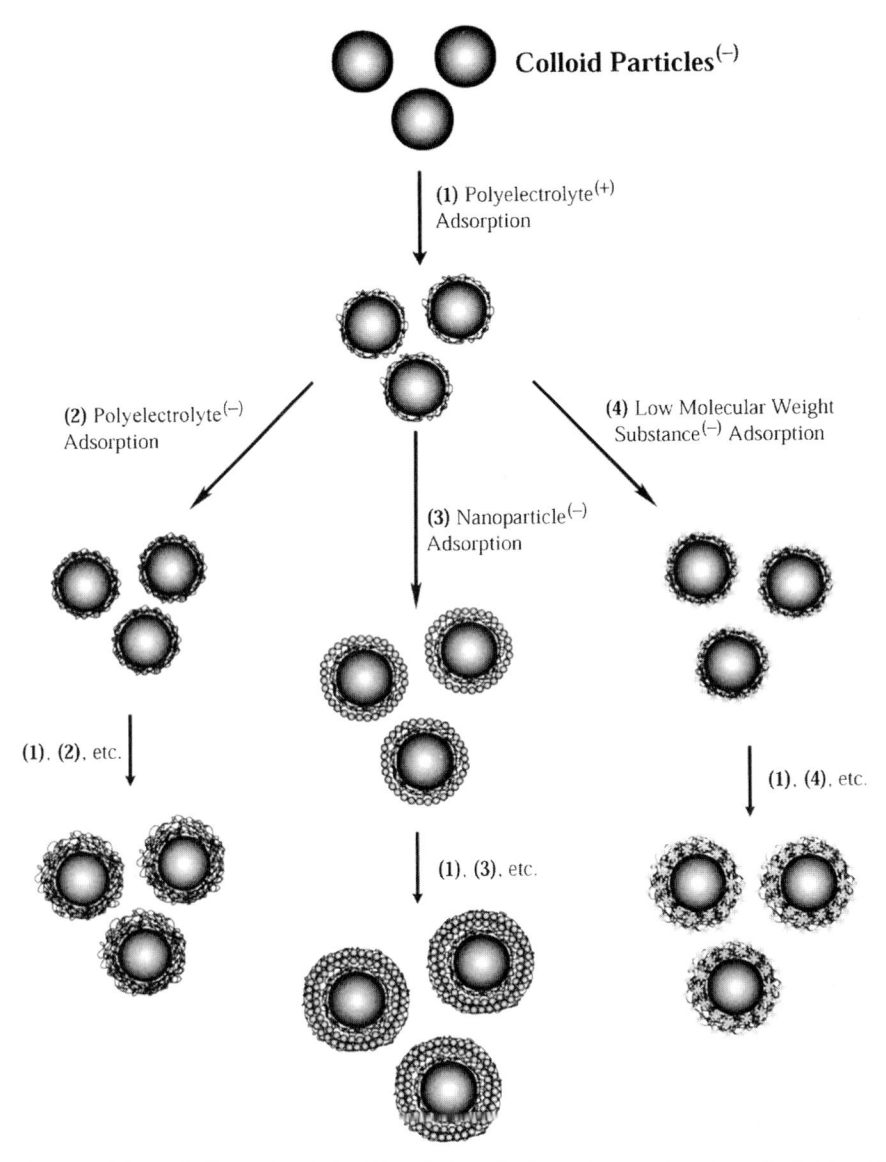

Fig. 8.1 Schematic illustration of the LbL technique for the surface modification of colloid particles. The procedure involves the sequential electrostatic assembly of oppositely charged species onto particles. Following deposition of each species, unadsorbed material is removed by repeated centrifugation/wash or filtration/wash cycles. The coatings can be polyelectrolytes (steps 1 and 2), or nanocolloids such as inorganic nanoparticles or proteins (steps 1 and 3), or low molecular weight substances such as dyes, lipids, or molecular precursors (steps 1 and 4). The templates can be of different composition, size, and shape (latex particles, metal nanorods and nanoparticles, proteins, and cells).

sembled with polyelectrolytes onto colloids to obtain multilayer coatings was reported [70–72]. Recently, Wang and coworkers [73] and Valtchev and Mintova [74, 75] reported the coating of PS particles with various types of zeolite nanoparticles. Dong et al. also deposited silver nanoparticles onto PS spheres by the LbL method [76].

This section will focus on our recent work on the LbL deposition of various charged materials on colloid particles of different size, shape and composition. In these studies, the layering process has been investigated step-by-step. Experimental techniques such as electrophoresis, single particle light scattering, fluorescence spectroscopy, electron microscopy, and atomic force microscopy provide evidence for the surface modification of particles through LbL nanolayering.

In the LbL colloid particle modification approach (Fig. 8.1), a polyelectrolyte solution is added to a colloidal suspension (step 1). The essential point is that the added polyelectrolyte has an opposite charge to that on the particles, thereby adsorbing to the particle surface through electrostatic interactions. The LbL assembly of polyelectrolytes onto colloid particles can be performed in two main ways: (i) the concentration of polyelectrolyte added at each step is equal to that required to form a saturated layer [47], or (ii) the concentration of polyelectrolyte is in excess so as to cause saturation adsorption [45–51]. The first method requires accurate determination of the exact amount of polyelectrolyte needed to saturate the particle surface. This involves separate adsorption experiments to determine the relative concentrations of polyelectrolyte and particles. The second method involves removing the excess polyelectrolyte prior to the addition of the next oppositely charged component to prevent the formation of polyelectrolyte complexes in bulk solution. The excess polyelectrolyte can be removed by several centrifugation/wash/redispersion cycles in water [45–51], or by filtration [77]. In the coating process, particle aggregation must be suppressed at all stages of the preparation. For example, when the centrifugation/wash/dispersion approach is used, care needs to be taken not to use too high centrifugation speeds, as strongly aggregated particles in the form of a pellet are difficult to redisperse. Regardless of the method used, template particle concentrations of up to a few weight percent are used in order to avoid particle aggregation upon addition of the coating species. After adsorption of a polyelectrolyte layer, the sign of the charge on the surface of the particles is reversed, which aids in the deposition of subsequent layers of a wide range of charged components. Subsequent stepwise adsorption of oppositely charged polyelectrolytes (step 2), preformed nanoparticles (step 3), or low molecular weight substances (step 4), and polyelectrolyte, results in the formation of nanocomposite multilayers. Excess materials are again removed either by centrifugation/wash/dispersion cycles or filtration. Generally, polyelectrolytes are used to separate layers of the same materials (e.g. nanoparticles or low molecular weight substances), as these polyelectrolyte interlayers act as molecular "glue". Additionally, they provide enhanced colloidal stability to the surface-modified particles. Hollow capsules, or capsular colloids, are obtained from the coated colloids by either chemical or thermal treatment. In the case of the polyelectrolyte-coated particles, the decomposable core is removed by exposure to an acid or organic solvent, leaving behind hollow polyelectrolyte capsules. For the nanoparticle/polyelectrolyte-coated particles, the elevated temperature treatment removes the organic matter (i.e. sacrificial colloid core and bridging polymer) and sinters the preformed inorganic nanoparticles, hence providing structural integrity for the resulting hollow inorganic capsules. Further details of hollow capsules can be found in Section 8.4.1.4 and Chapter 18.

There are various experimental parameters that need to be controlled and optimized in order to deposit layers uniformly onto colloid particles via the LbL method. These include particle concentration and size, polymer type, length, and concentration, and total salt concentration in the adsorbing solution. Of particular note is that it becomes progressively more challenging to deposit layers as the size of the core particles used is reduced from micrometers to nanometers. These issues will be discussed in Sections 8.2.1 and 8.2.2.

8.2.1
Materials for Surface Modification

As is the case with planar supports, a range of materials can be used to selectively modify colloid particles. Polymer coating of colloid particles is one method to modify surface properties, with the polymer type determining the final surface characteristics of the particles. Polyelectrolytes (charged polymers) are attractive for this purpose as they can be self-assembled and because they contain certain functional groups that may be directly introduced through adsorption onto the particles. This alleviates the need for chemical routes to modify particles, protocols that often lead to unwanted particle aggregation. Nanoparticles can also be used as materials for modifying larger particles. This represents a direct way to introduce the properties that are inherent to the nanoparticles (e.g. magnetic, optical, catalytic) to the resulting composite larger particles. Similarly, the use of low molecular weight dyes further expands the range of possibilities for imparting unique properties to particles. The following sub-sections present examples of these systems.

8.2.1.1 Polyelectrolytes

Polyelectrolyte-modified particles offer interesting prospects in applications ranging from catalysis to additives and pigments [18]. For centuries, colloid particles have been surface modified and stabilized by the direct adsorption of polyelectrolytes from solution onto their surface [4]. Particle stabilization arises from both electrostatic and steric (polymeric) effects. However, most studies in the past have focused on the adsorption of a single polyelectrolyte layer on particles. Under certain conditions, such polymers can also induce flocculation of colloidal materials – this phenomenon has been widely exploited in many industrial processes [78] – although in many cases it is undesirable. In the following, it is shown that multiple layers of polyelectrolytes of opposite charge can be assembled on particles via the LbL method. Importantly, this approach leads to modified particles that are colloidally stable and unaggregated.

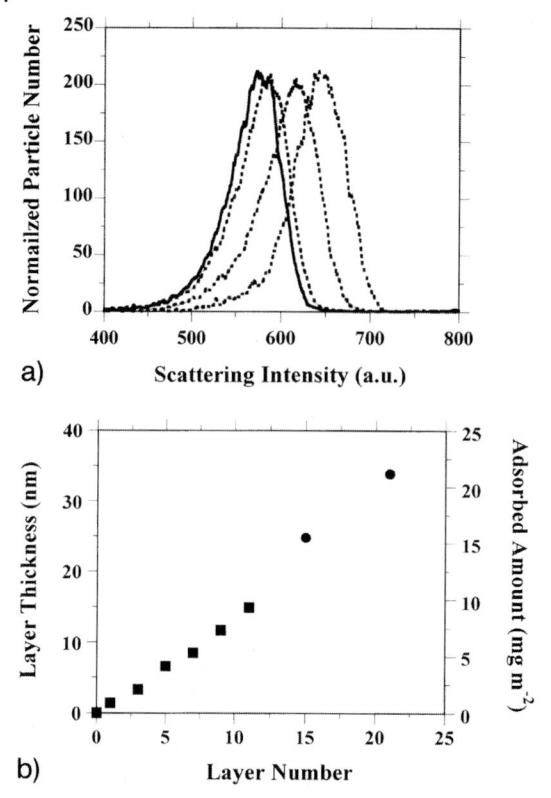

Fig. 8.2 a From left to right: Normalized SPLS intensity distributions of bare 640 nm diameter PS spheres, and identical PS spheres coated with a total of one, five, and nine layers of PDADMAC/PSS. The layers were assembled by the consecutive adsorption of PDADMAC and PSS. The systematic shift in the SPLS intensity distributions is indicative of an increase in thickness of the multilayer coating on the PS particles. The distributions shown correspond to single, coated particles (reproduced from [49] by permission of the American Chemical Society); b Thickness of polyelectrolyte multilayers assembled on 640 nm diameter PS spheres as a function of the number of deposited layers. The thicknesses were determined from SPLS data. The values are the same, within experimental error, for PDADMAC/PSS and PAH/PSS systems. The squares represent PDADMAC/PSS layers, and the circles PAH/PSS layers. The square data points were obtained for layers deposited and where centrifugation/water wash cycles were used to remove unadsorbed polyelectrolyte, while the circles correspond to layers adsorbed and where excess polyelectrolyte was removed by filtration/water wash cycles. No difference in the layer thickness was observed when using either of these two separation methods.

The technique of microelectrophoresis provides a qualitative indication of the deposition of polyelectrolyte layers on colloid particles [45–51]. In these experiments, the measured parameter, electrophoretic mobility (u) is converted to a ζ-potential via the Smoluchowski relation $\zeta = u\eta/\varepsilon$, where η and ε are the viscosity and permittivity of the solution, respectively. In the case of negatively charged (uncoated) polystyrene (PS) particles, a ζ-potential of ca. –65 mV is calculated, indicating that the particles are negatively charged [49]. Subsequent adsorption of a positively charged polyelectrolyte (e.g. poly(diallyldimethylammonium chloride), PDADMAC) onto the PS particles causes a reversal in sign of the ζ-potential (ca. +55 mV) [49]. This suggests that the particles are modified with an outermost layer of PDADMAC. Additional deposition of alternate layers of the polyanion poly(styrenesulfonate) (PSS) and PDADMAC onto the modified PS particles results in alternating negative and positive ζ-potentials, respectively [49]. This alternating trend in the data suggests a successful recharging of the particle surface with deposition of each polyelectrolyte layer. Similar data has been observed for a range of other polycation/polyanion combinations [45–48, 50, 51].

Single particle light scattering (SPLS) is an optical technique [79, 80] that enables determination of the thickness of adsorbed layers on colloid particles [49]. It is capable of discriminating between uncoated and coated particles, and, if present, the degree of aggregation of the coated colloids. The SPLS technique involves recording the light scattered from a single particle at a given moment in time, thus being sensitive to the amount of adsorbed material on the colloid particles. The resolution of SPLS is about 1–2 nm of adsorbed material (assuming the refractive index of the adsorbed component is ~ 1.5). By recording the light scattered from many individual particles, one obtains a histogram of particle number versus scattering intensity (or SPLS intensity distributions). The normalized SPLS intensity distributions for neat PS latices and those coated with one, five, and nine PDADMAC/PSS layers are shown in Fig. 8.2a. The deposition of the polyelectrolytes onto the particles is seen as a shift (on the x-axis) of the SPLS intensity distribution. This systematic shift with increasing polyelectrolyte layer number confirms polyelectrolyte layer growth, as suggested by the ζ-potential data. The SPLS intensity distributions shown correspond to those of single particles. No peaks at higher intensities, which are characteristic of aggregated particles (i.e. doublets, triplets, or higher order aggregates), were observed.

The thickness of the deposited polyelectrolyte layers can be derived from the SPLS data using the Rayleigh-Debye-Gans theory and a refractive index for the polymer layers (1.47) [49]. The calculated average PDADMAC/PSS layer thickness increases with the number of polyelectrolyte layers deposited, as depicted in Fig. 8.2b. The average thickness of each polyelectrolyte layer is about 1.5 nm (equivalent to approximately 1.0 mg m^{-2}) [49]. Similar values have also been measured for the polyelectrolyte pair poly(allylamine hydrochloride) (PAH) and PSS when assembled from solutions containing the same concentration of salt (typically 0.5 M NaCl) [49]. Increasing the salt concentration in the solution from which the polyelectrolytes are adsorbed typically leads to thicker layers [81]. Similar thicknesses are also obtained, regardless of whether the centrifugation or filtra-

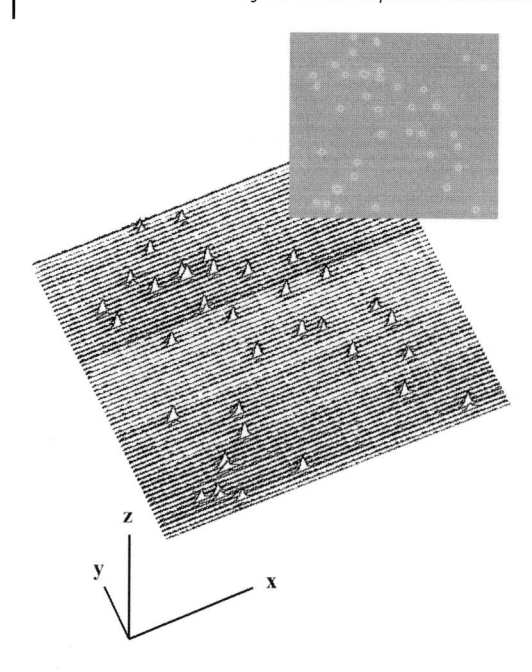

Fig. 8.3 Fluorescence microscopy image surface plot of PS spheres (640 nm diameter) modified with (PAH-FITC/PSS)$_5$ (10 layers total). The z-axis represents the fluorescence intensity of the particles, given as a pixel value. The insert is the corresponding light microscopy image. The diameter of the coated PS particles is approximately 680 nm (reproduced from [82] by permission of Academic Press).

tion methods were used in the preparation of the layers [49, 77]. The SPLS data show that the polyelectrolyte layer thickness can be controlled at the nanometer level (to within 2 nm).

When the polyelectrolytes contain absorbing or fluorescing chromophores, UV-vis spectrophotometry and fluorescence spectroscopy can be used to evidence polyelectrolyte adsorption onto the particles [50, 82]. An example is the fluorescent microspheres prepared by the LbL assembly of dye-labeled PAH and PSS onto PS particles [82]. By selecting PAH labeled with different fluorescent dyes (e.g. fluorescein isothiocyanate (FITC), RhBITC, Cy5), particles with three different emission maxima were prepared (526 nm, 586 nm, and 665 nm, respectively). Also of importance is tailoring the fluorescence intensity of the particles, which can be achieved by controlling the number of dye-labeled polyelectrolyte layers deposited. Fig. 8.3 shows a fluorescence microscopy image (FMI) of individual 640 nm diameter PS particles modified by deposition of FITC-PAH/PSS layers. It was found that the fluorescence intensity of the surface-modified particles increases linearly with increasing number of deposited FITC-PAH layers [82]. This shows that the LbL approach is promising for the preparation of tailored fluorescent particles.

The nanoscale surface modification of colloid particles with polyelectrolyte multilayers can be further verified by transmission electron microscopy (TEM). The same system presented in Fig. 8.3 is considered here. A TEM image of highly monodisperse 640 nm diameter PS particles coated with nine polyelectrolyte layers [(PAH-FITC/PSS)$_4$/PAH-FITC] is shown in Fig. 8.4. Uniformly modified

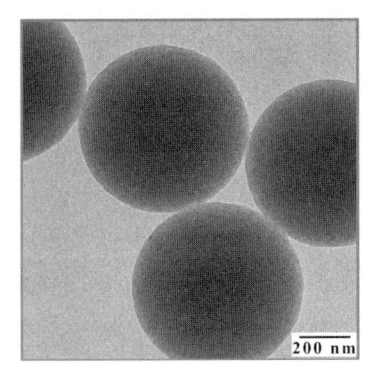

8.2 Sequential Electrostatic Assembly Applied to Particles

Fig. 8.4 TEM micrograph of [(PAH-FITC/PSS)$_4$/PSS]-modified PS microparticles. The multilayer coating cannot be seen, as the contrast of the polyelectrolytes is similar to that of the PS core particles in the electron beam. However, an increase in particle diameter of approximately 640 nm to 680 nm occurred as a result of polyelectrolyte multilayer modification. No aggregation of the modified particles was observed by TEM (reproduced from [82] by permission of Academic Press).

Fig. 8.5 TEM micrographs of unpolymerized hollow capsules made of 10 layers of diazoresin/PSS. The hollow capsules were prepared from diazoresin/PSS-coated 640 nm PS particles that were exposed to THF, which dissolves the PS core. The scale bar in the inset corresponds to 200 nm (reproduced from [84]).

and individual particles are obtained. The presence of the polyelectrolyte multilayers on the PS particles was confirmed by TEM. From the diameter differences between the polyelectrolyte-modified and the bare particles, a polyelectrolyte layer thickness of approximately 1.5–2.0 nm per layer is obtained [17], which is in good agreement with SPLS data for similar systems (see earlier).

It should be noted that the layer thickness corresponds to an *average* thickness. In terms of the internal structure of the polyelectrolyte multilayers, they comprise an overlapping network of polycations and polyanions. That is, the multilayer films are not composed of stratified polyelectrolyte layers, but rather the layers interpenetrate each other over about four neighboring layers [35, 83]. This has been verified by neutron reflectivity [35] and dye adsorption-desorption [49] experiments. Such interpenetration provides an even greater number of contact points for the polycations to electrostatically associate, thus enhancing film stability.

Unequivocal evidence that polyelectrolyte multilayers are formed on colloid particles is obtained after decomposition and removal of particle cores that have been surface-modified with polyelectrolytes [46, 48, 50, 84]. The polyelectrolyte multilayer coatings are semipermeable, thus allowing removal of the decomposed cores. An example of such hollow polyelectrolyte capsules made of diazoresin and PSS polyelectrolytes is shown in Fig. 8.5 [84]. The folds and creases seen are due to evaporation of the aqueous content upon drying the colloidal suspension of hollow capsules on a solid support. This image also shows the exceptionally strong electrostatic forces involved in the formation of these ultrathin polyelectrolyte layers. The wall thickness of the capsules can be determined from atomic force microscopy (AFM) where there is overlapping of the two walls. The layer thicknesses obtained in this way are generally in close agreement with those determined from SPLS and TEM. Details on the formation and potential applications of hollow polyelectrolyte capsules can be found in Chapter 18.

8.2.1.2 Nanoparticles

Nanoparticles, especially those made of metals, metal oxides, and semiconductors are of intense interest because they possess unique catalytic, electronic, magnetic, and optical properties. Depositing nanoparticles onto larger particles provides a pathway to impart new properties, originating from the nanoparticles, to the larger core particles. These systems, typically microsphere-nanoparticle composites, are of interest because of their potential use as heterogeneous catalysts and their relevance in electronic and optical sensor applications, as well as for surface-enhanced Raman scattering [40, 85–89].

The use of preformed nanoparticles has the advantages that a broad range of nanoparticles of different composition and crystal phase, as well as of predefined size, can be selected to modify (or coat) larger particles. The LbL deposition of nanoparticles and polyelectrolytes yield nanoparticle-based multilayer coatings on colloid particles [52–63]. The system of SiO_2 nanoparticle/polyelectrolyte multilayers on submicrometer-sized PS particles [52–55] will be considered first to exemplify the LbL method of coating particles and the methods used for characterizing the nanoparticle-modified microparticles. As is the case with polyelectrolyte multilayers (see earlier), microelectrophoresis, SPLS, and electron microscopy can be used to follow the surface modification of the PS particles. Microelectrophoresis experiments reveal that the ζ-potential alternates from negative to positive for deposition of SiO_2 nanoparticle and PDADMAC layers, respectively [52–54]. The measurements were conducted at pH=5.6, which is above the isoelectric point of SiO_2 and hence the SiO_2 nanoparticles have an overall negative charge at this pH. The alternating ζ-potentials are characteristic of stepwise modification of the PS particles upon deposition of each layer. It is important to note that prior to deposition of the SiO_2 nanoparticles, a precursor polyelectrolyte film, PDADMAC/PSS/PDADMAC, was deposited on the PS spheres in order to facilitate adsorption of the first layer of nanoparticles.

SPLS revealed stepwise surface modification of the PS particles, and that the modified particles were unaggregated. The average thickness of each SiO_2 nanoparticle/PDADMAC layer pair deposited onto the PS spheres, as determined from SPLS, was 36±10 nm [52–54], suggesting that on average approximately a monolayer of SiO_2 nanoparticles was deposited with each nanoparticle adsorption step. The PDADMAC layer thickness was estimated to be ca. 2 nm.

TEM provided further evidence for modification of the PS particles. TEM micrographs of bare PS particles (Fig. 8.6a) showed that the particles have a smooth surface, while those modified with SiO_2 nanoparticle/PDADMAC multilayers (Fig. 8.6b–d) showed increased surface roughness. Additionally, the presence of the

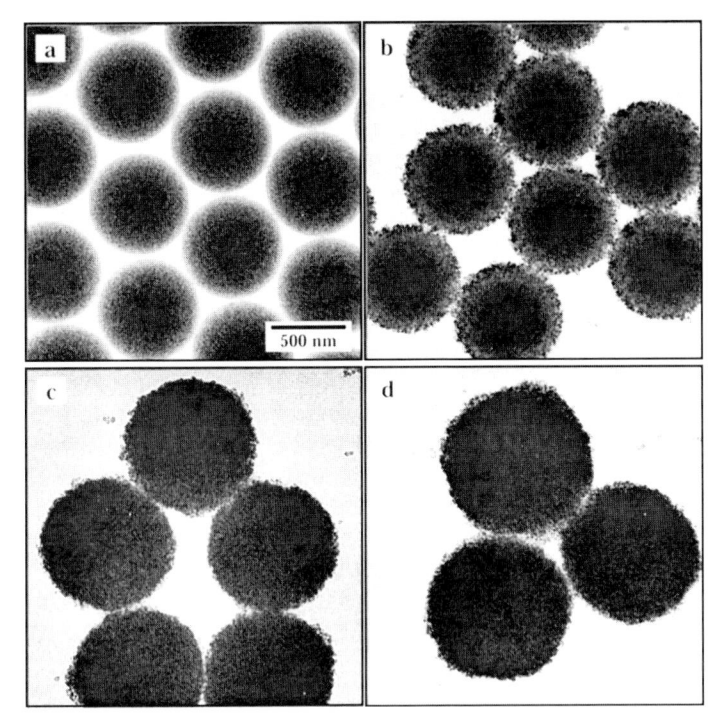

Fig. 8.6 TEM images of **a** bare PS spheres, and polyelectrolyte-modified PS spheres coated by the LbL method with **b** one, **c** three, and **d** five SiO$_2$ nanoparticle/PDADMAC layer pairs. The formation of SiO$_2$ nanoparticle/PDADMAC multilayers on the PS particles is confirmed by the increase in particle diameter with increasing layer number. The scale bar corresponds to all four TEM images shown (reproduced from [91]).

multilayers on the PS spheres resulted in a systematic increase in their diameter. The diameter increases of the coated PS spheres (relative to bare PS particles (a)) were (b) 60, (c) 200, and (d) 310 nm, yielding an average diameter increment of ca. 65 nm, corresponding to a layer thickness of approximately 32±5 nm for each SiO$_2$/PDADMAC layer pair. These data are in agreement with the SPLS results.

In the above system, it was essential to use processing conditions that sufficiently screen the silica surface charge on the nanoparticles, for example salt-containing solutions. This allows the nanoparticles to pack closely on the PS particle surface. Using these conditions, approximately one monolayer of nanoparticles was adsorbed with each deposition step [52–54]. The modified PS particles remain as individual particles in solution. In contrast, deposition of Au@SiO$_2$ nanoparticles directly from pure water solutions resulted in only sparse nanoparticle coverage on the PS spheres, which in turn caused particle clumping [61]. This example highlights the importance of controlling the solution conditions to achieve uniform nanoparticle surface modification of larger colloid particles by the LbL technique.

The versatility and applicability of the LbL strategy to prepare coated colloids is exemplified by the range of other nanoparticles that have been utilized to modify larger particles, spanning metals (gold), metal oxides (silica, zeolite, titania, iron oxide), luminescent semiconductors (CdTe, HgTe), and composite nanoparticles (Au@SiO$_2$, octa(3-aminopropyl)silsesquioxane-stabilized silver nanoparticles) [52–63]. For these nanoparticles, similarly to those shown for silica, regular surface coatings were formed on particles when they were deposited in alternation with oppositely charged polyelectrolytes. Recently, it was shown that a new class of nanoparticles, rare earth-doped lanthanum phosphates (LaPO$_4$), could be electrostatically assembled to form thin films on microspheres with the aid of oppositely charged polyelectrolyte interlayers [90]. The fluorescence from these nanoparticles originates from the bulk properties of the material, and is therefore independent of their size. Different colors are available from the various dopants used in their synthesis (e.g. Ce, Tb, Eu, Dy). By mixing the nanoparticles in different ratios and depositing them on microparticles, modified microparticles with tailored optical properties were prepared. In all of the aforementioned nanoparticle-based systems, the thickness of the multilayer coatings could be controlled on the nanoscale through variation of the number of deposited layers.

In the case where the surface coatings comprise nanoparticles, the nanoparticle/polyelectrolyte-coated particles prepared by the LbL method can be subjected to elevated temperatures (i.e. calcination) to form hollow inorganic capsules [52, 55, 60, 61, 63]. The calcination step removes the organic matter (core and polymer) during heating to 450 °C, as confirmed by thermogravimetric analysis. In contrast to the polyelectrolyte coatings, the incorporation of nanoparticles can introduce further structural stability to the coatings and the resulting hollow particles. Details of the preparation and application of hollow inorganic capsules can be found elsewhere [52, 55, 60, 61, 63, 91].

The surfaces of colloid particles have also been modified through the LbL deposition of proteins and oppositely charged polyelectrolytes [64–68]. In this context, the proteins can be treated as biological nanoparticles. PS particles have been modified through the deposition of multilayer films of various proteins, including bovine serum albumin (BSA) [64], immunoglobulin G (IgG) [64], glucose oxidase (GOD) [65, 66], horseradish peroxidase (POD) [66], β-glucosidase (β-GLS) [67], and urease [68]. The thicknesses of the protein layers were varied from several to hundreds of nanometers [64]. The actual thickness depends on the type of protein, the number of protein layers deposited, and the conditions of assembly (e.g. ionic strength and pH of the adsorption solution). For the FITC-labeled bovine serum albumin (FITC-BSA)/PDADMAC system, the coating thickness increases linearly with the number of protein layers deposited [64]. The average layer thickness for each FITC-BSA layer was ∼3 nm when adsorbed from pure water (pH ∼ 5.6), and ∼6 nm when deposited from a phosphate-buffered saline solution (pH 7). The difference in these values can be attributed to the higher percentage of charged groups on BSA at pH 7. For immunoglobulin G (IgG)/PSS layers, layer thicknesses of up to ∼37 nm were observed, which are equivalent to several monomolecular IgG layers [64].

TEM studies on the protein-modified particles showed both an increase in surface roughness and an increase in particle diameter compared with the bare PS spheres [64]. The TEM layer thicknesses are in agreement with those determined from SPLS. In the modification of the particles with the protein layers, careful selection of the processing conditions was required to prevent particle aggregation. Enhanced colloidal stability was imparted to the particles by the subsequent deposition of polyelectrolyte layers. On the whole, the LbL method allows the biofunctionalization of particles with various proteins according to specifically designed architectures. Examples of the potential uses of such protein-modified particles are provided in Section 8.4.1.1.

8.2.1.3 Low Molecular Weight Substances

The LbL deposition of low molecular weight substances in alternation with polyelectrolytes onto colloid particles provides an alternative means to introduce, step-by-step, functional materials (e.g. chromophores) into the coatings and hence specific properties to the particles [69, 92]. Multilayers of spermidine/DNA [92], Tb^{3+}/PSS and pyrenetetrasulfonic acid (4-PSA)/PAH [62] were assembled on latex particles. These multilayers can be deconstructed by the addition of salt or by altering the temperature, making them attractive for use in sustained-release experiments of low molecular weight species. When the core is removed and replaced with other materials to be subsequently released, these systems can be used as drug delivery systems (see Section 8.4.1.4 and Chapter 18).

In the context of this section, lipids will be considered as low molecular weight substances. Lipids have been LbL assembled in alternation with polyelectrolytes onto colloid particles. The lipids can be deposited onto the particles from methanol solution or by adsorption of preformed lipid vesicles from aqueous solution [69, 93–95]. Energy transfer measurements suggest a bilayer conformation of the lipids on the particles for charged lipids and multilamellar structures for non-charged lipids [93]. These lipid coatings have been shown to affect the permeability properties of the polyelectrolyte multilayers, making them impermeable to water-soluble dyes [94] and less permeable to ions [95]. Lipid coatings on polyelectrolyte capsules are attractive models for basic biophysical studies (see Chapter 18 for details).

The LbL deposition of inorganic molecular precursors and polyelectrolytes onto colloid particles can also be used to prepare coated particles with uniform inorganic-organic composite coatings of defined thickness. The precursor, through hydrolysis, can be converted to inorganic nanoparticle coatings on the spheres. This method avoids the separate preparation of nanoparticles prior to adsorption. Additionally, the coated particles may then be converted to hollow inorganic capsules upon removal of the core. The LbL assembly method has been used to sequentially deposit the inorganic molecular precursor titanium (IV) bis(ammonium lactato) dihydroxide (TALH) and PDADMAC onto PS spheres [70]. The success of this approach for forming uniform coatings (and subsequently hollow spheres) relies on the use of water-stable inorganic molecular precursors that can be assembled in alternation with polyelectrolytes. TALH is relatively stable and hydrolyzes slowly at ambient temperature in neutral solution [96–98], unlike most other titanium alkoxides, which hydrolyze rapidly in the presence of water. This relative stability prevents significant hydrolysis and condensation reactions, which would cause precipitation of titania in solution and possible aggregation of the coated particles. The precursor interacts with the polyelectrolyte coatings through electrostatic interactions [99]. Refluxing the coated particles at 100 °C resulted in titania nanoparticle/polyelectrolyte-coated PS particles (as observed by TEM), which is in agreement with the conversion of TALH to titania anatase nanoparticles at this temperature [96]. Calcination of the coated particles yielded hollow titania spheres as a result of hydrolysis and condensation of TALH [70]. For alkoxide precursors that are highly water sensitive, preformed polyelectrolyte multilayers can be used as nanoreactors to form inorganic-organic hybrid coatings on particles (see Section 8.3.2).

Low molecular weight dyes, 4-PSA and 6-carboxyfluorescein, were also incorporated into PAH/PSS and PDADMAC/PSS coatings preassembled onto latex particles through electrostatic binding [49]. Using fluorescence spectroscopy, probe binding was observed only when the outermost layer was oppositely (positively) charged to the probe, and the amount of probe bound was found to increase linearly with polyelectrolyte layer number up to a film thickness of ∼15 nm (11 layers), with saturation binding occurring beyond 11 layers. This study showed that a minimum of about 10–30% of the cationic charges of the polycations in the upper region of the multilayer films (where the probe binds) are not directly electrostatically utilized in the multilayer assembly process through ion-pair binding to oppositely charged sites on the polyanions. Further, through experiments on the removal of probe bound to multilayers of different thicknesses upon subsequent exposure to PSS, it was found that the polyelectrolyte multilayer films are long-range electrostatically coupled systems. The polyelectrolyte-dye modified particles prepared by this approach also represent interesting systems for use as dye-labeled particles, as the dye type and dye content per particle can be readily controlled.

8.2.2
Types of Particles

Particles of different sizes, shapes, and composition have been modified by the LbL technique. These include latex particles, metal nanorods, metal nanoparticles, enzyme crystals, crystals of low molecular weight substances, and biological cells. The following sub-sections detail the different types of particles that have been used and the processing conditions necessary for their modification.

8.2.2.1 Latex Particles
Typically, PS and melamine formaldehyde (MF) spherical particles in the 0.1 to 5 μm diameter range have been utilized for LbL modification. The main reasons

have been that PS particles of various sizes and high monodispersity can be easily prepared. They are also commercially available. These particles have been of great benefit in monitoring the LbL build-up of materials on their surface, especially because of their high monodispersity, which is highly useful for SPLS measurements. On the other hand, the (weakly cross-linked) MF particles have been used in the preparation of hollow polyelectrolyte capsules because they can be decomposed when exposed to acidic solution (pH < 1.6) [46, 48, 50, 92]. Similarly, PS particles can be decomposed upon exposure to tetrahydrofuran [84], but care needs to be taken as PS particles prepared with cross-linking agents are difficult to remove once modified (coated) with polyelectrolyte coatings. Therefore, PS particles with no or very low levels of cross-linking are highly preferable.

8.2.2.2 High Aspect Ratio Particles

The LbL method is essentially insensitive to template shape when polyelectrolytes are being used to modify particles that are larger than about 50 nm. The LbL method has been applied to modify the surface of nickel nanorods of average diameter ~65 nm and length ~1.5 μm (Fig. 8.7a). The nickel nanorods were coated with eight layers of PDADMAC and PSS (Fig. 8.7b) [71]. TEM images of the modified particles reveal that the nickel nanorods were coated with polyelectrolyte multilayers, the thickness of the coating being ca. 8 nm, corresponding to ca. 1 nm per layer. This thickness is comparable to those observed for similar polyelectrolyte multilayers on latex particles. Dissolving the nickel core particles by using dilute hydrochloric acid (pH = 2) resulted in polyelectrolyte nanotubes (Fig. 8.7c). The polymer nanotubes retain the shape of the nickel nanorod template. The original nanotube shape is preserved mainly because of the exceptionally strong electrostatic interactions between the polyelectrolyte layers. Fig. 8.7d shows a higher magnification TEM image of a polymer nanotube, the diameter of which is ca. 90 nm. An increase in diameter of ca. 10% of the polymer nanotubes was observed, compared with the coated nickel nanorods. This is likely to be due to the effect of drying, as a result of which the polymer nanotubes collapse to form a "sheet" with increased diameter. Similar observations have been made for hollow polymer capsules derived from coated polymer colloid particles [46, 48, 50, 84, 92]. Titania-polyelectrolyte coatings were also deposited on the nickel nanorods by depositing TALH alternately with PDADMAC [71]. Hollow titania-based nanotubes were prepared from these precursor particles [71]. Concentric nanotubes, that is, with a different composition on the inside of the nanotube from that of the surface, could also be prepared via this method.

8.2.2.3 Nanoparticles

The importance of modifying the surface properties of nanoparticles lies in modulating the unique size-dependent properties afforded by colloids in this size range, as well as their surface characteristics. Reducing the size of the particles to the

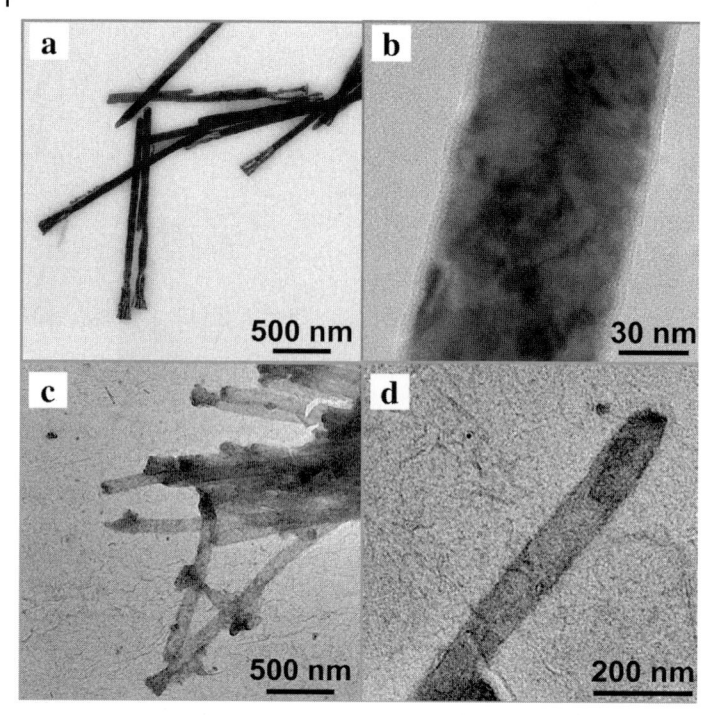

Fig. 8.7 TEM images of **a** bare nickel nanorods; **b** a nickel nanorod coated with eight layers of polyelectrolyte (PDADMAC/PSS); **c** PDADMAC/PSS nanotubes (eight layers of polyelectrolyte) obtained upon dissolution of the sacrificial nickel nanorod from the polyelectrolyte-coated nanorods; and **d** a higher magnification image of a nanotube of (PDADMAC/PSS)$_4$ (reproduced from [71] by permission of the American Chemical Society).

nanometer regime, however, presents two challenges: (i) establishing appropriate experimental protocols; and (ii) accounting for the effects of curvature of the small particles. The first point is related to the methods used to effectively and efficiently separate excess polyelectrolyte from particles that are less than 100 nm in diameter. This is straightforward for the typically employed latex particles of several hundreds of nanometers in diameter. Commercial, bench-top centrifugation systems can be used. This approach has also been used to separate latex particles down to about 70 nm in diameter [50]. However, centrifugal forces of $\sim 40\,000\,g$ were required, compared with $\sim 10\,000\,g$ for latex particles of 500 nm diameter. Moving to even smaller particle sizes represents an issue for centrifugation protocols to routinely separate polyelectrolytes and nanoparticles. This is particularly the case with respect to the time required, as a number of centrifugation/water wash cycles are required.

One way of overcoming the above challenge is to employ metal (or metal oxide) nanoparticles, which have a higher density than latex particles. These inorganic colloid particles can be readily separated by centrifugation. Additionally, gold nanoparticles have the added benefit that the optical properties of the particles can be exploited to investigate the deposition of material on their surface [100, 101]. As already mentioned, as the particles become smaller issues with regard to particle curvature become important. While the nature of the polymer, polymer length, polymer concentration, and total salt concentration in the adsorption solution are important general experimental parameters that affect the coating of particles with polyelectrolytes, these become increasingly important when coating nanoparticles. In order to modify the surfaces of gold nanoparticles 30 nm in diameter, while avoiding nanoparticle aggregation, it was necessary to take into account the polymer length, polymer stiffness, and added salt (leading to finite screening lengths) [100, 101]. Experimental conditions of 1 mM salt in the adsorption solution and a polyelectrolyte molecular weight of about 15–20 kDa were found to be suitable. These conditions are in general agreement with theoretical predictions on the model system of histone-DNA complexation [102]. The uniform deposition of PDADMAC and PSS (eight layers total) on the nanoparticles was confirmed by a systematic red shift in the spectral position of the surface plasmon absorption band (with little broadening) upon polyelectrolyte deposition [100, 101].

A recent study demonstrated that sub-10-nm diameter particles (7 nm diameter) can also be surface modified with polyelectrolytes via the LbL method [103]. As with previous work [100, 101], successful polyelectrolyte modification of gold nanoparticles was possible when the nanoparticles were stabilized using molecules that bound strongly to the nanoparticle surface. The gold nanoparticles were first capped with mercaptoundecanoic acid (MUA), where the thiol group is covalently attached to the nanoparticle surface, and the carboxylic acid moiety renders a negative surface charge at pH above its pK_a (~ 4.6). These nanoparticles exhibit a peak plasmon absorption band at 525 nm (Fig. 8.8a). The choice of ionic strength is made considering the inverse Debye-Hückel screening length (nm^{-1}) [104]. It is known that the ionic strength required to reduce the chain stiffness in order to coat particles with polyelectrolytes of a given length increases with decreasing particle size [102]. Recent calculations by Kunze and Netz on salt-induced DNA-histone complexation, where the size of the protein is similar to the size of the nanoparticles discussed here, have shown that for a polyelectrolyte with renormalized charge density of $\tau = 1/l_b$ (l_b = Bjerrum length, ~ 0.7 nm for water) and length 50 nm, full wrapping is obtained for salt concentrations corresponding to the inverse Debye Hückel screening length larger than typically $\kappa \sim 0.2$ nm^{-1} [102]. We chose similar conditions for the polyelectrolyte coating of gold nanoparticles, except that the polymer molecular weight was selected such that the polymer length (~ 15 nm) was approximately equal to the circumference ($2\pi \times 3.5$ nm) of the nanoparticles.

The 7 nm diameter MUA-modified gold nanoparticles were coated with a layer of PDADMAC by rapidly mixing the PDADMAC solution with the gold nanoparticle dispersion. The polyelectrolyte adsorption was carried out at a constant temperature of 20 °C and at an ionic strength of 20 mM (NaCl). The peak plasmon absorption was found to shift from 525 to 529 nm on adsorption of PDADMAC (Fig. 8.8a). This 4 nm red shift in the peak plasmon absorption is related to a

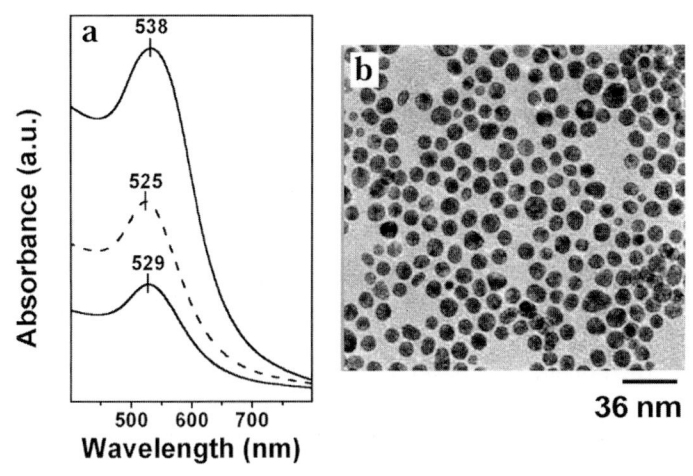

Fig. 8.8 a UV-visible spectra of MUA-capped gold nanoparticles (dashed spectrum), and MUA-modified gold nanoparticles coated with PDADMAC (lowest spectrum) or PDADMAC/PSS (top spectrum); b TEM image of MUA-capped gold nanoparticles coated with two polyelectrolyte layers (PDADMAC/PSS). The diameter of the gold nanoparticles was ~7 nm (adapted from [103]).

change in the local dielectric constant around the gold nanoparticles as a result of adsorption of PDADMAC [105]. A layer of PSS was subsequently deposited under the same conditions, resulting in a further red shift in the peak plasmon position to 538 nm (Fig. 8.8a). No broadening of the gold nanoparticle peak plasmon absorption occurred as a result of coating with polyelectrolytes, suggesting the absence of aggregated nanoparticles. The ζ-potential values were also in agreement with stepwise polyelectrolyte modification of the nanoparticles. The polyelectrolyte-modified gold nanoparticles were found to be colloidally stable for more than three months. In addition, analytical ultracentrifugation (AUC) measurements showed an increase of 4 nm in the mean diameter of the gold nanoparticles upon deposition of four [(PDADMAC/PSS)$_2$] polyelectrolyte layers [103]. The average polyelectrolyte layer thickness derived from the AUC data is approximately 0.5 nm, which is similar to layer thicknesses obtained for polyelectrolyte layers assembled under low ionic strength conditions [106]. Fig. 8.8b shows a TEM image of PDADMAC/PSS-modified MUA-capped gold nanoparticles. The nanoparticles are well separated, with no signs of aggregation.

The above studies demonstrate that the LbL method can be used to functionalize nanoparticles with a variety of polyelectrolytes, either positively or negatively charged. When generalized, this method has the potential to allow surface modification of a range of other nanoparticles, including semiconductor quantum dots, phosphors, and metal oxides. Such particles are of interest in, for example, biolabeling applications.

8.2.2.4 Bioparticles

Biocolloids have also been utilized as particles for coating with polyelectrolyte multilayers via the LbL method. These include enzyme crystals, amorphous particles of proteins, and biological cells. The main motivation for using biocolloids was to encapsulate them within thin coatings and to tailor their surface properties. The following will briefly outline the different systems investigated.

Microcrystals of the enzyme catalase (~ 10 µm in diameter) were coated via the sequential deposition of oppositely charged polyelectrolytes [107]. Deposition of the layers was confirmed by microelectrophoresis and by using dye-labeled fluorescent polyelectrolytes. The crystals are soluble in water at pH > 6 and pH < 4, so the polyelectrolytes were deposited from polyelectrolyte solutions containing 1 M potassium acetate at pH 5 and 4 °C to minimize enzyme dissolution. The rectangular shape of the crystals was maintained upon polyelectrolyte coating. The important features of these particles are that they represent a system with an extremely high enzyme loading in a thin polymer coating and that the activity of the encapsulated catalase is preserved. Additionally, the thin coatings act as protective layers for the catalase, shielding them from enzyme-degrading substances such as proteases [107]. These particles also represent useful building blocks for the formation of high-enzyme content films for biocatalytic applications (see Section 8.4.1.2).

Micron-sized amorphous particles of protein aggregates (lactate dehydrogenase [108] and chymotrypsin [109]) were also used as cores for polyelectrolyte multilayer assembly. The amorphous particles were prepared by precipitating them in saline solutions. Each aggregate consisted of numerous, smaller protein particles (primary aggregates), in the shape of small spheres with a diameter of the order of 100–300 nm. Variations in the protein and sodium chloride concentrations used permitted optimization of the enzyme precipitation reaction. Similar to the catalase crystals described above, the polyelectrolyte coatings result in encapsulation of the protein, and they also provide a selective barrier for the diffusion of different species (substrates, inhibitors) from the exterior [108]. Related work has shown that three polyelectrolyte layers can be deposited onto condensed DNA particles of size 50–100 nm, yielding stable particle suspensions [110, 111]. Modification of the DNA particles by the polyelectrolytes facilitates the uptake of the particles by biological cells with sequential gene expression [111, 112].

Biological cells such as glutaraldehyde-fixed echinocytes have also been employed as cores for encapsulation. These cells have been coated with polymer [113, 114] and polymer/nanoparticle [91] multilayers. Although the biocolloids have a jagged and highly structured surface, the coating mimics the original shape, including the secondary structure (spikes) of the biocolloid [91, 113, 114]. Coating of biological cells in this way not only encases them in ultrathin polyelectrolyte coatings, thus providing a protective barrier for them, but also allows modification of the surface properties of the cells.

8.2.2.5 Crystals of Low Molecular Weight Materials

The surface modification of uncharged particles is of intense interest, especially with regard to preparing stable colloidal dispersions of such particles, and also for encapsulation of low molecular weight, crystallized (or amorphous) materials, such as drugs. Crystallized, largely water-insoluble low molecular weight substances including pyrene and fluorescein diacetate [115, 116] have been coated by the sequential deposition of polyelectrolytes [115, 116]. A surfactant pre-charging step prior to polyelectrolyte multilayer coating was essential. The particles were exposed to a solution containing amphiphilic molecules (e.g. surfactants or phospholipids), which resulted in them becoming dispersible. The amphiphile adsorbs onto the particles and gives them a well-defined charge. The amphiphile-modified crystals were then sequentially coated with polyelectrolyte multilayers. These (model) and related systems are attractive for sustained-release applications, as the low molecular weight substances can be removed by exposure of the coated particles to organic solution, causing their solubilization and removal. The release rate of the encapsulated low molecular weight materials has been shown to be a function of the polymer coating thickness and the nature of the pre-charging layer [116].

8.3 Polyelectrolyte Multilayers on Particles as Matrices for Particle Modification

An alternative procedure to modify the surface of particles is to exploit preassembled polyelectrolyte multilayers on colloid particles as environments for the selective adsorption of various materials and for localizing chemical reactions. In this way, materials that would otherwise be difficult to directly assemble in a LbL process (or by other means) may be deposited onto particles. The following section considers both of these possibilities and gives examples of each. Polyelectrolyte multilayers on planar supports have already been utilized as nanoreactors for the synthesis of nanoparticles [117, 118].

8.3.1 Infiltration

The following example demonstrates the use of polyelectrolyte multilayers as matrices for infiltration of metal nanoparticles [62]. It shows how the challenge of forming densely packed layers of metal nanoparticles can be overcome through a single adsorption step. Typically, submonolayer nanoparticle coverage is achieved when metal (e.g. gold and silver) nanoparticles are adsorbed from aqueous solutions onto surfaces: surface coverage ratios of less than 0.3 are usually observed [119–125]. Several attempts have been made to increase the metal nanoparticle loading on surfaces by using an additional linker molecule to bind extra particles to the surface [120–122], by modification of the surface by pre-adsorption of polyelectrolytes [123], or by repeated washing and re-exposure of substrates to nano-

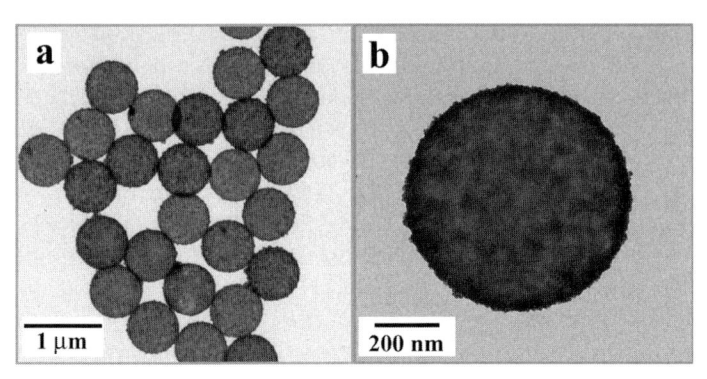

Fig. 8.9 a, b TEM micrographs of polyelectrolyte [(PAH/PSS)$_2$]-modified PS spheres coated with 4-dimethylaminopyridine capped-gold nanoparticles. The images show the uniformity of the gold nanoparticle coating. No aggregation was observed between the coated PS spheres a (adapted from [62]).

particle solutions [124]. By using polyelectrolyte-coated particles (or planar surfaces), dense films of nanoparticles were prepared by a *single* nanoparticle adsorption step [62]. The PS particle surface to be metallized was first coated with four polyelectrolyte layers to provide a surface with a negatively charged outermost layer. The polyelectrolyte multilayer-modified particles were then exposed to the nanoparticle dispersion for 12 h. A significant increase in surface coverage of the gold nanoparticles, compared with the unmodified substrates, was observed. Fig. 8.9 shows TEM images of the metal nanoparticle-coated PS spheres, illustrating the uniformity of the coatings. The modified particles formed were unaggregated. Complementary experiments on quartz crystal microbalance electrodes (planar supports) under identical conditions revealed surface loadings of between 1.1 and 1.3, which is much higher than the 0.3 reported in the absence of the polyelectrolyte support coatings [119–125]. Recent work has shown that the nanoparticle density can be controlled by the number of preassembled polyelectrolyte layers [126]. Tuning the film thickness by varying the polyelectrolyte layer number combined with subsequent metal plating approaches also allows the preparation of metal-coated spheres with tailored optical properties [62, 127]. Such particles are of interest in surface-enhanced Raman scattering (SERS) applications, in biological assays, as well as in electrochemistry or catalysis reactions.

8.3.2
Chemical Reactions

In this section, the use of sol–gel reactions coupled with the infiltration method (see Section 8.3.1) is described. Although metal oxide precursors are widely used to form thin inorganic coatings on surfaces, the vast majority of them are extremely water sensitive, hydrolyzing and condensing upon direct contact with water [128]. (An exception to this is the TALH precursor discussed in Section

8.2.1.3.) When applied directly to modifying colloids, it is often difficult to control the precipitation of these inorganic species in solution, which leads to non-uniform coatings and hence aggregation of the particles [17, 129, 130]. A further issue is that the water-sensitivity of the precursors makes their direct application in the LbL modification of colloid particles difficult because aqueous-based deposition solutions are typically used. LbL-preassembled polyelectrolyte multilayers on particles as nanoreactors can be used to overcome the main issue of precursor water sensitivity for modifying particles. The key points are that the particles are dispersed in organic solutions after coating and that the (sol–gel) reactions are localized within the thin polyelectrolyte coatings on the particles [131]. Here, the water-sensitivity of the precursor is exploited to conduct the sol–gel reaction within the multilayer films, which remain hydrated to a certain extent. Once the sol–gel reaction is completed, and after several washing cycles with alcohol, the particles can be readily dispersed in aqueous solution without aggregation.

Considering the case of lithium niobate ($LiNbO_3$), the following procedure was employed [131]. Polyelectrolyte (PE) multilayer-coated PS particles were centrifuged and redispersed in anhydrous alcohol (ethanol or isopropanol) ten times in order to replace water in the colloidal dispersion with alcohol. The precursor for $LiNbO_3$ ($LiNb(OC_2H_5)_6$) was then added to the polyelectrolyte-coated particles in alcohol, causing it to infiltrate the multilayer shell. The presence of adsorbed water contained in the polyelectrolyte shell results in hydrolysis and condensation of the precursor (i.e. *in situ* sol–gel reaction), causing its gelation, and the formation of thin, inorganic-polyelectrolyte coatings. Fig. 8.10a shows the TEM micrograph of a PS sphere coated with eighteen polyelectrolyte layers and infiltrated by the $LiNbO_3$ precursor (as described above). The presence of $LiNbO_3$ in the PE shell is indicated by the increase in surface roughness compared with the polyelectrolyte-coated particles, which appear smooth in comparison (data not shown). There is no significant change in diameter of the colloids after infiltration of the precursor. Energy-dispersive X-ray (EDX) spectra confirmed the presence of Nb in the shell, suggesting the formation of $LiNbO_3$. The modified particles formed were unaggregated. It was also found that the thickness of the inorganic-polyelectrolyte films is determined by the number of preassembled polyelectrolyte multilayers. Hollow metal oxide spheres were subsequently obtained by calcining the modified colloids at 500 °C [131]. During calcination, $LiNbO_3$ nanoparticles were formed and coalesced, forming the hollow spheres. The current approach was also extended to modify silica particles with titania [131]. The main reason for this is that titania-coated silica supports are widely utilized in catalysis applications [132, 133], but particles with uniform titania coatings are generally difficult to synthesize [134]. The silica particles (450 nm diameter) were coated with ten PAH/PSS layers (SiO_2-PE_{10}). The aqueous SiO_2-PE_{10} colloidal dispersions had the water replaced by isopropanol (IPA). They were then exposed to an IPA solution of the titania precursor, titanium (IV) isopropoxide (TIP). After hydrolysis and condensation of TIP within the PE layers, composite colloids comprising a silica core and a PE_{10}/TiO_2 hybrid shell were produced, as shown by TEM (Fig. 8.10b). EDX spectra demonstrated that the shells surrounding the silica cores comprise Ti, indicating

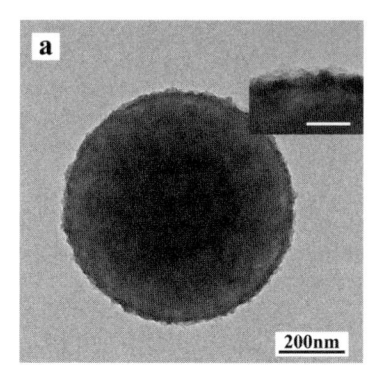

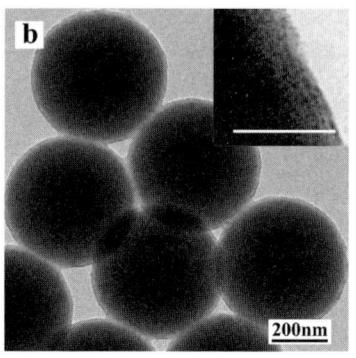

Fig. 8.10 a TEM micrographs of PS spheres coated with 18 polyelectrolyte multilayers and infiltrated with a lithium niobate (LiNbO$_3$) precursor, resulting in PS sphere-PE$_{18}$/LiNbO$_3$ core–shell colloids. The roughness seen on the particles is due to the presence of LiNbO$_3$ in the shell. The scale bar in the inset is 50 nm; **b** TEM micrographs of SiO$_2$ spheres coated with ten polyelectrolyte multilayers and infiltrated with titanium (IV) isopropoxide (TIP), giving rise to SiO$_2$ sphere-PE$_{10}$/TiO$_2$ core–shell colloids. The coating can be clearly visualized in the inset. The silica colloids used were 450 nm in diameter. The scale bar in the inset is 50 nm (adapted from [131], by permission of the American Chemical Society).

the formation of TiO$_2$. The TiO$_2$ at the surface of the spheres is seen as dark spots because of the higher electron contrast of TiO$_2$ compared with the polyelectrolyte (inset in Fig. 8.10b). This infiltration/sol–gel chemistry procedure is rather versatile, as precursor infiltration occurs regardless of the sign of the charge of the outermost layer (e.g. polycation or polyanion), and can be applied to various inorganic molecular precursors [131] to modify the surface of colloid particles.

Similar to what has been reported for polyelectrolyte multilayer systems of weak polyelectrolytes on planar supports [117], carboxylic acidic groups in films of weak polyelectrolytes deposited on particles [135] are potentially useful for metal ion binding and subsequent nanoparticle formation, thus using the multilayers as chemical nanoreactors. This should open up new synthesis routes for the preparation of nanoparticle-modified (micro)particles.

8.4
Potential Uses of Nanoscale Modified Particles

8.4.1
Bioscience

Colloid particles have long been used in biological applications [136, 137]. Particles with biological molecules such as enzymes, antibodies or antigens coupled to their surface can specifically react with antigens, target cells, or viruses, and can be used in both in vitro and in vivo applications [5, 137]. These include immunoassays, bioseparations, hybridization assays, biochemical or enzymatic reactions, affinity chromatography, clinical analysis and diagnostics, and localization and markers in electron or standard light microscopy [5, 136–138]. Recent interest has focused on the use of biofunctionalized nanoparticles for constructing nanoparticle-based assemblies through specific biomolecular interactions [11–14].

In the following sub-sections, examples are given of the use of LbL surface-modified colloid particles in the field of bioscience. These include enzymatic biocatalysis, with both carrier particles and immobilized particles, biological assays, and delivery vehicles.

8.4.1.1 Enzyme Carrier Particles

The inherently high surface area of colloid particles makes them attractive for use as enzyme nanoreactors. The preparation of colloids with organized enzyme-containing multilayer shells for exploitation as colloidal enzymatic nanoreactors has been reported [64–68]. It was demonstrated that particles with defined enzyme content and film architecture could be readily prepared by the LbL method [64–68]. The PS-core–enzyme-shell particles were prepared by assembling urease, GOD, POD, β-GLS, or preformed enzyme-polyelectrolyte complexes in alternation with oppositely charged polyelectrolytes onto PS spheres. Experiments utilizing the enzyme-coated carrier particles showed that the enzymatic activity (per particle) increased with the number of enzyme layers deposited for all of the systems studied. However, differences in the enzymatic activities for each system were observed. Particles coated with preformed enzyme-polyelectrolyte complexes showed approximately a 10-fold lower enzymatic activity than those prepared by the direct adsorption of uncomplexed enzyme. Differences in the relative increase in activity were dependent on the polyelectrolyte layer used to assemble the enzyme multilayer films. These studies demonstrate that enzyme multilayer-coated particles with tailored enzymatic yields could be prepared. Multi-component enzyme multilayer films on colloids were also prepared and examined for sequential enzymatic catalysis [66]. Additionally, particles pre-modified with magnetic nanoparticles and then coated with enzyme multilayers were repeatedly used as catalysts [66, 68]. The presence of the magnetite nanoparticle coating provided a magnetic function that allowed the biocolloids to be easily and rapidly separated with a permanent magnet after the enzymatic reactions were performed. These biocolloid systems are of importance because of their high surface areas for reaction, and because they can be prepared with tailored catalytic activities and with complex and multiple catalytic functions. Further, they can be easily separated and reused many times.

8.4.1.2 Biofunctional Particle-based Films

The LbL coating of crystallized enzyme particles, as outlined in Section 8.2.2.3, provides a system of high enzyme content in a thin polymer coating [107]. Fig. 8.11a shows a fluorescence micrograph of catalase crystals coated with eight [(PSS/FITC-PAH)$_4$] polyelectrolyte layers. Immobilizing these polyelectrolyte-encased microcrystals on solid supports via the LbL method provides a facile method to prepare high-enzyme-content films of controllable thickness. The films were prepared by LbL assembly with oppositely charged polyelectrolytes. Stable, relatively thin (few micrometers), high-enzyme-content multilayer films were prepared [139], as shown in the SEM micrograph (Fig. 8.11b). The image corresponds to a multilayer film that comprises four layers of PSS-coated catalase crystals, assembled in alternation with PAH, on polyelectrolyte-modified quartz crystal microbalance (QCM) electrodes. The catalase crystals are not densely packed on the surface. The image also shows the polydispersity of the micrometer-sized catalase crystals. The inset of Fig. 8.11b is a cross-sectional image of the film, showing the

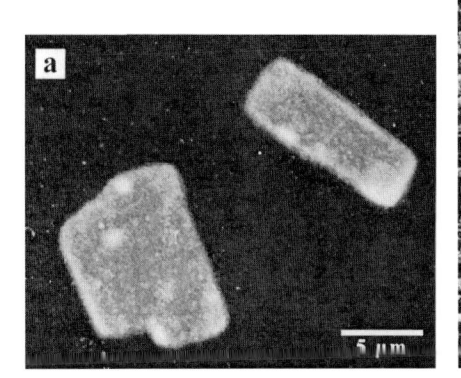

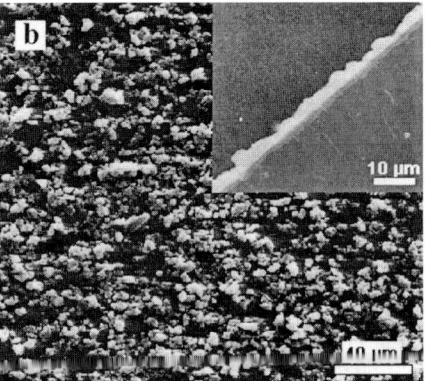

Fig. 8.11 **a** Fluorescence optical micrographs of catalase crystals coated with eight polyelectrolyte layers [(PSS/FITC-PAH)$_4$]. FITC-PAH was deposited in alternation with PSS. No noticeable change in the enzyme crystal morphology or activity of the enzyme was observed as a result of polyelectrolyte multilayer coating; **b** SEM micrographs of the surface and a cross-section (inset) of films composed of four alternating layers of (PSS/coated catalase crystals) and PAH. The films were deposited on polyelectrolyte-coated gold QCM electrodes. The enzyme crystals can be seen on the surface, and the film thickness is ∼2–5 µm (reproduced from [107, 139] by permission of the American Chemical Society).

film roughness and the film thickness (ca. 2–5 μm). The polyelectrolyte coating was essential to obtain high enzyme loadings: QCM studies revealed that each polyelectrolyte-encapsulated catalase crystal layer had a mass of about 120 mg m^{-2}, which is significantly higher than that of 12 mg m^{-2} for uncoated catalase crystals and about 3 mg m^{-2} for solubilized catalase. The films made of polyelectrolyte-encased microcrystals were found to be highly bioactive, displaying biocatalytic activities up to fifty times higher than those prepared by conventional LbL deposition of solubilized enzyme. These films can potentially also be used as biosensors because they are relatively thin, thereby providing fast response times. Related multicomponent films can also be prepared by depositing polyelectrolyte-encapsulated colloids of different enzymes, making them of interest for use in diverse bioapplications.

8.4.1.3 Bioanalytical Assays

Biofunctionalized particles are widely used in immunodiagnostics. Particle-based systems such as fluorophore-loaded latex beads [140], liposome-encapsulated fluorophores [141, 142], and chemiluminescent microemulsions [143] have been used in immunoassays. A different class of particles was prepared via the LbL modification of core PS beads. Biofunctional fluorescent microparticles were prepared by sequentially depositing layers of oppositely charged polyelectrolytes (one of which was fluorescently labeled) onto colloid particles, followed by the passive adsorption of mouse immunoglobulin G (IgG) [82]. Immunotests showed that goat anti-mouse IgG adsorbed on polystyrene culture dish surfaces recognized the mouse IgG-modified particles, as strong binding was observed between anti-mouse IgG and mouse IgG immobilized on the fluorescent PS microparticles. This was detected as fluorescence from the dish. Control experiments confirmed the specific interaction between the adsorbed mouse IgG on the particles and the immobilized antibodies or antigens on the dish. Similar studies were conducted with smaller particles (<100 nm). These studies showed that the sensitivity of the biofunctional particles was close to that observed with conventional immunoprobes.

Since the sensitivity of fluorescence assays is determined by the number of light quanta emitted per analyte molecule, increasing the dye/biomolecule ratio, while minimizing dye self-quenching, is highly desirable. This results in signal amplification and hence enhanced sensitivity. To this end, the preparation and utilization of a novel class of particulate labels based on nanoencapsulated organic microcrystals was reported [144]. The particulate labels were constructed by encapsulating microcrystalline fluorescein diacetate (FDA) (average size ∼500 nm) within polyelectrolyte layers of PAH and PSS via the LbL technique, and subsequently attaching anti-mouse antibodies through adsorption (Fig. 8.12a). Following the scheme outlined in Fig. 8.12b, the microcrystal-based label system was applied in a model sandwich immunoassay for mouse immunoglobulin G (IgG) detection. The important aspect here for obtaining highly amplified bioanalytical assays is the high molar ratio of fluorescent molecules present in the microcrystal

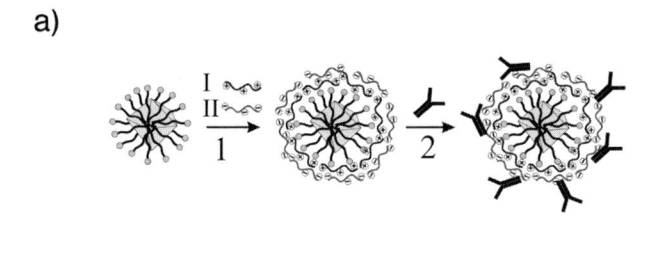

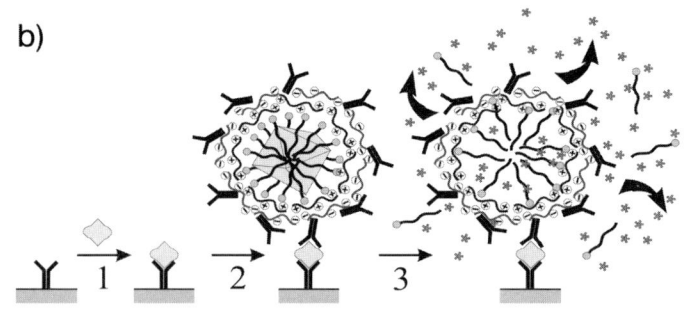

Fig. 8.12 a Schematic illustration of the preparation of biolabelled, polyelectrolyte-encapsulated fluorescein diacetate (FDA) microcrystals. FDA was ball milled into micrometer-sized crystals in an aqueous surfactant medium, followed by encapsulation with polyelectrolyte multilayers (1), and the attachment of a specific antibody (2); **b** Principle of a sandwich immunoassay using biomodified, polyelectrolyte-encapsulated FDA particulate labels. The analyte is first immobilized by the capture antibody preadsorbed on the solid phase (1), followed by exposure to antibody-labelled microparticle detectors (2). High signal amplification is achieved after solubilization, release, and conversion of the precursor FDA into fluorescein molecules by the addition of DMSO and NaOH (3) (adapted from [144], by permission of the American Chemical Society).

core to biomolecules on the particle surface. Following the immunoreaction (step 2), the FDA core was dissolved by exposure to organic solvent, leading to the release of the FDA molecules into the surrounding medium (step 3). Amplification rates of 70- to 2000-fold of the microcrystal label-based assay compared with the corresponding immunoassay performed with direct fluorescently labeled antibodies were observed. This approach provides a general and versatile means to prepare a novel class of highly sensitive bioassay labeling systems.

It is also noted that colloid particles modified with optically active materials such as semiconductor and rare earth-doped nanoparticles are potentially useful for bioanalytical applications (e.g. multiplexing, assays, etc.).

8.4.1.4 Delivery Vehicles

Removal of the cores following the surface modification of colloid particles with polyelectrolytes via the LbL technique affords novel ultrathin-walled hollow capsules that can be utilized for delivery applications [46, 48, 50, 52, 55, 60, 61, 63, 70, 71, 84, 91, 92, 145–150]. Hollow capsules prepared via the LbL self-assembly method on spherical templates comprise oppositely charged linear polyelectrolyte pairs, polyelectrolyte/dye, polyelectrolyte/polyvalent ion, polyelectrolyte/inorganic nanoparticle, and polyelectrolyte/dendrimer pairs, as well as solely inorganic materials (TiO_2, SiO_2, Fe_3O_4, and composites of these). The polyelectrolyte-based capsular colloids exhibit unique physicochemical properties: for example, ultrathin (nanometer-thick) capsular membranes; stimuli-responsive membrane permeability; membrane elasticity; the ability to encapsulate/entrap a range of chemical compounds, solvents, and biological macromolecules in the hollow core or membrane; and the possibility of surface modification and/or functionalization. These capsules can be used as containers for the encapsulation and release of various substances, including pharmaceutically relevant species such as doxorubicin and ibuprofen and enzymes. These species can be loaded after the formation of the hollow capsules via ionic exchange mechanisms with pre-encapsulated polyelectrolyte in the interior of the capsules [146], or via pH changes of the dispersing medium [146–149]. Alternatively, low molecular weight uncharged model drugs in crystalline form can be encapsulated in polyelectrolyte multilayers [116] (see Section 8.2.2.5). Depending on the systems, the release rate of the encapsulated substances can be tuned by varying the number of polyelectrolyte layers composing the hollow capsules, salt concentration, pH, and outer amphiphilic coatings (e.g. lipids). These systems are potentially useful as delivery vehicles for the sustained release of drugs (also see Chapter 18). The hollow capsules can be designed to be robust, deconstructible, biocompatible and with macropores, mesopores and/or micropores in their walls/membranes.

8.4.2
Organization and Templates for Macroporous Materials

8.4.2.1 Colloidal Crystals

Ordered assemblies made of micron- and submicron-sized monodisperse colloid spheres (e.g. silica or polystyrene), also known as colloidal crystals or artificial opals, can possess an optical stop band in the visible and near-IR spectral ranges (see Chapters 9 and 15). These ordered structures are routinely prepared by various methods, with the most common approach involving the spontaneous self-organization of the spheres into crystal structures with long-range periodicity [151–155]. In the case of silica, crystallization of the spheres occurs by slow sedimentation of the particles under controlled conditions. For polymer spheres, where the specific density is closer to that of the solvent (usually water), crystallization occurs by an interplay of drying and capillary forces. Most studies on the formation of photonic crystals from colloids have utilized uncoated colloid spheres. Recently, however, surface-modified colloids have been used [156, 157]. It is important to

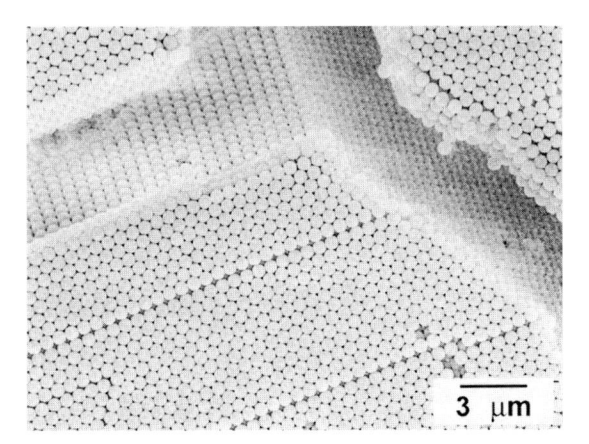

Fig. 8.13 SEM image of a crystalline array of 640 nm diameter PS spheres coated with four polyelectrolyte layers and an outermost PEO layer [(PAH/PSS)$_2$/PEO]. The crystalline arrays form as close-packed planes arranged along the (111) direction. Viewing of the cracks (caused by drying) shows the ordering in the vertical direction (adapted from [126]).

note that uniformly coated, isolated colloid spheres are required to form high-quality colloidal crystals, as the presence of a small number of aggregates can significantly influence or prevent their crystallization. Surface modification of colloid spheres by the LbL method provides control over the thickness and composition of the coating. This, coupled with independent control over the colloid particle size, makes it a versatile route to prepare colloidal crystals with tailored optical properties. Colloidal crystals made from LbL-modified colloid particles have recently been constructed [59, 126, 127, 158, 159]. Surface coatings of polyelectrolyte multilayers [126, 158], semiconductor nanoparticle/polyelectrolyte layers [59, 159], and gold nanoparticle/polyelectrolyte layers [126, 127] were deposited onto submicron-sized PS spheres. The modified particles were assembled into ordered arrays by sedimentation of the spheres or by using a specially designed filtration-flow cell. As an example, PS spheres of 640 nm diameter were coated with PAH/PSS multilayers. A maximum of eight layers [(PAH/PSS)$_4$] were deposited, although colloids with more layers can be readily prepared. The colloids were then further modified by adsorption of an amine-terminated random copolymer of 70% poly(ethylene oxide) (PEO) with 30% poly(propylene oxide), forming a brush layer and facilitating assembly of the colloids into ordered assemblies [126]. Fig. 8.13 shows an SEM image of colloidal crystals made of the corresponding PS particles coated with PAH/PSS multilayers and PEO. The samples crystallized with hexagonal 3D ordered packing with the (111) face parallel to the quartz substrates. Ordered areas extend over hundreds of square micrometers. The ordered arrays produced from PE/PEO multilayer-coated particles were found to be comparable in quality to those observed for uncoated PS colloids. The optical band gap of the colloidal crystals can be modulated by varying the thickness of the PAH/PSS multilayer coating on the spheres. The reflectance spectra of colloidal crystals made of uncoated PS spheres and those coated with [(PAH/PSS)$_x$] layers and terminated with PEO (i.e. PS-PE$_x$/PEO) show stop bands at 1358 ± 3 nm ($x=0$), 1364 ± 3 nm ($x=4$), 1383 ± 3 nm ($x=6$), and 1409 ± 3 nm ($x=8$) [126]. The red shift observed is consistent with Bragg's law, as 3D crystalline arrays of colloid spheres diffract light according to the Bragg equation [151–155]. In essence, the position of the stop band is defined by the particle size and the effective refractive index of the colloidal crystal. Using the relative difference between the lattice parameters (center-to-center distances) of the uncoated and coated particles from the SEM images, the Bragg equation yields an expected red shift in the spectral peak position of the optical stop band of 49 ± 7 nm for the colloidal crystals of PS-PE$_8$/PEO (relative to the pure PS particles). This is in close agreement with the measured peak position of 51 nm. The above data show that the position of the optical stop band of colloidal crystals prepared from LbL surface-modified colloid spheres can be tuned with nanoscale precision through variation of the coating thickness. The composition of the coating on the spheres can also be exploited to modulate the stop band. This was demonstrated for spheres coated with gold nanoparticle/polyelectrolyte layers, which were assembled to form metallodielectric colloidal crystals [126, 127], and also for spheres coated with HgTe semiconductor nanocrystal/polyelectrolyte layers [159]. Additionally, for the latter system, the photoluminescence properties of the HgTe semiconductor nanocrystals LbL assembled around colloid spheres used to form colloidal crystals are impacted by the stop band of the colloidal crystals, giving rise to modification of their emission properties [159]. This provides a means to combine electronic confinement, originating from the semiconductor nanocrystals, with photon confinement, due to the ordered dielectric structures, thus potentially opening new avenues in the design and construction of novel electro-optical devices based on photonic crystals. It is expected that as the level of sophistication for modifying colloid particles increases, a broad range of colloidal crystals with unique optical properties will be reported.

8.4.2.2 Macroporous Films

A diverse range of 3D macroporous materials have been constructed by using colloidal crystals as 3D ordered scaffolds for the infiltration or synthesis of various materials using wet chemistry techniques (e.g. sol–gel processes) [160, 161] (see Chapter 15). The colloid particles are subsequently removed by calcination or chemical means, giving a periodic and open pore structure, often termed inverse opals. Recently, inverse opals of metals [162, 163], metal oxides [164, 165], semiconductors [166], carbon and silicon [167, 168], polymers [169–175], and goldsilica composites [176] have been prepared. The templates used for the preparation of these macroporous materials have predominantly been colloidal assemblies made from uncoated spheres of silica or polystyrene. The use of surface-modified spheres as building blocks to construct tailored composite macroporous materials with new properties and structures has only recently been reported. LbL surface-modified spheres were used to prepare macroporous inorganic and inorganic-composite materials [56, 159, 177]. For example, an ordered silicalite (zeolite) macroporous monolith was prepared by centrifugation of zeolite nanoparticle (~ 50 nm)/polyelectrolyte-coated 640 nm diameter PS particles into close-packed arrays on filters, followed by calcination [56]. Shrinkage ($\sim 20\%$) of the structure occurred upon calcination. TEM images revealed that the pore size was ~ 500 nm

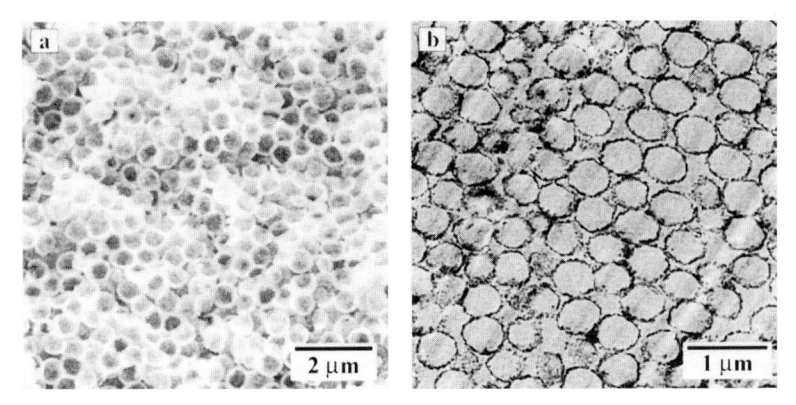

Fig. 8.14 a SEM and **b** TEM micrographs of cross-sections of a macroporous TiO_2/SiO_2 structure fabricated by templating close-packed arrays of PS spheres coated with SiO_2 nanoparticle/polyelectrolyte layers with titanium(IV) isopropoxide, followed by calcination (adapted from [177] by permission of the American Chemical Society).

and that the walls were ~200 nm thick. The wall structure consisted of a dense array of (microporous) nanoparticles, and they also contained disordered mesopores as a result of the interparticle voids. Electron diffraction confirmed the crystalline nature of the nanoparticles. The wall thickness and pore diameters of the hierarchically ordered monolith can be controlled by the coating thickness and template diameter, respectively.

Macroporous (composite) inorganic structures of hollow spheres were also prepared by infiltrating close-packed SiO_2 nanoparticle/polyelectrolyte-coated colloid spheres with a titanium dioxide precursor, followed by removal of the organic material by calcination (Fig. 8.14) [177]. SEM and TEM images of cross-sections of the resulting TiO_2/SiO_2 macroporous structure confirmed the hollow nature of the spheres in the material (Fig. 8.14). High-resolution TEM confirmed that two different types of particles with sizes of about 3 and 25 nm are present in the walls of the resulting material. The example described here has a "closed pore" structure, as opposed to inverse opals that comprise highly ordered air spheres interconnected to each other by small channels (i.e. "open pore" structure). The pore morphology (open or closed pores) of the macroporous structures derived from LbL-modified particles could be controlled by the nature of the multilayers deposited on the colloid particles [177]. Further, the wall thickness of the pores was tuned by varying the number of layers deposited on the particles [177].

Heterogeneous inverse opals can also be prepared from colloidal crystals of nanoscale-modified colloid spheres. A titanium(IV) isopropoxide (TIP)-isopropanol solution was infiltrated into the voids of the colloidal crystals made of submicrometer-sized PS spheres coated with hybrid films consisting of HgTe semiconductor nanocrystal and polyelectrolyte (PE) multilayers by the LbL method. After solidification of TIP and removal of the colloidal crystal template by calcination, heterogeneous TiO$_2$/HgTe inverse opals with HgTe embedded in the interior surface of the macroporous TiO$_2$ were obtained [159].

8.5
Summary and Outlook

This chapter has described the application of the LbL method for generating surface-modified colloid particles. The examples presented demonstrate the wide range of surface nanoengineered particles that can be prepared by utilizing this highly versatile technique. Compositionally complex particles with defined composition, shape, size, and wall thickness can be prepared as the components are deposited in a stepwise sequence via electrostatic assembly. It has also shown that the surface-modified particles could be exploited for subsequent particle modification by using the coatings as nanolayers for confining the infiltration of various species and/or chemical reactions near the particle surface. The particles can be exploited as dispersions or as building blocks to construct functional films. The nanoengineered colloid particles described are intriguing new materials with designed and specific properties for applications in various areas, ranging from biocatalysis to bioanalytical assays, drug delivery, photonics, and porous materials. Despite rapid advances in this area since the late 1990s, much remains to be accomplished with respect to engineering the surfaces of particles. Future work in this area is likely to continue to concentrate on designing novel colloid particles with tailored optical, magnetic, catalytic, and biological properties. Advances in nanotechnology and biotechnology are foreseen to rely on the availability of "smart" particles with multiple, complex functions. The rapidly growing interest in colloid particles and their application is certain to fuel excitement and stimulate multidisciplinary research in this area.

8.6
Acknowledgments

I extend my thanks to many outstanding colleagues, whose names are listed in the references, for their excellent contributions to the research reported in this chapter and for fruitful collaborations. J. Quinn is thanked for critical reading of the manuscript. Funding from the BMBF, DFG, Volkswagen Foundation, Alexander von Humboldt Foundation, the DAAD, and the Australian Research Council is gratefully acknowledged.

8.7
References

1. P. M. CLAESSON, T. EDERTH, V. BERGERON, M. W. RUTLAND, *Adv. Colloid Interface Sci.* **1996**, *67*, 119.
2. E. MATIJEVIC, in *Fine Particle Science and Technology* (Ed. E. PELIZETTI), Kluwer Academic, Dordrecht, **1996**, pp. 1–16.
3. J. N. ISRAELACHVILI, *Intermolecular and Surface Forces*, 2nd edn, Academic Press, New York, **1991**.
4. D. C. BLACKLEY, *Polymer Latices: Science and Technology*, 2nd edn, Vol. 2, Chapman & Hall, London, **1997**.
5. *Colloidal Gold: Principles, Methods and Applications*, Vol. 1 (Ed. M. A. HAYAT), Academic Press, New York, **1989**.
6. For recent reviews, see: (a) C. M. NIEMEYER, *Angew. Chem. Int. Ed.* **2001**, *40*, 4128. (b) I. WILLNER, E. KATZ, A. SHIPWAY, *Chem. Phys. Chem.* **2000**, *1*, 18. (c) P. V. Kamat, *J. Phys. Chem. B* **2002**, *106*, 7729.
7. W. C. W. CHAN, S. NIE, *Science* **1998**, *281*, 2016.
8. M. BRUCHEZ Jr, M. MORONNE, P. GIN, S. WEISS, A. P. ALIVISATOS, *Science* **1998**, *281*, 2013.
9. H. MATTOUSSI, J. M. MAURO, E. R. GOLDMAN, G. P. ANDERSON, V. C. SUNDAR, F. V. MIKULEC, M. G. BAWENDI, *J. Am. Chem. Soc.* **2000**, *122*, 12142.
10. M. HAN, X. GAO, J. Z. SU, S. NIE, *Nature Biotechnol.* **2001**, *19*, 631.
11. C. A. MIRKIN, R. L. LETSINGER, R. C. MUCIC, J. J. STORHOFF, *Nature* **1996**, *382*, 607.
12. P. ALIVISATOS, K. P. JOHNSSON, X. PENG, T. E. WILSON, C. J. LOWETH, M. BRUCHEZ, P. G. SCHULTZ, *Nature* **1996**, *382*, 609.
13. W. SHENTON, S. A. DAVIS, S. MANN, *Adv. Mater.* **1999**, *11*, 449.
14. S. CONNOLLY, D. FITZMAURICE, *Adv. Mater.* **1999**, *11*, 1202.
15. B. J. BATTERSBY, D. BRYANT, W. MEUTERMANS, D. MATTHEWS, M. L. SMYTHE, M. TRAU, *J. Am. Chem. Soc.* **2000**, *122*, 21.
16. M. H. HUANG, S. MAO, H. FEICK, H. YAN, Y. WU, H. KIND, E. WEBER, R. RUSSO, P. YANG, *Science* **2001**, *292*, 1897.
17. For a review, see: F. CARUSO, *Adv. Mater.* **2001**, *13*, 11, and references therein.
18. C. H. M. HOFMAN-CARIS, *New J. Chem.* **1994**, *18*, 1087, and references therein.
19. F. LIM, A. M. SUN, *Science* **1980**, *210*, 908.
20. *Hollow and Solid Spheres and Microspheres: Science and Technology Associated With Their Fabrication and Application*, Vol. 372 (Eds. D. L. WILCOX, M. BERG, T. BERNAT, D. KELLERMAN, J. K. COCHRAN), Materials Research Society Proceedings, Pittsburgh, **1995**.
21. S. J. SHUTTLEWORTH, S. M. ALLIN, P. K. SHARMA, *Synthesis* **1997**, *11*, 1217.
22. R. PARTCH, in *Materials Synthesis and Characterization* (Ed. D. PERRY), Plenum Press, New York, **1997**, pp. 1–17.
23. R. DAVIES, G. A. SCHURR, P. MEENAN, R. D. NELSON, H. E. BERGNA, C. A. S. BREVETT, R. H. GOLDBAUM, *Adv. Mater.* **1998**, *10*, 1264.
24. R. LANGER, *Acc. Chem. Res.* **2000**, *33*, 94.
25. S. M. MARINAKOS, J. P. NOVAK, L. C. BROUSSEAU, A. B. HOUSE, E. M. EDEKI, J. C. FELDHAUS, D. L. FELDHEIM, *J. Am. Chem. Soc.* **1999**, *121*, 8518.
26. C. L. HUANG, E. MATIJEVIC, *J. Mater. Res.* **1995**, *10*, 1327.
27. R. H. OTTEWILL, A. B. SCHOFIELD, J. A. WATERS, N. ST. J. WILLIAMS, *Colloid Polym. Sci.* **1997**, *275*, 274.
28. L. QUARONI, G. CHUMANOV, *J. Am. Chem. Soc.* **1999**, *121*, 10642.
29. R. K. ILER, US PATENT NO. 2, 885, 366, **1959**.
30. N. KAWAHASHI, E. MATIJEVIC, *J. Colloid Interface Sci.* **1991**, *143*, 103.
31. L. M. LIZ-MARZÁN, M. GIERSIG, P. MULVANEY, *Langmuir* **1996**, *12*, 4329.
32. R. K. ILER, *J. Colloid Interface Sci.* **1966**, *21*, 569.
33. G. DECHER, J.-D. HONG, *Ber. Bunsen.-Ges. Phys. Chem.* **1991**, *95*, 1430.
34. G. DECHER, J.-D. HONG, *Makromol. Chem., Macromol. Symp.* **1991**, *46*, 321.
35. For reviews, see: (a) G. DECHER, *Science* **1997**, *277*, 123, and references therein. (b) G. DECHER, in *Templating, Self Assembly and Self-Organisation*, Vol. 9 (Eds. J.-P. SAUVAGE, M. W. HOSSEINI), Pergamon, Oxford, **1996**, pp. 507–528.

36 P. T. Hammond, Curr. Opin. Colloid Interface Sci. **1999**, *4*, 430.
37 P. Bertrand, P. A. Jonas, A. Laschewsky, R. Legras, Macromol. Rapid Commun. **2000**, *21*, 319.
38 S. W. Keller, S. A. Johnson, E. S. Brigham, E. H. Yonemoto, T. E. Mallouk, *J. Am. Chem. Soc.* **1995**, *117*, 12879.
39 T. Chen, P. Somasundaran, *J. Am. Ceram. Soc.* **1998**, *81*, 140.
40 A. Dokoutchaev, J. T. James, S. C. Koene, S. Pathak, G. K. S. Prakash, M. E. Thompson, Chem. Mater. **1999**, *11*, 2389.
41 R. Pommersheim, J. Schrezenmeir, W. Vogt, Macromol. Chem. Phys. **1994**, *195*, 1557.
42 P. Rilling, T. Walter, R. Pommersheim, W. Vogt, *J. Membrane Sci.* **1997**, *129*, 283.
43 O. Gaserod, A. Sannes, G. Skjak-Braek, Biomaterials **1999**, *20*, 773.
44 A. Bartkowiak, D. Hunkeler, Chem. Mater. **2000**, *12*, 206.
45 F. Caruso, E. Donath, H. Möhwald, *J. Phys. Chem. B* **1998**, *102*, 2011.
46 E. Donath, G. B. Sukhorukov, F. Caruso, S. A. Davis, H. Möhwald, Angew. Chem. Int. Ed. **1998**, *37*, 2201.
47 G. B. Sukhorukov, E. Donath, H. Lichtenfeld, E. Knippel, M. Knippel, H. Möhwald, Colloids Surf A: Physicochem. Eng. Aspects **1998**, *137*, 253.
48 G. B. Sukhorukov, E. Donath, S. Davis, H. Lichtenfeld, F. Caruso, V. I. Popov, H. Möhwald, Polym. Adv. Technol. **1998**, *9*, 759.
49 F. Caruso, H. Lichtenfeld, E. Donath, H. Möhwald, Macromolecules **1999**, *32*, 2317.
50 F. Caruso, C. Schüler, D. G. Kurth, Chem. Mater. **1999**, *11*, 3394.
51 D. G. Kurth, F. Caruso, C. Schüler, Chem. Commun. **1999**, 1579.
52 F. Caruso, R. A. Caruso, H. Möhwald, Science **1998**, *282*, 1111.
53 F. Caruso, H. Lichtenfeld, H. Möhwald, M. Giersig, *J. Am. Chem. Soc.* **1998**, *120*, 8523.
54 F. Caruso, H. Möhwald, Langmuir **1999**, *15*, 8276.
55 R. A. Caruso, A. Susha, F. Caruso, Chem. Mater. **2001**, *13*, 400.

56 K. Rhodes, S. A. Davis, F. Caruso, B. Zhang, S. Mann, Chem. Mater. **2000**, *12*, 2832.
57 F. Caruso, A. S. Susha, M. Giersig, H. Möhwald, Adv. Mater. **1999**, *11*, 950.
58 A. Susha, F. Caruso, A. L. Rogach, G. B. Sukhorukov, A. Kornowski, H. Möhwald, M. Giersig, A. Eychmüller, H. Weller, Colloids Surf. A: Physicochem. Eng. Aspects **2000**, *163*, 39.
59 A. Rogach, A. Susha, F. Caruso, G. Sukhorukov, A. Kornowski, S. Kershaw, H. Möhwald, A. Eychmüller, H. Weller, Adv. Mater. **2000**, *12*, 333.
60 F. Caruso, M. Spasova, A. Susha, M. Giersig, R. A. Caruso, Chem. Mater. **2001**, *13*, 109.
61 F. Caruso, M. Spasova, V. Salgueiriño-Maceira, L. M. Liz-Marzán, Adv. Mater. **2001**, *13*, 1090.
62 D. I. Gittins, A. S. Susha, B. Schöler, F. Caruso, Adv. Mater. **2002**, *14*, 508.
63 T. Cassagneau, F. Caruso, Adv. Mater. **2002**, *14*, 732.
64 F. Caruso, H. Möhwald, *J. Am. Chem. Soc.* **1999**, *121*, 6039.
65 C. Schüler, F. Caruso, Macromol. Rapid Commun. **2000**, *21*, 750.
66 F. Caruso, C. Schüler, Langmuir **2000**, *16*, 9595.
67 F. Caruso, H. Fiedler, K. Haage, Colloids Surf A: Physicochem. Eng. Aspects **2000**, *169*, 287.
68 Y. Lvov, F. Caruso, Anal. Chem. **2001**, *73*, 4212.
69 L. Radtchenko, G. B. Sukhorukov, S. Leporatti, G. B. Khomutov, E. Donath, H. Möhwald, *J. Colloid Interface Sci.* **2000**, *230*, 272.
70 F. Caruso, X. Shi, R. A. Caruso, A. Susha, Adv. Mater. **2001**, *13*, 740.
71 S. Mayya, D. I. Gittins, A. M. Dibaj, F. Caruso, Nano Lett. **2001**, *1*, 727.
72 S. Mayya, D. I. Gittins, F. Caruso, Chem. Mater. **2001**, *13*, 3833.
73 X. D. Wang, W. L. Yang, Y. Tang, Y. J. Wang, S. K. Fu, Z. Gao, Chem. Commun. **2000**, 2161.
74 V. Valtchev, Chem. Mater. **2002**, *14*, 956.
75 V. Valtchev, S. Mintova, Micropor. Mesopor. Mater. **2001**, *43*, 41.
76 G. Dong, Y. J. Wang, Y. Tang, W. Ren, W. L. Yang, Z. Gao, Chem. Commun. **2002**, 350.

77 A. Voigt, H. Lichtenfeld, G. B. Sukhorukov, H. Zastrow, E. Donath, H. Bäumler, H. Möhwald, *Ind. Eng. Chem. Res.* **1999**, *38*, 4037.
78 C. Blackley, *Polymer Latices: Science and Technology*, 2nd edn, Vol. 3, Chapman & Hall, London, **1997**.
79 H. Lichtenfeld, L. Knapschinsky, H. Sonntag, V. Shilov, *Colloids Surf. A: Physicochem. Eng. Aspects* **1995**, *104*, 313.
80 H. Lichtenfeld, L. Knapschinsky, C. Dürr, H. Zastrow, *Progr. Colloid Polym. Sci.* **1997**, *104*, 148.
81 G. B. Sukhorukov, J. Schmitt, G. Decher, *Ber. Bunsenges. Phys. Chem.* **1996**, *100*, 948.
82 W. Yang, D. Trau, R. Renneberg, N. T. Yu, F. Caruso, *J. Colloid Interface Sci.* **2001**, *234*, 356.
83 M. C. Hsieh, R. J. Farris, T. J. McCarthy, *Macromolecules* **1997**, *30*, 8453.
84 I. Pastoriza-Santos, B. Schöler, F. Caruso, *Adv. Functional Mater.* **2001**, *11*, 122.
85 S. J. Oldenburg, R. D. Averitt, S. L. Westcott, N. J. Halas, *Chem. Phys. Lett.* **1998**, *288*, 243.
86 A. Henglein, *J. Phys. Chem.* **1993**, *97*, 5457.
87 L. Armelao, R. Bertoncello, M. D. Dominicus, *Adv. Mater.* **1997**, *9*, 736.
88 T. Sun, K. Seff, *Chem. Rev.* **1994**, *94*, 857.
89 M. Haruta, *Catal. Today* **1997**, *36*, 153.
90 P. Schuetz, F. Caruso, *Chem. Mater.* **2002**, *14*, 4509.
91 For a review, see: F. Caruso, *Chem. Eur. J.* **2000**, *6*, 413, and references therein.
92 C. Schüler, F. Caruso, *Biomacromolecules* **2001**, *2*, 921.
93 S. Moya, E. Donath, G. B. Sukhorukov, M. Auch, H. Bäumler, H. Lichtenfeld, H. Möhwald, *Macromolecules* **2000**, *33*, 4538.
94 G. B. Sukhorukov, E. Donath, S. Moya, A. S. Susha, A. Voigt, J. Hartmann, H. Möhwald, *J. Microencapsulation* **2000**, *17*, 177.
95 R. Georgieva, S. Moya, S. Leporatti, B. Neu, H. Bäumler, C. Reichle, E. Donath, H. Möhwald, *Langmuir* **2000**, *16*, 7075.
96 A. Hanprasopwattana, T. Rieker, A. G. Sault, A. K. Dayte, *Catal. Lett.* **1997**, *45*, 165.
97 H. Möckel, M. Giersig, F. Willig, *J. Mater. Chem.* **1999**, *9*, 3051.
98 S. Baskaran, L. Song, J. Liu, Y. L. Chen, G. L. Graff, *J. Am. Ceram. Soc.* **1998**, *81*, 401.
99 X. Shi, T. Cassagneau, F. Caruso, *Langmuir* **2002**, *18*, 904.
100 D. I. Gittins, F. Caruso, *Adv. Mater.* **2000**, *12*, 1947.
101 D. I. Gittins, F. Caruso, *J. Phys. Chem. B* **2001**, *105*, 6846.
102 K.-K. Kunze, R. R. Netz, *Phys. Rev. Lett.* **2000**, *85*, 4389.
103 K. S. Mayya, B. Schoeler, F. Caruso, *Adv. Functional Mater.* **2003**, *13*, 183.
104 The inverse Debye-Hückel screening length is given by $\kappa = c^{0.5}/0.304$, where c is the ionic strength of a monovalent salt solution. See reference [2].
105 P. Mulvaney, *Langmuir* **1996**, *12*, 788.
106 J. B. Schlenoff, S. T. Dubas, *Macromolecules* **2001**, *34*, 592.
107 F. Caruso, D. Trau, H. Möhwald, R. Renneberg, *Langmuir* **2000**, *16*, 1485.
108 M. E. Bobreshova, G. B. Sukhorukov, E. A. Saburova, L. I. Elfimova, B. I. Sukhorukov, L. I. Sharabchina, *Biophysics* **1999**, *44*, 813.
109 N. G. Balabushevitch, G. B. Sukhorukov, N. A. Moroz, D. V. Volodkin, N. I. Larionova, E. Donath, H. Möhwald, *Biotech. Bioeng.* **2001**, *76*, 207.
110 V. S. Trubetskoy, A. Loomis, J. E. Hagstrom, V. G. Budker, J. A. Wolff, *Nucleic Acids Res.* **1999**, *27*, 3090.
111 D. Finsinger, J. S. Remy, P. Erbacher, C. Koch, C. Plank, *Gene Therapy* **2000**, *7*, 1183.
112 R. Dallüge, A. Haberland, S. Zaitsev, M. Schneider, H. Zastrow, G. B. Sukhorukov, M. Böttger, *Biochim. Biophys. Acta* **2002**, *43*, 1376.
113 E. Donath, G. B. Sukhorukov, H. Möhwald, *Nach. Chem. Tech. Labor.* **1999**, *47*, 400.
114 B. Neu, A. Voigt, R. Mitlöhner, S. Leporatti, E. Donath, C. Y. Gao, H. Kiesewetter, H. Möhwald, H. J. Meiselman, H. Bäumler, *J. Microencapsulation* **2001**, *18*, 385.
115 F. Caruso, W. Yang, D. Trau, R. Renneberg, *Langmuir* **2000**, *16*, 8932.
116 X. Shi, F. Caruso, *Langmuir* **2001**, *17*, 2036.

117 S. Joly, R. Kane, L. Radzilowski, T. Wang, A. Wu, R. E. Cohen, E. L. Thomas, M. F. Rubner, *Langmuir* **2000**, *16*, 1354.

118 J. Dai, M. L. Bruening, *Nano Lett.* **2002**, *2*, 497.

119 R. G. Freeman, K. C. Grabar, K. J. Allison, R. M. Bright, J. A. Davis, A. P. Guthrie, M. B. Hommer, M. A. Jackson, P. C. Smith, D. G. Walter, M. J. Natan, *Science* **1995**, *267*, 1629.

120 M. D. Musick, C. D. Keating, M. H. Keefe, M. J. Natan, *Chem. Mater.* **1997**, *9*, 1499.

121 K. C. Grabar, K. J. Allison, B. E. Baker, R. M. Bright, K. R. Brown, R. G. Freeman, A. P. Fox, C. D. Keating, M. D. Musick, M. J. Natan, *Langmuir* **1996**, *12*, 2353.

122 M. D. Musick, C. D. Keating, L. A. Lyon, S. L. Botsko, D. J. Pea, W. D. Holliway, T. M. McEvoy, J. N. Richardson, M. J. Natan, *Chem. Mater.* **2000**, *12*, 2869.

123 J. Schmitt, G. Decher, W. J. Dressick, S. L. Brandow, R. E. Geer, R. Shashidhar, J. M. Calvert, *Adv. Mater.* **1997**, *9*, 61.

124 (a) A. Dokoutchaev, J. T. James, S. C. Koene, S. Pathak, G. K. S. Prakash, M. E. Thompson, *Chem. Mater.* **1999**, *11*, 2389; (b) S. Pathak, M. T. Greci, R. C. Kwong, K. Mercado, G. K. Surya Prakash, G. A. Olah, M. E. Thompson, *Chem. Mater.* **2000**, *12*, 1985.

125 S. J. Oldenburg, R. D. Averitt, S. L. Westcott, N. J. Halas, *Chem. Phys. Lett.* **1998**, *288*, 243.

126 Z. Liang, A. S. Susha, F. Caruso, *Adv. Mater.* **2002**, *14*, 1160.

127 Z. Liang, A. Susha, F. Caruso, *Chem. Mater.* **2003**, *15*, 3176.

128 C. J. Brinker, G. W. Scherer, *Sol–Gel Science: The Physics and Chemistry of Sol–Gel Processing*, Academic Press, San Diego, **1990**.

129 W. P. Hsu, R. Yu, E. J. Matijevic, *J. Colloid Interface Sci.* **1993**, *156*, 56.

130 R. Castillo, B. Koch, P. Ruiz, B. Delmon, *J. Mater. Chem.* **1994**, *4*, 903.

131 D. Wang, F. Caruso, *Chem. Mater.* **2002**, *14*, 1909.

132 P. V. Kamat, *Nanoparticles and Nanostructured Films*, Ch. 9 (Ed. J. H. Fendler), Wiley-VCH, Weinheim, **1998**, and references therein.

133 U. Stafford, K. A. Gray, P. V. Kamat, *J. Phys. Chem.* **1994**, *98*, 6343.

134 A. Imhof, *Langmuir* **2001**, *17*, 3579.

135 N. Kato, P. Schütz, A. Fery, F. Caruso, *Macromolecules* **2002**, *35*, 9780.

136 A. Rembaum, W. J. Dreyer, *Science* **1980**, *208*, 364.

137 G. T. Hermanson, A. K. Mallia, P. K. Smith, *Immobilized Affinity Ligand Techniques*, Academic Press, London, **1992**.

138 *Antibodies: A Practical Approach*, Vol. II. (Eds. D. Catty, C. Raykundalla), IRL Press, Oxford **1989**.

139 W. Jin, X. Shi, F. Caruso, *J. Am. Chem. Soc.* **2001**, *123*, 8121.

140 L. B. Bangs, *Pure Appl. Chem.* **1996**, *68*, 1873.

141 L. Truneh, P. Machy, P. K. Horan, *J. Immunol Methods* **1987**, *100*, 59.

142 H. Schott, D. von Cunow, H. Langhals, *Biochim. Biophys. Acta* **1992**, *1110*, 151.

143 A. Kamyshny, S. Magdassi, *Colloids Surf. B* **1998**, *11*, 249.

144 W. Yang, D. Trau, M. Lehmann, F. Caruso, N. T. Yu, R. Renneberg, *Anal. Chem.* **2002**, *74*, 5480.

145 A. J. Khopade, F. Caruso, *Nano Lett.* **2002**, *2*, 415.

146 A. J. Khopade, F. Caruso, *Biomacromolecules* **2002**, *3*, 1154.

147 G. B. Sukhorukov, A. A. Antipov, A. Voigt, E. Donath, H. Möhwald, *Macromol. Rapid Commun.* **2001**, *22*, 44.

148 G. Ibarz, L. Dahne, E. Donath, H. Möhwald, *Adv. Mater.* **2001**, *13*, 1324.

149 Y. Lvov, A. A. Antipov, A. Mamedov, H. Möhwald, G. B. Sukhorukov, *Nano Letters* **2001**, *1*, 125.

150 C. Gao, E. Donath, S. Moya, V. Dudnik, H. Möhwald, *Eur. Phys. J. A* **2001**, *5*, 21.

151 S. H. Park, Y. Xia, *Langmuir* **1999**, *15*, 266.

152 P. Jiang, J. F. Bertone, K. S. Hwang, V. L. Colvin, *Chem. Mater.* **1999**, *11*, 2132.

153 W. L. Vos, M. Megens, C. M. van Kats, P. Boeseche, *Langmuir* **1997**, *13*, 6004.

154 A. van Blaaderen, R. Ruel, P. Wiltzius, *Nature* **1997**, *385*, 321.

155 O. D. Velev, A. M. Lenhoff, E. W. Kaler, *Science* **2000**, *287*, 2240.
156 M. L. Breen, A. D. Dinsmore, R. H. Pink, S. B. Qadri, B. R. Ratna, *Langmuir* **2001**, *17*, 903.
157 K. P. Velikov, A. van Blaaderen, *Langmuir* **2001**, *17*, 4779.
158 G. Kumaraswamy, A. M. Dibaj, F. Caruso, *Langmuir* **2002**, *18*, 4150.
159 D. Wang, A. L. Rogach, F. Caruso, *Chem. Mater.* **2003**, *15*, 2724.
160 O. D. Velev, E. W. Kaler, *Adv. Mater.* **2000**, *12*, 531.
161 Y. Xia, B. Gates, Y. Yin, Y. Liu, *Adv. Mater.* **2000**, *12*, 693.
162 O. D. Velev, P. M. Tessier, A. M. Lenhoff, E. W. Kaler, *Nature* **1999**, *401*, 548.
163 K. M. Kulinowski, P. Jiang, H. Vaswani, V. L. Colvin, *Adv. Mater.* **2000**, *12*, 833.
164 J. E. G. J. Wijnhoven, W. L. Vos, *Science* **1998**, *281*, 802.
165 B. T. Holland, C. F. Blanford, A. Stein, *Science* **1998**, *281*, 538.
166 Y. A. Vlasov, N. Yao, D. J. Norris, *Adv. Mater.* **1999**, *11*, 165.
167 A. A. Zakhidov, R. H. Baughman, Z. Iqbal, C. X. Cui, I. Khayrullin, S. O. Dantas, I. Marti, V. G. Ralchenko, *Science* **1998**, *282*, 897.
168 A. Blanco, E. Chomski, S. Grabtchak, M. Ibisate, S. John, S. W. Leonard, C. Lopez, F. Meseguer, H. Miguez, J. P. Mondia, G. A. Ozin, O. Toader, H. M. van Driel, *Nature* **2000**, *405*, 437.
169 B. Gates, Y. Yin, Y. Xia, *Chem. Mater.* **1999**, *11*, 2827.
170 S. A. Johnson, P. J. Olivier, T. E. Mallouk, *Science* **1999**, *283*, 963.
171 P. Jiang, J. Cizeron, J. F. Bertone, V. L. Colvin, *J. Am. Chem. Soc.* **1999**, *121*, 11630.
172 D. Wang, F. Caruso, *Adv. Mater.* **2001**, *13*, 350.
173 T. Cassagneau, F. Caruso, *Adv. Mater.* **2002**, *14*, 34.
174 T. Cassagneau, F. Caruso, *Adv. Mater.* **2002**, *14*, 1629.
175 T. Cassagneau, F. Caruso, *Adv. Mater.* **2002**, *14*, 1837.
176 D. Wang, V. Salgueiriño-Maceira, L. M. Liz-Marzán, F. Caruso, *Adv. Mater.* **2002**, *14*, 908.
177 D. Wang, R. A. Caruso, F. Caruso, *Chem. Mater.* **2001**, *13*, 364.

9
Colloidal Crystals: Recent Developments and Niche Applications

Younan Xia, Hiroshi Fudouzi, Yu Lu, and Yadong Yin

9.1
Introduction

Colloids are usually referred to as small particles with at least one characteristic dimension in the range of a few nanometers to one micrometer, dispersed in a different phase [1]. This range of size has been defined to underscore the importance of Brownian motion [2] – the ceaseless diffusion resulting from incompletely averaged-out bombardments that each particle receives from the molecules of a dispersion medium. In broad terms, colloidal particles can also be considered as giant molecules, with their behaviors being well described by the laws of statistical mechanics (e.g. the Boltzman distribution). Since the pioneering work by Ostwald and Graham more than 140 years ago, colloids have become a subject of extensive research in the context of chemistry, biology, materials science, condensed matter physics, applied optics, and fluid dynamics [3–5]. The enormous importance of colloids can also be appreciated from their ubiquitous presence in a rich variety of commercial products such as foods, drinks, inks, paints, toners, coatings, papers, cosmetics, photographic films, rheological fluids, and magnetic recording media.

Fundamental studies on colloidal particles usually require the production of monodisperse samples that are uniform in size, shape, composition, and surface properties. It is no wonder that most advances in these studies were brought about by the elaboration of many synthetic strategies capable of generating monodisperse colloids in relatively copious quantities [6]. Thanks to many years of continuous efforts, a variety of colloids can now be synthesized as truly monodisperse samples in which the size and shape of the particles, and the charges chemically fixed on their surfaces, are all identical to within 1–2% [6–8]. Spherical colloids, in particular, have been the most successful, best-established example of monodisperse systems. Driven by the minimization of interfacial energy, the spherical shape may represent the simplest form that a colloidal particle can easily develop during the nucleation and growth processes. Over the past several decades, a wealth of chemical routes have been developed to synthesize spherical colloids from various materials including organic polymers and inorganic materials. In principle, the properties of a monodisperse sample of spherical colloids can be

Colloids and Colloid Assemblies. Edited by Frank Caruso
Copyright © 2004 Wiley-VCH Verlag GmbH & Co. KGaA, Weinheim
ISBN: 3-527-30660-9

conveniently controlled by varying their intrinsic parameters that include size, chemical composition, crystallinity (polycrystalline vs. amorphous), phase, substructure, and surface functional groups (and thus the density and polarity of charges and the interfacial free energy). It is also worth noting that many theoretical treatments of colloidal systems depend on the assumption of spherical symmetry. For example, the Mie scattering theory [9] that describes the scattering properties of colloidal particles and the Derjaguin-Landau-Vervey-Overbeek (DLVO) model [10] that deals with the interactions between colloidal particles are both derived from spherical colloids.

Experimental studies on spherical colloids have greatly enriched our understanding of their light scattering powers [11], the interactions between colloidal particles [12], and their hydrodynamic properties in various types of dispersion media [13]. On the other hand, spherical colloids may also represent a class of ideal building blocks that could be readily assembled into long-range ordered lattices such as colloidal crystals [14–16]. The ability to crystallize spherical colloids into spatially periodic structures allows for the observation of interesting and often useful functionality not only from the constituent material of the colloidal particles, but also from the long-range order exhibited by these periodic lattices. The beautiful, iridescent, and attractive colors of opals (both natural and synthetic), for instance, are caused by their three-dimensionally periodic lattices of silica and zirconia colloids that are colorless by themselves [17]. As a matter of fact, recent studies on the optical properties of these materials have now evolved into a new, active, and exciting field of research that is usually referred to as photonic crystals or photonic bandgap (PBG) structures [18]. The unique features associated with this new class of materials are technologically important because they might eventually lead to the fabrication of more efficient light sources, detectors, and waveguiding structures.

In addition to their use as PBG materials, colloidal crystals have been extensively explored for many other intriguing applications. For example, two-dimensional (2D) crystals of spherical colloids have been successfully demonstrated as arrays of microlenses in image processing [19] and photolithography [20]; as physical masks for evaporation or reactive ion etching to fabricate regular arrays of micro- or nanostructures [21] and as regular arrays of relief structures to cast elastomeric stamps for use in soft lithographic techniques [22]. Three-dimensional (3D) crystals have recently been exploited as sacrificial templates to fabricate highly ordered, macroporous materials [23] as diffractive elements to fabricate sensors, filters, switches, and many other types of optical and electro-optical devices [24] and as a directly observable (in real space) model system to study various fundamental phenomena such as crystallization, phase transition, melting, and fracture mechanics [25]. The success of all these investigations has strongly depended on the availability of spherical colloids with tightly controlled sizes and surface/bulk properties, and on the ability to assemble them into long-range ordered arrays with well-defined crystal structures, sufficiently large domain sizes, and well-controlled thickness or volume. When used as photonic crystals, it is also desirable to vary the refractive index and geometric shape of the colloidal particles in an effort to fully control their photonic band structures. In this case, tight control over the degree of perfection of the 3D periodic structure seems to be as necessary to the photonic exploitation of a crystalline lattice of spherical colloids as it has been in the microelectronic usage of a single crystalline semiconductor. Most of these issues will be addressed as we proceed. It is believed that all of these issues must be solved before colloidal crystals can reach their ultimate potential as a new class of industrial materials.

The objectives of this chapter are the following: (a) to briefly discuss a number of chemical methods commonly used for the facile synthesis of spherical colloids with well-controlled sizes and properties; (b) to fully demonstrate the potential of spherical colloids in generating three-dimensionally periodic lattices; (c) to address experimental issues related to the crystallization of nonspherical colloids into long-range ordered 3D lattices; and (d) to assess a range of intriguing applications associated with the periodic structure of a colloidal crystal. The materials of this chapter will focus on 3D lattices, with only some brief discussions on 1D and 2D crystals in the sections that deal with their applications.

9.2
Monodisperse Spherical Colloids: the Key Component

Like many other areas of colloid science, most advances related to colloidal crystals were brought about by the elaboration of simple, convenient, and reproducible methods that were able to generate spherical colloids as monodisperse samples and in relatively large quantities [6–8]. As we shall see in the following sections, the use of a monodisperse sample is also central to the self-assembly of these colloids into homogeneous, long-range ordered lattices characterized by large domain sizes. Thanks to many years of continuous effort, a variety of colloids can now be treated as truly monodisperse systems in which the size and shape of the particles and the net charges fixed on their surfaces are all identical to within 1–2%. The best established and most commonly used methods seem to be *emulsion polymerization* for polymer latexes and *controlled precipitation* for inorganic colloids (as well as polymer beads). Spherical colloids synthesized using these methods can be obtained in relatively large quantities from a number of commercial sources. Tab. 9.1 gives a partial list (in alphabetical order) of such companies whose products are usually supplied as stabilized suspensions (>1%) in water. Most of them can be directly used in crystallization without further purification and separation. Fig. 9.1 shows the transmission electron microscopy (TEM) images of two typical examples: polystyrene beads (~200 nm in diameter) that were obtained from Duke Scientific, and silica spheres (~600 nm in diameter) that were synthesized in our group using the Stöber method. The energy-dispersive X-ray (EDX) spectra indicate the elemental compositions on their surfaces.

Tab. 9.1 Some commercial sources of monodisperse spherical colloids [a]

Company	Contact information	General comments
Bangs Laboratories [b]	317 570 7020 (Tel) 317 570 7034 (Fax) www.bangslabs.com	Polystyrene (plain, fluorescent, magnetic, 0.02–5.0 μm) and silica spheres (0.3–5.0 μm)
Duke Scientific [b]	650 424 1177 (Tel) 650 424 1158 (Fax) www.dukesci.com	Polystyrene (certified particle size standards, plain, fluorescent, 0.02–1.0 μm) and silica spheres (0.5–1.6 μm)
Interfacial Dynamics [b]	503 684 8008 (Tel) 503 684 9559 (Fax) www.idclatex.com	Polystyrene spheres (fluorescent, 0.02–10.0 pm)
Nissan Chemicals	713 532 4745 (Tel) 713 532 0363 (Fax) www.snowtex.com	Colloidal silica (various dispersing media, 0.003–0.100 μm), antimony pentoxide
Polysciences [b]	215 343 6484 (Tel) 215 343 0214 (Fax) www.polysciences.com	Polystyrene (plain, fluorescent, 0.05–90 μm), silica, and glass spheres (0.05–0.45 μm)
Seradyn	317 266 2956 (Tel) 317 266 2991 (Fax) www.seradyn.com	Polystyrene spheres (plain, fluorescent, magnetic, 0.05–5.0 μm)

a) Information contained here was current at the time of writing.
b) Custom synthesis available from these companies.

9.2.1
Polymer Latexes through Emulsion Polymerization

Polymer latexes of various chemical compositions can be produced as exceedingly uniform spheres via the so-called emulsion polymerization process that involves at least four components: monomer, dispersion medium (in most cases, water), emulsifier (surfactant), and initiator (often water-soluble) [1, 11]. Most surfactant molecules aggregate into micelles ($\sim$ 10 nm in size) swollen by the monomer. The majority of monomer exists as an aqueous emulsion (1 $\sim$ 100 μm in size) stabilized by the emulsifier. The formation of polymer latexes begins with formation of a burst of primary free radicals through the decomposition of the water-soluble initiator. These radicals polymerize the small amount of monomer that is dissolved in the aqueous phase into oligomers (in the form of tiny particles). These tiny nuclei subsequently enter the micelles and eventually grow into larger particles until all monomers dissolved in each micelle have been polymerized. Meanwhile, monomers encapsulated in emulsion droplets act as the reservoir (through diffusion) to ensure continuous growth of the polymer chains. The growth of polymer latexes will stop at the point when all monomers have been exhausted.

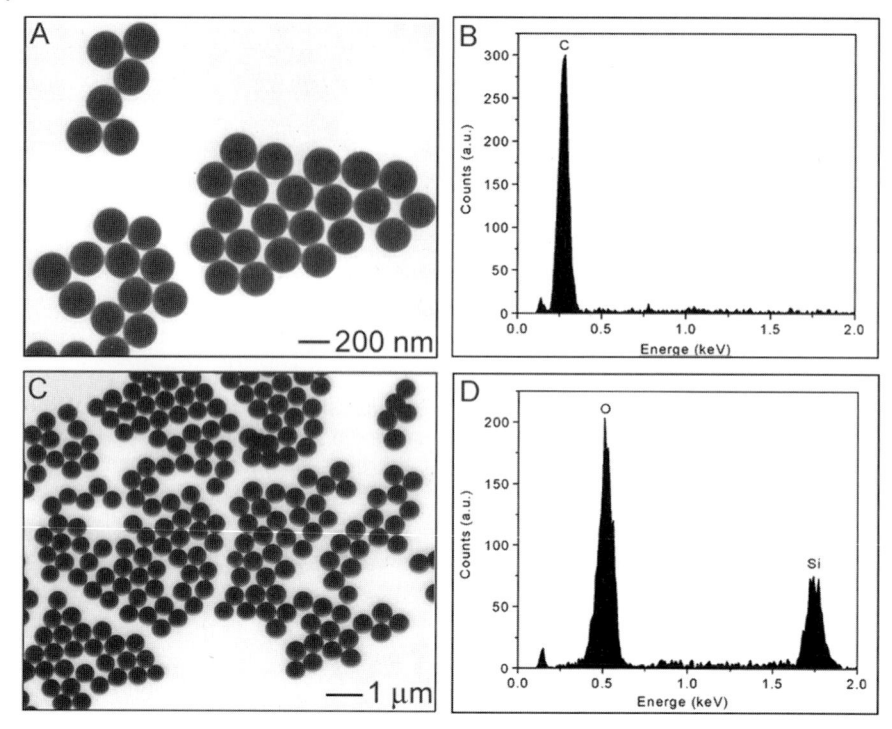

Fig. 9.1 The TEM images and EDX spectra of two representative colloidal systems that can be readily prepared as monodispersed samples in large quantities: **A, B** polystyrene beads ∼200 nm in diameter, **C, D** silica spheres ∼600 nm in diameter.

The surface properties of polymer latexes synthesized using emulsion polymerization are largely determined by the initiator. When potassium persulfate – a water-soluble initiator – is used, the surface of polymer latexes should be terminated by the negatively charged sulfate group [26]. Other acidic (e.g. –COOH) and basic (e.g. –NH$_2$) groups could also be introduced into the surface layer by adding the appropriate component to the monomer solution. The polarity of these surface groups can be changed by controlling the pH value of the dispersion medium. In some cases, both positively and negatively charged groups can be simultaneously placed on the surface of a single polymer latex particle. At a particular pH value, the negative and positive charges on such latex surface are balanced so that these particles behave like an amphoteric or zwitterionic species [27].

The emulsion polymerization process has been successfully applied to a number of polymer systems such as polystyrene and poly(methyl methacrylate) [11]. A similar approach has also been extended to synthesize monodisperse samples of inorganic colloidal particles [28]. The size of polymer latex spheres can be controlled at will to cover a broad range from 20 nm to ∼1 μm. Polymer spheres with diameters larger than 1 μm have to be prepared using a two-stage method developed by Ugelstad, El-Aasser, and their co-workers [29]. In this case, submicrometer-sized polystyrene beads were synthesized via the classical emulsion polymerization route, and then swollen with another monomer (or more of the original monomer) in the presence of an aprotic solvent miscible with water. Subsequent polymerization of the encapsulated monomer could lead to the formation of monodisperse latex spheres with diameters up to several hundred micrometers. A comprehensive treatment on the formation of latexes can be found in Chapter 1.

9.2.2
Inorganic Colloids through Controlled Precipitation

Colloids made of inorganic materials are usually prepared via precipitation, a process that involves at least two steps: nucleation and growth. To achieve monodispersity, these two steps must be separated into sequential stages: that is, nucleation should be strictly avoided once growth has started. In a typical process, the monomer (usually existing as a complex or a solid precursor) must be added or released slowly at a well-controlled rate in order to keep it from passing the critical supersaturation concentration during the period of growth. This criteria for obtaining monodisperse colloidal particles was summarized by La Mer et al. as a general rule (the La Mer diagram) in their study of sulfur colloids prepared by acidifying thiosulfate solutions [30] and later was successfully extended to a broad range of other inorganic systems by various groups.

In 1968, Stöber et al. took advantage of this diagram and demonstrated an extremely useful route to the preparation of monodisperse silica colloids [31]. They hydrolyzed a dilute solution of tetraethylorthosilicate (TEOS) in ethanol at high pH and obtained uniform spheres of amorphous silica. The diameters of these colloids could be varied from 50 nm to 2 µm simply by changing the precursor concentration. This method was later improved by many other groups and has now become the simplest and most effective route to monodisperse silica spheres [32]. In the following years, Matijevic and his co-workers further developed this method, and successfully applied it to the production of monodisperse spheres, cubes, rods, and ellipsoids essentially from all classes of metal oxides and carbonates [7, 33]. As a general rule, it is necessary to precisely control the reaction conditions – for example, the temperature, the pH value, the concentration of reactants, and the concentration of counterions – to ensure the formation of a single, short burst of nuclei and then let them grow uniformly without new nucleation events. It is worth noting that this simple strategy can also be extended to organic systems to synthesize uniform polymer latexes [34]. This route is commonly known as precipitation or emulsifier-free polymerization, a process that does not require the addition of any surfactant to the synthetic system.

As one of the best-characterized inorganic colloids, silica spheres can now be manufactured as monodisperse samples in industrial quantities [32]. The surfaces of as-synthesized silica colloids are usually terminated in silanol groups (–Si–OH) that will ionize to generate negatively charged interfaces when the pH value of dispersion medium is higher than 7. When pristine silica colloids are thermally treated at elevated temperatures, they will undergo a series of changes: the absorbed water (~5% by weight) will be released at ~150 °C; the silanol groups will be cross-linked via dehydration in the temperature range of 400–700 °C; and finally these particles will be fused into aggregates when the temperature is raised above the glass transition temperature of amorphous silica (~800 °C).

9.2.3
Core-Shell Spherical Colloids

The properties of spherical colloids can be conveniently varied by modifying their surfaces. For example, the wettability of silica colloids can be easily changed using siloxane chemistry to form self-assembled monolayers (SAMs) with the silanol groups [35]. In some cases, modification can also be accomplished by coating the surfaces of colloids with shells of a different chemical composition, as well as in varying thickness [36]. It has been demonstrated that the size, structure, and composition of such core-shell particles can all be varied in a controllable way to tailor their optical, electrical, thermal, mechanical, magnetic, and catalytic properties over a broad range [37]. In addition, the cores can also be removed in a subsequent step using procedures such as solvent extraction or calcination at elevated temperatures to generate hollow particles that might exhibit optical properties substantially different from those of solid ones. The empty interiors of hollow particles also make them particularly useful in several other niche applications. When employed as fillers (or pigments), for example, hollow particles definitely offer advantages over their solid counterparts as a result of their much lower densities. As demonstrated by computer simulations, the use of core-shell colloids (e.g. silica colloids coated with silver shells) may even lead to the formation of complete photonic gaps, an interesting feature that is still lacked by crystals made of most spherical colloids [38].

A wealth of methods have been demonstrated for coating spherical colloids with thin shells of various materials. Most of them involve the use of controlled adsorption and/or reaction (e.g. precipitation, grafted polymerization, or sol–gel condensation) that could be effectively confined to the surfaces of spherical colloids [39]. Although these methods look simple and straightforward, they used to have difficulties in controlling the homogeneity and thickness of the coatings, and sometimes even led to clumping and heterocoagulation. These problems have now been solved (at least, partially) by several groups, and it is possible to generate homogeneous, dense coatings of silica or polymers on various colloidal templates with well-controlled thickness. In the first method, the surfaces of inorganic colloids were grafted with an appropriate primer that could greatly enhance the coupling (and thus deposition) of silica monomers and/or oligomers to these surfaces [40]. This method allowed the incorporation of semiconductor nanocrystallites or metal sols into silica colloids to functionalize them with fluorescence or other useful properties [41]. More recently, Xia et al. demonstrated that the surfaces of metal and metal oxide colloids could be directly coated with silica shells of various thickness as long as the concentration of templates was sufficiently high to eliminate the homogeneous nucleation process [42–44]. Fig. 9.2 A shows the back-scat-

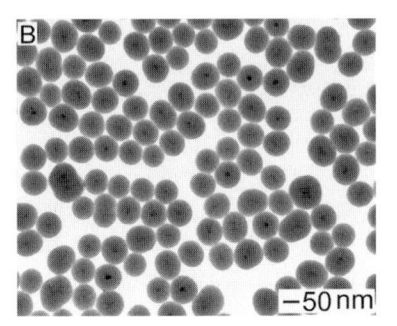

Fig. 9.2 The surfaces of metal and metal oxide particles could be directly coated with silica shells of various thicknesses to form core-shell colloids. **A** The back scattering SEM image of gold colloids whose surface had been coated with conformal shells of amorphous silica. **B** The TEM image of superparamagnetic magnetite nanoparticles whose surfaces had been coated with uniform shells of silica.

tering scanning electron microscopy (SEM) image of some gold colloids whose surface had been coated with conformal shells of amorphous silica. Fig. 9.2B shows the TEM image of superparamagnetic particles of magnetite whose surfaces had been coated with uniform shells of silica. Both samples could be prepared as stable dispersions (in alcohols or water) and in relatively copious quantities (also see Chapter 7). In the second method, electrostatic attractive adsorption of polyelectrolytes and charged nanoparticles was used to build a thin shell (layer-by-layer) around template cores – spherical colloids whose surfaces had been derivatized with appropriate charges [45] (see Chapter 8). In the third method, colloids could be coated with uniform shells of organic polymers by grafting their surfaces with radicals, followed by living polymerization [46]. All these methods have been successfully applied to form homogeneous and dense coatings of ceramic or organic materials on the surfaces of a variety of spherical colloids. Subsequent removal of the core particles yielded hollow spheres made of the ceramic or polymeric material.

9.3
Crystallization of Colloids under Physical Confinement

Monodisperse spherical colloids often spontaneously organize themselves into long-range ordered lattices when they are subjected to a physical confinement (provided, for example, by a pair of parallel substrates) [47]. On the basis of this observation, Xia et al. have demonstrated an effective approach that allowed the fabrication of colloidal crystals with domain sizes as large as several square centimeters [48]. In a typical procedure, spherical colloids (both polymer latexes and silica spheres, with diameters ranging from ~ 50 nm to ~ 5 µm) were injected into a specially designed fluidic cell, and crystallized into a cubic-close-packed (ccp) 3D

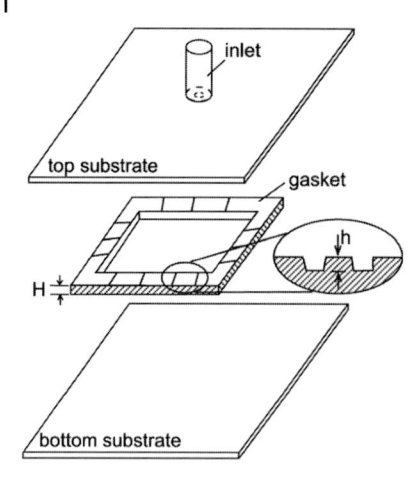

Fig. 9.3 A schematic illustration of the flow cell used to crystallize spherical colloids into 3D crystalline lattices. The gasket was cut from a plastic film whose thickness could vary in the range of a few to several hundred microns. A glass tube was attached to the hole drilled into the top glass substrate whose surface had been made hydrophilic by treating with oxygen plasma. The cell was assembled by sandwiching the gasket between the top and bottom glass substrates, and tightening with binder clips. The aqueous dispersion of monodispersed colloids was injected into the cell through the inlet.

lattice under constant agitation supplied by sonication or mechanical vibration. Fig. 9.3 shows a schematic illustration of the fluidic cell that could be conveniently fabricated in copious quantities using conventional microlithographic techniques. In our original demonstration [48], the fluidic cell was constructed by sandwiching a photoresist frame of uniform thickness (<12 μm) between two glass slides held together using binder clips. The surface of this photoresist frame could be patterned with parallel arrays of shallow channels (their depth had to be controlled below the diameter of the spherical colloids to be assembled) using a double-exposure procedure that involved overlapping of two photomasks. All components of each cell were assembled in a clean-room or under a relatively dust-free environment (e.g. a hood equipped with laminar flow and intake filtration). Most recently, the design of fluidic cells was extended to systems foregoing the requirement of clean-room facilities. A number of simple and convenient methods that rapidly generated fluidic cells were demonstrated by sandwiching a square frame (cut from commercial Mylar film) between two glass substrates [49]. Uniform Mylar films with thickness in the range of 20–100 μm could be obtained from Fralock (Canoga, CA) or Dupont. Small channels between the Mylar film and glass substrates could be easily fabricated using a number of non-photolithographic methods: for example, by scraping both sides of the Mylar film with a piece of soft paper to create channels via abrasion; by coating both surfaces of the Mylar film with colloids smaller than those to be crystallized in the cell; and by patterning the glass substrate(s) with arrays of channels in thin films of gold through the use of microcontact printing (μCP) and selective wet etching [50]. Although these methods are not as reproducible as conventional techniques based on photolithography, a yield of more than ∼90% could still be routinely accomplished.

The aqueous dispersion of spherical colloids is usually injected into the cell through a glass tube glued to the small hole (1–3 mm in diameter) drilled in the top glass substrate. As shown in Fig. 9.3, colloids with a diameter (D) larger than the depth (h) of channels in the gasket surfaces will be concentrated at the bottom of the fluidic cell and crystallized into a long-range ordered 3D lattice. The rate of crystallization (including both nucleation and growth) near the edge of this fluidic cell can be increased by applying a slightly positive pressure of N_2 gas through the glass tube. Constant vibration provided by a Branson 1510 sonicator (Danbury, CT) has to be used to mechanically agitate the system and thus to ensure that each spherical particle rests at the lattice site represented as a thermodynamic minimum so that no random packing occurs in the cell. We have been able to routinely assemble monodisperse spherical colloids (including polymer latexes and silica beads) into ccp lattices over several square centimeters that are characterized by a homogeneous structure, well-controlled thickness, and long-range ordering. Both scanning electron microscopy and optical diffraction studies indicate that the (111) planes of these crystalline lattices are oriented parallel to the substrates [48 b]. Colloidal crystals (made of polymer beads) several square centimeters in area and tens of micrometers in thickness could be conveniently obtained within a few days.

This method is simple, flexible, and relatively fast. It also provides good control over the surface morphology, as well as the exact number of (111) planes in the direction perpendicular to the glass substrates [51]. For any crystalline lattice fabricated using the fluidic cell shown in Fig. 9.3, the number of (111) planes is largely determined by the ratio between the thickness (H) of the cell and the diameter (D) of the spherical colloids. By using gaskets with different thicknesses, we were able to conveniently tune this number from 1 to more than 200. Fig. 9.4 shows the SEM images of several typical examples that were crystallized from 220 nm diameter polystyrene beads. The gaskets used in these demonstrations had thicknesses of 2, 10, and 22 µm, respectively. These SEM images indicated that long-range order was achieved in the direction perpendicular to the substrates by physically confining spherical colloids within two parallel substrates. Their cross-sections also suggest that these crystals had a ccp structure with an "ABC" stacking. The (111) planes of these crystals were oriented parallel to the surfaces of glass substrates. In addition to the number of (111) planes, tight control could also be reproducibly accomplished for the surface morphology and structural orientation of these 3D crystalline lattices. Since the lattices formed using this method are long-range ordered in all three dimensions, they could serve as optical notch filters to selectively reject a narrow band of light (as determined by the Bragg equation) in the spectral region that extends from ultraviolet to near-infrared [52]. In addition to polymer beads, this method has also been successfully applied to other colloidal systems such as silica spheres and core-shell particles [44]. Fig. 9.4 D shows the back-scattering SEM image of a crystalline lattice assembled from $Au@SiO_2$ spherical colloids, clearly indicating the existence of a gold core in the middle of each particle. The gold cores were ~ 50 nm in diameter, and the silica shells had a thickness of ~ 100 nm (as measured from TEM images). This three-dimensionally ordered lattice also exhibited a ccp structure, with its (111) planes oriented parallel to the supporting substrate. Because gold nanoparticles exhibit a strong plasmon resonance absorption, the optical properties of these

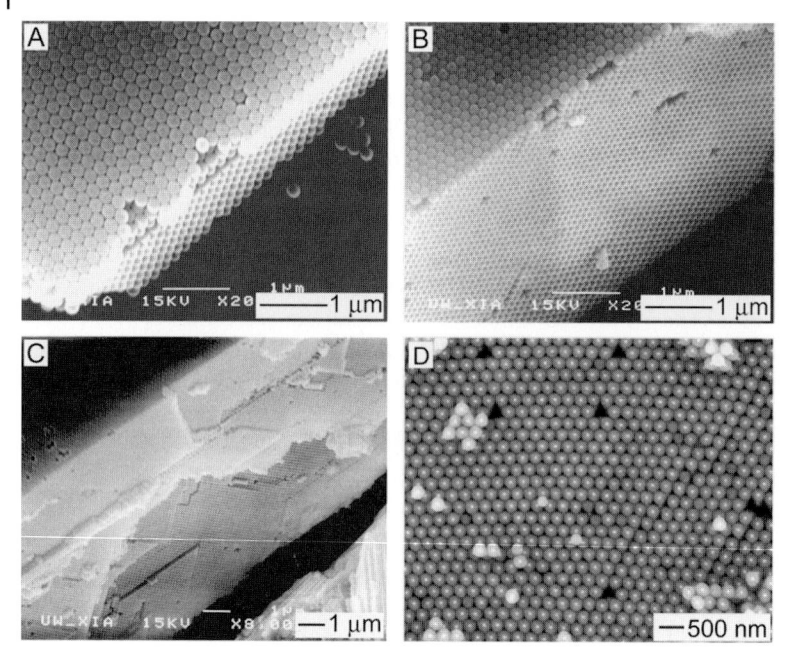

Fig. 9.4 **A–C** Oblique SEM images showing the cross-sections of 3D crystalline lattices of 220 nm polystyrene beads. They had different thicknesses, with **A** 13, **B** 56, and **C** 127 layers of (111) planes in the direction perpendicular to the substrate. Such highly ordered lattices could be readily obtained over areas of several square centimeters.
D The backscattering SEM image of a 3D lattice crystallized from colloids that had gold cores 50 nm in diameter and silica shells ~80 nm in thickness.

crystals are expected to be different from those of crystals assembled from pristine silica or polymer beads. For example, strong coupling between the plasmon modes of adjacent gold cores may lead to the creation of new and/or wider gaps in the photonic band structure [53].

9.4
Other Methods for Fabricating Colloidal Crystals

In addition to the above approach based on physical confinement, many other methods have also been reexamined or demonstrated for organizing monodisperse spherical colloids into long-range ordered 3D crystals. For reasons of space, we concentrate here on just three methods that have been employed by more than two research groups to successfully generate 3D colloidal crystals characterized by relatively large domain sizes. Fig. 9.5 shows the schematic diagrams of these methods, and more detailed descriptions will be provided in the following sections.

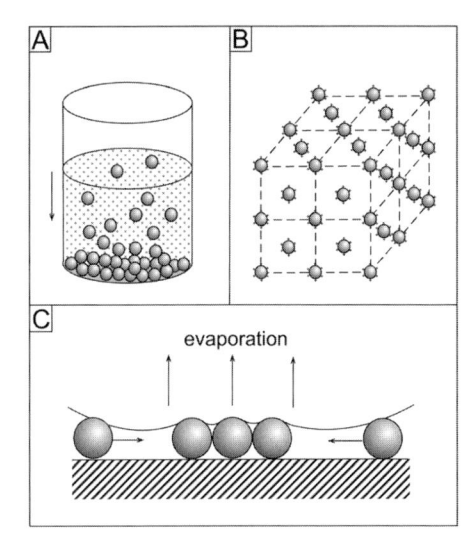

Fig. 9.5 Schematic outlines of three experimental procedures that have been used to assemble spherical colloids into 3D crystalline lattices: **A** sedimentation in the gravitational field; **B** ordering via repulsive electrostatic interactions; **C** crystallization via attractive capillary forces induced by solvent evaporation.

9.4.1
Crystallization via Sedimentation in a Force Field

Sedimentation in a force field seems to provide the crudest mechanism for concentrating a dilute suspension of colloidal particles. The application of gravitational force, for example, often leads to the formation of a dense sediment (either crystalline or amorphous) on the bottom and a clear liquid above (Fig. 9.5 A) [54]. Although it may look simple, sedimentation actually involves the coupling of several complex processes in a highly nonideal system that include gravitational settling, translational diffusion (or Brownian motion), and order-disorder phase transition (crystal nucleation and growth). Successful use of this method in generating colloidal crystals relies on tight control of several parameters – for example, the size and density of the colloids and the rate of sedimentation. Colloidal particles can always settle completely to the bottom of a dispersion as long as their size and density are sufficiently large. Only when spherical colloids are used and the settling process is slow enough, will the particles concentrated at the bottom undergo a hard-sphere disorder-to-order phase transition and form a long-range ordered lattice [55]. If the colloidal particles are too small (less than 100 nm) and/or their density is too close to that of the dispersion medium, they will exist as a well-dispersed equilibrium state, in which the number of spherical colloids per unit volume will vary with height according to the Boltzmann distribution function.

Natural opals are probably the best-known products of sedimentation, and they are usually formed within the cracks of a rock when spherical silica colloids are forced to settle by the gravitational field [56]. A similar protocol has been extended to synthesize equivalent lattices in the setting of a research laboratory. Monodisperse silica spheres are commonly employed in these sedimentation experiments owing to the relatively higher density of amorphous silica compared with those of organic polymers. Opalescent lattices (often referred to as synthetic or artificial opals) with large domains could be obtained under carefully controlled conditions [57]. It is generally accepted that the 3D crystals fabricated using this method have a ccp structure similar to that of a natural opal [58]. The preference for a ccp structure over the hexagonal close-packed (hcp) one has been suggested to result from the difference in entropy between these two structures [59].

There are a number of disadvantages associated with the sedimentation method. First of all, it usually takes very long periods of time (weeks to months) to completely settle submicrometer-sized colloidal particles. Second, this method provides very little control over the morphology of the top surface, as well as the number of layers contained in the 3D crystalline lattices. Third, layered sedimentation often occurs in practical operations, which usually leads to the formation of a number of layers of different densities and structural orders along the direction of the gravitational field. Fourth, the 3D lattices produced by the sedimentation method are often polycrystalline or amorphous in structure. Some of these problems have recently been solved (at least partially) by modifying the experimental procedure. For example, centrifugation could be used to increase the settling rate [1]. Lopez et al. demonstrated that the sedimentation of small particles could be greatly accelerated by applying an electric field along the direction of gravitational force [60]. Kumacheva et al. found that sedimentation under an oscillatory shear could greatly improve the crystallinity and ordering in the resultant lattices [61]. In addition to its major function as a means to concentrate colloidal particles, the gravitational field may also play an important role in determining the crystal structure. Chaikin et al. recently performed a space shuttle experiment showing that spherical colloids preferentially organized themselves into the random hexagonal close-packed (rhcp) structure at volume fractions up to 61.9% when the gravitational force was not present [62]. This observation suggested that the ccp structure obtained under normal gravity might be induced by gravity-related stresses.

9.4.2
Crystallization via Repulsive Electrostatic Interactions

The long-range (over distances >100 nm) interaction(s) between colloidal particles provides another driving force to organize spherical colloids into crystalline lattices. In general, the pair-wise potential energy of interaction between two hard spheres (that is, particles whose surfaces are electrically neutral) can be divided into two terms: the short-range, steric repulsive force and the moderately long-range, attractive force that is commonly known as the van der Waals (or London dispersion) interaction [10]. For highly charged colloidal particles suspended in a liquid containing dissociated electrolytes, a third term has to be added to the effective potential energy: that is, the much stronger and truly long-range Coulomb repulsion screened by electrolytes [63]. The sum of these three potentials gives the well-known Sogami-Ise potential. As documented in many previous studies, the

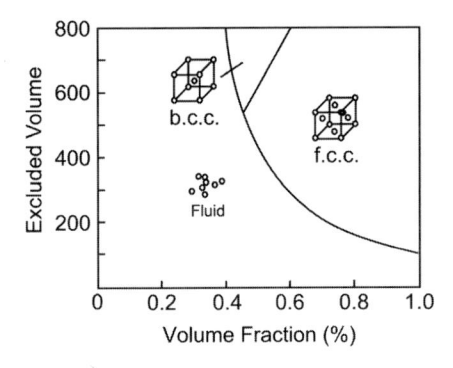

Fig. 9.6 The phase diagram for a colloidal system containing charged spherical colloids. These spheres can order into either bcc or fcc lattices as the volume fraction or the screening length increases. Note that the disorder-to-order (or Kirkwood-Alder) transition for such soft spheres occurs at much lower volume fractions than the hard-sphere system.

magnitude of electrostatic interaction decreased significantly with an increase in the concentration of stray electrolytes because of the shielding effect caused by the counterions presented in the double layer. At high levels of stray electrolytes, the van der Waals attraction exceeds the repulsive force at all separations and the colloidal particles tend to agglomerate. As the concentration of electrolytes is reduced to sufficiently low levels ($<10^{-5}$ M), the intensity of electrostatic repulsion becomes strong enough to stabilize an ordered array for these particles at separations greater than the diameter of a particle. It is also worth noting that electrostatic interactions between charged colloids are strongly dependent on the presence of other highly charged particles and/or external charged surfaces. In some cases, a long-range attractive interaction was also observed between similarly charged colloidal particles, which could be attributed to the redistribution of the electric double layers of ions and counterions around the colloidal particles [64].

When highly charged spherical colloids are suspended in a liquid medium containing very few stray ions, they can spontaneously organize themselves into a variety of crystalline structures driven by the minimization of electrostatic repulsive interaction (Fig. 9.5 B) [65]. In 1988, Robbins et al. noted that the phase diagram of such a system could be adequately described using two thermodynamic variables: that is, the volume fraction or solid content of colloidal particles in the dispersion and the electrostatic screening length (or the Debye-Hückel length, κ^{-1}) of the dispersion medium [66]. Colloidal crystals that have been observed in this system may exhibit a number of different structures such as body-centered cubic (bcc), face-centered cubic (fcc), rhcp and AB_2 [67]. Fig. 9.6 shows a typical phase diagram for this system. When the screening length is shorter than the center-to-center distance between two spheres, the spherical colloids behave like "hard balls" and they do not interact with each other until they are in physical contact. In this case, an fcc crystal structure is formed, and no heat or energy change is involved during the crystallization process. If the screening length is longer than the center-to-center distance, these spherical colloids act like "soft balls" and crystallize into a bcc lattice similar to that of a confined, one-component plasma system [68]. In both crystalline lattices, the spherical colloids are separated by a distance comparable to their dimensions. From the phase diagram, it is also clear that the disorder-to-order transition can be provoked either by increasing the volume fraction of spherical colloids or by extending the range of the screening length (by reducing the concentration of stray electrolytes). A synchrotron diffraction study indicated that charge-stabilized spherical colloids could form fcc structures at all volume fractions up to ∼60% [69].

The method based on repulsive electrostatic interactions may represent the most powerful and successful route to the fabrication of large colloidal crystals. This method, however, has a number of strict requirements on the materials and experimental conditions. For example, stray electrolytes have to be removed from the dispersion medium (usually by adding ion-exchange resins). It is also necessary to have good control over the temperature, monodispersity, surface charge density, and number density of colloids in the suspension. In some cases, a shear flow might be needed in order to achieve long-range translational ordering [70]. As a result of the repulsive interactions among colloidal particles, the lattice points in the crystalline arrays formed by this method are always separated by distances comparable to the sizes of these colloids.

9.4.3
Crystallization via Attractive Capillary Forces

This method was originally developed by Nagayama and his co-workers for generating 2D arrays of spherical colloids supported on various substrates [71]. In a typical procedure, a liquid dispersion of spherical colloids is spread onto the surface of a support. As the solvent evaporates slowly under well-controlled conditions, the spherical colloids are driven by attractive capillary forces to crystallize into a closely packed, hexagonal lattice (Fig. 9.5 C). Velev et al. followed this self-assembly process experimentally by using an optical microscope [72]. They found that a nucleus (i.e. an ordered region consisting of several spherical colloids) was first formed when the thickness of the liquid layer approached the diameter of the colloids. More and more colloids were driven toward this nucleus through convective transport, and eventually crystallized on the nucleus owing to the attractive capillary forces. Flat, clean, and chemically homogeneous surfaces have to be used in order to generate long-range ordered lattices over relatively large areas. Solid substrates such as glass slides and silicon wafers can be directly used in this technique. Liquids such as perfluorinated oil (F-oil) or mercury have also been explored by Lazarov et al. for use as supporting substrates in generating highly ordered 2D lattices of spherical colloids [73].

This method could also be modified in a number of ways to improve the process efficiency or the product quality. For example, the evaporation of solvent could be accelerated by carefully spin-coating a colloidal dispersion placed on a solid substrate. In this case, the formation of a uniform monolayer requires that the colloidal dispersion is able to wet completely the surface of this solid substrate and there exists a weak electrostatic repulsion between the spherical colloids and the solid substrate. As shown by several groups, the wetting could be greatly improved by adding a surfactant to the colloidal dispersion or simply by precoating the substrate with a thin layer of surfactant. As observed by Deckman et al. [74]

the surfactant might form residuals on the substrate, and the domain size of the highly ordered region often decreased with the diameter of spherical colloids (ordering was not observed when spherical colloids were smaller than 50 nm). Van Duyne et al. also explored the use of this technique in forming single- and double-layered structures from spherical colloids [75]. They further examined the potential of these 2D lattices as physical masks in generating ordered arrays of metal nanoparticles (such as silver) to be used for surface-enhanced Raman scattering. As demonstrated by Colvin et al., the capability of this method could be further extended to fabricate 3D crystalline lattices with their (111) planes oriented parallel to the substrates [76]. As a major advantage for this method, the number of (111) planes or the thickness of these colloidal crystals could be precisely controlled using a layer-by-layer approach [77]. More recently, van Blaaderen et al. further demonstrated that this layer-by-layer approach could be exploited to grow binary crystals containing spherical colloids of two different sizes [78]. Superlattices with various crystal structures or compositions (e.g. LS, LS_2 and LS_3) were obtained.

9.5
Control over the Spatial Orientation of Colloidal Crystals

When grown on the surface of a flat substrate, the 3D crystalline lattice of spherical colloids usually exhibits an fcc structure, with its (111) planes oriented parallel to the surface of the underlying substrate. To obtain 3D colloidal crystals with other orientations, it is necessary to use substrates whose surfaces have been patterned with appropriate arrays of relief structures. These relief patterns can serve as physical templates to directly control the nucleation and growth of colloidal crystals that will exhibit various spatial orientations. This process can be considered as a mesoscopic equivalent of epitaxial growth, a process that has been extensively explored at the atomic and molecular scales to generate thin solid films with well-controlled crystallographic orientations [79].

Van Blaaderen and Wiltzius first demonstrated the use of lithographically defined surfaces as templates to grow (via a slow sedimentation process) 3D colloidal crystals with either (100) or (110) planes oriented parallel to the supporting substrates [80]. Later, Yodh and his co-workers observed the formation of colloidal crystals with (100) orientations when periodically patterned substrates were used as the supports for entropically driven self-organization [81]. For templates involved in these two studies, the minimum feature sizes had to be kept on the same scale as the diameter of spherical colloids. To crystallize colloids with sizes smaller than 0.5 µm, one needs to use advanced nanolithographic techniques (e.g. e-beam writing) to fabricate templates. This difficulty could be overcome by using relief structures with tapered dimensions. For example, Ozin et al. pioneered the use of square pyramidal pits or V-grooves anisotropically etched in Si(100) wafers to fabricate small photonic crystals whose (100) planes were oriented parallel to the silicon substrates [82]. Xia et al. further extended the potential of these templates to assemble spherical colloids into discrete clusters characterized by a

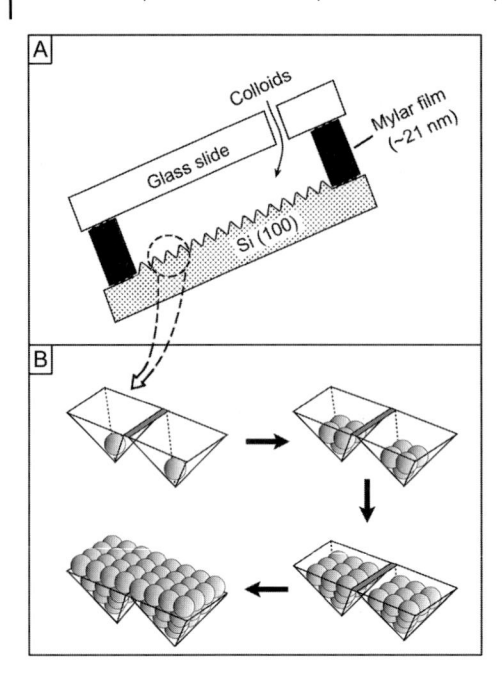

Fig. 9.7 **A** A schematic illustration of the flow cell that used confinement to crystallize monodispersed spherical colloids into 3D crystalline lattices with their (100) planes oriented parallel to the supporting substrates. **B** Step-by-step formation of a (100)-oriented crystal by templating against the 2D array of square pyramidal pits etched in the surface of a Si(100) wafer.

square pyramidal shape [83]. They also demonstrated that these templates could be adopted to form (100)-oriented colloidal crystals as large as several square centimeters in area [84]. As illustrated in Fig. 9.7, the polymer beads could be obliged by the square pyramidal pits in the Si(100) surface to grow into a 3D crystalline lattice with its (100) planes oriented parallel to the surface of the supporting substrate. In this template-directed process, the atomic scale symmetry of the underlying Si(100) substrate was faithfully transferred into the mesoscale colloidal crystal. Fig. 9.8 shows the SEM images of two samples fabricated by crystallizing 1.0 μm polystyrene beads against 2D arrays of pyramidal pits and V-grooves etched in the surfaces of Si(100) wafers. Fig. 9.8A shows the top view of a portion of the 3D lattice, clearly indicating the square symmetry of (100) planes parallel to the surface of supporting substrate. Fig. 9.8B shows an SEM of the backside of this crystal, whose surface is decorated by an array of nuclei formed in the pyramidal pits. Fig. 9.8C gives an SEM image of some colloidal crystals formed within a parallel array of V-grooves. If the crystals in adjacent grooves were spatially registered with no mismatching, they could merge and continue to grow into a large crystal without changing their spatial orientations (Fig. 9.8D).

These results strongly suggest that the 3D lattices generated using the present procedure also exhibit long-range ordering along both directions parallel and perpendicular to the surface of Si(100) template. The key to the success of this approach was to precisely control the lateral dimensions of and separations between the square pyramidal pits or V-grooves. When the size does not match well between spherical colloids and templates, (111)-oriented lattices will be formed outside the confine-

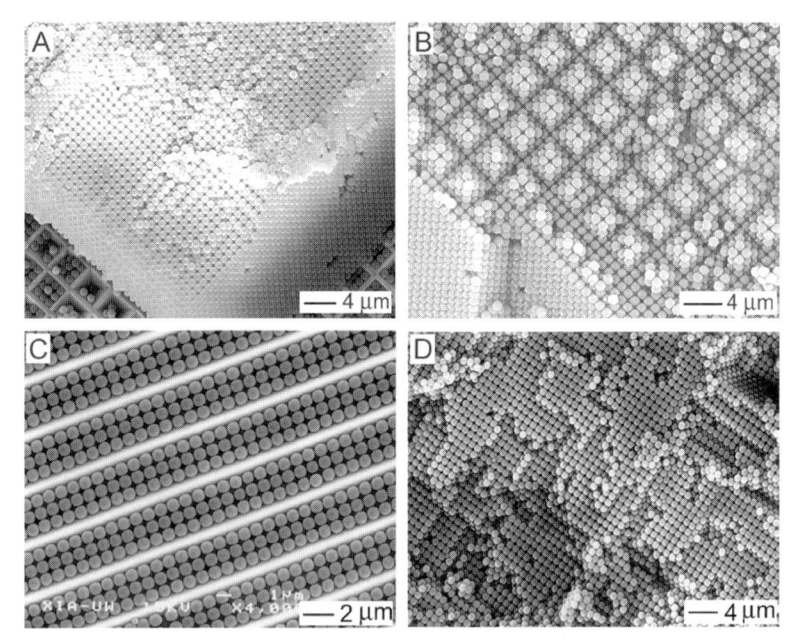

Fig. 9.8 A, B SEM images of (100)-oriented colloidal crystals generated by templating 1.0 μm polystyrene beads against a 2D array of 4 μm wide square pyramidal pits. The crystals belong to the same sample that happened to stick to the **A** bottom and **B** top substrate when the top glass substrate was separated from the cell. **C, D** SEM images of (100)-oriented crystals fabricated by self-assembling 1 μm polystyrene beads against an array of 4 μm wide V-shaped grooves. Crystals formed within the V-grooves could continue to grow without changing their spatial orientations when a relatively low concentration of polystyrene beads was used.

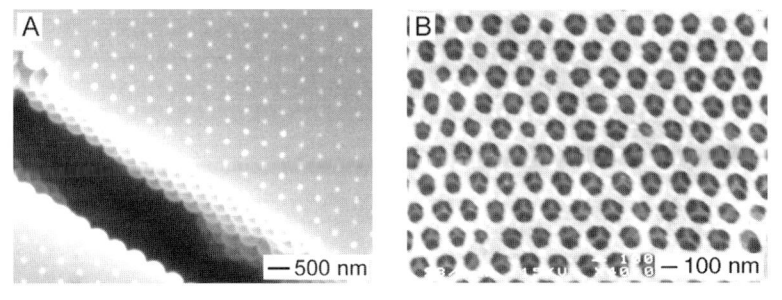

Fig. 9.9 A The SEM image of a crystalline lattice of 480 nm polystyrene beads infiltrated with UV-cured poly(acrylate methacrylate) copolymer. Bright spots correspond to the polystyrene beads underneath the surface. **B** The SEM image of a polyurethane inverse opal that was fabricated by templating against a 3D crystalline lattice of 200 nm polystyrene beads, followed by etching in toluene. Note that the second layer could also be observed through the holes in the top layer.

ment of pyramidal pits (or V-grooves). For templates with tapered dimensions, the critical feature sizes could be many times larger than the diameters of the spherical colloids to be crystallized; no access to advanced nanolithographic facility is needed. In another related work, Xia et al. also demonstrated that polygonal clusters of spherical colloids (formed through a template-assisted method [83]) could be used as templates to dictate the crystallization of spherical colloids into lattices with symmetries or orientations other than those most commonly observed for the spherical system. For example, when ~1% of square planar clusters were mixed with spherical colloids, they were able to direct spherical colloids to crystallize into an ccp lattice with its (100) planes oriented parallel to the surface of the supporting substrate [85]. In this case, the colloidal crystal tended to develop the same symmetry as the clusters during the self-organization process. This ability to control the crystallographic orientation of a colloidal crystal provides another way to change the photonic properties of this crystalline lattice.

9.6
Colloidal Crystals Embedded in Solid Matrices

Colloidal crystals are usually characterized by limited temperature and mechanical stability. Their ordered structures can be easily destroyed when they are subjected to thermal treatment or mechanical shock. This problem could be effectively solved by embedding the colloidal crystal in a solid matrix. Two different approaches have been demonstrated to accomplish this goal: (1) direct assembly of colloidal particles in a dispersion medium that can be cured thermally or by exposure to UV light; (2) infiltration of the voids between colloidal particles of a crystalline lattice with a liquid precursor that can be solidified in the next stage. Kamenetzky et al. demonstrated the first approach by embedding colloidal crystals of polystyrene beads in acrylamide-methylene-bisacrylamide monomers that were later polymerized in situ into gels [86]. This approach was later extended by Ford et al. and Asher et al. to prepare colloidal crystals embedded in other polymer matrices [87, 88]. The second approach was first demonstrated by Velev et al., who fabricated colloidal crystals of polystyrene beads embedded in a silica matrix by infiltrating the crystalline lattice with a sol–gel precursor [89]. Subsequent removal of these polystyrene beads led to the formation of highly ordered and interconnected 3D porous structures. This approach was later extended to many other systems that include ceramics, polymers, inorganic semiconductors, and metals [90–98]. These materials are also known as inverse opals, and all of them are characterized by an fcc lattice of air balls that are interconnected in three-dimensional space. In addition to the use of liquid precursors, the void spaces among colloids could also be filled with materials decomposed from gaseous precursors [99–101] or with suspensions of nanoparticles having smaller dimensions [102].

Embedding of colloidal crystals into solid matrices also opens the door to new applications, such as the fabrication of smart optical sensors. In general, a colloidal crystal diffracts light of a specific wavelength according to the Bragg equation:

$$m \cdot \lambda = 2 \cdot n \cdot d_{hkl} \cdot \sin \theta$$

where m is the diffraction order, λ is the wavelength of diffracted light, n is the mean refractive index of this crystal, d_{hkl} is the spacing along the [hkl] direction, and θ is the angle between the incident light and the normal to the (hkl) planes. This equation indicates that the wavelength of diffracted light is directly proportional to the interplanar spacing and mean refractive index. Any variation in the spacing and refractive index should result in an observable shift in the diffraction peak. As a result, a 3D colloidal crystal can serve as an optical sensor capable of displaying and measuring environmental changes. In one demonstration, Stein et al. have demonstrated the use of such colloidal crystals in differentiating various solvents [24a]. In a series of publications, Asher et al. described the fabrication and testing of a number of smart sensors [24b, 103]. They assembled highly charged spherical colloids into 3D colloidal crystals (with lattice constants in the range of 50–500 nm) and then embedded them into polymer matrices whose states were sensitive to the environmental change. When the matrix around the polymer beads were made of poly(N-isopropyl acrylamide) PNIPAM, the reversible shrinkage of this material between 10 and 35 °C changed the lattice spacing and subsequently led to a shift of the diffraction peak [24b]. When a polymer gel incorporated with a crown-ether – 4-acryloyl aminobenzo-18-crown-6 – was used as the matrix, the colloidal crystal was shown to exhibit a spectral response to Pb^{2+}, Ba^{2+}, and K^+ cations. The crown-ether selectively trapped these cations, drawing in counterions, increasing the Donnan osmotic pressure of the gel, and swelling the matrix. The magnitude of this reversible swelling depended on the number of charge groups covalently attached to the gel. A ~ 150 nm red-shift was observed when the concentration of Pb^{2+} was increased from ~ 20 ppb to ~ 2000 ppm [103].

9.7
Photonic Bandgap Properties of Colloidal Crystals

The long-range order of colloidal crystals has made them ideal candidates for use in many applications. For instance, they have been extensively exploited as a model system with which to study the mechanism for structural phase transition [25]. The relatively large sizes and long timescale of motion associated with colloids allows for real-time observation and analysis on various phenomena that might be difficult or impossible to accomplish at the atomic scale. Most recently, crystallization of colloids has also been explored by many groups as an alternative route to the fabrication of photonic bandgap (PBG) crystals [18].

A photonic bandgap crystal is a spatially periodic lattice made of materials having different dielectric constants [104]. The concept of this novel class of material was first proposed independently by Yablonovich [105] and John [106] in 1987, and since then a variety of applications has been suggested or demonstrated for these materials. For example, a PBG crystal can influence the propagation of elec-

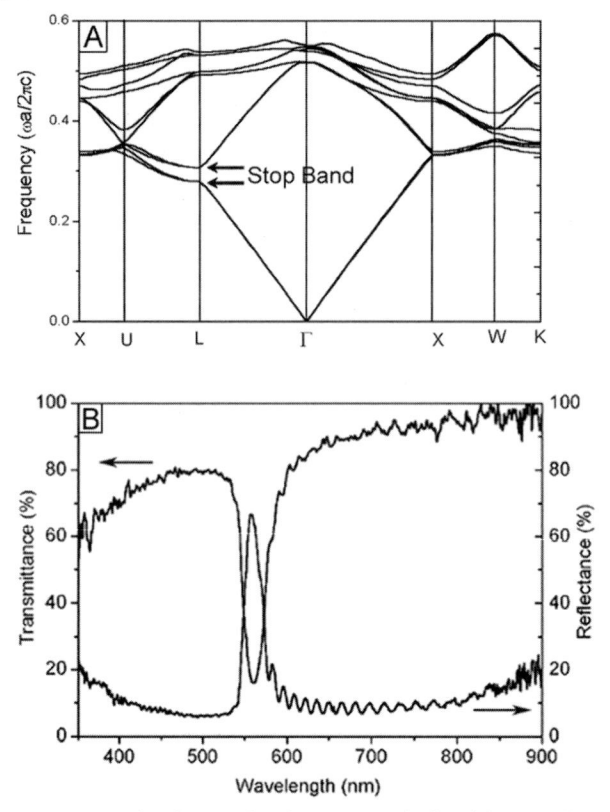

Fig. 9.10 **A** The photonic band structure calculated for a *ccp* lattice made of silicon spheres and surrounded by air. The filling fraction of spheres is 74%. It only exhibits pseudo bandgaps, such as the one between the 2nd and 3rd bands at the L-point (indicated by the arrows). **B** The UV-Vis transmission and reflectance spectra taken from a 3D crystalline lattice made of 208 nm polystyrene beads. Both spectra were measured at normal incidence relative to the (111) planes of the crystal.

tromagnetic waves in much the same way as a semiconductor does for electrons – that is, there exists a bandgap that can exclude the passage of photons of a chosen range of frequencies. It thus provides a convenient and powerful tool to confine, control, and manipulate photons in all three dimensions of space, and to produce more efficient optical, optoelectronic, and quantum electronic devices [107]. Research on PBG crystals has extended to cover all three dimensions, and the spectral regions range from ultraviolet to radio frequencies [108]. The real impetus in this area, however, has been the strong desire to obtain a complete bandgap around 1.55 µm – the wavelength now used in optical fiber communications. Although many advances have been made, it still remains a challenge to apply conventional microlithographic techniques to the fabrication of 3D periodic lattices in which the feature size is comparable to the wavelength of near-infrared or visible light. As shown in previous sections, colloidal crystals with 3D long-range ordering can be conveniently assembled from spherical colloids with sizes ranging from tens of nanometers to several micrometers. It is a natural choice to fully examine the potential use of these crystalline lattices as photonic crystals.

In principle, the band structure of a PBG crystal can be obtained (in advance) by solving the Maxwell equations where the dielectric constant is expressed as a spatially periodic function [104]. Because the Maxwell equations enjoy scale invariance, it is also possible to shift a bandgap theoretically to any frequency range simply by scaling the feature sizes of a periodic structure. Several numerical approaches have been demonstrated to calculate the photonic band structure of a periodic lattice. The plane wave expansion method (PWEM) seems to be the most commonly used tool for 3D systems, even though it cannot be applied to those systems whose dielectric constants contain large imaginary parts (due to absorption) [109]. It is generally accepted that the traditional scalar-wave approximation is inadequate in fully describing many of the important aspects of a 3D photonic crystal [110]. The photonic band structure calculated for an fcc lattice of spherical colloids indicate that this simple system does not possess a full bandgap (Fig. 9.10A); it only has a number of pseudo bandgaps (or stop bands, as marked by two arrows on the graph) [111]. The existence of a complete bandgap in such a system is mainly inhibited by the degeneracy at the W- or U-point that is induced by the spherical symmetry of the building blocks. It is impossible to completely eliminate it by increasing the refractive index contrast, or by changing the filling ratio of spheres from 74% (close-packed) to any other values. It is generally accepted that spherical symmetry is not well-suited for generating full gaps in the photonic band structures, although spherical colloids seem to be the simplest building blocks to synthesize as monodisperse samples and/or to crystallize into 3D periodic lattices. The stop band of an fcc lattice made of spherical colloids can be easily measured from its transmission or reflectance spectrum. Fig. 9.10B shows two such spectra, recorded from a 3D colloidal crystal assembled from polystyrene beads. The peaks observed in these transmission and reflectance spectra match very well in position and intensity. The position could also be calculated from the lattice constants and mean refractive index by using Bragg's diffraction equation.

Although fcc lattices of spherical colloids do not exhibit full bandgaps, they offer a simple and easily prepared model system to experimentally probe the photonic band diagrams of certain types of 3D periodic structures. They also serve as sacrificial templates to prepare inverse opals, which are expected to display complete bandgaps when the contrast in dielectric constant is high enough. Computational simulations indicate that the minimum contrast in the refractive index at which a complete bandgap will be observed (between the eighth and ninth bands) in an inverse opal is around 2.8 [111]. This contrast can be accomplished using a number of inorganic materials such as group IV or II–VI semiconductors (Si, Ge, or CdSe), rutile titania, and iron oxides. Fig. 9.11A shows the photonic band structure of an inverse opal made of ferric oxide, and the refractive index

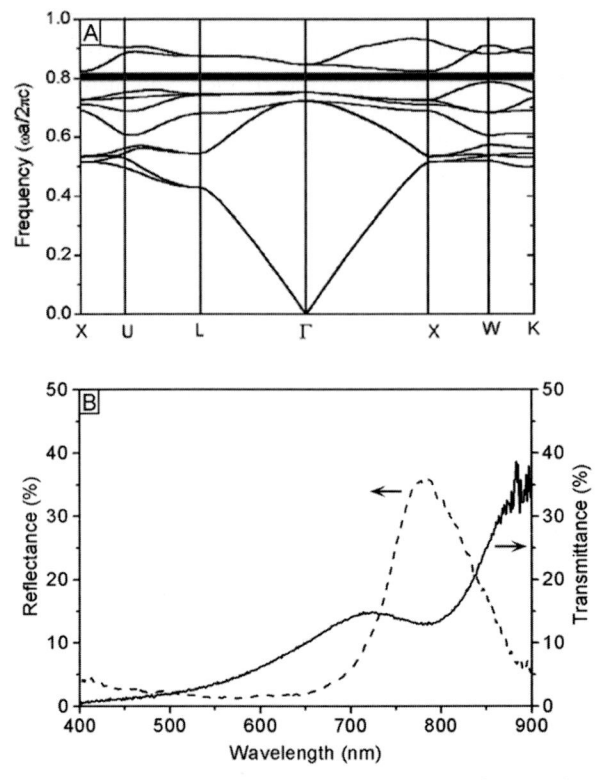

Fig. 9.11 **A** The photonic band structure calculated for a *ccp* lattice made of air balls in a silicon matrix. The filling fraction of air is 74%. A complete bandgap opens between the 8th and 9th bands (marked by the horizontal bar). **B** The UV-Vis transmission and reflectance spectra of an inverse opal that was fabricated by templating magnetite nanoparticles against polystyrene beads (480 nm in diameter). Both spectra were measured at normal incidence relative to the (111) planes of the inverse opal.

contrast is 3.01/1.00. This diagram suggests the existence of a complete bandgap (between the eighth and ninth bands) that extends over the entire Brillouin zone. Fig. 9.11 B shows the transmission and reflectance spectra taken from an inverse opal of magnetite. In this case, only a stop band exists (although its bandwidth is broader than that of an opal). Part of the reason lies in the fact that the filling of void spaces among spherical colloids is not complete and the resultant materials may not be sufficiently dense to acquire a refractive index close to that of the bulk solid. As shown by Vos et al., the refractive index of their anatase inverse opal was only 1.18–1.29, a value that is much lower than that of single crystalline bulk anatase (~2.6) [112]. By using semiconductors with higher refractive indices (such as Si and Ge), the definitive existence of the signature of a complete photonic band gap has been experimentally observed by a number of research groups [101].

Computational studies by John further demonstrated that an inverse opal could exhibit a fully tunable gap if the surface of the 3D porous structure could be coated with a few layers of an optically birefringent material such as a nematic liquid crystal [113]. In this case, the bandgap could be easily opened or closed by applying an external electric field to rotate the liquid crystal molecules with respect to the normal to the surface of the inverse opal.

9.8
Concluding Remarks

Monodisperse spherical colloids have emerged as the material of choice for a wide variety of niche applications that range from drug delivery to fabrication of photonic devices. They also represent the simplest class of building blocks that can be readily synthesized in large quantities and be organized into long-range ordered lattices – colloidal crystals. In addition, new types of 3D periodic structures (such as inverse opals) can be conveniently fabricated by templating various precursors against colloidal crystals. These two simple approaches provide a flexible and cost-effective route to three-dimensionally periodic structures with remarkably little of the investment required by conventional microfabrication techniques. More importantly, the feature sizes of these structures can be varied in a controllable fashion to cover a broad range of scales that spans from tens of nanometers to several micrometers. The ability to generate such periodic structures with well-controlled feature sizes has allowed useful functionality to be obtained, not only from the constituent materials but also from the long-range order that characterizes these lattices.

Monodisperse spherical colloids, and most of the applications derived from them, are still in an early stage of technical development. Many issues still need to be addressed before these materials can reach their potential in industrial applications. For example, the diversity of materials must be greatly expanded to include every major class of functional materials. At the moment, only silica and a few organic polymers (e.g. polystyrene and poly(methylmethacrylate)) can be prepared as *truly* monodisperse spherical colloids. These materials, unfortunately, do not exhibit any particularly interesting optical, nonlinear optical, or electro-optical functionality. In this regard, it is necessary to develop new methods to either dope currently existing spherical colloids with functional components or to directly deal with the synthesis of other materials. Second, formation of complex crystal structures other than ccp, bcc, and rhcp lattices has met with limited success. As a major limitation to the self-assembly procedures described in this chapter, all of them seem to lack the ability to form 3D lattices with arbitrary structures. Recent demonstrations based on the optical trapping method may provide a potential solution to this problem, albeit this approach seems to be too slow to be useful in practice [114]. Third, the density of defects in the crystalline lattices of spherical colloids must be well-characterized and kept below the level tolerated by the particular application. Some work has started to address this issue by studying the

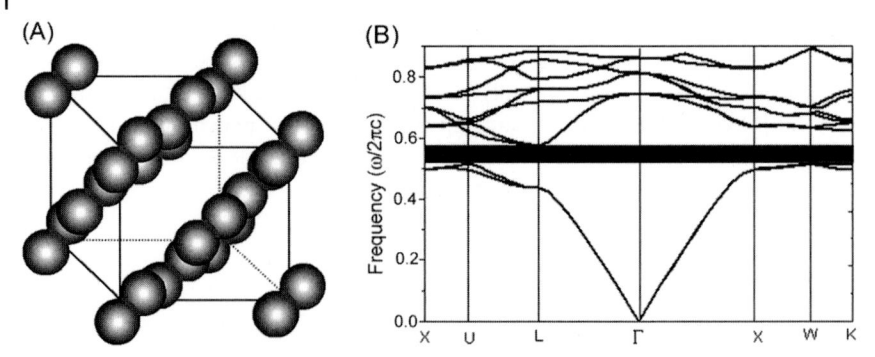

Fig. 9.12 **A** Schematic illustration of a diamond-type structure that could be produced by self-assembling dimers of monodispersed spherical colloids. **B** The photonic band structure of an *fcc* lattice self-assembled from the peanut-shaped building blocks made of ferric oxide. The filling ratio of these building blocks is 34%, and their longitudinal axes are directed towards the ⟨111⟩ direction. The refractive indexes of ferric oxide and surrounding medium are ∼3.01 and 1.00, respectively.

crystallization process of spherical colloids with mixed sizes [115]. In fact, the use of polydispersed spherical colloids as building blocks in self-assembly may provide another route to the fabrication of colloidal crystals with complex structures [116]. Last but not least, the compatibility of any crystallization techniques with the processes involved in the production of other types of integrated electro-optical devices (such as diode lasers) still needs to be defined and improved.

Spherical colloids with monodisperse sizes will certainly continue as the dominant subject of research in colloid science. They are, however, not necessarily the best and/or the only option for all fundamental studies and applications that involve the use of colloidal particles. They are not, for example, ideal building blocks for generating colloidal crystals with complete photonic bandgaps. As limited by the spherical symmetry of lattice points, the photonic band structures of these crystals are marked by degeneracy at the W- and U-point no matter how high the refractive index contrast might be. On the other hand, nonspherical colloids seem to offer some immediate advantages over their spherical counterparts in applications that require periodic structures with lower symmetries. Computational studies from several groups suggested that such a symmetry-induced degeneracy at the W- or U-point could be lifted by fabricating the 3D crystalline lattices with *nonspherical* colloids [117]. For example, a complete bandgap could develop between the second and third bands in the photonic structure when the building blocks were switched to dimeric types of units consisting of two interconnected dielectric spheres (Fig. 9.12 A) [118]. Such a full bandgap could have a gap-to-midgap ratio of 11.2% when the filling ratio was 30%, and a refractive index contrast approaching 3.60/1.00. Fig. 9.12 B shows the photonic band structure that we calculated for an fcc lattice constructed from peanut-shaped colloids of ferric oxide, a common inorganic substance with a refractive index of 3.01. As indicated by a shaded zone on this photonic band structure, such a 3D periodic lattice does

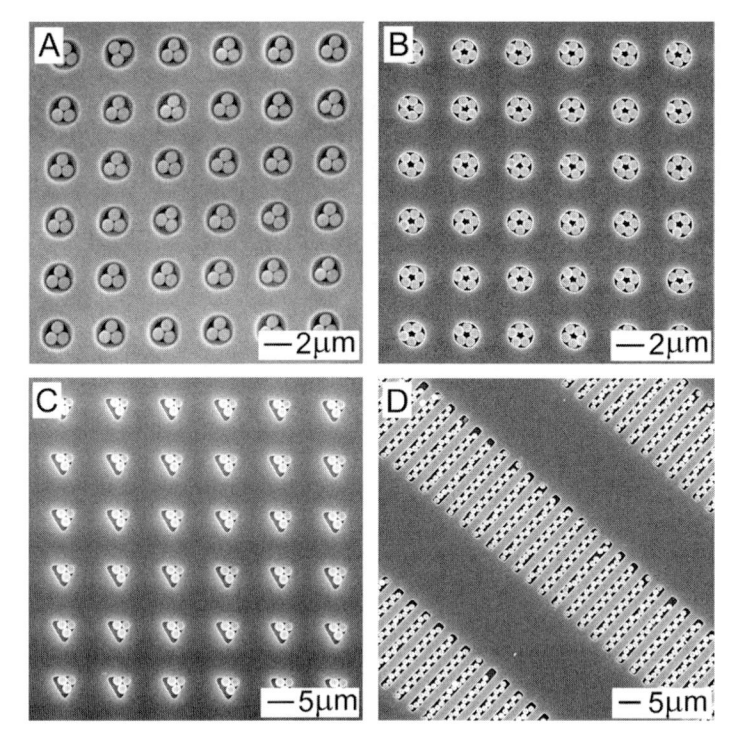

Fig. 9.13 SEM images of some typical examples of well-defined aggregates that were formed by templating spherical colloids against 2D arrays of templates with different geometric shapes: **A** a 2D array of trimers formed from 0.9 µm polystyrene beads in 2 µm cylindrical holes; **B** a 2D array of pentagons formed from 0.7 µm polystyrene beads in 2 µm cylindrical holes; **C** a 2D array of trimers formed from 1.75 µm silica spheres in triangular templates with a 5 µm edge length; **D** a 2D array of zigzag aggregates formed from 1.6 µm polystyrene beads in 2.0 µm wide and 20 µm long channels. All templates were etched in thin films of photoresist using photolithography.

exhibit a complete gap that can extend throughout the entire Brillouin zone. The minimum contrast in refractive index that is required for generating a complete bandgap in this new system is only ∼2.40, a relatively low value that could be easily achieved using a large number of inorganic materials (such as titania, II–VI semiconductors, and selenium). The major challenge is to synthesize them as monodisperse, *nonspherical* colloids, and to find ways to organize them into long-range ordered lattices.

A seemingly straightforward approach to colloidal crystals with nonspherical lattice points might involve the following two steps: (i) the synthesis of nonspherical colloids characterized by monodisperse sizes and shapes; and (ii) the self-assembly of these nonspherical building blocks into 3D lattices with both spatial and orientational orders. Thanks to the efforts of many research groups, ellipsoidal

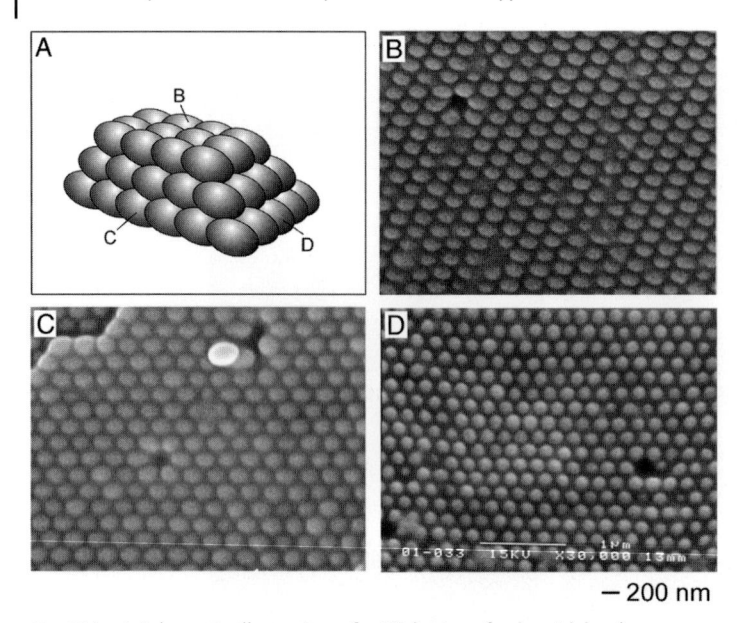

Fig. 9.14 **A** Schematic illustration of a 3D lattice of spheroidal polystyrene beads that was prepared using an approach based on viscoelastic deformation. The spherical polystyrene beads were 240 nm in diameter, the matrix was PDMS, and a uniaxial elongation of $\sim 1.3 \times$ was achieved by stretching the PDMS matrix film at a temperature around 190 °C. **B–D** The SEM images of three different (111) facets of the 3D lattice shown in **A**.

(and rod-shaped) colloids can now be directly synthesized from a number of inorganic materials [119]. Ellipsoidal colloids can also be fabricated by mechanically deforming spherical colloids using a number of different approaches [120–123]. Xia et al. have demonstrated a new approach that combined physical confinement and attractive capillary forces to assemble spherical colloids into discrete, uniform clusters with well-controlled sizes, shapes, and structures [84]. The capability and feasibility of this method have been demonstrated by assembling polystyrene beads and/or silica spheres (0.5–1 μm in diameter) into polygonal and polyhedral clusters that include, for example, dimers, triangles, squares, pentagons, hexagons, tetrahedrons, octahedrons, linear or zigzag chains, and spirals [84, 124]. Fig. 9.13 shows the SEM images of several typical examples that we have fabricated using this "*template-assisted self-assembly*" process. Once assembled, the beads contained in each cluster could be joined together through viscoelastic deformation at a temperature slightly higher than the glass transition temperature of the material. These clusters could then be released as stable dispersions in water by dissolving the photoresist film in ethanol under sonication. These well-defined clusters with nonspherical morphologies should provide complexity or new types of functionality that can not be offered by spherical colloids.

Although the above methods have been successfully applied to generate nonspherical colloids with well-controlled sizes and shapes and in relatively large quantities, it still remains a great challenge to organize these nonspherical building blocks into 3D crystalline lattices over large areas that can be further characterized using spectroscopic methods. As a matter of fact, very little is known about crystallization processes that involve nonspherical building blocks on the scale from 100 nm to 1 µm (the size range most interesting for photonic applications) [113–117]. It seems to be very difficult (or impossible) to simultaneously achieve both spatial and positional order when nonspherical colloids are used as the building blocks. We have partially overcome this difficulty by developing a procedure that formed 3D colloidal crystals containing spheroidal lattice points by mechanically stretching a crystalline lattice of polystyrene beads embedded in an elastomeric matrix [125]. Fig. 9.14 A gives a schematic illustration of such a 3D lattice, and the SEM images taken from three different facets (labeled as B, C, and D in Fig. 9.14 A) are shown in Fig. 9.14 B–D, respectively. The long-range order was essentially preserved along all three dimensions in this uniaxial elongation process, although the original fcc lattice of spherical beads had been transformed into a triclinic type. Due to the low contrast in refractive index, the colloidal crystal made of polymer spheroids only exhibited a stop band (rather than a complete band gap) in the optical regime. Nevertheless, such crystals can serve as a good model system to investigate the dependence of photonic band structures on the shape (or symmetry) of lattice points. Future work on this particular system will be directed toward colloidal particles with higher refractive indices.

9.9
Acknowledgment

This work has been supported in part by the AFOSR-MURI program at the UW, a Career Award from the National Science Foundation (DMR-9983893), a Research Fellowship from the Sloan P. Foundation; a Fellowship from the David and Lucile Packard Foundation, and a Camille Dreyfus Teacher Scholar Award from the Dreyfus Foundation. Y.Y. and Y.L. thank the Center for Nanotechnology at the UW for Student Fellowship Awards.

9.10 References

1. a) D. H. EVERETT, *Basic Principles of Colloid Science*, Royal Society of Chemistry, London, **1988**. b) W. B. RUSSEL, D. A. SAVILLE, W. R. SCHOWALTER, *Colloidal Dispersions*, Cambridge University Press, New York, NY, USA, **1989**. c) R. J. HUNTER, *Introduction to Modern Colloid Science*, Oxford University Press, Oxford, **1993**.
2. A. EINSTEIN, *Brownian Motion Investigations: on the Theory of the Brownian Movement*, Dover Publications, New York, NY, USA, **1956**.
3. a) A. HENGLEIN, *Chem. Rev.* **1989**, *89*, 1861–1873. b) G. SCHMID, *Chem. Rev.* **1992**, *92*, 1709–1727. c) G. SCHON, U. SIMON, *Colloid Polym. Sci.* **1995**, *273*, 101–117. d) Z. L. WANG, *Adv. Mater.* **1998**, *10*, 13–36. e) G. SCHMID, L. F. CHI, *Adv. Mater.* **1998**, *10*, 515–526.
4. a) M. L. STEIGERWALD, L. E. BRUS, *Acc. Chem. Res.* **1990**, *23*, 183–188. b) M. G. BAWENDI, M. L. STEIGERWALD, L. E. BRUS, *Annu. Rev. Phys. Chem.* **1990**, *41*, 477–496. c) Y. WANG, *Acc. Chem. Res.* **1991**, *24*, 133–139. d) H. WELLER, *Angew. Chem. Int. Ed. Engl.* **1993**, *32*, 41–53. e) A. P. ALIVISATOS, *Science* **1996**, *271*, 933–937.
5. a) J. H. FENDLER, *Chem. Rev.* **1987**, *87*, 877–899. b) G. A. OZIN, *Adv. Mater.* **1992**, *4*, 612–649. c) G. C. HADJIPANAYIS, R. W. SIEGEL, *Nanophase Materials: Synthesis, Properties, Applications*, Kluwer Academic Publishers, New York, NY, USA, **1994**. d) D. A. TOMALIA, *Adv. Mater.* **1994**, *6*, 529–539.
6. T. SUGIMOTO, *Fine Particles: Synthesis, Characterization, and Mechanisms of Growth*, Marcel Dekker, New York, NY, USA, **2000**.
7. E. MATIJEVIC, *Chem. Mater.* **1993**, *5*, 412–426.
8. a) I. PIIRMA, *Emulsion Polymerization*, Academic Press, New York, NY, USA, **1982**. b) G. W. POEHLEIN, R. H. OTTEWILL, J. W. GOODWIN, *Science and Technology of Polymer Colloids Vol. II*, Martinus Nijhoff Publishers, Hingham, MA, USA, **1983**.
9. G. MIE, *Ann. Phys.* **1908**, *25*, 377.
10. a) B. V. DERJAGUIN, L. LANDAU, *Acta Physicochimica (USSR)* **1941**, *14*, 633–652. b) E. J. VERWEY, J. T. G. OVERBEEK, *Theory of the Stability of Lyophobic Colloids*, Elsevier, Amsterdam, Netherlands, **1948**.
11. M. I. MISHCHENKO, J. W. HOVEENIER, L. D. TRAVIS, *Light Scattering by Nonspherical Particles: Theory, Measurements, and Applications*, Academic Press, San Diego, CA, USA, **2000**.
12. J. VISSER, IN *Surface and Colloid Science Vol. 8* (E. MATIJEVIC, ed.), John Wiley & Sons, New York, **1976**, pp. 3–84.
13. a) R. L. ROWELL, *NATO ASI Ser., Ser. C* **1990**, *303*, 187–208. b) T. RADEVA, *Curr. Top. Colloid Interface Sci.* **1997**, *2*, 131–142.
14. a) P. PIERANSKI, *Contemp. Phys.* **1983**, *24*, 25–73. b) W. VAN MEGAN, I. SNOOK, *Adv. Colloid Interface Sci.* **1984**, *21*, 119–194. c) A. P. GAST, W. B. RUSSEL, *Physics Today* **1998**, December, 24–30. d) D. G. GRIER, *From Dynamics to Devices: Directed Self-Assembly of Colloidal Materials*, a special issue in *MRS Bull.* **1998**, *23*, 21–50. e) A. K. ARORA, B. V. R. TATA, *Ordering and Phase Transitions in Colloidal Systems*, VCH, New York, NY, USA, **1996**.
15. Y. XIA, B. GATES, Y. YIN, Y. LU, *Adv. Mater.* **2000**, *12*, 693–713.
16. a) V. L. COLVIN, *MRS Bull.* **2001**, *26*, 637–641. b) A. D. DINSMORE, J. C. CROCKER, A. G. YODH, *Curr. Opin. Colloid Interface* **1998**, *3*, 5–11.
17. J. V. SANDERS, *Acta Crystallog.* **1968**, *24*, 427–434.
18. See, for example, a) special issue *Adv. Mater.* **2001**, *13*, 369–450. b) A. POLMAN, P. WILTZIUS, *Materials Science Aspects of Photonic Crystals*, special issue *MRS Bull.* **2001**, *26*, 608–646.
19. a) S. HAYASHI, Y. KUMAMOTO, T. SUZUKI, T. HIRAI, *J. Colloid Interface Sci.* **1991**, *144*, 538–547. b) Y. LU, Y. YIN, Y. XIA, *Adv. Mater.* **2001**, *13*, 34–37.
20. H. WU, T. W. ODOM, G. M. WHITESIDES, *J. Am. Chem. Soc.* **2002**, *124*, 7288–7289.
21. a) H. W. DECKMAN, J. H. DUNSMUIR, *Appl. Phys. Lett.* **1982**, *41*, 377–379. b) J. C. HULTEEN, D. A. TREICHEL, M. T. SMITH, M. L. DUVAL, T. R. JENSEN, R. P.

Van Duyne, *J. Phys. Chem. B* **1999**, *103*, 3854–3863.
22 Y. Xia, J. Tien, D. Qin, G. M. Whitesides, *Langmuir* **1996**, *12*, 4033–4038.
23 a) M. E. Turner, T. J. Trentler, V. L. Colvin, *Adv. Mater.* **2001**, *13*, 180–183. b) A. Stein, R. C. Schroden, *Curr. Opin. Solid State Mater. Sci.* **2001**, *5*, 553–564. c) P. V. Braun, P. Wiltzius, *Curr. Opin. Colloid Interface* **2002**, *7*, 116–123.
24 a) R. C. Schroden, M. Al-Daous, C. F. Blanford, A. Stein, Andreas, *Chem. Mater.* **2002**, *14*, 3305–3315. b) M. Weissman, H. B. Sunkara, A. S. Tse, S. A. Asher, *Science* **1996**, *274*, 959–960. c) P. L. Flaugh, S. E. O'Donnell, S. A. Asher, *Appl. Spectrosc.* **1984**, *38*, 847–850.
25 a) C. Murray, *MRS Bull.* **1998**, *23*, 33–38. b) C. A. Murray, D. G. Grier, *Am. Sci.* **1995**, *83*, 238–245.
26 E. S. Daniels, E. D. Sudol, M. S. El-Aasser, *Polymer Colloids: Science and Technology of Latex Systems*, American Chemical Society, Washington, DC, **2002**.
27 I. H. Harding, T. W. Healy, *J. Colloid Interface Sci.* **1982**, *89*, 185–201.
28 S. Y. Chang, L. Liu, S. A. Asher, *J. Am. Chem. Soc.* **1994**, *116*, 6739–6744.
29 H. R. Sheu, M. S. El-Aasser, J. W. Vanderhoff, *Polym. Mater. Sci. Eng.* **1987**, *57*, 911–915.
30 a) E. M. Zaiser, V. K. LaMer, *J. Colloid Interface Sci.* **1948**, *3*, 571–598. b) V. K. LaMer, R. H. Dinegar, *J. Am. Chem. Soc.* **1950**, *72*, 4847–4854. c) V. K. LaMer, *Ind. Eng. Chem.* **1952**, *44*, 1270–1277.
31 W. Stöber, A. Fink, *J. Colloid Interface Sci.* **1968**, *26*, 62–69.
32 R. K. Iler, *The Chemistry of Silica*, Wiley-Interscience, USA, **1979**.
33 a) E. Matijevic, *Acc. Chem. Res.* **1981**, *14*, 22–29. b) E. Matijevic, *Fine Particles* a special issue in *MRS Bull.* **1989**, *14*, 18–48. c) E. Matijevic, *Langmuir* **1994**, *10*, 8–16.
34 a) R. Arshady, *J. Microencapsulation*, **1988**, *5*, 101–114. b) M. Kumakura, *Eur. Polym. J.* **1995**, *31*, 1095–1098. c) W.-H. Li, H. D. H. Stoever, *Macromolecules* **2000**, *33*, 4354–4360.
35 M. Ueda, H.-B. Kim, K. Ichimura, *J. Mater. Chem.* **1994**, *4*, 883–889.

36 F. Caruso, M. Spasova, V. Saigueiriño-Maceira, L. M. Liz-Marzán, *Adv. Mater.* **2001**, *13*, 1090–1094.
37 a) A. L. Aden, *J. Appl. Phys.* **1951**, *22*, 1242–1246. b) T. Sugimoto, *MRS Bull.* **1989**, *14*, 23–28. c) M. Ohmori, E. Matijevic, *J. Colloid Interface Sci.* **1992**, *150*, 594–598. d) W. P. Hsu, R. Yu, E. Matijevic, *J. Colloid Interface Sci.* **1993**, *156*, 56–65. e) A. P. Philipse, M. P. B. van Bruggen, C. Pathmamanoharan, *Langmuir* **1994**, *10*, 92–99. f) D. Walsh, S. Mann, *Nature* **1995**, *377*, 320–323. g) R. D. Averitt, D. Sarkar, N. J. Halas, *Phys. Rev. Lett.* **1997**, *78*, 4217–4220. h) M. Giersig, L. M. Liz-Marzán, T. Ung, D. Su, P. Mulvaney, *Bunsenges. Phys. Chem.* **1997**, *101*, 1617–1620. i) H. Bamnolker, B. Nitzan, S. Gura, S. Margelo, *J. Mater. Sci. Lett.* **1997**, *16*, 1412–1415. j) S. J. Olderburg, R. D. Averitt, S. L. Westcott, N. J. Halas, *Chem. Phys. Lett.* **1998**, *288*, 243–247.
38 a) C. T. Chan, W. Y. Zhang, Z. L. Wang, X. Y. Lei, D. Zheng, W. Y. Tam, P. Sheng, *Physica B* **2000**, *279*, 150–154. b) Z. L. Wang, C. T. Chan, W. Zhang, N. Ming, P. Sheng, *Phys. Rev. B* **2001**, *64*, 113108/1–4.
39 See, for example, a) A. Garg, E. Matijevic, *J. Colloid Interface Sci.* **1988**, *126*, 243–250. b) N. Kawahashi, E. Matijevic, *J. Colloid Interface Sci.* **1990**, *138*, 534–542. c) X. C. Guo, P. Dong, *Langmuir* **1999**, *15*, 5535–5540.
40 a) M. A. Correa-Duarte, M. Giersig, L. M. Liz-Marzán, *Chem. Phys. Lett.* **1998**, *286*, 497–501. b) T. Ung, L. M. Liz-Marzán, P. Mulvaney, *Langmuir* **1998**, *14*, 3740–3748.
41 M. Giersig, *Mater. Sci. Forum* **1999**, *312–314*, 623–678.
42 Y. Yin, Y. Lu, Y. Sun, Y. Xia, *Nano Lett.* **2002**, *2*, 427–430.
43 Y. Lu, Y. Yin, Y. Xia, *Nano Lett.* **2002**, *2*, 183–186.
44 Y. Lu, Y. Yin, Y. Xia, *Nano Lett.* **2002**, *2*, 785–788.
45 a) F. Caruso, H. Lichtenfeld, M. Giersig, H. Möhwald, *J. Am. Chem. Soc.* **1998**, *120*, 8523–8524. b) F. Caruso, R. A. Caruso, H. Möhwald, *Science* **1998**, *282*, 1111–1114.

46 E. Kumacheva, O. Kalinina, L. Lilge, *Adv. Mater.* **1999**, *11*, 231–234.
47 P. Pieranski, L. Strzelecki, B. Pansu, *Phys. Rev. Lett.* **1983**, *50*, 900–903.
48 a) S. H. Park, D. Qin, Y. Xia, *Adv. Mater.* **1998**, *10*, 1028–1032. b) S. Park, Y. Xia, *Langmuir* **1999**, *15*, 266–273. c) B. Gates, D. Qin, Y. Xia, *Adv. Mater.* **1999**, *11*, 466–469.
49 a) S. H. Park, Y. Xia, *Chem. Mater.* **1998**, *10*, 1745–1747. b) S. H. Park, Y. Xia, *Adv. Mater.* **1998**, *10*, 1045–1048. c) Y. Lu, Y. Yin, B. Gates, Y. Xia, *Langmuir* **2001**, *17*, 6344–6350.
50 Y. Xia, G. M. Whitesides, *Angew. Chem. Int. Ed.* **1998**, *37*, 550–575.
51 B. Gates, Y. Lu, Z.-Y. Li, Y. Xia, *Appl. Phys. A* **2003**, *76*, 509–513.
52 B. Gates, S. H. Park, Y. Xia, *J. Lightwave Tech.* **1999**, *17*, 1956–1962.
53 a) W. Wang, S. A. Asher, *J. Am. Chem. Soc.* **2001**, *123*, 12528–12535. b) K. P. Velikov, A. Moroz, A. van Blaaderen, *Appl. Phys. Lett.* **2002**, *80*, 49–51. c) B. Rodriguez-Gonzalez, V. Salgueiriño-Maceira, F. Garcia-Santamaria, L. M. Liz-Marzan, *Nano Lett.* **2002**, *2*, 471–473. d) H. Miguez, F. Meseguer, C. Lopez, A. Mifsud, J. S. Moya, L. Vazquez, *Langmuir* **1997**, *13*, 6009–6011.
54 K. E. Davis, W. B. Russel, W. J. Glantschnig, *Science* **1989**, *245*, 507–510.
55 P. N. Pusey, W. Vanmegen, *Nature* **1986**, *320*, 340–342.
56 J. V. Sanders, *Nature* **1964**, *204*, 1151–1153.
57 a) T. C. Simonton, R. Roy, S. Komarneni, E. Breval, *J. Mater. Res.* **1986**, *1*, 667–674. b) A. P. Philipse, *J. Mater. Sci. Lett.* **1989**, *8*, 1371–1373. c) R. Mayoral, J. Requena, J. S. Moya, C. Lopez, A. Cintas, H. Miguez, F. Meseguer, L. Vazquez, M. Holgado, A. Blanco, *Adv. Mater.* **1997**, *9*, 257–260. d) V. N. Bogomolov, S. V. Gaponenko, I. N. Germanenko, A. M. Kapitonov, E. P. Petrov, N. V. Gaponenko, A. V. Prokofiev, A. N. Ponyavina, N. I. Silvanovich, S. M. Samoilovich, *Phys. Rev. E* **1997**, *55*, 7619–7625.
58 B. Y. Cheng, P. G. Ni, C. J. Jin, Z. L. Li, D. Z. Zhang, P. Dong, X. C. Guo, *Opt. Commun.* **1999**, *170*, 41–46.
59 L. V. Woodcock, *Nature* **1997**, *388*, 235–237.
60 M. Holgado, F. Garcia-Santamaria, A. Blanco, M. Ibisate, A. Cintas, H. Miguez, C. J. Serna, C. Molpeceres, J. Requena, A. Mifsud, F. Meseguer, C. Lopez, *Langmuir* **1999**, *15*, 4701–4704.
61 O. Vickreva, O. Kalinina, E. Kumacheva, *Adv. Mater.* **2000**, *12*, 110–112.
62 J. Zhu, M. Li, R. Rogers, W. Meyer, R. H. Ottewill, STS-73 Space Shuttle Crew, W. Z. B. Russel, P. M. Chaikin, *Nature* **1997**, *387*, 883–885.
63 I. Sogami, N. Ise, *J. Chem. Phys.* **1984**, *81*, 6320–6332.
64 See, for example, a) G. M. Kepler, S. Fraden, *Phys. Rev. Lett.* **1994**, *73*, 356–359. b) J. C. Crocker, D. G. Grier, *Phys. Rev. Lett.* **1994**, *73*, 352–355. c) J. C. Crocker, D. G. Grier, *Phys. Rev. Lett.* **1996**, *77*, 1897–1900.
65 a) A. Kose, S. Hachisu, *J. Colloid Interface Sci.* **1974**, *46*, 460–469. b) N. A. Clark, B. J. Ackerson, *Nature* **1979**, *281*, 57–60. c) D. H. Vanwinkle, C. A. Murray, *Phys. Rev. A* **1986**, *34*, 562–573. d) J. L. Harland, S. I. Henderson, S. M. Underwood, W. Vanmegen, *Phys. Rev. Lett.* **1995**, *75*, 3572–3575. e) T. Okubo, *Langmuir* **1994**, *10*, 1695–1702. f) T. Palberg, W. Monch, J. Schwarz, P. Leiderer, *J. Chem. Phys.* **1995**, *102*, 5082–5087.
66 M. O. Robbins, K. Kremer, G. S. Grest, *J. Chem. Phys.* **1988**, *88*, 3286–3312.
67 a) P. N. Pusey, W. Vanmegen, P. Bartlett, B. J. Ackerson, J. G. Rarity, S. M. Underwood, *Phys. Rev. Lett.* **1989**, *63*, 2753–2756. b) H. Versmold, P. Lindner, *Langmuir* **1994**, *10*, 3043–3045. c) S. M. Clarke, A. R. Rennie, R. H. Ottewill, *Langmuir* **1997**, *13*, 1964–1969.
68 a) T. B. Mitchell, J. J. Bollinger, D. H. E. Dubin, X. P. Huang, W. M. Itano, R. H. Baughman, *Science* **1998**, *282*, 1290–1293. b) T. M. O'Neil, *Physics Today* **1999**, February, 24–30.
69 W. L. Vos, M. Megens, C. M. van Kats, P. Bösecke, *Langmuir* **1997**, *13*, 6004–6008.
70 T. Okubo, *Langmuir* **1994**, *10*, 1695–1702.
71 a) N. D. Denkov, O. D. Velev, P. A. Kralchevski, I. B. Ivanov, H. Yoshimura, K.

Nagayama, *Langmuir* **1992**, *8*, 3183–3190. b) P. Kralchevski, V. Paunov, I. Ivanov, K. Nagayama, *J. Colloid Interface Sci.* **1992**, *151*, 79–94.

72 a) N. D. Denkov, O. D. Velev, P. A. Kralchevsky, I. B. Ivanov, H. Yoshimura, K. Nagayama, *Nature* **1993**, *361*, 26–26. b) P. A. Kralchevsky, V. N. Paunov, N. D. Denkov, I. B. Ivanov, K. Nagayama, *J. Colloid Interface Sci.* **1993**, *155*, 420–437.

73 G. S. Lazarov, N. D. Denkov, O. D. Velev, P. A. Kralchevsky, K. Nagayama, *J. Chem. Soc., Faraday Trans.* **1994**, *90*, 2077–2083.

74 H. W. Deckman, J. H. Dunsmuir, S. Garoff, J. A. McHenry, D. G. Peiffer, *J. Vac. Sci. Technol. B* **1988**, *6*, 333–336.

75 J. C. Hulteen, R. P. Van Duyne, *J. Vac. Sci. Technol.* **1995**, *A13*, 1553–1558.

76 a) P. Jiang, J. F. Bertone, K. S. Hwang, V. L. Colvin, *Chem. Mater.* **1999**, *11*, 2132–2140. b) Y. H. Ye, F. LeBlanc, A. Hache, V. V. Truong, *Appl. Phys. Lett.* **2001**, *78*, 52–54. c) Y. A. Vlasov, X. Z. Bo, J. C. Sturm, D. J. Norris, *Nature* **2001**, *414*, 289–293. d) Z. Z. Gu, A. Fujishima, O. Sato, *Chem. Mater.* **2002**, *14*, 760–765. e) L. M. Goldenberg, L. M. Goldenberg, J. Wagner, J. Stumpe, B. R. Paulke, E. Gornitz, *Langmuir* **2002**, *18*, 3319–3323.

77 J. F. Bertone, P. Jiang, K. S. Hwang, D. M. Mittleman, V. L. Colvin, *Phys. Rev. Lett.* **1999**, *83*, 300–303.

78 K. P. Velikov, C. G. Christova, R. P. A. Dullens, A. van Blaaderen, *Science* **2002**, *296*, 106–109.

79 J. R. Arthur, *Surf. Sci.* **2002**, *500*, 189–217.

80 A. van Blaaderen, R. Ruel, P. Wiltzius, *Nature* **1997**, *385*, 321–324.

81 K. H. Lin, J. C. Crocker, V. Prasad, A. Schofield, D. A. Weitz, T. C. Lubensky, A. G. Yodh, *Phys. Rev. Lett.* **2000**, *85*, 1770–1773.

82 a) S. M. Yang, G. A. Ozin, *Chem. Commun.* **2000**, 2507–2508. b) G. A. Ozin, S. M. Yang, *Adv. Funct. Mater.* **2001**, *11*, 95–104. c) V. Kitaev, S. Fournier-Bidoz, S. M. Yang, G. A. Ozin, *J. Mater. Chem.* **2002**, *12*, 966–969.

83 a) Y. Yin, Y. Lu, B. Gates, Y. Xia, *J. Am. Chem. Soc.* **2001**, *123*, 8718–8729. b) Y. Yin, Y. Xia, *Adv. Mater.* **2002**, *14*, 605–608.

84 Y. Yin, Y. Xia, *Adv. Mater.* **2001**, *13*, 267–271.

85 Y. Xia, B. Gates, Y. Yin, *Aust. J. Chem.* **2001**, *54*, 287–290.

86 E. A. Kamenezky, L. G. Magliocco, H. P. Panzer, *Science* **1994**, *263*, 207–210.

87 Y. Y. Chen, W. T. Ford, N. F. Materer, D. Teeters, *Chem. Mater.* **2001**, *13*, 2697–2704.

88 S. A. Asher, J. Holtz, J. Weissman, G. Pan, *MRS Bull.* **1998**, *23*, 44–50.

89 a) O. D. Velev, T. A. Jede, R. F. Lobo, A. M. Lenhoff, *Nature* **1997**, *389*, 447–448. b) O. D. Velev, T. A. Jede, R. F. Lobo, A. M. Lenhoff, *Chem. Mater.* **1998**, *10*, 3597–3602.

90 O. D. Velev, A. M. Lenhoff, *Curr. Opin. Colloid Interface Sci.* **2000**, *5*, 56–63.

91 J. Wijnhoven, L. Bechger, W. L. Vos, *Chem. Mat.* **2001**, *13*, 4486–4499.

92 B. Gates, Y. Yin, Y. Xia, *Chem. Mater.* **1999**, *10*, 2827–2836.

93 a) D. Wang, R. A. Caruso, F. Caruso, *Chem. Mater.* **2001**, *13*, 364–371. b) K. H. Rhodes, S. A. Davis, F. Caruso, B. Zhang, S. Mann, *Chem. Mater.* **2000**, *12*, 2832–2834.

94 a) J. F. Bertone, P. Jiang, K. S. Hwang, D. M. Mittleman, V. L. Colvin, *Phys. Rev. Lett.* **1999**, *83*, 300–303. b) P. Jiang, K. S. Hwang, D. M. Mittleman, J. F. Bertone, V. L. Colvin, *J. Am. Chem. Soc.* **1999**, *121*, 11630–11637. c) K. M. Kulinowski, P. Jiang, H. Vaswani, V. L. Colvin, *Adv. Mater.* **2000**, *12*, 833–838. d) M. E. Turner, T. J. Trentler, V. L. Colvin, *Adv. Mater.* **2001**, *13*, 180–183.

95 A. Stein, R. C. Schroden, *Curr. Opin. Solid State Mat. Sci.* **2001**, *5*, 553–564.

96 S. A. Johnson, P. J. Ollivier, T. E. Mallouk, *Science* **1999**, *283*, 963–965.

97 D. J. Norris, Y. A. Vlasov, *Adv. Mater.* **2001**, *13*, 371–376.

98 G. Subramanian, V. N. Manoharan, J. D. Thorne, D. J. Pine, *Adv. Mater.* **1999**, *11*, 1261–1265.

99 A. A. Zakhidov, R. H. Baughman, Z. Iqbal, C. Cui, I. Khayyrullin, O. Dantas, J. Marti, V. G. Ralchenko, *Science* **1998**, *282*, 897–901.

100 A. Blanco, E. Chomski, S. Grabtchak, M. Ibisate, S. John, S. W. Leonard, C. Lopez, F. Meseguer, H. Miguez, J. P.

Mondia, G. A. Ozin, O. Toader, H. M. van Driel, *Nature* **2000**, *405*, 437–440.
101 a) H. Yang, N. Coombs, G. A. Ozin, *Adv. Mater.* **1997**, *9*, 811–814. b) Y. A. Vlasov, X. Z. Bo, J. C. Sturm, D. J. Norris, *Nature* **2001**, *414*, 289–293.
102 a) A. Imhof, D. J. Pine, *Nature* **1997**, *389*, 948–951. b) A. Imhof, D. J. Pine, *Adv. Mater.* **1998**, *10*, 697–700. c) B. Gates, Y. Xia, *Adv. Mater.* **2001**, *13*, 1605–1608. d) O. D. Velev, E. W. Kaler, *Adv. Mater.* **2000**, *12*, 531–534.
103 a) G. Pan, R. Kesavamoorthy, S. A. Asher, *Phys. Rev. Lett.* **1997**, *78*, 3860–3863. b) G. Pan, R. Kesavamoorthy, S. A. Asher, *J. Am. Chem. Soc.* **1998**, *120*, 6525–6530. c) J. H. Holtz, S. A. Asher, *Nature* **1997**, *389*, 829–832. d) J. H. Holtz, J. S. W. Holtz, C. H. Munro, S. A. Asher, *Anal. Chem.* **1998**, *70*, 780–791.
104 J. D. Joannopoulos, R. D. Meade, J. N. Winn, *Photonic Crystals*, Princeton University Press, Princeton, NJ, USA, **1995**.
105 E. Yablonovitch, *Phys. Rev. Lett.* **1987**, *58*, 2059–2062.
106 S. John, *Phys. Rev. Lett.* **1987**, *58*, 2486–2489.
107 a) S. John, *Physics Today* **1991**, *May*, 32–40. b) E. Yablonovitch, *J. Opt. Soc. Am. B* **1993**, *10*, 283–295. c) J. D. Joannopoulos, P. R. Villeneuve, S. Fan, *Nature* **1997**, *386*, 143–149.
108 a) C. M. Soukoulis, *Photonic Band Gap Materials*, Kluwer Academic Publishers, New York, NY, USA, **1996**. b) A. Scherer, T. Doll, E. Yablonovitch, H. O. Everitt, J. A. Higgins, special issue *J. Lightwave Technol.* **1999**, *17*, 1928–2411.
109 a) K. M. Leung, Y. F. Liu, *Phys. Rev. Lett.* **1990**, *65*, 2646–2649. b) Z. Zhang, S. Satpathy, *Phys. Rev. Lett.* **1990**, *65*, 2650–2653. c) H. S. Sozuer, J. W. Haus, R. Inguva, *Phys. Rev. B* **1992**, *45*, 13962–13972.
110 E. Yablonovitch, T. J. Gmitter, *Phys. Rev. Lett.* **1989**, *63*, 1950–1953.
111 a) J. W. Haus, *J. Mod. Opt.* **1994**, *41*, 195–207. b) R. Biswas, M. M. Sigalas, G. Subramania, K.-M. Ho, *Phys. Rev. B* **1998**, *57*, 3701–3705.
112 a) J. E. G. J. Wijnhoven, W. L. Vos, *Science* **1998**, *281*, 802–804. b) M. S. Thijssen, R. Sprik, J. E. G. J. Wijinhoven, M. Megens, T. Narayanan, A. Lagendijk, W. L. Vos, *Phys. Rev. Lett.* **1999**, *83*, 2730–2733.
113 a) K. Busch, S. John, *Phys. Rev. E* **1998**, *58*, 3896–3908. b) A. Moroz, C. Sommers, *J. Phys.: Condens. Matter* **1999**, *11*, 997–1008.
114 a) M. M. Burns, J. M. Fournier, J. A. Golovchenko, *Science* **1990**, *249*, 749–754. b) H. Misawa, K. Sasaki, M. Koshioka, N. Kitamura, H. Masuhara, *Appl. Phys. Lett.* **1992**, *60*, 310–312. c) W. Hu, H. Li, B. Chang, J. Yang, Z. Li, J. Xu, D. Zhang, *Opt. Lett.* **1995**, *20*, 964–966. d) C. Mio, M. D. W. Mar, *Langmuir* **1999**, *15*, 8565–8568.
115 a) S. Hachisu, S. Yoshimura, *Nature* **1980**, *283*, 188–189. b) P. Bartlett, R. H. Ottewill, P. N. Pusey, *Phys. Rev. Lett.* **1992**, *68*, 3801–3804. c) B. Gates, Y. Xia, *Appl. Phys. Lett.* **2001**, *78*, 3178–3180.
116 a) S. Yoshimura, S. Hachisu, *Progr. Colloid Polymer Sci.* **1983**, *68*, 59–70. b) P. Bartlett, R. H. Ottewill, P. N. Pusey, *Phys. Rev. Lett.* **1992**, *68*, 3801–3804. c) C. J. Kiely, J. Fink, M. Brust, D. Bethell, D. J. Schiffrin, *Nature* **1998**, *396*, 444–446.
117 a) Z.-Y. Li, J. Wang, B.-Y. Gu, *Phys. Rev. B* **1998**, *58*, 3721–3729. b) Z.-Y. Li, J. Wang, B.-Y. Gu, *J. Phys. Soc. Jpn.* **1998**, *67*, 3288–3291.
118 Y. Xia, B. Gates, Z.-Y. Li, *Adv. Mater.* **2001**, *13*, 409–413.
119 a) N. Sasaki, Y. Murakami, D. Shindo, T. Sugimoto, *J. Colloid Interface Sci.* **1999**, *213*, 121–125. b) T. Sugimoto, M. M. Khan, A. Muramatsu, H. Itoh, *Colloids Surf. A* **1993**, *79*, 233–247.
120 a) M. Nagy, A. Keller, *Polym. Commun.* **1989**, *30*, 130–132. b) K. M. Keville, E. I. Franses, J. M. Caruthers, *J. Colloid Interface Sci.* **1991**, *144*, 103–126.
121 Y. Lu, Y. Yin, Y. Xia, *Adv. Mater.* **2001**, *13*, 415–420.
122 P. Jiang, J. F. Bertone, V. L. Colvin, *Science* **2001**, *291*, 453–457.
123 E. Snoeks, A. van Blaaderen, T. van Dillen, C. M. van Kats, M. L. Brongersma, A. Polman, *Adv. Mater.* **2000**, *12*, 1511–1514.
124 Y. Yin, Y. Lu, Y. Xia, *J. Mater. Chem.* **2001**, *11*, 987–989.
125 Y. Lu, Y. Yin, Z.-Y. Li, Y. Xia, *Langmuir* **2002**, *18*, 7722–7727.

10
Surface-directed Colloid Patterning: Selective Deposition via Electrostatic and Secondary Interactions

Paula T. Hammond

10.1
Introduction

A great deal of excitement has been generated around the idea of controlled and ordered arrays of micron- to nanometer-scale particles arranged on surfaces. The ability to direct the positions of such small "objects" onto a substrate provides a large advantage in the creation of molecular- to nanometer-scale devices, biological sensors, combinatorial arrays, and electronic and photonic devices. For these reasons, a number of approaches have been investigated which involve the manipulation of colloidal systems using electrostatics [1–6], lithography, dip-coating, physical confinement in etched or molded grooves [7], sedimentation [8–13], capillary forces [14], flow fields [15, 16], electrophoretics [17], and microfluidics to guide specific systems to different regions of a surface. An especially compelling approach to directing deposition is to use secondary interactions such as coulombic interactions, hydrogen bonding, hydrophobic interactions, and biospecific interactions, as a means of guiding different elements to a surface. This form of self-assembly provides the basis for positioning a broad range of elements based on the presence of complementary functional groups on planar and nonplanar surfaces; this concept of "surface sorting" has been explored in our group with polyelectrolytes [18–21], and more recently, colloidal particles [22–25], as a potential tool in microfabrication. The chemical basis of this approach is the use of surface functionality for templating and directing the assembly of nanometer to micron-sized materials systems.

The concept of surface sorting is illustrated schematically in Fig. 10.1. A chemically patterned surface containing different regions of charged, hydrogen bonding, hydrophobic, or biospecific ligand groups, acts as a template to which materials species of complementary functionality deposit onto their corresponding sites. This process can take place through a series of sequential adsorption steps, or, under highly controlled conditions with more specific interactions, it may be possible to achieve the desired effect in a single step. This concept is universal, and can be extended to a broad range of systems, including polyelectrolytes, colloidal particles, proteins, DNA, nanotubes, etc. Nonlithographic patterning methods, particularly contact printing techniques [26], allow us to create such systems on a

Colloids and Colloid Assemblies. Edited by Frank Caruso
Copyright © 2004 Wiley-VCH Verlag GmbH & Co. KGaA, Weinheim
ISBN: 3-527-30660-9

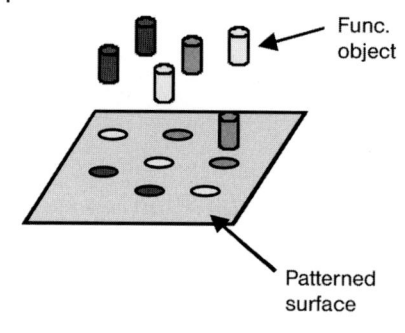

Fig. 10.1 Schematic representation of concept of surface sorting with multiple surface components during the adsorption process. Cylinders represent charged functional "objects" present in solution ranging from macromolecules to micron-sized particles with a specific binding group. Colored dots on the surface indicate surface regions with complementary functionality which act as regions for deposition.

broad range of surfaces, and directly position a number of chemical and biological functional groups, without the constraints of traditional photolithography.

10.2
Observation of Selective Deposition with Simple Polyamines

By understanding the interfacial interactions between the polyelectrolyte and the chemical functional group at the surface, it is possible to manipulate the deposition of polyion pairs based on electrostatic, hydrogen bonding, and hydrophobic interactions. We have now established a set of rules that define the conditions re-

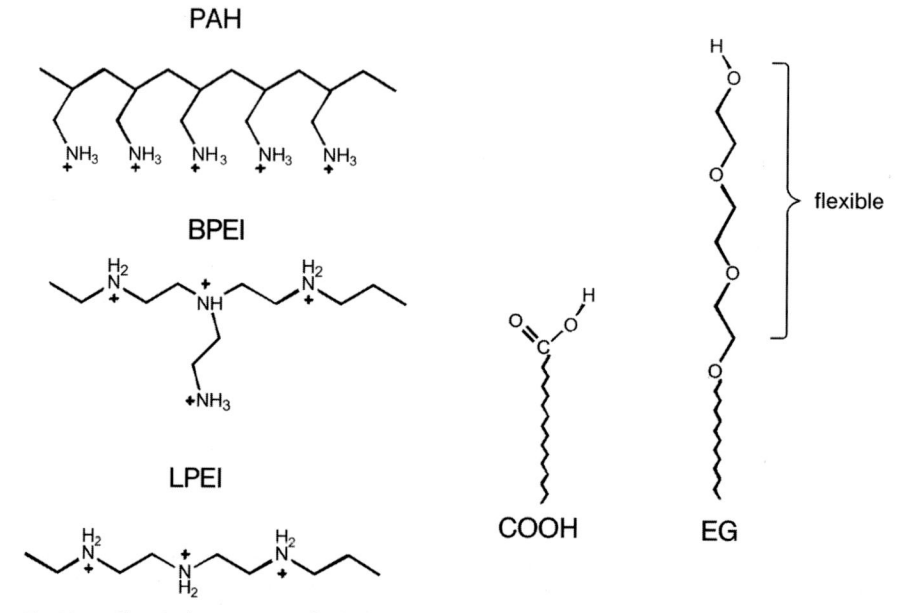

Fig. 10.2 Chemical structures of polyamines and surface functional groups used in selective adsorption studies.

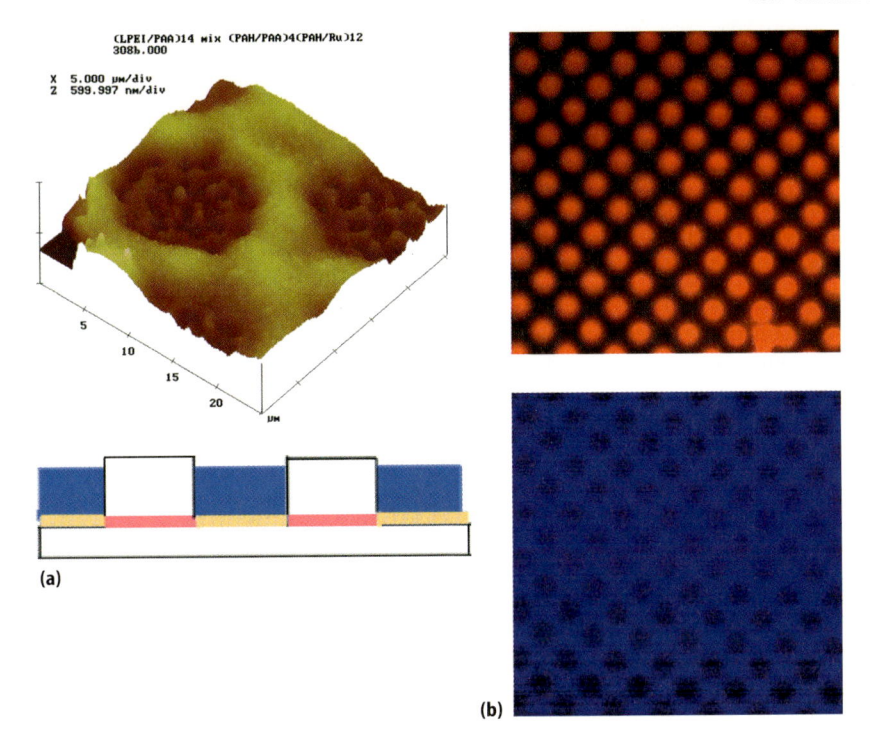

Fig. 10.3 a AFM topographic image of patterned (LPEI/PAA) multilayers adsorbed on a patterned COOH SAM (self-assembled monolayer) surface (outer regions) and (LPEI/PAA) 10(LPEI/Ru(phen)$^{4-}$) adsorbed on the EG surface respectively (circular dots). Schematic shows cross-sectional view [20]. **b** Laterally patterned bicolor OLED model structure, containing 10(PDAC/PPP(−)) on COOH surface and 5(PAH/PAA)10(PAH/Ru(phen)$^{4-}$) on the EG surface. Image on top is a fluorescence micrograph taken with the filter for red dye on the EG surface and the bottom image is the same sample taken with the fluorescence filter for the blue dye on the COOH surface [20].

quired for selective deposition on specific surface regions for both strong [27–29] and weak [19–21, 30] polyelectrolyte adsorption, utilizing systematic adsorption studies and chemical force microscopy. An example of this ability is demonstrated in the selectivity of polyamines with hydrophobic versus hydrophilic backbones to different surface regions based on secondary as well as electrostatic interactions. In previous studies, it was found that weak polyamines such as those shown in Fig. 10.2 deposit selectively on preferred regions of a surface as a function of pH [19, 21]. At intermediate pH ranges, the more hydrophobic species, poly(allylamine hydrochloride) (PAH), which has a hydrophobic backbone, will adsorb in larger quantities to an oligoethylene oxide-functionalized self-assembled monolayer (EG) than to an acid functionalized monolayer (COOH). In particular, the hydrophobic nature of the PAH backbone can induce dispersion interactions with the ethylene groups in PEO. These interactions do not occur with linear polyethyleneimine

LPEI, which is highly hydrated in aqueous solution, thus introducing strong steric repulsions with the similarly hydrated EG oligomer. The conclusion is that the role of secondary interactions in such adsorption processes is extremely important in determining where adsorption occurs. These preferences can be tuned with adsorption conditions such as ionic strength and pH; for example, the preferred region for deposition of strong, highly charged polyelectrolytes can be manipulated by shielding the charge of the backbone, thus making secondary interactions more important than ionic interactions in selective adsorption.

These concepts lead to an ability to guide different material elements to a surface based on these types of interactions. This understanding leads to the ability to direct different items to a surface, a concept we have termed "surface sorting" to indicate the ability of systems in solution to adsorb to predetermined regions based on specific and non-specific interactions. Fig. 10.1 contains a schematic illustrating this general concept.

The differences in adsorption behavior of these and other polymers have been used as a tool to effectively control the region of film deposition and create side-by-side polyelectrolyte multilayer films on a surface [20]. A demonstration of side-by-side multilayer adsorption using LPEI and PAH was accomplished by first adsorbing multilayers of LPEI with polyacrylic acid (PAA) onto the COOH surface, followed by the adsorption of PAH/polyanions-based multilayers onto the EG regions of the surface. The red sulfonated ruthenium dye, $Ru(phen')_3^{4-}$, was chosen as the multiply charged anionic species to be directed to the EG regions by PAH. The resulting composite structure, shown in the AFM micrograph in Fig. 10.3a, consists of the $PAH/Ru(phen')_3^{4-}$ surrounded by a continuous region of LPEI/PAA multilayers. This technique can be extended to demonstrate the adsorption of two independent dye systems on one substrate, as shown by the images in Fig. 10.3b. The permanently charged polycation, poly(diallyldimethylammonium chloride) (PDAC) was used to assemble patterned films with a sulfonated poly-p-phenylene (PPP(–)) blue dye on the COOH regions. PAH was then co-adsorbed with the ruthenium dye at pH 4.8, resulting in the adsorption of the red dye on the EG circular regions. The result was an alternating surface pattern of red and blue luminescent dye multilayers on a single surface [20].

10.2.1
Extension of Selective Deposition to Colloids

The principles introduced above can be applied to small-scale 3D objects over a range of sizes, from nanometers to micrometers, simply by using particle systems with modified surfaces. The advantages of being able to manipulate materials on this length scale are manifold. Colloidal particles arrayed on surfaces can act as elements in a microphotonic device, or as components in a microfluidic channel or reactor. Particles functionalized with biofunctional materials can guide cell growth and attachment, direct deposition and/or hybridization of DNA, or serve as assays for proteonomics or other bioassay approaches, and systems with electro-optical properties may serve as part of a display, sensor, or other functional device. The desire to gain control of the assembly of nano- to micro-scale objects on surfaces and in 3D has been led to a number of studies for which capillary forces, fluid flow fields [15, 16], sedimentation, and crystallization [8–13] have been used extensively. Further approaches include confinement within etched or molded topographical surfaces [8, 31], sometimes combined with capillary forces [14] or electrophoretics [17], to produce ordered colloidal structures. Many of these techniques rely primarily on a physical quality or effect, such as the presence of walls in topographic surfaces, or the presence of a fluid flow field. For these reasons, none of the above named approaches would allow for straightforward differentiation between particles of different types on a given surface. The concepts of surface-sorting can be applied to such systems to achieve the ability to guide materials systems in a more specific manner. The most versatile, and the strongest of these interactions are electrostatic, or charge-based, interactions. Electrostatic interactions have been used in colloid assembly approaches using uniform, charged surfaces [2–5] and patterned surfaces [1, 6, 32]. Patterned polyelectrolyte multilayers are a unique and versatile means to create electrostatic templates for colloid deposition [22–25], opening the field of colloid assembly to a broad range of new applications via the introduction of functional materials. An important new development that will be discussed below is the concept of using additional secondary interactions to achieve 2D arrays that can differentiate different components, leading to two or more component colloidal array systems [23]. The ability to employ "surface sorting" allows the integration of colloidal particle systems with a variety of functionalities onto a multiply patterned surface.

There are three primary advantages to the integration of polyelectrolyte multilayers as a template for the creation of patterned colloidal particle arrays, as opposed to the direct use of traditional low molar mass templates such as SAMs. First, the fact that polyelectrolytes can be adsorbed onto a broad range of substrates, including plastic, metal, metal oxide, and glass surfaces, and even textile fiber surfaces, allows one to create functional colloidal arrays on a much larger set of materials surfaces than the more traditional use of silanes or thiols on silicon or metal surfaces for the creation of chemically templates. This advantage is an important one, in that a number of new markets exist for inexpensive, plastic electronic and photonic devices that can be fabricated using common plastics such as Mylar, polystyrene, polymethacrylates, etc. A second advantage of the use of layer-by-layer thin films is that it enables the incorporation of functionality in the underlying film; this is a capability that is not easily achieved through the use of SAMs such as alkanethiols or alkoxysilanes. The polyelectrolyte multilayer process provides a means of incorporating multiple layers of functional materials systems, including conducting [33, 34] or luminescent polyelectrolyte multilayer films, non-linear optical dyes, or electrochemical systems [41–43], to name a few. By modifying adsorption conditions, as well as the components incorporated into these thin films, the functionality of these films can be tuned, as well as their thickness and their heterostructure. This aspect of multilayer/directed colloid assembly can lead to a number of promising sensors, photonic or electronic devices, bioactive systems, and a wealth of other composite systems. Finally, the presence of the under-

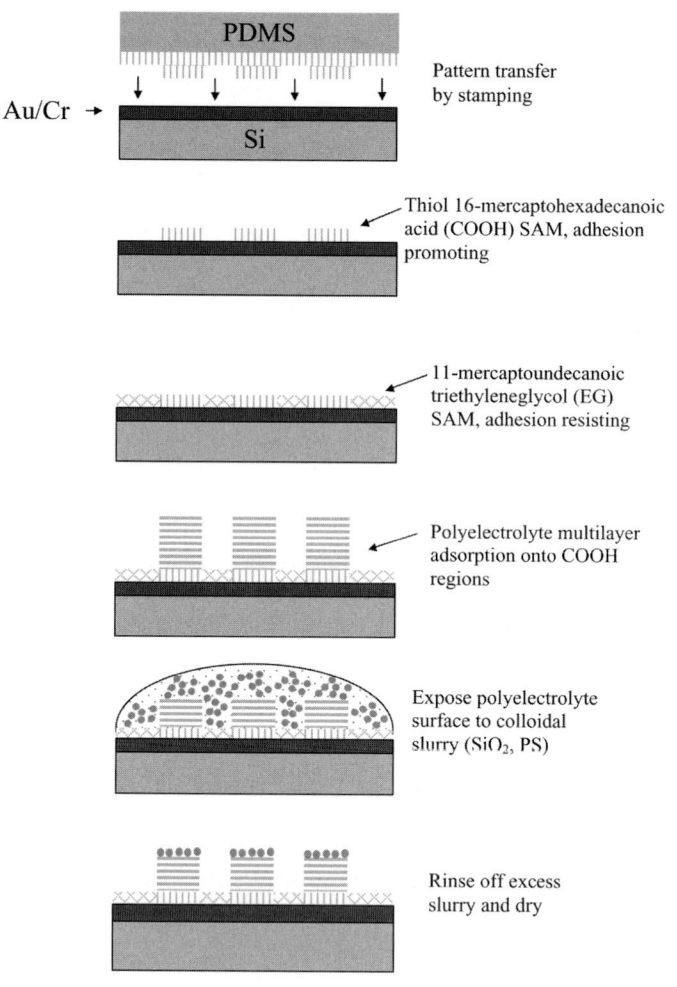

Fig. 10.4 Schematic of sample fabrication process: substrate patterning, polyelectrolyte adsorption, and colloid deposition [22].

lying multilayer provides important adhesion and stability to the adsorbed particle upon deposition, due to the presence of a highly functional, partially charged, polymer layer that can be further modified to enhance adhesion as needed by the incorporation of cross-links or increase in the thickness or density of functional groups at the film surface [44, 45].

Fig. 10.4 illustrates the sample fabrication process employed to form colloidal arrays on patterned polyelectrolyte surfaces [22]. The key to this approach to colloid assembly is to tune the interactions between the colloid, the underlying patterned polyelectrolyte surface, and the alternating surface region. For example, three different mechanisms were systematically explored [22], as shown schematically in Fig. 10.5, representing the effects of (a) variation of the surface charge of

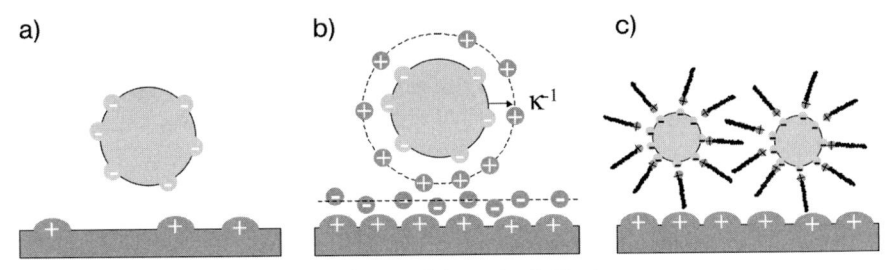

Fig. 10.5 Three mechanisms for regulation of charged colloid adsorption on an oppositely charged substrate region: **a** surface charge density regulated by pH, **b** charge screening by electrolytes, and **c** charge shielding by surfactant [22].

a weak polyelectrolyte top layer, (b) the effect of shielding the surface charge of both colloid and surface with additional electrolyte, and (c) the shielding of the colloid charged surface with small quantities of an oppositely charged surfactant molecule. The first two cases vary the electrostatic attraction between the particle and the surface through variation of net surface charge or introduction of shielding. The third case introduces additional secondary interactions between colloids in the form of hydrophobic attractions, while decreasing the overall charge of the particles and thus the electrostatic repulsions between particles.

The result of variation of the surface charge has been demonstrated with a polyelectrolyte multilayer consisting of a weak polyamine/polyacid pair, LPEI and polyacrylic acid (PAA). The adsorption of negatively charged 700 nm SiO_2 spheres onto patterned LPEI/PAA bilayers at different pH values is shown in Fig. 10.6. Based on the reported pK_a of LPEI of approximately 5.0, the surface charge is still relatively high, at 35 to 40% ionization of the amine groups, when the pH is 6.8 [46, 47]. Under these conditions, the surface charge of the polyelectrolyte multi-

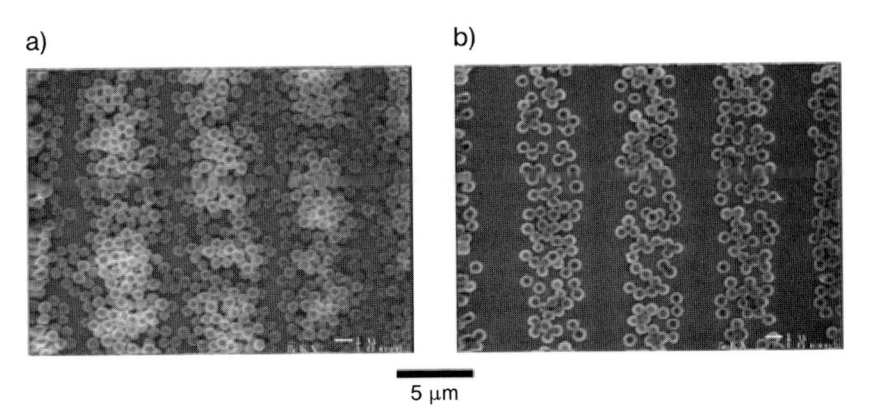

Fig. 10.6 SEM micrographs of SiO_2 microspheres bound by electrostatic attraction onto underlying 5.5(LPEI/PAA) bilayers at **a** pH = 6.8 and **b** pH = 8.1 [22].

layers is high enough to attract large numbers of particles, resulting in the deposition of multiple layers of particles. There is also a tendency of particles to "bridge" across the resist layer, presumably due to favorable particle–particle interactions at this pH range, and potential hydrogen bonding interactions with the EG resist region, resulting in reduced selectivity. At a higher pH of 8.1, the silanol particles become highly charged, and, despite the change in the degree of ionization for LPEI (20–30%), a controlled, selective deposition occurs, resulting in a monolayer of particles on the charged amine surface. It is evident from this experiment that the charge density of both the particles and the surface must be considered independently, and that interparticle attractions are also important in gaining selective deposition. When electrostatic interactions are optimized, higher levels of selectivity can be gained; however, when ionic interactions are lowered, secondary interactions become more important, an observation earlier made with selective polyelectrolyte deposition [18, 19].

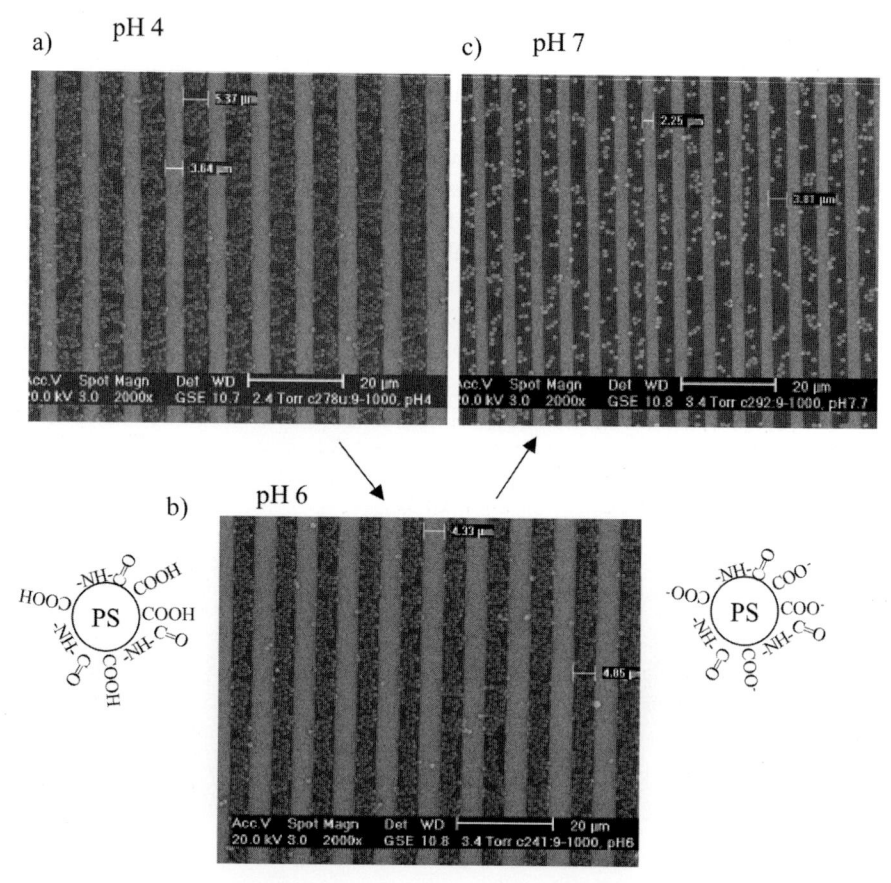

Fig. 10.7 pH effect on the adsorption of 9–1000 particles on patterned 5.5(PDAC/SPS) multilayers (SEM pictures): the pH value of 9–1000 particle suspension was adjusted to **a** pH 4, **b** pH 6 (unadjusted) and **c** pH 7.7.

It is also interesting to observe trends in selectivity with pH when the surface is permanently charged, but the particle contains weak ionizable groups such as acids or amines. In an examination of such effects, polystyrene latex particles containing carboxylamide and carboxylic acid surface groups were used in deposition experiments for which a sulfonated polystyrene (SPS)/PDAC multilayer was used as the template. In this case, the positively charged top PDAC surface acted as a region for deposition of the particles, whereas the EG surface acted as a resist. Results of adsorption of the colloidal particles at different pH values is shown in Fig. 10.7. At lower pH values of 4.0, the particles tended to adsorb to the surface in a somewhat loose-packed fashion; this packing is influenced by the degree of charge on the particle surface, which is low when the acid groups are below their pK_a. However, as the pH is increased above the pK_a of approximately 5.0, the particles begin to pack more densely onto the positively charged multilayer surface as the charge on the particle increases, and the electrostatic attraction increases correspondingly. Ultimately, at pH values of 7 and higher, the density of particles adsorbed on the multilayer surface is greatly decreased. This is likely to be due to the high particle–particle repulsion that occurs when approximately all of the acid groups on the surface are charged at high pH. It should be noted that the surface charge density of the particles in Fig. 10.7 are considerably larger than that of the silica particles discussed earlier.

Additional interactions may also play a role in how the particles self-arrange on the surface at low pH, when electrostatics are less important. In this case, it is likely that the particles undergo interparticle hydrogen bond dimerization, which could lead to clustering and aggregation. When the pH was increased to 7.7, hydrogen bonding was no longer possible for the highly charged carboxylate anions, and sparse adsorption on PDAC surface with high selectivity was observed (Fig. 10.7 c). It is interesting to note that for these particularly hydrophilic carboxyl amide/acid-functionalized particles, the resist region is effective regardless of pH. It appears that, just as hydrophilic backbones played an important role in determining selectivity for the weak polyamines and their multilayers, so it plays a role in colloid adsorption. The hydrophilic particles in this case may exhibit a shell of hydration that undergoes repulsion with a similar hydration layer in the EG brush surface. The result is a strong repulsive interaction, and high selectivity for the PDAC multilayer surface.

On the other hand, direct shielding of electrostatic charge via addition of salt also has a strong effect on the selective deposition of charged species [22]. When polyelectrolyte multilayer platforms of PDAC and SPS are used as a template, the top PDAC surface is permanently charged. Silica particles initially adsorb selectively onto this oppositely charged surface region when no additional ions are added to the adsorption solution; however, the shielding introduced with the addition of NaCl results in a loss of selectivity. As the colloidal particles are shielded, repulsive interactions between neighboring colloidal particles are reduced, and the attractive interactions between the PDAC and particle surfaces are lowered. Ultimately, with a decrease in the electrostatic driving forces, the shielded particles deposit on the EG surface as well as the PDAC surface in large quantities as shown

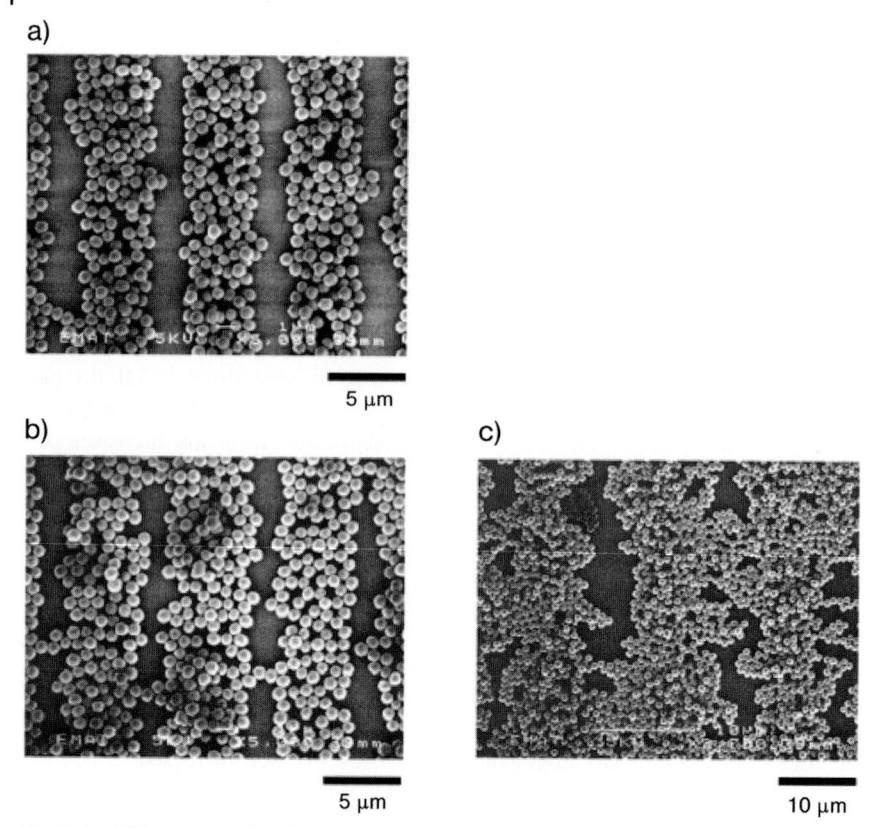

Fig. 10.8 SEM micrographs of SiO$_2$ microspheres bound by electrostatic attraction onto underlying 5.5(PDAC/SPS) bilayers in the presence of **a** no NaCl, **b** 0.01 M NaCl, and **c** 0.1 M NaCl [22].

by the SEM images in Fig. 10.8. As the ionic strength increases, the selectivity decreases. The interactions of the particles with the EG surface may include weak hydrogen bonding interactions; deposition on this resist region is also facilitated by the decreased repulsion between colloid particles at high ionic strength.

After the samples had been dried, the adhered colloids withstood repeated dipping cycles in DI water and blowing with compressed air, remaining adsorbed to the surface. Moreover, the colloids could not be stripped off by use of common adhesive tape. Thus, the use of polyelectrolyte multilayers results in well-adhered, stable films of patterned colloidal particles, as suggested previously. The strong adhesion may be attributed to the very high surface charge densities of polyelectrolytes versus those of the SAMs. Importantly, the strong adhesion of colloids to the patterned polyelectrolyte surface, coupled with adhesion tailorability, offers a robust and flexible platform for subsequent processing that is absent when SAM underlayers are used exclusively.

10.3
Use of Surfactants to Guide Colloidal Surface Assembly

A unique approach to the dual shielding and chemical modification of the colloid particle surface is the introduction of small amounts of surfactant to partially or fully compensate the charge of the colloidal particle surface. This approach has been used successfully to manipulate particles in three-dimensional assemblies [48–50]. We have used the same approach to control particle assembly on the surface to achieve significant differences in the positioning and ordering of particles on polyelectrolyte-templated surfaces [22]. When the cationic surfactant dodecyltrimethylammonium bromide (DTAB) is added to suspensions of negatively charged polystyrene latex particles, the result is the decoration of the latex particles with bound surfactant molecules at the surface. Ultimately, the surface charge is reduced and the surface is made more hydrophobic as more surfactant is added to the suspension. One of the goals in altering surface chemistry is to increase interparticle interactions and thus gain more closely packed particle assemblies on the surface. Fig. 10.9 illustrates the effects of adding surfactant to sulfate particle suspensions. The concentrations of surfactant added were gradually increased from 1×10^{-5} M to 5×10^{-4} M DTAB. The templated surface consists of a positively charged top surface of PDAC from a patterned SPS/PDAC multilayer; the alternating surface consists of the EG SAM resist surface. At low degrees of shielding, the particles actually pack somewhat closer on the surface due to reduced particle–particle repulsion and increased hydrophobic interactions (Fig. 10.9b). As the surfactant concentration was slowly increased, however, the overall selectivity approaches zero, as particles adsorb on both regions of the surface (not shown). With continued increases in the surfactant concentration, an unexpected change in the preferred region of selectivity occurred, resulting in a reversal in selectivity (Fig. 10.9c and d).

In the original paper, the degree of shielding was determined based on the surface charge density of particles reported by the particle manufacturer. More recent studies performed in our group have used ζ-potential analysis to determine the surface charge of particles treated with DTAB at the same concentrations; the values obtained from ζ-potential analysis are shown in Tab. 10.1 and in Fig. 10.10; it is clear from these values that the negative surface charge is ultimately fully compensated, and then reversed, by the binding of surfactant molecules at higher concentrations. Charge reversal was observed for sulfate-functionalized particles when the DTAB concentration was above 5×10^{-5} M. Similar charge reversal was observed in other work when 2.2×10^{-4} M HTAB (hexadecyltrimethylammonium bromide) was added to a similar sulfate PS latex sphere suspension [50]. This confirms the existence of surfactant bilayers at higher DTAB concentration. On the basis of this set of ζ-potential data, the calculated extents of charge shielding are listed in Tab. 10.1 as well, and the adsorbed surfactant structures on latex particles are modeled in Fig. 10.10b–d. Surfactant molecules form monolayers, bilayers, or multilayers with increasing surfactant concentrations. At more than 100% charge compensation, the surfactant molecules begin to form a double layer around the

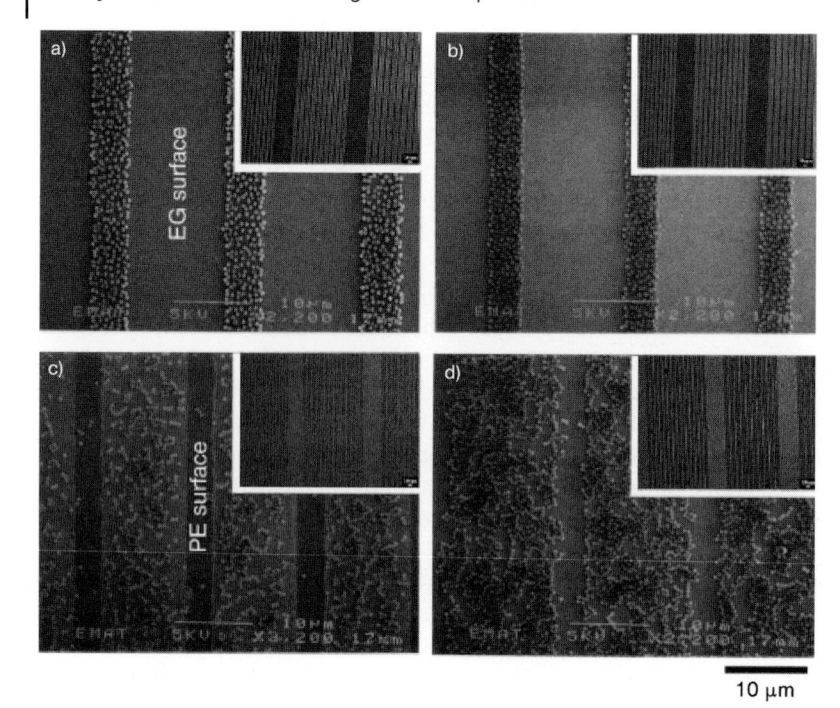

Fig. 10.9 SEM micrographs of patterned 5.5(PDAC/SPS) bilayers immersed in the 1 500 polystyrene latex sphere suspension with **a** 0 M, **b** 1×10^{-5} M, **c** 1×10^{-4} M, and **d** 5×10^{-4} M DTAB surfactant. The insets are optical micrographs [22].

Tab. 10.1 ζ-Potentials of virgin colloids and DTAB-modified colloid

DTAB concentration (M)	ζ-potentials of 1–500 sulfate particles (mV)	Extent of charge shielding for 1–500 sulfate particles
0	-71.95 ± 1.17	0
2.00×10^{-7}	-68.38 ± 0.91	5%
2.00×10^{-6}	-67.64 ± 0.66	6%
1.00×10^{-5}	-39.79 ± 0.87	44.5%
2.00×10^{-5}	-30.99 ± 0.52	57%
1×10^{-4}	29.22 ± 1.18	141%
5×10^{-4}	36.50 ± 0.56	151%
1×10^{-3}	40.59 ± 0.76	156%

All measurements were carried out at 0.1 mg/mL particle concentration.

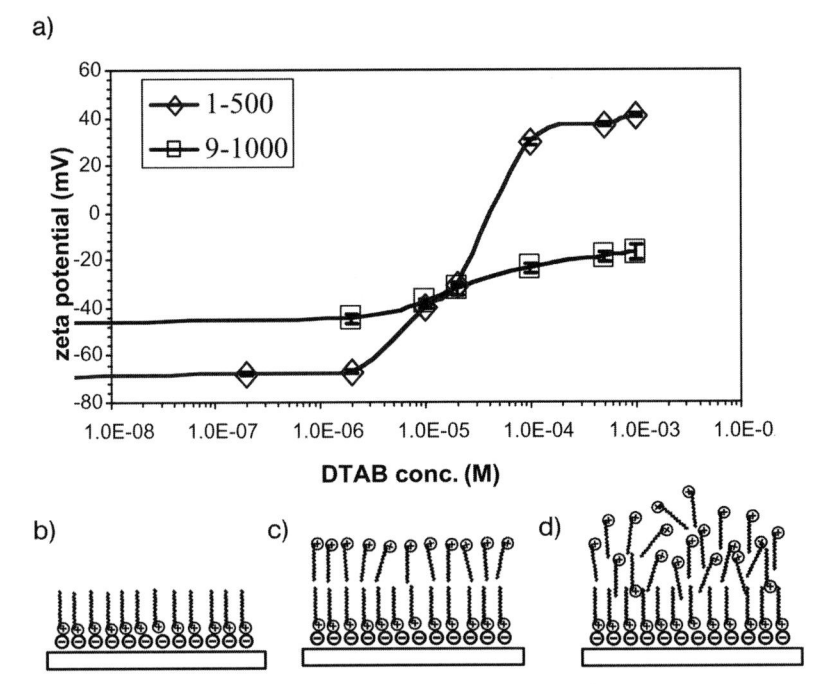

Fig. 10.10 The ζ-potential versus DTAB concentration plots **a** for both 1-500 particles and 9–1000 particles (all ζ-potential measurements were performed at 0.1 mg/mL particle concentrations) and the model structures of surfactant adsorption on charged surfaces; **b** submonolayer or monolayer adsorption at a low surfactant concentration; **c** double layer adsorption at a medium surfactant concentration; **d** multilayer adsorption at a high surfactant concentration.

colloidal particle, for which the hydrocarbon regions interpenetrate the original surfactant layer, and the charged surfactant "heads" face the water solution. In this case, the net charge of the particle is reversed, and the particle now acts as a positively charged entity. This charge reversal results in a repulsion between the top PDAC multilayer surface and the surfactant-modified colloidal particle, and deposition is prevented on the multilayer surface. On the other hand, the EG surface appears to undergo weak interactions with hydrophobic/hydrophilic molecular systems, as observed in studies with the weak polyamines. In this case, the surfactant-coated particles are attracted to the EG surface, and are able to form a uniform, ordered monolayer on that surface. It is also possible that small amounts of free surfactant molecules may have been attracted to the EG surface, thus making it a surface with additional hydrophobic character.

10.4
Surface Sorting with Two Different Colloid Types

The concepts addressed above lead to the possibilities of sorting particles on surfaces, much as we were able to sort polyelectrolytes. An initial set of particles can be attracted to the surface based on ionic interactions, and a second directed to an alternate surface region based on hydrophobic, hydrogen bonding, or other interactions. To demonstrate this concept, we have reported the transfer of particles of different sizes to different regions of a patterned polyelectrolyte multilayer-templated surface [23]. A sequential series of adsorption steps are used to accomplish this particle positioning, as illustrated schematically in Fig. 10.11. In this process, a series of polymer multilayers such as SPS/PDAC are adsorbed onto a chemically patterned surface of alternating COOH and EG SAMs, utilizing the chemistry-directed polyelectrolyte adsorption process developed in earlier work. A first set of colloidal particles is deposited onto the positively charged top surface of the multilayer based on ionic interactions. Following this deposition, the negatively charged colloidal particles are coated with a single layer of polycation. The EG resist surface remains free during each of these processes, as it acts to prevent the deposition of both polyions and particles. Finally, this EG resist is actually used as a region of deposition by the introduction of materials systems and adsorption conditions that promote deposition on the EG via secondary interactions. In the first case illustrated as the final step in Fig. 10.11 (left-hand arrow), the particles are deposited in the presence of surfactant at concentrations that exceed shieldings of

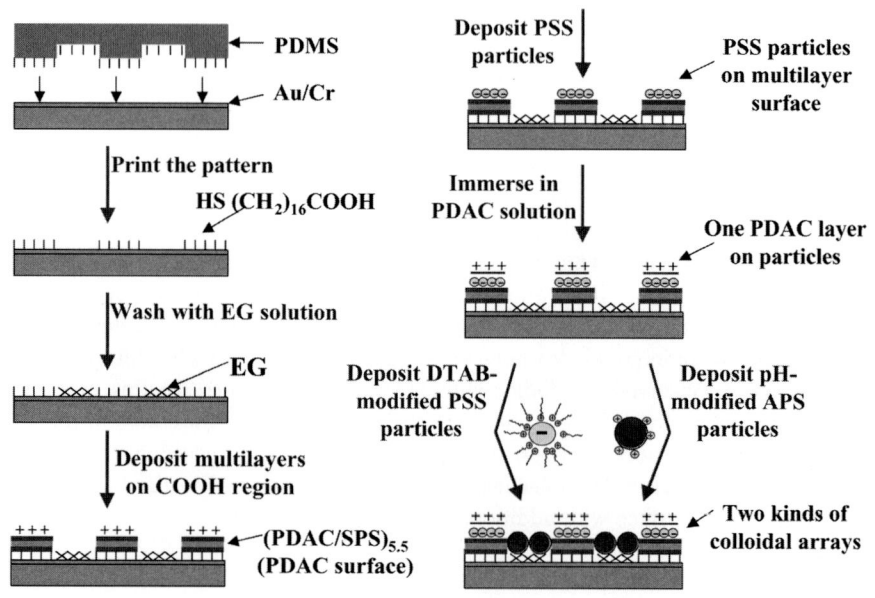

Fig. 10.11 Procedural schematic for deposition of two kinds of particle arrays on patterned templates [23].

100%, thus resulting in particles with a net positive charge, and an affinity for the EG layer, as discussed in the previous section. Such particles are repelled from the polycation blanket layer atop the original set of particles, and attracted only to the EG region. The blanket layer in this case plays a critical role; without its presence, the second set of particles adsorbs to both regions of the surface, and in many cases, the original set of particles desorbs in the presence of other positively charged particles in solution. The nature of the polymer blanket is such that the underlying particles are stabilized and held in place during the second adsorption step; thus the polycation adsorption step provides an important charge reversal as well as increased stability. Examples of the results of the first approach are shown in Fig. 10.12.

A second approach is to use hydrogen bonding as an attractive force for the EG layer. In this case, amidine-functionalized polystyrene latex particles are used [23], which feature a charged amidine group that can undergo hydrogen bonding with the lone pair electrons of the ether oxygens in oligoethylene glycol functional groups of the EG resist. As in the case of the surfactant-modified colloids, these particles are repelled by the positively charged polymer overlayer, and attracted to the EG at a pH of approximately 5.0. The two examples in Fig. 10.13 demonstrate the fact that the packing of particles can also be varied in this process through the use of particles with different charge densities, as was the case here, or the introduction of additional charge shielding or surface charge densities. The packing on the EG surface in all of these examples illustrates that the kinetic effects which affect the ordering of particles on surfaces are greatly decreased when the interactions on the surface are lowered. When charged particles are adsorbed on a surface of high charge density, the arrangement of the particles is limited by the lack of mobility of the colloids upon adsorption. On the EG surface, the particles are

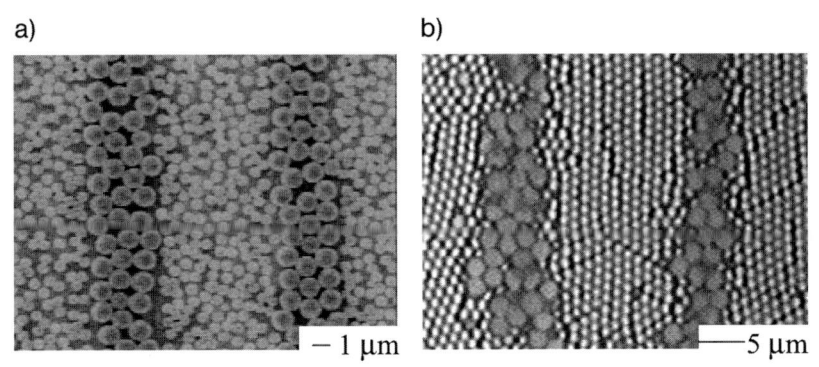

Fig. 10.12 Two different types of particles can be adsorbed to the charged multilayer surface of the alternating oligoethylene glycol monolayer based on ionic charge interactions combined with hydrophobic (left **a** – large particles are DTAB shielded) SEM **a** and OM **b** images of PSS particles on patterned (PDAC/SPS) 5.5/EG surfaces: **a** 0.5 µm particles on the PDAC surface and 1.0 µm particles on the EG surface; **b** 1.9 µm particles on the PDAC surface and 1.0 µm particles on the EG surface [23].

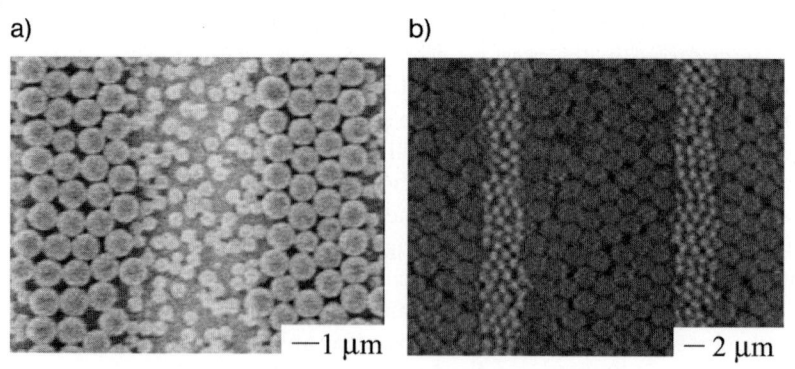

Fig. 10.13 Deposition of PSS and APS particles on patterned (PDAC/SPS) 5.5/EG surfaces: **a** SEM images of 0.5 μm PSS particles on the PDAC surface and 1.0 μm APS particles on the EG surface; **b** OM image of 1.0 μm PSS particles on the PDAC surface and 2.1 μm APS particles on the EG surface [23].

adhered to the surface via much weaker interactions, and are therefore able to rearrange on the surface to obtain ideal hexagonal close packing. These dynamics can actually be utilized to vary the range and density of colloidal elements on a surface for a variety of applications. The trade-off in this approach is that the adhesion to the surface is also lowered when weak interactions are used as the basis of attraction. The particles adsorbed onto the polymer multilayers are very strongly attached to the surface, and are difficult to remove using the adhesive tape test. Particle adhesion is considerably lower on the EG surface due to lowered interaction forces. For this reason, one might consider post-processing steps such as cross-linking or creation of covalent bonds to the particle surface after particle positioning.

The ability to create arrays of particles of different type, size, or functionality leads to a number of possible applications from electro-optical applications, such as the formation of multicolor luminescent display pixels and the formation of microphotonic devices using variation of refractive index, to biological and medical applications including sensors and combinatorial arrays based on biomolecular systems such as DNA or antibodies, the directed attachment of cells of different cell type, and the creation of viral array systems. This development has been demonstrated with nanometer-scale macromolecules and micrometer-scale particles; because the basis of surface sorting is the relative attraction *and adhesion* of materials systems to different chemical surface groups, rather than just the deposition of a given material on a surface, gravitational forces are not a controlling factor. The adsorption process is always followed by a rinse step, often with ultrasonics, which eliminates the presence of materials that have adventitiously deposited onto a given surface region, but have not bound to the surface through primary or secondary interactions to the surface. This principle is important to the concept of surface sorting – the primary driving forces for directed assembly are controlled using secondary interactions as tools to guide deposition.

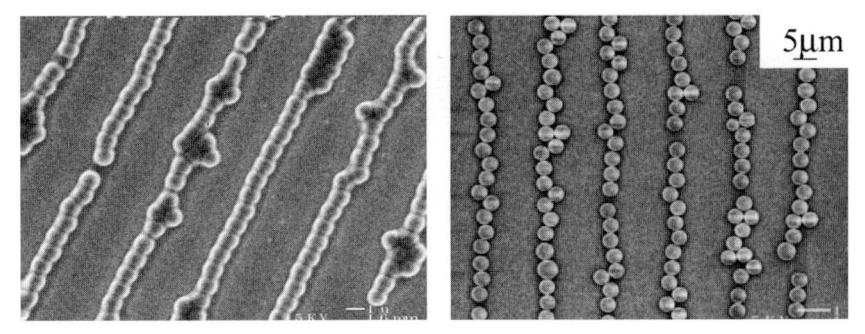

Fig. 10.14 Colloids directed to confined regions of a surface arranged in ordered patterns. **a** 1 μm particles on a 2 μm wide stripe [51], and **b** 5 μm particles on a 5 μm wide stripe.

10.5
Surface Confinement Effects – an Additional Tool

Surface chemistry has been shown to be an important tool for directing different elements to a surface; however, it can still be difficult to control the ordering of particles on the desired surface regions, depending on the interactions used for assembly, as illustrated in the examples in Figs. 10.12 and 10.13. Confinement of particles has been demonstrated as a means of controlling the placement of particles using topography, fluidics, and other approaches. By adding the concept of surface confinement to the secondary interactions and selective deposition tools described above, we can gain greater control of the arrangements of particles on surfaces. Such effects can be achieved without the introduction of etched or molded grooves simply by controlling the area of the chemical template used to guide particles to the surface. Confinement effects were initially observed with the deposition of particles on patterned polyelectrolyte multilayer lines with widths close to the diameter of the particle [22]. The results shown in Fig. 10.14a and b, are that single lines of particles can be observed for large portions of a given pattern. In the case of Fig. 10.14a, the particles are also thought to undergo strong hydrogen bonding interactions due to the carboxylamide functionalities on the surface exterior. It is interesting to note that when the linewidth is approximately twice the particle diameter, continuous single lines are observed, with occasional defects containing two or more rows. The relatively strong alignment of the particle is attributed in part to the strong particle–particle interactions. In Fig. 10.14b, sulfate-functionalized polystyrene particles, which lack the hydrogen bonding interactions between particles, arrange in lines along a multilayer line approximately the same width as the particle. Despite the greater degree of confinement in this case, clusters or groupings of particles are found containing two or three particles across the linewidth. In this case, the perfect alignment found in the hydrogen-bonded particles is not found in these particles, which simply contain highly charged groups on the surface. The effects observed in all cases are at least

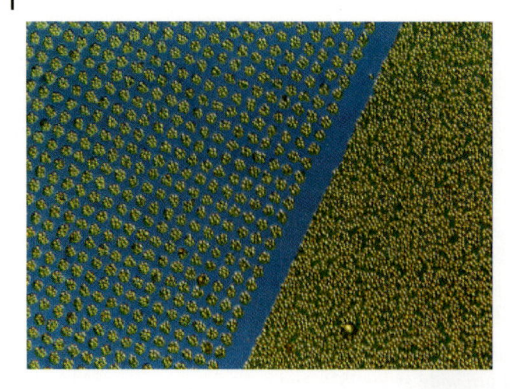

Fig. 10.15 Optical micrograph of colloids assembled on isolated patterned polyelectrolyte multilayer templates (left) and continuous, unpatterned multilayer region [25]; image © by Felice Frankel.

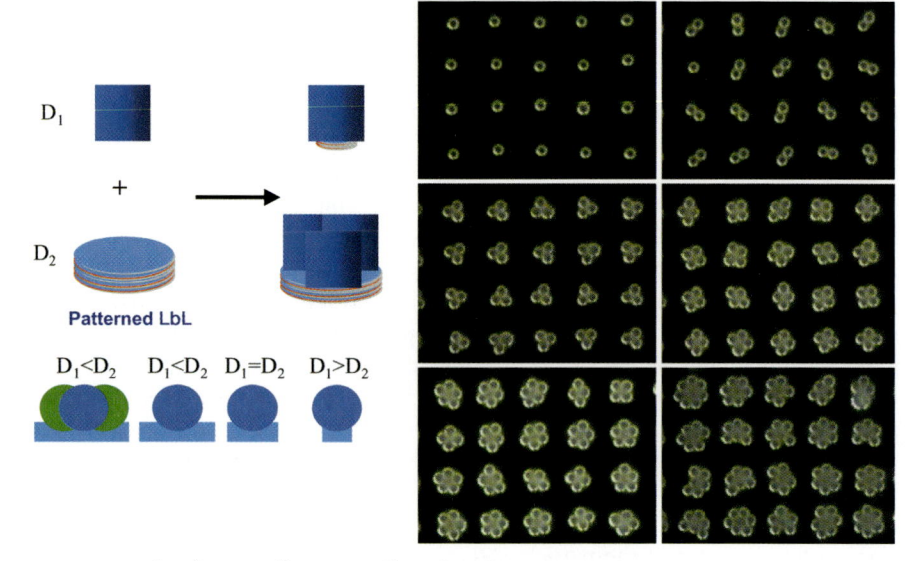

Fig. 10.16 Left: Schematic illustrating effect of confinement in a circular patterned layer-by-layer film in which the dot pattern is varied in size. Right: Controlled colloid cluster arrays adsorbed LbL templates of increasing size from top to bottom and left to right [25].

partially the result of capillary forces on the particles during the drying process; for this reason, the degree to which a particle is initially bound to the surface will affect the extent to which it can rearrange on removal from aqueous solution.

Because the capillary effects described above also result in drying patterns in which particles ultimately separate into isolated groupings, the arrangement of particles on isolated two-dimensional surface regions offers greater promise for the control of particle placement. Along these lines, patterned circular surface regions result in the clustering of particles into isolated patches [25], as shown to the left of the image in Fig. 10.15. The right of the image illustrates a continuous, unpatterned multilayer surface onto which particles have adsorbed, leaving a dense, relatively uniform covering of latex particles. By systematically controlling the size of the patterned polyelectrolyte multilayer, it is possible to control deposition to achieve regular arrays of single and multiple particle clusters [25], as illustrated in Fig. 10.16. In this case, it is possible to predict the numbers of particles that can fit onto a circle of given diameter, and thus extrapolate the cluster size with respect to the ratio of particle diameter to circle template diameter. In general, the particle diameter must be considerably larger than the template diameter to achieve single particle deposition; for example, in the case shown in Fig. 10.16, a 4 µm particle fits onto a 2.5 µm diameter multilayer template. It should be noted that for these systems, the patterned surface consists of alternating regions of positive charge and the EG SAMs resist surface. A different relationship would result from surfaces consisting of alternating charge, as will be discussed in the next section.

10.6
Use of Polymer Stamping as a Means to Functional Surfaces

It has been demonstrated that a range of surface chemistries can be used to direct the deposition of nanometer- to micrometer-scale materials systems to surfaces, and that one, two, or more such systems may be introduced to the surface to create ordered particle arrays. The approaches of directed assembly and the use of nonlithographic approaches to create such arrays become particularly interesting when a number of different surfaces, including low-cost, flexible plastic substrates, common glass and oxide surfaces, and even treated paper surfaces, can be utilized as the basis of devices, displays and sensors. Thus far, much of the work described above utilized alkanethiol-based self-assembled monolayers [26]; however, for many applications, microcontact printing of monolayers on gold and other metal or metal oxide surfaces can be limiting. There is a strong interest in the use of polymers as lightweight substrates for displays, sensors, electronic media, and other applications; however, most such surfaces are not accessible to patterning using silane or thiol chemistry. Further, the use of low molar mass systems often requires targeted synthesis to gain new functional systems. A new approach to the creation of chemical surface patterns has been developed which utilizes the large variety of functionalized polymers, polyions and block and graft copolymers which can be stamped directly onto other surfaces by careful selection of surface chemistry [52, 53].

The polymer-on-polymer stamping process is shown schematically in Fig. 10.17. In this process, which is an adaptation of microcontact printing, a PDMS stamp is inked with a polymer containing functional or charged groups complementary to the functionality of the substrate to be patterned. The substrate can be any surface containing functional groups, including acid, amine, alcohol, or other surface chemistries, or containing any charged group. An ideal surface for this approach is the polyelectrolyte multilayer. Layer-by-layer thin films contain highly functional

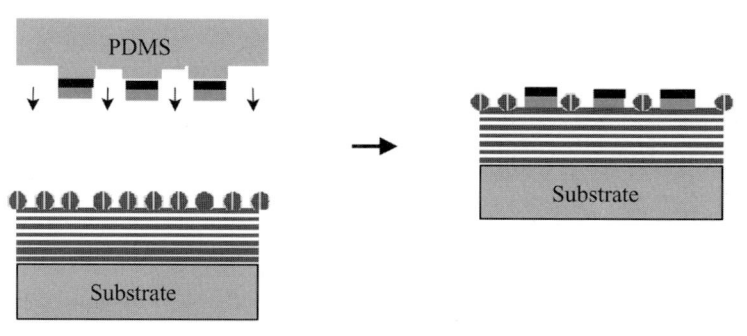

Fig. 10.17 Schematic illustrating the concept of polymer-on-polymer stamping.

top surfaces, and the density of functional groups on the surface can be modified by choice of deposition conditions as well as polyions; further, multilayers can be applied to many different surfaces, including common plastic surfaces, elastomer surfaces, metals, glasses, and metal oxides, thus allowing the patterning of each of these surface types. Upon contact of the stamp with the surface, the polymer is transferred directly to the top surface of a polyelectrolyte multilayer thin film based on covalent and/or non-covalent interactions.

This approach can be used to create surfaces with alternating positive and negative charge, which can be used for directed colloid deposition. A continuous multilayer can be adsorbed onto a desired substrate such that the top surface layer is negatively charged, and then the surface stamped with a polycation to create a positively charged pattern. When the substrate is later exposed to a colloidal suspension of oppositely charged particles, those particles adsorb selectively onto the oppositely charged regions, and are strongly repelled from the areas with the same charge. This process is shown schematically in Fig. 10.18, as well as an example of a polymer surface patterned in such a matter.

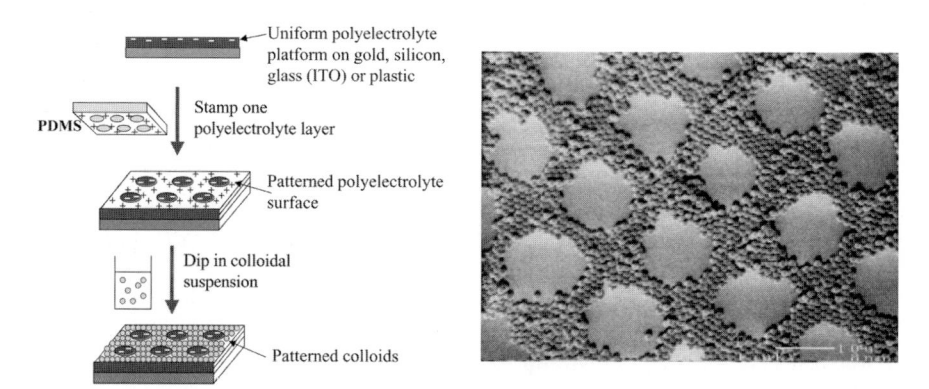

Fig. 10.18 Left: Schematic outline of the procedure for polymer-on-polymer stamping using polyelectrolyte as ink and fabrication of patterned colloidal arrays. Right: SEM pictures of PSS particles atop patterned PDAC surfaces on (PDAC/SPS) 10 bilayers [24].

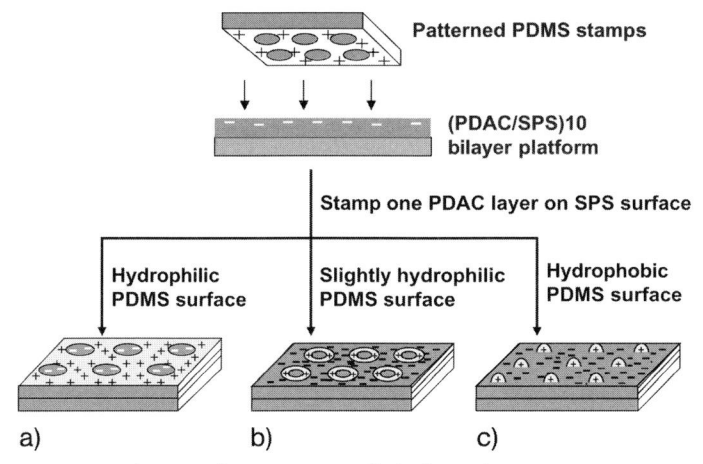

10.6 Use of Polymer Stamping as a Means to Functional Surfaces

Fig. 10.19 Schematic illustrating controlled inking of PDMS stamp following plasma treatment of stamp to achieve varying degrees of wettability [24].

The PDMS stamp is intrinsically hydrophobic, and must be plasma treated to achieve a wettable surface for the aqueous polycation-inking solution used to create alternating positive/negative surfaces. By varying the amount of exposure time of the PDMS stamp to oxygen or air plasma, it is possible to create heterogeneous stamp surfaces for which only the edges of the stamp are fully wettable by the ink [24]. In these cases, the ink wets only the edges of an inverse circular array pattern, and the end result is that a series of rings are created with dimensions considerably smaller than the original dot feature. If the stamp is not treated at all, it is possible to get single, uniform dots in a hexagonal array based on the dewetting of the ink to periodic regions on the stamp surface. Both of these effects are shown schematically in Fig. 10.19. When negatively charged particles are then adsorbed onto the surface, the ring and dot patterns are clearly templated (Fig. 10.20). The advantages of polymer-on-polymer stamping include the fact that

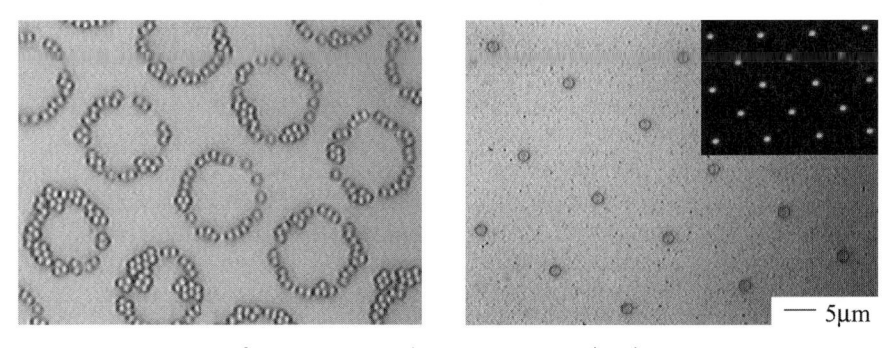

Fig. 10.20 OM images of 1.0 μm PSS particles on ring-patterned and dot-patterned PDAC layer atop the SPS outermost layer [24].

such templates can be easily and readily made, and applied to high-speed industrial processes such as reel-to-reel processing. Ultimately, multi-component arrays may be created utilizing the versatility of polymer chemistry in conjunction with alignment techniques to generate surfaces patterned with a variety of different functional groups that can act as templates for materials deposition.

10.7
Applications, Examples and Conclusions

Many functional polymers can be used in layer-by-layer assembly, including conducting and light-emitting polymers [33, 54], non-conjugated redox-active polymers, metal-polymer complexes [60–62], reactive polymers [63, 64], non-linear optically active polymers [64–66], liquid crystalline polyelectrolytes [67–69], dendrimers [70–72], and biologically derived polyelectrolytes such as nucleic acids, proteins, and polysaccharides [73–75]. The ability to pattern these systems onto numerous surfaces can lead to a large number of interesting applications. For example, inexpensive flexible plastic circuits can be fabricated using the polymer-on-polymer stamping approach to pattern conducting polymers on plastic substrates. Full-color LED displays could result from using the surface-sorting approach to make three-color pixel arrays with electroluminescent dyes. Incorporating both colloid particles and luminescent multilayers into the same microstructure would result in novel optical devices and photonic systems. Selective metallization using patterned polymer templates would be useful for the formation of electrodes and circuits in thin-film applications. The combination of selective metallization on glass or plastic and patterned polymeric multilayers can be used to create microreactors, microcolumns, and microfluidic channels in microelectrical mechanical systems (MEMS) and chemical plants on a chip. One example of a final complex structure device is depicted in Fig. 10.21. The colloid array and the underlying patterned multilayer form a 3D photonic system. Surface sorting can be used to make three multilayer-component arrays. The white block represents the multilayers that contain luminescent dye components and emit light under the colloids; the grid blocks represents metal electrodes formed by selective electroless metal plating and can be used to address the luminescent multilayers; the black blocks represent dielectric multilayer stacks, forming dielectric contrast with the environment. This structure might be used as a light source with incorporated waveguide.

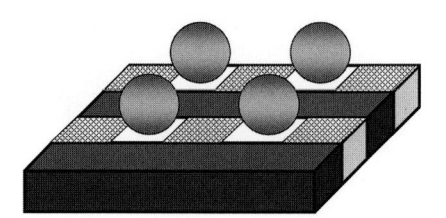

Fig. 10.21 Complex microfabricated structures consisting of luminescent multilayer arrays (white), metal electrodes (grid), dielectric multilayer blocks (black) and colloid arrays (gray).

Bioelectronics is a new emerging field, at the interface of molecular biology and submicrometer electronics [76]. Examples of bioelectronic devices include biosensors [77–79], bioactuators [80], photovoltaic cells [81, 82], DNA microchips [83, 84], and cellular automata [85]. The capabilities of biosensors and bioelectronic devices can be expanded if biomolecules can be deposited in microscopically defined arrays on surfaces [83, 86]. For example, information about spatial distribution, chemical gradient, multiple analytes, and simultaneously addressed internal control could be obtained. In addition, molecular electronic devices might consist of the supramolecular assembly of biomolecules, or interacting arrays of responsive molecules that are addressable using image process technology in real time. Undoubtedly, patterned polymeric multilayers have great potential in this area, as most biological molecules are charged and easily embedded into multilayers. Immobilization of proteins [74, 75], cholesterols [87], enzymes [88], and oligonucleotides [73] using layer-by-layer assembly and their sensing ability have been reported. Current collaborative efforts in our group with the Rubner, Cohen, Mayes, and Griffith groups involve using patterned multilayer templates and patterning approaches for cell co-culture and biosensing.

The efforts described here will have a large impact on the versatility and, ultimately, the commercialization of layer-by-layer techniques for a number of different applications, ranging from electro-optical thin films and devices to biological and chemical sensors, microelectromechanical systems (MEMS), and other areas in which patterned regions of polymer thin films are of interest. By investigating several new electroactive or optically functional systems, we have begun to explore new areas that show promise in the development of devices. For example, the use of this technique to allow the nanoscale manipulation and incorporation of components in a photorefractive film is novel, and shows a great deal of promise for the development of new photorefractive materials. The examination of electrochromic and photochromic properties in these films is particularly interesting for the development of smart windows and electrochromic displays. Finally, the ability to control the deposition of two or more sets of polymers with differing properties onto specific regions of a surface can provide simple and elegant approaches to the formation of complex organic systems or devices without expensive photolithographic steps for a broad range of applications.

10.8 References

1. Tien, J.; Terfort, A.; Whitesides, G. M. *Langmuir* **1997**, *13*, 5349–5355.
2. Serizawa, T.; Takeshita, H.; Akashi, M. *Langmuir* **1998**, *14*, 4088–4094.
3. Yonezawa, T.; Onoue, S.; Kunitake, T. *Chem. Lett.* **1998**, 689.
4. Schmitt, J.; Decher, G.; Dressick, W. J.; Brandow, S. L.; Geer, R. E.; Shashidar, R.; Calvert, J. M. *Adv. Mater.* **1997**, *9*, 61.
5. Schmitt, J.; Machtle, P.; Eck, D.; Möhwald, H.; Helm, C. A. *Langmuir* **1999**, *15*, 3256.
6. Aizenberg, J.; Braun, P. V.; Wiltzius, P. *Physical Review Letters* **2000**, *84*, 2997–3000.
7. Park, S. H.; Qin, D.; Xia, Y. *Adv. Mater.* **1999**, *10*, 1028–1032.

8 van Blaaderen, A.; Ruel, R.; Wiltzius, P. *Nature* **1997**, *385*, 321–323.
9 Lu, Y.; Yin, Y. D.; Gates, B.; Xia, Y. N. *Langmuir* **2001**, *17*, 6344–6350.
10 Donselaar, L. N.; Philipse, A. P.; Suurmond, J. *Langmuir* **1997**, *13*, 6018–6025.
11 Miguez, H.; Meseguer, F.; Lopez, C.; Mifsud, A.; Moya, J. S.; Vazquez, L. *Langmuir* **1997**, *13*, 6009–6011.
12 Rakers, S.; Chi, L. F.; Fuchs, H. *Langmuir* **1997**, *13*, 7121–7124.
13 Dimitrov, A. S.; Nagayama, K. *Langmuir* **1996**, *12*, 1303–1311.
14 Yin, Y. D.; Lu, Y.; Gates, B.; Xia, Y. N. *J. Am. Chem. Soc.* **2001**, *123*, 8718–8729.
15 Kim, E.; Xia, Y.; Whitesides, G. M. *Nature* **1995**, *376*, 581.
16 Denkov, N. D.; Velev, O. D.; Kralchevsky, P. A.; Ivanov, I. B.; Yoshimura, H.; Nagayama, K. *Nature* **1993**, *361*, 26.
17 Kumacheva, E.; Golding, R. K.; Allard, M.; Sargent, E. H. *Adv. Mater.* **2002**, *14*, 221.
18 Clark, S. L.; Montague, M. F.; Hammond, P. T. *ACS Symp. Ser.* **1998**, *695*, 206–219.
19 Clark, S. L.; Hammond, P. T. *Langmuir* **2000**, *16*, 10206–10214.
20 Jiang, X. P.; Clark, S. L.; Hammond, P. T. *Adv. Mater.* **2001**, *13*, 1669–1673.
21 Jiang, X. P.; Ortiz, C.; Hammond, P. T. *Langmuir* **2002**, *18*, 1131–1143.
22 Chen, K. M.; Jiang, X.-P.; Kimerling, L. C.; Hammond, P. T. *Langmuir* **2000**, *16*, 7825–7834.
23 Zheng, H. P.; Lee, I.; Rubner, M. F.; Hammond, P. T. *Adv. Mater.* **2002**, *14*, 569–572.
24 Zheng, H. P.; Rubner, M. F.; Hammond, P. T. *Langmuir* **2002**, *18*, 4505–4510.
25 Lee, I.; Zheng, H. P.; Rubner, M. F.; Hammond, P. T. *Adv. Mater.* **2002**, *14*, 572–577.
26 Kumar, A.; Biebuyck, H. A.; Whitesides, G. M. *Langmuir* **1994**, *10*, 1498–1511.
27 Hammond, P. T.; Whitesides, G. M. *Macromolecules* **1995**, *28*, 7569–7571.
28 Clark, S. L.; Montague, M. F.; Hammond, P. T. *Macromolecules* **1997**, *30*, 7237–7244.
29 Clark, S. L.; Montague, M.; Hammond, P. T. *Supramol. Sci.* **1997**, *4*, 141–146.
30 Jiang, X.-P.; Ortiz, C.; Hammond, P.T. *Langmuir* **2002**, *18*, 1131–1143.
31 Braun, P. V.; Zehner, R. W.; White, C. A.; Weldon, M. K.; Kloc, C.; Patel, S. S.; Wiltzius, P. *Adv. Mater.* **2001**, *13*, 721–724.
32 Brandow, S. L.; Dressick, W. J.; Dulcey, C. S.; Koloski, T. S.; Shirey, L. M.; Schmidt, J.; Calvert, J. M. *J. Vac. Sci. Tech. B* **1997**, *15*, 1818–1824.
33 Cheung, J. H.; Fou, A. F.; Rubner, M. F. *Thin Solid Films* **1994**, *244*, 985–989.
34 Fou, A. C.; Onitsuka, O.; Ferreira, M. S.; Howie, D.; Rubner, M. F. in *ACS Meeting Proceedings, Spring Meeting*: Anaheim, **1995**.
35 Liu, Y. J.; Wang, Y. X.; Claus, R. O. *Chem. Phys. Lett.* **1998**, *298*, 315–319.
36 Gao, M.; Richter, B.; Kirstein, S. *Adv. Mater.* **1997**, *9*, 802–805.
37 Hong, H.; Davidov, D.; Avny, Y.; Chayet, H.; Faraggi, E. Z.; Neumann, R. *Adv. Mater.* **1995**, *7*, 846–849.
38 Hong, H.; Davidov, D.; Tarabia, M.; Chayet, H.; Benjamin, I.; Faraggi, E. Z.; Avny, Y.; Neumann, R. *Synth. Met.* **1997**, *85*, 1265–1266.
39 Balanda, P. B.; Ramey, M. B.; Reynolds, J. R. *Macromolecules* **1999**, *32*, 3970–3978.
40 Li, D.; Ratner, M. A.; Marks, T. J. *J. Am. Chem. Soc.* **1990**, *112*, 1789–1790.
41 Bergstedt, T. S.; Hauser, B. T.; Schanze, K. S. *J. Am. Chem. Soc.* **1994**, *116*, 8380–8381.
42 Hanken, D. G.; Corn, R. M. *Israel J. Chem.* **1997**, *37*, 165–172.
43 Wu, A. P.; Lee, J.; Rubner, M. F. *Thin Solid Films* **1998**, *329*, 663–667.
44 Yoo, D.; Shiratori, S. S.; Rubner, M. F. *Macromolecules* **1998**, *31*, 4309–4318.
45 Shiratori, S.; Rubner, M. F. *Macromolecules* **2000**, *33*, 4213–4219.
46 Mandel, M. *Euro. Polym. J.* **1970**, *6*, 807–822.
47 Weyts, K. F.; Goethals, E. J. *Makromol. Chem., Rap. Comm.* **1989**, *10*, 299–302.
48 Velikov, K. P.; Durst, F.; Velev, O. D. *Langmuir* **1998**, *14*, 1148–1155.
49 Denkov, G. S.; Velev, O. D.; Kralchevsky, P. A.; Ivanov, I. B.; Yoshimura, II.; Nagayama, K. *Langmuir* **1992**, *8*, 3183–3190.
50 Velev, O. D.; Furusawa, K.; Nagayama, K. *Langmuir* **1996**, *12*, 2374–2384 and 2385–2391.

51 Chen, K.; Jiang, X.-P.; Kimerling, L.C.; Hammond, P.T. *Langmuir* **2000**, *16*, 7825–7834.
52 Jiang, X.-P.; Hammond, P.T. *Langmuir* **2000**, *20*, 8501–8509.
53 Jiang, X.P.; Zheng, H.P.; Gourdin, S.; Hammond, P.T. *Langmuir* **2002**, *18*, 2607–2615.
54 Cheung, J.H.; Stockton, W.B.; Rubner, M.F. *Macromolecules* **1997**, *30*, 2712–2716.
55 Fou, A.C.; Onitsuka, O.; Ferreira, M.; Rubner, M.F.; Hsieh, B.R. *J. Appl. Phys.* **1996**, *79*, 7501–7509.
56 Ferreira, M.; Rubner, M.F. *Macromolecules* **1995**, *28*, 7107–7114.
57 Ferreira, M.; Cheung, J.H.; Rubner, M.F. *Thin Solid Films* **1994**, *244*, 806–809.
58 Stockton, W.B.; Rubner, M.F. *Macromolecules* **1997**, *30*, 2717–2725.
59 Kaschak, D.M.; Lean, J.T.; Waraksa, C.C.; Saupe, G.B.; Usami, H.; Mallouk, T.E. *J. Am. Chem. Soc.* **1999**, *121*, 3435–3445.
60 Wu, A.; Yoo, D.; Lee, J.K.; Rubner, M.F. *J. Am. Chem. Soc.* **1999**, *121*, 4883–4891.
61 Laschewsky, A.; Mayer, B.; Wischerhoff, E.; Arys, X.; Jonas, A. *Ber. Bunsenges. Phys. Chem.* **1996**, *100*, 1033–1038.
62 Sun, Y.; Sun, J.; Zhang, X.; Sun, C.; Fang, Y.; Zhang, X. *Thin Solid Films* **1998**, *327/9*, 730–733.
63 Laschewsky, A.; Mayer, B.; Wischerhoff, E.; Arys, X.; Bertrand, P.; Delcorte, A.; Jonas, A. *Thin Solid Films* **1996**, *284/5*, 334–337.
64 Koetse, M.; Laschewsky, A.; Mayer, B.; Rolland, O.; Wischerhoff, E. *Macromolecules* **1998**, *31*, 9316–9327.
65 Roberts, M.J.; Lindsay, G.A.; Herman, W.N.; Wynne, K.J. *J. Am. Chem. Soc.* **1998**, *120*, 11202–11203.
66 Balasubramanian, S.; Wang, X.; Wang, H.C.; Yang, K.; Kumar, J.; Tripathy, S.K. *Chem. Mater.* **1998**, *10*, 1554–1560.
67 Fischer, P.; Laschewsky, A.; Wischerhoff, E.; Arys, X.; Jonas, A.; Legras, R. *Macromol. Symp.* **1999**, *137*, 1–24.
68 Passmann, M.; Wilbert, G.; Cochin, D.; Zentel, R. *Macromol. Chem. Phys.* **1998**, *199*, 179–189.
69 DeWitt, D.M. in *Chemical Engineering*; MIT, Cambridge, MA, **2002**.
70 He, J.A.; Valluzzi, R.; Yang, K.; Dolukhanyan, T.; Sung, C.M.; Kumar, J.; Tripathy, S.K.; Samuelson, L.; Balogh, L.; Tomalia, D.A. *Chem. Mat.* **1999**, *11*, 3268–3274.
71 Tsukruk, V.V.; Rinderspacher, F.; Bliznyuk, V.N. *Langmuir* **1997**, *13*, 2171–2176.
72 Anzai, J.; Kobayashi, Y.; Nakamura, N.; Nishimura, M.; Hoshi, T. *Langmuir* **1999**, *15*, 221–226.
73 Sukhorukov, G.B.; Montrel, M.M.; Petrov, A.I.; Shabarchina, L.I.; Sukhorukov, B.I. *Biosens. Bioelectron.* **1996**, *11*, 913–922.
74 Lvov, Y.; Ariga, K.; Ichinose, I.; Kunitake, T. *J. Am. Chem. Soc.* **1995**, *117*, 6117–6123.
75 Lvov, Y.M.; Lu, Z.; Schenkman, J.B.; Zu, X.; Rusling, J.M. *J. Am. Chem. Soc.* **1998**, *120*, 4073–4080.
76 Nicolini, C. *Thin Solid Films* **1996**, *284/285*, 1–5.
77 Adami, M.; Sartore, M.; Nicolini, C. *Biosens. Bioelectron.* **1995**, *10*, 155.
78 Nicolini, C.; Adami, M.; Erokhin, V.; Facci, P.; Sartore, M. *Sens. Actuators B* **1995**, *24*, 121–128.
79 Nicolini, C.; Erokhin, V.; Facci, P.; Guerzoni, S.; Ross, A.; Paschkevitsch, P. *Biosens. Bioelectron.* **1997**, *12*, 613–618.
80 Antolini, F.; Paddeu, S.; Nicolini, C. *Langmuir* **1995**, *11*, 2719–2725.
81 Janzen, A.F.; Seibert, M. *Nature* **1980**, *286*, 584.
82 Miyake, J.; Majima, T.; Asada, Y.; Sugino, H.; Ajiki, S.; Toyotama, H. *Mat. Sci. Eng. C* **1994**, *2*, 63–67.
83 Fodor, S.P.A.; Read, J.L.; Pirrung, M.C.; Stryer, L.; Lu, A.T.; Solas, D. *Science* **1991**, *251*, 767–773.
84 Mirzabekov, A.D. *Trends Biotechnol.* **1994**, *12*, 27–32.
85 Nicolini, C. *Molecular Bioelectronics*; World Scientific Publishing, London, **1996**.
86 Britland, S.; Perez-Arnaud, E.; Clark, P.; McGinn, B.; Connoly, P.; Moores, G. *Biotechnol. Prog.* **1992**, *8*, 155–160.
87 Ram, M.K.; Bertoncello, P.; Ding, H.; Paddeu, S.; Nicolini, C. *Biosens. Bioelectron.* **2001**, *16*, 849–856.
88 Ram, M.K.; Adami, M.; Paddeu, S.; Nicolini, C. *Nanotechnology* **2000**, *11*, 112–119.

11
Evolving Strategies of Nanomaterials Design

Sean Davis

11.1
Introduction

The complexity of form of biominerals has fascinated scientists for decades [1]. Even comparatively simple single-celled organisms such as protozoa produce intricate exoskeletons, with species-specific structures that imply an active role of the cell in the mineralization process (Fig. 11.1). The control of structure and morphology is exerted on length scales ranging from the nanometer to the micrometer. The realization that organisms can regulate properties such as the size, shape, phase, orientation and assembly of particles with high fidelity under ambient conditions has provided the impetus for multidisciplinary studies on the process of biomineralization [2]. These investigations have highlighted the importance of the organic/inorganic interface, in particular the influence of biomolecules and biostructures on the formation and deposition of mineral phases. They have also provided inspiration for the development of new strategies in materials production (biomimetics) [3].

At one extreme biomimetic studies are concerned with the direct mimicking of biomineralization. In biological systems the inorganic component is generally limited to insoluble salts of the commonly available elements. Thus, although over sixty different minerals have been identified, phases such as crystalline calcium

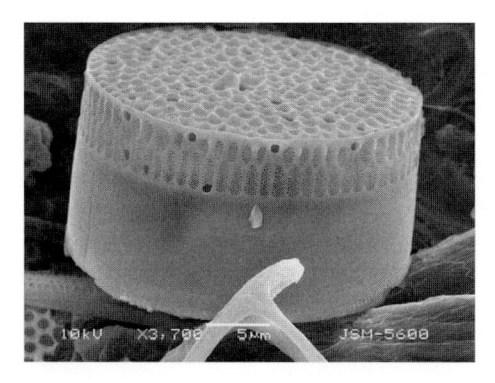

Fig. 11.1 Scanning electron microscopy image of the patterned silica exoskeleton of a diatom.

Colloids and Colloid Assemblies. Edited by Frank Caruso
Copyright © 2004 Wiley-VCH Verlag GmbH & Co. KGaA, Weinheim
ISBN: 3-527-30660-9

salts (carbonate and phosphate), iron oxides and amorphous silica predominate [4]. With these restricted inorganic resources organisms have evolved strategies to optimize the form of these materials for specific functions, such as skeletal support, protection and ion storage. The diversity of form of biominerals is a reflection of the abundance and variation of organic structure-directing agents rather than of the inorganic components.

The basic premise that a given material may have an improved range of properties and applications (function) if its form can be controlled has led to the development of bioinspired routes for introducing structural complexity into more technologically relevant materials [5]. In particular there has been increased interest in the use of preformed nanoparticles as building blocks for the construction of materials with higher-order architectures [6]. One of the main reasons for using colloidal particles as building blocks is that the physical and chemical properties can be predefined. Some of the most frequently used particles are gold (metallic), cadmium sulfide (semi-conductor), iron oxide (magnetite, magnetic), titania (photocatalytic) and silica (ceramic). New innovations in colloid synthesis and protocols for assembling them mean this is an active and continually evolving area of research. A number of generic strategies have been identified (e.g. self-, programmed and templated assembly) to form meso- and macroscopic nanostructured materials, which have collectively been termed nanotectonics [7].

In this chapter some of these approaches to the synthesis and assembly of nanoparticles are reviewed, in particular those which utilize biomolecules (lipids, carbohydrates, peptides, and nucleotides) and biomolecular structures. Section 11.2 discusses the conceptually simplest strategy; the use of a pre-organized template or scaffold to spatially define particle formation and/or deposition. Biotemplates have been used to produce a wide variety of nanostructured materials from discrete particles to hierarchically porous monoliths. Section 11.3 reviews how biomolecular interactions can be used to produce ordered structures. Specific molecular recognition properties can be used to subtly program the aggregation/assembly of nanoparticles or catalyze the formation of mineral phases.

11.2
Biotemplates

11.2.1
Nanohybrids

The mineralization of the interior of capsules is a common strategy used by biomineralizing organisms for ion storage, concentration and transport. Colloid and materials chemists have used self-assembled supramolecular structures extensively as templates to control the formation of nanostructured materials [8]. In general these approaches are used to control the size and/or morphology of the hybrid colloids produced. Aggregate structures of synthetic surfactants and copolymers such as micelles, microemulsions and lyotropic liquid crystalline phases have

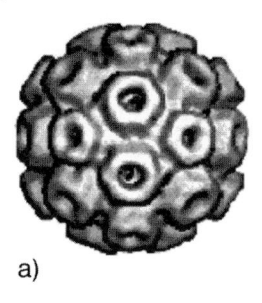

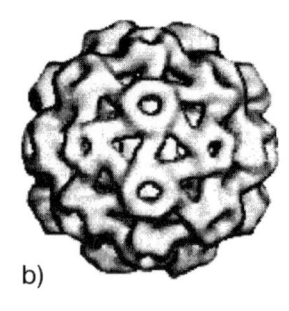

Fig. 11.2 Cryo-electron microscopy and image reconstruction of the cowpea chlorotic mottle virus: **a** Unswollen condition at low pH; **b** swollen condition at high pH (reprinted from [19a] by permission of *Nature*. Copyright 1998 Macmillan Publishers Ltd).

been used in the synthesis of nanoparticles and mesostructured materials. In natural systems the major components of membranes are proteins and lipids. Self-assembled structures of these biomolecules provide a diverse set of templates that can be utilized for the formation and deposition of nanoparticles, to produce bio-inorganic colloids.

For example a number of minerals have been precipitated in the interior of phospholipid vesicles. The physicochemical environment on the interior of these closed membranes can be different to the exterior. This was used as a means to preferentially synthesize magnetite inside the cavity [9]. Other materials such as aluminum oxide [10] and cadmium sulfide have also been prepared within vesicles [11]. The membrane does not act as a rigid barrier to control particle size, but regulates the volume of liquid, and hence the number of ions that can crystallize. Thus inorganic colloids with narrow size distribution can be produced provided that monodisperse unilamellar vesicles are used.

Similarly, the iron storage protein ferritin has been studied extensively as an archetypal biomimetic system. The protein consists of 24 subunits that form a spherical shell (external diameter 12.5 nm; internal diameter=8 nm) [12]. The protein shell contains hydrophilic ion channels and specific enzymatic sites which oxidize Fe(II) to Fe(III), thus facilitating the movement and concentration of iron ions. Iron is stored in the internal cavity in the form of the weakly crystalline iron oxyhydroxide ferrihydrite. The size to which the iron oxide precipitate can grow is restricted, thus very monodisperse ferrihydrite nanoparticles, encased in a biocompatible protein shell are produced under physiological conditions. This biosystem has been exploited with the aim of producing non-native cores of more technological interest. Iron sulfide cores are produced simply by in situ reaction of the native iron oxide core [13]. Alternatively the native core can be removed by reductive dissolution to form the apoprotein, allowing a more general approach to the production of a variety of inorganic phases within this nanoreactor, such as semiconductor particles [14] and uranium salts for neutron capture therapy [15].

The formation of magnetic phases within the biocompatible protein shell has been a specific goal. Much work has been devoted to the formation of magnetic iron oxide phases. Under appropriate conditions a crystalline magnetite core can be produced [16]. The protein shell acts to limit the maximum size to which the crystals grow (4500 Fe atoms). Nanoparticles of the inverse spinel magnetite of this size display superparamagnetic properties, and have potential applications in medical imaging [17]. More recently it has been demonstrated that cobalt ferrites can be synthesized within the shell, and these are currently being investigated for use in ultra-high-density storage media, amongst other things [18].

Although proving very successful both commercially and as a model system, ferritins provide a very specific example of nanoreactors. A greater variety of sizes and shapes are available using other, albeit not naturally mineralized, biocapsules. For example, the removal of nucleic acids from virus capsids leaves protein shells that can also function as templates for the formation of inorganic nanoparticles [19]. Although they have no specific channels for concentrating metal cations within their interior, many virions such as the cowpea chlorotic mottle virus (CCMV) can be chemically induced to undergo reversible structural changes (Fig. 11.2) [20]. At low pH (<6.5) CCMV virions have an external diameter of 28 nm and an internal diameter of 18 nm. Normally the cavity contains viral RNA, but the viral coat protein subunits can be purified and reassembled into empty capsids, with a positively charged interior cavity [21]. At pH>6.5 these shells swell causing the formation of 2 nm diameter pores in the wall which allow molecular exchange between the interior and exterior of the capsule. Thus, exposure of these capsids to tungstate ions at high pH, followed by lowering of the pH results in mineralization of the cavity. The decrease in pH causes the pores in the wall to close and also initiates the oligomerization of tungstate to the less-soluble paratungstate. However, the particles formed are larger in size (6.7 nm) than that predicted based only on the crystallization of entrapped ions, implying that the positively charged interior cavity also acts to induce polyoxometallate crystallization.

Complete structural rearrangements can also occur during mineralization as physicochemical conditions dictate the way in which subunits assemble. The bacterial enzyme complex lumazine synthase is normally spherical and similar in size to apoferritin (inner and outer diameters 7.8 and 14.7 nm) [22]. The 60 subunits form a structure in which there are positive and negative charged channels through the shell, and concentrations of negative charge on the interior surface, which make it susceptible to mineralization. Under non-physiological conditions the structure can rearrange into a larger (30 nm diameter) icosahedron containing 240 subunits [23]. This rearrangement occurs during iron oxide mineralization [24] although crystal growth does remain restricted to the interior of the capsids rather than the bulk (Fig. 11.3). At loadings of 17 iron atoms per subunit, lepidocrocite particles 10–15 nm in diameter are produced within some, but not all the capsids. Although these initial mineralization reactions are less specific than in ferritins, lumazine synthase has two types of wall channel (positive and negative), and it is anticipated that further studies will enable a greater range of chemical reactions to be conducted within the core.

Most of the aforementioned strategies result in the formation of isotropic hybrid particles (inorganic core/bioshell). The presence of a biological interface provides additional functionality to the colloid. The bioshell can confer stability and biocompatibility, and can also be modified with other biomolecules for targeting

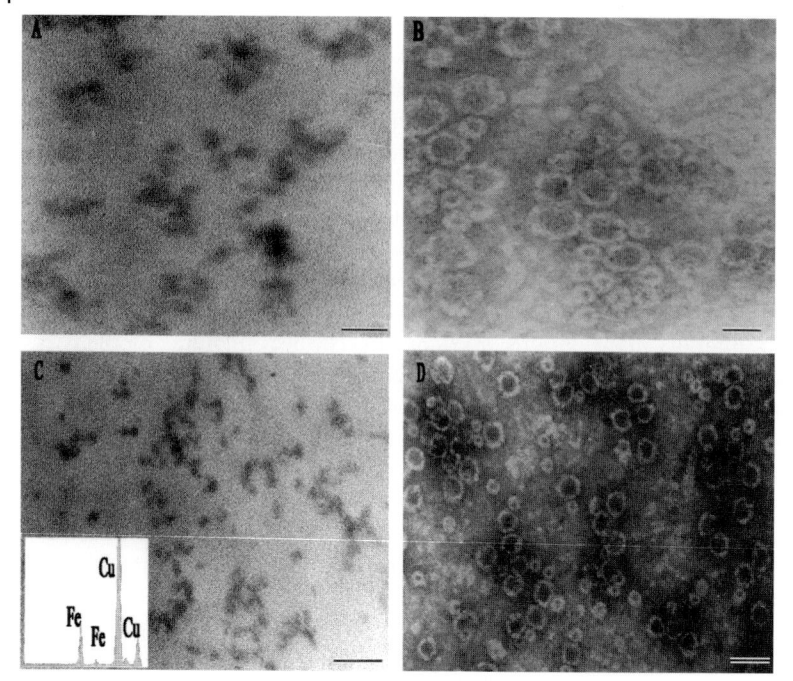

Fig. 11.3 Transmission electron micrographs of iron oxide-mineralized lumazine synthase: **A** and **C** Unstained samples prepared with a loading of 5 and 17 Fe atoms, respectively, per subunit showing the presence of iron oxide nanoparticles, scale bars = 30 and 50 nm respectively; **B** and **D** corresponding negatively stained images of the samples shown in **A** and **C**, respectively, showing the presence of capsids with a bimodal size distribution, scale bars = 30 and 50 nm (reprinted with permission of [24]).

and conjugation. The primary properties of the hybrid are dictated by the habit of the mineral core, in essence the type and arrangement of the atoms. Other properties such as magnetic, catalytic, semiconducting and optical are governed by the size and shape of the particles. Some of these aspects, and other approaches to producing such particles are discussed further in Section 11.3.1.2.

The self-assembly of biomolecules can also lead to the formation of structures with anisotropic shapes such as tubes and fibers. If sufficiently charged these structures provide a suitable interface for mineral formation or adsorption. Such high-aspect-ratio hybrid nanocolloids represent amongst other things potential precursors for higher-order assembly. For example the chirality of certain phospholipid headgroups means that under appropriate conditions the bilayer sheets can curl up to produce tubules or helical ribbons [25]. Ceramic fibers were produced by exposing pre-assembled lipid tubules to dispersions of silica colloids, which aggregated on the surface to form a coherent silica coating [26]. Calcination removed the template to produce closed hollow silica tubes. These were considered for various potential applications such as fiber reinforcements. Subsequently various other mineral coatings were produced on similar templates, such as magnetic and non-magnetic iron oxides on self assembled tubules from galactocerebrosides [27]. In these systems although the charged external surface provides a suitable interface for mineralization the 10–30 nm central lumen remains unmineralized. In general it is difficult to control the texture of the coating, with mineralization occurring non-specifically across the membrane surface. However, patterned arrays of gold nanoparticles have been produced on helical coiled ribbons of phospholipid bilayers [28]. Deposition of the gold colloid occurred preferentially on the headgroups exposed along the helical pitch.

The negatively charged phosphate backbone of DNA can also be used as a non-specific interface for mineralization. Favorable electrostatic interactions are used to concentrate ions or nanoparticles at the surface. For example, CdS nanoparticles are produced by reacting suspensions of DNA and Cd^{2+} ions with Na_2S [29]. The size of the semiconductor particles produced is sensitive to the sequence composition [30]. Semiconductor/nucleotide hybrids have also been produced using preformed CdS (3 nm diameter) nanoparticles [31]. The particles were stabilized with thiocholine, and therefore had the required positive surface charge, to enable deposition. Similarly positively charged lysine-modified gold particles have also been assembled onto DNA [32]. Addition of DNA to nanoparticle dispersions produces close packed linear arrays. No such structures are observed when negatively charged gold particles are used.

Virus capsids have been used as substrates for the formation of bioinorganic nanotubes. A large diversity of virus structures exist, including those based on the helical assembly of protein subunits producing tubules [33]. Tobacco mosaic virus (TMV) is a classic virus particle and its structure has been extensively studied by electron microscopy. The virion has a tubular morphology and is 300×18 nm in size with a 4 nm lumen. 2130 protein subunits assemble around a single strand of RNA to produce the structure. The stability of the capsid enables a variety of chemical reactions to be carried out on the charged external surface [34]. The viral particles have been coated with cadmium sulfide, lead sulfide, silica and iron oxides. Although no evidence was found for mineralization of the interior channel, this remains a goal as it is hoped this will lead to the production of protein-coated nanowires, suitable for further assembly. The fact that the interior channel can be stained demonstrates that small inorganic clusters can access the lumen. Interestingly the conditions of silica and iron oxide precipitation also led to hybrid structures with larger aspect ratios than TMV alone. This is due to "head to tail" assembly of TMV particles under the reaction conditions. Other higher-order hybrid structures were reported when high concentrations of the TMV template were mineralized [35]. Under these conditions a nematic liquid crystal phase is formed. Either extended mesostructures or mesostructured nanoparticles are produced depending on the relative concentration of silane precursor added.

11.2.2
Patterned Templates

The porous exoskeleton of the diatom represents a biological archetype for the formation of patterned materials. In particular the species-specific patterns often show hierarchical pore structures. From a structural materials perspective there is considerable interest in preparing framework materials with controlled pore size and distribution. The use of templates is pervasive throughout structural inorganic as well as bioinorganic chemistry. The simplest strategy for producing materials with extended structure is to use pre-formed templates which themselves have a higher-order architecture; in essence all the structural information is pre-programmed and just has to be replicated in inorganic form. This approach has proved particularly useful for producing macroporous materials (pore size > 50 nm) that have potential applications such as in catalysis [36] and separation [37]. In addition synthetic protocols have recently been established to prepare nanophase dispersions of crystalline zeolites [38]. These molecularly templated silicas and aluminosilicates can be used in the preparation of materials with hierarchical porosity.

Recently colloidal crystals have attracted particular attention as templates for the preparation of three-dimensionally ordered macroporous (3DOM) materials (inverse opals) [39]. This class of materials also shows interesting optical and photonic crystal properties. The general strategy involves the concentration of monodisperse colloidal spheres (polystyrene, poly(methyl methacrylate)) and silica into close-packed structures, which resemble natural opals. Various strategies have been used to infiltrate the voids in these static template structures with inorganic materials. However, using preformed nanoparticles offers some advantages, in particular reducing shrinkage on template removal. Metallic [40], ceramic [41], semiconducting [42] and microporous [43] nanoparticles have all successfully been used to produce 3DOM structures in this way. Although the inverse opals are very regular, the actual microarchitecture of the pore structure is limited. Hence, extended templates of a biological origin have attracted attention because of the variety of structures available.

11.2.2.1 Bacteria
Bacterial cells have a number of properties that make them amenable as templates. Most biological surfaces, including bacteria [44], have a predominantly negative surface charge. Thus the bacterial surface can act as a site for nucleation and mineral deposition, albeit poorly controlled and unspecific. Indeed, the formation of fossilized bacteria is associated with the mineralization of the external surface of the cells.

For example, *Bacillus subtilis* has a cylindrical-shaped cell 0.8 µm in diameter and 4 µm in length. This cell can be engineered in order to suppress cell separation. Thus the cells can be grown, such that they assemble end on end to produce multicellular filaments 0.5 µm in diameter. The high aspect ratio of these filaments results in them close packing together pseudo hexagonally on drawing through a fluid/air interface, producing a bacterial thread [45].

Electrostatic interactions can be used to infiltrate this microstructured template with nanoparticles. Good infiltration is achieved with dispersions of nanoparticles with a negative surface charge (silica [46], zeolite [47], magnetite [48]). Under conditions of high pH the bacterial thread swells owing to repulsive electrostatic interactions, increasing the void volume within the structure. On drying the thread, the component multicellular filaments regain the original close-packed structure, with the inorganic particles replicating the interfilament spaces in the form of a continuous inorganic framework. In contrast, low pH dispersions of titania (positive surface charge) cause no swelling of the bacterial superstructure, and deposition is restricted to the surface of the macroscopic thread [48].

Removal of the bacterial template by calcination produces stable macroporous replicas for silica-based materials. Indeed a hierarchical fibrous silica containing ordered macropores and micropores was produced from colloidal zeolite (TPA-silicalite-1) [47]. Furthermore, a similar hierarchical structure, consisting of mesopores and micropores, was obtained by calcination of a bacterial structure infiltrated with an MCM-41 precursor solution [46].

11.2.2.2 Plants

Natural multicellular structures are also potential templates for ordered materials. An almost limitless number of patterned microstructures are available as templates and they are also environmentally benign. For example, wood has attracted recent attention as a preorganized ordered array of plant cells. The structure of both hardwood (poplar) and softwood (pine) has been replicated using surfactant-templated sol–gel mineralization [49]. A positive replica is produced as the silicate species penetrate and mineralize the cell walls rather than the void space, presumably aided by the surfactant. The action of the surfactant as a porogen is also critical, as it allows pathways for decomposition products to leave during calcination. Attempts to mineralize without surfactant resulted in structural collapse on calcination.

Hierarchical porous zeolitic materials have also been produced from cedar wood tissue [50]. Preformed silicalite-1 nanoparticles were absorbed onto the wood structure, which had been premodified with a cationic polyelectrolyte (Fig. 11.4). The layer of zeolite particles was then used to seed secondary overgrowth from clear zeolite precursor solution to produce the composite. Calcination produced arrays of hollow zeolite fibers or rods, depending on the processing conditions, with diameters of 10–20 µm commensurate with the original template cells. Finer microstructural details of the cells were also replicated. Similar results were also obtained with bamboo grass although the template has polydisperse pores (2–20 µm).

Interestingly plant tissue, such as grasses often contain significant natural silica deposits. Ultimately it may be possible to engineer plants to over silicify or tolerate higher silica concentrations such that natural composites could be grown then

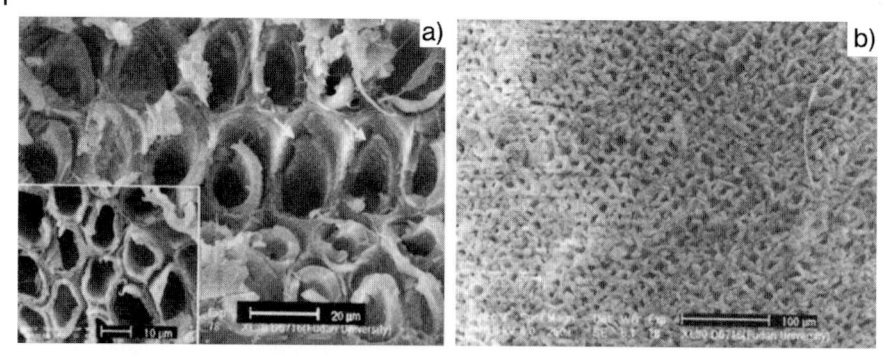

Fig. 11.4 Scanning electron microscopy images of: **a** Cedar/zeolite composite, inset original cedar cells; **b** cross-section of self-standing porous zeolite replica (reproduced with permission of [50]).

processed. For example the diatoms are unicellular plants encased in a patterned silica shell. These porous shells, which are the major component in certain earths and oozes, have been used as preformed and patterned inorganic templates. The natural shells, which are amorphous silica, have been recrystallized into zeolites, providing a comparatively cheap route to hierarchical zeolites [51].

11.2.2.3 Biominerals

Similarly other inorganic bioskeletons have been used in the production of ordered, macroporous structures [52]. The skeletal plates of sea urchins are single crystals of magnesium calcite. The defect-free bicontinuous structure is reminiscent of certain binary surfactant/water phases [53]. They offer the advantage over colloidal crystal templates of a larger void space available for filling (50% versus

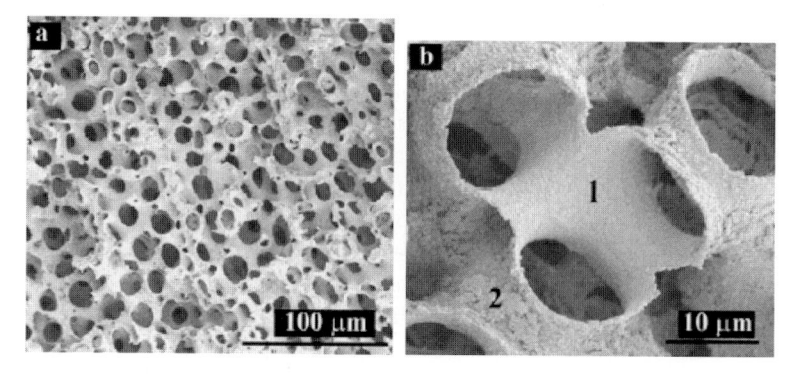

Fig. 11.5 Scanning electron micrographs of a gold replica of an echinoid skeletal plate. 1 and 2 in **b** indicate smooth inner and rough outer surfaces of the templated gold structure (reprinted with permission of [52a]).

26%). Also, unlike the majority of synthetic templates, the inorganic phase can be removed by comparatively mild acid treatment rather than calcination. The deposition of gold colloid onto the bioinorganic template can be controlled between a thin coating and full void space infiltration (Fig. 11.5). As well as affecting the macro structure, with either a positive or negative replica being produced, the variation of wall thickness also influences mechanical stability.

Cuttlefish bone has also been used as a template in the preparation of macroporous materials. In this case the calcium carbonate component was dissolved by acid treatment to leave the chitinous scaffold. This was used as a substrate for the deposition of sol–gel-derived silica [54].

11.2.2.4 Starch Sponges

Various synthetic polymer gels with complex macroporous structures and reversible swelling properties have been used as templates for mineral deposition. For example bicontinuous polymeric gels have been used as templates for the in situ mineralization of magnetite [55] and polymerization of titanium alkoxide [56]. Functionalized copolymer gels, with tunable chemical properties have also been infiltrated with preformed magnetite (Fe_3O_4) or titanium dioxide (TiO_2) nanoparticles [57]. These provide a generic strategy in that both the micro architecture and surface properties can be tailored to optimize loading with a range of colloidal particles. Subsequent calcination removes the polymer, producing macroporous monoliths. However, attempts to produce hierarchical porous silica materials using zeolite particles with these templates was less successful [58].

This led to the development of another macroporous biopolymer template, more amenable to zeolite infiltration, namely starch sponges [59]. These materials offer other advantages such as being readily available, low cost and environmentally benign. The biopolymer sponges (Fig. 11.6a) are prepared by freezing and thawing a starch gel. The pore sizes and structure can be controlled by varying the starch content of the initial gel, with porosity decreasing with increased starch content. Immersion of these sponges in zeolite suspensions did not result in loss of structure of the template (Fig. 11.6b). However, calcination of the composite to remove the molecular and polymer templates successfully produced a microporous/macroporous silica monolith albeit with some disorganization with respect to the template (Fig. 11.6c).

Monoliths and thin films can also be prepared by in situ blending of the nanoparticles with starch gel solutions. These can then be cast as thin films or shaped monoliths. Again calcination can be used to produce structures with porosity which depended on the initial starch content.

The use of suitably sized and structured biotemplates facilitates the production of a wide spectrum of nanostructured materials. In general non-specific electrostatic interactions drive the formation or assembly of the nanophase. This means that often the controlled deposition is produced by manipulating reaction conditions rather than being an inherent property of the template. For example, the nanoparticle aggregates are "amorphous", in that they are randomly arranged with

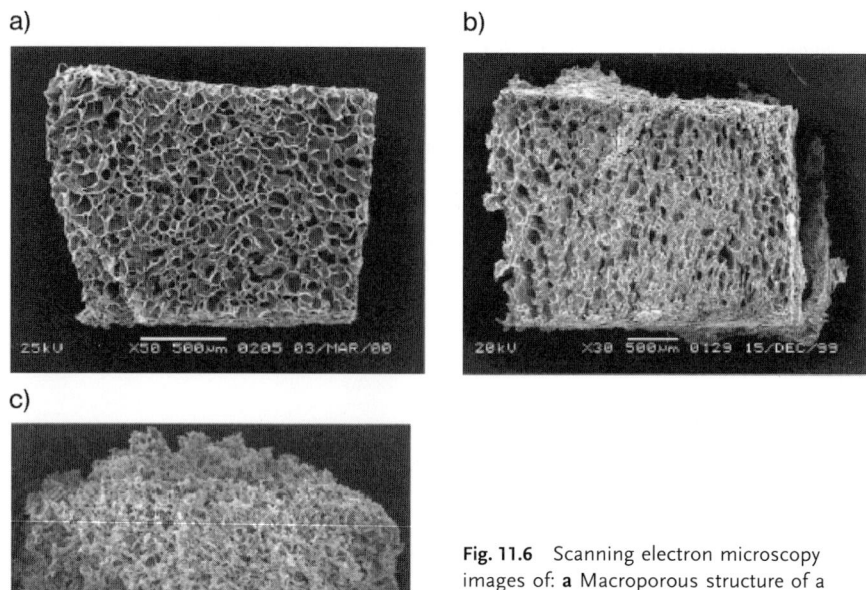

Fig. 11.6 Scanning electron microscopy images of: **a** Macroporous structure of a starch gel sponge; **b** starch-silicalite porous composite; **c** macroporous monolith of silicalite prepared by calcination of a starch-silicalite composite (reprinted by permission of [59]. Copyright 2002 American Chemical Society).

respect to each other. This is resolved to some extent using suitable engineered surfaces that promote regiospecific binding.

11.2.2.5 Structured Interfaces

Prepatterned substrates can also be used to produce organized 2D arrays of nanoparticles (substrate engineering). For example the technique of microcontact printing pioneered by the group of Whitesides can produce structured 2D interfaces with well-ordered micrometer-scale domains of hydrophilic functional groups [60]. These have been used to control the regiospecific crystallization of aligned $CaCO_3$ crystals [61]. The self-assembly of S-layer proteins from bacteria can be used to produce analogous interfaces at the mesoscale [62].

Again as with other biological templates there is a tremendous diversity of structures available. The protein monomers self-assemble on various substrates to produce crystalline lattices with oblique, square or hexagonal lattice symmetry and unit cell spacings of 3–30 nm. Arrays of CdS [63], Au [64] and Pt [65] nanoparticles have been produced by site-specific nucleation, with the symmetry of the nanoparticle array dictated by the bacterial species. Recently it has been demonstrated that preformed gold particles can be regiospecifically absorbed onto these templates [66]. The S-layer protein of *Deinococcus radiodurans* forms a hexagonal crystal lattice

(18 nm), with regularly arranged cationic binding sites. Well-ordered arrays are only produced using negatively charged gold particles smaller in size than the lattice constant. Positively charged or larger particles (20 nm) cannot bind effectively to the template, and show no periodic structure. The arrays formed from both 5 and 10 nm particles show the same periodic structure and interparticle spacing, confirming that registry with the template is the driving force for pattern formation rather than other effects. The studies also indicate that in any population of particles the smaller ones tend to bind preferentially at the expense of larger particles.

In biomineralizing organisms even more specific molecular scale interactions are generally involved in the process of mineral formation and assembly. In the next section molecular recognition at the organic/inorganic interface and programmed assembly are discussed.

11.3
Molecular Recognition

There is considerable interest in the formation of well-ordered arrays of nanoparticles. Crystalline arrays of nanoparticles can be formed by the simple concentration of hydrophobic nanoparticle dispersions to form superlattices analogous to the formation of colloidal crystals from silica and latex colloids [67]. Alternatively bifunctional coupling ligands can be used to cross-link particles [68]. Thus well-ordered arrays can be prepared from monodisperse nanoparticles given favorable interparticle interactions. Such approaches to highly symmetric, single-component structures can be useful for preparing materials such as high-density magnetic storage media. However, there is a desire to be able to produce more complicated 2D and 3D topologies of hybrid multicomponent materials from more flexible strategies for nanoelectronic applications. As such most studies on these "bottom up" approaches have concentrated on metallic and semiconducting particles.

11.3.1
Programmed Assembly

The most significant advances in nanoparticle organization have been made in the field of programmed assembly. Rather than directly mimicking biostructures this involves the exploitation of the specific molecular recognition properties of biomolecules to induce nanoparticle assembly. For example mesostructured materials have been prepared using antibody–antigen coupling [69, 70], biotin-streptavadin conjugation [70–72] and DNA hybridization [73–75].

11.3.1.1 **Proteins**
The potential of using specific antibody/antigen reactions to cross-link nanoparticles was demonstrated by Shenton et al. [69]. The procedure is relatively simple as it involves well-established techniques routinely used in, for example immunocyto-

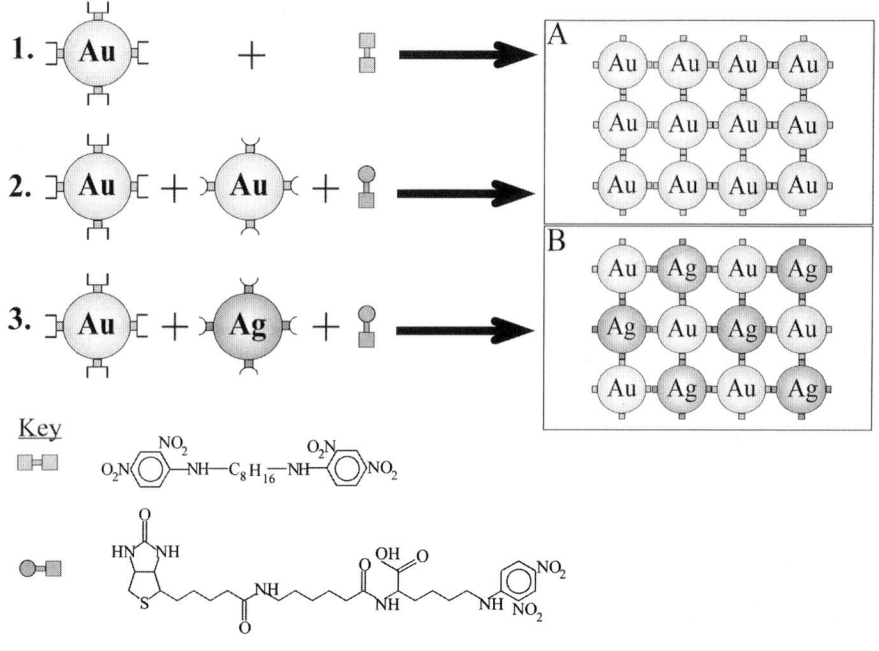

Fig. 11.7 Schematic diagram showing approaches to the antibody–antigen cross-linking of inorganic nanoparticles. (1) Gold particles coated with anti-DNP IgE antibodies linked by DNP-DNP antigen. (2) Gold particles coated with either anti-DNP IgE or anti-biotin IgG antibodies linked by DNP-biotin bivalent antigen. (3) 1:1 mixture of Au/anti-DNP IgE and Ag/anti-biotin IgG linked by DNP-biotin antigen (reprinted with permission of [69]).

chemistry. In particular gold nanoparticles have been extensively used as markers to allow the investigation of biomolecular interactions by microscopy. Antibodies are routinely adsorbed onto gold particles by isoelectric focusing to produce stable dispersions of conjugates.

Adding a suitable bifunctional, antigenic cross-linker can induce aggregation of the particles. For example, the addition of a synthetic antigen (bis-N-2,4-dinitrophenyloctamethylene diamine) to gold particles coated with anti-DNP IgE (DNP, dinitrophenyl) results in the formation of close-packed aggregates (Fig. 11.7). A key advantage of these biomolecularly based systems is the ability to form heterostructures. This can be done to a limited extent with colloids with a suitable relative size ratio [76]. However, the structural diversity is limited to simple close packing and it is difficult to prepare more complex architectures or those containing more than two components. Silver particles were coated with anti-biotin IgG, then mixed 1:1 with anti-DNP gold conjugates. Aggregation only occurs on addition of a hetero-Janus antigen, consisting of biotin and DNP on either end of a flexible C_{19} backbone.

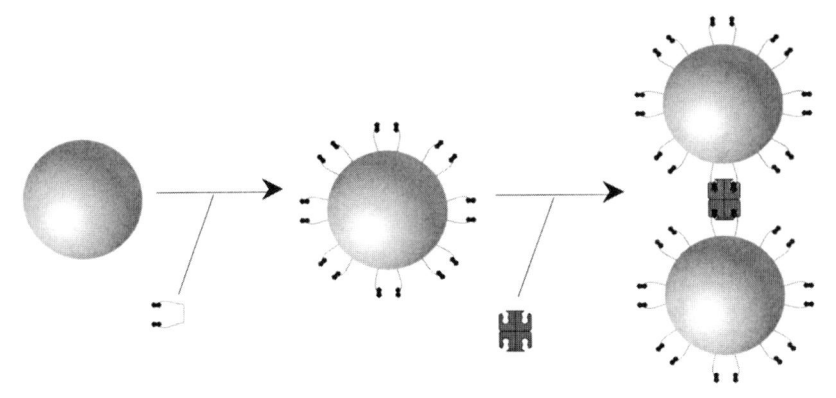

Fig. 11.8 Schematic diagram of the streptavidin-biotin-induced cross-linking of gold nanoparticles (reprinted with permission of [70]).

The biotin/streptavidin coupling system has a very high binding constant ($K_a > 10^{14}$ mol^{-1} dm^3), which can be used to drive particle assembly. Connolly and Fitzmaurice studied gold nanoparticles conjugated with a disulfide biotin analog (DSBA) [71]. The addition of the protein streptavidin caused diffusion-limited aggregation, as indicated by dynamic light scattering measurements (Fig. 11.8). Evidence was also provided by a visible color change in the dispersion (red→purple) and electron microscopy imaging of aggregate structures in which the interparticle distance of 5 nm was commensurate with the use of streptavidin.

One of the limitations with these systems is that conjugate formation is driven by thiol adsorption onto the particle surface. This works well with gold particles but for other colloids it is often necessary to tailor the coupling group or the particle interface. Such an example is provided by the protein ferritin, where the protein surface acts as a generic interface that can easily be modified with conjugates using standard procedures. For example the lysine residues on the external surface of ferritin can be biotinylated, and the particles cross-linked by adding streptavidin [72]. In fact there is an optimum ratio of ferritin to streptavidin of 1:6. Below this ratio conjugate formation is slower, and above these problems occur, with streptavidin effectively binding to all biotin on ferritin and preventing interparticle aggregation.

Various factors need to be addressed such as size and shape of particles as well as length, flexibility, number and orientation of binding molecules per particle to enable the rational design of ordered superstructures, with architectures not limited to simple close-packed geometries. Such problems are beginning to be addressed, most notably in nucleotide-driven assembly.

11.3.1.2 Nucleic Acids

Currently, some of the most elegant studies are those driven by DNA hybridization. The self-assembly of complementary DNA strands into a double helix is one

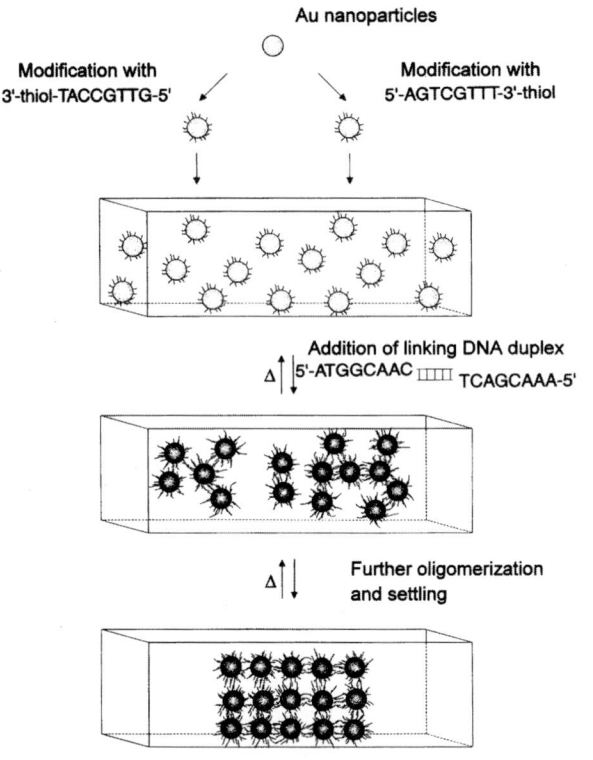

Fig. 11.9 Schematic diagram showing the DNA-based aggregation of gold colloids into extended structures (reprinted [73] by permission from *Nature*. Copyright 1996 Macmillan Publishers Ltd).

of the most important biomolecular recognition processes known. The production of complex organic DNA structures has been pioneered by Seeman [77]. Factors such as the specificity of the base pairing and its reversibility (self annealing structures) have contributed to the interest in DNA-mediated assembly of nanoparticles. In addition, recent advances in biotechnology that allow oligonucleotides to be prepared synthetically should enable a variety of multicomponent periodic structures to be engineered. However, unlike antibody-coated nanoparticles for which there was extensive "prior art", little work had been done until comparatively recently on preparing nanoparticle DNA conjugates. Various approaches have now been developed to use DNA as an intelligent cement to enable the processing of inorganic nanoparticle building blocks into periodic mesostructures.

The classic approach adopted by Mirkin and co-workers was to prepare two sets of gold particles (13 nm) coated with non-complementary oligonucleotides [73]. The oligonucleotides prepared contained a thiol group at one end, which facilitated binding to the gold surface. The coating of oligonucleotides was sufficient to stabilize the particles against irreversible, non-specific interactions as well as providing enough strands projecting from the surface to enable duplex formation.

11.3 Molecular Recognition

Thus, mixtures of these suspensions remained stable until a DNA duplex molecule with sticky ends, each of which was complementary to one of the functionalized nanoparticles, was added. This caused a slow aggregation due to DNA hybridization, with a corresponding color change of the suspension (red→blue) due to a shift in the plasmon band. Ultimately 3D aggregates were produced with an interparticle separation of 6 nm, commensurate with the length of the duplex linker (Fig. 11.9). Significantly it was also demonstrated that the aggregation process is reversible for this system. Heating to 80 °C, i.e. above the melting temperature of the duplex, causes disaggregation and a switch back to the characteristic red color of a stable dispersed gold colloid suspension [78]. These systems have been developed as probes for detecting DNA [79].

This system has also been used to prepare extended periodic heteroaggregates. The heterostructure can be composed of particles of different size [80] and/or composition [81]. For example, two sets of non-complementary oligonucleotide-gold particle conjugates with different sizes (31 and 8 nm diameter) were prepared. In the absence of cross-linker no regular structures are produced. When the appropriate DNA cross-linker is added to a mixture (eightfold molar excess of smaller particles) extended structures with regularly alternating particle sizes are produced. Interestingly, with a 120-fold excess of the 8 nm particles they tend to just coat the 31 nm particles and extended aggregates are not observed.

Analogous to the antibody/antigen system, binary networks have been produced using gold and CdSe/Zns semiconductor particles [81]. The key step was the coupling of the thiol-functionalized oligonucleotides to the quantum dot surface. The nature of the particle synthesis results in semiconductor particles that are capped with ligands, which impart a hydrophobic surface. Thus to prepare DNA-QD conjugates involved changing the interfacial properties so they were hydrophilic in nature. This was accomplished by reaction with 3-mercaptopropionic acid, followed by deprotonation with 4-(dimethylamino)pyridine and dispersion in water. Surface modification by thiol-modified oligonucleotides was then facile, and the particles retained size-dependent fluorescent properties in the conjugate. Hybrid assemblies of non-complementary gold (13 nm) and quantum dot (3.2 nm CdSe core) conjugates were produced only when the appropriate complementary cross-linker was added to mixtures. Another recently reported method to confer water-soluble properties involves the silanization of the semiconductor nanoparticle surface [82]. As well as the potential use in the construction of nanomaterials these DNA/QD conjugates also have the flexibility to be useful fluorescent probes to target and mark various biomolecules.

Rather than producing extended networks an alternative approach has been developed to produce oligomers of nanoparticles (Fig. 11.10) [74]. Again initial studies were concerned with preparing gold oligonucleotide conjugates. To enable the production of "nanocrystal molecules" small gold clusters (1.2 nm) were conjugated with single thiol-functionalized oligonucleotides, as opposed to the large number necessary for extended network formation. In this regard the oligonucleotide is used solely to direct site-specific hybridization events, with other molecules conferring the colloid stability. These were then assembled onto a single-stranded

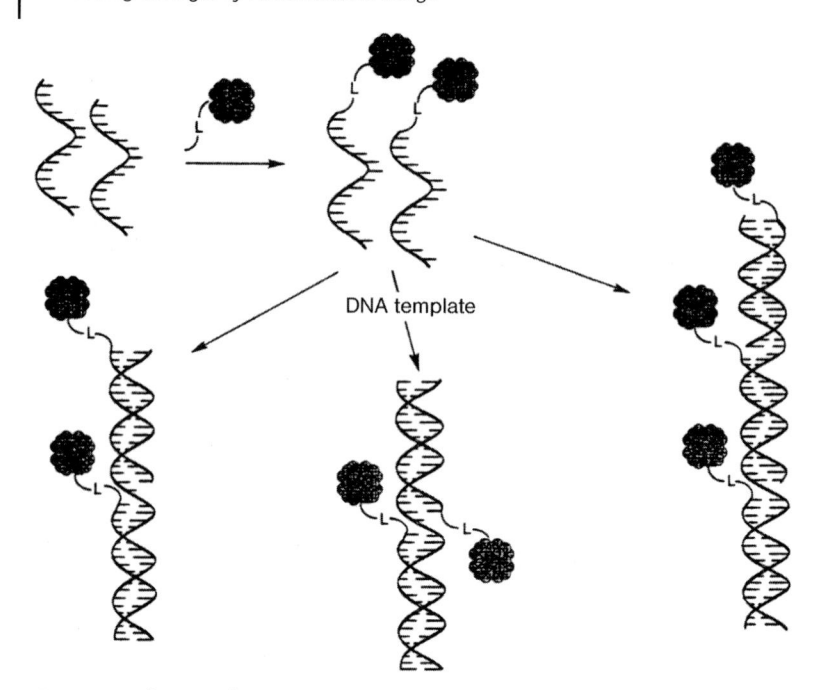

Fig. 11.10 Schematic diagram showing gold nanoparticle assembly into dimers and trimers based on Watson-Crick base-pairing (reprinted [74] by permission of *Nature*. Copyright 1996 Macmillan Publishers Ltd).

template, specifically where complementary sequences exist, to produce gold nanoparticle dimers and trimers.

Subsequently, various simple heteroaggregates have been produced using the same methodology from 5 and 10 nm diameter gold particles [83]. DNA/gold conjugates could be directly hybridized into dimers and trimers by combination with either free single strands containing complementary sections or alternatively complementary strands conjugated to another nanoparticle. Alternatively, two gold particles were attached at either end of a thiolated double-stranded DNA molecule. Reaction with a complementary gold/single-stranded DNA conjugate produced a trimer with all particles aligned along one face of the double helix. The interparticle separation determined from TEM images correlated with the modifications in number of base pairs introduced to space particles. However, there was some variation in spatial control due to fluxionality of the molecules induced by the non-rigidity of the interconnects.

The potential of combining protein and DNA interactions to produce a variety of novel systems has been demonstrated [75]. Six conjugates of single-stranded DNA and the tetrameric binding protein streptavidin with six different base sequences were prepared by covalent coupling. Biotinylated gold clusters (1.4 nm diameter) were then aggregated with the streptavidin component and the metal protein structures could then be site-specifically assembled onto a single RNA strand containing all six complementary base sequence sections. Furthermore, it was demonstrated that other functionality could also be introduced in the hybrid structure by coupling the DNA–streptavidin to biotinylated antibodies.

It seems reasonable to expect that more complicated architectures may be available from building blocks with discrete numbers of potential interconnects [84]. In the case of oligonucleotides and gold particles electrophoretic methods have been used to separate conjugates with precise amounts of oligonucleotides attached per crystal. This presents the possibility of using these particles as discrete structural units for building higher-order structures. For this to be fully realized the orientation of the connects with regards to the nanoparticle will also need to be considered. For example, to create linear arrays one needs the connectors to be at 180° to each other.

The shape of the nanoparticles is also a key feature, which can dictate the structure of the aggregates. Recently there has been much interest in the formation of high aspect ratio colloidal particles. Rod-like particles have a natural tendency to form liquid crystalline phases. Amphiphilic molecules such as surfactants and block-copolymers form lyotropic liquid crystalline structures at high concentrations. Such mesostructures have been used to template the deposition of silica to form mesoporous materials such as MCM-41 [8].

New syntheses have recently been developed that allow the production of monodisperse metallic and semiconducting nanorods in particular [85]. These are interesting for potential uses as interconnects in nanodevices. For example semiconducting particles show both size- and aspect ratio-dependent variation in properties [85b]. Thus in much the same way that surfactant-capped isotropic semiconductor particles form crystalline structures when concentrated, CdSe nanorods form liquid crystalline structures when concentrated suspensions are produced by solvent (cyclohexane) evaporation [86]. Similarly colloids prepared in microemulsions spontaneously self-assemble into linear arrays and superlattices depending on shape and concentration of particles [87].

Anisotropic nanoparticles have also been assembled using more specific interactions such as DNA hybridization [88]. Aqueous suspensions of gold nanorods (with residual surfactant double layers) were functionalized with oligonucleotides to produce either complementary or non-complementary conjugates. The duplexation of mixtures of complementary gold–DNA conjugates was thermally reversible. However, aggregation of the non-complementary conjugates induced by addition of a strand half complementary to each was non-reversible, because non-specific interactions dominated in the structure. The initial structures formed in both systems consisted of chains of rods packed side by side, which were stacked in three dimensions.

For many applications, including nanoelectronics, it is necessary to assemble particles onto 2D substrates. Nanostructured thin films have been prepared by various techniques such as layer-by-layer assembly of polyelectrolytes [89], self-assembled monolayers [90], Langmuir-Blodgett films [91] and using bifunctional organic molecules [92]. Larger gold wires (up to 2 µm in width and 6 µm in length), produced by evaporation into anodized alumina [93], can be modified with single-stranded DNA and assembled onto surfaces bearing complementary strands [94]. This synthesis procedure has the particular advantage that controlling the processing allows regiospecific modification of the nanowires. For example, unprotected tips of gold nano wires can be specifically functionalized with single-stranded DNA or for bimetallic wires competitive adsorption can be used to derivatize gold domains.

DNA-based assembly of nanoparticles has also been extended to form mono- and multilayered structures analogous to the layer-by-layer assembly of polyelectrolytes on substrates [95]. Thus, stepwise assembly involves hybridization of strands tethered to gold particles each half complementary with a linker molecule, rather than oppositely charged colloids or polymers. These surface structures show similar temperature-dependent reversible aggregation properties to the extended networks and have been used as a model system for understanding the sharp melting transition of the solution-based structures. The strategy has also been shown to be viable for regiospecific binding of nanoparticles to patterned solid substrates [97]. In this case dip pen nanolithography was used to pattern the substrate with two sets of domains of oligonucleotides with different sequences. When exposed to a dispersion of two sets of oligonucleotide-derivatized gold particles, each set specifically binds to the domains of the corresponding complementary strands.

Other strategies for synthesizing and assembling inorganic colloids have been inspired by biomolecular recognition properties at the inorganic interface, as discussed next.

11.3.2
Biomolecular/Inorganic Interfaces

Biominerals are generally associated with biomolecules such as proteins and polysaccharides. One of the most successful strategies in biomimetic studies has been to directly utilize the soluble macromolecules extracted from biominerals for in vitro investigations of mineral formation. Such studies provide insight into the mechanisms of biomineral formation as well as potential new routes to control mineral properties such as size, shape and phase. Many biomimetic investigations have been made on the crystallization of calcium carbonate minerals [97, 98].

For example, molecules have been extracted from the $CaCO_3$ phases produced by various organisms. The most thermodynamically stable form of $CaCO_3$ is calcite. However, the other phases (aragonite, vaterite and the intrinsically unstable amorphous) are also common components of biomineral structures. Indeed biominerals often contain mixtures of $CaCO_3$ phases laid down at specific stages of biomineralization, such as in the abalone shell where a thin layer of calcite crystals is overgrown with oriented aragonite crystals [97]. The molecular recognition properties of the polyanionic proteins extracted from each component of the shell are such that they initiate phase-specific nucleation in vitro.

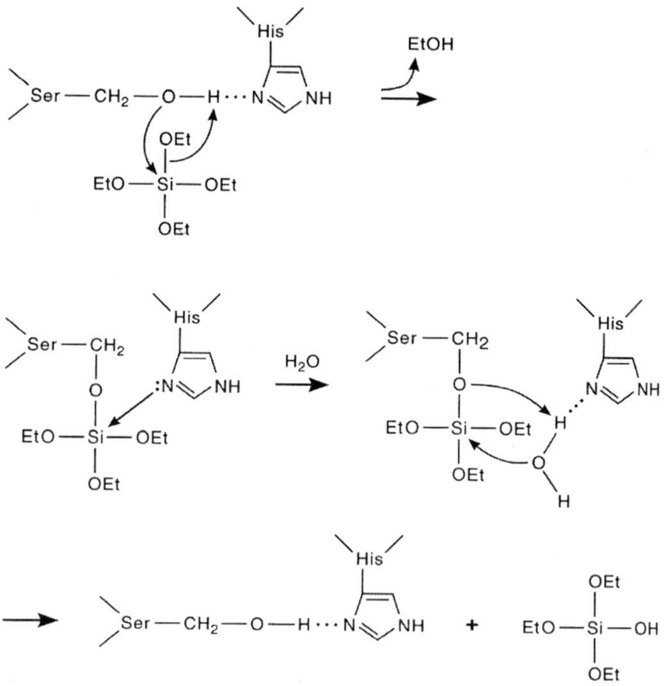

Fig. 11.11 Schematic diagram showing the role of the hydroxy group of serine and the imidazole group of histidine in the catalysis of alkoxysilane polycondensation at neutral pH by silicatein (reprinted with permission of [103]).

Recently there has been considerable interest in isolating and characterizing the molecules associated with biosilica formation [99, 100]. Synthetic routes to silica-containing materials often involve the use of relatively harsh reaction conditions such as high temperature and extremes in pH. Ceramic sols for instance are commonly produced under conditions of high pH. For example the classical Stöber synthesis of silica affords charge-stabilized colloidal particles [101]. The size of the amorphous spherical particles is sensitive to the reaction conditions and can be varied by adjusting alkoxysilane precursor concentration. These silica precursors are typically less reactive than other metal alkoxides such as titanium and aluminum, hence the need for acid or base catalysis of the hydrolysis and condensation reactions. In contrast, silica biominerals, such as diatom exoskeletons, sponge spicules and plant crystoliths, are precipitated at ambient temperatures and neutral pH.

The spicules produced by the marine sponge *Tethya aurantia* consist of a protein fiber up to 2 mm in length and 2 µm in diameter encased in a 30 µm thick silica shell. These filaments consist of three protein subunits, which have been termed silicateins by Morse and co-workers [99]. They found that these biomolecules catalyze the hydrolysis and condensation of silicon alkoxides at neutral pH and low temperature in vitro [102]. Site-directed mutagenesis studies indicate that the specific activity of these substrates is associated with the spatial arrangement of both the hydroxy and imidazole side chains of the serine and histidine residues of the protein (Fig. 11.11) [103]. Thus, whilst protein subunit structure governs the molecular-scale interactions the macroscopic form of the spicules is templated by the filaments formed by the assembly of subunits. Other fibrous templates such as cellulose and silk, although rich in surface hydroxy groups, do not initiate silica polymerization at neutral pH.

These studies have inspired attempts to design new catalysts for silica precipitation. It was discovered that of the simple polymers of amino acids poly(L-cysteine) was the most catalytically active at neutral pH, but showed no control over silica form [104]. To enable the production of silica species with controlled morphology it was necessary for the polypeptides to form well-defined aggregate structures, to serve as templates for silica deposition. Thus synthetic block copolypeptides were developed, with water-soluble and insoluble components. In particular the self-assembled aggregate structures of cysteine-lysine block copolypeptides were found to be the best mimics of natural silicateins, as they catalyzed the deposition of silica with controlled morphology. Under an inert atmosphere silica spheres were produced from the reaction of tetraethoxysilane with suspensions of the copolypeptides. In contrast the same experiment conducted in air results in the formation of silica rods owing to the oxidation of the poly(L-cysteine) component, and consequent change in the aggregate structure.

Nanoparticulate hollow shells have been prepared by a number of techniques including the layer-by-layer assembly of nanoparticles onto sacrificial templates including biological cells [105] and the self-assembly of gold colloids conjugated with phospholipids [106]. Recently these poly(L-lysine-L-cysteine) block copolymers have been used to prepare micrometer-sized multicomponent nanoparticulate shells [107]. Preformed gold particles (10–12 nm) are reacted with aqueous suspensions of the block copolypeptide, presumably interacting with the cysteine residues. Subsequent addition of silica particles (10–12 nm), which may interact with the lysine component, resulted in the formation of hollow shells with an inner layer of gold and a 250 nm thick outer layer of silica. Other morphologies are available by changing, for example, block length, whilst changing the number and types of blocks should allow numerous shell compositions to be produced.

In related work the molecules associated with diatom formation have also been investigated [100]. Initial studies demonstrated that polycationic polypeptides (termed silaffins) extracted from the diatom *Cylindrotheca fusiformis* catalyzed the deposition of aggregates of silica spheres from silicic acid solution [100]. Subsequent investigations on a number of different species confirmed the presence of silaffins and also long-chain (N-methyl-propylamine) polyamines [108]. Each diatom has a species-specific set of silaffins and polyamines. In vitro studies showed that the polyamines catalyze the formation of silica particles. Increasing pH or decreasing polyamine molecular weight can be used to reduce particle size (Fig. 11.12). Interestingly the most monodisperse particles form at around neutral pH. In contrast the silaffin proteins produce monolithic porous aggregates of ir-

Fig. 11.12 Silica structures produced using polyamines extracted from the shell of a diatom (*N. angularis*). Precipitates formed in the presence of polyamines of molecular masses 1000 to 1250 **A** and 600 to 750 Da **B**. Natural mixtures of polyamines show a pH-dependent effect on silica form **C** pH 5.4, **D** pH 6.3, **E** pH 7.2, **F** pH 8.3 (reprinted with permission of *Proceedings of the National Academy of Sciences* [108]. Copyright 2000 National Academy of Sciences, USA).

regular small particles under similar conditions. Combinations of the two macromolecular components afford monolithic aggregates of regular silica spheres. Thus these structure-directing agents allow fast precipitation of controlled silica morphologies at ambient conditions, implying a role in the species-specific patterning. Synthetic mimics of these molecules should enable new approaches to silica formation under ambient conditions, similar to those currently under investigation by Morse and co-workers. For example, there have been many studies on the encapsulation of biological molecules and structures within silica sol–gel matrices, for applications such as sensors. The catalysis of silica-gel production at neutral pH and low temperature should further extend the range of sensitive biomolecules and structures that can be immobilized in silica whilst retaining activity.

Based on these observations Sumper has proposed a phase-separation model for diatom morphogenesis [109]. In this mechanism the methylated polyamines initially form a close-packed array of emulsion-like droplets that delineate the space and initial pattern, with the polyamines catalyzing and co-precipitating with the silica. This consumption of the polyamine causes repeated phase separation into smaller droplets which again template silica deposition. This simple model adequately explains the hierarchical structures. The variation in morphology between species of the same genus is rationalized based on the pore-to-pore distance in the initial close-packed structure.

The studies outlined above elegantly demonstrate how biomineral-derived macromolecules and synthetic analogues can be used directly in biomimetic materials synthesis to control the atomic arrangement of the inorganic phase as well as the form, at all length scales up to macroscopic. However, one of the key aims is to build structures with more technologically relevant materials e.g. magnetic, semiconducting and metallic particles. Iron oxide phases are relatively common components of biominerals, for example ferritins and the linear chains of magnetic nanoparticles found formed by magnetotactic bacteria [110]. However, the production of semiconducting and metallic colloidal particles in biological systems is rare.

Belcher and co-workers have pioneered an alternative, indirect strategy to screen molecules for recognition properties at non-biological mineral interfaces [111]. They have used combinatorial phage display libraries to select peptide sequences that show specific affinities for desired mineral phases. Essentially this enables approximately 10^9 peptides to be reacted at the mineral surface. Those that bind non-specifically are washed off, before eluting and amplifying the peptides that show a stronger binding. This process is repeated 3–5 times and then the peptide sequence which has the strongest interaction is determined. This has allowed peptides to be identified that can differentially bind to certain minerals such as GaAs rather than Si and can show crystalline-face specificity. Peptides can also be selected that control semiconductor size, shape and, hence, optical properties [112]. Thus, as well as providing an approach to the assembly of preformed nanoparticles, based on synthesizing peptides with multiple recognition sites, it could also enable the synthesis and specific assembly of nanomaterials from molecular precursors.

11.4
Summary and Outlook

In this chapter a number of methodologies for preparing nanostructured materials using biomolecules have been outlined. Biotemplates can be used to control mineral formation, deposition and extended structure formation. The use of these templating strategies means nanoparticle aggregate form can be pre-defined to a certain extent, and a diverse range of hybrid or purely inorganic structures can be produced. In particular non-specific interfacial electrostatic forces commonly dictate structure formation. The strategies are therefore generic in the sense that any mineral that can be prepared in nanoscale form, with suitable surface properties can be used. Thus the combination of new building blocks and/or biomolecular scaffolds will continue to extend the catalog of nanoscale materials.

More complex, multicomponent structures are beginning to emerge from "bottom up" approaches based on molecular scale interactions. These "bricks and mortar" routes challenge the colloid chemist to produce an array of potential building blocks of controlled size, shape, composition and interfacial properties. The range of available specific biomolecular coupling systems can then be used to produce hybrid materials ordered at the mesoscale. Biotechnological approaches should also allow the inorganic molecular recognition properties to be engineered to attach the biomolecules which direct aggregation specifically to the appropriate particles or even crystallographic faces. Thus, further advances in colloid chemistry and biotechnology are expected to yield more complex building blocks pre-programmed to self-assemble into complicated, hybrid, aggregate topologies.

11.5
Acknowledgments

I would like to thank S. Mann for his help and advice during the preparation of this manuscript and the EPSRC and University of Bristol for financial support.

11.6
References

1 D'A. THOMSON, *On Growth and Form*, Cambridge University Press, Cambridge, **1961**.
2 E. BAUERLEIN, *Biomineralization*, Wiley-VCH, Weinheim, **2000**.
3 S. MANN, *Biomimetic Materials Chemistry*, Wiley-VCH, New York, **1996**.
4 H. A. LOWENSTAM, S. WEINER, *On Biomineralization*, Oxford University Press, New York, **1989**.
5 (a) I. A. AKSAY, M. TRAU, S. MANNE, I. HUNMA, N. YAO, L. ZHOU, P. FENTER, P. M. EISENBERGER, S. M. GRUNER, *Science* **1996**, *273*, 892. (b) G. A. OZIN, *Acc. Chem. Res.* **1997**, *30*, 17–27. (c) S. MANN, *J. Chem. Soc., Dalton Trans.* **1997**, 3953. (d) E. DUJARDIN, S. MANN, *Adv. Mater.* **2002**, *14*, 775–788.
6 (a) A. P. ALIVISATOS, *Endeavor* **1997**, *21*, 56–60. (b) C. M. NIEMEYER, *Angew. Chem. Int. Ed.* **2001**, *40*, 4128–4158. (c) C. A. MIRKIN, *Inorg. Chem.* **2000**, *39*, 2258–2272. (d) S. MANN, S. A. DAVIS, S. R. HALL, M. LI, K. H. RHODES, W. SHENTON, S. VAUCHER, B. ZHANG, *J. Chem. Soc., Dalton Trans.* **2000**, 3753.
7 S. A. DAVIS, M. BREULMANN, K. H. RHODES, B. ZHANG, S. MANN, *Chem. Mater.* **2001**, *13*, 3218–3226.
8 (a) C. GÖLTNER in: *Reactions and Synthesis in Surfactant Systems* (J. TEXTER, ed.), Marcel Dekker, New York, **2001**, pp. 797–818. (b) W. MEIER, *Curr. Opin. Colloid Interface Sci.* **1999**, *4*, 6–14.
9 S. MANN, J. P. HANNINGTON, R. J. P. WILLIAMS, *Nature* **1986**, *324*, 565–567.
10 S. BHANDARKER, A. BOSE, *J. Colloid Interface Sci.* **1990**, *135*, 531–538.
11 M. T. KENNEDY, B. A. KORGEL, H. G. MONBOUQUETTE, J. A. ZASADZINSKI, *Chem. Mater.* **1998** *10*, 2116–2119.
12 N. D. CHASTEEN, P. M. HARRISON, *J. Struct. Biol.* **1999**, *126*, 182–194.
13 T. DOUGLAS, D. P. E. DICKSON, S. BETTERIDGE, J. CHARNOCK, C. D. GARNER, S. MANN, *Science* **1995**, *269*, 54–57.
14 K. K. W. WONG, S. MANN, *Adv. Mater.* **1996**, *8*, 928–932.
15 (a) F. C. MELDRUM, V. J. WADE, D. L. NIMMO, B. R. HEYWOOD, S. MANN, *Nature* **1991**, *349*, 684–687. (b) J. F. HAINFELD, *Proc. Natl. Acad. Sci. USA* **1992**, *89*, 11064–11068.
16 (a) F. C. MELDRUM, B. R. HEYWOOD, S. MANN, *Science* **1992**, *257*, 522–523. (b) K. K. W. WONG, T. DOUGLAS, S. GIDER, D. D. AWSCHALOM, S. MANN, *Chem. Mater.* **1998**, *10*, 279–285.

17 J.W.M. Bulte, T. Douglas, S. Mann, R.B. Frankel, B.M. Moskowitz, R.A. Brooks, C.D. Baumgarner, J. Vymazal, J.A. Frank, *Invest. Radiol.* **1994**, *29*, S214–S216.
18 B. Warne, K.I. Oksana, E.L. Mayes, J.A.L. Wiggins, K.K.W. Wong, *IEEE Trans. Magn.* **2000**, *36*, 3009–3011.
19 (a) T. Douglas, M. Young, *Nature* **1998**, *393*, 152–155. (b) T. Douglas, M. Young, *Adv. Mater.* **1999**, *11*, 679–681.
20 J.A. Speir, S. Munshi, G. Wang, T.S. Baker, J.E. Johnson, *Structure* **1995**, *3*, 63–78.
21 X. Zhao, J.M. Fox, N.H. Olson, T.S. Baker, M.J. Young, *Virology* **1995**, *207*, 486–494.
22 R. Ladenstein, M. Schneider, R. Huber, H.D. Bartunik, K. Wilson, K. Schott, A. Bacher, *J. Mol. Biol.* **1988**, *203*, 1045–1070.
23 K. Schott, R. Ladenstein, A. König, A. Bacher, *J. Biol. Chem.* **1990**, *265*, 12686–12689.
24 W. Shenton, S. Mann, H. Cölfen, A. Bacher, M. Fischer, *Angew. Chem. Int. Ed.* **2001**, *40*, 442–445.
25 P. Yager, P.E. Schoen, *Mol. Cryst. Liquid Cryst.* **1984**, *106*, 371–381.
26 S. Baral, P. Schoen, *Chem. Mater.* **1993**, *5*, 145–147.
27 D.D. Archibald, S. Mann, *Nature* **1993**, *364*, 430–433.
28 S.L. Burkett, S. Mann, *Chem. Commun.* **1996**, *3*, 321–322.
29 J.L. Coffer, S.R. Bigham, R.F. Pinizzotto, H. Yang, *Nanotechnol.* **1992**, *3*, 69–76.
30 S.R. Bigham, J.L. Coffer, *Colloids Surf. A* **1995**, *95*, 211–219.
31 T. Torimoto, M. Yamashita, S. Kuwabata, T. Sakata, H. Mori, H. Yoneyama, *J. Phys. Chem. B* **1999**, *103*, 8799–8803.
32 A. Kumar, M. Pattarkine, M. Bhadbhade, A.B. Mandale, K.N. Ganesh, S.S. Datar, C.V. Dharmadhikari, M. Sastry, *Adv. Mater.* **2001**, *13*, 341–344.
33 A. Klug, *Phil. Trans. R. Soc. London B* **1999**, *354*, 531–535.
34 W. Shenton, T. Douglas, M. Young, G. Stubbs, S. Mann, *Adv. Mater.* **1999**, *11*, 253–256.
35 C.E. Fowler, W. Shenton, G. Stubbs, S. Mann, *Adv. Mater.* **2001**, *13*, 1266–1269.
36 P.T. Tanev, M. Chibwe, T.J. Pinnavaia, *Nature* **1994**, *368*, 321.
37 Y.N. Yun, D.M. Dabbs, I.A. Aksay, S. Erramilli, *Langmuir* **1994**, *10*, 3377.
38 (a) A.E. Persson, B.J. Schoeman, J. Sterte, J.E. Otterstedt, *Zeolites* **1994**, *14*, 557. (b) R. Ravishankar, C. Kirschock, B.J. Schoeman, P. Vanoppen, P.J. Grobet, S. Storck, W.F. Maier, J.A. Martens, F.C. De Schryver, P.A. Jacobs, *J. Phys. Chem. B* **1998**, *102*, 2633. (c) E.R. Geus, J.C. Jansen, H. van Bekkum, *Zeolites* **1994**, *14*, 82.
39 A. Stein, R.C. Schroden, *Curr. Opin. Solid State Mat. Sci.* **2001**, *5*, 553–564.
40 (a) O.D. Velev, P.M. Tessier, A.M. Lenhoff, E.W. Kaler, *Nature* **1999**, *401*, 548. (b) O.D Velev, E.W. Kaler, *Adv. Mater.* **2000**, *12*, 531.
41 G. Subramanian, V.N. Manoharan, J.D. Thorne, D.J. Pine, *Adv. Mater.* **1999**, *11*, 1261.
42 Y.A. Vlasov, N. Yao, D.J. Norris, *Adv. Mater.* **1999**, *11*, 165–169.
43 (a) Y.J. Wang, Y. Tang, Z. Ni, W.M. Hua, W.L. Yang, X.D. Wang, W.C. Tao, Z. Gao, *Chem. Lett.* **2000**, 510–511. (b) L. Huang, Z. Wang, J. Sun, L. Miao, Q. Li, Y. Yan, D. Zhao, *J. Am. Chem. Soc.* **2000**, *122*, 3530–3531.
44 T.J. Beveridge, *Metal Ions and Bacteria*, Wiley, New York, **1989**.
45 (a) N.H. Mendelson, J.J. Thwaites, *MRS Symp. Proc.* **1990**, *174*, 171. (b) J.J. Thwaites, N.H. Mendelson, *Proc. Natl. Acad. Sci. USA* **1985**, *82*, 2163. (c) N.H. Mendelson, J.J. Thwaites, *J. Bacteriol.* **1989**, *171*, 1055.
46 S.A. Davis, S.L. Burkett, N.H. Mendelson, S. Mann, *Nature* **1997**, *385*, 420.
47 B. Zhang, S.A. Davis, N.H. Mendelson, S. Mann, *Chem. Commun.* **2000**, 781.
48 S.A. Davis, H.M. Patel, E.L. Mayes, N.H. Mendelson, G. Franco, S. Mann, *Chem. Mater.* **1998**, *10*, 2516.
49 Y. Shin, J. Liu, J.H. Chang, Z. Nie, G.J. Exarhos, *Adv. Mater.* **2001**, *13*, 728–732.
50 A. Dong, Y. Wang, Y. Tang, N. Ren, Y. Zhang, Y. Yue, Z. Gao, *Adv. Mater.* **2002**, *14*, 926–929.

51 M. W. Anderson, S. M. Holmes, N. Hanif, C. S. Cundy, *Angew. Chem. Int. Ed.* **2000**, *39*, 2707.

52 (a) R. Seshadri, F. C. Meldrum, *Adv. Mater.* **2000**, *12*, 1149–1151. (b) F. C. Meldrum, R. Seshadri, *Chem. Commun.* **2000**, 29.

53 K. M. McGrath, *Adv. Mater.* **2001**, *13*, 989–992.

54 W. Ogasawara, W. Shenton, S. A. Davis, S. Mann, *Chem. Mater.* **2000**, *12*, 2835–2837.

55 M. Breulmann, H. Coelfen, H.-P. Hentze, M. Antonietti, D. Walsh, S. Mann, *Adv. Mater.* **1998**, *10*, 237.

56 R. A. Caruso, M. Giersig, F. Willig, M. Antonietti, *Langmuir* **1998**, *14*, 6334.

57 M. Breulmann, S. A. Davis, S. Mann, H. P. Hentze, M. Antonietti, *Adv. Mater.* **2000**, *12*, 502.

58 B. Zhang, PhD thesis, University of Bristol, **2001**.

59 B. Zhang, S. A. Davis, S. Mann, *Chem. Mater.* **2002**, *14*, 1369–1375.

60 A. Kumar, A. Biebuyck, G. M. Whitesides, *Langmuir* **1994**, *10*, 1498–1511.

61 J. Aizenberg, A. J. Black, G. M. Whitesides, *Nature* **1999**, *398*, 495.

62 U. B. Sleytr, P. Messner, D. Pum, M. Sara, *Crystalline Bacterial Cell Surface Proteins*, Landes Co., Austin and Academic, San Diego, **1996**.

63 W. Shenton, D. Pum, U. B. Sleytr, S. Mann, *Nature* **1997**, *389*, 585–587.

64 S. Dieluweit, D. Pum, U. B. Sleytr, *Supramol. Sci.* **1998**, *5*, 15–19.

65 M. Mertig, R. Kirsch, W. Pompe, H. Engelhardt, *Eur. Phys. J.* **1999**, *D9*, 45–48.

66 S. R. Hall, W. Shenton, H. Engelhardt, S. Mann, *Chem. Phys. Chem.* **2001**, *3*, 184.

67 (a) C. B. Murray, C. R. Kagan, M. G. Bawendi, *Science* **1995**, *270*, 1335. (b) T. Vossmeyer, G. Reck, L. Katsikas, E. T. K. Haupt, B. Schulz, H. Weller, *Science* **1995**, *267*, 1476. (c) C. Petit, A. Taleb, M.-P. Pileni, *Adv. Mater.* **1998**, *10*, 259. (d) M.-P. Pileni, *Adv. Funct. Mater.* **2001**, *11*, 323–336.

68 (a) M. Brust, D. Bethell, D. J. Schiffrin, C. J. Kiely, *Adv. Mater.* **1995**, *7*, 795. (b) R. P. Andres, J. D. Bielefeld, J. I. Henderson, D. B. Janes, V. R. Kolagunta, C. P. Kubiak, W. J. Mahoney, R. G. Osifchin, *Science* **1996**, *273*, 1690. (c) B. A. Korgel, D. Fitzmaurice, *Adv. Mater.* **1998**, *10*, 661.

69 W. Shenton, S. A. Davis, S. Mann, *Adv. Mater.* **1999**, *11*, 449.

70 S. Mann, W. Shenton, M. Lei, S. Connolly, D. Fitzmaurice, *Adv. Mater.* **2000**, *12*, 147–150.

71 S. Connolly, D. Fitzmaurice, *Adv. Mater.* **1999**, *11*, 1202.

72 M. Li, K. K. W. Wong, S. Mann, *Chem. Mater.* **1999**, *11*, 23.

73 C. A. Mirkin, R. L. Letsinger, R. C. Mucic, J. J. Storhoff, *Nature* **1996**, *382*, 607.

74 P. Alivisatos, K. P. Johnsson, X. Peng, T. E. Wilson, C. J. Loweth, M. Bruchez, P. G. Schultz, *Nature* **1996**, *382*, 609.

75 C. M. Niemeyer, W. Bürger, J. Peplies, *Angew. Chem. Int. Ed.* **1998**, *37*, 2265–2268.

76 (a) C. J. Kiely, J. Fink, M. Brust, D. Bethell, D. J. Schiffrin, *Nature* **1998**, *396*, 444. (b) C. J. Kiely, J. Fink, J. G. Zheng, M. Brust, D. Bethell, D. J. Schiffrin, *Adv. Mater.* **2000**, *12*, 640.

77 N. C. Seeman, *Acc. Chem. Res.* **1997**, *30*, 357–363.

78 R. Elghanian, J. J. Storhoff, R. C. Mucic, R. L. Letsinger, C. A. Mirkin, *Science* **1997**, *277*, 1078–1081.

79 J. J. Storhoff, R. Elghanian, R. C. Mucic, C. A. Mirkin, R. L. Letsinger, *J. Am. Chem. Soc.* **1998**, *120*, 1959.

80 R. C. Mucic, J. J. Storhoff, C. A. Mirkin, R. L. Letsinger, *J. Am. Chem. Soc.* **1998**, *120*, 12674–12675.

81 G. P. Mitchell, C. A. Mirkin, R. L. Letsinger, *J. Am. Chem. Soc.* **1999**, *121*, 8122–8123.

82 W. J. Parak, D. Gerion, D. Zanchet, A. S. Woerz, T. Pellegrino, C. Micheel, S. C. Williams, M. Seitz, R. E. Bruehl, Z. Bryant, C. Bustamante, C. R. Bertozzi, A. P. Alivisatos, *Chem. Mater.* **2002**, *14*, 2113–2119.

83 C. J. Loweth, W. B. Caldwell, X. Peng, A. P. Alivisatos, P. G. Schultz, *Angew. Chem. Int. Ed.* **1999**, *38*, 1808–1812.

84 D. Zanchet, C. M. Micheel, W. J. Parak, D. Gerion, A. P. Alivisatos, *Nano Lett.* **2001**, *1*, 32–35.

85 (a) X. Peng, L. Manna, W. Wang, J. Wickham, E. Scher, A. Kadavanich, A. P. Alivisatos, *Nature* **2000**, *404*, 59. (b) J. Hu, L.-S. Li, W. Yang, L. Manna, L.-W. Wang, A. P. Alivisatos, *Science* **2001**, *292*, 2060. (c) Y. Y. Yu, S. S. Chang, C. L. Lee, C. R. C. Wang, *J. Phys. Chem. B* **1997**, *101*, 6661. (d) S. S. Chang, C. W. Shih, C. D. Chen, W. C. Lai, C. R. C. Wang, *Langmuir* **1999**, *15*, 701. (e) C. S. Yang, D. D. Awschalom, G. D. Stucky, *Chem. Mater.* **2002**, *14*, 1277–1284. (f) C. J. Johnson, E. Dujardin, S. A. Davis, C. J. Murphy, S. Mann, *J. Mater. Chem.* **2002**, *12*, 1765–1770.

86 L.-S. Li, J. Walda, L. Manna, A. P. Alivisatos, *Nano Lett.* **2002**, *2*, 557–560.

87 M. Li, H. Schnablegger, S. Mann, *Nature* **1999**, *402*, 393.

88 E. Dujardin, L.-B. Hsin, C. R. C. Wang, S. Mann, *Chem. Commun.* **2001**, 1264–1265.

89 (a) J. Schmitt, G. Decher, W. J. Dressik, S. L. Brandow, R. E. Geer, R. Shashidar, J. M. Calvert, *Adv. Mater.* **1997**, *9*, 61. (b) T. Cassagneau, J. H. Fendler, T. E. Mallouk, *Langmuir* **2000**, *16*, 241.

90 V. L. Colvin, A. N. Goldstein, A. P. Alivisatos, *J. Am. Chem. Soc.* **1992**, *114*, 5221–5230.

91 N. A. Kotov, F. C. Meldrum, J. H. Fendler, *J. Phys. Chem.* **1994**, *98*, 8827.

92 T. Baum, D. Bethell, M. Brust, D. J. Schiffrin, *Langmuir* **1999**, *15*, 866–871.

93 B. R. Martin, D. J. Dermody, B. D. Reiss, M. Fang, L. A. Lyon, M. J. Natan, T. E. Mallouk, *Adv. Mater.* **1999**, *11*, 1021.

94 J. K. N. Mbindyo, B. D. Reiss, B. R. Martin, C. D. Keating, M. J. Natan, T. E. Mallouk, *Adv. Mater.* **2001**, *13*, 249–254.

95 A. T. Taton, R. C. Mucic, C. A. Mirkin, R. L. Letsinger, *J. Am. Chem. Soc.*, **2000**, *122*, 6305–6306.

96 L. M. Demers, S-J Park, T. A. Tatan, Z. Li, C. A. Mirkin, *Angew. Chem. Int. Ed.* **2001**, *40*, 3071–3073.

97 A. M. Belcher, X. H. Wu, R. J. Christensen, P. K. Hansma, G. D. Stucky, D. E. Morse, *Nature* **1996**, *381*, 56.

98 J. Aizenberg, J. Hanson, T. F. Koetzle, S. Weiner, L. Addadi, *J. Am. Chem. Soc.* **1997**, *119*, 881–886.

99 K. Shimizu, J. Cha, G. D. Stucky, D. E. Morse, *Proc. Natl. Acad. Sci. USA* **1998**, *95*, 6234–6238.

100 N. Kröger, R. Dautzmann, M. Sumper, *Science* **1999**, *286*, 1129–1132.

101 W. Stöber, A. Fink, E. Bohn, *J. Colloid Interface Sci.* **1968**, *26*, 62–69.

102 J. N. Cha, K. Shimizu, Y. Zhou, S. C. Christiansen, B. F. Chmelka, G. D. Stucky, D. E. Morse, *Proc. Natl. Acad. Sci. USA* **1999**, *96*, 361–365.

103 Y. Zhou, K. Shimizu, J. N. Cha, G. D. Stucky, D. E. Morse, *Angew. Chem. Int. Ed.* **1999**, *38*, 780–782.

104 J. N. Cha, G. D. Stucky, D. E. Morse, T. J. Deming, *Nature* **2000**, *403*, 289–292.

105 F. Caruso, *Chem. Eur. J.* **2000**, *6*, 413–419.

106 J. F. Hainfeld, F. R. Furuya, R. D. Powell, *J. Struct. Biol.* **1999**, *127*, 152–160.

107 M. S. Wong, J. N. Cha, K.-S. Choi, T. J. Deming, G. D. Stucky, *Nano Lett.* **2002**, *2*, 583–587.

108 N. Kröger, R. Dautzmann, C. Bergsdorf, M. Sumper, *Proc. Natl. Acad. Sci. USA* **2000**, *97*, 14133–14138.

109 M. Sumper, *Science* **2002**, *295*, 2430–2433.

110 D. Schuler in *Biomineralization* (E. Bauerlein, ed.), Wiley-VCH, Weinheim, **2000**, pp. 47–60.

111 S. R. Whaley, D. S. English, E. L. Hu, P. F. Barbara, A. M. Belcher, *Nature* **2000**, *405*, 665–668.

112 N. C. Seeman, A. M. Belcher, *Proc. Natl. Acad. Sci. USA* **2002**, *99*, 6451–6455.

12
Nanoparticle Organization at the Air-Water Interface and in Langmuir-Blodgett Films

Murali Sastry

12.1
Introduction

Impressive strides are being made in the synthesis of nanomaterials covering a range of chemical compositions, sizes and shapes [1–5] as well as in understanding their unusual physicochemical and optoelectronic properties [5, 7–10]. An important step en-route to commercial exploitation of nanotechnology concerns the assembly of nanomaterials into topologically predefined superstructures and their packaging. While the properties of individual nanoparticles are no doubt exciting, the collective properties of ensembles of nanoparticles interacting with each other can be radically different from their individual counterparts, with interesting application potential. This is best illustrated by the classic example of gold hydrosols that change color from ruby red/pink to blue during aggregation of the nanoparticles in solution [11–13]. Mirkin and co-workers have used this phenomenon to develop a colorimetric method for detection of the base sequence in DNA tagged onto the surface of colloidal gold [14, 15]. Using controlled hybridization of DNA bound to gold nanoparticles in solution, this group has shown that the collective properties of gold nanoparticle ensembles can be tuned by varying both the inter-nanoparticle separation (the interaction strength) and the size of the ensembles [16]. The collective properties of nanoparticle assemblies are also critically dependent on the shape of the superstructures. Linear assemblies of gold nanoparticles exhibit optical properties very different from spherical aggregates [17]. Very recently, Atwater and co-workers have discussed the exciting possibility of using linear arrays of closely spaced metal nanoparticles in the fabrication of waveguides for electromagnetic energy for integration in optical devices operating below the diffraction limit of light [18]. This may be achieved by the creation of coupled plasmon modes ("plasmonics", as the authors termed this approach) between neighboring particles in the array that would then lead to energy transmission [18].

While controlled assembly of nanoparticles in solution is important [14–16], application of nanomaterials, particularly in the optoelectronics industry [9, 18], would require their immobilization on solid surfaces in either monolayer or multilayer form. This may conveniently be done by synthesis of nanoparticles by the

Colloids and Colloid Assemblies. Edited by Frank Caruso
Copyright © 2004 Wiley-VCH Verlag GmbH & Co. KGaA, Weinheim
ISBN: 3-527-30660-9

colloidal route, surface modification of the nanoparticles and then self-assembly of the nanoparticles using a variety of interactions between the nanoparticles and the surface on which immobilization is to be carried out. Nanoparticle monolayers may be obtained by assembly on ω-functionalized self-assembled monolayers [19–22], surface-modified polymers [23, 24] and by solvent evaporation of hydrophobic nanoparticles dispersed in organosols [25–27]. Multilayers of nanoparticles may be assembled in a layer-by-layer fashion using electrostatic interactions involving polyelectrolytes [28, 29] and oppositely charged nanoparticles [30, 31] as well as via bifunctional linker molecules [32, 33].

While the beginnings of assembly of amphiphilic monolayers on the surface of water can be traced to Babylonian times [34], it was the pioneering work of Pockels [35, 36] and Langmuir [37] that laid the foundation for understanding and exploiting the excellent organizational capability of the air-water interface. A major advance in this direction was provided by the work of Blodgett [38] who showed that monolayers of amphiphilic molecules organized at the air-water interface (termed Langmuir monolayers) could be transferred onto suitable solid surfaces in a layer-by-layer fashion. This process results in highly ordered lamellar multilayers of amphiphilic molecules and is referred to as the Langmuir-Blodgett method in honor of the two remarkable scientists who established this field. Their seminal work has opened up a very active branch of interfacial chemistry with exciting applications in the synthesis of advanced materials, particularly nanomaterials. One may now routinely assemble small and more complex inorganic ions such as polyoxometallates (Keggin salts) [39, 40] and biomacromolecules such as proteins/enzymes [41, 42] and DNA [43, 44] at the air-water interface. The synthesis and assembly of inorganic nanoparticles at the air-water interface is a relatively new area of research and the remainder of this chapter will focus on this aspect and on how one may realize multilayer films of nanoparticles on different solid surfaces by the versatile Langmuir–Blodgett (LB) method.

12.2
Nanoparticle Assembly at the Air-Water Interface

One may arrive at inorganic nanoparticles assembled at the air-water interface by a variety of routes that may be broadly classified into physical and chemical assembly methods. This classification and the underlying techniques in each of the sub-groups is illustrated schematically in Fig. 12.1. In the physical assembly method, the nanoparticles are synthesized separately (commonly as colloidal solutions) and depending on the mode of organization, the nanoparticle surface is modified by chemisorption of suitable surfactant molecules. If surface-modification of the nanoparticles renders them hydrophobic and soluble in non-polar organic solvents, such nanoparticles can be spread on the surface of water like conventional amphiphilic molecules (stearic acid, for example), compressed to a close-packed configuration and transferred by the LB method onto different substrates (Fig. 12.1). On the other hand, if the nanoparticles are water-soluble and capped

12.2 Nanoparticle Assembly at the Air-Water Interface

```
                    Langmuir-Blodgett films of nanoparticles
                                      |
                    ┌─────────────────┴─────────────────┐
              Physical methods                    Chemical methods
              ╱         ╲                         ╱          ╲
Assembly of hydrophobic    Chemical reaction of ions
nanoparticles at the air-  immobilized at the air-
water interface            water interface

Electrostatic attachment of nano-    Chemical reaction of ions
particles with charged Langmuir      in Langmuir-Blodgett films
monolayers at the air-water interface  of metal salts of fatty lipids
```

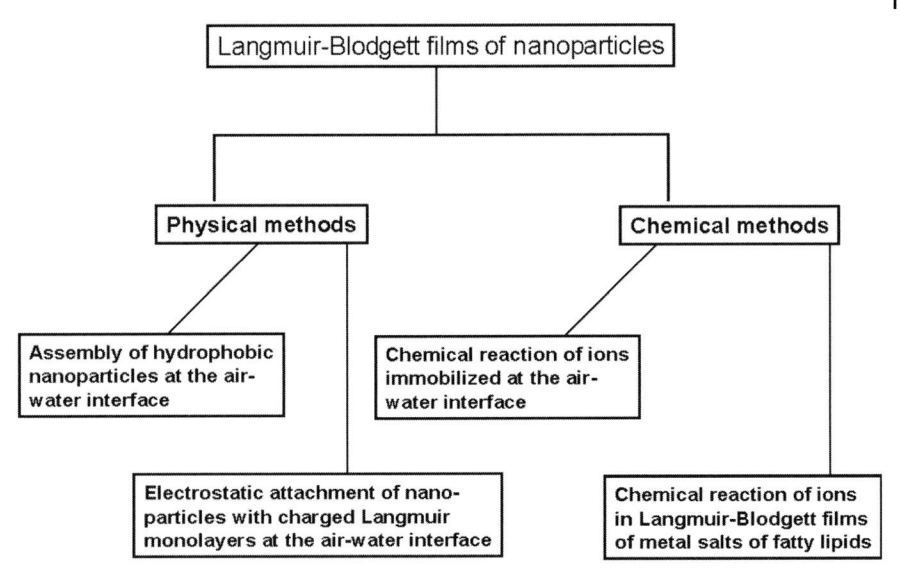

Fig. 12.1 Schematic outlining the classification of different experimental protocols for realizing Langmuir-Blodgett films of inorganic nanoparticles.

with hydrophilic functional groups, they may be taken in the aqueous subphase and their surface functionality used to complex with Langmuir monolayers at the air-water interface via electrostatic or hydrogen bonding and other interactions (Fig. 12.1).

The second (and chemical) method requires assembly of ions at the air-water interface followed by chemical reaction with the ions either in situ at the air-water interface or after formation of LB films of the ion-amphiphile complex (Fig. 12.1). A number of reactions can be envisaged leading to the formation of metal, metal sulfide and metal oxide nanoparticles in lamellar superstructures. Some representative examples belonging to the physical and chemical nanoparticle assembly domains and the formation of their LB films are presented and discussed below.

12.2.1
Physical Methods for Generation of Nanoparticle Langmuir-Blodgett Films

12.2.1.1 Organization of Hydrophobic Nanoparticles

The pioneering work of Fendler's group on the growth and assembly of nanoparticles at the air-water interface has, in many ways, set the tone for research in this area [45]. The demonstration that surfactant-stabilized hydrophobic silver [46], CdS [47], magnetite [48] and ferroelectric lead zirconium titanate nanoparticles [49] could be dispersed on the surface of water and thereafter transferred as close-packed nanoparticle multilayers by the LB method has led to the development of an extremely simple and elegant method for the organization of nanoparticles at the air-water interface. Many groups have since enlarged the scope of this protocol to include assembly of hydrophobized CdSe [50], buckyballs (C_{60}) [51], TiO_2, barium ferrite and Y-Ba-Cu superconductors [52], Pt [53, 54], Q-state CdS [55–57], alkanethiol/alkylamine-capped gold [58–64], silver [58, 65], alkane chalcogenide (Se, Te)-capped gold [66], SnO_2 [67], Fe_2O_3 [68], CdTe [69, 70] and MoS_2 nanoparticles [71].

The first stage in this method consists of surface modification of the nanoparticles to render them hydrophobic and soluble in non-polar organic solvents. A number of capping agents may be used depending on the chemical composition of the nanoparticles and on whether they are synthesized in an aqueous or non-polar organic environment. Once the particles are rendered hydrophobic, they are dispersed in a non-polar organic solvent and carefully spread on the surface of water as illustrated in Fig. 12.2. On evaporation of the solvent, the nanoparticles are known to self-assemble into monolayer rafts floating on the surface of water that on compression, form a two-dimensional array of close-packed nanoparticles. These nanoparticle monolayers may then be compressed using a suitable barrier (Fig. 12.2) and under constant surface pressure conditions [72], they may be transferred onto solid substrates to yield multilayers of the nanoparticles in a manner identical to that adopted for classical amphiphiles [38].

The report of Meldrum, Kotov and Fendler on the assembly of oleic acid-stabilized silver nanoparticles on the surface of water is interesting not only from the point of view that it was one of the first papers in this direction but also because it describes the phase transfer of silver nanoparticles synthesized in water to hexane by complexation with oleic acid molecules [46]. Binding of oleic acid molecules to silver nanoparticles at the interface between water and hexane resulted in the hydrophobization of the nanoparticles and their transfer into hexane [46]. The silver nanoparticles in hexane (diameter 103 ± 38 Å) were then spread on the surface of water and the surface pressure (π) versus surface area of the monolayer (A) isotherm was measured. Fig. 12.3 shows the π–A isotherms for both the oleic acid-stabilized silver nanoparticles (left) and pure oleic acid on water (right). While the π–A isotherm for oleic acid does not show much structure, at least five phases under different stages of compression have been identified for the silver

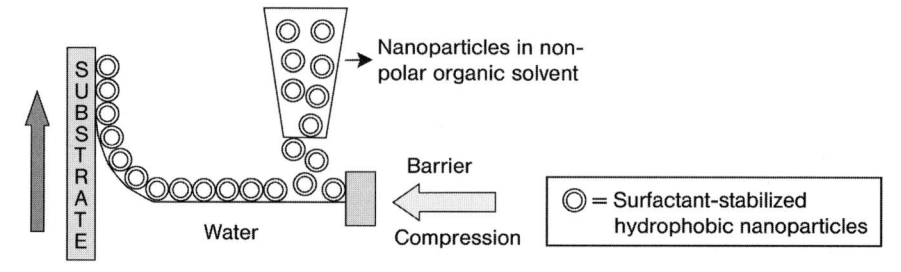

Fig. 12.2 Diagram illustrating the steps involved in the assembly of hydrophobized nanoparticles on the surface of water and transfer of the nanoparticle monolayers by the Langmuir-Blodgett method.

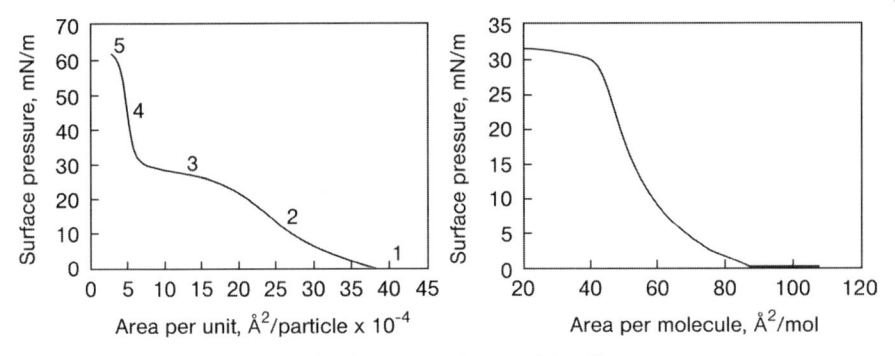

Fig. 12.3 Surface pressure vs molecular area isotherms of thin films on distilled water subphases of (**a**, left) 0.6 mL of Ag/oleic acid solution where labels 1–5 refer to the different phases of the monolayer structure, as discussed in the text and (**b**, right) oleic acid (reprinted with permission from [46], © 1994 American Chemical Society).

nanoparticle monolayer (Fig. 12.3, left). Together with transmission electron microscopy (TEM) studies of the silver nanoparticle monolayer transferred at different surface pressures, Fendler and co-workers concluded that the features in the π–A isotherms could be rationalized in terms of the existence of two separate phases consisting of oleic acid-stabilized silver nanoparticles and uncoordinated oleic acid molecules. The plateau region 3, which occurs at a pressure of ca. 30 mN/m, represents the collapse of oleic acid domains while the large rise in surface pressure in region 4 is due to compression of silver nanoparticle domains. At this stage, the authors observed that the silver nanoparticle monolayer had a distinct shiny, yellow color. LB films of the silver nanoparticles were grown on silanized glass plates at a surface pressure of 27 mN/m and the UV-vis absorption spectra were recorded as a function of film thickness. Fig. 12.4 shows a plot of the surface plasmon absorbance at 470 nm [73] as a function of number of immersion cycles of the substrate in the silver nanoparticle monolayers. The linearity in the plot clearly indicates lamellar growth of the nanoparticle LB film with good uniformity within the plane of the film [46].

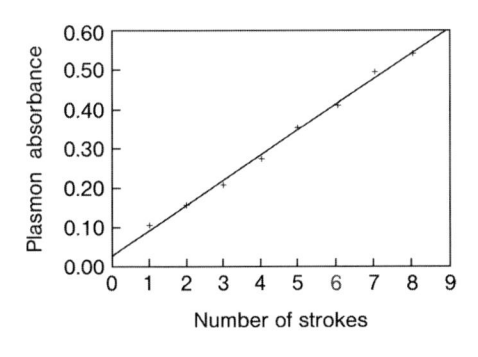

Fig. 12.4 Graph showing the intensity of the plasmon absorption of silver particles in an LB film with respect to the number of strokes (n) of a silanized glass plate through the silver particulate film (reprinted with permission from [46], © 1994 American Chemical Society).

As briefly alluded to above, the air-water interface provides an excellent medium for studying the phase behavior of nanoparticles. A detailed study of the phase behavior of gold and silver nanoparticles of different sizes and size distributions as well as different surface passivants was presented by Heath, Knobler and Leff [58]. By a systematic investigation of the π–A isotherms of the silver/gold nanoparticles and TEM images of LB films of the nanoparticles, the authors have constructed the pressure/temperature phase diagrams of the metal nanocrystal monolayers. Generality in the phase behavior was observed that depended on the excess conical volume available to the capping ligand as it extends from the particle surface.

Bawendi and co-workers presented an elegant study of the manipulation of size-selected CdSe quantum dots capped with trioctylphosphine oxide on the surface of water [50]. They observed that the particles assembled into hexagonal close-packed structures on water and that the optical properties of the CdSe nanoparticles in the LB films resembled those of the individual nanocrystallites [50]. However, even though highly monodisperse hydrophobic nanoparticles do form hexagonally organized close-packed structures under suitable compression, the temporal and temperature stability of the nanoparticle LB films is not high. In order to circumvent this problem, Yoneyama et al. and Chen have demonstrated that cross-linking of CdS [56] and gold nanoparticles [61, 62] respectively may be accomplished at the air-water interface with suitable cross-linking agents. In the case of gold nanoparticles, place exchange of alkanethiols capping the particles with a dithiol such as 4,4′-thiobisbenzenethiol (TBBT) added to the surface of water along with the nanoparticles results in the covalent cross-linking of the gold nanoparticles [61, 62]. Indirect evidence of the TBBT-induced networking of the gold nanoparticles (diameter = 4.7 nm) at the air-water interface is provided by monitoring the pressure-area isotherms under different conditions (Fig. 12.5) [62]. While there is a small increase in the takeoff area of the monolayer after addition of TBBT (compare curves a and b in Fig. 12.5), the slope of the isotherms in the low molecular area regime is almost identical indicating similar compressibility of the isotherms. However, on maintaining the barrier in the closed position for 15 h, the isotherm changed significantly showing a much more incompressible monolayer in a more compact, close-packed configuration (curve c, Fig. 12.5).

More direct evidence of gold nanoparticle network formation by TBBT cross-linking is shown in TEM pictures of monolayers of the particles transferred onto SiO_2-coated Cu grids (Fig. 12.6) [62]. From Fig. 12.6a and b, it is seen that transfer of the octanethiol-capped gold nanoparticles (C8Au) at different surface pressures without addition of TBBT does not result in nanoparticle assembly with translational order. However, addition of the cross-linking agent at the air-water interface during compression of the particles results in the formation of a beautiful network whose long-range order increases with increasing surface pressure (Fig. 12.6c and d) [62]. In addition to stabilizing the two-dimensional nanoparticle network, the covalent coupling strategy of Chen may be used to electrically connect the nanoparticles that otherwise possess an insulating barrier between the particles.

Yoneyama and co-workers adopted a conceptually slightly different approach to achieve the cross-linking of surface-modified CdS nanoparticles on the surface of

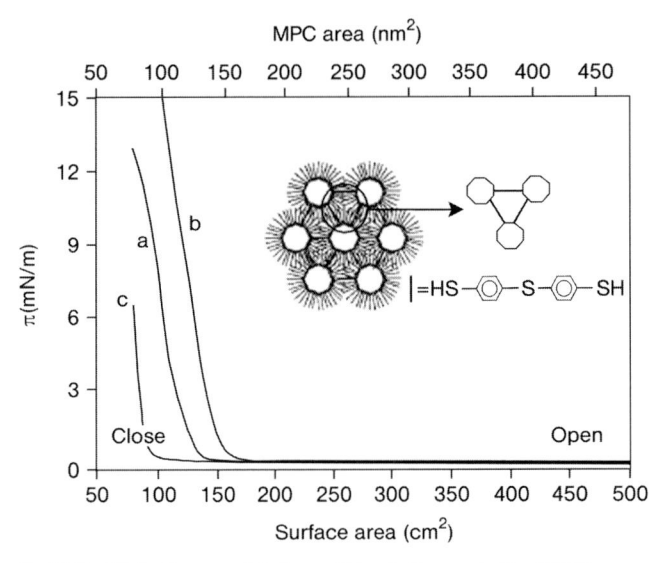

Fig. 12.5 LB isotherms of C_8Au particles before and after TBBT cross-linking. Initially, 150 μL of the C8Au particle solution in hexane (1 mg/mL) was spread onto the water surface. The TBBT concentration was 4 mM in CH_2Cl_2 and 75 μL was added to the water surface. **a** The isotherm of the C_8Au particles prior to addition of TBBT; **b** the first compression after the introduction of TBBT; **c** the isotherm of the resulting particle networks after holding the barrier in the closed position for 15 h. The inset shows the schematic of the particle cross-linked networks (reprinted with permission from [62], © 2001 American Chemical Society).

water [56]. This process is illustrated schematically in Fig. 12.7. First, the 2-aminoethanethiol-capped CdS nanoparticles (2-AET/Q-CdS, 2.7 ± 0.4 nm diameter) were spread on the surface of water at 0 °C containing a known amount of glutaraldehyde (Step 1, Fig. 12.7). The 2-AET/Q-CdS monolayer was then held at different constant areas at a temperature of 25 °C for 1 h to enable cross-linking to proceed to completion (Step 2, Fig. 12.7). The glutaraldehyde molecules bind to the free amine groups surrounding neighboring CdS nanoparticles thereby cross linking them. Thereafter, the monolayer of cross-linked CdS nanoparticles were transferred at a pressure of 17 mN/m onto Au substrates previously modified with 2-aminoethanethiol by the LB method (Step 3, Fig. 12.7) for further studies.

Fig. 12.8 shows the STM images recorded from the Au (111) substrate covered with a monolayer of 2-AET (a), the 2-AET/Q-CdS nanoparticle film without the glutaraldehyde cross-linker (b) and 2-AET/Q-CdS film cross-linked with glutaraldehyde (c). While the surface is essentially featureless in Fig. 12.8a, the CdS nanoparticles aggregate to form islands of dimensions greater than 10 nm on the gold substrate in the absence of glutaraldehyde. The cross-linked CdS nanoparticle monolayer, however, shows an extremely uniform and regular arrangement of the

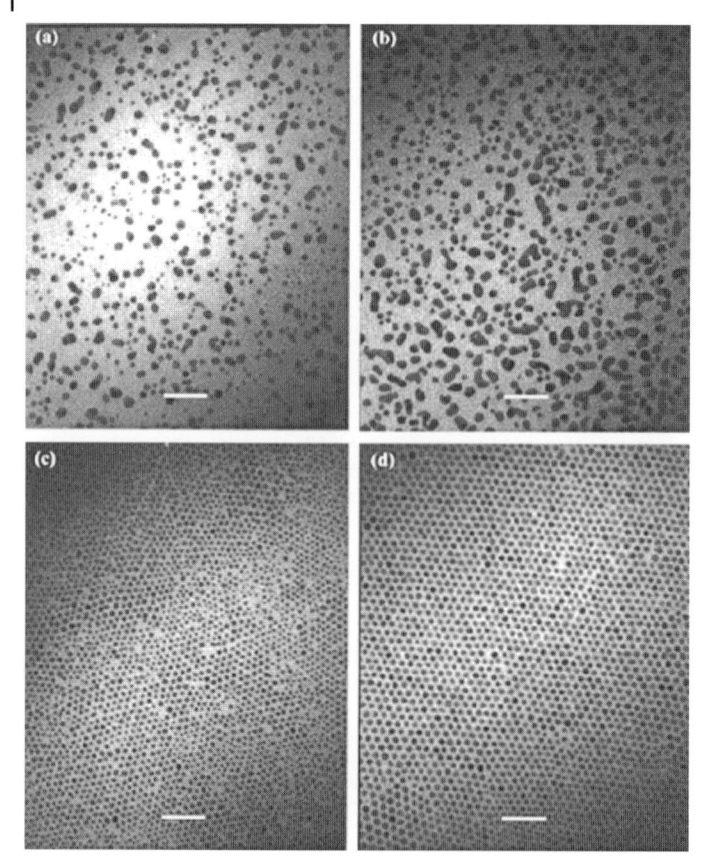

Fig. 12.6 TEM micrographs of the C_8Au particles deposited onto a SiO_x-coated Cu grid at varied surface pressures before and after TBBT cross-linking: **a** no TBBT, 4 mN/m; **b** no TBBT, 7 mN/m; **c** after TBBT networking, 4 mN/m; **d** after TBBT networking, 7 mN/m. The scale bars are all 33 nm (reprinted with permission from [62], © 2001 American Chemical Society).

nanoparticles on the gold substrate (Fig. 12.8c) thereby clearly attesting to the stabilizing nature of the cross-linking agent, glutaraldehyde. Yoneyama and co-workers then went on to investigate the photoelectrochemical properties of the cross-linked CdS nanoparticle films [56].

A potentially exciting application of monolayer films of hydrophobized gold nanoparticles deposited by the LB technique has been recently shown by Bourgoin and co-workers [64]. This group has shown that sub-50 nm wide nanowires can be created by electron beam lithography of monolayers of gold nanoparticles capped with alkanethiols. After electron beam irradiation and pattern writing, the films are immersed in toluene. The nanoparticle film behaves as a negative resist, i.e. the area exposed to the electron beam is rendered toluene insoluble. Cross-linking

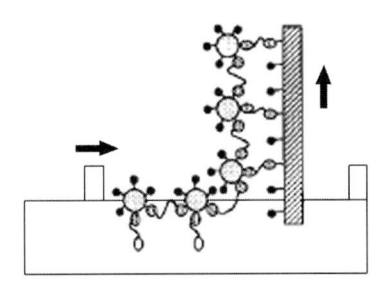

Fig. 12.7 Schematic illustrations of two-dimensional cross-linking of 2-aminoethanethiol-modified CdS nanoparticles spread at the air-water interface (reprinted with permission from [56], © 1999 American Chemical Society).

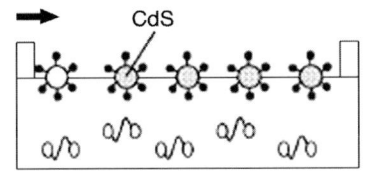

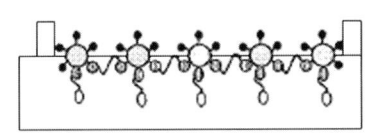

of the gold nanoparticles during electron beam irradiation was proposed as the mechanism behind immobilization of the nanoparticles [64]. The gold nanowires are one particle diameter in thickness and robust enough to withstand lithography aimed at making contacts with electrodes [64].

12.2.1.2 Complexation of Aqueous Nanoparticles with Langmuir Monolayers

The second method for the physical assembly of nanoparticles at the air-water interface is based on the attachment of suitably surface-modified nanoparticles from the aqueous subphase with Langmuir monolayers. In this method, the nanoparticles are derivatized by binding to their surface molecules bearing appropriate terminal functional groups that would enable their complexation with complementary functionality present in Langmuir monolayers on the nanoparticle subphase. The use of coulombic or electrostatic interactions in nanoparticle assembly at the air-water interface has been until now the most popular approach. Here again, Fendler and co-workers published the first report on using electrostatic interactions for the assembly of size-quantized CdS nanoparticles [74]. Nanoparticles of metals such as silver [75–82] and gold [77, 80, 83, 84]; semiconductor quantum dots such as CdS [74, 85] and CdTe [86]; oxides such as SiO_2, TiO_2 [87] and Fe_3O_4 [88] and latex spheres [89] have since been assembled from solution at the air-water interface using electrostatic interactions.

The basic ingredients of the electrostatic assembly of nanoparticles with oppositely charged Langmuir monolayers are illustrated in Fig. 12.9A. In this particular example, nanoparticles in the aqueous subphase are derivatized with carboxylic

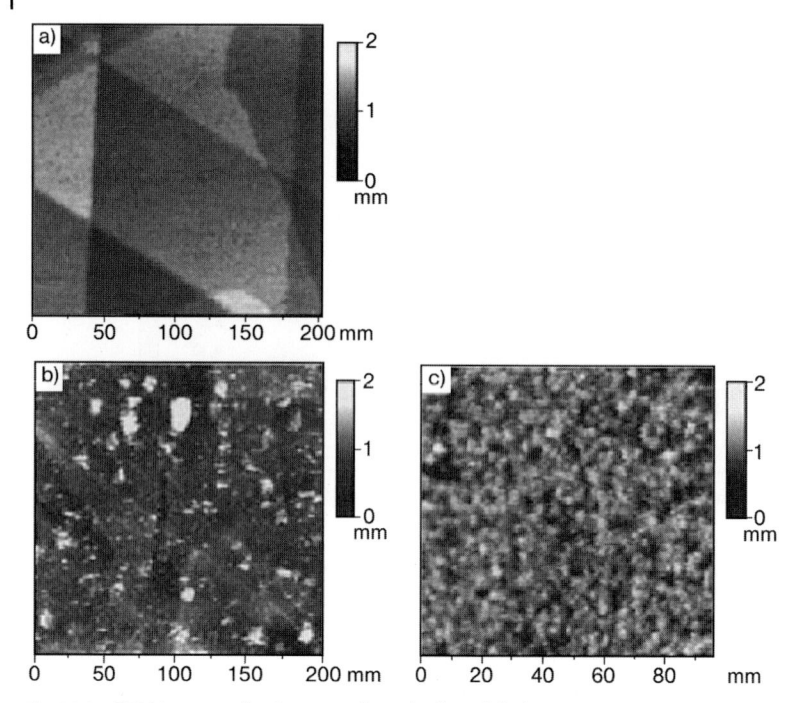

Fig. 12.8 STM images of **a** 2-aminoethanethiol-modified Au (111) surface (bias voltage 0.10 V; current 30 pA) and CdS monoparticulate films formed on 2-aminoethanethiol-modified Au (111) surface (1.5 V, 10 pA) **b** without addition of glutaraldehyde and **c** with two-dimensional cross-linking of CdS nanoparticles by glutaraldehyde in the subphase under conditions of $A_{link}/A_{CdS}=1.0$. The number of cumulations of CdS monoparticle layer was 1 in both cases (reprinted with permission from [56], © 1999 American Chemical Society).

acid functionality. Upon spreading a Langmuir monolayer of a cationic lipid such as octadecylamine (ODA) on the subphase surface held at a pH where both the carboxylic acid groups on the nanoparticle surface and the amine groups in the Langmuir monolayer are fully charged, the nanoparticles are attracted to the Langmuir monolayer through coulombic interactions. This process results in the assembly of the nanoparticles at the interface in the form of a monolayer that can be easily transferred by the LB technique onto solid substrates (Fig. 12.9 A). We have used this procedure to immobilize CdS quantum dots in the manner discussed below [85].

CdS nanoparticles of dimensions 3.5 nm were synthesized in water and capped with 4-carboxythiophenol (4-CTP) molecules. It is well known that thiol groups bind to CdS nanoparticles and, thus, this process leads to the presence of free carboxylic groups on the nanoparticle surface. By suitably adjusting the pH of the CdS hydrosol, the carboxylic acid groups may be ionized and the particles stabilized in solution electrostatically. Fig. 12.10 shows the surface pressure versus mo-

12.2 Nanoparticle Assembly at the Air-Water Interface

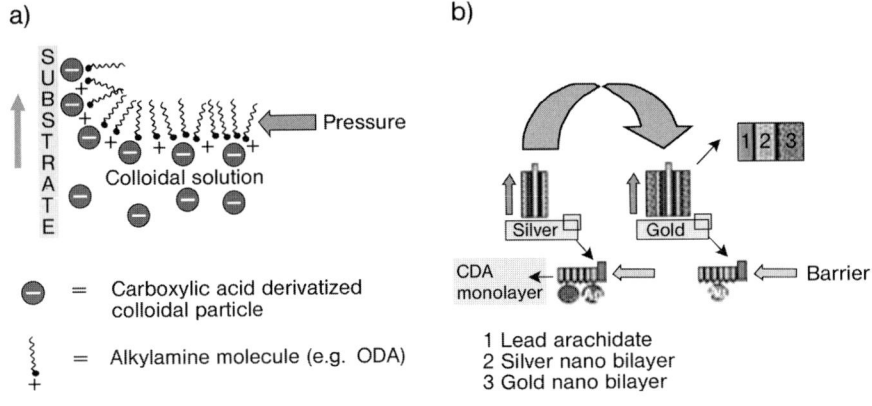

Fig. 12.9 **a** Electrostatic immobilization of carboxylic acid-derivatized colloidal particles at the air-water interface with fatty amine Langmuir monolayers and vertical lifting of the nanoparticle monolayers onto solid supports by the LB method; **b** assembly of colloidal gold and silver particles in an alternating layer-by-layer fashion on solid substrates by the LB technique (reprinted with permission from [82], © 2002 American Chemical Society).

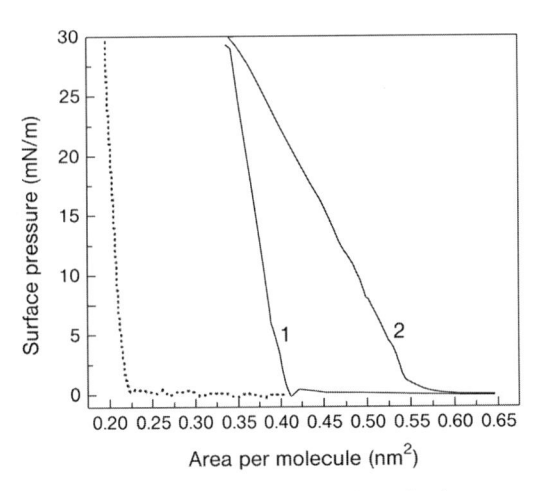

Fig. 12.10 Pressure-area isotherms of octadecylamine on 4-CTP-stabilized CdS hydrosol (pH 9) after 15 min of equilibration (curve 1) and 120 min of equilibration (curve 2). The dotted line is the isotherm of octadecylamine on water at pH 9 (reprinted with permission from [85], © 1998 Elsevier Science).

lecular area (π–A) isotherms recorded from ODA monolayers spread on the surface of 4-CTP-capped CdS nanoparticle solution at pH 9 as a function of time of spreading the monolayer. At this pH, the ODA monolayer is protonated (pK_B of ODA is 10.5) while the carboxylic acid groups on the CdS nanoparticles are completely ionized leading to maximum attractive electrostatic interaction between them. It is observed that there is a large expansion in the ODA monolayer with time and the liftoff area increases from ca. 0.41 to ca. 0.55 nm^2/molecule within 2 h of spreading the monolayer. This expansion is clearly an indication that the CdS nanoparticles are being immobilized at the ODA monolayer surface and is to be contrasted with the takeoff area of 0.2 nm^2/molecule recorded for the ODA monolayer on plain water at pH 9 (dotted line, Fig. 12.10). The kinetics of complexation of the nanoparticles is slow, often requiring 15 h for nanoparticle equilibration at the air-water interface as observed in the case of 4-CTP-capped silver particles [76]. The fact that the interaction between the CdS nanoparticles and the ODA monolayer is electrostatic in nature was established by the fact that no nanoparticle complexation was observed when the CdS subphase was maintained at pH 12. At this pH, the ODA monolayer is uncharged and consequently the driving force for surface assembly is absent [85]. Similar variation in the pH-dependent complexation of silver [76, 78], gold [84] and small SiO_2 and TiO_2 nanoparticles with ODA monolayers [87] have been made, thereby strengthening the electrostatic interaction model proposed.

The ability to modulate the electrostatic interaction between nanoparticles in the subphase and Langmuir monolayers is nicely illustrated in the work of Muramatsu et al. who have carefully recorded not only the variation in the π–A isotherms of ODA Langmuir monolayers during complexation of negatively charged SiO_2 and TiO_2 nanoparticles, but also scanning electron microscopy (SEM) images of the nanoparticle monolayer deposited under different pH conditions of the nanoparticle subphase [87]. Fig. 12.11 shows SEM images recorded from ODA-SiO_2 nanoparticle (diameter 110 nm) monolayers deposited from SiO_2 subphases at different pH values. It is clear that the SiO_2 nanoparticle density in the monolayer is highest at pH 7 and decreases progressively as the pH is increased to 11.6 [87]. The charge on the ODA monolayer reduces significantly above pH 10 (pK_B of ODA = 10.5) and would lead to a weaker electrostatic interaction between the silica nanoparticles and the Langmuir monolayer as observed in the SEM images (Fig. 12.11). The SiO_2 nanoparticles pack into a hexagonal close-packed

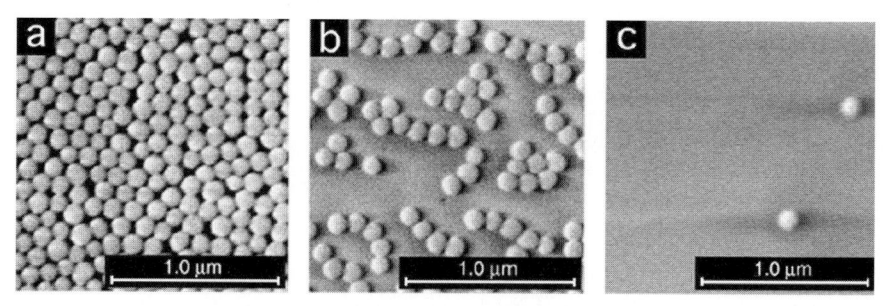

Fig. 12.11 SEM images of the ODA-SiO_2 (d=110 nm) LB monolayers deposited at various pH values: **a** 7.0, **b** 10.0, **c** 11.6 (reprinted with permission from [87], © 1998 Academic Press).

monolayer structure at pH 7 (Fig. 12.11a) indicating that the electrostatic repulsive interactions between the negatively charged silica spheres is to a large extent neutralized by the cationic ODA monolayer. These authors also performed a detailed analysis of SEM pictures obtained from SiO_2 nanoparticle monolayers transferred at different pH values as a function of the silica particle size. The data obtained is shown in Fig. 12.12 for 110 nm (open circles), 300 nm (open triangles) and 560 nm (open squares) diameter SiO_2 particles complexed with ODA Langmuir monolayers [87]. It can be seen that while the nanoparticles assemble into a hexagonal close-packing configuration for the 110 nm silica spheres (ca. 80% surface coverage), the surface coverage is much less for the larger SiO_2 particles (Fig. 12.12). Muramatsu et al. speculated that the reduced surface coverage of the large SiO_2 particles in the Langmuir monolayers could be due to detachment of the particles during the process of substrate withdrawal from the subphase or to a reaction of ODA molecules with atmospheric CO_2 leading to the formation of a carbamate [87].

After equilibration of the CdS nanoparticle density at the air-water interface, the amine-CdS nanoparticle monolayers were transferred onto different hydrophobized substrates at a surface pressure of 25 mN/m by the LB technique [85]. Close to unity transfer ratio was observed during both the upward and downward strokes of the substrate indicating Y-type growth of the CdS nanoparticle superlattice. The UV-vis spectra recorded from the amine-CdS nanoparticle LB films grown on quartz substrates are shown in Fig. 12.13. An absorption threshold of 450 nm was determined from the figure which corresponds to an average particle diameter of ~3.5 nm [90]. This absorption edge arises from the first excitonic transition and is dependent on the size of the nanoparticles [90]. An increase in

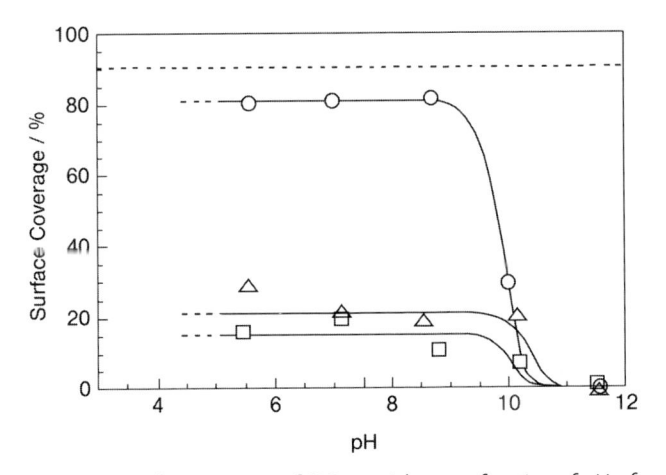

Fig. 12.12 Surface coverage of SiO_2 particles as a function of pH of the colloidal subphase. The dotted line at a surface coverage of 91% is for perfect hexagonal packing of spherical particles. d: O, 110 nm; △, 300 nm; □, 560 nm (reprinted with permission from [87], © 1998 Academic Press).

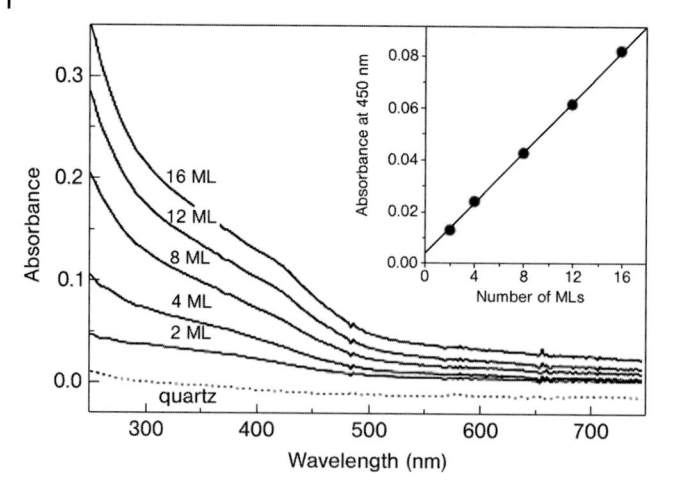

Fig. 12.13 Optical absorption spectra of several multilayer amine-CdS nanoparticle LB films deposited on hydrophobized quartz substrates at CdS subphase pH=9. The dashed line is the absorption spectrum of the bare quartz substrate. The inset shows the variation of the absorption intensity at 450 nm with film thickness while the solid line is a least-squares fit to the data (reprinted with permission from [85], © 1998 Elsevier Science).

the absorption intensity at 450 nm with increasing thickness of the LB film can clearly be seen in the figure. The inset of Fig. 12.13 shows a plot of the absorbance at 450 nm plotted against the number of monolayers in the different CdS nanoparticle films. It can be seen that there is a good linear increase in the absorbance at 450 nm with LB film thickness, indicating that lamellar growth of the CdS nanoparticle monolayers occurs without significant variation in the nanoparticle density in the different layers of the film. This is an important requirement for the formation of ordered superlattice structures. Ellipsometry and quartz crystal microgravimetry measurements were also performed on LB films of the CdS nanoparticles and in agreement with the optical absorption spectroscopy measurements, lamellar growth of the CdS nanoparticle monolayers was established [85].

It is well known that LB films of ODA complexed with ions such as $PtCl_6^{2-}$ show a lamellar, c-axis-oriented structure akin to those of LB films of metal salts of fatty acids [91]. This is understandable given that the dimensions of the ion are considerably smaller than the dimensions of the ODA bilayer (ca. 5 nm). On the other hand, the size of the CdS nanoparticles (3.5 nm) is larger than the molecular dimensions of ODA (2.5 nm) and, therefore, the presence of the nanoparticles in the bilayer stacks would lead to a significant distortion to the bilayer structure in LB films of the nanoparticles. An artist's representation of the possible structure of the CdS nanoparticle-ODA LB film is shown in Fig. 12.14 [85]. It is expected that the ODA monolayer in the vicinity of the nanoparticle would be distorted as shown in the figure. Another possibility is the formation of surface mi-

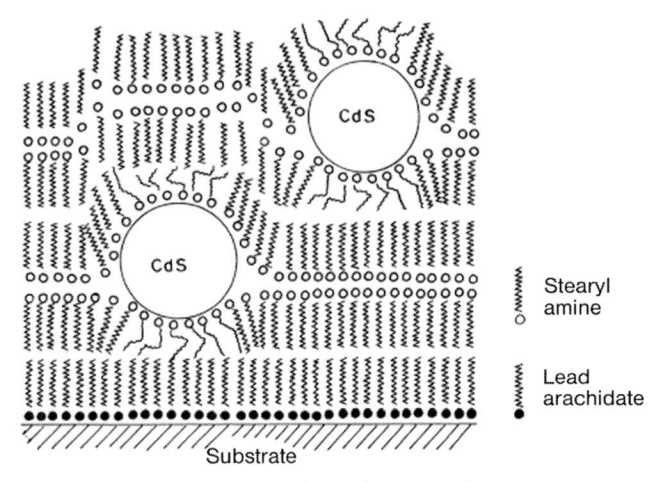

Fig. 12.14 Probable structure of the CdS-octadecylamine Langmuir-Blodgett film (reprinted with permission from [85], © 1998 Elsevier Science).

cellar assemblies of the CdS nanoparticles capped with ODA resulting in hydrophobic nanoparticles floating on the water surface.

The versatility of the Langmuir-Blodgett technique arises from the fact that superlattice structures consisting of amphiphilic molecules [92] or nanoparticles [80] in any desired sequence can be easily deposited on a solid substrate by using multiple troughs in the deposition process. This process is illustrated in Fig. 12.9 B for the formation of alternating layers of silver and gold nanoparticles by the LB technique [80]. Sequential immersion of substrates alternately in two troughs containing silver (7 nm diameter) and gold (13 nm diameter) colloidal solutions results in the transfer of a bilayer of the nanoparticles per dip and formation of a hetero-nanoparticle assembly. This process may be extended to films of larger thickness and more complicated nanoparticle sequences.

12.2.2
Chemical Methods for Generation of Nanoparticle Langmuir-Blodgett Films

As briefly mentioned in the introduction, organized assembly of nanoparticles can be achieved by a chemical process involving ions bound to Langmuir monolayers at the air-water interface or ions entrapped in LB films of metal salts of fatty lipids (Fig. 12.1). This is a potentially exciting approach, with many distinctive advantages over the physical assembly methods discussed earlier, and details are provided below.

12.2.2.1 Nanoparticle Formation by Reaction of Metal Ions at the Air-Water Interface

Historically, the synthesis of nanoparticles in LB films of metal salts of fatty lipids (the so-called chemical insertion method) preceded the first demonstration of nanoparticle formation at the air-water interface. However, I address the latter method first since, in many ways, understanding of the special features of the chemical formation of nanoparticles at the air-water interface would lead to a logical appreciation of the technique for nanoparticle synthesis in metal salt LB films.

The problem of electrostatic complexation of metal ions with oppositely charged Langmuir monolayers has been studied in great detail (see Ulman's book, in [72]). This process leads to a very high concentration of the metal counterions at the interface, often five orders of magnitude higher than the nominal concentration in the bulk subphase [93]. The metal ions are immobilized in a monolayer of atomic dimensions in crystalline assemblies, the crystallography of which can be modulated by changing a wide spectrum of experimental parameters such as the nature of insoluble surfactant in the Langmuir monolayer, the subphase pH, temperature, etc. Products formed by the reaction of ions immobilized at the interface would be expected to show a degree of anisotropy arising from constrained lateral diffusion of reactants and products within the interface. It is clear that these features taken together provide ample scope for chemical manipulation of ions bound at the air-water interface. Recognizing these special attributes of the air-water interface, researchers in the field of biomineralization have used the air-water interface as a model system and have shown that highly oriented crystals of minerals such as $BaSO_4$ [94–96], $CaCO_3$ [94, 97, 98] etc. can be grown at the air-water interface in the presence of cationic lipid Langmuir monolayers. Face-selective nucleation of the mineral crystals and crystal morphology control are some of the exciting features of this approach [94–98] that follow from the nature of ion assembly at the air-water interface.

The formation of metal nanocrystals such as gold [99] and silver [100] and semiconductor quantum dots such as CdS [101, 102] and PbS [103] by chemical treatment of the Langmuir monolayer-metal ion precursor has been demonstrated. The first step consists of formation of the complex of the metal ion with the Langmuir monolayer, this organo-inorganic complex serving as the precursor for chemical treatment. Thereafter, the precursor is exposed to the chemical agent. If metal nanoparticles are required, reducing agents such as hydrazine, CO and ultraviolet radiation [99] may be used while treatment with H_2S gas is indicated for formation of quantum dots of CdS [101, 102] and PbS [103]. The nanoparticles are formed at the air-water interface and, consequent to their strong interaction with the Langmuir monolayer, may be transferred in a layer-by-layer fashion by the LB method onto different supports.

The chemical formation of nanoparticles at the air-water interface can be illustrated with the first report on the formation of CdS nanoparticles by Yi and Fendler [101]. The formation of CdS nanoparticles at the air-water interface was accomplished in the following manner. A 2.5×10^{-4} M aqueous $CdCl_2$ solution was taken and a Langmuir monolayer of the molecule n-$C_{16}H_{33}C(H)[(CON(H)$-

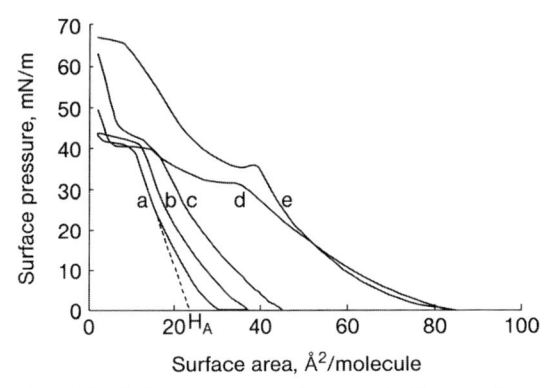

Fig. 12.15 Surface pressure–surface area isotherms of monolayers of 1 on aqueous 2.5×10^{-4} M $CdCl_2$. pH of the subphase was adjusted to **a** 6.1, **b** 7.4, **c** 8.4, **d** 8.8, and **e** 9.1 by addition of 0.10 M NaOH. Headgroup areas, indicated on the figure by H_A, were calculated to be **a** 24, **b** 27, **c** 35, **d** 38, and **e** 44 $Å^2$/molecule (reprinted with permission from [101], © 1990 American Chemical Society).

$(CH_2)_2$-$NH_2]_2$ (1) was spread on the surface. Thereafter, infusion of H_2S gas into the $CdCl_2$ solution led to the formation of CdS nanoparticles at the air-water interface. The process of complexation of cadmium ions with Langmuir monolayers of 1 as a function of subphase pH is shown in the π–A isotherms of Fig. 12.15. It is seen that as the pH of the subphase increases the monolayer area also increases, indicating more binding of the cadmium ions in the cavity of molecule 1 at higher pH. The formation of CdS nanoparticles was studied by UV-vis spectroscopy and TEM measurements. Fig. 12.16 shows the absorption spectra recorded from monolayers of the CdS nanoparticles complexed with 1 after different times of reaction with H_2S. It can be seen that as the time of exposure of the Cd^{2+}-1 monolayer complex at the air-water interface to H_2S increases, the onset of absorption

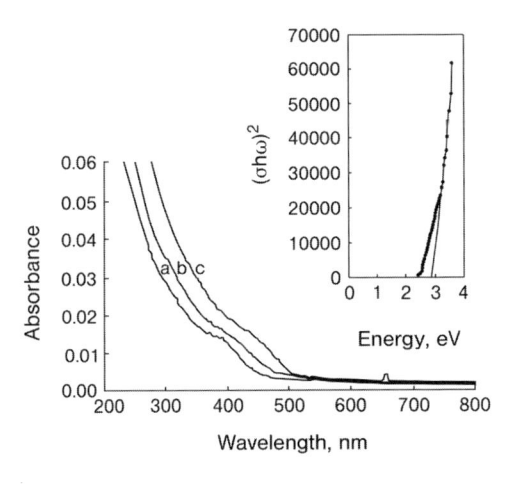

Fig. 12.16 Absorption spectra of CdS particles, generated in situ at 1 monolayers floating on the surface of 2.5×10^{-4} M $CdCl_2$ at pH 9.1 by **a** 3, **b** 5, and **c** 10 min exposure to H_2S and subsequently transferred to quartz substrates by horizontal lifting. The insert shows a plot of the data according to the equation $(\sigma h\omega)^2 = h\omega C - E_g C$ where σ is the absorption coefficient, $h\omega$ is the photon energy and E_g is the direct bandgap (reprinted with permission from [101], © 1990 American Chemical Society).

in the LB films shifts to larger wavelengths. The onset of absorption is due to the creation of excitons in the quantum dots of CdS, whose wavelength is a strong function of the size of the nanoparticles [90]. The shift to larger wavelengths with increasing reaction time indicates formation of larger CdS nanoparticles at the air-water interface. The inset of Fig. 12.16 shows a plot of the absorption data shown in Fig. 12.16, curve a, according to the equation given in the caption. This plot enables determination of the direct bandgap for the CdS quantum dots and was estimated to be 2.82, 2.76 and 2.74 eV for the films exposed to H_2S gas for 3, 5 and 10 min respectively. These bandgap values correspond to particle diameters in the range 30–50 Å [90]. Thus, the size of the quantum dots may be controlled by varying the reaction time [101].

TEM images recorded from monolayers of the CdS nanoparticles formed in situ after reaction of the metal ions with H_2S gas for 30 min at different pH values of the aqueous $CdCl_2$ subphase are shown in Fig. 12.17. Higher pH leads to the formation of smaller CdS nanoparticles [101].

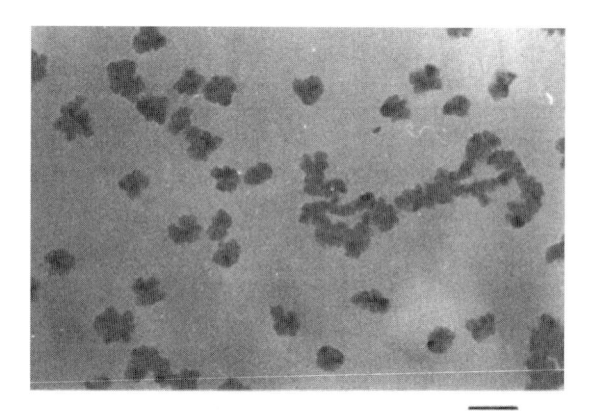

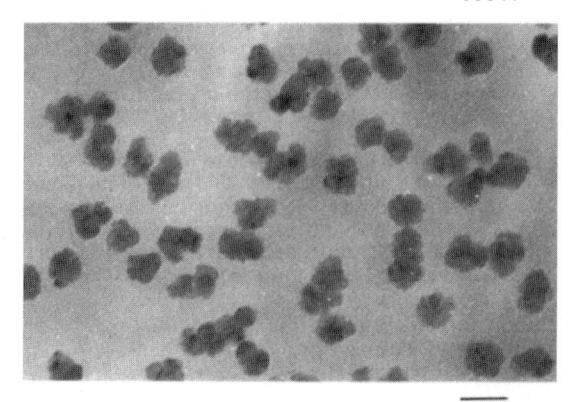

Fig. 12.17 Transmission electron micrographs of CdS particles, generated in situ at 1 monolayers floating on 2.5×10^{-4} M $CdCl_2$ at **a** pH 9.1 and **b** pH 8.6 by exposure to H_2S for 30 min. Particles were subsequently transferred to 200-mesh, celluloid-coated copper grids (reprinted with permission from [101], © 1990 American Chemical Society).

Yi and Fendler also demonstrated the formation of CdS nanoparticles in LB films of 1 complexed with Cd^{2+} ions by reaction with H_2S gas [101]. Formation of CdS particles smaller than 50 Å diameter was indicated in these experiments.

12.2.2.2.2 Nanoparticle Formation by Reaction of Metal Ions in Langmuir-Blodgett Films of Salts of Fatty Lipids

Lamellar LB films of metal salts of fatty lipids wherein the metal ions are entrapped within nanoscale containers possess features that provide excellent opportunities for controlling the size, orientation and shape of nanoparticles synthesized in situ chemically. The concentration of ions within the hydrophilic regions of the bilayers of LB films can be easily controlled during preparation of the film. These hydrophilic regions are of nanoscale dimensions (along a direction normal to the bilayers) and highly anisotropic. Furthermore, the lamellar structure of the precursor organic-inorganic hybrid LB film could result in the formation of a highly periodic assembly of nanoparticles in the reaction matrix. The periodicity of the nanoparticle layers may be easily varied by varying experimental parameters such as the hydrocarbon chain length of the LB host, thereby modulating the interaction between the nanoparticle layers. Realizing such periodic organic–inorganic nanoparticle hybrid structures is an important goal in advanced materials engineering.

Barraud and co-workers were the first to realize the implications of using LB films of metal salts of lipids in nanoparticle synthesis and showed that silver nanoparticles could be chemically inserted into the LB films by reduction of Ag^+ ions entrapped within the lipid lamellae [104, 105]. Subsequently, a number of groups have investigated this approach from different angles and have reported on the in situ synthesis of nanoparticles of gold [106], CdI_2 [107], CdS [101, 108–113], CdSe [114], PbS [115–118], TiO_2 [119], copper and copper sulfide [118, 120, 121] to name just some of the chemical compositions investigated. While not always leading to the formation of nanoparticles, the chemical insertion method in LB films has also been used with considerable success in realizing ultrathin films of oxides such as TiO_2 [122], Y_2O_3 [123], ZnO [124, 125] and Er_2O_3 [126].

Controlling the morphology of inorganic nanoparticles is an area of topical interest. Development of a photoinduced method for the large-scale conversion of silver nanoparticles into nanoprisms has led to the first measurement of two distinct quadrupole plasmon resonances in the particles [6]. Nanorods of variable aspect ratios are exciting from the point of view of synthesis of nanowires in electronic circuitry and many solution-based methods have begun to address this issue [8, 127]. The synthesis of nanoparticles of different morphologies in the form of a highly ordered thin film would, as mentioned earlier, have exciting potential. Talham and co-workers have addressed this problem in their study of the growth of gold nanoparticles in LB films of different organic matrices [106]. They have systematically studied the complexation of $AuCl_4^-$ ions with Langmuir monolayers of amphiphiles such as octadecylamine, 4-hexadecylaniline (HDA), benzyldimethylstearylammonium chloride (BDSAC) and the phospholipids dipalmitoyl-

DL-α-phosphatidyl-L-serine (DPPS) and dipalmitoyl-L-α-phosphatidylcholine (DPPC) by monitoring the π–A isotherms [106]. Thereafter, LB films of the $AuCl_4^-$-lipid complexes were formed on different substrates as precursors for the in situ synthesis of gold nanoparticles by UV reduction of the metal ions. They observed that while the gold cations complexed readily with the cationic lipids ODA, HDA and BDSAC and resulted in the formation of gold nanoparticles after UV treatment, no nanoparticles were observed in the phospholipids [106]. As a representative example, the UV-vis spectra recorded from a 13-monolayer ODA film complexed with gold ions as a function of time of UV irradiation is shown in Fig. 12.18. It is observed that as the irradiation proceeds, an absorption at 550 nm grows steadily. This growth of absorption (due to excitation of surface plasmon oscillations in the nanoparticles [73]) is related to the creation of gold nanoparticles and is accompanied by a fall in the absorption at 330 nm due to $AuCl_4^-$ species [106]. The width of the resonance at 550 nm decreases with time of irradiation indicating that the gold particles grow in size.

Talham et al. observed that the morphology of the gold nanoparticles was strongly dependent on the nature of the organic host. Figures 12.19, 12.20 and 12.21 show TEM pictures of gold nanoparticles synthesized in 9-monolayer LB films of BDSAC, ODA and HDA respectively.

In the case of the BDSAC matrix (Fig. 12.19), the particles ranged in size from 20 to 300 nm and could be classified into two categories. The first class includes multiply-twinned particles (MTPs) and in Fig. 12.19a, decahedral and icosahedral MTPs have been identified. A number of ⟨111⟩-oriented triangular single crystals of gold were also observed by the authors (Fig. 12.19b). Clearly, the face-selective nucleation and growth of the gold nanocrystals indicates epitaxy with the underlying BDSAC organic template. Close examination of this image shows that the gold nanocrystal edges are truncated, which is a feature observed by Mirkin and co-workers in their silver nanoprism samples [6]. The number density of gold nanoparticles in the ODA LB film is much higher than in both BDSAC (Fig. 12.19a) and HDA (Fig. 12.21a) matrices indicating that ODA is an effective matrix for

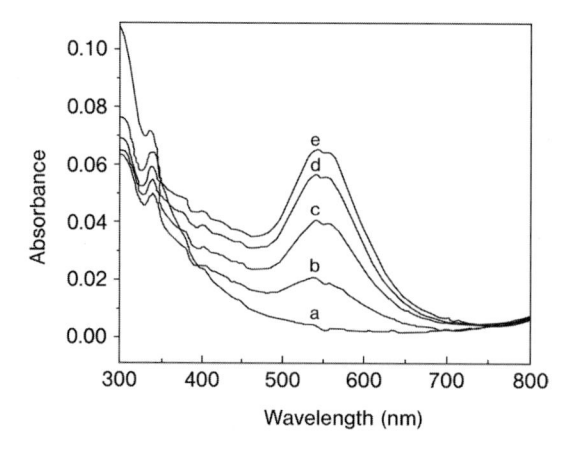

Fig. 12.18 UV-visible absorption spectra of 13-layer ODA LB film after exposure to UV light for **a** 0, **b** 15, **c** 20, **d** 25, and **e** 30 min. The small feature between 530 and 550 nm is an instrumental artifact (reprinted with permission from [106], © 1998 American Chemical Society).

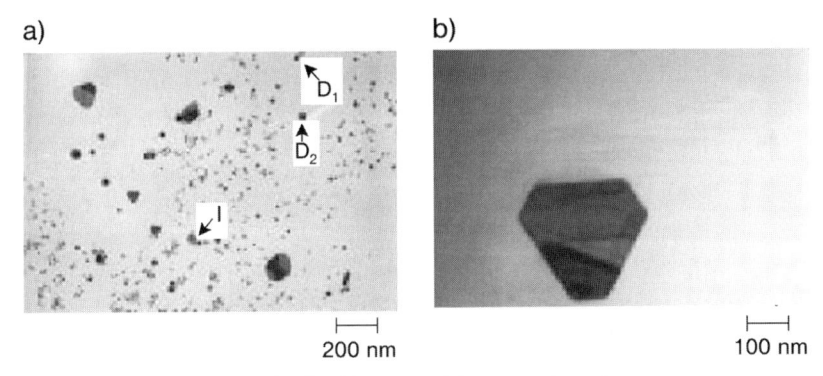

Fig. 12.19 a TEM micrograph of gold nanoparticles grown in a nine-layer BDSAC LB film exposed to UV irradiation for 120 min. The arrows indicate MTPs of decahedral (D_1 and D_2) and icosahedral (I) morphology. **b** TEM micrograph of a triangular gold single crystal with truncated corners, grown in a nine-layer BDSAC LB film after exposure to UV irradiation for 120 min (reprinted with permission from [106], © 1998 American Chemical Society).

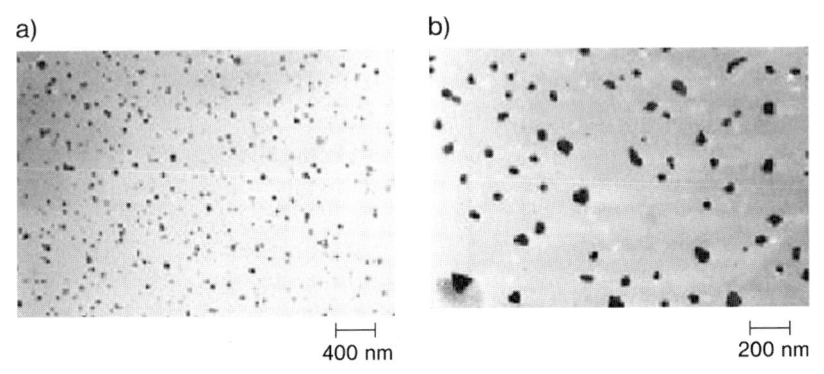

Fig. 12.20 a, b TEM micrographs of gold nanoparticles grown in a nine-layer ODA LB film irradiated for 120 min (reprinted with permission from [106], © 1998 American Chemical Society).

gold nanoparticle growth [106]. The nanoparticle size was also much more uniform, possibly due to the homogeneous distribution of gold ions in the precursor LB film. On the other hand, gold nanoparticles in HDA were much larger and far more irregular (Fig. 12.21). Their size ranged from 200 to 800 nm and their platelike morphology resembled those of gold nanoparticles synthesized under Langmuir monolayers of octadecyl mercaptan by Fendler and co-workers [99].

From the results of Talham et al. and those of others, it is clear that there are two factors that would control the size and morphology of nanoparticles grown chemically within LB precursor matrixes. They are the transport of the inorganic ions within the anisotropic spaces provided by the LB bilayers during reaction and

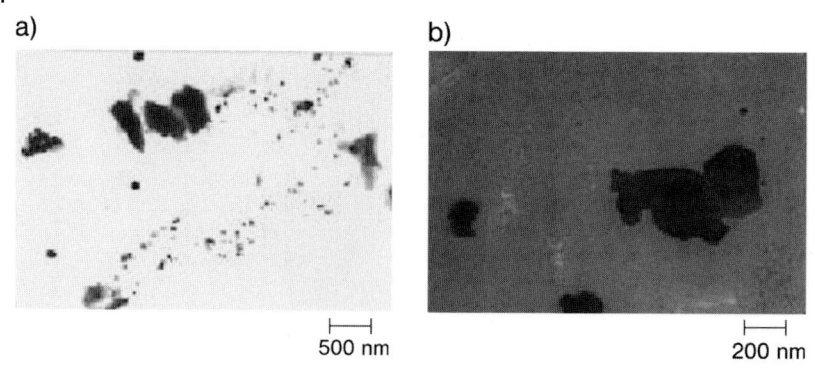

Fig. 12.21 a, b TEM micrographs of gold nanoparticles grown in a nine-layer HDA LB film irradiated for 120 min (reprinted with permission from [106], © 1998 American Chemical Society).

nucleation of the particles at specific sites in the reaction phase. A rational approach to address these factors was attempted by Lahav and co-workers in their study of growth of CdS nanoparticles in both 3D crystals and LB films of Cd alkanoates [111]. Their strategy was to form mixed bilayers of cadmium alkanoates and thioalkanotes as illustrated in Fig. 12.22. The presence of thioalkanoates is designed to act as nucleation centers for CdS nanoparticles thus preventing their uncontrolled growth during treatment of the crystals/LB films with H_2S.

Lahav et al. observed that in LB films in which the percentage of cadmium thioalkanoates was high relative to that of cadmium alkanoates, the onset of absorption in the UV-vis spectra of LB films of CdS nanoparticles shifted to the blue, indicating the formation of much smaller CdS particles. TEM images of LB films of CdS nanoparticles formed in a mixed matrix (50% thioalkanoate) and in a pure Cd alkanoate matrix are shown in Fig. 12.23. The image in Fig. 12.23a shows the side view of the LB film from the 50% thioalkanoate sample caused by folding in a region of the sample. Periodic order in the CdS nanoparticles can clearly be seen in this film, which is absent in the CdS nanoparticle LB film formed from the pure fatty acid alone (Fig. 12.23b). This important result clearly underlines the role played by the thioalkanoate nucleation centers in controlling the size of the nanoparticles as well as in retention of lamellar order of the LB precursor [111].

Richardson et al. have approached the problem of nanoparticle size control in a slightly different manner [110]. They have studied the formation of CdS nanoparticles in calixarene Cd salt LB films by reaction with H_2S. It was observed that extremely small (1.5±0.3-nm diameter) CdS nanoparticles were formed in the calixarene LB films and this was attributed to reduction in the lateral motion of CdS molecules and small aggregates owing to their capture by the calixarene cavities [110].

An interesting application of such nanoparticle-LB film composites has been recently shown by Erokhin and co-workers [118]. They have shown that in-plane pat-

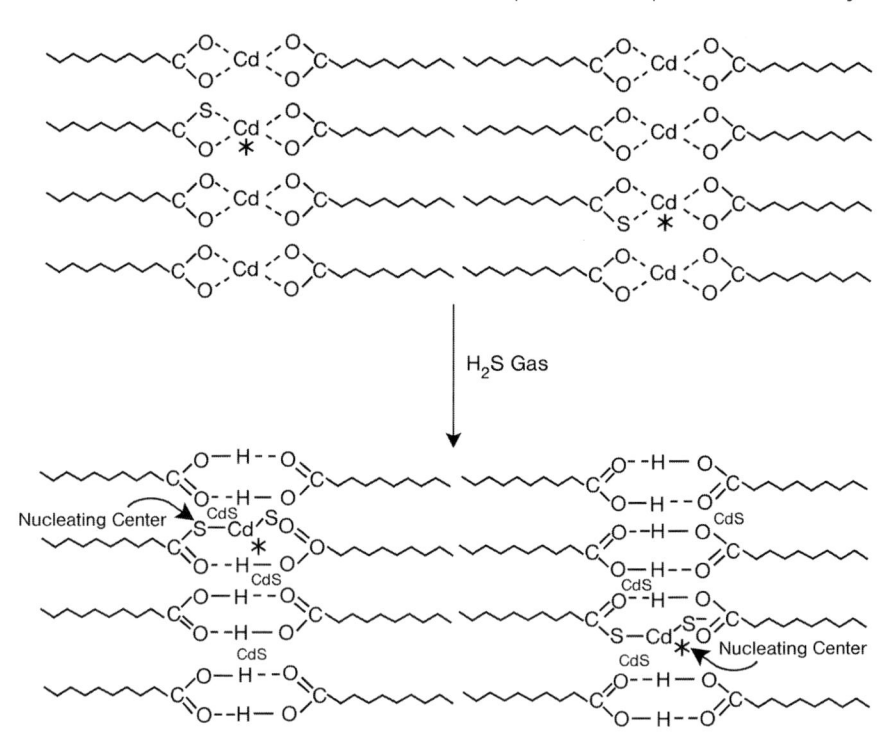

Fig. 12.22 Schematic packing arrangement of mixed crystals of cadmium alkanoates with thioalkanoates and the proposed nucleation centers for CdS during reactions of these mixed crystals with dry H_2S (reprinted with permission from [111], © 1999 American Chemical Society).

terning of nanoparticle films may be accomplished by a simple electron irradiation process as illustrated in Fig. 12.24. Exposure of an LB film of the metal salt of a fatty acid such as stearic acid (the metal ions used were Cd^{2+}, Pb^{2+} and Cu^{2+}) through a mask resulted in the cross-linking of stearic acid molecules in the exposed regions (Fig. 12.24a). The exposed regions thereby become insoluble in an organic solvent. After electron beam irradiation, the LB films were reacted with H_2S to form metal sulfide nanoparticles throughout the film (Fig. 12.24b). After formation of the nanoparticles, the film was immersed in a solvent such as chloroform and the nanoparticles aggregated in the shadow regions as shown in Fig. 12.24c. The authors also studied the electrical conductivity of the different regions in the aggregated nanoparticle film and showed that the conductivity could vary by as much as five orders of magnitude across irradiated and non-irradiated regions of the film surface [118].

This section is concluded by mentioning a new method developed in my laboratory for realizing nanoparticles in lipid bilayer structures very similar to that obtained by chemical reaction of LB films of metal salts of fatty lipids. We have

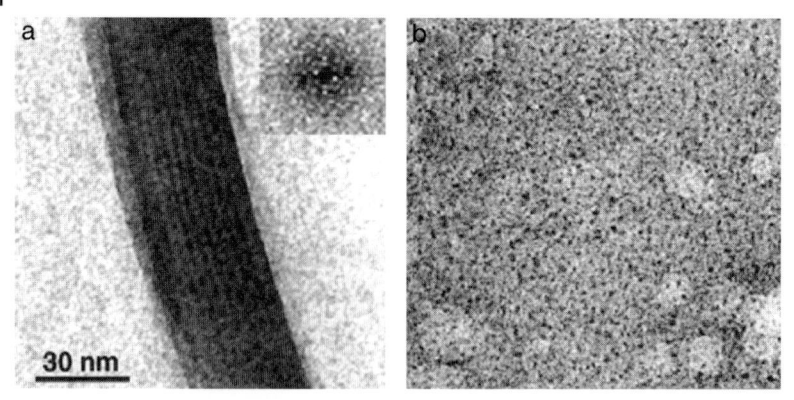

Fig. 12.23 TEM images of CdS particles formed in LB films of C_{22} alkanoate **a** containing 50% of the corresponding thioalkanoate and **b** pure Cd C_{22} alkanoate; **a** shows a folding in the LB film, revealing both top and side views. The inset shows FFT from the side view region. All images were taken at the same magnification (reprinted with permission from [111], © 1999 American Chemical Society)

shown that thermally evaporated films of fatty acids (arachidic acid, for example) on solid supports when immersed in suitable electrolyte solutions such as $CdCl_2$/$PbCl_2$ resulted in electrostatic entrapment of the metal ions in the lipid matrix and their spontaneous organization into a lamellar c-axis-oriented structure very similar to that observed for LB films of cadmium/lead arachidate [128]. Chemical treatment of these lamellar lipid films with entrapped metal ions would lead to the formation of nanoparticles as outlined above for LB films with the important advantage that patterned lipid films can be deposited by thermal evaporation, such patterning being almost impossible to achieve by the LB method. We have

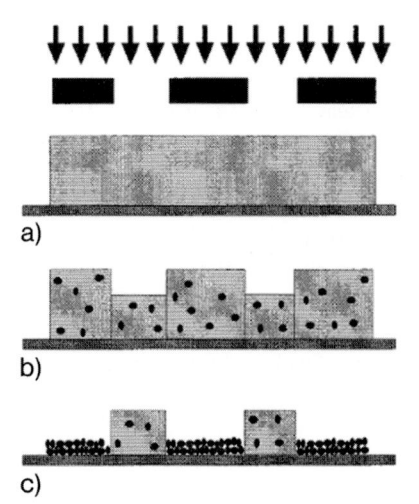

Fig. 12.24 Scheme of the patterning process of the aggregated layer: **a** irradiation of the sample by an electron beam through the mask; **b** formation of the particles in the sample by the chemical reaction; **c** aggregation of the particles by selective removal of fatty acid molecules in the non-irradiated regions (reprinted with permission from [118], © 2002 American Chemical Society)

used this approach to form patterned assemblies of gold [129] and CdS nanoparticles [130] and have also recently investigated the low-temperature alloying of nanoparticles produced within the thermally evaporated lipid matrices [131–134].

12.3 Summary

The use of the air-water interface as an organizational medium for nanoparticles has been outlined. Two main approaches based on physical organization of nanoparticles and chemical treatment of ions bound to Langmuir monolayers and in Langmuir-Blodgett films have been described and illustrated with specific examples from the literature. The versatility of the air-water interface in nanoparticle assembly lies in the fact that multilayer films with tunable interaction between nanoparticle layers may be easily deposited by the LB technique. This field is relatively new and there are many interesting issues yet to be resolved and modifications possible. Some of them include the formation of LB films containing simultaneously different ions for chemical treatment to yield doped semiconductor nanoparticles, ternary oxides or alloys; highly localized reduction of metal ions at the air-water interface to yield nanosheets/nanoribbons and the immobilization of biological molecules in LB films as templates for the hierarchical assembly of nanoparticles.

12.4 Acknowledgements

I am grateful to my former doctoral student, Dr. K. Subramanya Mayya and my current students, Mr. Ashavani Kumar and Ms. Anita Swami who have carried out most of the experimental work from my group discussed in this chapter. This work owes much to the generous financial support of the Department of Science and Technology, Government of India; the Indo-French Centre for the Promotion of Advanced Scientific Research (IFCPAR) and the Council of Scientific and Industrial Research, Government of India, and is gratefully acknowledged. I apologize for any omissions and oversights – they are unintentional.

12.5 References

1 D.A. HANDLEY, *Colloidal Gold: Principles, Methods and Applications* (HAYAT, M.A., ed.), Academic Press, San Diego, **1989**, Vol. 1, Chapter 2.
2 G. SCHMID, *Clusters and colloids – from theory to applications*, VCH, Weinheim, **1994**.
3 D.V. LEFF, P.C. OHARA, J.C. HEATH, W.M. GELBART, *J. Phys. Chem.* **1995**, *99*, 7036.
4 P. MUKHERJEE, A. AHMAD, D. MANDAL, S. SENAPATI, S.R. SAINKAR, M.I. KHAN, R. RAMANI, R. PARISCHA, P.V. AJAYAKU-

MAR, M. ALAM, M. SASTRY, R. KUMAR, *Angew. Chem. Int. Ed.* **2001**, *40*, 3585.
5 X. PENG, L. MANNA, W. WANG, J. WICKHAM, E. SCHER, A. KADANAVICH, A. P. ALIVISATOS, *Nature* **2000**, *404*, 59.
6 R. JIN, Y. CAO, C. A. MIRKIN, K. L. KELLY, G. C. SCHATZ, J. G. ZHENG, *Science* **2001**, *294*, 1901.
7 A. HENGLEIN, *J. Phys. Chem.* **1993**, *97*, 5457.
8 M. A. EL-SAYED, *Acc. Chem. Res.* **2001**, *34*, 257.
9 A. P. ALIVISATOS, *Science* **1996**, *271*, 933.
10 R. TURTON, *The quantum dot: a journey into the future of microelectronics*, Oxford University Press, New York, **1995**.
11 W. S. WEISBECKER, M. V. MERRITT, G. M. WHITESIDES, *Langmuir* **1996**, *12*, 3763.
12 K. S. MAYYA, V. PATIL, M. SASTRY, *Langmuir*, **1997**, *13*, 3944.
13 M. SASTRY, N. LALA, V. PATIL, S. P. CHAVAN, A. G. CHITTIBOYINA, *Langmuir* **1998**, *14*, 4138.
14 R. ELGHANIAN, J. J. STORHOFF, R. C. MUCIC, R. L. LETSINGER, C. A. MIRKIN, *Science* **1997**, *277*, 1078.
15 J. J. STORHOFF, R. ELGHANIAN, R. C. MUCIC, C. A. MIRKIN, R. L. LETSINGER, *J. Am. Chem. Soc.* **1998**, *120*, 1959.
16 J. J. STORHOFF, A. L. LAZARIDES, R. C. MUCIC, C. A. MIRKIN, R. L. LETSINGER, G. C. SCHATZ, *J. Am. Chem. Soc.* **2000**, *122*, 4640.
17 T. SAWITOWSKI, Y. MIQUEL, A. HEILMANN, G. SCHMID, *Adv. Funct. Mater.* **2001**, *11*, 435.
18 S. A. MAIER, M. L. BRONGERSMA, P. G. KIK, S. MELTZER, A. A. G. REQUICHA, H. A. ATWATER, *Adv. Mater.* **2001**, *13*, 1501.
19 V. L. COLVIN, A. N. GOLDSTEIN, A. P. ALIVISATOS, *J. Am. Chem. Soc.* **1992**, *114*, 5221.
20 A. GOLE, S. R. SAINKAR, M. SASTRY, *Chem. Mater.* **2000**, *12*, 1234.
21 F. AUER, M. SCOTTI, A. ULMAN, R. JORDAN, B. SELLERGREN, J. GARNO, G.-Y. LIU, *Langmuir* **2000**, *16*, 7554.
22 E. W. L. CHAN, L. YU, *Langmuir* **2002**, *18*, 311.
23 K. C. GRABAR, K. J. ALLISON, B. E. BAKER, R. M. BRIGHT, K. R. BROWN, R. G. FREEMAN, A. P. FOX, C. D. KEATING, M. D. MUSICK, M. J. NATAN, *Langmuir* **1996**, *12*, 2353.
24 S. PESCHEL, G. SCHMID, *Angew. Chem. Int. Ed. Engl.* **1995**, *34*, 1442.
25 Z. L. WANG, *Adv. Mater.* **1998**, *10*, 13.
26 K. VIJAYA SARATHY, G. RAINA, R. T. YADAV, G. U. KULKARNI, C. N. R. RAO, *J. Phys. Chem. B* **1997**, *101*, 9876.
27 L. O. BROWN, J. E. HUTCHINSON, *J. Phys. Chem. B* **2001**, *105*, 8911.
28 G. DECHER, *Science*, **1997**, *277*, 1232 and references therein.
29 A. MAMEDOV, J. OSTRANDER, F. ALIEV, N. A. KOTOV, *Langmuir* **2000**, *16*, 3941.
30 R. K. ILER, *J. Colloid Interface Sci.* **1966**, *21*, 569.
31 A. KUMAR, A. B. MANDALE, M. SASTRY, *Langmuir* **2000**, *16*, 6921.
32 M. D. MUSICK, C. D. KEATING, M. H. KEEFE, M. J. NATAN, *Chem. Mater.* **1997**, *9*, 1499.
33 K. V. SARATHY, P. J. THOMAS, G. U. KULKARNI, C. N. R. RAO, *J. Phys. Chem. B* **1999**, *103*, 399.
34 D. TABOR, *J. Colloid Interface Sci.* **1980**, *75*, 240.
35 A. POCKELS, *Nature* **1891**, *43*, 437.
36 A. POCKELS, *Nature* **1894**, *50*, 223.
37 I. LANGMUIR, *J. Am. Chem. Soc.* **1917**, *39*, 1848.
38 K. B. BLODGETT, *J. Am. Chem. Soc.* **1935**, *57*, 1007.
39 P. GANGULY, D. V. PARANJAPE, M. SASTRY, *J. Am. Chem. Soc.* **1993**, *115*, 793.
40 M. CLEMENTE-LEON, E. CORONADO, P. DELHAES, C. J. GOMEZ-GARCIA, C. MIGNOTAUD, *Adv. Mater.* **2001**, *13*, 574.
41 A. RICCIO, M. LANZI, F. ANTOLINI, C. DE NITTI, C. TAVANI, C. NICOLINI, *Langmuir* **1996**, *12*, 1545.
42 J. M. R. PATINO, M. R. R. NINO, C. C. SANCHEZ, *Ind. Eng. Chem. Res.* **2002**, *41*, 2652.
43 Y. EBARA, K. MIZUTANI, Y. OKAHATA, *Langmuir* **2000**, *16*, 2416.
44 M. SASTRY, V. RAMAKRISHNAN, M. PATTARKINE, A. GOLE, K. N. GANESH, *Langmuir* **2000**, *16*, 9142.
45 J. H. FENDLER, F. MELDRUM, *Adv. Mater.* **1995**, *5*, 607 and references therein.
46 F. C. MELDRUM, N. A. KOTOV, J. H. FENDLER, *Langmuir* **1994**, *10*, 2035.

47 Y. Tian, J. H. Fendler, *Chem. Mater.*, **1996**, *8*, 969.
48 F. C. Meldrum, N. A. Kotov, J. H. Fendler, *J. Phys. Chem.* **1994**, *98*, 4506.
49 N. A. Kotov, G. Zavala, J. H. Fendler, *J. Phys. Chem.* **1995**, *99*, 12375.
50 B. O. Dabbousi, C. B. Murray, M. F. Rubner, M. G. Bawendi, *Chem. Mater.* **1994**, *6*, 216.
51 P. Ganguly, D. V. Paranjape, K. R. Patil, S. K. Chaudhari, S. T. Kshirsagar, *Ind. J. Chem. A* **1992**, *31*, F42.
52 T. Nakaya, Y.-J. Li, K. Shibata, *J. Mater. Chem.* **1996**, *6*, 691.
53 M. Sastry, V. Patil, K. S. Mayya, D. V. Paranjape, P. Singh, S. R. Sainkar, *Thin Solid Films* **1998**, *324*, 239.
54 H. Perez, R. M. L. de Sousa, J.-P. Pradeau, P.-A. Albouy, *Chem. Mater.* **2001**, *13*, 1512.
55 C. Damle, A. Gole, M. Sastry, *J. Mater. Chem.* **2000**, *10*, 1389.
56 T. Torimoto, N. Tsumura, M. Miyake, M. Nishizawa, T. Sakata, H. Mori, H. Yoneyama, *Langmuir* **1999**, *15*, 1853.
57 T. Torimoto, N. Tsumura, H. Nakamura, S. Kuwabata, T. Sakata, H. Mori, H. Yoneyama, *Electrochim. Acta* **2000**, *45*, 3269.
58 J. R. Heath, C. M. Knobler, D. V. Leff, *J. Phys. Chem. B* **1997**, *101*, 189.
59 X. Y. Chen, J. R. Li, L. Jiang, *Nanotechnol.* **2000**, *11*, 108.
60 S. Huang, G. Tsutsui, H. Sakaue, S. Shingubara, T. Takahagi, *J. Vac. Sci. Technol. B* **2001**, *19*, 2045.
61 S. Chen, *Adv. Mater.* **2000**, *12*, 186.
62 S. Chen, *Langmuir* **2001**, *17*, 2878.
63 M. Sastry, A. Gole, V. Patil, *Thin Solid Films* **2001**, *384*, 125.
64 M H V Werts, M. Lambert, J.-P. Bourgoin, M. Brust, *Nano Lett.* **2002**, *2*, 43.
65 S. Kuwajima, Y. Okada, Y. Yoshida, K. Abe, N. Tanigaki, T. Yamaguchi, H. Nagasawa, K. Sakurai, K. Yase, *Coll. Surf. A* **2002**, *197*, 1.
66 M. Brust, N. Stuhr-Hansen, K. Norgaard, J. B. Christensen, L. K. Nielsen, T. Bjornholm, *Nano Lett.* **2001**, *1*, 189.
67 L. Cao, L. Huo, G. Ping, D. Wang, G. Zeng, S. Xi, *Thin Solid Films* **1999**, *347*, 258.
68 J. Yang, X.-G. Peng, Y. Zhang, H. Wang, T.-J. Li, *J. Phys. Chem.* **1993**, *97*, 4484.
69 D. G. Kurth, P. Lehmann, G. Lesser, *Chem. Commun.* **2000**, 949.
70 G. K. Zhavnerko, V. S. Gurin, *Interface Sci.* **2002**, *10*, 83.
71 P. Y. Zhang, Q. J. Xue, Z. L. Du, Z. J. Zhang, *Wear* **2000**, *242*, 147.
72 During compression, the surface tension of the amphiphilic monolayer-covered water surface changes and the monolayer undergoes a number of phase transitions. A plot of the surface pressure (defined as the difference in surface tension of pure water and the monolayer-covered surface) enables one to identify these phase transitions and identify the optimum surface pressure for transfer of amphiphiles/nanoparticles in a close-packed configuration for LB film growth (A. Ulman, *An introduction to ultrathin organic films: from Langmuir-Blodgett to self-assembly*, Academic Press, San Diego, **1991**).
73 Silver (bright yellow) and gold colloidal solutions (ruby red/pink) possess beautiful colors due to excitation of surface plasmon vibrations in the nanoparticles (see [7] for details).
74 Y. Tian, C. Wu, J. H. Fendler, *J. Phys. Chem.* **1994**, *98*, 4913.
75 V. Patil, K. S. Mayya, S. D. Pradhan, M. Sastry, *J. Am. Chem. Soc.* **1997**, *119*, 9281.
76 M. Sastry, K. S. Mayya, V. Patil, D. V. Paranjape, S. G. Hegde, *J. Phys. Chem. B* **1997**, *101*, 4954.
77 K. S. Mayya, M. Sastry, *J. Phys. Chem. B* **1997**, *101*, 9790.
78 K. S. Mayya, M. Sastry, *Langmuir* **1998**, *14*, 74.
79 M. Sastry, K. S. Mayya, V. Patil, *Langmuir* **1998**, *14*, 5291.
80 M. Sastry, K. S. Mayya, *J. Nano. Res.*, **2000**, *2*, 183.
81 Y. Tran, S. Bernard, P. Peretti, *Eur. Phys. J. Appl. Phys.* **2000**, *12*, 201.
82 M. Sastry, M. Rao, K. N. Ganesh, *Acc. Chem. Res.* **2002**, *35*, 847.
83 K. S. Mayya, V. Patil, M. Sastry, *Langmuir* **1997**, *13*, 2575.

84 K. S. Mayya, V. Patil, M. Sastry, *J. Chem. Soc., Faraday Trans.* **1997**, *93*, 3377.
85 K. S. Mayya, V. Patil, P. M. Kumar, M. Sastry, *Thin Solid Films* **1998**, *312*, 300.
86 G. K. Zhavnerko, V. E. Agabekov, M. O. Gallyamov, I. V. Yaminsky, A. L. Rogach, *Coll. Surf. A* **2002**, *202*, 233.
87 K. Muramatsu, M. Takahashi, K. Tajima, K. Kobayashi, *J. Colloid Interface Sci.* **2001**, *242*, 127.
88 D. K. Lee, Y. S. Kang, C. S. Lee, P. Stroeve, *J. Phys. Chem. B* **2002**, *106*, 7267.
89 H. Du, Y. B. Bai, Z. Hui, L. S. Li, Y. M. Chen, X. Y. Tang, T. J. Li, *Langmuir* **1997**, *13*, 2538.
90 A. Henglein, *Chem. Rev.* **1989**, *89*, 1861.
91 P. Ganguly, D. V. Paranjape, K. R. Patil, M. Sastry, *J. Am. Chem. Soc.* **1993**, *115*, 891.
92 P. Ganguly, M. Sastry, S. K. Chaudhary, D. V. Paranjape, *Langmuir* **1997**, *13*, 6582.
93 J. N. Israelachvili, *Intermolecular and surface forces*, Academic Press, San Diego, **1993**.
94 B. R. Heywood, S. Mann, *Adv. Mater.* **1994**, *6*, 9 and references therein.
95 B. R. Heywood, S. Mann, *Langmuir* **1992**, *8*, 1492.
96 B. R. Heywood, S. Mann, *J. Am. Chem. Soc.* **1992**, *114*, 4681.
97 A. L. Litvin, S. Valiyaveettil, D. L. Kaplan, S. Mann, *Adv. Mater.* **1997**, *9*, 124.
98 P. J. J. A. Buijnsters, J. J. J. M. Donners, S. J. Hill, B. R. Heywood, R. J. M. Nolte, B. Zwanenburg, N. A. J. M. Sommerdijk, *Langmuir* **2001**, *17*, 3623.
99 K. C. Yi, V. S. Mendieta, R. L. Castanares, F. C. Meldrum, C. Wu, J. H. Fendler, *J. Phys. Chem.* **1995**, *99*, 9869.
100 S. X. Ji, C. Y. Fan, F. Y. Ma, X. C. Chen, L. Jiang, *Thin Solid Films* **1994**, *242*, 16.
101 K. C. Yi, J. H. Fendler, *Langmuir* **1990**, *6*, 1519.
102 J. Yang, F. C. Meldrum, J. H. Fendler, *J. Phys. Chem.* **1995**, *99*, 5500.
103 J. Yang, J. H. Fendler, *J. Phys. Chem.* **1995**, *99*, 5505.
104 A. Ruadel-Texier, J. Leloup, A. Barraud, *Mol. Cryst. Liq. Cryst.* **1986**, *134*, 347.
105 A. Barraud, J. Leloup, P. Maire, A. Ruadel-Texier, *Thin Solid Films* **1985**, *133*, 133.
106 S. Revaine, G. E. Fanucci, C. T. Seip, J. H. Adair, D. R. Talham, *Langmuir* **1998**, *14*, 708.
107 J. K. Pike, H. Byrd, A. A. Morrone, D. R. Talham, *J. Am. Chem. Soc.* **1993**, *115*, 8497.
108 H. S. Mansur, F. Grieser, M. S. Marychurch, S. Biggs, R. S. Urquhart, D. N. Furlong, *J. Chem. Soc., Faraday Trans.* **1995**, *91*, 665.
109 R. S. Urquhart, C. L. Hoffmann, D. N. Furlong, N. J. Geddes, J. F. Rabolt, F. Grieser, *J. Phys. Chem.* **1995**, *99*, 15987.
110 A. V. Nabok, T. Richardson, F. Davis, C. J. M. Stirling, *Langmuir* **1997**, *13*, 3198.
111 S. Guo, L. Konopny, R. Popovitz-Bior, H. Cohen, H. Porteanu, E. Lifshitz, M. Lahav, *J. Am. Chem. Soc.* **1999**, *121*, 9589.
112 L. S. Li, L. H. Qu, R. Lu, X. G. Peng, Y. Y. Zhao, T. J. Li, *Thin Solid Films* **1998**, *329*, 408.
113 B. M. D. O'Driscoll, I. R. Gentle, *Langmuir* **2002**, *18*, 6391.
114 R. S. Urquhart, D. N. Furlong, T. Gegenbach, N. J. Geddes, F. Grieser, *Langmuir* **1995**, *11*, 1127.
115 X. Peng, S. Guan, X. Chai, Y. Jiang, T. Li, *J. Phys. Chem.* **1992**, *96*, 3170.
116 L. S. Li, L. Qu, L. Wang, R. Lu, X. Peng, Y. Zhao, T. J. Li, *Langmuir* **1997**, *13*, 6183.
117 Y. Savin, T. Pak, A. Tolmachev, *Mol. Cryst. Liq. Cryst.* **2001**, *361*, 223.
118 V. Erokhin, V. Troitsky, S. Erokhina, G. Mascetti, C. Nicolini, *Langmuir* **2002**, *18*, 3185.
119 D. N. Furlong, R. S. Urquhart, F. Grieser, K. Matsuura, Y. Okahata, *J. Chem. Soc., Chem. Commun.* **1995**, 1329.
120 D. J. Elliot, D. N. Furlong, F. Grieser, *Coll. Surf. A* **1998**, *141*, 9.
121 G. Hemakanthi, A. Dhathathreyan, *Langmuir* **1999**, *15*, 3317.
122 D. V. Paranjape, M. Sastry, P. Ganguly, *Appl. Phys. Lett.* **1993**, *63*, 18.
123 D. T. Amm, D. Johnson, T. Laursen, S. K. Gupta, *Appl. Phys. Lett.* **1992**, *61*, 522.

124 D. M. Taylor, J. N. Lambi, *Thin Solid Films* **1994**, *243*, 384.
125 M. Seidl, M. Schurr, A. Brugger, E. Volz, H. Voit, *Appl. Phys. A* **1999**, *68*, 81.
126 D. T. Amm, D. J. Johnson, N. Matsuura, T. Laursen, G. Palmer, *Thin Solid Films* **1994**, *242*, 74.
127 N. R. Jana, L. Gearheart, C. J. Murphy, *J. Phys. Chem. B* **2001**, *105*, 4065.
128 P. Ganguly, M. Sastry, S. Pal, M. N. Shashikala, *Langmuir* **1995**, *11*, 1078.
129 S. Mandal, S. R. Sainkar, M. Sastry, *Nanotechnol.* **2001**, *12*, 358.
130 S. Mandal, C. Damle, S. R. Sainkar, M. Sastry, *J. Nanosci. Nanotechnol.* **2001**, *1*, 281.
131 A. Kumar, C. Damle, M. Sastry, *Appl. Phys. Lett.* **2001**, *79*, 3314.
132 C. Damle, K. Biswas, M. Sastry, *Langmuir* **2001**, *17*, 7156.
133 C. Damle, A. Kumar, M. Sastry, *J. Phys. Chem. B* **2002**, *106*, 297.
134 C. Damle, A. Gopal, M. Sastry, *Nano Lett.* **2002**, *2*, 365.

13
Layer-by-layer Self-assembly of Metal Nanoparticles on Planar Substrates: Fabrication and Properties

Thierry P. Cassagneau

13.1
Introduction

Self-assembly processes involve non-covalent interactions based on electrostatics, charge-transfer, hydrophilicity-hydrophobicity, H-bonding, chelation, metal coordination, $\pi-\pi$ coupling, etc. The term "self" in self-assembly accounts for the fact that each interactive building block unit carries information that inherently defines its binding properties and is relevant enough to determine a spontaneous interaction at a receptive site or surface. The building block under consideration can be any molecule, ion, cluster, particle, polymer, etc., that participates in the film growth. The idea of using the binding capabilities (especially through electrostatic interactions) of micrometric objects to sequentially build-up structures on a support was first presented in 1966 [1]. Since then micrometer-sized objects have been designed and functionalized in such a way that they can self-organize into complex patterns [2]. Sequential electrostatic self-assembly was further extended to the build-up of biphosphonate anions with polycationic species (1988) [3] and to the adsorption of polyelectrolyte multilayers (1991) [4] as an alternative to the Langmuir-Blodgett monolayer transfer technique for film formation. The layer-by-layer assembly technique relies on the alternate immersion of a derivatized substrate, S, into a solution of building blocks A and a solution of building blocks B, both bearing chemical functions directing their mutual spontaneous interaction. Upon repeating adsorption cycles a periodic film grows at the substrate surface, $S-(AB)_n$. The consecutive adsorption of particles with a binder on a planar substrate defines layers whose properties integrate over the entire film. The sequences of layering along with the choice of building blocks to be used determine a certain synergy between the layers. The versatility of this technique is not limited to the build-up of binary systems, and complex arrangements can be designed to form three-dimensional nanostructures with interesting properties [5]. Very often one building block is used as an inert binder to facilitate the adsorption of an element that brings the desired functionality upon assembly (Fig. 13.1). A given functionality is not only "packaged" within a discrete entity, say a nanoparticle, with specific magnetic, optical or electrical properties, but also localized within a three-dimensional nanostructure. Ideally, the layer-by-layer self-assembly

Colloids and Colloid Assemblies. Edited by Frank Caruso
Copyright © 2004 Wiley-VCH Verlag GmbH & Co. KGaA, Weinheim
ISBN: 3-527-30660-9

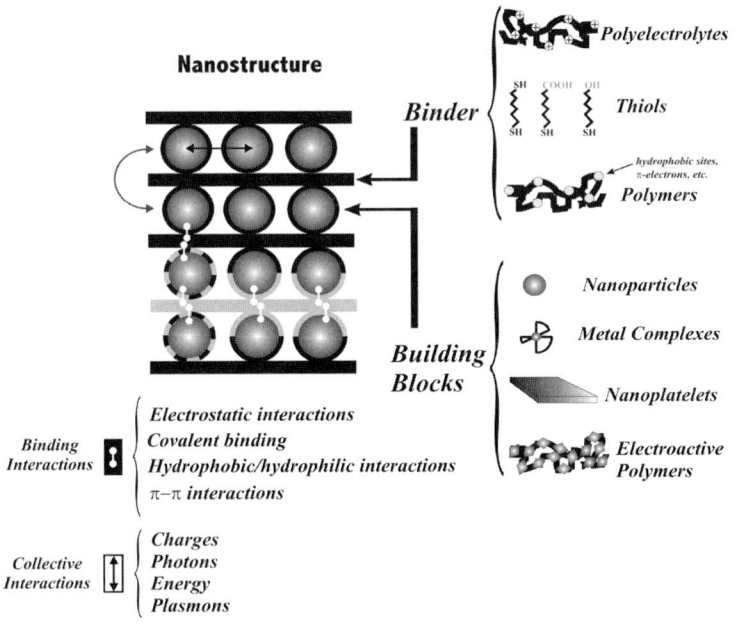

Fig. 13.1 Abandoned, unfinished, the "little" tower of Babel (1563) by Pieter Bruegel the Elder, symbolizes the difficulty of creating a piece of work when everyone tries to impose their own style, their own ideas. Building blocks must not only be commensurate with the type of construction needed but also possess binding properties that do not disrupt the stability of the edifice. Somehow a direction, an "agreement", must be promoted to raise a viable construction. In a beaker, nanosized objects can be handled through chemistry by surface functionalization. The layer-by-layer self-assembly technique reckons with the necessity of functionalizing both interacting surfaces, and exploits chemical interactions to direct the assembly process of a whole pool of building blocks. It aims to reconstruct a functional film capable of exhibiting novel properties derived from the collective interactions developed throughout the entire film by the assembly of building blocks (and sometimes their binder). The resulting nanostructure may interact with charges, plasmons, electrons, as well as exchange energy, to form a solid-like system that the chemist can tune for specific purposes.

technique provides the tools needed by an experimentalist aiming at forming three-dimensional structures in which useful objects are distributed in a controllable fashion.

The synthesis of nanoparticles often involves the use of solvents, which are also used as storage media. Means of manipulating such small objects rests on their ability to form stable dispersions in liquids and interact with a surface acting as a "collector". Hence, surface functionalization of such particles by chemical means allows programming the nature of the interactions at a given surface. In addition, the intimate chemical binding to a receptive surface leads to more uniform coatings and is amenable to deposition on large surface areas. Furthermore, the surface loading can be tuned by varying the concentration of the sol used for deposition and/or by altering the amount of available binding sites at a chosen surface. Such a control of the surface functionalization of both a substrate and a particle has fostered the processing of metal nanoparticles into films.

Metallic nanoparticles exhibit unique optical, electrical, catalytic or magnetic properties stemming from their high surface-to-volume ratio and their ability to couple with surface plasmons of neighboring metal particles or with electromagnetic waves. Integration of such preformed particles into heterostructures necessitates developing protocols of stabilization, functionalization and organization of the particles. Efforts toward the fabrication of nanostructures based on nanosized objects have benefited from the most recent advances in surface chemistry, supramolecular chemistry and physical chemistry. At a scale where a particle has dimensions comparable to a supramolecule, chemical interactions play a dominant role in directing the assembly of inorganic nanosized building blocks and the layer-by-layer self-assembly technique is one of the most appropriate layering technique to form multilayer assemblies of such objects.

In the following sections, different methods of functionalization of metal particles are described prior to discussing their manipulation in self-assembly deposition. The formation of ordered arrays is considered in a separate section emphasizing the need for ordered nanostructures when high-density storage information and sensing are of special interest. Self-assembled into films, the collective properties of metal particles are shown to be controllable through a variety of parameters determining their optical, electrical or magnetic characteristics. This chapter exclusively considers the use of layer-by-layer self-assembly in the build-up of nanostructured films incorporating metal particles. The reader is also invited to refer to recent reviews [6] and books [7] describing the use of alternative methods to form thin films of metal particles.

13.2
Layer-by-layer Self-assembly from Preformed Building Blocks

13.2.1
Functionalization of Nanoparticles

As-prepared nanoparticles often develop interactions with a substrate; however, understanding and controlling these interactions necessitate functionalizing these particles with well-defined capping molecules or organic or inorganic coatings. The choice of the protective layer can then be determined by the type of interaction of value at a given solid interface. In the following, a distinction is made between several procedures of particle functionalization that are well adapted to the preparation of layer-by-layer self-assembled metallodielectric films. Fig. 13.2 summarizes the main achievements in that field.

13.2.1.1 Covalent Capping

Nanosized particles naturally exhibit a surface charge imparted by the existence of dangling bonds and surface defects (introduced through chemical reduction via borohydride, citrate, hydrazine, etc.). However, most of them require stabilization prior to incorporation into films. When the nanoparticles contain metallic atoms (i.e. metal particles or II–VI semiconducting nanoparticles), thiolated molecules are commonly used to covalently attach an organic protective capping layer bearing specific functional moieties (for example, amine, carboxylic acid or alkyl groups). The synthesis of metal nanoparticles with a covalently bound protective capping layer is based on the growth of metallic nuclei undergoing chemical reduction with the concomitant attachment of thiolated molecules on the growing clusters [8]. Preformed particles, ionically stabilized (for example with citrate or consecutive to a borohydride reduction) can be covalently functionalized in the presence of thiolate ligands [9, 10]. These organic groups determine the solubility of the nanoparticles in different solvents as well as the sol stability through repulsive interactions. These two aspects are important in that they allow processing the particle adsorption under controllable conditions. When a hydrophobic monolayer surrounds the particles, the spontaneous evaporation of a suspension of particles from an apolar solvent leads to self organized structures at the solid–solvent interface (due to inter-particle hydrophobic interactions) [11]. Two- and three-dimensional superlattices of metal particles can spontaneously form when long alkyl chain thiolates are used [12–14]. However, such regular arrays are difficult to obtain on a large area with controlled assembly structure, thickness and uniform surface coverage. When ionizable moieties are present at the particle surface, an alternative consists in using polar solvents in order to tune the electrostatic interactions from the ionic strength, pH, additives and complexing metal ions. As the layer-by-layer self-assembly technique favors the use of polar solvents to electrostatically adsorb particles, functionalization of the particle surface with ionizable organic moieties is usually preferred. This is realized either by stabilizing a metal

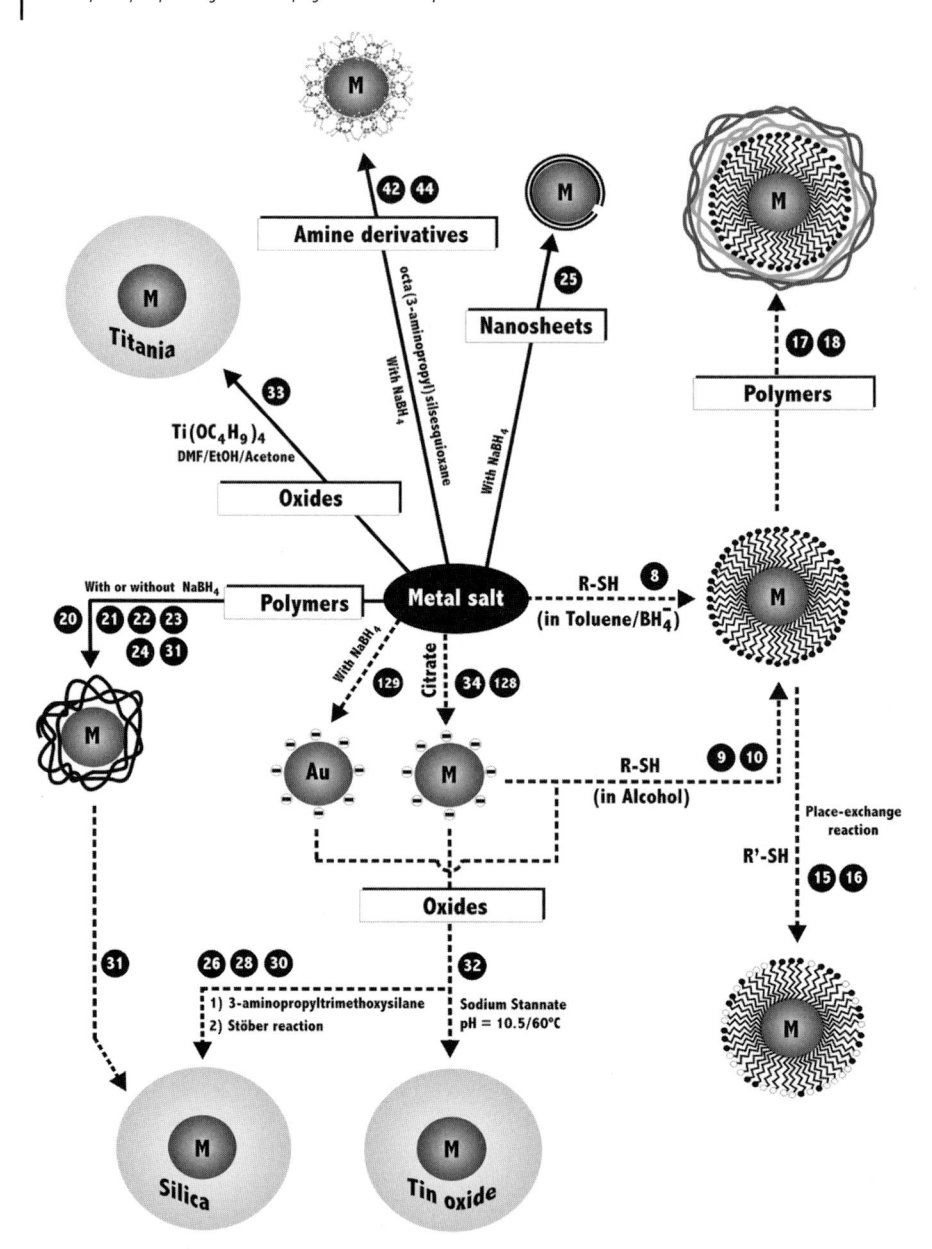

Fig. 13.2 Schematic overview of the different methods of particle functionalization employed in layer-by-layer self-assembly. The solid lines correspond to synthetic methods in which the particles are formed in the presence of the protecting layer. The dashed lines correspond to the case where the particles are first prepared and then modified with the appropriate protective coating. Reference numbers have been added on every path.

core with the appropriate thiol in a two-phase transfer synthesis or by "place-exchange" reaction upon exposure of the thiol-capped particles to an excess solution of another thiolate ligand for partial functionalization of the surface [15, 16]. This method offers the advantage of functionalizing the particle surface with any type of organic moieties, provided that the corresponding thiol derivative can be synthesized. This allows the isolation of particles in a solid form.

13.2.1.2 Electrostatic Encapsulation

The surface charge of particles can be exploited to attract soft matter electrostatically and form a protective coating and vice versa if the soft matter can be used to nucleate the particle formation. Two types of materials have been used to encapsulate metal particles: polymer chains and nanosized sheets.

13.2.1.2.1 Polymer Coating

Layer-by-layer Coating Gold particles (15–35 nm) capped with mercaptodecanesulfonate have been layer-by-layer coated with polyions of molecular weight ranging from 15 000 to 20 000 [17]. The presence of salt is mandatory to induce, through screening of the repulsive charges on the polymer chain, the occurrence of chain conformations favorable to spherical colloid encapsulation. Charge reversal allows alternate coating of the particle with thin, well-defined protective multilayers. The coating of 5-nm mercaptoundecanoic acid-capped gold particles has also been shown to be practically possible by choosing polyelectrolytes of length comparable to the diameter of the particle and by fine adjustment of the ionic strength at which the coating is carried out [18]. This coating protocol allows varying the thickness of the dielectric shell while introducing a surface charge ensuring the stability of the sol. It is most appropriate when a thin dielectric film (below 10 nm) is needed.

An alternative to sequential polymer coating consists in grafting a polymer at the surface of the particles. Polymer grafting on citrate-stabilized gold particles has been achieved by exposure of thiolated and non-thiolated polymers based on N-[tris-(hydroxymethyl)methyl]acrylamide or N-(isopropyl)acrylamide. All of these polymers exhibit comparable affinity for gold surface (although it is slightly larger when disulfide groups are present in the chain). The coating efficiently prevents particle aggregation by providing steric stabilization. It also ensures a good water solubility and can survive lyophilization. However, under certain treatments (high polymer concentration, use of other thiols, presence of proteins capable of unspecific binding) the polymer coating can be partially or fully displaced [19].

In situ Encapsulation In situ encapsulation takes place in a one-step reaction when metal ions are directly in contact with the polymer and undergo reduction either by the polymer itself or by subsequent addition of a reducing agent. For example, the addition of poly(N-vinyl-2-pyrrolidone), PVP, to a H_2PdCl_4 water/alcohol mixture leads, under reflux, to the formation of Pd particles with sizes ranging from 1 to 5 nm depending on the type and volume of alcohol used and the initial polymer concentration [20]. When π-conjugated polymers exhibit strong electron-donating properties, in situ reduction of metal ions may occur by electron transfer from the conjugated polymer to the neighboring ions. For example, such a reaction is favored by incorporation of dithiafulvene moieties in the polymer backbone. Dissolving π-conjugated poly(dithiafulvene) in dimethyl sulfoxide in the presence of metal salts spontaneously results in the formation of stable 5–6-nm metal nanoparticles (Au, Pd or Pt) with a positively charged coating of conjugated polymer [21]. When the reducing capabilities of the polymer are limited, addition of a reducing agent may be necessary to form the particles. For example, silver nanoparticles can be prepared in the presence of poly(acrylic acid), PAA, acting as a protective layer, when $NaBH_4$ is added to an aqueous solution of silver perchlorate and PAA [22]. Gold nanoparticles have been prepared from methanolic solutions of PVP [23] or aqueous solutions of PDDA [24] in the presence of $HAuCl_4$ salt upon addition of $NaBH_4$.

13.2.1.2.2 Encapsulation with Nanosized Sheets

Ultrathin charged nanosized sheets may provide sufficient nucleation sites to assist the formation of metal particles. It has been reported that circularly shaped 20–100-nm diameter graphite oxide, GO, sheets can be used for the in situ reduction of silver ions leading to GO-encapsulated 10-nm silver nanoparticles [25]. GO must be previously exfoliated to such an extent that ultrathin (4–6-Å thick) sheets are capable of bending around the incipient metallic nuclei to properly encapsulate a cluster as it expands. Thicker sheets can still act as nucleation supports but lead to the formation of larger unevenly sized particles due to incomplete encapsulation (Fig. 13.3). It should be noted that in contrast to polymer-coated particles, the metal core is very resistant to chemical or electrochemical oxidation owing to the efficient GO barrier, making this type of electrostatic capping particularly stable and not suited to the formation of hollow organic capsules by conventional methods.

13.2.1.3 Inorganic Coating

Inorganic coating of metal particles offers a means of not only stabilizing the particles but also tuning their optical properties from an inert and optically transparent coating (owing to the effect of the dielectric layer on the core absorption). Most studies have focused on using silica as coating material. However, the surface of some metals (e.g. gold) does not spontaneously oxidize in solution or air and presents little affinity for silica owing to the absence of an oxide layer. Besides, any chemical synthesis of metal particles involves the use of an organic stabilizer that increases the vitreophobic character of the metal. Vitreophilic modifi-

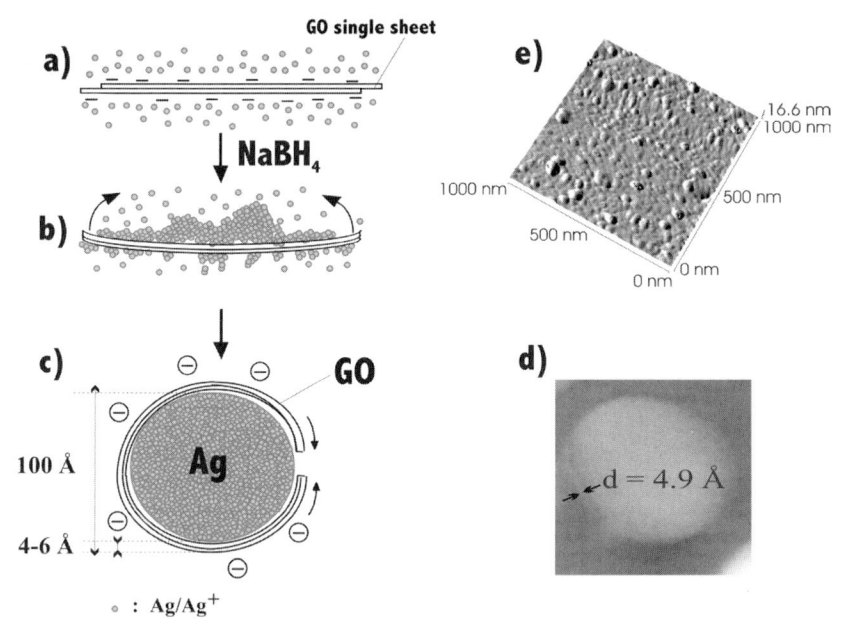

Fig. 13.3 Sequences (**a–c**) describe the assisted graphite oxide, GO, encapsulation of silver particles upon in situ reduction of the metal salt. The thickness and size of the graphite oxide should match the size of the particles. **d** If it is too thick and not wide enough, the GO sheet acts as a nucleation site but does not bend around the particle. **e** Properly encapsulated particles are extremely stable and massively adsorb on positively charged polyelectrolyte-coated surfaces.

cation of the surface is necessary to activate the growth of a silica shell. This can be done by using a silane coupling agent, (3-aminopropyl)trimethoxysilane, to form a monolayer with silane thiols pointing to the solution. Further addition of silicate at pH 8–10 in water/ethanol mixtures promotes the formation of a thin silica layer around the metal core (typically 4-nm thick) used to grow larger silica shells by the Stöber method (from TEOS as the silica source) [26]. A combination of silicate and TEOS deposition is known to lead to homogeneous silica coatings. When a surfactant (cetyltrimethylammonium bromide) is added to the Stöber mixture of NH_4OH and TEOS in water, a MCM-41 coating forms at an appropriate incubation temperature [27].

Silica encapsulation has been extended to gold clusters [28], Fe_3O_4 clusters [29] and silver nanoparticles [30, 31]. Metal oxide coating with SnO_2 and TiO_2 have also been proposed to coat gold and silver particles, respectively. A SnO_2 shell is formed by contacting preformed Au particles (citrate method) with a sodium stannate solution at pH 10.5 upon heating at 60 °C [32]. Titania-coated silver particles are produced by concomitant reduction of silver in a dimethylformamide/ethanol/acetylacetone mixture and condensation of $Ti(OC_4H_9)_4$ [33]. This coating procedure complements well the technique based on layer-by-layer self-assembly of polyelectrolyte in that it can also apply to large core particles and allows the formation of much thicker spherical metal oxide shells acting as spacers. An extra benefit is that the porosity of the shell can be modified through addition of organic additives during the shell growth [27].

13.2.2
Layer-by-layer Self-assembly of Metal Nanoparticles

13.2.2.1 Polymer-mediated Assembly

13.2.2.1.1 Electrostatic Assembly
Particles sterically stabilized and/or encapsulated with charged polymers can be sequentially adsorbed with polyions of opposite charge. In some cases, the polymer coating around the particles is thick enough to prevent the interparticle plasmon coupling from occurring. The case of silver particles stabilized with poly(acrylic acid) is given as an example (Fig. 13.4). Layering metal particles with soft matter is commonly achieved using the layer-by-layer self-assembly technique, provided that the particles are adequately functionalized. The following sections survey the different approaches developed to form multilayers by electrostatic assembly. Fig. 13.5 compiles the structures of the polymers used in that context.

From Citrate-stabilized Particles Metal particles prepared by reduction of chloroauric acid in the presence of citrate ions bear a negative charge originating from the citrate carboxylate groups adsorbed at the particle surface upon reduction [34]. This surface charge allows the particles to form stable colloidal suspensions and can also be exploited to electrostatically guide the particles to a positively charged substrate. The sequential layering of citrate-reduced gold particles with poly(allylamine hydrochloride), PAH, was one of the first studies conducted on multilayers incorporating metal particles [35]. Earlier, it was shown that the principles governing the layer-by-layer self-assembly of polyelectrolytes remain valid when

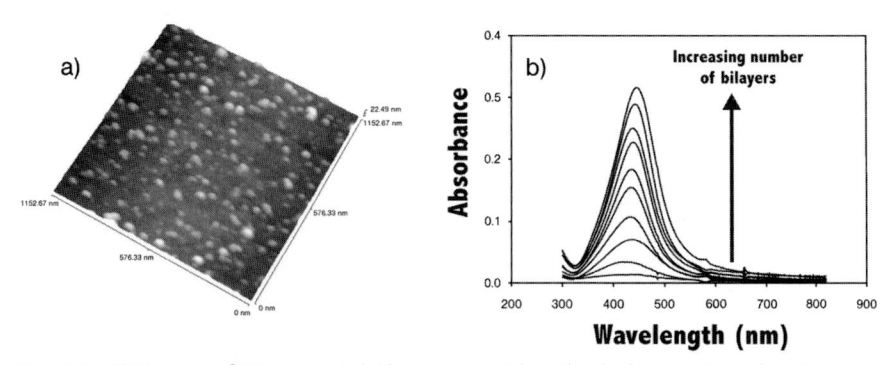

Fig. 13.4 AFM image of PAA-protected silver nanoparticles adsorbed on a mica substrate pre-coated with PDDA. A steady increase of the plasmon absorption band is observed upon sequential layering of both components. As the film grows, there is almost no red shift of the band because the large separation between particles prevents interparticle plasmon coupling.

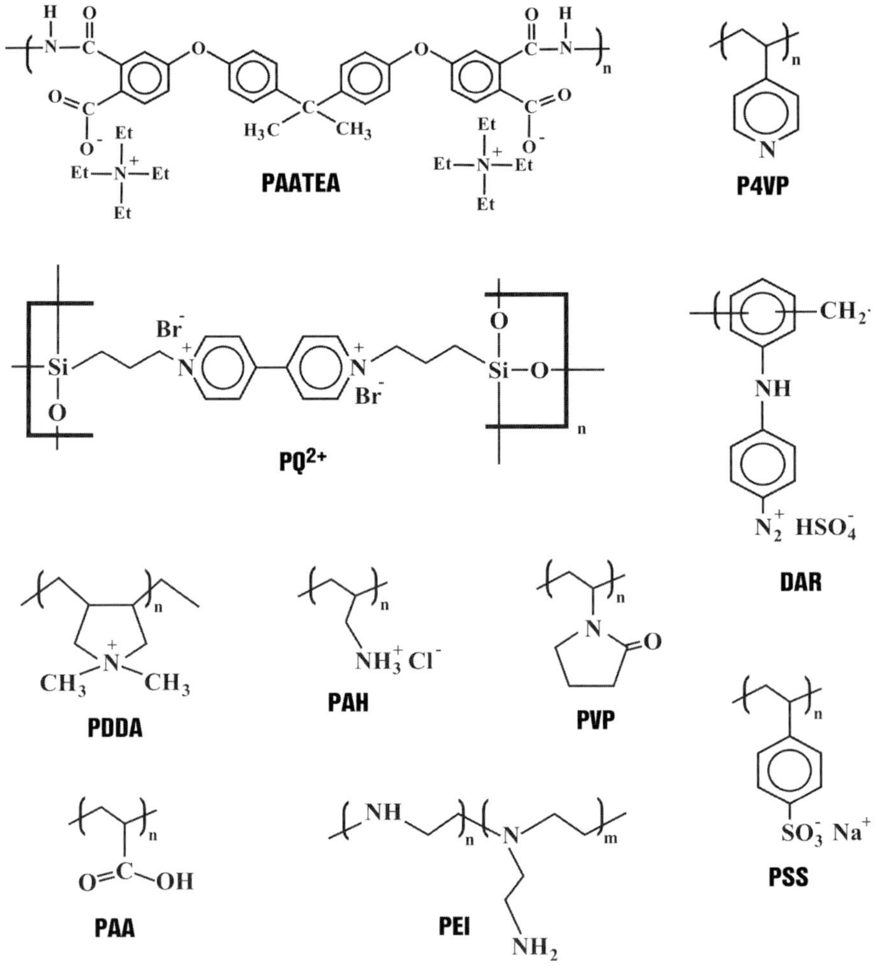

Fig. 13.5 Abbreviations and chemical structures of the different polymers utilized in the next sections. Polyamic acid, **PAATEA**; poly(4-vinylpyridine), **P4VP**; a polymer containing bipyridinium units, **PQ^{2+}**; diazoresin, **DAR**; poly(diallyldimethylammonium chloride), **PDDA**; poly(allylamine hydrochloride), **PAH**; poly(N-vinyl-2-pyrrolidone), **PVP**; poly(acrylic acid), **PAA**; poly(ethyleneimine), **PEI**; poly(sodium 4-styrenesulfonate), **PSS**.

hard nanosized spherical objects are alternately deposited with a polyelectrolyte [36]. In particular, the charge overcompensation of the outermost polyelectrolyte layer is still effective, allowing the adsorption of a next particle layer. With particle coverage of nearly 30% (1 particle per 1000 nm^2), X-ray reflectivity studies have indicated a periodic structure when a polyelectrolyte spacer (made of PSS/PAH bilayers) separates the particle layers. Variations in the optical properties of such films are due to the distance-dependent plasmon coupling of metal particles [35, 37]. The local dielectric environment influences the position of the plasmon band as well as its intensity. The metallic particles are essentially diluted in a dielectric medium (Maxwell Garnett model) and interact with each other via dipole plasmon coupling. Depending on the separation length between the particle layers the proportion of isolated and closely packed particles may vary within the film. At short separation length, two absorption bands appear at 510 nm and 600–650 nm. The former corresponds to the dipole plasmon band of single particles; the latter is broader and generated when interparticle plasmon coupling takes place (i.e. particles are closely packed). The contribution of the aggregates absorption vanishes as the spacing between adjacent particle layers is increased (causing an apparent hypsochromic shift of the whole plasmon absorption) while the plasmon band at 510 nm becomes dominant (for a spacer comprising at least 3–5 polyelectrolyte bilayers). In addition, it has been shown that the dipole plasmon intensity of one particle layer adsorbed on a PEI-coated substrate increases upon subsequent deposition of polyelectrolyte bilayers until the polyelectrolyte coating thickness matches the particle diameter [37].

Citrate-stabilized particles may be modified by reaction with thiolated derivatives to impart different surface functionality and additional means of film formation. For example, gold particles derivatized with 3-mercapto-1-propanesulfonic acid can be layer-by-layer self-assembled with diazoresin (DAR) and covalently bound to the polymer layer by UV light curing [38]. Such films may be of interest when the composite coating must withstand special conditions or exposure to strong solvents.

From Core-shell Particles The distance between particles is critical in determining the optical features of metal nanoparticulate films. For best control of these properties, it has been proposed to vitrophilize the metal particles in order to impart a hydrophilic character to the surface. To this aim, synthetic procedures of silica and titania coating of metal particles have been recently established (see Section 13.2.1.3). The ensuing core-shell particles exhibit a surface charge that depends on the isoelectric point of the oxide shell. Silica-coated particles are naturally negatively charged in air-equilibrated water (above the isoelectric point of silica). Hence, the sequential layering of such particles is achieved by coating the substrate with a polycation (i.e. PDDA) layer prior to each exposure to the particle sol [39]. Similarly, it has been shown that titania-coated particles can be alternately deposited with a polyanion, PAA, when working at a pH below the isoelectric point of titania [33]. Films incorporating oxide-coated metal particles exhibit optical properties controllable through tuning of the shell thickness (acting as a spacer), which determines the magnitude of particle–particle interactions (see below, Section 13.5.1).

Silver nanoparticles encapsulated with nanosheets of graphite oxide have been shown to electrostatically adsorb to PDDA-coated substrates owing to ionic pairing between the carboxylate or phenolate moieties present at the surface of the sheets and the ammonium groups of PDDA. Multilayer structures have been fabricated, in which the encapsulated silver particles exhibit a strong resistance toward oxidation as compared to particles stabilized with a polyelectrolyte coating only [25].

13.2.2.1.2 Hydrogen-mediated Assembly

Build-up Based on H-Bonding Hydrogen-bonding is predominantly an electrostatic interaction where the H atom is not shared but remains closer to its parent atom while interacting with another electronegative atom. The possibility of forming polyelectrolyte multilayers from poly(4-vinylpyridine), PVP, and poly(acrylic acid), PAA, via H-bonding has been demonstrated [40]. Obviously, using gold particles functionalized with carboxylic acid moieties (or pyridine groups) to promote H-bonding with PVP (or PAA, respectively) offers another alternative to metal particle/polymer multilayer film build-up. For example, gold particles (2.6 ± 0.9 nm) stabilized with 4-mercaptobenzoic acid have been successfully layer-by-layer assembled with PVP [23]. In addition, PVP-coated gold nanoparticles prepared by in situ reduction (see above) could be layer-by-layer self-assembled with PAA from the same type of interaction.

Build-up Based on Acid–Base Interaction When a proton is shared between two organic groups exhibiting opposite affinities towards protons (i.e. acid and base), the resulting proton exchange generates ionization. Typically the interaction involves amine-derivatized surfaces and acidic surfaces (sulfonated or carboxylated). If one of these surfaces is acidified (for example, a sulfonated surface) and allowed to interact with the complementary surface (an aminated surface), which is not acidified, in an appropriate solvent (preferably free of water to not de-acidify the target surface), binding develops. Such conditions can be utilized when the metal particle bears basic or acidic moieties. It should be noted that in that case, both interacting surfaces are initially neutral or negligibly charged. An alternative consists of adjusting the pH conditions in such a way that ionized moieties form prior to self-assembly. For example, mixed monolayer-protected clusters, MPCs, consisting of 1.6-nm Au nanoparticles capped with 4-aminothiophenol or mercaptoundecanoic acid can be sequentially assembled with PSS or PAH, respectively, when pre-ionized at appropriate pH. A pronounced plasmon band enhancement is observed for multilayers of PSS/aminothiophenol-capped Au clusters, probably because of a stronger interparticle electronic interaction involving the surrounding aromatic medium (more effective in relaying electron coupling) and/or the closer proximity of the nanoparticles [41]. However, several monolayers of particles adsorb at each step (3.7 ± 1.3 monolayers of MPCs on both sides of the substrate) at optimized pH conditions, suggesting that the polymeric binder may rearrange, increasing the permeability to MPCs. The loading of particles at the polyelectrolyte surface is pH-dependent and can be brought to a value below monolayer coverage. The acidity or basicity retained in the multilayer films is thought to affect the state of protonation on the MPCs at the interface where they are being incorporated into the films [41].

A case of binding via acid–base interaction has been also demonstrated under pH and solvent conditions where both interacting surfaces are neutral. Octa(3-aminopropyl)silsesquioxane is an efficient capping molecule for metal particles by virtue of its amine groups. Although non-covalent, the amine–metal interaction is

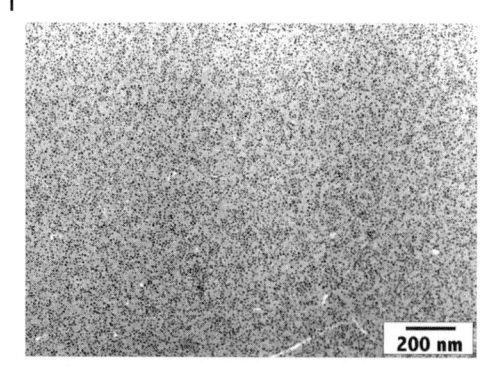

Fig. 13.6 Transmission electron microscope (TEM) image of 9 nm silver particles capped with NSi$_8$, Ag-NSi$_8$, and adsorbed on a bilayer PAH/PSS (rinsed at pH = 2 prior to adsorption with Ag-NSi$_8$). Although the coverage is significant, there is no long-range close packing assembly.

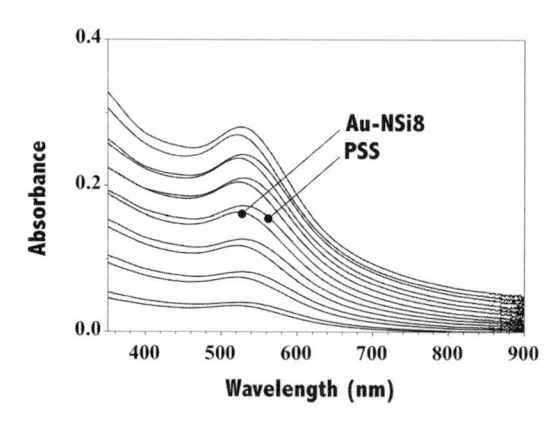

Fig. 13.7 Absorption spectra of films composed of 2 nm gold nanoparticles capped with NSi$_8$, Au-NSi$_8$, sequentially layered with PSS (pre-protonated prior to exposure to the gold sol). Acid–base interactions can be used to deposit quantum dots on a polyelectrolyte layer.

stable enough to resist manipulation under different pH conditions. Such molecules are synthesized in a mixture of ethanol/water (trace amount) from 3-aminopropylethoxysilane, with an apparent pH of 12. Under these conditions, the octamers (neutral) can be layer-by-layer self-assembled with poly(4-styrenesulfonate), PSS, by a judicious rinsing procedure involving the protonation of sulfonate groups. Particles of noble metal are prepared by in situ reduction of metal salts (AgClO$_4$, HAuCl$_4 \cdot 4\,H_2O$, CoCl$_2 \cdot 6\,H_2O$, etc.) with NaBH$_4$ in a concentrated ethanolic solution of octa(3-aminopropyl)silsesquioxane. Selecting the nature of the salt and the ratio of metal salt to capping molecule (i.e. large excess of amine groups), allows the formation of particles with sizes ranging from 1–2 nm to 10 nm and more [42]. Further adsorption on solid surfaces is made possible by precoating the surface with PSS and rinsing at pH = 2 prior to exposure to the sol (Fig. 13.6). This procedure allows the consecutive assembly of PSS with capped Au clusters (Fig. 13.7) based on the acid–base interaction taking place between sulfonic acid groups (at the PSS surface) and amine groups (from the capping monolayer) [43]. Metallodielectric coatings on spheres, giving novel properties, have been prepared using this layering procedure [44].

13.2.2.2 Molecular Cross-linking

13.2.2.2.1 Based on Electrostatic Interactions
Citrate-stabilized gold particles can be layer-by-layer assembled with molecular cross-linkers of positive charge such as N,N'-bis(2-aminoethyl)-4,4'-bipyridinium, tetracationic cyclophanes, $[(en)Pd(4,4'-bipy)]_4(NO_3)_8$ (en, ethylenediamine; bipyridine, bipyridinium) or $[Cd(4,4'-bipy)]_4(NO_3)_8$ [45–48]. A regular growth was observed as evidenced from the electrochemical oxidation and reduction waves of the gold particle surface. Particle densities of 0.2 to 1.0×10^{11} particles cm^{-2} and about 450 bridging molecules per particle were estimated. Any multifunctional molecule, bearing aurophilic moieties and/or positive charges, can actually be used to cross-link citrate-stabilized gold particles. Thionine, which is both a photoactive species and a charge carrier in photovoltaic cells, can also act as a cross-linker in the build-up of gold particle multilayers. In that case, both covalent and electrostatic binding are thought to be involved [49]. An average of 120 molecules per adsorbed particle was deduced from coulometric analysis. One drawback of this layering method is the absence of two-dimensional ordering; however, the resulting films are particularly stable. These assemblies are useful in the design of electrochemical–sensoric interfaces (see Section 13.5.2).

13.2.2.2.2 Based on Covalent Binding
Bifunctional molecules such as dithiols [50, 51] or mercaptoalkylamines [52] with binding groups at both ends can be used to consecutively bind gold particles on a substrate by exposure to a molecular cross-linker solution prior to a gold sol. Using mercaptoethylamine or hexanedithiol to cross-link 12-nm gold particles, films with metallic conductivity are obtained after seven or eight deposition cycles [50, 52], while the conduction proceeds from activated electron hopping events when the particle size decreases (i.e. 6 nm) with similar or slightly longer molecular binders [51]. Thiolated dyes have also been used to elaborate photosensitive multilayer assemblies with gold particles (see Section 13.5.2) [53].

13.2.2.3 DNA-mediated Assembly
Metal nanoparticles functionalized with thiolated oligonucleotides can be sequentially adsorbed onto surfaces on which complementary nucleic acid strands are immobilized. First the substrate (gold or glass) is derivatized with an x-mer (x corresponds to the number of constitutive bases) oligonucleotide to provide a branching point to a complementary target $2x$-mer DNA-analyte. Half of the bound analyte remains available for further hybridization with nanoparticles derivatized with x-mer oligonucleotides. If there is no (or very little) base mismatch between the analyte and the probe, a significant amount of particle adsorbs to the surface. Repeating this layering process a second time enables a dramatic amplification of the probe signal owing to the existence of numerous branching points at the surface of the first particle sub-monolayer, after which the surface is saturated with particles and the same amount of particles is added at each step.

This method can be used to identify the sequence of a target DNA-sequence when the analyte has a sequence complementary to both the surface of the derivatized substrate and the probe. The detection is achieved microgravimetrically [54] or optically [55]. Microgravimetric sensing of the target sequence with concentrations as low as 10^{-10} M has been reported upon attachment of the first sub-monolayer of particles [54]. The hybridization of the surface with the particles also induces a deeper red color with increasing layer number, ultimately leading to a reflective surface for thicker films [55]. The amplification efficiency is size-dependent when water washing is applied on the substrates. The binding strength between complementary oligonucleotides is weakened when the particle size becomes larger than 4–5 nm (for $x=8$ or 17) [56]. One salient feature of these assemblies is their sharp melting properties at temperatures lower than the corresponding pure nucleotide duplexes.

13.2.2.4 Ionic Assembly of Ligand-protected Particles

Electrostatic self-assembly requires the exploitation of coulombic attractive forces to govern the assembly onto a substrate. Multilayered systems (mainly based on polyelectrolytes) in which the interaction of ionic ammonium–carboxylate or ammonium–sulfonate pairs dominates the binding have been extensively studied. Such ionizable groups may be introduced on nanosized objects to direct their assembly electrostatically. One simple means to functionalize metal nanoparticles consists of using thiol derivatives to convey a specific chemical functionality through a judicious choice of organic moieties. 4-Aminothiophenol-capped gold particles (13 nm) have been sequentially assembled with 4-carboxythiophenol-capped silver particles (7 nm) when ionized under appropriate pH conditions. Estimates of the filling fraction for each adsorbed particle layer revealed that upon adsorption of a first layer of positively charged gold particles an average filling fraction of 30% was obtained, responsible for a surface charge reversal. As the silver particles were about half the size of the gold particles, the filling fraction of the next silver layer was close to 60%. However, due to surface inhomogeneities generated by the incomplete packing of particles, filling fractions decreased by about 25% upon successive adsorption of particle layers without affecting the film growth [57].

Multilayers of mixed MPCs $Au_{145}[S(CH_2)_{10}COO^-]_{17}[S(CH_2)_5CH_3]_{33}$, 4.4 nm in size, and gold clusters capped with N,N-trimethyl(undecylmercapto)ammonium ligand, >1.7 nm in size, have been constructed using the same principles of electrostatic binding. It was found that ca. 9×10^{-11} mol cm^{-2} of the former adsorb on 1.6×10^{-11} mol cm^{-2} of the latter after each exposure cycle on a mercaptan-functionalized substrate [58].

Monolayer-protected clusters with mixed hexanethiolate/mercaptohexanoic acid capping layers can attach to a COOH-functionalized surface by means of carboxylate/metal(II) (Zn^{2+}, Cu^{2+}) bridges formed during consecutive dipping and rinsing cycles in a metal salt solution and a suspension of clusters [59]. However, the amount of adsorbed MPCs at each step exceeds the monolayer coverage owing to migration of excess surface-coordinated metal ions promoting multiple adsorption growth. Using this procedure to concentrate MPCs on a conductive substrate allows the formation of ultrathin films with well-defined single-electron charging properties (see below).

13.2.2.5 π–π Interactions

Gold clusters have been prepared by the in situ reduction of $HAuCl_4 \cdot 4H_2O$ with KBH_4 in the presence of 2,6-bis(1′-(8-thiooctyl)benzimidazol-2-yl)pyridine [60]. The latter covalently attaches to the gold cluster surface exposing benzimidazole and pyridine groups to the solution. Such aromatic moieties promote interparticle π–π interactions responsible for the two-dimensional self-organization of the clusters upon drying of a suspension on a flat substrate. It has also been shown that tetrapyridylporphine-derivatized gold nanoparticles form two-dimensional hexagonally packed arrays at interfaces due to π–π interactions between pyridine moieties [61]. Although close-packed arrays of particle could be obtained as a monolayer, the formation of multilayers has not been yet attempted.

13.3 Alternate Polymer/Metal Particle Films Prepared by Salt Incubation into Polymer Multilayers

13.3.1 Polyelectrolyte/Metal Particle System

The synthesis of metal nanoparticles within preformed polymer multilayers relies on the availability of binding sites from which metal ions can diffuse and fix throughout the layer. In the case of polyelectrolyte multilayers, the formation of ion pairs between oppositely charged moieties induces charge neutrality within the film. Such electrostatically assembled structures should be unfavorable to the penetration of foreign ions by ion-exchange reaction owing to the poor availability of the ionized sites within the films. To circumvent this limitation two approaches have been envisaged. One involves using weak polyelectrolytes (having a pH-dependent degree of ionization) to generate a certain fraction of ionic sites not involved in direct electrostatic pairing with adjacent oppositely charged polyelectrolyte layers. Another method involves fabricating multilayers of strong polyelectrolytes with acidifying washing steps after each polyelectrolyte adsorption to modify the charge overcompensation at the outermost layer.

13.3.1.1 Multilayers Incorporating Weak Polyelectrolytes

The charge of weak polyelectrolytes originates from the presence of ionizable moieties such as carboxylic acid or amine, and depends on their degree of ionization, which is itself pH-dependent. In solution, poly(acrylic acid), PAA, has a pK_a value ranging from 5 to 6. Incorporation of cationic transition metal complexes, such as $Ru(NH_3)_6^{3+}$, within a PAA film can be accomplished by fixing the pH of the metal salt solution above this pK_a [62]. However, in a polyelectrolyte multilayer structure assembled at a pH higher than 5, all of the PAA carboxylate groups form ionic pairs with the adjacent polycationic layers, preventing any further ion exchange with cationic species. Adjusting the pH of a PAA solution to a value lower than its pK_a value is critical in generating films in which a majority of nonionized carboxylic acid groups remains present. These nonionized groups do not contribute to the electrostatic layer-by-layer self-assembly of PAA with polycations if the layering is carried out at pH conditions that preserve their integrity (typically at pH values varying from 2.5 to 4.5). It has been estimated that about 70% of carboxylic acid groups remain nonionized at pH 2.5, while this value decreases to 30% at pH 4.5 [63]. The presence of nonionized groups does not hinder the adsorption of polycations [64] and can be used to subsequently ion-exchange protons for metal ions in the vicinity of carboxylic acid groups [65, 66].

Using silver acetate solutions to incubate multilayers of PAA/PAH prior to exposure to a reducing hydrogen atmosphere, nanoparticle volume fractions of 4 and 8% were obtained with multilayers self-assembled at pH 4.5 and 2.5, respectively. As the metal ion loading is proportional to the number of displaceable protons within the PAA layers, the particle diameter increases from 2.1 nm to 3.8 nm when the pH at which the multilayers are fabricated passes from 4.5 to 2.5 (Fig. 13.8). It should be noted that regeneration of carboxylic acid groups is made possible after silver particle formation and permits a regular growth of the particles upon repeating incuba-

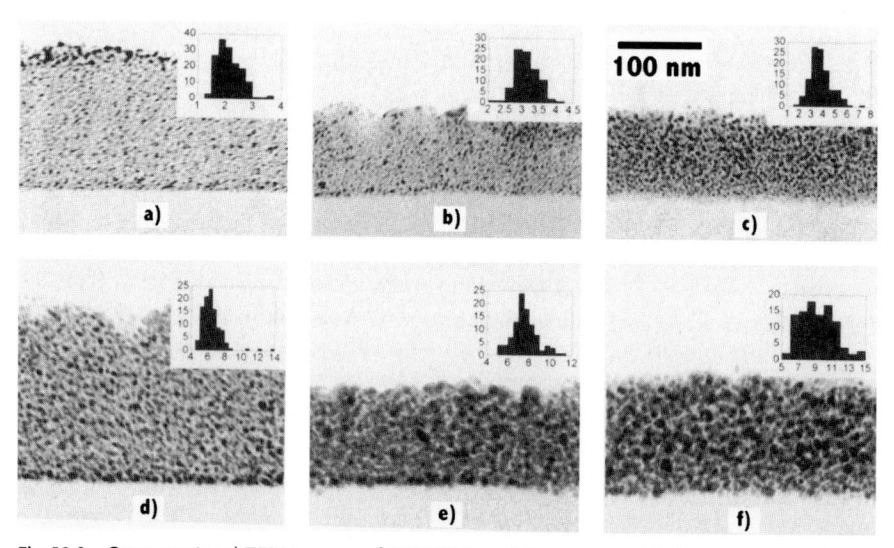

Fig. 13.8 Cross-sectional TEM images of PAH/PAA multilayers assembled at pH = 4.5 (**a,d**); 3.5 (**b,e**) and 2.5 (**c,f**) prior to be subjected to one (**a–c**) or five (**d–f**) cycles of silver cation exchange and reduction. Histograms of the particle size distribution are shown in insets (reproduced from [66]).

tion and reduction cycles. For example, after five cycles, the nanoparticle volume fraction increases to 14% and 24% when the reprotonation of the films is performed at pH 4.5 and 2.5, respectively. A significant splitting of the dipole plasmon mode was observed after three incubation/reduction cycles.

An alternative to controlling the ionization of the polyelectrolyte solutions through pH adjustment may consist in reversibly blocking a fraction of ionic sites by complexation or chemical functionalization. For example, PAH and PAA can be derivatized with photolabile 2-nitrobenzyl groups upon reaction with 2-nitrobenzyl chloroformate and 2-nitrobenzyl bromide, respectively. Further layering necessitates adjusting the pH of the polyelectrolyte solutions above the pK_a of PAA. Hence, a fixed charge is introduced within the film by photolysis of the 2-nitrobenzyl groups [67]. Varying the fixed charge of such multilayers may be achieved by first complexing the carboxyl groups with mono- or divalent metal ions. Soluble low molecular weight Cu^{2+}-complexed PAA can be formed at pH close to 5.5 (most carboxylic groups are deprotonated and the polymer does not precipitate) and layer-by-layer self-assembled with PAH from solutions at the same copper salt concentration and pH (to prevent leaching of the copper ions from the adsorbed Cu^{2+}-complexed PAA). Regeneration of the carboxyl moieties is made possible by simple washing at pH 3.5 to displace Cu^{2+} with H^+, followed by immersion in a pH 5–6 solution to ionize PAA [68]. This procedure allows the preparation of films with enhanced anion-transport selectivity and permits a seven-fold increase of ionization within the films as compared to films prepared by direct pH adjustment of the polyelectrolytes [64] and subsequent incubation in copper salt solutions. Although these procedures have been developed to fabricate anion-selective multilayer membranes, they are deemed to be very useful for incorporation and reduction of metal ions within polyelectrolytes.

13.3.1.2 Multilayers Incorporating Strong Polyelectrolytes

The use of strong polyelectrolytes to form multilayers capable of receiving metal ions has also been explored to form semiconducting and magnetic crystals by in situ nucleation [69–71]. In that case, alternate layers of PDDA and PSS were built up from their aqueous solutions with rinsing steps in water and in pH 4.5 HCl solutions after each immersion. When PSS is the outermost layer, acidifying the surface diminishes the magnitude of the charge overcompensation by protonation of sulfonate groups. The isoelectric point of adsorbed PSS can reach values much higher than PSS in aqueous solutions, owing to surface charge regulation. For example, it was found that PSS alternately deposited (with acidifying washing steps at pH 2) with an octaamine molecule has an isoelectric point as high as a weak polyelectrolyte [43]. Partial reprotonation of the surface screens the access to a fraction of nonionized moieties to the ammonium groups of PDDA. These protons can be leached out upon appropriate washing, liberating sulfonate groups capable of binding with metal ions. Using that procedure Fe^{2+}, Pb^{2+} and Co^{2+} ions were infiltrated into PDDA/PSS multilayers to undergo oxidation or sulfidation and form crystals of β-FeOOH [69a], γ-FeOOH [69b], α-Co(OH)$_2$ [70] or PbS [71].

13.3.1.3 Multilayers Incorporating Transition Metal Complexes

An alternative approach to metal ion incorporation utilizes the displacement of weakly bound counter-anions by anionic metal complexes. The incorporation of negatively charged metal complexes within cationic polyelectrolyte films deposited at an electrode surface has been the subject of numerous studies aiming to elucidate the thermodynamics and kinetics of interfacial ion diffusion and binding. Poly(4-vinylpyridine), PVP, has been used to persistently bind anionic complexes such as $Fe(CN)_6^{4-}$ and $IrCl_6^{2-}$ under acidic conditions [72]. The incorporation of these multiply-charged complexes takes place by anion exchange with $CF_3CO_2^-$ initially present within PVP at low pH. Other studies have emphasized the usefulness of ion-exchange reactions to incorporate anionic metal complexes within electroactive polycations. In that case, the binding behaviors were clarified by probing the electrochemical characteristics of both the host and the guest at the electrode surface [73, 74]. Using an electrode coated with a polymer containing bipyridinium repeat units, $(PQ^{2+})_n$, it has been shown that the complex $IrCl_6^{2-}$ can be introduced (without potentiostating the electrode) by equilibrating the electrode to an aqueous solution of this anion. The binding to $IrCl_6^{2-}$ occurs to an extent that the positive charge of the polymer is fully compensated. Initially, the polymer charge is compensated with monovalent anions (e.g. Br^-, Cl^-), readily exchanged upon equilibration. When a second multicharged anion is added to the solution, such as SO_4^{2-}, the polymer retains both anions. However, holding the electrode at potential negative to the $IrCl_6^{2-}/IrCl_6^{3-}$ redox couple leads to concentration of $IrCl_6^{3-}$ species in the film. Thermodynamics controls the distribution of anions present in the surface-bound polymer when the electrolyte includes two or more anions. Further examination of the electrostatic binding of other negatively charged metal complexes has revealed the existence of preferential binding properties hinging on the nature of both the metal ion and the ligands. It was shown that the binding increases as the anion changes from Cl^- to $IrCl_6^{2-/3-}$, $Fe(CN)_6^{4-}$, $Ru(CN)_6^{4-}$ and $Mo(CN)_6^{4-}$. The transition metal complex anions are much more firmly bound than other types of anions such as bromide, chloride, iodide, sulfate, perchlorate, etc. These thermodynamically weakly bound anions are labile and rapidly displaced, and do not affect the kinetics of the polymer redox properties. On the other hand, the strongly bound metal complex anions are not kinetically labile and do affect the kinetics of the polymer redox properties [73]. The possibility of ion-exchanging weakly bound anions with anionic metal complexes has been exploited for surface metallization of semiconducting photocathodes to enhance the performance for H_2 evolution. In that case, $PtCl_6^{2-}$ is first loaded prior to potentiostating at –0.1 V vs. SCE to induce the formation of Pt particles and introducing the desired catalytic nuclei [74].

Similarly, a quaternized poly(4-vinylpyridine) complexed with $[Os(bpy)_2Cl]^{2+/+}$ was layer-by-layer self-assembled with $PdCl_4^{2-}$. Subsequent electrochemical reduction of the metal complex at –0.8 V vs. Ag/AgCl leads to a composite film in which 3–7-nm Pd particles are homogeneously distributed. Catalytic films of metallic Pd suffer from dissolution upon oxidation–reduction cycling in acidic electrolyte. Embedding such particles within a polymer may provide a means of im-

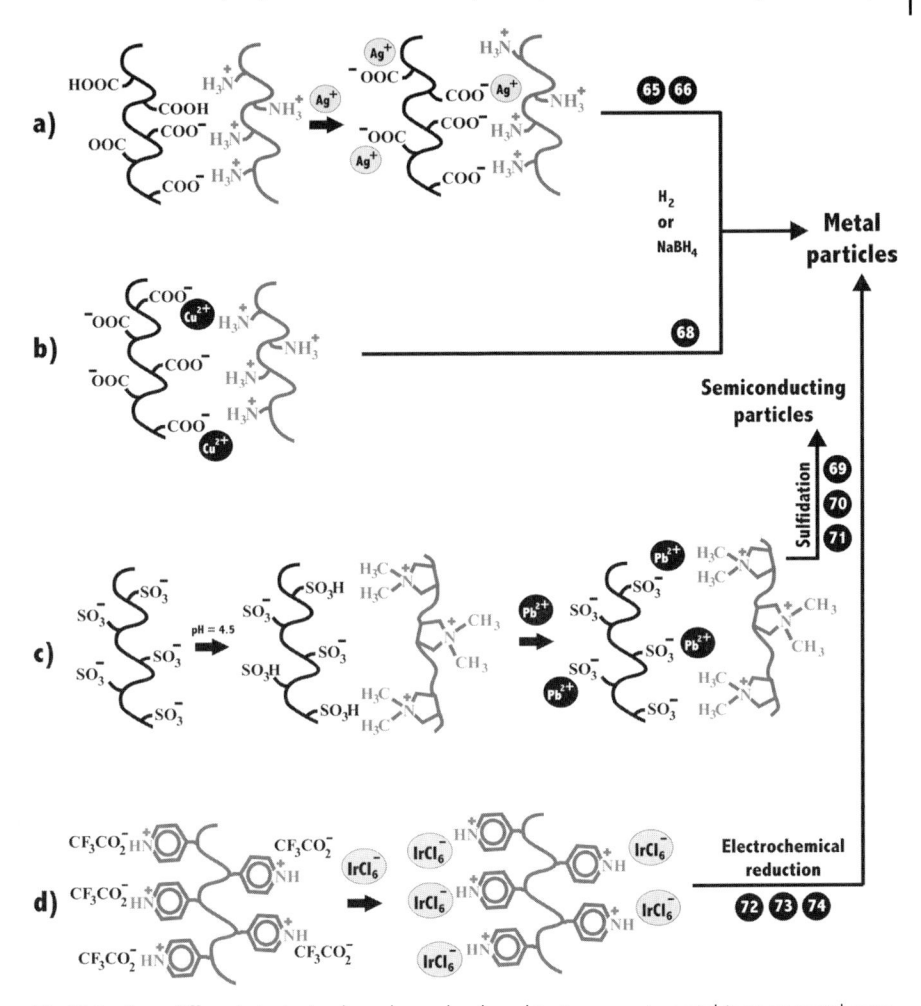

Fig. 13.9 Four different strategies have been developed to incorporate metal ions or metal complexes within polyelectrolytes layers. In **a**, the pH at which the multilayers are grown is adjusted in such a way that a fraction of non-ionized carboxylic groups remains and can be used to further exchange with metal ions. In **b** the multilayer growth is promoted in the presence of bound metal ions. In **c** the polyanionic layer is treated with an acidic solution during the build-up with a polycationic layer to favor the subsequent exchange with metal ions. In **d** the polycationic layer, whose charge is compensated with displaceable negative ions, can be loaded with anionic metal complexes.

proving the particle stability for catalytic uses (e.g. O_2 reduction) [75]. Fig. 13.9 proposes a schematic summary of the different methods of polyelectrolyte film infiltration with metal ions.

13.3.2
Diblock Copolymer/Metal Particle System

Symmetric diblock copolymers have the property of forming multilayers of macroscopic lamellae parallel to a substrate exhibiting a preferential interaction for one of the blocks. Surface-induced ordering spontaneously occurs upon annealing of the film at moderate temperatures, whether the final film thickness is commensurate with the bulk period of the lamellae L or not. Evidences of such periodic morphologies have been reported using poly(styrene-β-methyl methacrylate) [76], poly(styrene-β_2-vinylpyridine) [77] or poly(ethylene-propylene-b-ethylethylene) [78] diblock copolymers. Unlike layer-by-layer self-assembled films, for which the thickness can be varied from a few nanometers to a few micrometers by repeated adsorption cycles, the thickness of each layer is intrinsically determined by the molecular weight of the diblock copolymer. X-ray reflectivity studies have shown that a periodic structure can be seen up to 12.5 L for poly(styrene-β-methyl methacrylate) films [79]. Utilizing a block bearing functional groups with affinity to metal ions can yield in situ particle formation upon chemical reduction. This was recently demonstrated using a poly(styrene-β_4-vinylpyridine) where Au^{3+} ions selectively coordinate to the pyridine units and generate 3-nm gold nanoparticles upon subsequent exposure to a $NaBH_4$ solution. Annealing of the film prior to salt incubation led to the formation of a lamellar morphology as a result of the preferential affinity of poly(4-vinylpyridine) to the polar substrate and poly(styrene) to the air interface. Using a copolymer with a molecular weight of 42 000 g mol^{-1}, the

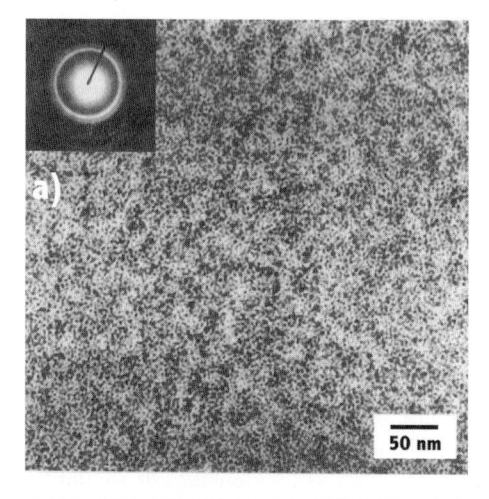

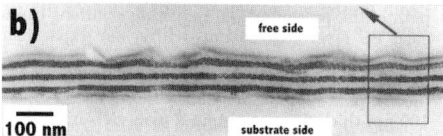

Fig. 13.10 a Plan-view TEM image of a thin polystyrene-block-poly(4-vinylpyridine) film of thickness equals 3.1x the bulk period of the lamellae, L, and containing gold particles. b Cross-sectional TEM images of the same film: the dark layers contain vinylpyridine groups in contact with gold particles, the gray layers only contain the styrene moieties (reproduced from [80]).

average bulk period was found to be 33 nm (16 nm for the PS layer and 17 nm for the P4VP layer). Since a poly(styrene) block has no affinity towards metal ions, a laminated alternate structure of metal-loaded layers with metal-free layers was obtained (Fig. 13.10). Cross-sectional transmission electron microscope imaging indicated that diffusion of the precursors and reduction solutions throughout the film was not limited [80].

13.4
Organization of Metal Nanoparticles on Planar Surfaces

Superlattices of metal nanoparticles can spontaneously form when using thiolate ligands bearing long alkyl chain (12 carbons or more) [12]. Hydrophobic interactions develop between protected particles as the solvent evaporates, leading to ordered close-packed arrays. Several approaches have also been proposed to adsorb *charged* metal particles on solid surfaces (see Section 13.2). However, devising the conditions favorable to two-dimensional or three-dimensional ordering of this type of particles brings another level of complexity. First, it should be noted that in most applications the organization of particles is not mandatory. Films containing metal particles are intrinsically interesting if the nature of the binding with the surrounding dielectric medium, the thickness of the film, the degree of mixing of the composite and the average spacing between particles are readily controllable. Ordered arrays of particles within a film would greatly improve the control of the film properties as well as their study. Nevertheless, *averaged* over an entire film, useful properties emerge. From this point of view, particle functionalization remains at the foreground of film fabrication. Second, the organization of preformed building blocks in a multilayer structure supposes a fine control of the interfacial properties at both the receptive surface and the particle surface. Such a control would involve varying the strength of binding interactions at the interface with the substrate and/or between neighboring particles. Therefore, designing the substrate (i.e. its topology) and/or surface chemistry are as critical as functionalizing the particles.

If ordered films can be constructed from non-covalent interactions between a binder and an ensemble of particles, these interactions should equally be sensitive to any change affecting them. In other words, such films should respond to a local chemical modification of their environment provided that they contain responsive functional groups. Alteration of the particle binding within the film transmutes into new optical, electrical or magnetic features that can then be used for sensing purposes. Hence, well-ordered structures should be particularly advantageous for designing accurate (selective) and highly sensitive sensors. It should be noted that the need for periodic sensing nanostructures becomes less stringent as the binding governing the interaction between a particle and a surface becomes extremely specific. A DNA-hybridized particle is by itself an encoding building block that does not necessitate a peculiar arrangement prior to pairing with a complementary strand of hybridized surface or particle (for colorimetric sensing).

The long-range order of particles (in particular their alignment) is of potential interest in miniaturizing electronic circuitry to a nanoscale range. Metal clusters are capable of undergoing single electron charging at room temperature and may be used as electroactive building blocks in nanoelectronics. In addition, well-ordered ensembles of larger particles may be sintered to form addressable two- or three-dimensional conductive networks. When particles are seen as storage information elements, their arrangement becomes critical in determining the storage density in a supporting material. This is particularly true for magnetic particles used to fabricate quantum magnetic disks (QMDs). QMDs consist of discrete magnetic elements uniformly distributed in a nonmagnetic support (e.g. SiO_2). Control parameters such as the shape and the location within the support are critical in maximizing the storage density [81]. The superparamagnetic limit of a magnetic particle corresponds to the size limit below which thermal fluctuations are sufficient to change the magnetization direction. For cobalt particles this limit is reached at a size of 10 nm [82], which implies that the ability to store information is lost below that size. At that scale, new nanolithographic techniques such as nanoimprint lithography [83–85] still allow the generation of highly resolved patterns, allowing storage densities as high as 40 Gbits cm^{-2} [81]. An alternative may consists of using pre-embedded elements, for example cobalt particles coated with a silica shell, to form ordered three-dimensional arrays of magnetic particles. For these prospects (in both sensing and nanoelectronics) ordered arrays of particle are highly desirable.

Below, two series of approaches are considered. One of them uses the adjustment of the adsorptive processes by altering parameters such as pH, ionic strength, solvent, temperature, etc., and effecting the non-covalent interactions through appropriate chemical functionalization of the two interacting surfaces. The other one concentrates on using patterned surfaces to direct the particle adhesion to specific domains.

13.4.1
Electrostatic Self-organization

Dispersed in a solution (isotropic medium), functionalized particles can assemble into networks or spheres when interacting with each other or with an added molecular body. For example, H-bonding interactions, derived from the pairing of diaminotriazine and thymine moieties, have been shown to be quite powerful in generating oblate aggregates or microspheres of nanoparticles [86, 87]. In most cases, shapeless aggregates are obtained when electrostatic interactions [88, 89] or DNA-hybridization [90] are used to drive the assembly of two distinct families of particles. Lowering the symmetry of these ensembles implies directing the adsorptive processes to solid surfaces.

Large-scale ordering of particles supposes cooperative interactions at an interface both with a substrate and between the particles, and is comparable to a crystallization process. The relative positioning of particles at the interface and their interaction with the surrounding medium (i.e. neighboring particles, polymers, solvent, etc.) is determined within a thermodynamic equilibrium. Ideally, the kinetics (i.e. the rate at which particles are adsorbed) of deposition should be slow enough to allow the particles to "see" their surrounding environment. Strong substrate–particle interactions may be detrimental to organization if they surpass interparticle interactions. Driving the organization of charged particles on charged substrates necessitates balancing the coulombic attractive substrate–particle interactions with the repulsive particle–particle interactions at the substrate/solution interface in such a way that ordered arrays of close-packed particles form. Two-dimensional organization of charged particles was demonstrated by exposure of PEI-coated substrates to a solution of 1.2-nm gold clusters, [Au_{55}-$(Ph_2PC_6H_4SO_3H)_{12}Cl_6$] (Fig. 13.11) [91]. This study revealed that the crystallinity of the adsorbed PEI influences the organization by acting as a structure-determining template (cubic and hexagonal packing could be obtained depending on the crystalline phase of PEI). In addition, it was indirectly proved that a matching between the particle size and the polymer molecular weight must be found to direct a preferential interaction. Most studies on the adsorption of charged particles onto charged flat substrates have revealed the scarcity of systems for which ordered long-range packing is possible. This probably originates from the number of parameters involved and the lack of theoretical modeling to direct the experimentalists. One general approach to ordering of nanoparticles may consist of coating the particle surface with a layer sensitive to parameters directly affecting the interac-

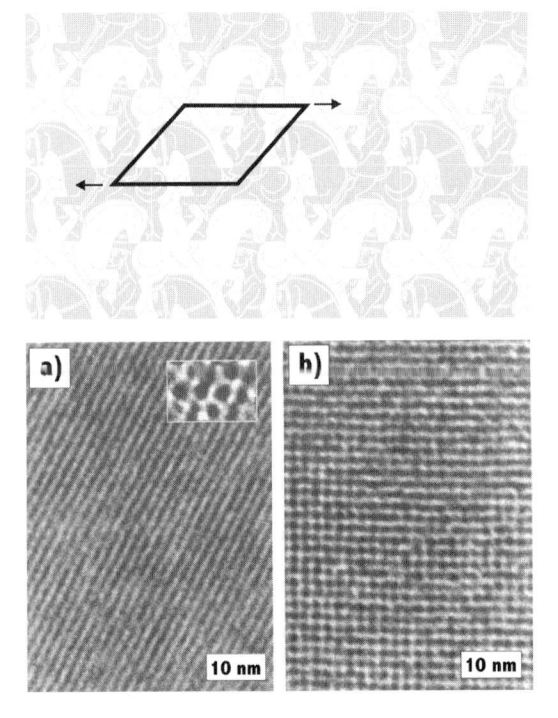

Fig. 13.11 Artistic tessellations (here from Escher) are based on a repeating design built around a basic geometric shape. They demonstrate that the order and the shape of the assembled building blocks serve to define, as in this drawing, complementary patterns oriented in specific directions. In the physical world, such patterns may be important to impart special polarization or electronic properties to an assembled film. Below is shown the TEM image of 1.2-nm [$Au_{55}(Ph_2PC_6H_4SO_3H)_{12}Cl_6$] clusters adsorbed as **a** a hexagonal packed monolayer and **b** a cubic packed monolayer on PEI. This is the first example of metal clusters arrayed through acid–base interactions (TEM images reproduced from [91]).

tions at the substrate/solution interface (i.e. pH, ionic strength, temperature, etc.), thereby modifying the adsorptive behavior of the particle during deposition. Preliminary studies have shown that the coverage of polyelectrolyte-coated nanoparticles on polyelectrolyte-primed substrates is tunable through adjustment of the ionic strength [92]. Polymer-encapsulated particles are promising entities, in that environmentally responsive groups can be introduced in the polymer chain, while the polymer coating itself may exhibit crystalline domains from which new recognition pathways can be generated. Another research direction may involve using polymerization-induced epitaxy (PIE) to grow polymer chains with two-dimensionally ordered binding loci. The epitaxial adsorption of a number of functional monomers on graphitic surfaces has been demonstrated [93–95]. Upon polymerization, the chains keep an alignment commensurate with the substrate lattice [95]. Such coated substrates could offer a simple means of directing the adsorption of pre-defined objects onto specified binding sites at a planar lattice.

13.4.2
Template-assisted Organization

Although template-directed colloidal crystallization or crystal nucleation are now well established techniques [96–98], innovative nanolithographic techniques (nanoimprint lithography) have not yet been used to direct the crystallization of nanosized objects. Instead, alternative imprinting protocols have been developed to mold the surface of substrates in such a way that nanoparticles adsorb in ordered arrays. For such small objects, surface patterning is technically challenging, as the topological features have to be commensurate with the size of the nanoparticle (typically 2–6 nm). Highly oriented patterns of thin polymers can be generated by friction transfer of poly(tetrafluoroethylene), PTFE, under controlled temperature, pressure and speed [99]. Another approach consists of fabricating faceted templates from NaCl (110) surfaces sublimed to form (100) and (010) terraces [100]. For both substrates, the resulting ridge-and-valley topology is molded by depositing a carbon layer, which is further used as a replica upon detachment from its pattern either by simple dissolution of the underlying NaCl or lifting off when PTFE is used. In both cases, the grooved topology allows the nanosized objects (thiolate-stabilized gold particles) to be lined up on solvent evaporation [101, 102]. The interplay of entropic gain and capillary forces near a groove explains this preferential adsorption [103, 104]. Effects due to surface functionalization as well as the electrical properties of these alignments have not yet been elucidated.

13.5
Properties of Metal Nanoparticle-containing Multilayers

13.5.1
Optical Properties

Dilute suspensions of metal particles in a transparent medium exhibit striking colors due to a strong absorption peak in the visible range [105]. Metal particles exposed to incident light radiation experience a collective oscillation of electrons (termed plasmon). As the wavelength of light is much larger than the size of the particles, retardation effects of the electromagnetic field over the particle diameter are negligible. The generally multipolar excitations [106] are restricted to the *dipolar* electric mode (homogeneous polarization). Each particle can be seen as a confining cavity in which electrons undergo resonant oscillations with the incident light. This resonance appears at a value of the metal complex dielectric function that depends on the shape of the particle and the dielectric constant of the surrounding medium.

The optical properties of metal nanoparticles have been reported in several studies on composite metal particle/polymer multilayers [25, 35–61]. However, few systematic investigations on the parameters influencing these properties have been achieved [37, 39]. Valuable information has been collected from layering silica-coated gold particles on planar substrates. Using such particles, the interparticle spacing could be programmed by choosing the silica shell thickness, which in re-turn allowed controlling the core–core coupling interactions within the assembled films [39]. One major difference between dilute dispersions of metal particles in solution and metal particle films originates from the high-volume fraction of particles found in the films. Such films can be seen as "suspensions" of identical spherical particles of dielectric constant $\varepsilon_2(\omega)$ embedded in a uniform medium with dielectric constant $\varepsilon_1(\omega)$ [107]. A field $E_\omega(r)$ is generated for any particular configuration, of local dielectric constant $\varepsilon(r, \omega)$, subjected to an applied field at frequency ω. The corresponding displacement current, originating from the dielectric nature of the dispersion in which charges move due to the electric field, is given by:

$$D_\omega(r) = \varepsilon(r, \omega) E_\omega(r) \qquad (1)$$

A satisfactory description of the optical properties of a "suspension" of particles has to take into account the polarization field due to the embedding matrix itself. The induced polarization, $P_\omega(r)$, relative to the medium in the absence of particles, is introduced within the relation:
Averaging over statistical ensemble of configurations Eq. (2) rewrites as:

$$\langle D_\omega(r) \rangle = \varepsilon_1(\omega) \langle E_\omega(r) \rangle + 4\pi \langle P_\omega(r) \rangle \qquad (2)$$

$$D_\omega(r) = \varepsilon_1(\omega) E_\omega(r) + 4\pi P_\omega(r) \qquad (3)$$

At sufficiently long wavelengths (as compared to the particle diameter and the interparticle distance) and without distinguishing between longitudinal and transverse fields, an effective dielectric constant $\varepsilon_{\text{eff}}(\omega)$ may be considered, so that:

$\varepsilon_{\text{eff}}(\omega)$ may then be estimated from solving the first and third Maxwell equations (on the microscopic level), which allow determination of $E_\omega(r)$ and $D_\omega(r)$, respectively:

$$\langle D_\omega(r) \rangle = \varepsilon_{\text{eff}}(\omega) \langle E_\omega(r) \rangle \tag{4}$$

$$\nabla \cdot E_\omega(r) = 0, \ \nabla \cdot D_\omega(r) = 4\pi \rho_{0\omega}(r) \tag{5}$$

The mathematical solution, proposed by Maxwell Garnett (1904) [108], has been shown to lead to good estimates of the effective dielectric constant for films consisting of particles isotropically dispersed in a non-absorbing medium [37, 39].

In multilayer assemblies, the optical properties are entirely due to dipole coupling of the plasmon modes. Using silica-coated gold particles as model entities, it has been shown, in accordance with the Maxwell Garnett model, that an in-plane coupling of plasmon takes place for particles coated with a thin silica shell, which causes a red shift and a broadening of the film plasmon absorption proportionate to the layer number (plateauing off after eight deposition cycles for shell thicknesses varying from 0.5 to 3 nm). Upon forming a thicker shell, particles do not experience dramatic changes from passing from solution to film and the peak position of the plasmon absorption becomes insensitive to the layer number [39].

Not surprisingly, injecting electrons within a metal particle causes an increase in the free electron concentration that affects the bulk plasma frequency, ω_p.

$$\omega_p = \frac{2\pi c}{\lambda_p} = \sqrt{\frac{4\pi^2 c^2 m \varepsilon_0}{N e^2}} \tag{6}$$

where m is the effective mass of the electron, λ_p is the bulk plasma wavelength and N is the concentration of free electrons. The plasmon band position of a metal particle obeying the Drude model depends itself on the refractive index of the surrounding medium according to [109]: where ε^∞ is the high-frequency dielectric constant due to interband and core transitions and ε_m is the real dielectric function of the surrounding medium. (The Drude model assumes that metals contain free electrons experiencing viscous damping due to the positive lattice, and no other forces except the applied electromagnetic field.) As a result,

$$\lambda^2 = \lambda_p^2(\varepsilon^\infty + 2\varepsilon_m) \tag{7}$$

the plasma absorption peak is blue-shifted for increasing values of N (i.e. upon electron injection). Particles placed in a reductive medium exhibit blue-shifted plasmon absorption (e.g. citrate-stabilized metal particles in a borohydride solution) [110] and direct evidence was reported by using pulse radiolysis [111]. In addition, the double layer capacitance of a particle plays an important role in regulating the charge within a particle (see next section). A simple electrical polarization of a film composed of metal nanoparticles can reversibly modify the position of the plasmon band absorption (typically 20 nm with silver particles) [112]. When the size of a particle becomes very small (quantum-size effect), electrons may also be allocated to the conduction band consisting of discrete energy levels, causing an increase in the Fermi level. The absorption spectra of metal nanoparticles of size below 2 nm present distinct steplike bands characteristic of more molecule-like properties and transitions to discrete LUMOs of the conduction band.

13.5.2
Electrical Properties

13.5.2.1 Electron Transfer

Metal nanoparticles adsorbed at an electrode surface are suitable elements to enhance the electronic conductivity of a composite coating (for example, in an electrochemical device) as well as to design microelectrode assemblies with unique diffusion behavior (fast electron kinetics studies) [113]. The implementation of selective ionic or electronic conductivity may be achieved using spatially organized metallic particles on a substrate to form ensembles of microelectrodes [114].

Electron transfer properties at a multilayer-supporting electrode are usually studied in an electrolyte solution containing a redox probe to evaluate the electron transfer resistance and the change of capacitance at the electrode/electrolyte interface. Electrochemical impedance spectroscopy allows determination of the charge transfer rate constant and charge transfer resistance of the probe at the electrode interface as well as the double-layer capacitance after coating [115, 116]. The presence of a coating at the electrode surface is expected to retard interfacial charge transfer kinetics, which manifests itself by an increase of the electron transfer resistance. Measurement of this resistance is derived from the faradaic impedance, Z_f, which is classically separated into the charge transfer resistance (kinetically controlled), R_{et}, and the Warburg impedance, Z_W, which is mass-transfer controlled. Charge transfer reactions at a planar electrode/film surface can be described by an equivalent circuit (see Scheme 13.1).

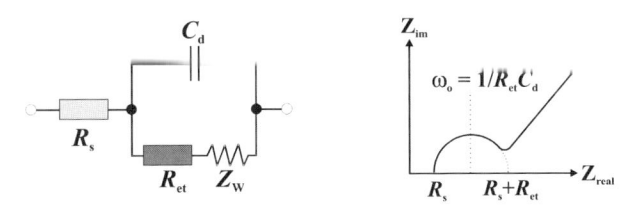

Scheme 13.1

The bulk electrolyte solution resistance and diffusion features of the redox probe in solution (both unaffected by the chemical transformations at the electrode surface) are represented by R_s and Z_W, respectively. While the double-layer capacitance, C_d, and the electron transfer resistance, R_{et}, sensitive to the electrode/electrolyte interface, are variable quantities, the term $1/C_d$ can be expressed as a sum of two terms, $1/C_{Au}$ (the capacitance of the uncoated electrode), and $1/C_{coating}$ (capacitance due to the coating). Similarly R_{et} decomposes into $R_{Au} + R_{coating}$.

A graphical determination of these two quantities is made possible from the Nyquist plots [$Z_{im}(\omega)$ vs. $Z_{real}(\omega)$] comprising a semicircle (corresponding to the electron transfer-limited process) followed by a straight line (corresponding to the diffusion-limited electron transfer process), as in Scheme 13.1. The maximum value of the imaginary impedance in the semicircle corresponds to $Z_{im} = R_{et}/2$ and is achieved at the characteristic frequency, ω_0:

$$\omega_0 = \frac{1}{C_d R_{et}} \tag{8}$$

Even though the net current is zero at equilibrium, a balanced faradaic activity takes place and can be expressed in terms of the exchange current, i_0, which is proportional to the rate constant of electron transfer, k_{et}, according to:

$$i_0 = nFAk_{et}C \tag{9}$$

where C is the bulk concentration of the redox probe, A is the electrode area and F the Faraday constant. The exchange current is also related to the electron transfer resistance by Eq. (10):

$$R_{et} = \frac{RT}{nFi_0} \tag{10}$$

where R is the gas constant and T the temperature (K). Eq. (10) allows the electron transfer rate constant to be deduced from Eq. (9).

Using the above model, the electrical properties of metal nanoparticle/hexanedithiol (HDT) multilayers have been evaluated. It was found that the dithiol layer generates an insulating layer partially blocking interfacial electron transfer [50]. However, when gold nanoparticles (12 nm in diameter) were interposed between the dithiol layers by stepwise self-assembly, the electron transfer resistance at the electrode was seen to diminish continuously with the number of adsorbed nanoparticle layers. Upon increasing the number of adsorbed HDT/Au nanoparticle bilayers, electron transfer changed from being kinetically limited to being a diffusion-limited process, indicating that the coating behaved nearly as a microelectrode array for films made of at least seven adsorbed bilayers. This trend was confirmed by cyclic voltammetry, showing that the peak-to-peak separation of the probe redox wave decreased with increasing layer number. However the capacitance of the coating (16 µF cm^{-2}) significantly differed from the capacitance of a bare gold electrode (40–60 µF cm^{-2}) [50].

13.5.2.1.1 Electrochemical Sensing

Incorporated within a film, conductive nanoparticles can be seen as small electrical relays facilitating communication between the redox loci (from enzymes, redox polymers, etc.) and the bulk electrode. Their petite size enables such entities to approach the surface, eliminating the need for diffusional electron mediators. The layer-by-layer self-assembly of sensing building blocks with metal nanoparticles has been especially studied to fabricate films with enhanced sensitivity to specific analytes.

Multilayer assemblies incorporating gold particles (12 ± 1 nm) and π-acceptor receptors such as cyclobis(1,1'-dimethyl-4,4'-bipyridinium-π-phenylene) have been constructed on conductive substrates for specific complexation of π-donor substrates (e.g. dialkoxybenzenes, hydroquinones) [47]. At low concentrations (e.g. 10 µM), these analytes are not electrochemically detectable on unmodified electrodes. The build-up of sensing multilayers allows not only the tuning of the sensitivity by varying the number of receptor/gold particle bilayers but also the introduction of a certain degree of selectivity towards analytes by altering the chemical nature of the cross-linking receptors. Non-specific interactions with the cyclophane/particle are not involved in the sensing behavior, although the porosity of the array is essential for satisfactory diffusion of the analyte throughout the film. Interestingly, regeneration of the sensing interface can be achieved by washing off the bound analyte by immersion in an appropriate buffer solution.

Several neurotransmitters (dopamine, adrenaline) possess an ω-hydroquinone moiety and may be detected within such nanostructured films. The possibility of detecting adrenaline was demonstrated by forming similar films on the PEI-precoated Al_2O_3 surface of an ion-sensitive field-effect transistor (ISFET). The accumulation of analyte within the bispyridinium cyclophane cavities could be detected by two means: either by measuring the source-drain current at different concentrations of adrenaline maintaining constant the gate-source potential or by monitoring the gate-source potential while keeping the source-drain current and the source-drain potential at fixed values. A detection limit of 1 µM was obtained using one bilayer Au particle/cyclophane on the PEI-coated sensing interface [117].

13.5.2.1.2 Photosensitization

Metal-nanoparticle arrays with three-dimensional conductivity offer a means to tailor photoelectrochemically active electrodes when cross-linking of these arrays with different photosensitizer-electron acceptor dyads is made possible. Such layered architectures are capable of enhancing the photocurrent by concentrating the photoactive species at the electrode surface (owing to the presence of receptive molecules) while maintaining effective electrical contact with the bulk electrode (via the metal nanoparticle layers). This was achieved by using Ru(II)-tris-(2,2'-bipyridine)-cyclobis(paraquat-π-phenylene) catenane or Zn(II)-protoporphyrin IX-bis(N-methyl-N'-undecanoate-4,4'-bipyridinium) as molecular cross-linkers for the fabrication of layer-by-layer self-assembled Au-particle (13±1 nm) arrays [118]. Intramolecular electron transfer in the photoactive dyads generates intermediate redox species leading to the generation of photocurrent that reaches the bulk electrode through the three-dimensional electrical contact path. A parallel protocol, sequentially layering gold particles with a sulfur-containing ruthenium (II) tris(2,2'-bipyridine) derivative, was developed to monitor the photocurrent response in the presence of triethanolamine as a sacrificial reducing molecule [53]. In that case, a photocurrent was generated from direct electron transfer from the photoexcited state of the photosensitizer to the gold particle. However, the measured responses were well below those expected from the amounts of immobilized photosensitizer in the film.

This type of film architecture is detrimental to the photocurrent quantum yield in that the metal particles cause a partial energy transfer quenching of the cross-linking photosensitizers. It is likely that the feasibility of high-quantum yield photosensitizing films necessitates a careful choice of the photosensitizer and the type of binding developed with the gold.

13.5.2.2 Single Electron Charging and Stoichiometric Electron Storage

As the size of a metallic particle becomes smaller, comparable to 2 nm, unusual electrical behaviors are promoted. Single electron conductivity between metal clusters becomes observable when the stepwise charging energy, E_c ($=e^2/C$), is much larger than the thermal energy, $k_B T$. Such clusters undergo single-electron charging events observable at room temperature when their capacitance becomes sufficiently small (sub-attofarad range). Quantized double layer capacitance charging of MPCs in the vicinity of an electrode has been detected in solutions using pulse, cyclic and steady-state voltammetry [119]. Impedance spectroscopic measurements (using the equivalent circuit shown in Scheme 13.1) have ruled out the interpretation that R_{et} might be due to the presence of a linker (thiol molecule) between the electrode and the metal core. Instead, it has been proposed that R_{et} represents the resistance to charge transfer of the MPC interface with the solution or with neighboring MPCs, and the capacitance C_d represents the double layer capacitance of the anchored MPCs (denoted C_{CLU}) [120].

Electrochemical studies (differential pulse voltammetry) of the MPCs at an interface electrode/electrolyte solution have shown that equally spaced current peaks are seen upon varying the potential applied to the electrode, suggesting that charging of these metal-like cores is controlled by electrostatic principles (double layer). During the double layer charging of MPCs an ionic space charge balancing the electronic charge on the core is generated. It is a kinetically fast process, so that the profile of working electrode current vs. average potential of MPCs near the electrode surface is determined by a combination of the Nernst equation and Fick's law of diffusion. A model has been proposed to account for these phenomena, in which the distribution of electrons between the working electrode and the MPCs near the surface is thought to be dependent on the applied potential, E_{app}, the potential of the MPC, E_p, and the capacitance, C_{CLU}, of an individual MPC [121]. Accordingly, the potential of an anchored MPC is given by:

$$E_p = E_{PZC} = \frac{ze}{C_{CLU}} \qquad (11)$$

where E_{PZC} is the potential of zero charge of the MPC, z is the number of electronic charges on the particle (signed positive or negative if the particle is oxidized or reduced, respectively) and e is the elementary charge (1.602×10^{-19} C). The capacitance of a spherical nanoparticle, C_{CLU}, uniformly coated with a dielectric layer (of dielectric constant ε) in an electrolyte solution is given by:

$$C_{CLU} = A_{CLU} \frac{\varepsilon \varepsilon_0}{r} \frac{r+d}{d} = 4\pi\varepsilon\varepsilon_0 \frac{r}{d}(r+d) \qquad (12)$$

where A_{CLU} is the surface area of the MPC core of radius r and d is the thickness of the dielectric coating. Upon charging, the initially uncharged core accumulates z electrons at an electrode potential E_{app}, whereas its potential changes from E_{PZC} to E_p according to Eq. (11). When considering a mixture of MPCs experiencing charging at the electrode surface, an analogy with a redox species can be made for which a nanoparticle is formally regarded as a *multivalent* redox system exhibiting equally spaced formal potentials. Within this analogy, two adjacent formal potentials can be seen as the result of passing from an initial charged state to a subsequent charged state differing by only one elementary charge. For example, for monolayer-protected gold clusters of 1.6 nm and capacitance $C_{CLU} = 0.5$ aF, the storage of one electron on the MPC core modifies its potential by $e/C_{CLU} = \Delta V = 0.32$ V (ΔV corresponding to the potential difference between two adjacent formal potentials). Thus, a mixture of MPCs undergoing charging is (formally) equivalent to a mixture of "redox couples" with formal potentials $E^0_{Z,Z-1}$. Taking into account this analogy, Eq. (11) can be rewritten as:

$$E^0_{Z,Z-1} = E_{PZC} + \frac{(z-1/2)e}{C_{CLU}} \qquad (13)$$

It should be noted that for conventional multivalent redox complexes the electrochemical formal potentials are generally not evenly spaced for the first versus subsequent electron transfers owing to the molecular nature of the systems. Provided that C_{CLU} is independent of the accumulated charge, z, the formal potentials of the capacitance charging steps vary linearly with the valence states of the MPC, which allows determination of the average nanoparticle capacitance. Fig. 13.12 presents some examples of quantized double layer charging observed by using gold clusters capped with octa(3-aminopropyl)silsesquioxane.

One of the most remarkable properties of these nanoparticles is their tolerance to double layer charging, making them comparable to nanocapacitors. The stored charge is retained upon drying and redispersion in solution [122]. Such entities can be seen as redox reagents and may be applicable to quantitative redox proce-

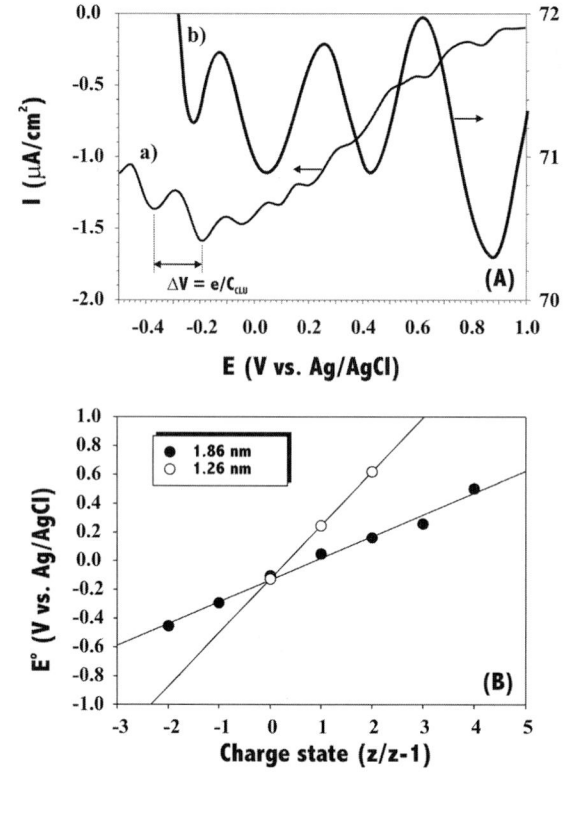

Fig. 13.12 In **A**, differential pulse voltammetry of Au-NSi$_8$ clusters with 1.86-nm **a** and 1.26-nm **b** core size in a LiBF$_4$ electrolyte ethanol/water solution (0.2 M) taken at a pulse amplitude of 55 mV, step potential of 50 mV, pulse time of 5 ms and interval time of 0.2 s. The corresponding formal potentials of the quantized capacitance charging versus the valence states are represented in **B**. The linear regression fit reveals that Eq. (13) satisfactorily describes the quantization phenomenon

dures. Stoichiometric redox capacities (i.e. oxidizing or reducing equivalents per mol) can be controllably assigned by tuning the potential at which the particles are charged (from differential pulse voltammetry). In this respect, charged MPCs constitute a promising source of electronic charge for stoichiometric electron transfer reactions.

Core–shell particles consisting of 15-nm gold cores protected with a 11-nm thick SnO$_2$ coating have been prepared to accumulate large quantities of electrons. The trapping energy corresponds to the difference of energy, about 1 eV, separating the Fermi level of the gold core from the conduction band edge of the semiconductor. In such a configuration, the core exhibits a higher electroaffinity than the shell and acts as an electron trap with a potential, E_{core}, given by a combination of Eqs. (11) and (12):

$$E_{core} = \frac{ze}{4\pi\varepsilon_{Shell}\varepsilon_0(r+d)}\frac{d}{r} \qquad (14)$$

where ε_{Shell} corresponds to the dielectric constant of SnO$_2$ (about 24). Cathodic polarization of the particles allows charging of the core (with loading of 1400±200 electrons per particle) to a limit determined by the trapping energy. A capacitance of

160 aF may be deduced from Eq. (14). Electrons can be stored for up to 10 h and used to reduce oxygen or electron acceptor molecules (viologen derivatives) [32].

Multilayered films of MPCs have been deposited on flat substrates by exploiting the complexing properties of divalent metal ions to form assemblies of bridged functionalized MPCs (see Section 13.2.2.4) that exhibit the same charging properties as MPCs dispersed in solution [59]. The quantized double layer charging of MPCs propagates throughout the entire film. By adjusting the length of the molecular bridges between MPC cores to slow the electron-hopping rate to a measurable range, it has been experimentally shown that the charge state change initiates at the electrode/nanoparticle interface and propagates outward into the bulk of the MPC film, until the entire film comes into equilibrium with the applied potential. The electron self-exchange was measured as a diffusion-like electron-hopping process (by analogy with redox polymers) and was characterized by very high rate constants (2×10^8 M^{-1} s^{-1}) [123]. When MPCs are functionalized with carboxylate or amine moieties to electrostatically adsorb on PAH or PSS, respectively, a few more factors have to be taken into consideration [41]. Polyelectrolytes add an additional tunneling barrier to electron hopping between MPC cores. Electron transport through the entire film supposes both a mobility of electrolyte ions and MPCs (thermally activated) within the film. It has been reported that the electronic conductivity of solid-state MPC films proceeds by bimolecular electron self-exchange reaction, whose rate constant is controlled by the core-to-core electron tunneling along alkanethiolate chains and the mixed valency of the MPC cores [124]. Upon application of a potential, the electronic charge is expected to propagate much more slowly to maintain an equilibrated Fermi level at the interface electrode/MPCs. The film potential follows the Nernst equation:

$$E_{\text{film}} = E^0_{Z,Z-1} + 0.059 \log \frac{[\text{MPC}^Z]}{[\text{MPC}^{Z-1}]} \tag{15}$$

from which the relative concentrations of the charge state formal "redox couple" can be calculated.

Driving the deposition of metal cores by using arylthiolate ligands and PSS is beneficial to the double layer capacitance charging compared with binders such as PAH and alkylthiolates. The residual content of charge-compensating counterions within the films plays a critical role in the charging processes and seems to be larger for films incorporating PSS. Although not yet attempted, the sequential layering of core-shell (Au-SnO_2) particles with polycations should also be a simple means of fabricating solid-state nanocapacitor thin films.

13.5.3
Magnetic Properties

The possibility of embedding magnetic nanoparticles into dielectric or metallic (and non-magnetic) matrices may be of technological interest for high-density information storage media (for example in QMDs). In addition, a judicious combination of magnetic and supporting material may lead to unusual giant magnetoresistive properties by tuning of the electrical resistance. In such applications, particles are considered as discrete magnetic elements with quantified magnetization that has only two states and allows the definition of a bit of binary information. Soft magnetic films can be prepared by the LbL self-assembly technique, using polyelectrolytes to hold the particles within the films electrostatically. Preliminary studies were conducted with nanosized iron oxide particles, Fe_3O_4, stabilized in the presence of poly(diallyldimethylammonium chloride), PDDA, to impart a positive charge to the particles prior to adsorption on a polyamic acid layer (PAATEA) [125]. It was shown that linear deposition occurred for at least 17 adsorbed Fe_3O_4-PDDA/PAATEA bilayers. An average magnetic flux density per bilayer of 850 nT was measured at a distance of 2 cm from the specimen film. It has also been demonstrated that other types of metallic nanoparticles, including Co, Fe or binary FePt and CoPt, can form densely packed layers by a polymer-mediated self-assembly process. The magnetic nanoparticles were first stabilized with oleic acid/ oleyl amine that is further displaced on approaching a poly(vinylpyrrolidone)- or poly(ethyleneimine)-coated surface. The particles attach to the surface through direct binding with the polymer film by a ligand-exchange reaction [126]. It should be noted that the deposition is carried out in non-aqueous solvents (chloroform, hexane). The magnetization pattern of these thin films is controlled using a laser diode-pulsed beam under a magnetic field. The laser power induces a local annealing of the particles reducing their coercivity and allowing thermally assisted magnetization reversal.

It should be noted that magnetic fields can be used to couple magnetic particles through dipole interactions and direct their assembly. Cobalt particles suspended in toluene and stabilized with a resorcinarene-based surfactant spontaneously form bracelet-like structures or chains upon drying on planar surfaces. The interplay of the kinetic stabilization due to the presence of surfactant and the magnetic moments of the particles is thought to be responsible for the formation of such structures [127].

13.6
Conclusions

Metal particles can be seen as building blocks in which electrons are confined within a discrete volume and at specific energy levels. Manipulation and organization of these objects on a surface are two particularly important objectives in thin film fabrication, as they offer a mean of tuning the solid-state collective properties (optical, electrical or magnetic) of metal particles. The layer-by-layer self-assembly technique allows the adsorptive processes to be directed in such a way that the coverage, the thickness and, in some cases, the arrangement of the particle layers can be controlled. This technique of deposition has been widely used to grow composite films in which synergisms, novel or similar to those attained with conventional film fabrication technologies, emerge from the spatial integration of self-assembled entities. This film construction approach is highly versatile in that a variety of tunable parameters can be identified, closely related to the type of chemical functionalization assigned to each building block unit. Although quite useful in designing hierarchical systems, this versatility is cumbersome when perfectly arrayed particles are needed, because the content of information placed on each particle is sufficient to determine its binding ability independently of the presence of neighboring particles. Extra binding forces (capillary forces, magnetic forces, etc.) have to be used to overcome the set of binding interactions present on each individual particle. Epitaxial-like deposition is, however, possible when attention is paid to patterning the substrate surface through crystallization processes. Most systems designed using layer-by-layer self-assembly techniques do not require highly ordered layers to be efficient. The way particles spatially distribute within the film only becomes critical when optimization of the film properties (e.g. density) are concerned and/or coherent couplings over a large surface area are essential to observe a property (e.g. amplification, accuracy of an optical signature, etc.).

13.7
References

1 R. K. ILER, *J. Coll. Interf. Sci.* **1966**, *21*, 569.
2 For a review see: G. L. WHITESIDES, B. BRZYBOSKI, *Science* **2002**, *295*, 2418.
3 H. LEE, L. J. KEPLEY, H.-G. HONG, T. E. MALLOUK, *J. Am. Chem. Soc.* **1988**, *110*, 618.
4 G. DECHER, J. D. HONG, *Makromol. Chem. Macromol. Symp.* **1991**, *110*, 618.
5 For a review see: T. P. CASSAGNEAU, J. H. FENDLER, in *Electrochemistry of Nanomaterials*, Chapter 9 (G. HODES, ed.), VCH-Wiley, Weinheim, **2001**, pp 247–282.
6 A. N. SHIPWAY, E. KATZ, I. WILLNER, *Chem. Phys. Chem.* **2000**, *1*, 18.
7 Metal nanoparticles: Synthesis, Characterization, and Applications (D. L. FELDHEIM, C. A. FOSS, eds.), Marcel Dekker, New York, **2001**.
8 M. BRUST, M. WALKER, D. BETHELL, D. J. SCHIFFRIN, R. WHYMAN, *J. Chem. Soc., Chem. Commun.* **1994**, 801.
9 K. S. MAYYA, V. PATIL, M. SASTRY, *Langmuir* **1997**, *13*, 3944.
10 M. SASTRY, K. S. MAYYA, K. BANDYOPADHYAY, *Coll. Surf. A* **1997**, *127*, 221.
11 C. B. MURRAY, C. R. KAGAN, M. G. BAWENDI, *Annu. Rev. Mater. Sci.* **2000**, *30*, 545.
12 R. P. ANDRES, J. D. BIELEFELD, J. I. HENDERSON, D. B. JAMES, V. R. KOLAGUNTA, C. P. KUBIAK, W. J. MAHONEY, R. G. OSIFCHIN, *Science* **1996**, *273*, 1690.
13 X. M. LIN, H. M. JAEGER, C. M. SORENSEN, K. J. KLABUNDE, *J. Phys. Chem. B* **2001**, *105*, 3353.
14 S. STOEVA, K. J. KLABUNDE, C. M. SORENSEN, I. DRAGIEVA, *J. Am. Chem. Soc.* **2002**, *124*, 2305.
15 R. S. INGRAM, M. J. HOSTETLER, R. W. MURRAY, *J. Am. Chem. Soc.* **1997**, *119*, 9175.
16 A. C. TEMPLETON, M. J. HOSTETLER, C. T. KRAFT, R. W. MURRAY, *J. Am. Chem. Soc.* **1998**, *120*, 1906.
17 D. I. GITTINS, F. CARUSO, *Adv. Mater.* **2000**, *12*, 1947.
18 K. S. MAYYA, B. SCHEELER, F. CARUSO, *Adv. Func. Mater.* **2003**, *13*, 183.
19 C. MANGENEY, F. FERRAGE, I. AUJARD, V. MARCHI-ARTZNER, L. JULLIEN, O. OUARI, E. D. RÉKAI, A. LASCHEWSKY, I. VIKHOLM, J. W. SADOWSKI, *J. Am. Chem. Soc.* **2002**, *124*, 5811.

20 T. Teranishi, M. Miyake, *Chem. Mater.* **1998**, *10*, 594.
21 (a) Y. Zhou, H. Itoh, T. Uemura, K. Naka, Y. Chujo, *Langmuir* **2002**, *18*, 277. (b) Y. Zhou, H. Itoh, T. Uemura, K. Naka, Y. Chujo, *Chem. Commun.* **2001**, 613.
22 T. Cassagneau, unpublished.
23 E. Hao, T. Lian, *Chem. Mater.* **2000**, *12*, 3392.
24 Y. Liu, Y. Wang, R. O. Claus, *Chem. Phys. Lett.* **1998**, *298*, 315.
25 T. Cassagneau, J. H. Fendler, *J. Phys. Chem. B* **1999**, *11*, 1789.
26 L. M. Liz-Marzán, M. Giersig, P. Mulvaney, *Langmuir* **1996**, *12*, 4329.
27 R. I. Nooney, T. Dhanasekaran, Y. Chen, R. Josephs, A. E. Ostafin, *Adv. Mater.* **2002**, *14*, 529.
28 P. Mulvaney, L. M. Liz-Marzán, M. Giersig, T. Ung, *J. Mater. Chem.* **2000**, *10*, 1259.
29 F. G. Aliev, M. A. Correa-Duarte, A. Mamedov, J. W. Ostrander, M. Giersig, L. M. Liz-Marzán, N. A. Kotov, *Adv. Mater.* **1999**, *11*, 1006.
30 T. Ung, L. M. Liz-Marzán, P. Mulvaney, *Langmuir* **1998**, *14*, 3740.
31 V. V. Hardikar, E. Matijevic, *J. Coll. Interf. Sci.* **2000**, *221*, 133.
32 G. Oldfield, T. Ung, P. Mulvaney, *Adv. Mater.* **2000**, *12*, 1519.
33 I. Pastoriza-Santos, D. S. Koktysh, A. A. Mamedov, M. Giersig, N. A. Kotov, L. M. Liz-Marzán, *Langmuir* **2000**, *16*, 2731.
34 J. Turkevich, P. C. Stevenson, J. Hillier, *Discuss. Faraday Soc.* **1951**, *11*, 55.
35 J. Schmitt, G. Decher, W. J. Dressick, S. L. Brandow, R. E. Geer, R. Shashidhar, J. M. Calvert, *Adv. Mater.* **1997**, *9*, 61.
36 N. A. Kotov, I. Dekány, J. H. Fendler, *J. Phys. Chem.* **1995**, *99*, 13065.
37 J. Schmitt, P. Mächtle, D. Eck, H. Möhwald, C. A. Helm, *Langmuir* **1999**, *15*, 3256.
38 Y. Fu, H. Xu, S. Bai, D. Qiu, J. Sun, Z. Wang, X. Zhang, *Macromol. Rapid Commun.* **2002**, *23*, 256.
39 T. Ung, L. M. Liz-Marzán, P. Mulvaney, *J. Phys. Chem. B* **2001**, *105*, 3441.
40 L. Wang, Z. Wang, X. Wang, J. Shen, L. Chi, H. Fuchs, *Macromol. Rapid Commun.* **1997**, *18*, 509.
41 J. F. Hicks, Y. Seok-Shon, R. W. Murray, *Langmuir* **2002**, *18*, 2288.
42 T. Cassagneau, F. Caruso, in preparation.
43 T. Cassagneau, F. Caruso, *J. Am. Chem. Soc.* **2002**, *124*, 8172.
44 T. Cassagneau, F. Caruso, *Adv. Mater.* **2002**, *14*, 732.
45 R. Blonder, L. Sheeney, I. Willner, *Chem. Commun.* **1998**, 1393.
46 M. Lahav, A. N. Shipway, I. Willner, M. B. Nielsen, J. F. Stoddart, *J. Electroanal. Chem.* **2000**, *482*, 217.
47 M. Lahav, A. N. Shipway, I. Willner, *J. Chem. Soc., Perkin Trans.* **1999**, *2*, 1925.
48 M. Lahav, R. Gabai, A. N. Shipway, I. Willner, *Chem. Commun.* **1999**, 1937.
49 W. Cheng, J. Jiang, S. Dong, E. Wang, *Chem. Commun.* **2002**, 1706.
50 M. Lu, X. H. Li, B. Z. Yu, H. L. Li, *J. Coll. Interf. Sci.* **2002**, *248*, 376.
51 M. Brust, D. Bethell, C. J. Kiely, D. J. Schiffrin, *Langmuir* **1998**, *14*, 5425.
52 M. D. Musick, D. J. Pea, S. L. Botsko, T. M. McEvoy, J. N. Richardson, M. J. Natan, *Langmuir* **1999**, *15*, 844.
53 Y. Kuwahara, T. Akiyama, S. Yamada, *Thin Solid Films* **2001**, *393*, 273.
54 F. Patolsky, K. T. Ranjit, A. Lichtenstein, I. Willner, *Chem. Commun.* **2000**, 105.
55 T. A. Taton, R. C. Mucic, C. A. Mirkin, R. L. Letsinger, *J. Am. Chem. Soc.* **2000**, *122*, 6305.
56 T. Liu, J. Tang, H. Zhao, Y. Deng, L. Jiang, *Langmuir* **2002**, *18*, 5624.
57 A. Kumar, A. B. Mandale, M. Sastry, *Langmuir* **2000**, *16*, 6921.
58 D. E. Cliffel, F. P. Zamborini, S. M. Gross, R. W. Murray, *Langmuir* **2000**, *16*, 9699.
59 F. P. Zamborini, J. F. Hicks, R. W. Murray, *J. Am. Chem. Soc.* **2000**, *122*, 4514.
60 T. Teranishi, M.-a. Haga, Y. Shiozawa, M. Miyake, *J. Am. Chem. Soc.* **2000**, *122*, 4237.
61 I. Šloufová-Srnová, B. Vlčková, *Nano Lett.* **2002**, *2*, 121.
62 N. Oyama, F. C. Anson, *J. Electrochem. Soc.* **1980**, *127*, 247.
63 D. Yoo, S. S. Shiratori, M. F. Rubner, *Macromolecules* **1998**, *31*, 4309.
64 S. S. Shiratori, M. F. Rubner, *Macromolecules* **2000**, *33*, 4213.

65 S. Joly, R. Kane, L. Radzilowski, T. Wang, A. Wu, R. E. Cohen, E. L. Thomas, M. F. Rubner, *Langmuir* **2000**, *16*, 1354.
66 T. C. Wang, M. F. Rubner, R. E. Cohen, *Langmuir* **2002**, *18*, 3370.
67 J. Dai, A. M. Balachandra, J. I. Lee, M. L. Bruening, *Macromolecules* **2002**, *35*, 3164.
68 A. M. Balachandra, J. Dai, M. L. Bruening, *Macromolecules* **2002**, *35*, 3171.
69 (a) S. Dante, Z. Hou, S. Risbud, P. Stroeve, *Langmuir* **1999**, *15*, 2176. (b) A. K. Dutta, G. Jarero, L. Zhang, P. Stroeve, *Chem. Mater.* **2000**, *12*, 176.
70 L. Zhang, A. K. Dutta, G. Jarero, P. Stroeve, *Langmuir* **2000**, *16*, 7095.
71 A. S. Dutta, T. Ho, L. Zhang, P. Stroeve, *Chem. Mater.* **2000**, *12*, 1042.
72 N. Oyama, F. C. Anson, *Anal. Chem.* **1980**, *52*, 1192.
73 J. A. Bruce, M. S. Wrighton, *J. Am. Chem. Soc.* **1982**, *104*, 74.
74 R. N. Dominey, N. S. Lewis, J. A. Bruce, D. C. Bookbinder, M. S. Wrighton, *J. Am. Chem. Soc.* **1982**, *104*, 467.
75 J. Liu, L. Cheng, Y. Song, B. Liu, S. Dong, *Langmuir* **2001**, *17*, 6747.
76 G. Coulon, T. P. Russell, V. R. Deline, P. F. Green, *Macromolecules* **1989**, *22*, 2581.
77 J. Heier, E. J. Kramer, S. Walheim, G. Krausch, *Macromolecules* **1997**, *30*, 6610.
78 M. D. Foster, M. Sikka, N. Singh, F. S. Bates, S. K. Satija, C. F. Majkrzak, *J. Chem. Phys.* **1992**, *96*, 8605.
79 T. P. Russell, *Physica B* **1996**, *221*, 267.
80 S. H. Sohn, B. H. Seo, *Chem. Mater.* **2001**, *13*, 1752.
81 S. Y. Chou, P. R. Krauss, L. Komg, *J. Appl. Phys.* **1996**, *79*, 6101.
82 S. H. Charap, P. L. Lu, Y. He, *IEEE Trans. Magn.* **1997**, *33*, 9/8.
83 H. Tuan, A. Gilbertson, S. Y. Chou, *J. Vac. Sci. Technol. B* **1998**, *16*, 3926.
84 S. Y. Chou, P. R. Krauss, P. J. Renstrom, *Science* **1996**, *272*, 85.
85 S. Y. Chou, C. Keimel, J. Gu, *Nature* **2002**, *417*, 835.
86 L. Cusack, R. Rizza, A. Gorelov, D. Fitzmaurice, *Angew. Chem. Int. Ed. Engl.* **1997**, *36*, 848.
87 A. K. Boal, F. Ilhan, J. E. DeRouchey, T. Thurn-Albrecht, T. P. Russell, V. M. Rotello, *Nature* **2000**, *404*, 746.
88 T. H. Galow, A. K. Boal, V. M. Rotello, *Adv. Mater.* **2000**, *12*, 576.
89 J. Kolny, A. Kornowski, H. Weller, *Nano Lett.* **2002**, *2*, 361.
90 C. A. Mirkin, R. L. Letsinger, R. C. Mucic, J. J. Storhoff, *Nature* **1996**, *382*, 607.
91 G. Schmid, M. Bäumle, N. Beyer, *Angew. Chem. Int. Ed.* **2000**, *39*, 181.
92 K. S. Mayya, B. Scheffer, F. Caruso, *Adv. Func. Mater.* **2003**, *13*, 183.
93 M. Sano, D. Y. Sasaki, T. Kunitake, *J. Chem. Soc. Chem. Commun.* **1992**, 1326.
94 M. Sano, D. Y. Sasaki, T. Kunitake, *Macromolecules* **1992**, *25*, 6961.
95 M. Sano, *Adv. Mater.* **1996**, *8*, 521.
96 A. van Blaaderen, R. Ruel, P. Wiltzius, *Nature* **1997**, *385*, 321.
97 Y. Yin, Y. Lu, B. Gates, Y. Xia, *J. Am. Chem. Soc.* **2001**, *123*, 8718.
98 J. Aizenberg, A. J. Black, G. M. Whitesides, *Nature* **1999**, *398*, 495.
99 J. C. Wittmann, P. Smith, *Nature* **1991**, *352*, 414.
100 A. Sugarawa, G. G. Hembree, M. R. Scheinfein, *J. Appl. Phys.* **1997**, *82*, 5662.
101 T. O. Hutchinson, Y.-P. Liu, C. Kiely, C. J. Kiely, M. Brust, *Adv. Mater.* **2001**, *13*, 1800.
102 T. Teranishi, A. Sugarawa, T. Shimizu, M. Miyake, *J. Am. Chem. Soc.* **2002**, *124*, 4210.
103 P. D. Kaplan, J. L. Rouke, A. G. Yodh, D. J. Pine, *Phys. Rev. Lett.* **1994**, *72*, 582.
104 P. A. Kralchevsky, V. N. Paunov, I. B. Ivanov, K. Nakayama, *J. Coll. Interf. Sci.* **1992**, *151*, 79.
105 U. Kreibig, M. Vollmer, *Optical Properties of Metal Clusters*, Springer, Berlin, **1995**.
106 G. Mie, *Annal. Phys* **1908**, *25*, 377.
107 B. U. Federhof, R. B. Jones, *Z. Phys. B – Condensed Matter* **1985**, *62*, 43.
108 J. C. Maxwell Garnett, *Philos. Trans. R. Soc. London* **1904**, *203*, 385.
109 I. N. Shklyarevskii, E. Anachkova, G. S. Blyashenko, *Opt. Spectrosc. (USSR)* **1977**, *43*, 427.
110 C. G. Blatchford, O. Siiman, M. Kerker, *J. Phys. Chem.* **1983**, *87*, 2503.
111 A. Henglein, P. Mulvaney, T. Linnert, *J. Chem. Soc., Faraday Discuss.* **1991**, *92*, 31.

112 P. Mulvaney, Langmuir **1996**, *12*, 788.
113 E. Sabatini, I. Rubinstein, *J. Phys. Chem.* **1987**, *91*, 6663.
114 H. Reller, E. Kirowa-Eisner, E. Gileadi, *J. Electroanal. Chem.* **1982**, *138*, 65.
115 A. J. Bard, L. R. Faulkner, *Electrochemical Methods: Fundamentals and Applications*, 1st edn, Wiley, New York, **1980**.
116 Z. B. Stoynov, B. M. Grafov, B. S. Savova-Staynov, V. V. Elkin, *Electrochemical Impedance*, Nauka Publisher, Moscow, **1991**.
117 A. B. Kharitonov, A. N. Shipway, I. Willner, *Anal. Chem.* **1999**, *71*, 5441.
118 M. Lahav, V. Heleg-Shabtai, J. Wasserman, E. Katz, I. Willner, H. Durr, Y.-Z. Hu, S. H. Bossmann, *J. Am. Chem. Soc.* **2000**, *122*, 11480.
119 S. Chen, R. S. Ingram, M. J. Hostetler, J. J. Pietron, R. W. Murray, T. G. Schaaff, J. T. Khoury, M. M. Alvarez, R. L. Whetten, *Science* **1998**, *280*, 2098.
120 S. Chen, R. W. Murray, *J. Phys. Chem. B* **1999**, *103*, 9996.
121 S. Chen, R. W. Murray, S. W. Feldberg, *J. Phys. Chem.* **1998**, *102*, 9898.
122 J. J. Pietron, J. F. Hicks, R. W. Murray, *J. Am. Chem. Soc.* **1999**, *121*, 5565.
123 J. F. Hicks, F. P. Zamborini, A. J. Osisek, R. W. Murray, *J. Am. Chem. Soc.* **2001**, *123*, 7048.
124 W. P. Wuelfing, S. J. Green, J. J. Pietron, D. E. Cliffel, R. W. Murray, *J. Am. Chem. Soc.* **2000**, *122*, 11465.
125 Y. Liu, A. Wang, R. O. Claus, *Appl. Phys. Lett.* **1997**, *71*, 2265.
126 S. Sun, S. Anders, H. F. Hamann, J.-U. Thiele, J. E. E. Baglin, T. Thomson, E. E. Fullerton, C. B. Murray, B. D. Terris, *J. Am. Chem. Soc.* **2002**, *124*, 2884.
127 S. L. Tripp, S. V. Pusztay, A. E. Ribbe, A. Wei, *J. Am. Chem. Soc.* **2002**, *124*, 7914.
128 P. C. Lee, D. Meisel, *J. Phys. Chem.* **1982**, *86*, 3391.
129 V. V. Vuković, J. M. Nedeljković, *Langmuir* **1993**, *9*, 980.

14
Assembly of Electrically Functional Microstructures from Colloidal Particles

Orlin D. Velev

14.1
Introduction: Electrical Forces in Colloidal Suspensions

Electric fields provide a powerful and convenient tool for the assembly of colloidal particles into structures that serve as microscopic elements in electrical circuits, sensors and displays, or act as photonic or electronic devices. This chapter presents an overview of colloidal structures that can be assembled by electric fields, the methods for fabrication of electrical circuits from colloidal particles and examples of devices made by these techniques. In general, the particles in a colloidal suspension can be manipulated by applying constant (DC) or alternating (AC) voltage to electrodes in contact with the liquid [1]. The forces operating on the particles in such systems are:

(1) The electrophoretic (EP) force F_{EP} between a particle of charge q and a DC electric field E,

$$\vec{F}_{EP} = q\vec{E} \qquad (1)$$

which makes the particle move along the direction of the field lines towards the electrode of opposite sign. The speed of the particle can be controlled via the applied voltage, and the particle charge can be adjusted via pH or addition of surfactant. A common complication in applying DC fields in ionic media (such as water) is the electrophoretic mobility of the liquid adjacent to the walls, which can drag the particles in an arbitrary direction or distort the assembled structures.

(2) Dielectrophoretic (DEP) attraction or repulsion along the *gradient* of an AC electric field. The use of alternating voltage allows high field strengths without water electrolysis and electro-osmotic currents. The DEP force acting on each particle, F_{DEP}, is [2–6]

$$\vec{F}_{DEP} = 2\pi\varepsilon_1 \text{Re}|\underline{K}(w)|r^3 \nabla E_{rms}^2 \qquad (2)$$

where r is the radius of the particle and K is the Clausius-Mossotti factor. The resultant force is dependent on the gradient, ∇E, and the particle can be attracted to inhomogeneities in the field. The sign and magnitude of the DEP force are de-

Colloids and Colloid Assemblies. Edited by Frank Caruso
Copyright © 2004 Wiley-VCH Verlag GmbH & Co. KGaA, Weinheim
ISBN: 3-527-30660-9

pendent on the real part of the Clausius-Mossotti function, K, i.e. the effective polarizability of the particle,

$$\mathrm{Re}|\underline{K}| = \frac{\varepsilon_2 - \varepsilon_1}{\varepsilon_2 + 2\varepsilon_1} + \frac{3(\varepsilon_1\sigma_2 - \varepsilon_2\sigma_1)}{\tau_{MW}(\sigma_2 + 2\sigma_1)^2(1 + w^2\tau_{MW}^2)} \qquad (3)$$

where ε_1 and σ_1 are the dielectric permittivity and conductivity of the media and ε_2 and σ_2 the same quantities for the particles. For dielectric particles this function changes sign (i.e. the force changes from attractive to repulsive) at a crossover frequency of $w_C = (\tau_{MW})^{-1}$, where τ_{MW} is the Maxwell-Wagner charge relaxation time $\tau_{MW} = \frac{\varepsilon_2 + \varepsilon_1}{\sigma_2 + 2\sigma_1}$. Frequency-dependent change of the sign of the interactions is commonly observed with latex microspheres in water [4–6] and allows a high degree of control. An example of the operation and possible directions of the DEP and EP forces generated by planar electrodes on a chip surface is presented in Fig. 14.1.

(3) A "chaining" force due to particle–particle dipolar attraction. This force is dependent on the field strength, E and operates in both AC and DC fields. A generalized expression for the force between adjacent particles is

$$F_{chain} = -C\pi\varepsilon r^2 K^2 E^2 \qquad (4)$$

where, depending on the distance between the particles and the length of the chain, the coefficient C ranges from 3 to $>10^3$. This force can provide much needed directional orientation during colloidal assembly as the particle chains become aligned along the field lines.

Due to the intense interest in photonic crystals from 2D and 3D arrays of colloidal spheres, a large body of research on the assembly of colloidal crystals by elec-

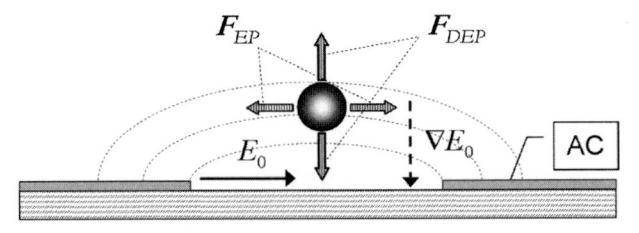

Fig. 14.1 Summary of the possible forces exerted on a charged dielectric particle by AC or DC fields applied to two planar electrodes. The electrophoretic (EP) force in DC fields moves the particle in the direction of the field E, towards the electrode of opposite charge. Its direction can be reversed by changing polarity. The dielectrophoretic (DEP) force caused by AC fields moves the particle along the gradient of the field, ∇E. Its direction can be reversed by increasing the frequency. Note that in this configuration the DEP and EP forces act in perpendicular directions.

tric fields has been reported. The techniques used in the formation of colloidal crystals by electric fields are classified and reviewed in Section 14.2. Colloidal assembly also has potential for direct fabrication of electrically functional structures, such as wires, sensors, switches, and logical and memory-element circuits. The state of the art in this area is reviewed in Section 14.3. Examples of actual devices made from structures of colloidal particles, including chemical and biological sensors, displays and electronic elements are presented in Section 14.4. In many of these devices the colloidal structures are both made by electric fields and serve as elements in electrical circuits themselves.

14.2
Electrical Assembly of Colloidal Crystals

Photonic crystals, structured materials with periodicity comparable to the wavelength of light, promise to revolutionize the way light is manipulated (see Chapter 9). The fabrication of such crystals by self-assembly of colloidal spheres is still far from maturity, as the rapid formation of long-range crystals with specific orientation for photonic applications remains a challenge. Many of the processes used in making colloidal crystals, such as sedimentation, slow drying, convective motion of particles or restriction of available volume by long-range electrostatic repulsion [7, 8], are slow and difficult to control. External electric fields can be a tool for rapid and controllable assembly. An outline of the four different geometries used in cells for colloidal crystal assembly is presented in Fig. 14.2. Each of these four methods is briefly discussed below.

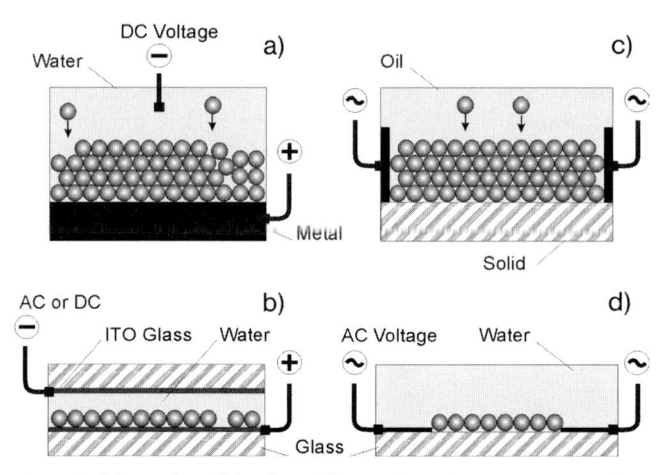

Fig. 14.2 Schematics of the four different electrode geometries used to assemble colloidal crystals by AC or DC fields. The forces operating in each system are discussed in the text.

14.2.1
Assembly by Electrophoretic Particle Attraction to Surfaces

Electrophoretic attraction can be used efficiently for deposition of multicrystalline layers of particles onto conductive solid surfaces (Fig. 14.2a). The process can be viewed as speeding up the deposition of dense particle crystals that would otherwise occur under gravity. Giersig and Mulvaney [9] have reported that ordered arrays of gold and latex particles deposit onto the surfaces of carbon-coated copper grids under the action of an electric field. Conductive grids have been immersed in the suspensions and polarized at a field of 1–5 V/cm. Alkanethiol-capped gold particles from 3.5 to 200 nm in diameter and 440-nm latex particles have been deposited in ordered arrays on the grid surface. Adsorption, nucleation and growth of polycrystalline arrays are observed when the grids are immersed for consecutively increasing periods of time. Multilayer latex arrays form at higher field strengths.

Direct electrophoretic control of the deposition speed of silica spheres for the fabrication of high-quality crystals has been reported by Holgado et al. [10]. The DC field is applied in the vertical direction in cylinders with sedimenting spheres, controlling the speed at which the spheres deposit on the bottom. The vertical electrophoretic mobility of the spheres can add to the Stokes sedimentation velocity so crystals can be assembled rapidly from small spheres. When the field is applied in the opposite direction it can slow the downward mobility of large spheres, whose Stokes sedimentation speed is so high that it inhibits the growth of well-ordered crystals. Electrophoretic deposition has been used to prepare crystals and multilayered deposits with a variety of potential applications [11–15].

14.2.2
Assembly in Thin Liquid Films Between Electrode Surfaces

The confinement of the particle suspension in thin film between electrodes few micrometers to tens of micrometers apart (Fig. 14.2b) can be used for quick assembly of 2D crystalline structures. The high intensity of the field in the thin gap leads to new effects, namely, the appearance of electrohydrodynamic currents around the particles, and lateral dipole–dipole interactions. The first observation of particle clustering in thin cells was reported by Richetti et al. [16] who studied the behavior of polyvinyl-toluene latex spheres sandwiched between conductive transparent slides as a function of the frequency and magnitude of an applied AC voltage. Above a certain threshold magnitude of the alternating voltage (1.7 V for 1 kHz in the 15-μm thick cell used) the particles are attracted to the surfaces of the electrodes and exhibit lateral attraction, forming ordered 2D aggregates. The ordered structure "melts" at frequencies higher than ≈ 5 kHz.

The dynamics of electrophoretic (DC) particle deposition and ordering onto conductive surfaces has been reported by Trau et al. [17, 18] and Böhmer [19]. The motion and crystallization of micrometer-sized latex spheres is followed directly by optical microscopy through transparent electrodes from indium tin oxide

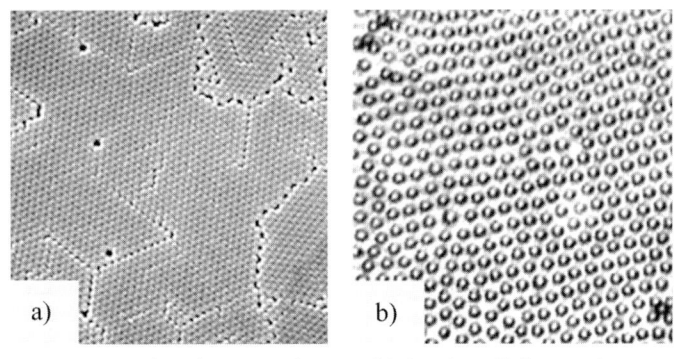

Fig. 14.3 Examples of 2-D crystals assembled in the cell illustrated in Fig. 14.2b. **a** Cell thickness many times larger than particle diameter, DC field. The crystal forms via electrohydrodynamic attraction [20]. **b** Cell thickness only slightly larger than particle diameter, AC field. The particles exhibit long range dipolar repulsion [26] (reprinted with permission from [20] and [26]. Copyright American Chemical Society, 1997 and 2001).

(ITO)-covered glass. When the potential applied between the electrodes exceeds ca. 0.5 V DC, the microspheres move towards the oppositely charged electrode and begin to form ordered 2D clusters. The microspheres clearly exhibit a lateral attraction force, which becomes more pronounced at higher potentials (1–2 V) where the particles attract each other at distances exceeding a few particle diameters. The microspheres form large 2D-ordered aggregates that can attract to, and merge with, other aggregates until the electrode is covered with a complete monolayer or even a multilayer of crystalline array (Fig. 14.3a). Thus the state of 2D particle aggregation and ordering can be reversibly controlled via the applied potential from "gaseous" through "liquid" to "crystalline" solid. At higher applied voltages the lateral attraction force may cause irreversible coagulation of the microspheres into fixed sheets. These ordered sheets can be removed from the surface by applying reverse potential or by liquid flow. The method, however, has its limitations, as the ordered layer is multicrystalline, similar to the ones obtained by direct particle adsorption.

Electrophoretic 2D colloidal crystallization can be observed with a large variety of particles under a wide range of conditions: the phenomenon has been observed with latex, silica and gold particles, suspended in deionized water or in moderate electrolyte, in DC and AC fields. The insensitivity of the assembly to the type of particles and the electrolyte concentration indicates that the ordering is not driven by the conventional Derjaguin-Landau-Verwey-Oberbeek (DLVO) forces between the particles. The forces behind this assembly have been theoretically interpreted as originating from electrohydrodynamic flows [18, 20–23]. These flows are generated because the counterionic atmosphere around the colloidal particles disturbs the concentration polarization layer on the electrode. The electroosmotic mobility of these counterionic layers leads to emergence of local flow streams perpendicular to the cell wall. The hydrodynamic interaction of the flows around the microspheres leads to attraction between the particles and their aggregation into ordered patterns. The electrohydrodynamic models explain the reversibility of the process and the "softness" of the incurred attractive forces that allows the lateral rearrangement of the particles into ordered arrays.

The electrophoretic method, concurrently with the photosensitivity of semiconductor ITO layers, has been used to make optically tunable micropatterns from colloidal crystals [24]. The ITO anodic layer absorbs light in the UV-region, which increases the rate of charge transfer at the electrode/solution interface and the current density in the illuminated regions. When a UV light pattern is applied to the surface through a lithographic mask, the illuminated ITO regions generate higher electrohydrodynamic currents that attract the particles and assemble them in the illuminated areas. The microspheres slowly migrate to the illuminated regions to form ordered patterns and diffuse back to an unordered state when the voltage is turned off. If the particles in the illuminated patterns are pushed beyond the coagulation threshold by increased current the assembled patterned arrays are permanently immobilized on the electrode. This study outlines the richness of structures that can be created by using combinations of the electrical and optical assembly techniques. Electrophoretic patterning has also been used to deposit live cells on surfaces [25].

Field-induced particle interactions in thin films (Fig. 14.2b) are not attractive in all possible combinations of cell thickness and particle sizes. Gong and Marr [26] have studied the "strictly confined" system where the thin gap between the electrodes is ca. 20% larger than the particle size. The proximity of the opposing solid electrodes suppresses the electrohydrodynamic currents and the related microsphere attraction. High-intensity fields, however, induce strong dipoles in the microspheres, and as these dipoles are parallel, they repel each other. The strongly repelling microspheres organize in 2D crystals with long-range separation between the particles (as compared to the closely packed crystals formed by electrohydrodynamic attraction). The lack of friction, the relatively high mobility and the long-range repulsion lead to rapid formation of relatively large crystal domains (Fig. 14.3b).

14.2.3
Annealing and Alignment of Three-dimensional Crystals by AC Fields

This method uses particle chaining caused by interaction between dipoles directed along the electric field (Eq. 4). The effect is used in electrorheological and magnetorheological fluids [27, 28]. The field-driven structure formation can also be used to assemble and align colloidal crystals. The particles in electrorheological fluids commonly arrange into three-dimensional body-centered tetragonal (bct) crystals [28–31]. Various stable and metastable crystalline phases formed in electrorheological suspension of silica spheres have been reported by Dassanayake et al. [31]. The formation of bct crystals is of particular interest; such symmetry cannot be obtained by conventional crystallization techniques based on phase transitions by increasing the volume fraction of the spheres.

The chaining effect can also be used to induce annealing and large-scale orientation of colloidal crystals formed by conventional techniques such as sedimentation. Electrodes situated on the two sides of the crystal can orient one of the principal axes along the field by the chaining force (Fig. 14.2c). This possibility has been discussed in the literature [32]. More conclusive data and application of the method in the practical fabrication of photonic materials can be expected in the future. Structures other than colloidal crystals can be fabricated by modifications of the method [33]. The chaining of millimeter-sized metallic particles, suspended in organic media, can be used to form electrical connections (discussed in detail in the following section).

14.2.4
Dielectrophoretic Assembly of Two-dimensional Crystals Between Planar On-chip Electrodes

Electrophoretic mobility at frequencies below 100 Hz can cause particle compression and long-range crystallization in highly deionized systems [34, 35]. At higher frequencies DEP forces provide powerful tools for controlled assembly. We have recently studied the formation of structures in suspensions of latex or silica particles subjected to an alternating electric field in a gap above planar electrodes on a surface (Fig. 14.2d) [36]. In this geometry, positive DEP forces attract the particles to the gap as shown in Fig. 14.1. The chaining forces assemble the particles in the direction of the field as explained above. Formation of switchable 2D crystals with areas above 25 mm^2 has been observed. These crystals are specifically oriented without the need for prior templating by microlithography or micromolds. The particles align into horizontal rows along the direction of the field and then crystallize into hexagonal arrays. The transitions between ordered and disordered states take place within seconds and can be repeated tens of times. The laser diffraction patterns of all consecutively formed crystals are identical (Fig. 14.4). Electrically tunable diffraction can be used to make rudimentary optical switches. The model of crystallization driven by a combination of dielectrophoresis and induced dipole chaining is supported by direct observation and diffraction. The assembly is driven by gradient-dependent forces and is not influenced by gravity. Electrostatic repulsion can be used to tune the lattice spacing in the crystal to a precision of about 10 nm. Various combinations of DEP, EP and chaining forces are likely to find applications in more elaborate techniques for precise assembly of ordered structures.

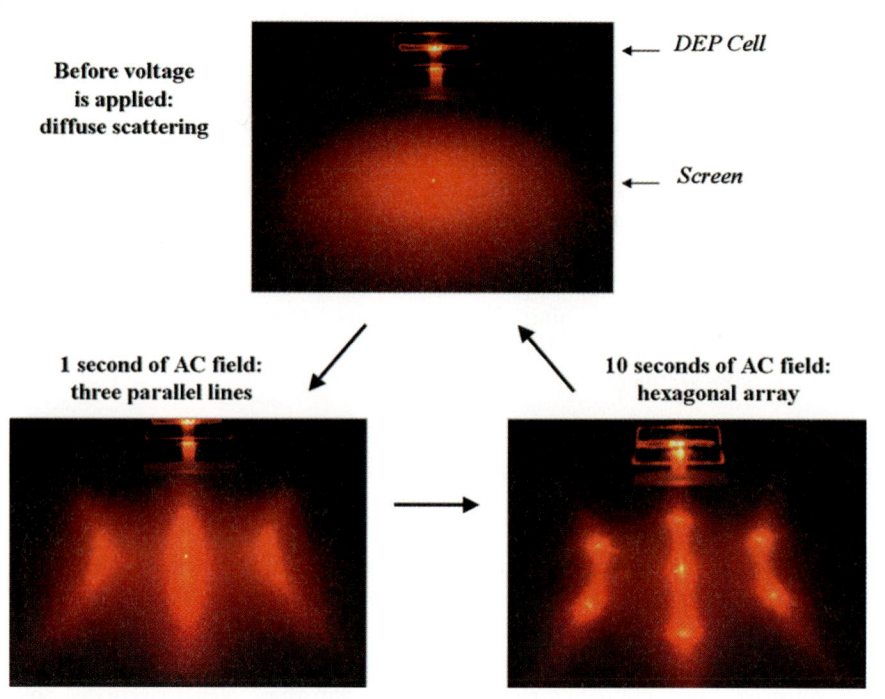

Fig. 14.4 Diffraction pattern of a laser beam directed through a 2-D crystallization cell with planar electrodes similar to the one in Fig. 14.2d [36]. **a** Diffuse scattering from a suspension of 0.7-μm latex particles before the electric field is applied; **b** 1 s after AC voltage is applied the particles align in chains diffracting in parallel lines; **c** 10 s after voltage is applied the particles crystallize into hexagonal close packed array. The process is fully reversible and can be used in beam splitters.

14.3
Fabrication of Electrical Circuits via Colloidal Assembly

14.3.1
Assembly of Electrical Circuits by Capillary Forces

One of the major requirements for self-assembly of electrical circuits from prefabricated components is the use of forces that could bring together the elements of the circuit in pre-defined, directional patterns. Most of the colloidal forces, such as electrostatic and van der Waals forces, do not provide directionality and may be too weak to align electronic components. However, the "capillary forces" generated by the action of interfacial tension on bodies wetted by liquid can provide the long-range attraction and alignment required to assemble objects of up to a millimeter scale [37–41]. This technique has been pioneered by Whitesides' group, originally for the self-assembly of topologically complex, millimeter-sized objects

Fig. 14.5 Example of an electrical circuit assembled via capillary forces from prefabricated components. The building blocks are millimeter-sized polyhedrons made from flexible plastic. The pattern of contact pads on the surface and the connection diagram for the internal LED are shown on top (excerpted from [53] with permission; copyright American Association for the Advancement of Science, 2000).

[42]. Crosses, hexagons, squares and "lock-and-key" polygons have been made from polydimethylsiloxane (PDMS) by a process that allows selective hydrophilization and hydrophobization of surfaces. When these objects are suspended on a water–oil interface they assemble into designed close-packed, or loose, ordered structures. Structures fabricated by this method include arrays and gratings with different symmetry, "host–guest" templated structures, or even analogues of biorecognition and DNA doublets [42–47].

Present technologies for the fabrication of electronic circuits and chips operate intrinsically in two dimensions and extending them into three dimensions promises higher densities and increased complexity. Recognizing that capillary forces can also be used to form 3D assemblies [48–51], Whitesides' group has extended their technique to the fabrication of 3D electrical circuits. The assembly blocks in this method are prefabricated octagons or prisms of millimeter size. Some of the facets of these blocks are fabricated with a pattern of solder patches that match the pattern on the opposing units only in certain orientation(s). Electronic elements are mounted in the objects and connected to the patches. The capillarity-based assembly is done by the attraction of liquefied low-melting-point solder for the connector patches. The process is carried out by tumbling the elements in heated liquid whose density matches the density of the blocks. Only configurations of complementary match of the patterns are held strongly enough by the capillary forces of the molten solder to stay together during the tumbling and thus the elements self-assemble into the designed circuit, rejecting wrong orientations during the process. Three-dimensional electrical networks and electronic devices have been demonstrated (Fig. 14.5) [52–55]. Other researchers have shown that similar forces can be used to raise and rotate electronic circuit components in semiconductor fabrication [56].

The capillary assembly method has also been used to fabricate arrays of small light-emitting diodes (LEDs) on the surfaces of cylindrical substrates with rows of addressable electrodes [57]. The process involves two steps of self-assembly and self-alignment: first the LED chips attach to arrays of patches on the bottom electrode, after which the top electrode self-aligns by capillarity action to the other side of the chips. The complex cylindrical displays obtained in this way illustrate the potential of self-assembly for the fabrication of 3D circuits.

Methods based on capillary forces are elegant and conceptually simple; however, so far they operate on complex prefabricated objects of millimeter size and assemble structures from special types of building blocks. Simpler planar LED arrays can be made by fluidic deposition [58].

14.3.2
Formation of Electrical Circuits by Chemical and Electrochemical Deposition

Direct metal deposition, by either chemical or electrochemical reduction, is one of the simplest techniques to make electrical connections in liquids. The reduced metal typically will not grow one-dimensional wires spontaneously, so the challenge in this approach is to modify the reduction process to make connectors and structures. There are two general solutions to this challenge: the first is to use templates that guide the metal nucleation and/or growth, and the second is to apply external electrical fields to steer the deposited structure in the desired direction.

One versatile method for liquid synthesis of metallic and semiconductor rods is based on electrochemical reduction of metals inside uniform cylindrical micropores in alumina membranes. The method, originally reported by Martin and Moskovits [59, 60], has been used to make the most comprehensive electrical microelements produced without microfabrication to date [61–67]. The process begins by vapor deposition of a metal electrode on the branched side of the membrane, which closes the pore openings on that side and serves as a substrate for metal growth inside the pores. The metal is reduced from electroplating solution(s) infiltrating the pores. The diameters of the metallic rods formed are equal to the diameter of the membrane micropores and typically are in the range of 200–300 nm. The rods are released into suspension by dissolving the templating alumina membrane in basic aqueous solution. A variety of precisely structured rods from different metals, semiconductors and polymers can be fabricated and their ends can be functionalized to invoke directional self-assembly.

The method can be used to sequentially deposit different metals in the pores to make striped rods of varying metal composition [61]. Seven different metals, Au, Ag, Pt, Pd, Ni, Co and Cu, have been deposited in controlled subsequent layers. The complex structures created in this way can serve as "submicrometer metallic barcodes", where the distinctive patterns of rings on the rods allow thousands of different types of rods to be recognized under the microscope [62]. More important for the focus of this chapter is the formation of rods that incorporate in their middle rectifying and other electrically non-linear metal–semiconductor junctions [63]. Rods containing "in-wire" diode junctions have been made by a combination of electrodeposition and nanoparticle assembly. After the bottom layer of metal is deposited, TiO_2 or ZnO semiconductor nanoparticles are assembled and bound with alternating polyelectrolyte layers of opposite charge. The self-assembled semiconductor layer is sandwiched with a second semi-rod of metal reduced on the other side. Semiconductor layers can also be deposited around the rods to make diodes with electrical contact pads on the side of the microscopic cylinders. Finally, non-linear in-wire junctions have been created by sandwiching monolayers of 16-mercaptohexanoic acid in the middle of the rods, illustrating the potential of interfacing the wires with molecular diodes and switches [64].

The suspensions of complex microscopic rods can be assembled in specific patterns and interfaced with electrical circuits. Chemical functionalization of the tips of the microrods during the process can be used to bind them to surfaces or to each other. Attachment of the molecular layer can be done by using agents with a terminal cyanide group (binding to Au and Pt) or thiol group (very strong binding to Au) [61, 64]; either the whole surface or only the tips of the rods can be functionalized [65]. The attachment of DNA and biomolecules is of specific importance [66], as it allows the assembly of rod–particle structures, or the application of the barcode rods in highly selective bioassays [62]. The rods can also be oriented and manipulated to interface electrical circuits by external electric and magnetic fields [67, 68]. Application of alternating electric fields in 2D and 3D electrode microchambers allows both aligning the rods in the direction of the field and positioning them in the area of high field density. Once the rods are positioned onto the electrical circuits, they can be connected to the probe pads by metals evaporated on top, which makes possible precise measurement of their current-voltage characteristics [67].

Nanowires and nanorods can be synthesized via direct chemical reduction in solution, by directional deposition onto specific crystallographic interfaces or by filling the channels of porous materials [69–73]. Enormous attention has recently been paid to the synthesis of electrically conductive carbon nanotubes and semiconductor nanowires via chemical deposition from vapor or liquid phases. Such nanotubes and nanowires make excellent building blocks for novel nanodevices [74–78], but since such syntheses are extensively described elsewhere [79–84] we will not review them in detail here. The major problem in assembling functional microstructures from prefabricated nanowires is the manipulation and positioning of the wires on the substrate. One elegant technique for aligning the tubes and interfacing them to electrical circuits uses parallel flows in microfluidic channels [76]. Arrays of microwires have then been overlaid by this technique on top of another array crossing the first, to generate homo- and heterojunctions. These junctions can serve as diodes, transistors and other microscopic electronic elements on chips [77]. It is likely that similar structures will grow in complexity and will find a variety of applications in electrical microcircuits, chemical and biological sensors, etc. [78].

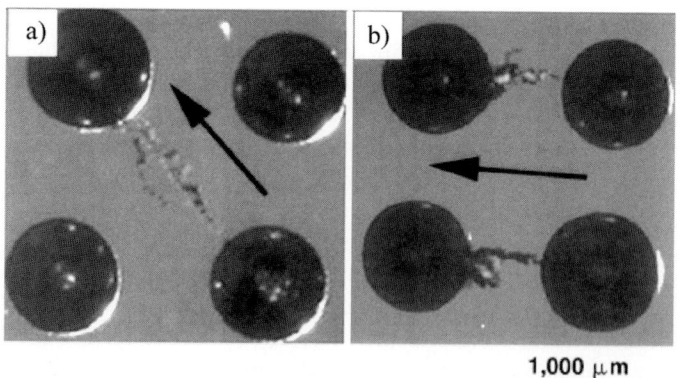

Fig. 14.6 Controlled connection of metallic particles via electrochemical deposition of copper wires. The direction of the field is indicated by the arrows (a) diagonal, (b) horizontal (reprinted from [88] with permission; copyright Macmillan Publishers, 1997).

The electrochemical deposition of metal has also been utilized for the fabrication of microscopic devices via growth of electrically conductive wires in electrolyte solutions. Direct electrical connections between electrodes have been created via electrodeposition [85] and electropolymerization [86, 87]. An original extension of these techniques has been the use of electrochemical deposition to directionally connect leads to metallic particles [88]. Copper particles of millimeter size have been placed in the gap between platinum electrodes, and constant electric fields higher than ca. 30 V cm^{-1} have been applied to induce the deposition of metallic Cu from the cupric ions dissolved in the water. The sides of any particle facing the electrode pairs become polarized with an effective voltage of opposite charge, thus serving as small electrodes in an electrochemical cell. This mechanism of electrochemical polarization, where each particle serves as both anode and cathode, has been named bipolar electrochemistry. This effect can be used to grow copper wires between pairs of nearby particles, one of which is closer to the positive electrode and the other closer to the negative electrode [88]. The growth starts from the particle situated closer to the cathode and continues until the gap is completely bridged and the particles are electrically connected.

The same group has demonstrated how to grow directly electrical circuits between specific particle pairs, by placing them onto an array of electrodes and then energizing the electrodes nearest to the desired particle pair, which makes the voltage-drop between them the highest so the metal is reduced there (Fig. 14.6). Such controllably deposited wires can be used to make connections on circuit boards and electrical circuits [89, 90]. Other interesting proof-of-principle experiments have been based on growth of copper connectors to silicon chips [91]. This has allowed preparation of the equivalents of Schottky and rectifying diodes. The bipolar electrochemistry method, however, has its disadvantages. In order to grow connectors between the particles, rather than between the particles and the electrodes, the interparticle distance has to be much smaller than the distance between the cathode and the nearest particle. The size of the particles can hardly be scaled down to the true microscopic regions, as the intensity of the bipolar polarization and the corresponding electrochemical driving voltage will also decrease. Although the wire electrodeposition technique can not possibly substitute for well-developed dry technologies for making electrical circuits, it has the potential advantage of being able to make connections in non-planar and 3D circuits.

Electrodeposition techniques have also been used to generate parallel arrays of wire-like conductive fingers. An earlier study [92] has produced the surprising result that electrodeposition in a thin-layer horizontal cell containing aqueous electrolyte prepared with cupric sulfate and sodium sulfate leads to the formation of branched metallic fingers pointing in the general direction of the electric field. The authors discuss the similarity between this mode of deposited growth and the theories for hydrodynamic instability and fingering in liquid fronts penetrating thin channels. The growth of many parallel copper fingers in thin films of $CuSO_4$ has been demonstrated [93, 94]. The technique is relatively complex, utilizing a thin cell where directional freezing of water creates a layer of saturated electrolyte solution of a thickness of ca. 200 nm, where the fingers grow by a relatively low field of a few volts applied to the electrodes. The structures formed consist of thousands of branched fingers of length of up to a few millimeters and thickness of hundreds of micrometers. Oscillations in the concentration of the electrolyte ions lead to periodic corrugations, with alternating nanoscopic regions rich in Cu_2O and metallic Cu. Such parallel electrodeposited fingers could find applications in electronic fabrication processes, subject to the limitations discussed in the previous paragraph. Microfilament connectors between two electrodes on a chip have also been fabricated by a process where the source of Cu^{2+} ions for the electrodeposition is not the salt dissolved in solution, but a massive copper anode deposited nearby on the chip surface [95].

Another electrodeposition technique used for fabrication of metallic nanowires is selective metal reduction on natural 2D templates provided by the step edges of materials such as graphite. Procedures for fabrication of nanowires from molybdenum and palladium have been developed by Penner et al. [96, 97]. The Mo wires are created by a two-step chemical process in which molybdenum oxides (MoO_x) are first electrodeposited on the step edges of cleaved molecularly smooth graphite and then converted to metal by hydrogen reduction at 500 °C. The Pd meso-wires are directly reduced on the graphite steps from a solution containing Pd^{2+} ions. The thickness of these hemicylindrical metallic nano- and meso-wires ranges from tens to hundreds of nanometers; their length can be larger than hundreds of micrometers. After the metallic structures are synthesized, they can be embedded in polymers such as polystyrene and cyanoacrylate, lifted off the graphite substrate, and their resistance and electrical response can be measured after metallic electrodes are deposited on the two sides of the gap spanned by the wires. The conductivity and mechanical resilience of the Mo wires have been found to be similar to those of bulk molybdenum. The palladium microwires consist of polycrystalline grains of sizes from 50 to 300 nm. Arrays of these Pd meso-wires have shown remarkable sensitivity to gaseous H_2, acting as switches that short-circuit the gap in the presence of hydrogen and disconnect the current in its absence. Hydrogen sensors with characteristics comparable, or superior, to the ones based on alternative principles have been made [97]. The hydrogen-sensing effect is based on the swelling and shrinking of the grains making or breaking the contacts between them. The simple fabrication principle will possibly be extended to many other metals and devices. The major disadvantage is the lack of capability to create the structures directly in the desired conformations in the electrical circuits.

Growth of complex electrical microwires can be achieved by purely chemical deposition of metal guided by liquid flow in microfluidic channels. This method was originally demonstrated by Whitesides' group [98] who directly grew silver wires of complex controlled shape. The method, termed "fabrication using laminar flow" uses the property of fluid flows in microfluidic channels that always occurs at low Reynolds numbers, i.e. under conditions of strictly laminar flow [99, 100]. Two solutions injected in parallel flow alongside each other and mixing of the chemical solutes occurs mostly via diffusion through the interface. By injecting the two components of a commercial electroless plating solution a silver microwire is deposited in the middle of the thin capillary channel. The metallic wire exactly follows the conformation of the templating channel and is significantly smaller in size than its cross-section, slowly broadening with the progression of the molecular diffusion fronts in the opposing liquid flux. This principle can be applied directly to the formation of a more complex functional device for electrochemical analysis by neatly positioning the wire in the gap between two electrodes on the surfaces [98]. The Ag microwire is used as a microscopic reference electrode after AgCl is deposited on its surface via treatment with HCl solution. The method holds promise for directly creating electrical leads and circuits inside microfluidic chips, a natural marriage between fluidic and electrical circuits.

Templated metal deposition into conductive microwires has also been achieved by the use of DNA as a substrate on which silver crystals are grown [101]. The bridge is assembled from three complementary DNA fragments in the 12–16 μm gap between two gold electrodes. Two oligonucleotides are attached to the gold surface by thiol-type chemistry. The third λ-DNA segment has "sticky" ends that are complementary to the two immobilized segments and attaches to them to complete the bridge. Metallic silver is deposited on the DNA surface via electroless reduction of a silver salt. The Ag^+ ions attracted to the surface of the DNA molecule nucleate metallic nanocrystals, so the polynucleotide bridge becomes encrusted with a dense shell of 30–50-nm sized nanoparticles. The electrical current through the nanoparticle shell is originally a non-linear function of the voltage, but the structure can be "annealed" by applying a higher voltage, resulting in wires of ohmic properties and of resistance in the megaohm region. The method has the potential to prepare electric wires connecting on-chip and three-dimensional structures. The technique has later been used for making DNA-guided metallic microwires from palladium [102] and platinum [103] and the properties of these structures have been characterized in detail [104–106]. In general, the complementarity of DNA pairing has been one of the most efficient tools for the designed assembly of various nanostructures from nanoparticles [107–112], or in prototype nanomechanical devices [113, 114]. It could be expected that more complex electrically functional nanodevices will soon be created via DNA assembly.

14.3.3
Assembly of Electrical Circuits from Nanoparticles

The fabrication of electrical wires by metal ion reduction in solution can pose problems because of the chemical reactivity of the reagents, or the relatively slow speed of deposition. An alternative technique for making such connectors is to assemble the wires from metallic particles. The possibility of making such connections arose from the directional chaining of particles in electrorheological and magnetorheological fluids. Simple electrical connections can be made via particle chaining similar to that found in electrorheological fluids. If the particles in such fluids are conductive, the chain will form a kind of conductive wire between the electrodes. Dueweke et al. have studied the assembly of millimeter-sized steel spheres in castor oil under the action of DC voltages of the order of 20 kV [115]. The spheres converge to build a chain in a few tens of seconds, bridging the gap between the electrodes submerged in the oil. The average growth time of the chains has been found to be proportional to the distance between the electrodes. The authors have also noted that these connections are self-repairing as new spheres are attracted to the ends of open chains, and have discussed the "Hebbian learning" in such systems, i.e. the easier restoration of previously built chains due to the presence of agglomerated particles in the gaps [116].

Chaining and column formation of conductive particles have been studied experimentally by Wen et al. [117, 118] for a system of 65-µm copper particles suspended in silicone oil. The chains have been grown between planar and coaxial electrodes and between vertical plates. The particles form tree-like agglomerates whose structure can be assumed fractal. The fractal dimension of the branched chains varies between 1.2 and 2, increasing with the volume fraction of the Cu particles. In analogy to the diffusion-limited aggregation (DLA), responsible for the formation of fractal aggregates in colloidal suspensions, the concept of electric-field-induced diffusion-limited aggregation (EDLA) has been extended. This concept however has been indirectly questioned by the theoretical study by Kun and Pál [119], who have performed lattice simulations of the growth of aggregates under similar conditions and have found that the Brownian diffusion of the particles does not play a crucial role in structure formation. This theoretical study confirms the power-law dependence of the fractal dimensions on the sphere concentration and points to the previously neglected importance of particle–particle interactions for structure growth. The above papers do not specifically discuss the assembly of electrical circuits, but the potential to use the structures as conductive wires is obvious. The particle chains, however, come apart after the electric field is turned off, so the wires assembled by chaining of metal beads are not permanent structures. The use of large particles and organic oils does not permit the application of such connectors in devices that are of microscopic size or include biological molecules or cells.

To overcome the above restrictions, a truly versatile technique for direct wire assembly should be based on dispersion of conductive nanoparticles in water. We have recently described a new class of electrically functional microwires assembled by dielectrophoresis from suspensions of metallic nanoparticles in water [120]. Microwires form following the introduction of a suspension of 15–30-nm gold nanoparticles into a thin chamber above planar metallic electrodes on a glass surface. When an alternating voltage of 50–250 V and frequency of 50–200 Hz is applied to the planar electrodes (resulting in a field intensity of ca. 250 V/cm), thin metallic fibers begin to grow on the electrode edge facing the gap. The wire has a diameter of the order of a micrometer and can grow longer than a centimeter with a speed that is typically ≈ 50 µm/s, but in some cases can be up to an order of magnitude higher. When the wire is completely assembled the electrodes become effectively short-circuited.

The finding that dielectrophoretic forces can readily assemble nanometer-sized metallic particles from aqueous suspensions into long, electrically conductive microwires is not directly predicted by theory. The dipole–dipole interaction energy between 15–30-nm particles separated by a gap of a few nanometers is smaller than $10^{-2}\,kT$ (kT being the thermal energy) and therefore insignificant. Instead of direct chaining, the assembly is driven by the DEP force arising from the interaction of the particle dipoles with the nonuniform AC field. The tip of a growing microwire creates local fields of high intensity and gradient, attracting and concentrating the particles at the end. The concentrated particles aggregate at the tip of the fibers, thereby extending them towards the opposite electrode. Purple coronas of highly concentrated areas in front of the growing wire ends and depletion zones behind them are clearly observed at low nanoparticle concentrations (Fig. 14.7a). Thus microwire formation is a collective effect whereby the nanoparticles are highly concentrated at the end of the tip and subsequently aggregate to extend the wire in the direction of the field gradient (Fig. 14.7b).

The microwire growth direction can be steered by inhomogeneities in the electric field obtained by introducing objects of higher dielectric permittivity in the gap between the electrodes. For example, when small islands of (conductive) carbon paint are deposited in the gap, the wires grow in their direction and spontaneously connect these islands to both electrodes. More complex structures involving multiple connections between multiple conductive pads can be formed. The field strength and particle concentration are also key variables, as they must exceed a threshold value in order for the wire to start growing. This threshold behavior emphasizes the collective nature of the assembly, as it shows that aggregation occurs only when the concentration of particles accumulated near the end of the wire is sufficiently high.

The microwires have linear (Ohmic) current-to-voltage response for both AC and DC voltages, with specific resistances that depend on the conditions of assembly and ranges from $3-20\times 10^{-6}$ Ωm (Fig. 14.7d). This resistivity is two to three orders of magnitude higher than that of bulk solid gold because of the porosity and

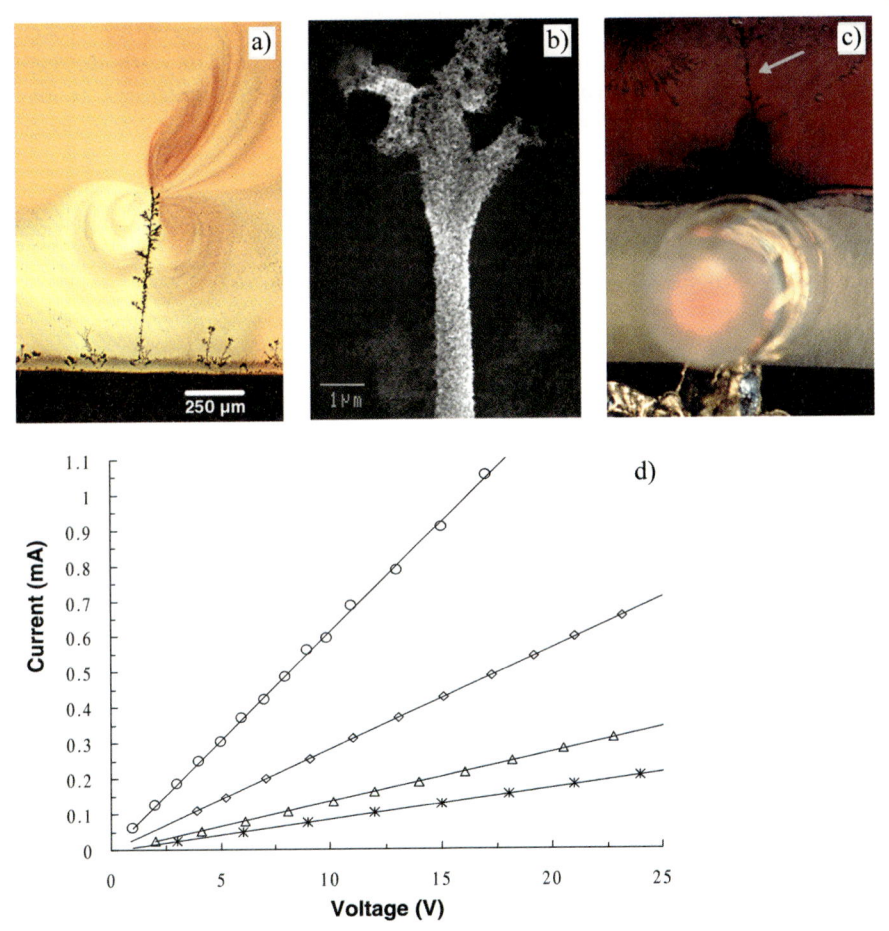

Fig. 14.7 Direct DEP assembly of electrically functional microwires from gold nanoparticles [120]. **a** Optical micrograph of a growing wire illustrating the area of high nanoparticle concentration in front of the wire where the DEP forces are strongest. **b** SEM photograph of a microwire end showing that the nanoparticles aggregate into a uniform cylindrical porous body. **c** Microwire connection of an electrical circuit through a droplet of nanoparticle suspension. The LED on the bottom is powered and lights up. **d** Current to voltage response of microwires formed in different conditions. The linear relation demonstrates Ohmic behavior (copyright American Association for the Advancement of Science, 2001).

small interparticle contact areas in the microwires, yet it is comparable to that of good electrical conductors. The self-assembled structures can quickly and simply connect electrical circuits in water, such as wiring and energizing light-emitting diodes through water gaps (Fig. 14.7c). Similarly to the large-particle chains mentioned above, these self-assembling electrical connections are also self-repairing as a consequence of the large field intensity in the small gap between the broken wires, which attracts new particles that restore the connection. The method is capable of assembling more complex core-shell metallo-dielectric structures. It may have potential for bioelectronic interfacing, by providing a tool for in situ connecting living cells to on-chip electrodes.

14.4
Electrically Functional Microstructures from Colloidal Particles

In the previous sections we have discussed a variety of generic methods to assemble colloidal crystals and electrical connectors by electrical fields and currents. In this section we present a few examples that demonstrate the potential of nanoparticle assemblies as microscopic devices interfaced to electrical circuits.

14.4.1
Chemical Sensors

The enormous surface area of nanoparticles and nanowires makes the electrical properties of nanoparticle structures susceptible to changes in the environment that affect the particle size or surface composition. One example of a sensor from polycrystalline palladium nanowires that drastically changes its conductivity in the presence of hydrogen is presented in Section 14.3.2 [97]. Similar "chemiresistance" effects can be observed in both gaseous and liquid media. The phenomenon has been extensively studied theoretically and experimentally for the cases of ultrathin metallic films and wires [121–125]. For example, the conductance of nanoscopic gold films and wire junctions has been shown to be sensitive to adsorbates ranging from N_2 to ethanol, pyridine, mercaptans and thiols [121, 122]. Similar sensors can be also be made from metallic nanoparticles [126].

The microwire assembly technique described above [120] also presents a simple and efficient way to make electrically interfaced nanoparticle structures. Microwires were formed in a thin flow chamber and their properties were measured in bridge mode to subtract the effect of the bulk solution conductivity. An easily resolvable and concentration-dependent increase of the resistivity of the microwires was observed after exposure to analytes that adsorb and modify the gold surface such as 2-(dimethylamino)ethanethiol hydrochloride and sodium cyanide. For example, the resistivity increased by $\sim 10\%$ after treatment with 2.5×10^{-4} M thiol solution and $\sim 7\%$ after treatment with NaCN at pH 11 [120]. This signal possibly results from a decrease of the surface conductivity of the gold nanoparticles in the wire structure as they become covered by an organic layer. No signal has been measured with solutions containing the protein lysozyme, which adsorbs on the gold surface without significantly affecting its surface conductivity, which demonstrates that the signal is specific to analytes intimately modifying the counterionic layer around the gold nanoparticles. The microwires are not expected to possess a lower limit of detection than other gold nanoparticle assemblies, but they can provide a quick response by virtue of their very high surface-to-volume ratio and easy analyte adsorption from the bulk of the surrounding solution.

14.4.2
Biological Sensors

Biosensors and bioarrays with direct electrical interfacing of electronic chips and particle assemblies have a number of potential advantages compared to the present technologies based on optical detection. The particle structures in such sensors interface the wet 3D biological environment and the dry 2D electronic circuits. The first example of how microscopic electrically readable sensors can be assembled in situ from micro- and nanoparticles has been presented by Velev and Kaler [127]. The method is schematically shown in Fig. 14.8. The sensors are assembled on glass substrates with four pairs of photolithographically fabricated gold electrodes separated by gaps of 30×12 µm. The particle patches are assembled in the gaps from the widely available latex particles used in traditional agglutination assays. The particles, suspended in a low-electrolyte aqueous medium, are collected in the gaps between the electrodes under the action of positive dielectrophoresis at relatively low frequencies (1–5 KHz). The mere collection of the microspheres is, however, not enough to produce sensor patches, as the latexes are stabilized either by electrostatic or by steric repulsion, and so disassemble when the field is turned off. The key step in particle immobilization into sensor patches is to decrease the repulsive interactions to a degree where the van der Waals and hydrophobic attractive forces coagulate the particles (Fig. 14.8). Particles stabilized by electrostatic repulsion from charged groups can be coagulated by adsorption of an oppositely charged surfactant on the surface. Particles stabilized by non-ionic surfactant are easily bound together by washing away the adsorbed steric protective layer. The assembly procedure can be repeated many times by exchanging the particle suspension and addressing different gaps so that different microscopic sensors can be assembled on the same "chip".

The model target molecule for the sensors was human immunoglobulin (IgG). Detection is carried out by tagging the IgG molecules with colloidal gold conjugated to a secondary antibody, anti-human IgG. The gold particles in the shells are enlarged and fused together by a silver layer deposited from a metastable solution of silver salts. Thus, the electrodes become short circuited and the electronic readout is carried out by simply measuring the resistance (Fig. 14.9). The size of these sensors is a state-of-the-art miniaturization and the experimentally estimated lower limit of detection (LOD) is of the order of the better IgG agglutination assays and immunosensors available. By further miniaturization, the sensitivity of the microscopic active elements can become much higher than that of agglutination assays, and may approach the limit of a few tens or hundreds of molecules.

The electrical detection method utilizing nanoparticle short-circuiting of electrodes has recently been extended to miniaturized DNA assays by Möller et al. [128] and Park et al. [129]. In both papers, oligonucleotide strands are chemically immobilized on the surface between the electrodes. Gold nanoparticles become

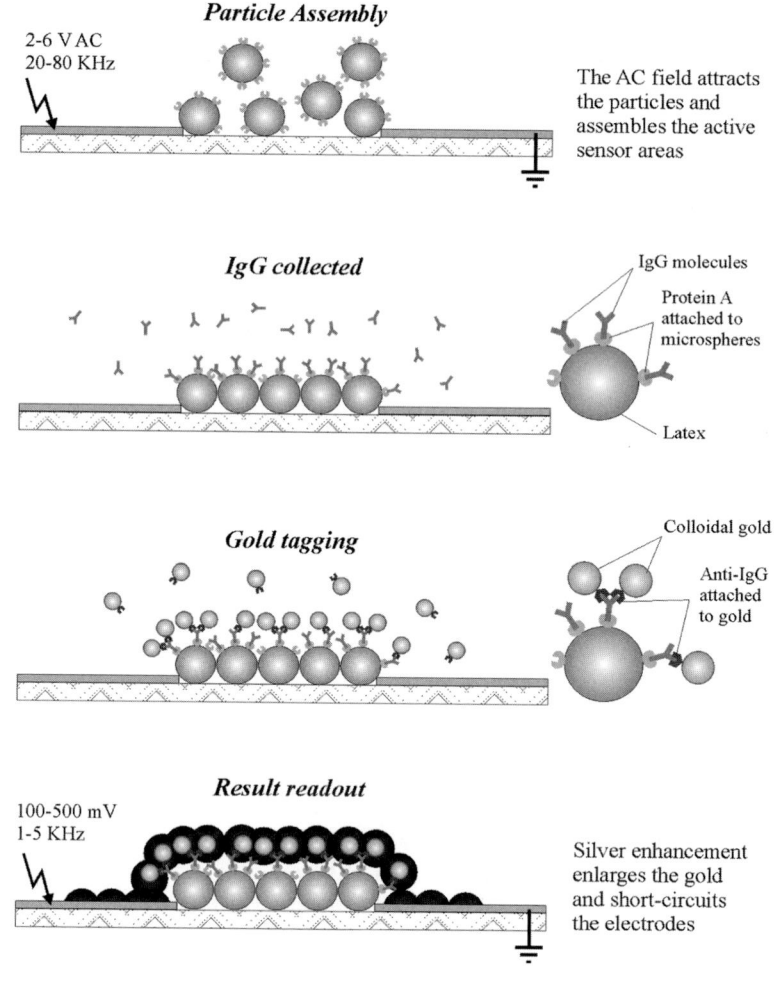

Fig. 14.8 Schematics of formation of microscopic biosensor patches via DEP assembly of latex particles and direct electric readout of the results via gold tagging and silver enhancement [127]. The procedure is illustrated on the basis of IgG detection (copyright American Chemical Society, 1999).

bound only if a complementary match between the DNA fragments is found; the complementary strands were either bound to the gold particles [128], or bridge the strands on the surface and the ones on the gold particles [129]. Some degree of quantification of the DNA concentration in the sample has been possible by measuring the time required to short-circuit the electrodes after the enhancer is introduced. The ability to fabricate parallel microscopic arrays of detectors for specific genetic markers is promising for the biotechnology and medical diagnostics industries.

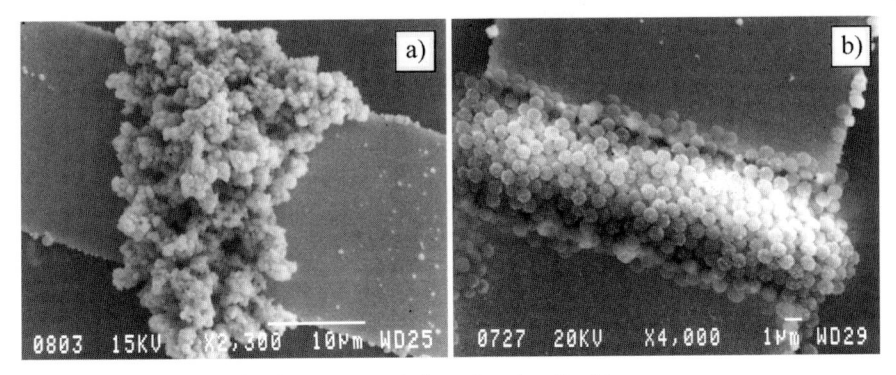

Fig. 14.9 SEM micrograph of a 10-μm patch from functionalized latex spheres in the miniature immunosensors with direct electronic read-out [127]. **a** Functionalized latexes heavily coated by deposited metal that short-circuits the electrodes. **b** The non-functionalized particles in the negative control patches are only marginally tagged by deposited metal (copyright American Chemical Society, 1999).

14.4.3
Displays

In the last few years technologies based on colloidal particles have provided an alternative to liquid crystal displays as a means for displaying graphical information. The breakthrough has been the development of electrophoretic ink for electronic "paper" [130]. This "ink" is made from thousands of capsules 30–300 μm in diameter, incorporating suspensions of white and black microparticles in the size range of 1–5 μm. The white and black particles carry opposite charges and can be separated to the opposite sides of the capsule by application of an electric field. The charged particles are protected from heterocoagulation by a polymeric adsorption layer around their surfaces. Alternatively, only white particles suspended in dark liquid can be used. The microcapsules are spread in a thin layer and addressable electrodes from ITO and silver-doped polymer are applied from top and bottom respectively.

In order to switch the color on and off, a potential field of a few volts per micrometer is applied across the electrodes. The particles move to the oppositely charged electrodes, separating into a black mass on one side of the capsule and white on the other side (Fig. 14.10). If the electrode potential is reversed, the position of the black and white particles changes, reversing the color of the capsule. An addressable resolution of 1200 dots per inch has been reported, suggesting that small features can be "printed" onto the surface. Such devices can be flexible and can retain the displayed information after the electric field is turned off. Recent commercial development of this technology has led to the fabrication of color displays of relatively low cost and low energy consumption. The commercialization of such displays is intertwined with the development of technologies for fab-

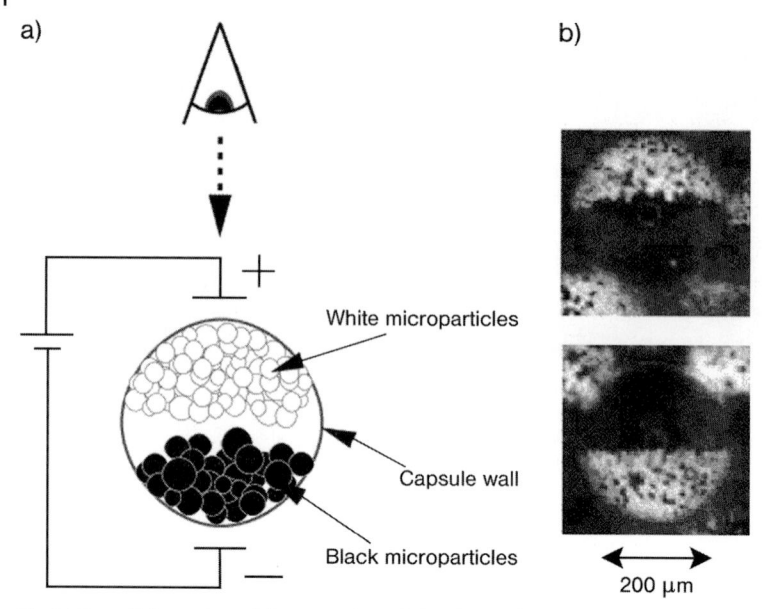

Fig. 14.10 a Schematics of the operation of the electrophoretic microcapsules for "electronic ink" applications. The attraction of the white and black particles to the oppositely charged electrodes leads to change of the visible color on the side of the electrodes. **b** Microphotograph of an individual e-ink capsule (reprinted from [130] with permission; copyright Macmillan Publishers, 1998).

rication of inexpensive organic transistors and electronic circuits by printing on flexible plastic films [131, 132].

14.4.4
Photovoltaic Cells and Light-emitting Diodes

Various types of semiconductor and metallic nanoparticles can be assembled in layer-by-layer structures by adsorption onto oppositely charged polyelectrolyte [133–138]. The delicate balance of the forces in layers of alternating charge allows sequential deposition of particle layers of various types. Ultrathin structures of semiconductor nanoparticles can form electronic elements similar to the way heterojunctions of doped semiconductors are used in solid-state electronics. The particles/polyelectrolyte layers are assembled on a metallic surface acting as the first electrode, and a layer of another metal is coated on top of the structure to form the second electrode. This technique has allowed the fabrication of nanoparticle-based rectifying and Zenner diodes [139, 140], LEDs [141–149] and photovoltaic cells [150–152]. The advantages offered by the nanoparticle technique are simple processing, precise control of the layer thickness, choice of a wide range of semiconductor nanoparticle types (e.g. CdSe, CdS, Si, TiO_2, ZnO, etc.) and ability to tune and improve the electrical characteristics of the devices via polyelectrolytes and self-assembled monolayers [136, 137, 146, 147].

Very interesting effects are observed when the nanoparticles are scaled down to a size of less than a few nanometers, and insulated from the electrodes (and each other) by a layer of surfactant. The tunneling current increases in discrete steps as a function of the number of electrons injected inside the nanoparticle [153–159]. The incremental electron tunneling manifests itself in the so-called "Coulomb staircase" current–voltage dependence. These quantum-based effects can, in principle, be used to store bits of information within a single nanoparticle and to make miniature lasers [137, 138, 157–159]. The assembly of such quantum dots in larger-scale structures is likely to produce devices bridging the domains of molecular electronics and semiconductor technologies.

14.5
Future Trends

The diverse scientific results described here are intended to give a brief presentation of the state of the art in the area of electrically functional colloidal assemblies. They are by no means inclusive of all related developments and trends. For example, carbon nanotubes and semiconductor nanowires form an enormous research area not reviewed in detail here. The same is also true for biomolecular signaling and bioelectronic interfacing. Equally, the research trends and techniques presented here cannot be viewed as encompassing the full range of the things to come in this rapidly evolving area. Many virtually unexplored fields for creative research exist at the intersection of colloidal assembly and other areas of nanoscience. For example, the combination of colloidal structures and molecular electronics has high potential for miniaturization of electrical circuits and their stacking in the third dimension. The marriage of colloidal assembly and microfluidics is likely to bring forward novel generations of sensors and bioelectronic devices. We should expect new and exciting discoveries and applications to emerge in this field in the near future.

14.6
Acknowledgements

The author's research reviewed here was carried out in collaboration with Simon Lumsdon, Eric Kaler, Kevin Hermanson and Jacob Williams. Financial support from the Camille and Henry Dreyfus Foundation, and from Oak Ridge Associated Universities, and the assistance of Ketan Bhatt in the preparation of the manuscript are gratefully acknowledged.

14.7 References

1. W. B. Russell, D. A. Saville, W. R. Schowalter, *Colloidal Dispersions*, Cambridge University Press, Cambridge, **1999**.
2. H. A. Pohl, *Dielectrophoresis*, Cambridge University Press, Cambridge, **1978**.
3. T. B. Jones, *Electromechanics of Particles*, Cambridge University Press, Cambridge, **1995**.
4. R. Pethig, Y. H. Huang, X. B. Wang, J. P. H. Burt, *J. Phys. D: Appl. Phys.* **1992**, *25*, 881–888.
5. T. Müller, A. Gerardino, T. Schnelle, S. G. Shirley, F. Bordoni, G. DeGasperis, R. Leoni, G. Fuhr, *J. Phys. D: Appl. Phys.* **1996**, *29*, 340–349.
6. G. Fuhr, T. Müller, T. Schnelle, R. Hagedorn, A. Voigt, S. Fiedler, W. M. Arnold, U. Zimmermann, B. Wagner, A. Heuberger, *Naturwissenschaften* **1994**, *81*, 528–535.
7. O. D. Velev, *Handbook of Surfaces and Interfaces of Materials*, Vol. 3 (H. S. Nalwa, ed.), Academic Press, San Diego, **2001**.
8. O. D. Velev, A. M. Lenhoff, *Curr. Opin. Colloid Interface Sci.* **2000**, *5*, 56–63.
9. M. Giersig, P. Mulvaney, *Langmuir* **1993**, *9*, 3408–3413.
10. M. Holgado, F. García-Santamaría, A. Blanco, M. Ibisate, A. Cintas, H. Míguez, C. J. Serna, C. Molpeceres, J. Requena, A. Mifsud, F. Meseguer, C. López, *Langmuir* **1999**, *15*, 4701–4704.
11. A. L. Rogach, N. A. Kotov, D. S. Koktysh, J. W. Ostrander, G. A. Ragoisha, *Chem. Mater.* **2000**, *12*, 2721–2726.
12. R. C. Bailey, K. J. Stevenson, J. T. Hupp, *Adv. Mater.* **2000**, *12*, 1930–1934.
13. J. Q. Sun, M. Y. Gao, J. Feldmann, *J. Nanosci. Nanotech.* **2001**, *1*, 133–136.
14. M. Y. Gao, J. Q. Sun, E. Dulkeith, N. Gaponik, U. Lemmer, J. Feldmann, *Langmuir* **2002**, *18*, 4098–4102.
15. Z. Z. Gu, S. Hayami, S. Kubo, Q. B. Meng, Y. Einaga, D. A. Tryk, A. Fujishima, O. Sato, *J. Am. Chem. Soc.*, **2001**, *123*, 175–176.
16. P. Richetti, J. Prost, P. Barois, *J. Phys. Lett. Paris* **1984**, *45*, 1137–1143.
17. M. Trau, D. A. Saville, I. A. Aksay, *Science* **1996**, *272*, 706–709.
18. M. Trau, D. A. Saville, I. A. Aksay, *Langmuir* **1997**, *13*, 6375–6381.
19. M. Böhmer, *Langmuir* **1996**, *12*, 5747–5750.
20. Y. Solomentsev, M. Böhmer, J. L. Anderson, *Langmuir* **1997**, *13*, 6058–6068.
21. Y. Solomentsev, S. A. Guelcher, M. Bevan, J. L. Anderson, *Langmuir* **2000**, *16*, 9208–9216.
22. P. J. Sides, *Langmuir* **2001**, *17*, 5791–5800.
23. J. Kim, S. A. Guelcher, S. Garoff, J. L. Anderson, *Adv. Colloid Interface Sci.* **2002**, *96*, 131–142.
24. R. C. Hayward, D. A. Saville, I. A. Aksay, *Nature* **2000**, *404*, 56–59.
25. V. Brisson, R. D. Tilton, *Biotechnol. Bioeng.* **2002**, *77*, 290–295.
26. T. Gong, D. W. M. Marr, *Langmuir* **2001**, *17*, 2301–2304.
27. A. P. Gast, C. F. Zukoski, *Adv. Colloid Interface Sci.* **1989**, *30*, 153–202.
28. M. Parthasarathy, D. J. Klingenberg, *Mat. Sci. Eng. R.* **1996**, *17*, 57–103.
29. R. Tao, J. M. Sun, *Phys. Rev. Lett.* **1991**, *67*, 398–401.
30. J. E. Martin, J. Odinek, T. C. Halsey, R. Kamien, *Phys. Rev. E* **1998**, *57*, 756–775.
31. U. Dassanayake, S. Fraden, A. J. van Blaaderen, *J. Chem. Phys.* **2000**, *112*, 3851–3858.
32. A. J. van Blaaderen, K. P. Velikov, J. P. Hoogenboom, D. L. J. Vossen, A. Yethiraj, R. Dullens, T. van Dillen, A. Polman, *Photonic crystals and light localization in the 21st century* (C. M. Soukoulis, ed.), Kluwer, Amsterdam, **2001**.
33. M. Trau, S. Sankaran, D. A. Saville, I. A. Aksay, *Nature* **1995**, *374*, 437–439.
34. A. E. Larsen, D. G. Grier, *Phys. Rev. Lett.* **1996**, *76*, 3862–3865.
35. A. E. Larsen, D. G. Grier, *Nature* **1997**, *385*, 230–233.
36. S. O. Lumsdon, J. P. Williams, E. W. Kaler, O. D. Velev, *Appl. Phys. Lett.*, **2003**, *82*, 949–951.
37. V. N. Paunov, P. A. Kralchevsky, N. D. Denkov, K. Nagayama, *J. Colloid Interface Sci.* **1993**, *157*, 100–112.

38 O. D. Velev, N. D. Denkov, V. N. Paunov, P. A. Kralchevsky, K. Nagayama, *Langmuir* **1993**, *9*, 3702–3709.

39 V. N. Paunov, *Langmuir* **1998**, *14*, 5088–5097.

40 P. A. Kralchevsky, K. Nagayama, *Particles at Fluid Interfaces and Membranes*, Elsevier, Amsterdam, **2001**.

41 P. A. Kralchevsky, N. D. Denkov, *Curr. Opin. Colloid Interface Sci.* **2001**, *6*, 383–401.

42 N. Bowden, A. Terfort, J. Carbeck, G. M. Whitesides, *Science* **1997**, *276*, 233–235.

43 I. S. Choi, N. Bowden, G. M. Whitesides, *J. Am. Chem. Soc.* **1999**, *121*, 1754–1755.

44 I. S. Choi, N. Bowden, G. M. Whitesides, *Angew. Chem. Int. Ed.* **1999**, *38*, 3078–3081.

45 I. S. Choi, M. Weck, B. Xu, N. L. Jeon, G. M. Whitesides, *Langmuir* **2000**, *16*, 2997–2999.

46 M. Weck, I. S. Choi, N. L. Jeon, G. M. Whitesides, *J. Am. Chem. Soc.* **2000**, *122*, 3546–3547.

47 N. B. Bowden, M. Weck, I. S. Choi, G. M. Whitesides, *Acc. Chem. Res.* **2001**, *34*, 231–238.

48 A. Terfort, N. Bowden, G. M. Whitesides, *Nature* **1997**, *386*, 162–164.

49 J. Tien, T. L. Breen, G. M. Whitesides, *J. Am. Chem. Soc.* **1998**, *120*, 12670–12671.

50 T. L. Breen, J. Tien, S. R. J. Oliver, T. Hadzic, G. M. Whitesides, *Science* **1999**, *284*, 948–951.

51 G. M. Whitesides, M. Boncheva, *Proc. Natl. Acad. Sci. USA* **2002**, *99*, 4769–4774.

52 A. Terfort, G. M. Whitesides, *Adv. Mater.* **1998**, *10*, 470–473.

53 D. H. Gracias, J. Tien, T. L. Breen, C. Hsu, G. M. Whitesides, *Science* **2000**, *289*, 1170–1172.

54 D. H. Gracias, M. Boncheva, O. Omoregie, G. M. Whitesides, *Appl. Phys. Lett.* **2002**, *80*, 2802–2804.

55 M. Boncheva, D. H. Gracias, H. O. Jacobs, G. M. Whitesides, *Proc. Natl. Acad. Sci. USA* **2002**, *99*, 4937–4940.

56 R. R. A. Syms, E. M. Yeatman, *Electron. Lett.* **1993**, *29*, 662–664.

57 H. O. Jacobs, A. R. Tao, A. Schwartz, D. H. Gracias, G. M. Whitesides, *Science* **2002**, *296*, 323–325.

58 H. J. J. Yeh, J. S. Smith, *IEEE Photonic. Tech. L.* **1994**, *6*, 706–708.

59 C. R. Martin, *Chem. Mater.* **1996**, *8*, 1739–1746.

60 D. Almawlawi, C. Z. Liu, M. Moskovits, *J. Mater. Res.* **1994**, *9*, 1014–1018.

61 B. R. Martin, D. J. Dermody, B. D. Reiss, M. M. Fang, L. A. Lyon, M. J. Natan, T. E. Mallouk, *Adv. Mater.* **1999**, *11*, 1021–1025.

62 S. R. Nicewarner-Pena, R. G. Freeman, B. D. Reiss, L. He, D. J. Pena, I. D. Walton, R. Cromer, C. D. Keating, M. J. Natan, *Science* **2001**, *294*, 137–141.

63 N. I. Kovtyukhova, B. R. Martin, J. K. N. Mbindyo, P. A. Smith, B. Razavi, T. S. Mayer, T. E. Mallouk, *J. Phys. Chem. B* **2001**, *105*, 8762–8769.

64 J. K. N. Mbindyo, T. E. Mallouk, J. B. Mattzela, I. Kratochvilova, B. Razavi, T. N. Jackson, T. S. Mayer, *J. Am. Chem. Soc.* **2002**, *124*, 4020–4026.

65 J. S. Yu, J. Y. Kim, S. Lee, J. K. N. Mbindyo, B. R. Martin, T. E. Mallouk, *Chem. Commun.* **2000**, *24*, 2445–2446.

66 J. K. N. Mbindyo, B. D. Reiss, B. R. Martin, C. D. Keating, M. J. Natan, T. E. Mallouk, *Adv. Mater.* **2001**, *13*, 249–254.

67 S. A. Sapp, D. T. Mitchell, C. R. Martin, *Chem. Mater.* **1999**, *11*, 1183–1185.

68 P. A. Smith, C. D. Nordquist, T. N. Jackson, T. S. Mayer, B. R. Martin, J. Mbindyo, T. E. Mallouk, *Appl. Phys. Lett.* **2000**, *77*, 1399–1401.

69 B. H. Hong, S. C. Bae, C. W. Lee, S. Jeong, K. S. Kim, *Science* **2001**, *294*, 348–351.

70 Y. D. Li, J. W. Wang, Z. X. Deng, Y. Y. Wu, X. M. Sun, D. P. Yu, P. D. Yang, *J. Am. Chem. Soc.* **2001**, *123*, 9904–9905.

71 B. Mayers, Y. N. Xia, *J. Mater. Chem.* **2002**, *12*, 1875–1881.

72 Y. G. Sun, B. Gates, B. Mayers, Y. N. Xia, *Nano Lett.* **2002**, *2*, 165–168.

73 B. H. Hong, S. C. Bae, C. W. Lee, S. Jeong, K. S. Kim, *Science* **2001**, *294*, 348–351.

74 S. J. Tans, A. R. M. Verschueren, C. Dekker, *Nature* **1998**, *393*, 49–52.

75 Y. Huang, X. Duan, Y. Cui, L. J. Lauhon, K. H. Kim, C. M. Lieber, *Science* **2001**, *294*, 1313–1317.
76 Y. Huang, X. Duan, Q. Wei, C. M. Lieber, *Science* **2001**, *291*, 630–633.
77 Y. Cui, C. M. Lieber, *Science* **2001**, *291*, 851–853.
78 Y. Cui, Q. Wei, H. Park, C. M. Lieber, *Science* **2001**, *293*, 1289–1292.
79 J. Hu, T. W. Odom, C. M. Lieber, *Acc. Chem. Res.* **1999**, *32*, 435–445.
80 J. T. Devreese, R. P. Evrard, V. E. Van Doren (eds.) *Highly Conducting One-Dimensional Solids*, Plenum, New York, **1979**.
81 M. Terrones, W. K. Hsu, H. W. Kroto, D. R. M. Walton, *Top. Curr. Chem.* **1999**, *199*, 189–234.
82 P. M. Ajayan, O. Z. Zhou, *Top. Appl. Phys.* **2001**, *80*, 391–425.
83 Y. Kondo, K. Takayanagi, *Science* **2000**, *289*, 606–608.
84 T. M. Whitney, J. S. Jiang, P. C. Searson, C. L. Chien, *Science* **1993**, *261*, 1316–1319.
85 C. Gurtner, M. J. Sailor, *Adv. Mater.* **1996**, *8*, 897–899.
86 M. J. Sailor, C. L. Curtis, *Adv. Mater.* **1994**, *6*, 688–692.
87 C. L. Curtis, J. E. Ritchie, M. J. Sailor, *Science* **1993**, *262*, 2014–2016.
88 J. C. Bradley, H. M. Chen, J. Crawford, J. Eckert, K. Ernazarova, T. Kurzeja, M. D. Lin, M. McGee, W. Nadler, S. G. Stephens, *Nature* **1997**, *389*, 268–271.
89 J. C. Bradley, J. Crawford, K. Ernazarova, M. McGee, S. G. Stephens, *Adv. Mater.* **1997**, *9*, 1168–1171.
90 J. C. Bradley, Z. M. Ma, E. Clark, J. Crawford, S. G. Stephens, *J. Electrochem. Soc.* **1999**, *146*, 194–198.
91 J. C. Bradley, Z. M. Ma, S. G. Stephens, *Adv. Mater.* **1999**, *11*, 374–378.
92 M. Q. Lòpez-Salvans, P. P. Trigueros, S. Vallmitjana, J. Claret, F. Sagués, *Phys. Rev. Lett.* **1996**, *76*, 4062–4065.
93 M. Wang, S. Zhong, X. B. Yin, J. M. Zhu, R. W. Peng, Y. Wang, K. Q. Zhang, N. B. Ming, *Phys. Rev. Lett.* **2001**, *86*, 3827–3830.
94 S. Zhong, M. Wang, X. B. Yin, J. M. Zhu, R. W. Peng, Y. Wang, N. B. Ming, *J. Phys. Soc. Jpn.* **2001**, *70*, 1452–1455.
95 M. W. Wu, L. L. Sohn, *IEEE Electr. Device L.* **2000**, *21*, 277–279.
96 M. P. Zach, K. H. Ng, R. M. Penner, *Science* **2000**, *290*, 2120–2123.
97 F. Favier, E. C. Walter, M. P. Zach, T. Benter, R. M. Penner, *Science* **2001**, *293*, 2227–2231.
98 P. J. A. Kenis, R. F. Ismagilov, G. M. Whitesides, *Science* **1999**, *285*, 83–85.
99 L. Bousse, C. Cohen, T. Nikiforov, A. Chow, A. R. Kopf-Sill, R. Dubrow, J. W. Parce, *Annu. Rev. Bioph. Biom.* **2000**, *29*, 155–181.
100 A. D. Stroock, S. K. W. Dertinger, A. Ajdari, I. Mezic, H. A. Stone, G. M. Whitesides, *Science* **2002**, *295*, 647–665.
101 E. Braun, Y. Eichen, U. Sivan, G. Ben-Yoseph, *Nature* **1998**, *391*, 775–778.
102 J. Richter, R. Seidel, R. Kirsch, M. Mertig, W. Pompe, J. Plaschke, H. K. Schackert, *Adv. Mater.* **2000**, *12*, 507–510.
103 W. E. Ford, O. Harnack, A. Yasuda, J. M. Wessels, *Adv. Mater.* **2001**, *13*, 1793–1797.
104 J. Richter, M. Mertig, W. Pompe, I. Monch, H. K. Schackert, *Appl. Phys. Lett.* **2001**, *78*, 536–538.
105 J. H. Gu, S. Tanaka, Y. Otsuka, H. Tabata, T. Kawai, *Appl. Phys. Lett.* **2002**, *80*, 688–690.
106 A. Rakitin, P. Aich, C. Papadopoulos, Y. Kobzar, A. S. Vedeneev, J. S. Lee, J. M. Xu, *Phys. Rev. Lett.* **2001**, *86*, 3670–3673.
107 A. P. Alivisatos, K. P. Johnsson, X. G. Peng, T. E. Wilson, C. J. Loweth, M. P. Bruchez, P. G. Schultz, *Nature* **1996**, *382*, 609–611.
108 C. J. Loweth, W. B. Caldwell, X. G. Peng, A. P. Alivisatos, P. G. Schultz, *Angew. Chem. Int. Ed.* **1999**, *38*, 1808–1812.
109 S. Connolly, D. Fitzmaurice, *Adv. Mater.* **1999**, *11*, 1202–1205.
110 Y. Maeda, H. Tabata, T. Kawai, *Appl. Phys. Lett.* **2001**, *79*, 1181–1183.
111 I. Willner, F. Patolsky, J. Wasserman, *Angew. Chem. Int. Edit.* **2001**, *40*, 1861–1864.
112 D. M. Hartmann, M. Heller, S. C. Esener, D. Schwartz, G. Tu, *J. Mater. Res.* **2002**, *17*, 473–478.

113 C. D. Mao, W. Q. Sun, Z. Y. Shen, N. C. Seeman, *Nature* **1999**, *397*, 144–146.
114 N. C. Seeman, *Trends Biotechnol.* **1999**, *17*, 437–443.
115 M. Dueweke, U. Dierker, A. Hubler, *Phys. Rev. E* **1996**, *54*, 496–506.
116 M. Sperl, A. Chang, N. Weber, A. Hubler, *Phys. Rev. E* **1999**, *59*, 3165–3168.
117 W. Wen, K. Lu, *Phys. Rev. E*, **1997**, *55*, R2100–R2103.
118 W. Wen, D. W. Zheng, K. N. Tu, *Phys. Rev. E* **1998**, *58*, 7682–7685.
119 F. Kun, K. F. Pál, *Phys. Rev. E* **1998**, *57*, 3216–3220.
120 K. D. Hermanson, S. O. Lumsdon, J. P. Williams, E. W. Kaler, O. D. Velev, *Science* **2001**, *294*, 1082–1086.
121 C. Z. Li, H. Sha, N. J. Tao, *Phys. Rev. B* **1998**, *58*, 6775–6778.
122 Y. Zhang, R. H. Terrill, P. W. Bohn, *J. Am. Chem. Soc.* **1998**, *120*, 9969–9970.
123 H. X. He, N. J. Tao, *Adv. Mater.* **2002**, *14*, 161–164.
124 C. Z. Li, H. X. He, A. Bogozi, J. S. Bunch, N. J. Tao, *Appl. Phys. Lett.* **2000**, *76*, 1333–1335.
125 R. G. Tobin, *Surf. Sci.* **2002**, *502–503*, 374–387.
126 H. Wohltjen, A. W. Snow, *Anal. Chem.* **1998**, *70*, 2856–2859.
127 O. D. Velev, E. W. Kaler, *Langmuir* **1999**, *15*, 3693–3698.
128 R. Möller, A. Csaki, J. M. Kohler, W. Fritzsche, *Langmuir* **2001**, *17*, 5426–5430.
129 S. J. Park, T. A. Taton, C. A. Mirkin, *Science* **2002**, *295*, 1503–1506.
130 B. Comiskey, J. D. Albert, H. Yoshizawa, J. Jacobson, *Nature* **1998**, *394*, 253–255.
131 J. A. Rogers, Z. Bao, K. Baldwin, A. Dodabalapur, B. Crone, V. R. Raju, V. Kuck, H. Katz, K. Amudson, J. Ewing, P. Drazic, *Proc. Natl. Acad. Sci. USA* **2001**, *98*, 4835–4840.
132 C. D. Dimitrakopoulos, P. R. L. Malenfant, *Adv. Mater.* **2002**, *14*, 99–117.
133 Y. Lvov, H. Haas, G. Decher, H. Möhwald, A. Kalachev, *J. Phys. Chem.* **1993**, *97*, 12835–12841.
134 S. W. Keller, H. N. Kim, T. E. Mallouk, *J. Am. Chem. Soc.* **1994**, *116*, 8817–8818.
135 J. H. Fendler, *Chem. Mater.* **1996**, *8*, 1616–1624.
136 G. Decher, M. Eckle, J. Schmitt, B. Struth, *Curr. Opin. Colloid In.* **1998**, *3*, 32–39.
137 J. H. Fendler, *Chem. Mater.* **2001**, *13*, 3196–3210.
138 A. N. Shipway, E. Katz, I. Willner, *ChemPhysChem* **2000**, *1*, 18–52.
139 T. Cassagneau, T. E. Mallouk, J. H. Fendler, *J. Am. Chem. Soc.* **1998**, *120*, 7848–7859.
140 B. Sweryda-Krawiec, T. Cassagneau, J. H. Fendler, *Adv. Mater.* **1999**, *11*, 659–664.
141 V. L. Colvin, M. C. Schlamp, A. P. Alivisatos, *Nature* **1994**, *370*, 354–357.
142 M. Y. Gao, M. L. Gao, X. Zhang, Y. Yang, B. Yang, J. C. Shen, *J. Chem. Soc. Chem. Comm.* **1994**, *24*, 2777–2778.
143 O. Onitsuka, A. C. Fou, M. Ferreira, B. R. Hsieh, M. F. Rubner, *J. Appl. Phys.* **1996**, *80*, 4067–4071.
144 A. C. Fou, O. Onitsuka, M. Ferreira, M. F. Rubner, B. R. Hsieh, *J. Appl. Phys.* **1996**, *79*, 7501–7509.
145 N. D. Kumar, M. P. Joshi, C. S. Friend, P. N. Prasad, R. Burzynski, *Appl. Phys. Lett.* **1997**, *71*, 1388–1390.
146 S. A. Carter, J. C. Scott, P. J. Brock, *Appl. Phys. Lett.* **1997**, *71*, 1145–1147.
147 Y. J. Liu, R. O. Claus, *J. Am. Chem. Soc.* **1997**, *119*, 5273–5274.
148 M. Y. Gao, C. Lesser, S. Kirstein, H. Mohwald, A. L. Rogach, H. Weller, *J. Appl. Phys.* **2000**, *87*, 2297–2302.
149 M. Eckle, G. Decher, *Nano Lett.*, **2001**, *1*, 45–49.
150 B. Oregan, M. Gratzel, *Nature* **1991**, *353*, 737–740.
151 M. Lahav, T. Gabriel, A. N. Shipway, I. Willner, *J. Am. Chem. Soc.* **1999**, *121*, 258–259.
152 M. Lahav, V. Heleg-Shabtai, J. Wasserman, E. Katz, I. Willner, H. Durr, Y. Z. Hu, S. H. Bossmann, *J. Am. Chem. Soc.* **2000**, *122*, 11480–11487.
153 R. P. Andres, T. Bein, M. Dorogi, S. Feng, J. I. Henderson, C. P. Kubiak, W. Mahoney, R. G. Osifchin, R. Reifenberger, *Science* **1996**, *272*, 1323–1325.
154 G. Markovich, D. V. Leff, S. W. Chung, H. M. Soyez, B. Dunn, J. R. Heath, *Appl. Phys. Lett.* **1997**, *70*, 3107–3109.

155 D. L. Feldheim, K. C. Grabar, M. J. Natan, T. E. Mallouk, *J. Am. Chem. Soc.* **1996**, *118*, 7640–7641.
156 T. Sato, H. Ahmed, *Appl. Phys. Lett.* **1997**, *70*, 2759–2761.
157 D. L. Klein, R. Roth, A. K. L. Lim, A. P. Alivisatos, P. L. McEuen, *Nature* **1997**, *389*, 699–701.
158 D. L. Feldheim, C. D. Keating, *Chem. Soc. Rev.* **1998**, *27*, 1–12.
159 D. J. Schiffrin, *MRS Bull.* **2001**, *26*, 1015–1019.

15
3D Ordered Macroporous Materials

Rick C. Schroden and Andreas Stein

15.1
Introduction

Materials exhibiting a uniform arrangement of pores offer a wide variety of applications that are based on both the chemical properties of the solid matrix and properties specific to the pore size and arrangement. The International Union of Pure and Applied Chemistry (IUPAC) has classified porous materials into three distinct categories according to their pore diameters [1]. "Micropores" are those with diameters less than 2 nm, "mesopores" range from 2 to 50 nm, and "macropores" are greater than 50 nm in diameter [1]. Interest in the synthetic creation of ordered porous materials began with the desire to replicate or improve upon the unique properties that result from the crystalline microporous structure of natural zeolites. With small pore sizes, open frameworks, high specific surface areas, and extremely well-defined structures, zeolites have found commercial applications as adsorbents, molecular sieves, and size- or shape-selective catalysts [2]. However, zeolites are limited to applications involving small-molecule substrates. For chemical applications involving larger guest molecules, such as biological compounds or polymers, materials with larger pores are required.

The development of laboratory techniques for the preparation of ordered porous materials with well-defined structures and pore sizes began in the 1950s with the creation of the first commercially significant synthetic zeolites, and has continued to the present [2]. The general procedure for the preparation of porous materials involves the use of sacrificial templates, space fillers, or structure directors around which a solid wall structure forms. Upon removal of the template by chemical or thermal methods, porous materials are produced. For zeolites, the sacrificial species is often a single molecule such as an alkyl-amine [2]. The size of the micropores of zeolites can be increased through the use of larger molecular templates, reaching a maximum pore size of approximately 2 nm [2]. While methods for preparing ordered microporous materials have been known for over fifty years, techniques for the fabrication of ordered materials with larger pores have only been developed since the early 1990s.

A new era in porous materials synthesis began with the development of techniques based on the use of self-assembled molecular arrays, instead of single mol-

---

Colloids and Colloid Assemblies. Edited by Frank Caruso
Copyright © 2004 Wiley-VCH Verlag GmbH & Co. KGaA, Weinheim
ISBN: 3-527-30660-9

ecules, as sacrificial species for the production of mesoporous materials [3–6]. By combining inorganic precursors with surfactants capable of forming micelles, ordered mesostructured materials are produced through cooperative self-assembly. Upon surfactant removal, mesoporous materials with cubic or hexagonal symmetry, and pore dimensions ranging from 2 to 5 nm, are produced. The use of longer-chain surfactants, swelling agents, and block copolymer templates further extends the range of achievable pore sizes to about 30 nm [7, 8]. Mesoporous materials offer the benefits of having extremely high specific surface areas, in excess of $1000 \text{ m}^2 \text{ g}^{-1}$, and the ability to tailor the surface reactivity by grafting various functionalities to the walls of the material [9]. However, to date there have been relatively few viable commercial applications for mesoporous materials. Some of the more interesting potential applications include adsorbents for toxic metal ions [10] and reactors for the production of high molecular weight crystalline polymers [11].

The latest size regime of ordered porous solids to be synthesized, and the primary focus of this chapter, is the class of macroporous materials. These materials are prepared by utilizing colloidal crystals, consisting of ordered arrays of polymer or silica microspheres with sub-micrometer to micrometer size diameters, as templates. Three-dimensionally ordered macroporous (3DOM) materials are produced by combining colloidal crystals with suitable structure-forming precursors, followed by removal of the colloidal crystal template from the solid composite by chemical or thermal methods. The resulting products are inverse replicas of the microsphere arrays. Since the colloidal crystal templates consist of solid spheres in a close-packed array reminiscent of the opal structure (Fig. 15.1 A), the macroporous products can be referred to as "inverse opals" (Fig. 15.1 B). The structure of inverse opals is that of a close-packed array of "air spheres" in the former location of the solid-sphere template, surrounded by solid walls that were formerly tetrahedral and octahedral voids in the opal template. Windows connecting the air spheres, or macropores, are typically present in the former location of contact points between neighboring spheres. The result is a structure of interconnected macropores with a regular, three-dimensional spacing and long-range order.

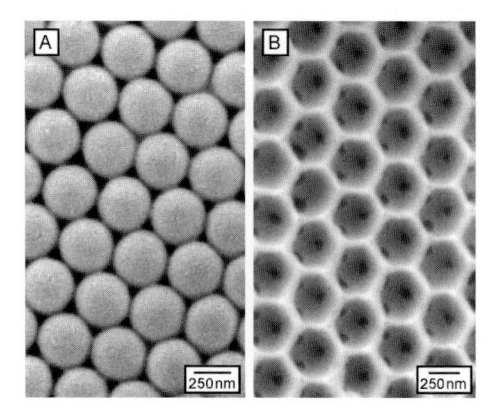

Fig. 15.1 **A** SEM image of a colloidal crystal of PMMA. The monodisperse polymer spheres have the face-centered cubic structure reminiscent of opals. **B** SEM image of three-dimensionally ordered macroporous silica, which has an "inverse opal" structure. The lighter regions in the image represent the solid matrix, and the gray circles are air spheres that were previously occupied by polymer spheres. Former contact points between neighboring polymer spheres appear as dark windows between the macropores.

In comparison to microporous and mesoporous materials, the large interconnected voids of macroporous materials potentially allow easier mass transport through the structure and less diffusional resistance to active sites [12]. These properties may enhance the activities of macroporous materials for processes such as catalysis, adsorption, chromatography, and sensing, especially if macropores are utilized in combination with the selectivity and high specific surface area of micropores or mesopores. In addition, the synthetic flexibility of the colloidal crystal templating technique allows the preparation of nearly any imaginable composition of macroporous materials.

Beyond their potential applications as chemically functional porous materials, the property that makes macroporous materials stand out from their smaller-pore counterparts is their behavior as photonic crystals [13, 14]. Diffraction of electromagnetic radiation from the uniform spacing of the dielectric walls leads to the formation of photonic stop bands, for which a specific range of light wavelengths is partially forbidden from propagating through the material in certain directions [15]. When the modulation of the refractive index is sufficiently large, complete photonic band gaps (PBGs) are formed where certain wavelengths of light are completely forbidden to exist within, or pass through, the material. The development of materials with complete PBGs would allow control over the propagation of light in a manner analogous to the control of electrons by the electronic band gaps of semiconductors. Numerous technological applications may result from PBG materials, including optical waveguides, optical circuits, and low-threshold telecommunications lasers [15–17].

This chapter reviews the various methods that have been developed to prepare macroporous materials by colloidal crystal templating. Compositions including metal oxides, metals, polymers, semiconductors, and many others have been prepared by innovative modifications of the general colloidal crystal templating technique. The discussion is organized by the method of preparation, and in many cases several methods are available for the preparation of a given composition. Numerous potential applications for macroporous materials have been cited in the literature, but relatively few have actually been experimentally demonstrated. A discussion of these demonstrated applications, including photonic crystals, surface-enhanced Raman spectroscopy, catalysis, adsorption, sensing, battery materials, bioactive materials, and templates for colloid synthesis, follows Section 15.2 on "Methods of Preparation." Reference is also made to several reviews devoted to the subjects of colloidal crystals [18, 19] and macroporous materials [20–25].

15.2
Methods of Preparation

15.2.1
Colloidal Crystal Templates

Colloidal crystals composed of silica, polystyrene (PS), or poly(methyl methacrylate) (PMMA) spheres are the most commonly utilized templates to direct the formation of ordered macroporous materials, because it is possible to prepare nearly monodisperse spherical particles of these materials with sizes ranging from tens of nanometers to several micrometers [18]. Numerous methods, including gravitational settling, centrifugation, convective assembly, and electrophoretic deposition may be exploited to order the spherical colloidal particles into colloidal crystals [18]. Polymer colloidal crystals can be annealed at temperatures slightly above their glass transition, and silica colloidal crystals can be sintered at high temperatures to form necks between the spheres. This process strengthens the colloidal crystal and prevents the spheres from separating during the infiltration of precursors, leading to fewer defects and more robust structures. In addition, the necks ensure interconnection of the pores by the formation of windows, thus allowing complete removal of the templates, and access of potential guests to the internal surfaces.

The choice of template material is often governed by the processing conditions required to form the walls of the macroporous solid and the ability to later remove the template from the composite structure without destroying the walls at the same time. Polymer templates are limited to low-temperature infiltration processes, and can be removed from the composite structures by calcination (thermal decomposition), or extraction with solvents such as THF, acetone, or toluene. Silica colloidal crystals may be used for high- or low-temperature infiltrations and can be removed from the composite structures by chemical etching with aqueous hydrofluoric acid solutions or alkaline metal hydroxide solutions. For many compositions of macroporous materials, either polymer or silica colloidal crystals may be used as templates, provided that the wall composition is amenable to the template removal technique.

15.2.2
Sol–Gel Chemistry

Metal alkoxides or mixtures of metal alkoxides and metal salt solutions are the typical precursors for sol–gel chemistry, where hydrolysis and condensation reactions lead to the formation of large metal oxide networks [26]. Macroporous oxides of a large variety of elements can be prepared by combining sol–gel processing with colloidal crystal templating. The sol–gel method has been used to prepare macroporous oxides of Si [27–39], Ti [31, 32, 38–49], Zr [31, 32, 39, 40, 46], Al [31, 34, 40], W [31], Fe [31], Sb [31], Sn [46], Eu [50], Nd [50], Sm [50], mixed oxides of Zr/Y [31], Co/Ti [42], Ba/Ti [51–53], Pb/Ti [32, 38], Pb/Ti/Zr [32], as well as hybrid

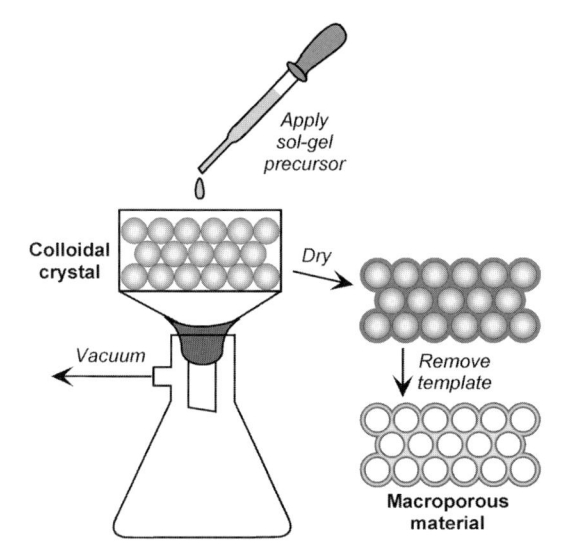

Fig. 15.2 Procedure for the preparation of macroporous materials by vacuum-assisted infiltration of colloidal crystals with sol–gel precursors. The sol–gel precursor penetrates the colloidal crystal, solidifying in the interstitial spaces between the spheres. After drying, the composite material is calcined to burn away the template, leaving behind a three-dimensionally ordered macroporous material that is an inverse replica of the polymer array.

organosilicates [31, 39, 54], aluminophosphates [31], and vanadium phosphates [55, 56]. The techniques used for the preparation of different materials by the sol–gel method are all very similar, each involving infiltration of the interstitial spaces of the colloidal crystal with a fluid precursor, solidification of the precursor, and removal of the template. In spite of this general similarity, modifications of each step can affect the structure on multiple levels, from the order of the pores, to the thickness, structure, and phase of the walls.

The first step in the fabrication of macroporous materials is to infiltrate fluid precursors into the voids between the close-packed spherical colloidal particles. Infiltration can be allowed to occur naturally by capillary forces [41, 45, 50, 51], or aided by the application of a vacuum [31, 34, 40, 43, 49]. An example of the latter method of vacuum-assisted infiltration is presented schematically in Fig. 15.2. In this technique, thin layers of a polymer colloidal crystal are deposited on a filter paper in a Büchner funnel [31, 40]. Sol–gel precursors are then applied dropwise to completely wet the polymer spheres, with vacuum applied to help draw the fluid precursor through the template. After solidification of the precursor and drying of the solid, the composite material is calcined to burn away the polymer colloidal crystal template, leaving a porous inverse replica in its place.

Improved penetration of sol–gel precursors through colloidal crystal templates and better wetting of the spheres is achieved by diluting the metal alkoxides with solvents, such as alcohols [31, 45, 50]. This is especially important for viscous and moisture-sensitive metal alkoxides, since alcohols decrease the viscosity and reactivity of the precursor and prevent the formation of a solid crust on the top surface of the polymer spheres [31, 40]. Dilution of the precursors also affects the wall structure, yielding products with thinner walls and larger windows between pores [31]. Wetting the templates with solvents prior to the addition of alkoxides

[40, 43, 48, 49], or functionalizing the surface of the polymer colloidal particles with various organic groups [29, 31, 40] can both be used to improve the interaction between the template and precursor solution. These processes facilitate penetration of the colloidal crystal template by the fluid precursor, leading to the formation of a continuous network.

After deposition of the sol–gel precursor in the voids between the spherical colloidal particles, the metal alkoxide undergoes hydrolysis and condensation to form a solid matrix. Water may either be added to the precursor solution [32, 45, 46], or the composite may be exposed to moisture from the air [31, 40, 41, 45], to facilitate hydrolysis of the alkoxide. The hydrolysis and condensation reactions can be catalyzed by an added acid [29, 34, 45] or surfactant [27, 28], or by acidic surface groups present on the microspheres [31, 40]. Although a close-packed arrangement of spheres contains an available void space of about 26% by volume, the macroporous products formed by sol–gel methods often have a solid fraction much lower than this value. Less than complete filling with solid is due to the loss of solvent upon hydrolysis and contraction of the structure upon condensation of the metal alkoxide. Multiple infiltration and drying cycles can be performed to increase the wall thickness [41, 45, 50, 51]. The pore size of macroporous materials can be predictably altered between approximately 100 nm and 1 μm by the use of templates with different diameters. However, significant linear shrinkage ranging from 10–40% is observed for sol–gel methods with respect to the diameter of the spheres and the spacing between macropores, producing cracks and defects in the structure. In spite of this large decrease in volume, the shrinkage tends to be very uniform, and materials with nearly monodisperse pore sizes can be prepared [45].

The degree of wetting of the spherical templates by the sol–gel precursor influences the manner in which the precursor solidifies and affects the filling of the interstitial spaces between the spheres. Fluid precursors that do not adhere to the template exhibit a volume-templating [57] behavior, in which the precursors completely fill the void volume between the spheres prior to solidification. As a result, these materials have solid walls at the intersection of air spheres, as can be seen in the SEM image of a representative material in Fig. 15.3 A. Alternatively, for precursors that strongly adhere to the template, a surface-templating [57] mechanism takes place in which condensation occurs at the surface of the spheres, blocking the introduction of additional precursor solution into the interstitial spaces between the spheres. The result is the observation of vacancies at the center of each triangular intersection of three air spheres [41, 43, 45, 49], as is apparent in the SEM image of a representative material in Fig. 15.3 B.

For most macroporous metal oxides prepared by the sol–gel method, polymer colloidal crystals are the preferred templates, and calcination is the most common method of template removal. For these materials, thermal processing serves the multiple roles of removing the template and controlling the microstructure of the walls with respect to the phase and grain size [58]. Macroporous silica remains amorphous regardless of the calcination temperature; however, compositions such as alumina, titania, zirconia, and many other non-oxide compositions have crystal-

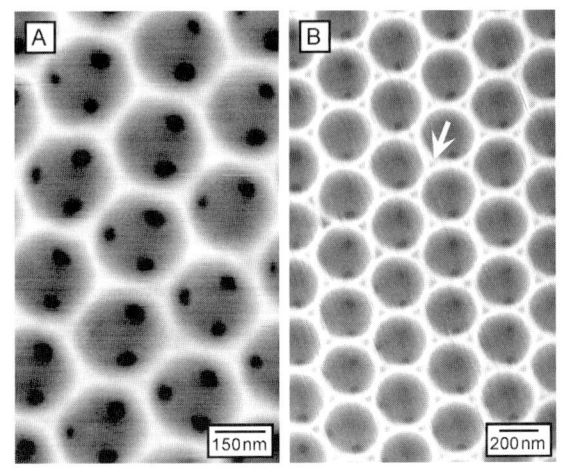

Fig. 15.3 SEM images of macroporous materials that are examples of **A** volume templating and **B** surface templating. Volume templating occurs with precursors that do not adhere very strongly to the template, resulting in the filling of the voids between spheres. Surface templating occurs when the precursor fluid strongly adheres to the spherical template, leading to the formation of voids at the center of each set of three intersecting macropores (indicated by an arrow). Sample **A** is 3DOM silica, and **B** is 3DOM mercaptopropyl-silica. Both were prepared with a PS template.

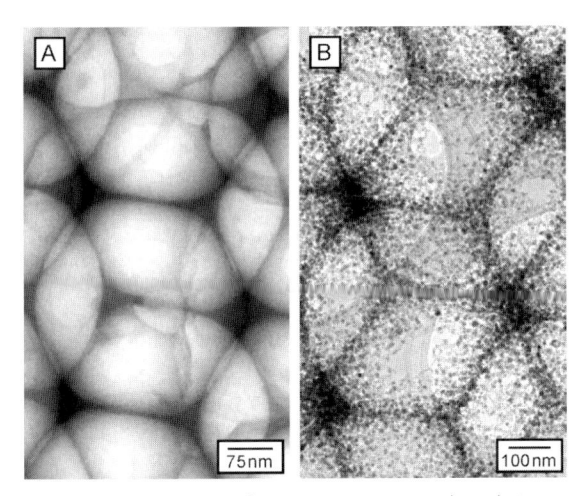

Fig. 15.4 TEM images of macroporous materials with **A** amorphous and **B** nanocrystalline walls. The walls are the dark regions in the images. Amorphous walls are smooth and continuous, whereas nanocrystalline walls are composed of nanometer-sized grains fused together to form the larger wall structure. Sample **A** is 3DOM silica, and **B** is 3DOM nickel.

line phases that form at elevated temperatures [27, 31, 39–41, 45, 58]. TEM images of macroporous materials with amorphous and nanocrystalline walls are given in Fig. 15.4. The high magnification images clearly show the smooth, continuous nature of amorphous walls, and the individual grains that are fused together to form the walls of nanocrystalline materials. Considerable effort has been directed to the preparation of macroporous materials containing nanocrystalline walls with controlled grain sizes [31, 39, 45, 58]. This is especially important for photonic crystal applications, because the grain size must be very small in comparison to the void dimensions in order to have the uniform periodicity required for optical diffraction [31, 39]. Well-ordered macroporous materials with small grains exhibit brilliant colorful reflections even if they are white as dense bulk materials, whereas those with larger grains remain white [39]. Excessive grain growth, where grain sizes approach void dimensions, results in a loss of periodicity and eventual destruction of the interconnected network of pores.

Materials with hierarchically ordered pore structures can be prepared by combining colloidal crystal templating techniques with methods for the synthesis of mesoporous or microporous materials. The addition of alkylammonium [31, 54] or triblock copolymer [30, 38, 59, 60] surfactants to precursors for silica, combined with colloidal crystal templating, allows the preparation of macroporous materials with mesoporous walls. In a similar manner, the addition of molecular species to the silica precursor allows the preparation of macroporous materials with microporous zeolitic walls [61]. The hierarchical porosity of these materials combines the selectivity and high specific surface area of the smaller pores with the improved transport through the macropores. Additional levels of order may also be created by using patterned poly(dimethylsiloxane) (PDMS) stamps to generate features on the micrometer scale [30, 38], or by templating within suspension droplets to control the geometry of macroporous particles [62].

The most extensively studied macroporous metal oxide prepared by sol–gel chemistry has been titania because of its high refractive index, which is a prerequisite for the formation of photonic crystals with complete photonic band gaps. Titania undergoes a transition from amorphous to the crystalline phases of anatase and rutile with increasingly higher processing temperatures. For example, removal of the polymer template from a titania/colloidal crystal composite by solvent extraction leads to the formation of macroporous titania with amorphous walls, whereas calcination at 450 °C produces macroporous titania with the anatase phase [31]. The preparation of macroporous titania with the rutile phase has been widely pursued because of its higher refractive index of 2.9, compared to 2.5 for anatase [31, 45, 46]. However, most attempts to prepare macroporous rutile have either failed to produce rutile or led to the formation of large crystals and a loss of the periodic structure [31, 45, 46]. For example, heating macroporous anatase to 1000 °C produces crystalline rutile, but excessive grain growth completely destroys the porous structure [31, 45]. Additional methods for the preparation of macroporous rutile have also been explored, including the templated assembly of pre-formed nanoparticle precursors (discussed in Section 15.2.5), emulsion templating (discussed in Section 15.2.7), and the employment of chemical additives to sol–gel precursors that control grain growth during calcination [63]. The latter approach seems promising, with preliminary results demonstrating the preparation of ordered macroporous rutile with relatively small grain sizes (~ 25 nm) [63].

15.2.3
Salt-precipitation and Chemical Conversion

An alternative to the use of sol–gel precursors for the fabrication of macroporous materials by colloidal crystal templating employs metal salt precursors in conjunction with in situ chemical reactions [64–67]. In this method, a metal salt is first deposited in the voids of a colloidal crystal by precipitation from a saturated solution. This salt may be further converted to a form capable of undergoing high-temperature decomposition to a structure-forming solid. This approach is especially suitable for the preparation of macroporous metal oxides where the corresponding metal alkoxide precursor is viscous, moisture-sensitive, or unavailable [58, 66]. Macroporous metal oxides prepared from metal salt precursors undergo shrinkage comparable to that observed with sol–gel precursors, although the products often have smaller grain sizes and more uniform pore structures [58, 66].

Fabrication of macroporous materials by this method begins with soaking a polymer colloidal crystal in a metal salt solution [64–67]. The metal salt may be an acetate, oxide, oxalate, or nitrate, and the solvent an alcohol, alcohol/water solution, or dilute acetic acid. Metal acetates are the most commonly used salts, since their high solubility in the desired solvents allows high loadings to be achieved. After a few minutes of soaking, the polymer colloidal crystal is removed from the salt solution, excess fluid is removed by vacuum filtration, and the sample is allowed to dry. Evaporation of the solvent leads to precipitation of the metal salt in the colloidal crystal voids. Direct calcination of these composites may lead to phase separation, owing to the low melting points of many metal acetate hydrates. To prevent phase separation, the composites may be soaked in ethanolic solutions of oxalic acid, followed by vacuum filtration and drying. This process converts the metal salts in the colloidal crystal voids to metal oxalates, which are desirable precursors because they undergo high-temperature decomposition instead of melting. Calcination of these composites burns away the polymer colloidal crystal, producing macroporous materials.

The atmosphere of calcination dictates the composition of macroporous materials prepared from metal salt precursors. Calcination of metal oxalate/polymer colloidal crystal composites in air has allowed the production of macroporous oxides of Mg, V, Cr, Mn, Fe, Co, Ni, Zn, and Zr, as well as calcium carbonate [34, 58, 64, 66]. Calcination in inert or reducing atmospheres has produced macroporous metals of Fe, Co, and Ni, and the alloys $Ni_{1-x}Co_x$ and Mn_3Co_7, from mixed metal salt precursors [64, 65, 67]. Macroporous Au and Pt metals can be prepared in a similar manner by reduction of the metal chloride salts in the voids of silica colloidal crystals by heating in hydrogen, followed by HF etching to remove the silica template [68]. The electrochemical potential of the metal must be considered when choosing a calcination atmosphere for the preparation of macroporous metals from salt precursors [67]. For example, both cobalt and nickel oxalates decompose to the metals upon heating in nitrogen, whereas iron oxalate decomposes to iron oxide in this atmosphere. However, macroporous iron metal can be produced by calcination in the reducing atmosphere of hydrogen [67]. Macroporous metals prepared in a nitrogen atmosphere contain a significant amount of residual carbon, typically >10 wt.%, present as thin layers surrounding the metal grains [58, 67]. In contrast, macroporous metals prepared in hydrogen have relatively little carbon, usually <2 wt.% [67].

Macroporous materials can also be prepared from metal salt precursors that do not undergo changes in composition upon template removal [41, 45]. For example, soaking a polymer colloidal crystal with a heated saturated solution of sodium chloride in water allows precipitation of the salt in the interstitial spaces of the template upon evaporation of the solvent. Calcination of the composite burns away the polymer, leaving a macroporous sodium chloride product [41, 45]. No shrinkage occurs for this material because sodium chloride precipitates from solution with its final structure and density.

15.2.4
Oxide Reduction

Certain macroporous metals and semiconductors can be prepared from the corresponding macroporous oxides by high-temperature reduction in an atmosphere of hydrogen. Macroporous Fe, Co, and Ni are formed after heating the macroporous oxides at 300 °C in hydrogen for 2 h [64, 67]. Significant shrinkage of >30% occurs upon reduction, and the macroporous metals prepared by this method generally have larger grain sizes and less regular order than metals prepared by direct conversion of metal oxalates to metals [64, 67]. In a similar manner, macroporous Ge is formed via high-temperature reduction of the oxide [69]. In this technique, germanium oxide is first formed in the voids of a silica colloidal crystal by the sol–gel method, followed by reduction of the oxide to Ge by heating at 550 °C in hydrogen. Several infiltration and reduction cycles are required to completely fill the voids. HF etching removes silica from the composite, leaving behind the macroporous semiconductor.

15.2.5
Nanoparticle Assembly

As an alternative to liquid structure-forming precursors that solidify in the voids of colloidal crystals, solid particles may be used as precursors provided that they are significantly smaller than the sphere templates. Macroporous materials can be prepared by introducing nanoparticle precursors (with sizes of about 10 nm) into the voids of colloidal crystals (with sphere sizes of hundreds of nanometers), followed by removal of the colloidal crystal template. This is a relatively flexible technique in that it can be applied to any composition for which nanoparticles can be

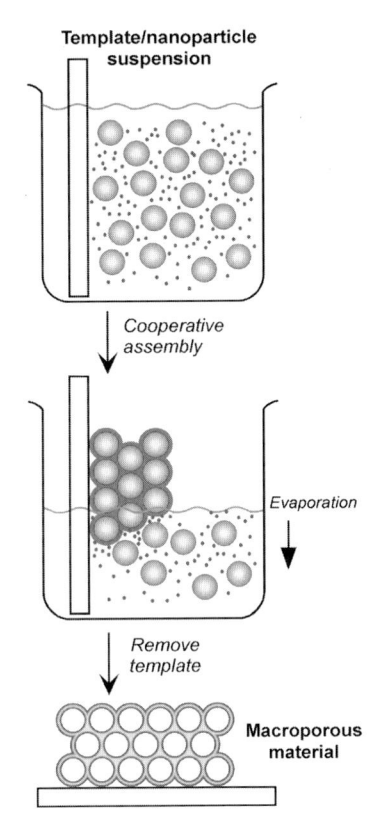

Fig. 15.5 Procedure for the preparation of macroporous materials from nanoparticle precursors by the vertical deposition method. A slurry of the nanoparticle precursors and silica or polymer spheres undergoes cooperative assembly to form a colloidal crystal of spheres filled with nanoparticles. Vertical deposition allows growth to occur at the meniscus as the solvent evaporates. Removal of the sphere templates by thermal or chemical methods leads to the formation of macroporous materials with nanoparticle walls.

synthesized. In addition, structural shrinkage is minimized by this technique because the nanoparticles are already in their final chemical form. However, macroporous materials produced by this method are generally limited to thin films. To date, the nanoparticle assembly method has been employed to prepare macroporous metals, metal oxides, hierarchically ordered porous silicates, and semiconductors, including Au [70–72], TiO_2 [73–77], SiO_2 [74], TiO_2/SiO_2 composites [78], magnetite (Fe_3O_4) [79], zeolites [80–82], and CdSe [83].

Two general methods can be used to prepare macroporous materials from nanoparticle precursors. In one method, pre-formed colloidal crystals are infiltrated with a suspension of nanoparticles [70, 83]. The nanoparticles can then either be allowed to settle into the voids of the colloidal crystal by slow evaporation of the solvent [83], or they can be drawn into the voids of the colloidal crystal by a flux of water by performing the infiltration on a permeable membrane [70]. In an alternative method (presented schematically in Fig. 15.5), the nanoparticles can be introduced into the composite while ordering of the spheres occurs by cooperative assembly of the two components from a mutual suspension [71–79]. Drying the mixture, for example by controlled evaporation of the solvent in a chamber with high relative humidity [73, 75, 76], leads to the formation of a colloidal crystal of spheres containing nanoparticles in the voids. After either method of assembly, removal of the template produces a macroporous material composed of the nanoparticle precursor.

Various processing methods can be used to change the physical characteristics of these macroporous materials. For example, thermal treatment of CdSe/silica sphere composites prior to template removal produces macroporous CdSe with sintered crystalline grains, whereas removal of the template without prior thermal treatment produces a macroporous network of CdSe nanoparticles held together only by van der Waals forces [83]. Additionally, heating macroporous titania prepared from anatase nanoparticles above 850 °C produces crystalline rutile; however, extensive grain growth destroys the macroporous order [74].

15.2.6
Solvothermal Synthesis

A solvothermal synthesis technique has been developed for the preparation of macroporous CdS [84]. Sintered silica colloidal crystals are first loaded with CdS nanoparticles, which serve as seeds for further growth of CdS from a liquid-phase reaction. The seeded colloidal crystal is then placed in an autoclave with cadmium chloride and thiourea precursors, along with carbon disulfide solvent. Heating the sample at 160 °C for 24 h, followed by HF etching to remove the template, produces macroporous CdS.

15.2.7
Emulsion Templating

While most methods for the preparation of macroporous materials rely on solid colloidal particles as templates, emulsion templating utilizes monodisperse liquid droplets as templates. Nearly monodisperse emulsions composed of a suspension of liquid oil droplets (isooctane), stabilized by surfactants, in an immiscible polar liquid (formamide) can be prepared by fractionation or shearing techniques [85–87]. To synthesize macroporous materials by this technique, various sol–gel precursors are first added to the polar phase, and then the emulsion is centrifuged to increase the droplet volume fraction above 50% and induce spontaneous ordering of the droplets into a close-packed structure. Addition of ammonia causes the emulsion to gel, after which the oil-droplet template is removed by washing in ethanol, and the gel is dried and calcined to solidify the walls and remove residual organics. The oil-in-formamide emulsion templating technique has been used to prepare macroporous materials with compositions of TiO_2, SiO_2, and ZrO_2 [85–87]. Related oil-in-water emulsion templating may also be used to prepare macroporous SiO_2 and polyacrylamide [86].

The primary benefits of the emulsion templating technique are that the oil droplet colloidal crystals are very deformable and they can be easily removed from the composite by a simple ethanol wash [85–87]. The deformability of the droplets allows the structure to undergo shrinkage during gelation without cracking, and the ability to remove the template prior to drying and heating allows the structure to withstand the stress of phase changes during calcination without loss of the porous structure. These aspects have allowed the preparation of macroporous rutile titania by calcination of the gels at 1000 °C [85–87]. Macroporous rutile produced by this method exhibits optical reflections; however, the large grain sizes (70 nm) and less than optimal size monodispersity of the emulsion droplets prevent the formation of a complete photonic band gap [87].

In a related approach, water droplets can be used as templates for the preparation of macroporous polymers [88]. Flowing moist air over the surface of a dilute solution of polystyrene in a volatile solvent results in the formation of water droplets on the surface of the solution due to evaporative cooling of the solvent. Hexagonally packed water droplets sink to the bottom of the lower density solvents (benzene or toluene). Successive layers of droplets stack on top of each other, producing a 3D close-packed array. As the solvent and water droplets evaporate, a 3DOM polystyrene film is formed [88]. The pore size of these macroporous polymers can be predictably altered between 0.20 and 20 µm by changing the velocity of airflow, with smaller pores produced from faster airflows.

15.2.8
Electrochemical Deposition

In many of the methods previously described, complete filling of the interstitial spaces of the colloidal crystal is not achieved owing to factors such as the loss of solvent during sol–gel processing, decomposition of the precursor to a smaller structure-forming species, or the presence of void spaces between nanoparticle precursors. A method that solves this problem is template-directed electrochemical deposition [89] (presented schematically in Fig. 15.6). This technique typically begins with the growing of a colloidal crystal on a conductive substrate, followed by sintering to improve the mechanical stability of the template. The substrate is then immersed in an electrobath along with a counter electrode, and a potential is applied. Electrodeposition fills the colloidal crystal sequentially from the bottom to the top, allowing the film thickness to be tailored by altering the electrodeposition time. The filled colloidal crystal is then processed to remove the template, yielding a macroporous material. Electrochemical deposition methods have been employed to prepare macroporous chalcogenides: CdS and CdSe [89, 90]; metals: Au [91–94], Ni [92, 95, 96], Co [97], Pt [94, 95, 97], Pd [97]; an alloy: SnCo [95]; metal oxides: ZnO [98] and WO_3 [99]; and conducting polymers: polypyrrole [100–102], polyaniline [101], polybithiophene [101], polythiophene [102], and polyaniline/carbon composites [103].

Electrochemical deposition typically produces volume-templated structures, in which the wall material grows in the void spaces between spheres and also fills any cracks that may be present within the template [91]. This behavior, combined with walls that already exhibit their final chemical form, allows nearly complete filling of the voids to be achieved. As a result, removal of the template leads to little or no structural shrinkage. The dense and continuous nature of the walls also

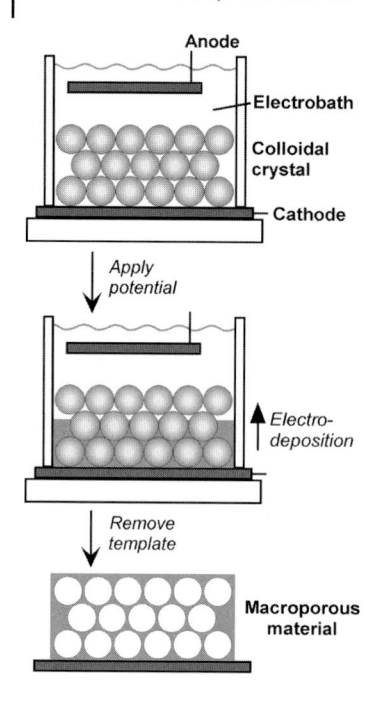

Fig. 15.6 General scheme for the preparation of macroporous materials by electrochemical deposition. Colloidal crystals are grown on an electrode and then immersed in an electrobath along with a counter electrode. A potential is applied, causing electrodeposition to occur from the bottom up. Upon completion of filling, the sample is withdrawn from the electrobath and the template is removed to produce a macroporous material.

increases the mechanical stability of the material and maximizes its refractive index. One notable exception in which complete filling was not observed by electrochemical deposition has been reported [101]. In this case, surface templating occurred when conductive polymers were grown around polystyrene spheres [101]. Triangular voids were observed at the intersections of air spheres in polypyrrole and polyaniline, whereas polybithiophene was so strongly surface-templated that hollow spheres were formed instead of a macroporous solid [101].

15.2.9
Electroless Deposition

A number of macroporous metals, including Ni, Cu, Ag, Au, and Pt, can be prepared by electroless deposition around colloidal crystals modified with catalytic particles [104]. Colloidal silica spheres are first functionalized with thiol groups via siloxane coupling agents, and are then organized into a colloidal crystal by convective self-assembly. The colloidal crystal is then immersed in a toluene solution containing gold nanoparticles, followed by drying and high-temperature treatment to remove the organic material and sinter the spheres. The resulting colloidal crystal consists of close-packed silica spheres with nanocrystalline gold particles attached to the surface, which serve as catalysts for electroless deposition. Upon immersion of these templates in electroless deposition baths, metal is deposited around the spheres. Removal of the silica spheres by etching in aqueous HF produces the macroporous metals.

15.2.10
Chemical Vapor Deposition

Perhaps the most promising method for the creation of materials with complete photonic band gaps is template-directed chemical vapor deposition (CVD). This method can be used for the synthesis of the macroporous semiconductors Si [105–107], Ge [108], and SnS_2 [109, 110], which as bulk materials have the extremely high refractive index that is required for the formation of a complete PBG. Absorption of semiconductors in the visible region, however, generally limits the PBG to infrared wavelengths. In addition to semiconductors, CVD can also be employed to prepare the macroporous carbon allotropes graphite and diamond [57, 111].

Disilane gas is used as the precursor to deposit Si around sintered silica colloidal crystals by CVD at an elevated temperature [105, 106]. Sintering of the silica spheres prior to deposition produces small necks between the spheres that provide mechanical stability and allow a continuous network to be formed. The amorphous Si framework deposited by CVD is transformed to polycrystalline Si by annealing at 600 °C, followed by HF etching to remove the silica template and leave behind macroporous Si [105–107]. Macroporous Ge is prepared by the same approach, using digermane gas as the precursor [108].

Macroporous SnS_2 is prepared by room temperature CVD around PMMA colloidal crystals by employing $SnCl_4$ vapor and H_2S gas as precursors [109, 110]. Removal of the template by dissolving in THF produces macroporous SnS_2. The refractive index of the SnS_2 framework is significantly lower than that of the bulk material, owing to the low density of the walls.

Macroporous graphitic carbon is prepared from propylene gas by CVD around silica colloidal crystals, with subsequent HF etching [57, 111]. Macroporous diamond is prepared by a plasma-enhanced CVD technique, in which a silica colloidal crystal is first seeded with diamond nanoparticles, and then carbon is deposited from a plasma of hydrogen and methane [57]. Etching of the silica with HF produces the macroporous product.

15.2.11
Pyrolysis and Spraying Techniques

In addition to CVD methods for the preparation of macroporous carbon, various pyrolysis techniques may also be used [57, 112, 113]. Either a phenolic resin [57] or sucrose [112, 113] may serve as the carbon precursor for these syntheses. In the first case, silica colloidal crystals are filled with a phenolic resin, which is thermally cured at a low temperature [57]. The template is subsequently removed by HF etching, and the sample is pyrolyzed at 1000 °C to form macroporous glassy carbon [57]. In the other method, silica colloidal crystals are filled with an aqueous solution of sucrose, then the sample is dried [112, 113]. Pyrolysis is subsequently carried out, followed by removal of the template by HF etching [112, 113]. Macroporous carbon prepared by this method can exhibit a specific surface area in excess of 500 $m^2 g^{-1}$ [112, 113].

Thin films of macroporous TiO$_2$ can be prepared by a spray-pyrolysis technique, in which titanyl acetylacetonate in ethanol is sprayed onto a 2D silica sphere array, followed by heating at 450 °C to convert the precursor to TiO$_2$, and HF etching to remove the silica template [114, 115]. Macroporous Au can be prepared by ionic spraying and macroporous Si by laser spraying of polystyrene spheres, followed by removal of the template by calcination or solvent extraction [95]. Macroporous materials prepared by spraying techniques are generally limited to thin films because of the limited penetration of the sprayed material through the colloidal crystal and pore blockage as the material is deposited.

15.2.12
Melt-imbibing

Low-melting metals and semiconductors, including Sb [116] and Se [117, 118], can be employed as molten precursors for the preparation of macroporous materials. Using silica colloidal crystals as templates, the melt is infiltrated into the voids under elevated temperature and pressure. The melt solidifies upon cooling, allowing the template to be removed by HF etching. This technique is generally limited to the preparation of macroporous films using highly ordered templates, since bulk material will solidify in any defects or spaces between colloidal crystal domains.

15.2.13
Organic Polymerization

In situ polymerization of organic precursors within the voids of a colloidal crystal can be employed for the preparation of a large variety of macroporous polymers [119]. Both silica and polymer colloidal crystals may be utilized as templates, provided that the macroporous polymer is insoluble in the extracting fluid. Macroporous polymers fabricated by this method include polyurethane [120–123], poly(acrylate-methacrylate) [120, 122], epoxy [120, 123], PS [123–125], PMMA [123], poly(methyl acrylate) [123], poly(allyl methacrylate) [126], poly(phenylene vinylene) [127], polyaniline [128], and epoxy-resin/gold-nanoparticle composites [129]. The polymerizations can be initiated by thermal treatment [120, 123], UV irradiation [120, 121, 123], or catalysis [125, 127], or may be performed by chemical oxidation polymerization [128]. Macroporous films prepared by this method may be either rigid or flexible, depending upon the glass transition temperature (T_g) of the polymer. For example, macroporous PS and PMMA (which both have a T_g near 100 °C) are rigid, whereas polyurethane (with a T_g near room temperature) is flexible [123].

Ordered mesoporous polymers may be prepared in a similar manner by utilizing colloidal crystals composed of 35-nm silica spheres [130]. Mesoporous poly(divinylbenzene) prepared with these templates is rigid and does not undergo shrinkage upon template removal, whereas mesoporous poly(ethyleneglycol dimethacrylate) is flexible and experiences over 50% shrinkage. By using mixtures of the two precursors, the pore size of the polymer can be continuously varied between 15 and 35 nm [130].

As an alternative to in situ polymerization, high-temperature solution infiltration may also be used for the preparation of macroporous polymers. In this method, concentrated solutions of pre-formed polymers at elevated temperatures are infiltrated into silica colloidal crystals, followed by evaporation of the solvent to precipitate the polymer and fill the colloidal crystal voids. Removal of the template by HF etching completes the synthesis. This method has been utilized for the preparation of several macroporous polymers, including PS [131], poly(alkyl-thiophenes) [132], poly(alkoxy-phenylene vinylenes) [132], and poly(vinylidene fluoride-trifluoroethylene) [133].

15.2.14
Coordination Polymerization

In addition to various organic polymerization techniques and inorganic polymerization of sol–gel precursors, colloidal crystal templating is also amenable to coordination polymerization. This method has been demonstrated for the preparation of macroporous cobalt hexacyanoferrate, $K_xCo_4[Fe(CN)_6]_y$ ($x=y=4$ for the stoichiometric compound) [134]. This material is prepared by successive immersion of silica or polystyrene colloidal crystals in aqueous solutions of cobalt(II) chloride and potassium ferricyanide. After four cycles of immersion, and template removal by solvent extraction or HF etching, a macroporous transition metal coordination polymer is produced.

15.2.15
Core-Shell Assembly and Rearrangement

Core-shell assembly (shown in Fig. 15.7) is an approach to the preparation of macroporous materials that combines layer-by-layer deposition and colloidal self-assembly. In this method, core-shell structures are formed by sequential adsorption of several layers of oppositely charged polyelectrolytes and nanoparticles on the surfaces of polymer spheres, followed by assembly of the spheres into a colloidal crystal. Subsequent calcination yields a macroporous material composed of sintered nanoparticles. Using this or related methods, macroporous materials composed of zeolites [135, 136], TiO_2 [137], TiO_2/SiO_2 composites [137], and silver [138] have been prepared. The primary advantage of this technique is that the wall thickness can be increased by depositing additional layers on the spheres prior to assembly into a colloidal crystal. A similar layer-by-layer deposition process can also be applied to macroporous materials post-synthesis, to tailor the composition and functionality of the surface [139].

Another method employing core-shell structures as precursors for macroporous materials is core-shell rearrangement [140, 141]. In this method, colloidal crystals composed of latex spheres with PS-rich cores and poly(2-hydroxyethyl methacrylate)-rich shells (i.e. polyHEMA) are exposed to styrene or toluene vapor. Since

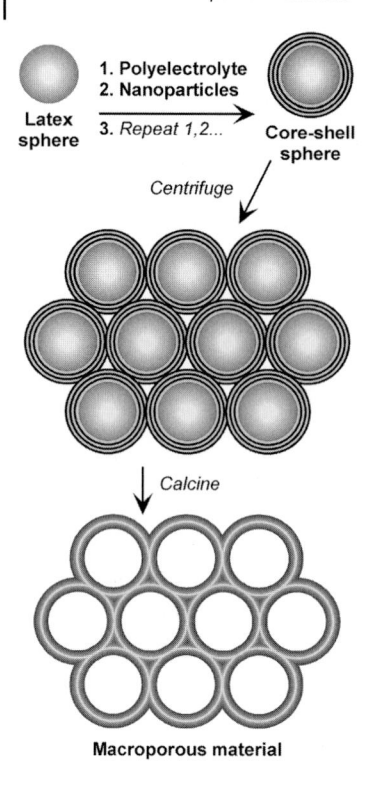

Fig. 15.7 General method used for the preparation of macroporous materials from core-shell spheres. Latex spheres are coated with several layers of polyelectrolytes and nanoparticles by sequential adsorption of the oppositely charged species. The core-shell spheres are then centrifuged to form a colloidal crystal and calcined to burn away the organics, leaving behind a macroporous material composed of sintered nanoparticles.

these fluids are good solvents for PS and poor solvents for polyHEMA, the vapors permeate and swell the PS cores, causing the PS to engulf the polyHEMA shell. The latex spheres are effectively turned inside out to form a porous polymer net.

15.3
Applications

15.3.1
Photonic Crystals

Photonic crystals with complete photonic band gaps (PBGs) are the most highly prized, and perhaps the most promising, application for 3DOM materials. The realization of materials exhibiting a complete PBG would be of great scientific and technological significance [13–17]. For example, PBG materials would be capable of localizing or directing the flow of light, which might lead to the development of optical waveguides and optical circuits. They would also be capable of inhibiting spontaneous emission, which could be used to improve the efficiency of photocatalytic reactions and reduce the threshold for telecommunications lasers.

15.3 Applications

These topics, and others in the field of photonic crystals, have been the subjects of reviews [142–145].

Several requirements must be met for the development of materials with a complete PBG. These include a low solid fraction (about 20% solid by volume), a minimum refractive index (RI) contrast of 2.8 (for materials with the fcc structure), negligible absorption in the spectral region of interest, and a uniform periodicity with a lattice spacing comparable to the desired PBG wavelengths [146]. The requirement of a low solid fraction immediately eliminates the opal structure of colloidal crystals from exhibiting a complete PBG, regardless of how large the RI contrast is. This aspect complicates the pursuit of PBG materials because not only is it very difficult to fabricate colloidal crystals that are free from defects, but most methods for the preparation of macroporous materials fail to yield a completely faithful replica (or more precisely, an inverse replica) of the template. Unwanted disorder is introduced into these systems from several sources, including incomplete filling of the voids and structural shrinkage upon template removal, which lead to cracks and other various defects. Moreover, the growth of bulk material in cracks between crystalline domains or on the surface of colloidal crystals provides additional forms of unwanted disorder. Defects such as these are detrimental to the formation of complete PBGs [147, 148]. In fact, theoretical calculations have shown that variations of as little as 2% of the lattice constant can close a PBG, even for materials with very high RI contrasts [149, 150].

While unwanted defects may be significantly reduced by process optimization, many of the macroporous materials that have been prepared to date do not have the requisite RI contrast for the formation of a complete PBG. In spite of this, many of these materials exhibit photonic crystal properties in the form of optical stop bands, which are manifested as colorful reflections. The spectral position of these stop bands, and therefore the color of the reflections, can be predictably altered over the entire visible spectrum by numerous methods, including tailoring the pore size [34, 45, 73, 77], altering the refractive index of the wall material [34, 78, 131], and filling the pores with fluids of various refractive indices [34, 39]. In addition, the stop bands are capable of modifying the emission characteristics of luminescent species intrinsic to [35, 127] or embedded in [151, 152] the macroporous material. Even with only a partial PBG, these materials remain potentially useful for applications as ultra-reflective pigments, optical filters, and chemical sensors [34, 39, 123, 131, 153].

Macroporous titania is one of the few photonic crystals that does not absorb in the visible region of the spectrum, yet has an RI contrast near the threshold value. With an RI contrast of about 2.5, macroporous titania in the anatase phase does not have a complete PBG, but it has stop bands that overlap for a wide range of angles covering over 55% of all directions [41, 154]. Macroporous titania with the rutile phase, on the other hand, meets the threshold RI contrast requirement. As a result, numerous attempts have been made to prepare this material, including sol–gel [31, 45, 46], nanoparticle [74], and emulsion templating [85–87] methods. Most attempts to prepare macroporous rutile have led to excessive grain growth and a loss of order. However, grain growth has been significantly limited by the employment of chemical additives [63], and stop bands have been observed for samples prepared by emulsion templating [87]. In general, the large grain sizes, structural disorder, and/or insufficient wall density of these rutile samples have prevented the formation of a complete PBG.

Macroporous semiconductors may hold the greatest promise for the development of photonic crystals with complete PBGs because of their extremely high RI contrast. These materials are generally limited to the near-IR region of the spectrum, however, owing to their strong absorption in the visible region. In spite of this, macroporous semiconductors remain technologically viable photonic crystals because the wavelength used in optical communications is 1.5 µm. A variety of macroporous semiconductors have been prepared, including Si (by CVD [105–107]), Ge (by CVD [108] and oxide reduction [69]), CdS (by solvothermal synthesis [84] and electrochemical deposition [89, 90]), CdSe (by nanoparticle assembly [83] and electrochemical deposition [89, 90]), Se (by melt-imbibition [117, 118]), and SnS_2 (by CVD [109, 110]). Of these, only macroporous Si and Ge prepared by CVD have had the requisite high RI contrast, along with the dense and homogeneous walls required for a complete PBG [105–108]. In fact, optical reflectance spectra of macroporous Si have exhibited near unity reflectance over two crystallographic directions, which is an important signature of the presence of a complete PBG [107].

A challenge that remains for the development of PBG technology from macroporous materials is the integration of these materials into devices. This process is complicated by the fact that self-assembly is used to fabricate the colloidal crystal templates for macroporous materials, thus making the design of specific features difficult. Several techniques that may aid this process have already been developed. These include growing macroporous materials on planar [107] or lithographically patterned [117, 144] silicon wafers or within PDMS molds [30, 38] to make the materials compatible with existing technology [107] and to transfer designed features into the product [30, 38, 117, 144]. In addition, further patterning has been performed after the structure was made, using photolithography and reactive ion etching [107].

While most types of defects in photonic crystals are unwanted, certain others are required for the localization and guiding of light. For example, point defects could serve as optical cavities to localize light, line defects as waveguides to direct light in lines or even around corners, and planar defects as mirrors [16]. Point defects have been produced in macroporous Si by doping the colloidal crystal template with a small amount of spheres of a different diameter prior to deposition of Si in the voids [107]. While the location of these defects was random, this approach was successful in creating additional air cavities in the photonic crystal. Multi-photon polymerization has been used to fabricate specific point and line defects within silica colloidal crystals [155]. When combined with templating to form a macroporous material, this method may allow the formation of optical cavities and waveguides in the structure.

An important requirement for the development of PBG devices is the ability to reversibly switch the band gap on and off. This has been accomplished by filling a macroporous material with an optically birefringent nematic liquid crystal [156, 157]. As an electric field was applied, the refractive index of the liquid crystal changed, causing a shift in the spectral position of the stop band. The reversibility of this shift was limited, however, and the peak position reverted only part of the way to its initial location after removal of the applied voltage [157]. A completely reversible method of tuning the stop band position of macroporous materials has also been demonstrated. By filling the voids of macroporous materials with fluids of various RI, the position of the stop band can be predictably shifted over a spectral range of several hundred nanometers [34, 39]. A potential switching mechanism may involve alternate filling of a photonic crystal in a flow cell with fluids of different RI to move the PBG in or out of a desired spectral region.

15.3.2
Surface-enhanced Raman Spectroscopy

Porous metallic films prepared by colloidal crystal templating may find an analytical application as substrates for surface-enhanced Raman spectroscopy (SERS). This application has been successfully demonstrated for thin macroporous gold films [71, 72]. These films were prepared by the co-assembly of gold nanoparticles and latex spheres on a flat substrate, followed by immersion of the sample in toluene to dissolve the polymer template. Coupling of surface plasmons in the gold nanoparticles that make up the wall structure gives rise to locally strong electric fields. As a result, molecules close to the surface of the film exhibit an enhancement in their Raman spectra. On model compounds, an average enhancement of 10^4 in the Raman signal was observed for spectra taken on the macroporous gold film relative to glass. This performance is similar to that observed for silver gratings produced by e-beam lithography, and an order of magnitude greater than results for gold nanoparticles randomly arranged on glass. These materials may be useful for the determination of trace contaminants or pollutants.

15.3.3
Chemically Functional Materials

While the majority of the research on macroporous materials has been directed to photonic and optical applications, these materials may also find chemical applications in areas traditionally occupied by zeolites or mesoporous materials. The large pores of macroporous materials may be beneficial for transformations or host–guest interactions involving chemical species that are too large for the smaller-pore materials. In addition, macropores are sufficiently large that mass transport is not diffusion limited [12]. Hierarchical porous materials that combine macropores with micro- or mesopores are expected to provide the benefits of each pore size. The macropores would provide efficient transport of guests, while the micro- or mesopores would provide selectivity and a high specific surface area [12]. While chemical transformations have yet to be performed using this type of bimodal porous material, numerous methods for their preparation have been developed. These methods include sol–gel chemistry combining colloidal crystals with molecular [61] or surfactant [30, 31, 38, 54, 59, 60] structure directors, and nanoparticle assembly [80–82] or core-shell [135, 136] methods using zeolitic precursors.

One method that has been utilized for the development of hybrid macroporous materials with adsorption or catalytic activities involves the attachment of chemically functional species to macroporous supports via linking groups. In one example, a direct synthesis (or "one-pot") approach combining sol–gel precursors with colloidal crystal templating was used to attach thiol functional groups to macroporous titania and zirconia supports via siloxane or sulfonate linking groups [158]. These materials were then used as adsorbents to remove heavy metal ions from aqueous solutions. The thiol groups provided sites for binding the metal ions, and the solid support allowed easy removal of the adsorbent from the solution. Leaching the metal ions with an acid wash allowed the material to be reused.

In another direct synthesis approach, polyoxometalate (POM) clusters were attached to macroporous silica via bifunctional siloxane linking groups [159]. The hybrid material contained a high loading of POMs that were nearly molecularly dispersed throughout the walls. The catalytic activity of the POM/silica hybrid was demonstrated by the epoxidation of cyclooctene. In a related approach, POMs were attached to macroporous silica by a post-synthesis grafting technique [160]. Macroporous silica was first functionalized with amines via siloxane linking groups, and then with transition metal-substituted polyoxometalate (TMSP) clusters, which attached datively to the amines. This material was an active catalyst for the epoxidation of cyclohexane, with conversions comparable to similarly prepared mesoporous silicates. Catalytically active thin films of macroporous titania have also been prepared [114, 115]. The photocatalytic activity of this material was confirmed by the conversion of silver ions in solution to metallic silver on the surface of the film upon UV irradiation.

Certain macroporous materials may find applications as chemical sensors. This has been demonstrated for macroporous SnO_2, which was tested as a gas sensor [161]. Semiconducting metal oxides are known to respond to reactive gases at elevated temperatures. The response is based on an increase in the conductivity of the oxide when exposed to a reactive gas, due to a release of trapped electrons when the gas reacts with adsorbed oxygen species. In tests with macroporous SnO_2, rapid and near ideal sensing behavior was observed when the sample was exposed to various gases at 400 °C. With process optimization, macroporous gas sensors are expected to give devices with behaviors that approach the all-surface limit.

Electrochemical power systems are another technology that may benefit by structuring their components into an interconnected macroporous arrangement, since this may facilitate diffusion between interfaces. As a cathode for a solid-oxide fuel cell, macroporous $Sr_{0.5}Sm_{0.5}CoO_3$ was prepared by templated precipitation of a stoichiometric solution of the metal salt precursors, followed by calcination to remove the template and form the mixed oxide [162]. When the macroporous cathode was combined with a gadolinia-doped ceria (GDC) electrolyte and a

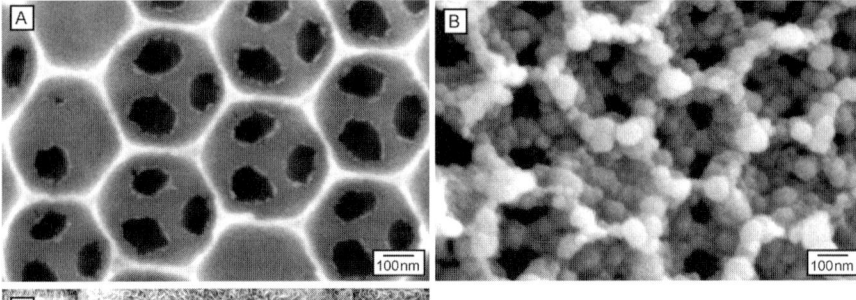

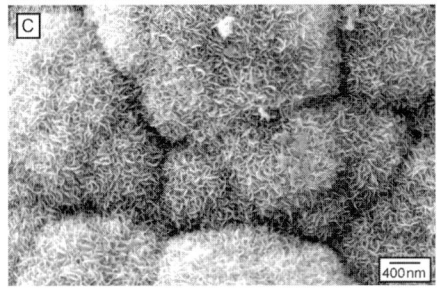

Fig. 15.8 SEM images illustrating the growth of hydroxycarbonate apatite (HCA) from macroporous bioactive glass in simulated body fluid (SBF) [164]. **A** Original sample of macroporous bioactive glass; **B** sample after immersion in SBF for 3 h; **C** sample after immersion in SBF for 4 days. The macroporous bioactive glass nucleates the growth of HCA, and is eventually completely replaced with bulk HCA.

GDC-NiO anode, the cell generated a maximum power density of 267 mW/cm^2 at 600 °C. An assortment of additional conducting macroporous metal oxides (e.g. LiMn$_2$O$_4$, LiCoO$_2$, LiNiO$_2$) has also been prepared for potential use as cathodes and anodes for lithium ion batteries [163].

Macroporous materials may also have medicinal or biological applications. An example is bioactive glass, which is a material that can bond to bone when placed in simulated body fluid through the growth of biologically compatible hydroxycarbonate apatite (HCA). This behavior may allow bioactive glass to be used as a bone graft substitute. A macroporous bioactive glass has been prepared using sol–gel chemistry and colloidal crystal templating to produce a CaO/SiO$_2$ composite [164]. The material was immersed in simulated body fluid to test its ability to nucleate the growth of HCA (Fig. 15.8). After 3 h, particle growth completely covered the walls, and after four days the material was completely replaced with bulk HCA. Tests on nontemplated control samples exhibited slower HCA growth. The improved performance of HCA growth on the macroporous sample was attributed to better access of the fluid to its surface. A related macroporous material with the desired medicinal application of a drug delivery agent has also been prepared [165]. Macroporous hydroxyapatite/calcium phosphate composites were prepared by templated precipitation of a precursor solution of calcium nitrate and phosphoric acid, followed by calcination in air [165]. The macroporous product was capable of adsorbing antibiotics and then releasing them in simulated body fluid. However, the release occurred too rapidly for the material to be used in a controlled drug delivery application.

15.3.4
Templates for Colloid and Hollow Sphere Synthesis

In a reverse of the colloidal crystal templating process, macroporous materials may be used to template the formation of colloids. This method provides for a physical, rather than chemical, means of controlling the size and shape of the colloid. This is an especially valuable method for the preparation of uniform colloidal particles for compositions that are difficult to prepare by chemical methods. This method of templating has been used to prepare colloidal particles or hollow spheres of SiO_2 [130, 166, 167], mesoporous SiO_2 [166], TiO_2 [166–168], ZrO_2 [166, 168], SnO_2 [161, 166], Al_2O_3 [168], Nd_2O_3 [167], polypyrrole [168], PPV [168], CdS [168], AgCl [168], Au [168], Ni [168], and Au [169]. Colloidal particles form when the precursors do not strongly wet the template, whereas hollow spheres form when they do. Macroporous PS templates typically yield hollow spheres, while PMMA yields solid spheres [168].

15.4
Summary and Outlook

In only a short period of time, colloidal crystal templating has far surpassed most other templating techniques with respect to the compositional diversity of ordered porous materials that can be attained. Macroporous materials with nearly any imaginable composition now appear within reach by the methods that have been presented in this chapter. This may lead to the development of new chemical applications for porous materials that were not thought possible before. Going beyond this, macroporous materials have created a new avenue for the pursuit of photonic band gap materials. This unique combination of chemical and optical properties that macroporous materials offer has attracted the interest of a wide variety of sciences, including chemistry, physics, and engineering.

While the methods are in place to prepare many different classes of macroporous materials, numerous challenges remain before these materials find viable applications. For chemical applications, the primary focus will be to take advantage of the large pore size and improved accessibility to active sites offered by macroporous materials. Applications in the areas of catalysis, adsorption, and separation would benefit from the efficient transfer of fluids through the macropores if they could be combined with the selectivity and high specific surface area of micro- or mesoporous walls. Battery materials, sensors, and even biological or medicinal agents are also expected to benefit from a three-dimensional macroporous arrangement. Since methods have already been demonstrated for the preparation of macroporous semiconductors, the primary challenge that remains for the creation of photonic band gap materials is to achieve a greater control over the structure of the material. More specifically, methods for the elimination of unwanted defects and the introduction of controlled defects will need to be developed. Considering that so much has already been accomplished with 3D ordered macroporous materials in only five years of existence, much more is surely yet to come.

15.5
Acknowledgments

Portions of the work described here were funded by 3M, Dupont, the David & Lucile Packard Foundation, the NSF (DMR-9701507), the MRSEC program of the NSF (DMR-9809364), the U.S. Army Research Laboratory and the U.S. Army Research Office (DAAD 19-01-1-0512), and the Office of Naval Research (grant number N00014-01-1-0810, subcontracted from NWU).

15.6
References

1. K.S.W. Sing, D.H. Everett, R.A.W. Haul, L. Moscou, R.A. Pierotti, J. Rouquérol, T. Siemieniewska, *Pure Appl. Chem.* **1985**, *57*, 603–619.
2. H.G. Karge, J. Weitkamp, *Molecular Sieves: Science and Technology*, Springer, Berlin Heidelberg, **1999**.
3. T. Yanagisawa, T. Shimizu, K. Kuroda, C. Kato, *Bull. Chem. Soc. Jpn.* **1990**, *63*, 988–992.
4. C.T. Kresge, M.E. Leonowicz, W.J. Roth, J.C. Vartuli, J.S. Beck, *Nature* **1992**, *359*, 710–712.
5. J.S. Beck, J.C. Vartuli, W.J. Roth, M.E. Leonowicz, C.T. Kresge, K.D. Schmitt, C.T.-W. Chu, D.H. Olson, E.W. Sheppard, S.B. McCullen, J.B. Higgins, J.L. Schlenker, *J. Am. Chem. Soc.* **1992**, *114*, 10834–10843.
6. S. Inagaki, Y. Fukushima, K. Kuroda, *J. Chem. Soc., Chem. Commun.* **1993**, 680–682.
7. M. Templin, A. Franck, A. Du Chesne, H. Leist, Y. Zhang, R. Ulrich, V. Schädler, U. Wiesner, *Science* **1997**, *278*, 1795–1798.
8. D. Zhao, J. Feng, Q. Huo, N. Melosh, G.H. Fredrickson, B.F. Chmelka, G.D. Stucky, *Science* **1998**, *279*, 548–552.
9. A. Stein, B.J. Melde, R.C. Schroden, *Adv. Mater.* **2000**, *12*, 1403–1419.
10. X. Feng, G.E. Fryxell, L.-Q. Wang, A.Y. Kim, J. Liu, K.M. Kemner, *Science* **1997**, *276*, 923–926.
11. K. Kageyama, J. Tamazawa, T. Aida, *Science* **1999**, *285*, 2113–2115.
12. U.A. El-Nafaty, R. Mann, *Chem. Eng. Sci.* **1999**, *54*, 3475–3484.
13. E. Yablonovitch, *Phys. Rev. Lett.* **1987**, *58*, 2059–2062.
14. S. John, *Phys. Rev. Lett.* **1987**, *58*, 2486–2489.
15. E. Yablonovitch, *J. Opt. Soc. Am. B* **1993**, *10*, 283–295.
16. J.D. Joannopoulos, P.R. Villeneuve, S. Fan, *Nature* **1997**, *386*, 143–149.
17. S.-Y. Lin, E. Chow, V. Hietala, P.R. Villeneuve, J.D. Joannopoulos, *Science* **1998**, *282*, 274–276.
18. Y. Xia, B. Gates, Y. Yin, Y. Lu, *Adv. Mater.* **2000**, *12*, 693–713.
19. A.D. Dinsmore, J.C. Crocker, A.G. Yodh, *Curr. Opin. Colloid Interface Sci.* **1998**, *3*, 5–11.
20. O.D. Velev, E.W. Kaler, *Adv. Mater.* **2000**, *12*, 531–534.
21. K.M. Kulinowski, P. Jiang, H. Vaswani, V.L. Colvin, *Adv. Mater.* **2000**, *12*, 833–838.
22. D.J. Norris, Y.A. Vlasov, *Adv. Mater.* **2001**, *13*, 371–376.
23. A. Stein, *Micropor. Mesopor. Mat.* **2001**, *44/45*, 227–239.

24 O. D. Velev, A. M. Lenhoff, Curr. Opin. Colloid Interface Sci. **2000**, *5*, 56–63.
25 A. Stein, R. C. Schroden, Curr. Opin. Solid State Mat. Sci. **2001**, *5*, 553–564.
26 C. J. Brinker, G. W. Scherer, *Sol-Gel Science: The Physics and Chemistry of Sol-Gel Processing*, Academic Press, San Diego, **1990**.
27 O. D. Velev, T. A. Jede, R. F. Lobo, A. M. Lenhoff, Nature **1997**, *389*, 447–448.
28 O. D. Velev, T. A. Jede, R. F. Lobo, A. M. Lenhoff, Chem. Mater. **1998**, *10*, 3597–3602.
29 M. Antonietti, B. Berton, C. Göltner, H. P. Hentze, Adv. Mater. **1998**, *10*, 154–159.
30 P. Yang, T. Deng, D. Zhao, P. Feng, D. Pine, B. F. Chmelka, G. M. Whitesides, G. D. Stucky, Science **1998**, *282*, 2244–2246.
31 B. T. Holland, C. F. Blanford, T. Do, A. Stein, Chem. Mater. **1999**, *11*, 795–805.
32 G. Gundiah, C. N. R. Rao, Solid State Sci. **2000**, *2*, 877–882.
33 Z. Zhong, Y. Yin, B. Gates, Y. Xia, Adv. Mater. **2000**, *12*, 206–209.
34 C. F. Blanford, R. C. Schroden, M. Al-Daous, A. Stein, Adv. Mater. **2001**, *13*, 26–29.
35 R. C. Schroden, M. Al-Daous, A. Stein, Chem. Mater. **2001**, *13*, 2945–2950.
36 S. Vaudreuil, M. Bousmina, S. Kaliaguine, L. Bonneviot, Adv. Mater. **2001**, *13*, 1310–1312.
37 A. N. Khramov, M. M. Collinson, Chem. Commun. **2001**, 767–768.
38 P. Yang, A. H. Rizvi, B. Messer, B. F. Chmelka, G. M. Whitesides, G. D. Stucky, Adv. Mater. **2001**, *13*, 427–431.
39 R. C. Schroden, M. Al-Daous, C. F. Blanford, A. Stein, **2002**, Chem. Mater. **2002**, *14*, 3305–3315
40 B. T. Holland, C. F. Blanford, A. Stein, Science **1998**, *281*, 538–540.
41 J. E. G. J. Wijnhoven, W. L. Vos, Science **1998**, *281*, 802–804.
42 J. S. Yin, Z. L. Wang, Adv. Mater. **1999**, *11*, 469–472.
43 A. Richel, N. P. Johnson, D. W. McComb, Appl. Phys. Lett. **2000**, *76*, 1816–1818.
44 A. F. Koenderink, M. Megens, G. van Soest, W. L. Vos, A. Lagendijk, Phys. Lett. A **2000**, *268*, 104–111.
45 J. E. G. J. Wijnhoven, L. Bechger, W. L. Vos, Chem. Mater. **2001**, *13*, 4486–4499.
46 M. E. Turner, T. J. Trentler, V. L. Colvin, Adv. Mater. **2001**, *13*, 180–183.
47 P. Ni, P. Dong, B. Cheng, X. Li, D. Zhang, Adv. Mater. **2001**, *13*, 437–441.
48 D. W. McComb, B. M. Treble, C. J. Smith, R. M. De La Rue, N. P. Johnson, J. Mater. Chem. **2001**, *11*, 142–148.
49 N. P. Johnson, D. W. McComb, A. Richel, B. M. Treble, R. M. De La Ru, Synth. Metals **2001**, *116*, 469–473.
50 Y. Zhang, Z. Lei, J. Li, S. Lu, New J. Chem. **2001**, *25*, 1118–1120.
51 Z. Lei, J. Li, Y. Zhang, S. Lu, J. Mater. Chem. **2000**, *10*, 2629–2631.
52 I. Soten, H. Miguez, S. M. Yang, S. Petrov, N. Coombs, N. Tetreault, N. Matsuura, H. E. Ruda, G. A. Ozin, Adv. Funct. Mater. **2002**, *12*, 71–77.
53 P. Harkins, D. Eustace, J. Gallagher, D. W. McComb, J. Mater. Chem. **2002**, *12*, 1247–1249.
54 B. Lebeau, C. E. Fowler, S. Mann, C. Farcet, B. Charleux, C. Sanchez, J. Mater. Chem. **2000**, *10*, 2105–2108.
55 M. A. Carreon, V. V. Guliants, Chem. Commun. **2001**, 1438–1439.
56 M. A. Carreon, V. V. Guliants, Chem. Mater. **2002**, *14*, 2670–2675.
57 A. A. Zakhidov, R. H. Baughman, Z. Iqbal, C. Cui, I. Khayrullin, S. O. Dantas, J. Marti, V. G. Ralchenko, Science **1998**, *282*, 897–901.
58 C. F. Blanford, H. Yan, R. C. Schroden, M. Al-Daous, A. Stein, Adv. Mater. **2001**, *13*, 401–407.
59 J. S. Yin, Z. L. Wang, Appl. Phys. Lett. **1999**, *74*, 2629–2631.
60 Q. Luo, L. Li, B. Yang, D. Zhao, Chem. Lett. **2000**, 378–379.
61 B. T. Holland, L. Abrams, A. Stein, J. Am. Chem. Soc. **1999**, *121*, 4308–4309.
62 G.-R. Yi, J. H. Moon, S.-M. Yang, Chem. Mater. **2001**, *13*, 2613–2618.
63 M. Al-Daous, A. Stein, unpublished results.
64 H. Yan, C. F. Blanford, B. T. Holland, M. Parent, W. H. Smyrl, A. Stein, Adv. Mater. **1999**, *11*, 1003–1006.
65 H. Yan, C. F. Blanford, W. H. Smyrl, A. Stein, Chem. Commun. **2000**, 1477–1478.

66 H. Yan, C. F. Blanford, B. T. Holland, W. H. Smyrl, A. Stein, *Chem. Mater.* **2000**, *12*, 1134–1141.

67 H. Yan, C. F. Blanford, J. C. Lytle, C. B. Carter, W. H. Smyrl, A. Stein, *Chem. Mater.* **2001**, *13*, 4314–4321.

68 G. L. Egan, J. Yu, C. H. Kim, S. J. Lee, R. E. Schaak, T. E. Mallouk, *Adv. Mater.* **2000**, *12*, 1040–1042.

69 H. Míguez, F. Meseguer, C. López, M. Holgado, G. Andreasen, A. Mifsud, V. Fornés, *Langmuir* **2000**, *16*, 4405–4408.

70 O. D. Velev, P. M. Tessier, A. M. Lenhoff, E. W. Kaler, *Nature* **1999**, *401*, 548.

71 P. M. Tessier, O. D. Velev, A. T. Kalambur, J. F. Rabolt, A. M. Lenhoff, E. W. Kaler, *J. Am. Chem. Soc.* **2000**, *122*, 9554–9555.

72 P. M. Tessier, O. D. Velev, A. T. Kalambur, A. M. Lenhoff, J. F. Rabolt, E. W. Kaler, *Adv. Mater.* **2001**, *13*, 396–400.

73 G. Subramanian, K. Constant, R. Biswas, M. M. Sigalas, K.-M. Ho, *Appl. Phys. Lett.* **1999**, *74*, 3933–3935.

74 G. Subramanian, V. N. Manoharan, J. D. Thorne, D. J. Pine, *Adv. Mater.* **1999**, *11*, 1261–1265.

75 G. Subramanian, K. Constant, R. Biswas, M. M. Sigalas, K. Ho, *Adv. Mater.* **2001**, *13*, 443–446.

76 G. Subramanian, K. Constant, R. Biswas, M. M. Sigalas, K.-M. Ho, *Synth. Metals* **2001**, *116*, 445–448.

77 Q.-B. Meng, C.-H. Fu, Y. Einaga, Z.-Z. Gu, A. Fujishima, O. Sato, *Chem. Mater.* **2002**, *14*, 83–88.

78 Z.-Z. Gu, S. Kubo, W. Qian, Y. Einaga, D. A. Tryk, A. Fujishima, O. Sato, *Langmuir* **2001**, *17*, 6751–6753.

79 B. Gates, Y. Xia, *Adv. Mater.* **2001**, *13*, 1605–1608.

80 L. Huang, Z. Wang, J. Sun, L. Miao, Q. Li, Y. Yan, D. Zhao, *J. Am. Chem. Soc.* **2000**, *122*, 3530–3531.

81 Y. J. Wang, Y. Tang, Z. Ni, W. M. Hua, W. L. Yang, X. D. Wang, W. C. Tao, Z. Gao, *Chem. Lett.* **2000**, 510–511.

82 G. Zhu, S. Qiu, F. Gao, D. Li, Y. Li, R. Wang, B. Gao, B. Li, Y. Guo, R. Xu, Z. Liu, O. Terasaki, *J. Mater. Chem.* **2001**, *11*, 1687–1693.

83 Y. A. Vlasov, N. Yao, D. J. Norris, *Adv. Mater.* **1999**, *11*, 165–169.

84 Z. Lei, J. Li, Y. Ke, Y. Zhang, H. Wang, G. He, *J. Mater. Chem.* **2001**, *11*, 1778–1780.

85 A. Imhof, D. J. Pine, *Nature* **1997**, *389*, 948–951.

86 A. Imhof, D. J. Pine, *Adv. Mater.* **1998**, *10*, 697–700.

87 V. N. Manoharan, A. Imhof, J. D. Thorne, D. J. Pine, *Adv. Mater.* **2001**, *13*, 447–450.

88 M. Srinivasarao, D. Collings, A. Philips, S. Patel, *Science* **2001**, *292*, 79–83.

89 P. V. Braun, P. Wiltzius, *Nature* **1999**, *402*, 603–604.

90 P. V. Braun, P. Wiltzius, *Adv. Mater.* **2001**, *13*, 482–485.

91 J. E. G. J. Wijnhoven, S. J. M. Zevenhuizen, M. A. Hendriks, D. Vanmaekelbergh, J. J. Kelly, W. L. Vos, *Adv. Mater.* **2000**, *12*, 888–890.

92 L. Xu, W. L. Zhou, C. Frommen, R. H. Baughman, A. A. Zakhidov, L. Malkinski, J.-Q. Wang, J. B. Wiley, *Chem. Commun.* **2000**, 997–998.

93 M. C. Netti, S. Coyle, J. J. Baumberg, M. A. Ghanem, P. R. Birkin, P. N. Bartlett, D. M. Whittaker, *Adv. Mater.* **2001**, *13*, 1368–1370.

94 P. N. Bartlett, J. J. Baumberg, P. R. Birkin, M. A. Ghanem, M. C. Netti, *Chem. Mater.* **2002**, *14*, 2199–2208.

95 Q. Luo, Z. Liu, L. Li, S. Xie, J. Kong, D. Zhao, *Adv. Mater.* **2001**, *13*, 286–289.

96 T. Sumida, Y. Wada, T. Kitamura, S. Yanagida, *Langmuir* **2002**, *18*, 3886–3894.

97 P. N. Bartlett, P. R. Birkin, M. A. Ghanem, *Chem. Commun.* **2000**, 1671–1672.

98 T. Sumida, Y. Wada, T. Kitamura, S. Yanagida, *Chem. Lett.* **2001**, 38–39.

99 T. Sumida, Y. Wada, T. Kitamura, S. Yanagida, *Chem. Lett.* **2002**, 180–181.

100 T. Sumida, Y. Wada, T. Kitamura, S. Yanagida, *Chem. Commun.* **2000**, 1613–1614.

101 P. N. Bartlett, P. R. Birkin, M. A. Ghanem, C. Toh, *J. Mater. Chem.* **2001**, *11*, 849–853.

102 T. Cassagneau, F. Caruso, *Adv. Mater.* **2002**, *14*, 34–38.

103 Z. Lei, H. Zhang, S. Ma, Y. Ke, J. Li, F. Li, *Chem. Commun.* **2002**, 676–677.

104 P. Jiang, J. Cizeron, J. F. Bertone, V. L. Colvin, *J. Am. Chem. Soc.* **1999**, *121*, 7957–7958.

105 A. Blanco, E. Chomski, S. Grabtchak, M. Ibisate, S. John, S.W. Leonard, C. Lopez, F. Meseguer, H. Miguez, J.P. Mondia, G.A. Ozin, O. Toader, H.M. van Driel, *Nature* **2000**, *405*, 437–440.

106 E. Chomski, G.A. Ozin, *Adv. Mater.* **2000**, *12*, 1071–1078.

107 Y.A. Vlasov, X.-Z. Bo, J.C. Sturm, D.J. Norris, *Nature* **2001**, *414*, 289–293.

108 H. Míguez, E. Chomski, F. García-Santamaría, M. Ibisate, S. John, C. López, F. Meseguer, J.P. Mondia, G.A. Ozin, O. Toader, H.M. van Driel, *Adv. Mater.* **2001**, *13*, 1634–1637.

109 M. Müller, R. Zentel, T. Maka, S.G. Romanov, C.M. Sotomayor Torres, *Adv. Mater.* **2000**, *12*, 1499–1503.

110 S.G. Romanov, T. Maka, C.M. Sotomayor Torres, M. Müller, R. Zentel, *Synth. Metals* **2001**, *116*, 475–479.

111 H. Kajii, Y. Kawagishi, H. Take, K. Yoshino, A.A. Zakhidov, R.H. Baughman, *J. Appl. Phys.* **2000**, *88*, 758–763.

112 G. Gundiah, A. Govindaraj, C.N.R. Rao, *Mater. Res. Bull.* **2001**, *36*, 1751–1757.

113 Z. Lei, Y. Zhang, H. Wang, Y. Ke, J. Li, F. Li, J. Xing, *J. Mater. Chem.* **2001**, *11*, 1975–1977.

114 S. Matsushita, T. Miwa, A. Fujishima, *Chem. Lett.* **1997**, 925–926.

115 S.I. Matsushita, T. Miwa, D.A. Tryk, A. Fujishima, *Langmuir* **1998**, *14*, 6441–6447.

116 N. Eradat, J.D. Huang, Z.V. Vardeny, A.A. Zakhidov, I. Khayrullin, I. Udod, R.H. Baughman, *Synth. Metals* **2001**, *116*, 501–504.

117 P.V. Braun, R.W. Zehner, C.A. White, M.K. Weldon, C. Kloc, S.S. Patel, P. Wiltzius, *Adv. Mater.* **2001**, *13*, 721–724.

118 P.V. Braun, R.W. Zehner, C.A. White, M.K. Weldon, C. Kloc, S.S. Patel, P. Wiltzius, *Europhys. Lett.* **2001**, *56*, 207–213.

119 H.-P. Hentze, M. Antonietti, *Curr. Opin. Solid State Mat. Sci.* **2001**, *5*, 343–353.

120 S.H. Park, Y. Xia, *Adv. Mater.* **1998**, *10*, 1045–1048.

121 S.H. Park, Y. Xia, *Chem. Mater.* **1998**, *10*, 1745–1747.

122 B. Gates, Y. Yin, Y. Xia, *Chem. Mater.* **1999**, *11*, 2827–2836.

123 P. Jiang, K.S. Hwang, D.M. Mittleman, J.F. Bertone, V.L. Colvin, *J. Am. Chem. Soc.* **1999**, *121*, 11630–11637.

124 P. Jiang, G.N. Ostojic, R. Narat, D.M. Mittleman, V.L. Colvin, *Adv. Mater.* **2001**, *13*, 389–393.

125 H. Míguez, F. Meseguer, C. López, F. López-Tejeira, J. Sánchez-Dehesa, *Adv. Mater.* **2001**, *13*, 393–396.

126 J.F. Bertone, P. Jiang, K.S. Hwang, D.M. Mittleman, V.L. Colvin, *Phys. Rev. Lett.* **1999**, *83*, 300–303.

127 M. Deutsch, Y.A. Vlasov, D.J. Norris, *Adv. Mater.* **2000**, *12*, 1176–1180.

128 D. Wang, F. Caruso, *Adv. Mater.* **2001**, *13*, 350–353.

129 B. Rodríguez-González, V. Salgueiriño-Maceira, F. García-Santamaría, L.M. Liz-Marzán, *Nano Lett.* **2002**, *2*, 471–473.

130 S.A. Johnson, P.J. Ollivier, T.E. Mallouk, *Science* **1999**, *283*, 963–965.

131 W. Qian, Z.-Z. Gu, A. Fujishima, O. Sato, *Langmuir* **2002**, *18*, 4526–4529.

132 K. Yoshino, Y. Kawagishi, S. Tatsuhara, H. Kajii, S. Lee, A. Fujii, M. Ozaki, A.A. Zakhidov, Z.V. Vardeny, M. Ishikawa, *Microelec. Eng.* **1999**, *47*, 49–53.

133 T.-B. Xu, Z.-Y. Cheng, Q.M. Zhang, R.H. Baughman, C. Cui, A.A. Zakhidov, J. Su, *J. Appl. Phys.* **2000**, *88*, 405–409.

134 S. Vaucher, E. Dujardin, B. Lebeau, S.R. Hall, S. Mann, *Chem. Mater.* **2001**, *13*, 4408–4410.

135 K.H. Rhodes, S.A. Davis, F. Caruso, B. Zhang, S. Mann, *Chem. Mater.* **2000**, *12*, 2832–2834.

136 S.A. Davis, M. Breulmann, K.H. Rhodes, B. Zhang, S. Mann, *Chem. Mater.* **2001**, *13*, 3218–3226.

137 D. Wang, R.A. Caruso, F. Caruso, *Chem. Mater.* **2001**, *13*, 364–371.

138 S. Han, X. Shi, F. Zhou, *Nano Lett.* **2002**, *2*, 97–100.

139 D. Wang, F. Caruso, *Chem. Commun.* **2001**, 489–490.

140 Y. Chen, W.T. Ford, N.F. Materer, D. Teeters, *J. Am. Chem. Soc.* **2000**, *122*, 10472–10473.

141 Y. Chen, W.T. Ford, N.F. Materer, D. Teeters, *Chem. Mater.* **2001**, *13*, 2697–2704.

142 Y. Xia, B. Gates, Z. Li, *Adv. Mater.* **2001**, *13*, 409–413.
143 A. Birner, R. B. Wehrspohn, U. M. Gösele, K. Busch, *Adv. Mater.* **2001**, *13*, 377–388.
144 G. A. Ozin, S. M. Yang, *Adv. Funct. Mater.* **2001**, *11*, 95–104.
145 T. F. Krauss, R. M. De La Rue, *Prog. Quant. Elec.* **1999**, *23*, 51–96.
146 K. Busch, S. John, *Phys. Rev. E* **1998**, *58*, 3896–3908.
147 R. Biswas, M. M. Sigalas, G. Subramanian, C. M. Soukoulis, K.-M. Ho, *Phys. Rev. B* **2000**, *61*, 4549–4553.
148 V. Yannopapas, N. Stefanou, A. Modinos, *Phys. Rev. Lett.* **2001**, *86*, 4811–4814.
149 Z.-Y. Li, Z.-Q. Zhang, *Phys. Rev. B* **2000**, *62*, 1516–1519.
150 Z. Li, Z. Zhang, *Adv. Mater.* **2001**, *13*, 433–436.
151 H. P. Schriemer, H. M. van Driel, A. F. Koenderink, W. L. Vos, *Phys. Rev. A* **2000**, *63*, 011801/1–011801/4.
152 S. G. Romanov, T. Maka, C. M. Sotomayor Torres, M. Müller, R. Zentel, *Appl. Phys. Lett.* **2001**, *79*, 731–733.
153 E. Yablonovitch, *Sci. Am.* **2001**, *285*, 46–55.
154 M. S. Thijssen, R. Sprik, J. E. G. J. Wijnhoven, M. Megens, T. Narayanan, A. Lagendijk, W. L. Vos, *Phys. Rev. Lett.* **1999**, *83*, 2730–2733.
155 W. Lee, S. A. Pruzinsky, P. V. Braun, *Adv. Mater.* **2002**, *14*, 271–274.
156 K. Busch, S. John, *Phys. Rev. Lett.* **1999**, *83*, 967–970.
157 M. Ozaki, Y. Shimoda, M. Kasano, K. Yoshino, *Adv. Mater.* **2002**, *14*, 514–518.
158 R. C. Schroden, M. Al-Daous, S. Sokolov, B. J. Melde, J. C. Lytle, M. C. Carbajo, J. T. Fernández, E. E. Rodríguez, A. Stein, *J. Mater. Chem.* **2002**, *12*, 3261–3267.
159 R. C. Schroden, C. F. Blanford, B. J. Melde, B. J. S. Johnson, A. Stein, *Chem. Mater.* **2001**, *13*, 1074–1081.
160 B. J. S. Johnson, A. Stein, *Inorg. Chem.* **2001**, *40*, 801–808.
161 R. W. J. Scott, S. M. Yang, G. Chabanis, N. Coombs, D. E. Williams, G. A. Ozin, *Adv. Mater.* **2001**, *13*, 1468–1472.
162 F. Chen, C. Xia, M. Liu, *Chem. Lett.* **2001**, 1032–1033.
163 H. Yan, S. Sokolov, J. C. Lytle, A. Stein, F. Zhang, W. H. Smyrl, *J. Electrochem. Soc.* **2003**, *150*, A1102–A1107.
164 H. Yan, K. Zhang, C. F. Blanford, L. F. Francis, A. Stein, *Chem. Mater.* **2001**, *13*, 1374–1382.
165 B. J. Melde, A. Stein, *Chem. Mater.* **2002**, *14*, 3326–3331.
166 S. M. Yang, N. Coombs, G. A. Ozin, *Adv. Mater.* **2000**, *12*, 1940–1944.
167 Z. Lei, J. Li, Y. Ke, Y. Zhang, H. Zhang, F. Li, J. Xing, *J. Mater. Chem.* **2001**, *11*, 2930–2933.
168 P. Jiang, J. F. Bertone, V. L. Colvin, *Science* **2001**, *291*, 453–457.
169 L. Xu, W. Zhou, M. E. Kozlov, I. I. Khayrullin, I. Udod, A. A. Zakhidov, R. H. Baughman, J. B. Wiley, *J. Am. Chem. Soc.* **2001**, *123*, 763–764.

16
Semiconductor Quantum Dots as Multicolor and Ultrasensitive Biological Labels

Warren C. W. Chan, Xiaohu Gao, and Shuming Nie

16.1
Introduction

Colloidal metal and semiconductor particles on the nanometer scale have unique optical, electronic, magnetic, and/or structural properties that are not available in either isolated molecules or bulk solids [1–8]. These properties are currently under intense study for potential use in microelectronics, quantum dot lasers, data storage, and a host of other applications. Recent advances have led to large-scale preparation of highly monodisperse nanoparticles [9–11], characterization of their lattice structures [3–5] and fabrication of nanoparticle arrays [12–15] and light-emitting diodes [16, 17]. These studies have mainly focused on the fundamental aspects of nanostructured materials and their applications in electronics and optics. However, recent developments indicate that the first *practical* applications of quantum dots (QDs) are occurring in biology and medicine [18, 19]. Key advances that have enabled these biological applications include (1) the synthesis of highly luminescent quantum dots in large quantities [20], (2) a reasonable understanding of the surface chemistry [21], (3) the preparation of water-soluble and biocompatible nanocrystals [22, 23], and (4) the incorporation of multicolor quantum dots into micro- and nanobeads for multiplexed optical encoding of biomolecules [24].

In this chapter, we discuss recent advances in luminescent semiconductor quantum dots and their emerging applications in biology and medicine. In comparison with organic dyes and fluorescent proteins, semiconductor quantum dots represent a new class of fluorescent labels with unique advantages and applications. For example, the fluorescence emission spectra of quantum dots can be continuously tuned by changing the particle size, and a single wavelength can be used for simultaneous excitation of all different-sized quantum dots [24]. Also, surface-passivated QDs are highly stable against photobleaching and have narrow, symmetric emission peaks (25–30 nm full width at half maximum (FWHM)). It has been estimated that CdSe quantum dots are about 20 times brighter and 100 times more stable than single rhodamine 6G molecules [23]. These properties are expected to open new possibilities in several research areas, such as single-molecule biophysics, multiplexed medical diagnostics, and high-throughput drug screening.

Colloids and Colloid Assemblies. Edited by Frank Caruso
Copyright © 2004 Wiley-VCH Verlag GmbH & Co. KGaA, Weinheim
ISBN: 3-527-30660-9

16.2
Synthesis and Surface Chemistry

Earlier attempts to synthesize quantum dots were conducted in aqueous environments with stabilizing agents such as thioglycerol and polyphosphate [1–5]. However, the resulting quantum dots showed poor quantum yields (<10%) and broad size distributions (relative standard deviation (RSD) >15%). In 1993, Bawendi and coworkers [25] reported a high-temperature organometallic procedure for quantum-dot synthesis. This method was later improved by three independent research groups [26–28], yielding nearly perfect nanocrystals with quantum yields as high as 50% at room temperature, and a particle size distribution as narrow as 5%.

To prepare type II–VI quantum dots, a metal precursor (such as dimethyl cadmium) and a chalcogenide compound (such as selenium) are first dissolved in tri-n-butyl phosphine (TBP) or tri-n-octyl phosphine (TOP), and then injected into a hot coordinating solvent such as tri-n-octylphosphine oxide (TOPO) at 340 to 360 °C. Recent work by Peng and coworkers showed that high-quality nanocrystals could also be prepared using CdO as an inexpensive starting material [20, 29, 30]. The nanocrystal size can be tuned by heating quantum dots in TOPO at 300 °C for an extended period of time (ranging from two minutes to two weeks, depending on the desired particle size), in which the quantum dots grow by Ostwald ripening. In this process, smaller nanocrystals are broken down, and the dissolved atoms are transferred to larger nanocrystals. The rate of growth is dependent upon temperature and the amount of limiting reagents [31, 32]. Alternatively, continuous injection of organometal/chalcogenide precursors at 300 °C can be used to increase the size of quantum dots [33]. Fig. 16.1 shows the fluorescence emissions observed from a series of 10 different-sized ZnS-capped CdSe quantum dots excited with a handheld UV-lamp.

In addition, the nanocrystal shape can be controlled by manipulating the kinetics of particle growth. This can be achieved by changing the chemical composition of the solvent (e.g. addition of hexylphosphonic acid and tetradecylphosphonic acid to pure TOPO). Successful synthesis of rod-shaped CdSe quantum dots (with aspect ratios of 1:1 to 1:30) has recently been reported using these modifications [34, 35]. For improved optical properties, the quantum dots are often coated and passivated by a thin layer of a higher bandgap material. For example, the fluorescence quantum yields of CdSe quantum dots increase from 5% to 50% with 1–2 monolayers of ZnS capping [26–28]. At present, ZnS and CdS are most commonly used to cap CdSe quantum dots. The bandgap energy of bulk CdS is about 0.9 eV higher than that of CdSe, while the ZnS and CdSe bond lengths are similar; these conditions lead to the epitaxial growth of a smooth ZnS layer on the surface of CdSe core particles. A similar procedure has been used to synthesize group III–V nanocrystals such as InP and InAs [36–39].

The surface of quantum dots has been studied by NMR and XPS spectroscopies [40, 41]. As revealed by high-resolution TEM, the quantum dots are not smooth spheres, but are faceted crystals with planes and edges. They are often negatively charged by adsorbed surface species. Two methods have been reported for convert-

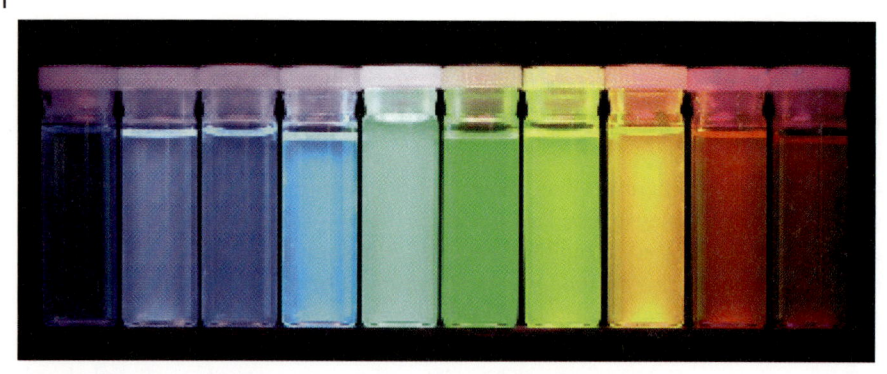

Fig. 16.1 Ten distinguishable emission colors of ZnS-capped CdSe quantum dots excited with a near-UV lamp. From left to right (blue to red), the emission maxima are located at 443, 473, 481, 500, 518, 543, 565, 587, 610, and 655 nm (adapted with permission from Han, Gao, Su, and Nie, Nat. Biotechnol. 19, 631, 2001).

ing hydrophobic quantum dots to water-soluble and biocompatible nanocrystals. In one procedure, mercaptopropyl trimethoxysilane (MPS) adsorbs on the quantum dot surface, and displaces the surface-bound TOPO molecules [22]. A silica shell is formed on the surface by introduction of a base and then hydrolysis of the MPS silanol groups. The polymerized silanol groups help stabilize nanocrystals against flocculation, and render the dots soluble in intermediately polar solvents such as methanol and dimethyl sulfoxide. Further reaction of bifunctional methoxy molecules, such as aminopropyltrimethoxysilane and trimethoxysilyl propyl urea, makes the dots more polar and soluble in aqueous solution.

Direct adsorption of bifunctional molecules such as mercaptoacetic acid and dithiothreitol also leads to water-soluble quantum dots [23]. Mercapto compounds and organic bases are added to TOPO-quantum dots dissolved in organic solvents. The base deprotonates the mercapto functional group and carboxylic acid (in the case of mercaptoacetic acid), which leads to a favorable electrostatic binding between negatively-charged thiols and the positively charged metal atoms. The quantum dots precipitate out of solution and can be redissolved in aqueous solution (pH > 5). The presence of highly polar functional groups, such as –COOH, –OH, or –SO$_3$Na (from bifunctional mercapto molecules) makes the nanocrystals soluble in water.

Polymerized silane quantum dots are stable against flocculation, but only small quantities of water-soluble quantum dots (in the milligram range) can be prepared per batch. Also, residual silanol groups on the nanocrystal surface can cause precipitation and gel formation at neutral pH. In comparison, the mercaptoacetic acid procedure yields large quantities of water-soluble dots, but slow desorption of the mercaptomolecules often causes dot aggregation and precipitation. We have solved this problem by using chemically modified BSA (bovine serum albumin) for surface passivation and bioconjugation of water-soluble quantum dots

[21]. This simple procedure not only yields highly stable quantum dots, but also restores the fluorescence quantum yields to the original values (as measured in chloroform). Furthermore, the BSA layer contains functional groups for covalent conjugation to other biomolecules.

Another approach for linking biomolecules onto the particle's surface is to use an exchange reaction, in which mercapto-coated quantum dots are mixed with thiolated biomolecules (such as oligonucleotides and proteins). After overnight incubation at room temperature, a chemical equilibrium is reached between the thiolated molecules in solution and on the quantum dot surface. This method has been used to adsorb oligonucleotides and biotinylated proteins onto the surface of quantum dots [42, 43]. A similar approach has been used by Mattoussi et al., in which engineered proteins are adsorbed onto the negatively charged surface of quantum dots through electrostatic binding [44].

16.3
Optical Properties

The optical properties of semiconductor nanoclusters arise from interactions between electrons, holes, and their local environments. Semiconductor quantum dots absorb photons when the energy of excitation exceeds the bandgap energy. During this process, electrons are promoted from the valence band to the conduction band. Measurements of UV-vis spectra reveal a large number of energy states in quantum dots. The lowest excited energy state is shown by the first observable peak (also known as the quantum-confinement peak), at a shorter wavelength than the fluorescence emission peak. Excitation at shorter wavelengths is possible because multiple electronic states are present at higher energy levels. In fact, the molar extinction coefficient gradually increases toward shorter wavelengths (Fig. 16.2). This is an important feature for biological applications because it allows simultaneous excitation of multicolor quantum dots with a single light source.

Light emission arises from the recombination of mobile or trapped charge carriers. The emission from mobile carriers is called excitonic fluorescence, and is observed as a sharp peak. The emission spectra of single ZnS-capped CdSe quantum dots are as narrow as 13 nm (FWHM) at room temperature [23]. Defect states in the crystal interior or on its surface can trap the mobile charge carriers (electrons or holes), leading to a broad emission peak that is red-shifted from the excitonic peak. Nanocrystals with a large number of trap states generally have low quantum yields, but surface capping or passivation can remove these defect sites and improve the fluorescence quantum yields (as discussed in the previous section).

Excitonic fluorescence is dependent on the nanocrystal size. Research in several groups has demonstrated an approximately linear relationship between the particle size and the bandgap energy [26, 45]. This quantum-size effect is similar to that observed for a "particle in a box." Outside of the box, the potential energy is

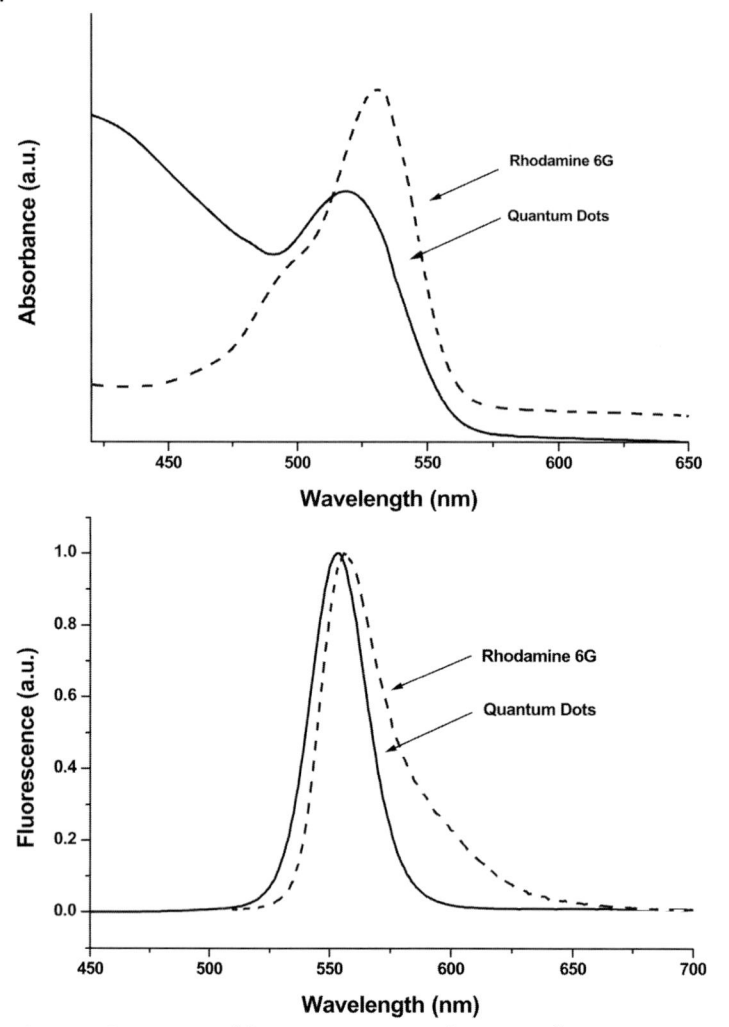

Fig. 16.2 Comparison of the excitation (top) and emission (bottom) profiles between rhodamine 6G (red curves) and CdSe quantum dots (black curves) (data taken from W. Chan, PhD Thesis, Department of Chemistry, Indiana University, 2001).

considered to be infinitely high. Thus, mobile carriers (similar to the particle) are confined within the dimensions of the nanocrystal (similar to the box) with discrete wavefunctions and energy levels. As the physical dimensions of the box become smaller, the bandgap energy becomes higher. For CdSe nanocrystals, the sizes of 2.5 nm and 5.5 nm correspond to fluorescence emission at 500 nm and 620 nm, respectively. In addition to size, the emission wavelength can be varied by changing the semiconductor material. For example, InP and InAs quantum dots usually emit in the far-red and near-infrared [36–39], while CdS and ZnSe dots often emit in the blue or near-UV [46]. It is also interesting to note that elongated quantum dots (called quantum rods) show linearly polarized emission [47], whereas the fluorescence emission from spherical CdSe dots is circularly polarized or not polarized [48, 49].

In comparison to organic dyes such as rhodamine 6G and fluorescein, CdSe nanocrystals show similar or slightly lower quantum yields at room temperature. The lower quantum yields of nanocrystals are compensated by their larger absorption cross-sections and much reduced photobleaching rates. Bawendi and coworkers [25, 26] estimated that the molar extinction coefficients of CdSe quantum dots are about 10^5 to 10^6 M^{-1} cm^{-1}, depending on the particle size and the excitation wavelength. These values are 10–100 times larger than those of organic dyes, but are similar to the absorption cross-sections of phycoerythrin, a multi-chromophore fluorescent protein. Chan and Nie [23] have estimated that single ZnS-capped CdSe quantum dots are ~20 times brighter than single rhodamine 6G molecules. Similarly, phycoerythrin is estimated to be 20 times brighter than fluorescein [50].

Another attractive feature of using quantum dots as biological labels is their high photostability. Gerion et al. examined the photobleaching rate of silica-coated ZnS-capped CdSe quantum dots against that of rhodamine 6G [51]. The quantum-dot emission stayed constant for 4 h, while rhodamine 6G was photobleached after only 10 min. It has been suggested that capped CdSe nanocrystals are 100–

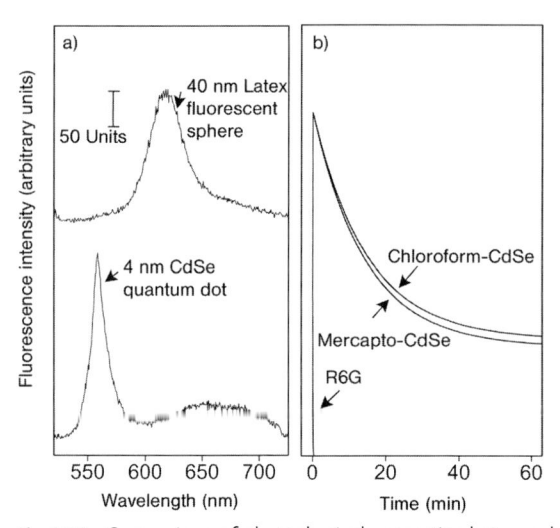

Fig. 16.3 Comparison of photophysical properties between luminescent quantum dots and organic dyes. **a** Wavelength-resolved spectra obtained from a single 40-nm fluorescent latex sphere and a single mercapto-quantum dot. The broad emission peak around 660 nm is believed to arise from surface defects on the quantum dot. **b** Time-resolved photobleaching curves for the original quantum dots, the solubilized quantum dots, and the dye rhodamine 6G (reproduced with permission from Chan and Nie, Science 281, 2016, 1998).

200 times more stable than organic dyes and fluorescent proteins (Fig. 16.3) [23]. Under intense UV excitation, single phycoerythrin molecules are found to photobleach after 70 s, while the fluorescence emission of quantum dots remains unchanged after 600 s [33]. The photobleaching of quantum dots is believed to arise from a slow process of photo-induced chemical decomposition. Henglein and co-workers speculated that CdS decomposition is initiated by the formation of S or SH radicals upon optical excitation [52, 53]. These radicals can react with O_2 from the air to form a SO_2 complex, resulting in slow particle degradation.

Single quantum dots have been shown to emit photons in an intermittent on-off fashion [54, 55], similar to a "blinking" behavior reported for single fluorescent dye molecules, proteins, polymers, and metal nanoparticles. The fluorescence of single quantum dots turns on and off at a rate that is dependent on the excitation power. This phenomenon has been suggested to arise from a light-induced process involving photoionization and slow charge neutralization of the nanocrystal [54]. When two or more electron–hole pairs are generated in a single nanocrystal, the energy released from the combination of one pair could be transferred to the remaining carriers, one of which is preferentially ejected into the surrounding matrix. Subsequent photogenerated electron–hole pairs transfer their energy to the resident, unpaired carrier, leading to non-radiative delay and dark periods. The luminescence is restored only when the ejected carrier returns to neutralize the particle. Banin et al. believe that thermal trapping of electrons and holes is also a contributing factor because they observed a dependence of the blinking rate on temperature [56]. A further finding is that single dots exhibit random fluctuations in the emission wavelength (spectral wandering) over time [55, 57]. This effect is attributed to interactions between excitons with optically induced surface changes.

16.4
Applications in Biology and Medicine

A number of biological and medical applications have been demonstrated including DNA hybridization, immunoassays, cell biology, and fluorescent labeling of cancer cells. For example, quantum dots conjugated with antibodies have been used for biomarker detection and for molecular profiling of cultured cancer cells and clinical tissue specimens (Fig. 16.4). In more fundamental studies, Nie and coworkers have observed single quantum dots in a single living cell [21, 23], in which quantum dots are conjugated with folic acid (a vitamin whose receptor is over-expressed in more than 80% of human cancers) or transferrin (a protein that is involved in iron transport). The bioconjugated dots are transported into living cells by receptor-mediated endocytosis.

Alivisatos and coworkers [22] have used green and red emitting nanocrystals for labeling cellular structures. In their work, the nucleus of mouse fibroblast cells are labeled with green quantum dots that are coated with trimethoxy silylpropyl urea and acetate groups. These organic molecules have high binding affinities to the cell's nucleus. Furthermore, red quantum dots coated with biotin bind to the

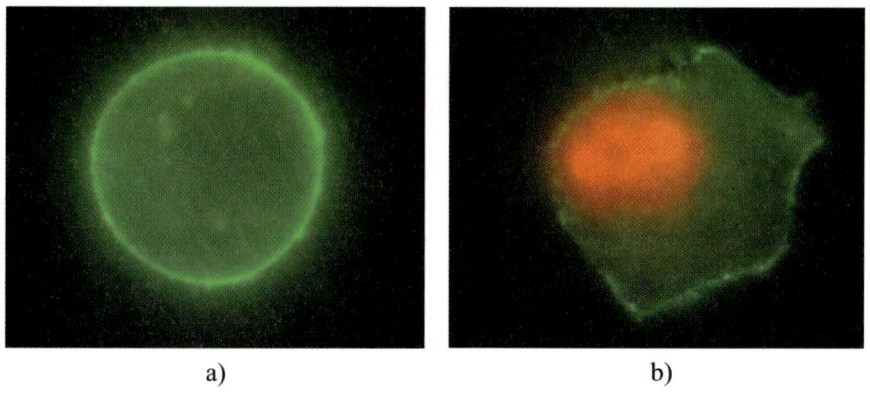

Fig. 16.4 Immunofluorescence images of human breast cancer cells (BT-474) labeled with green quantum dots (excitation wavelength = 490 nm and emission wavelength = 530 nm). **a** Detailed image of a cancer cell with overexpressed Her-2 receptors and labeled with Her-2 antibody-QD conjugates. **b** Detailed image of a cancer cell with its Her-2 receptors labeled with QD-antibody conjugates and its nucleus (orange) labeled with propidium iodide (PI) (Xiaohu Gao and Shuming Nie, Department of Biomedical Engineering, Emory University School of Medicine, unpublished data).

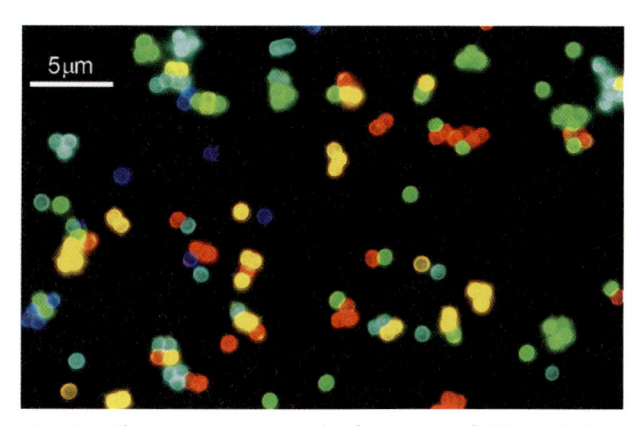

Fig. 16.5 Fluorescence micrograph of a mixture of CdSe/ZnS QD-tagged beads emitting single-color signals at 484, 508, 547, 575, and 611 nm. The beads were spread and immobilized on a polylysine-coated glass slide, which caused a slight clustering effect (reproduced with permission from Han, Gao, Su, and Nie, Nat. Biotechnol. 19, 631, 2001).

fibroblast's actin structures through a phalloidin-biotin (phalloidin is a peptide with a high binding affinity to the protein subunits of actin) and then with a streptavidin molecular bridge. A single excitation source can be used to observe the dual-labeling of quantum dots on the fibroblast cells.

Quantum dots can also be used to develop homogeneous bioassays. The detection sensitivity of heterogeneous assays is limited by non-specific binding of biomolecules or fluorophores to surfaces. Fluorescence correlation spectroscopy (FCS) can be used in conjunction with quantum dot bioconjugates to detect various analytes such as proteins and nucleic acids. FCS measures fluctuations in the fluorescence signal in a small probe volume ($\sim 10^{-15}$ L), and when multi-color quantum dots are brought together by an analyte molecule, the fluorescence signals will be correlated with each other. In addition, analyte binding could lead to fluorescence resonance energy transfer (FRET), in which a donor dot transfers its energy to a nearby acceptor dot [43].

The unique optical properties of quantum dots have been used to overcome the autofluorescence problem associated with most cells and tissues. For most biological and biomedical samples, the autofluorescence exhibits a decay time (excited state lifetime) of nanoseconds and a broad spectrum with a maximum at ~ 520 nm. Recently, Weiss and coworkers used quantum-dot probes in time-gated fluorescence microscopy [58, 59]. In contrast to organic dyes, quantum dots have excited-state lifetimes that are 10–100 times longer. Another method to overcome the autofluorescence of tissue is to use red or near-infrared emitting quantum dots, in which the fluorescence is shifted from that of the tissue or cell. Akerman and coworkers detected the binding of red-emitting peptide-coated quantum dots to liver, spleen, and lung tissues, as well as to tumor cells [60].

A further application of quantum dots is for multiplexed optical encoding and high-throughput analysis of genes and proteins, as reported by Nie and co-workers [24]. Polystyrene beads are embedded with multicolor CdSe quantum dots in various color and intensity combinations (Fig. 16.5). The use of 6 colors and 10 intensity levels can theoretically encode one million protein or nucleic acid sequences. Specific capturing molecules such as peptides, proteins, and oligonucleotides are covalently linked to the beads and are encoded by the bead's spectroscopic signature. A single light source is sufficient for reading all the quantum-dot-encoded beads. To determine whether an unknown analyte is captured or not, conventional assay methodologies (similar to direct or sandwich immunoassay) can be applied. This so-called "bar-coding technology" can be used for gene profiling and high-throughput drug and disease screening. Based on entirely different principles, Natan and coworkers [61] reported a metallic nanobarcoding technology for multiplexed bioassays. Together with quantum-dot-encoded beads, these "barcoding" technologies offer significant advantages over planar chip devices (e.g. improved binding kinetics and dynamic range), and are likely to find use in various biotechnological applications (also see Chapter 17).

16.5
Conclusions

We anticipate that luminescent quantum dots will be combined with other technologies such as microfluidics and microarrays for massively parallel biosensing and analytical detection. Using quantum dots as fluorescent labels, compact and portable devices could also be developed for screening diseases or analyzing complex biological mixtures. In addition, we believe that the multiplexing capabilities of quantum dots will allow the early detection of cancer, AIDS, and other diseases. Furthermore, near-IR emitting quantum dots provide a new class of optical probes for in-vivo molecular imaging. Finally, quantum dots should find broad applications in basic molecular biology, especially in probing the dynamics and functions of single macromolecules and single viruses inside living cells.

16.6
Acknowledgements

This work was supported by grants from the National Institutes of Health (R01 GM58173 and R01 GM60562) and the Department of Energy (DOE FG02-98ER14873).

16.7
References

1 HALPERIN, W. P. Quantum size effects in metal particles. *Rev. Mod. Phys.* **1986**, *58*, 533–606.
2 (a) ALIVISATOS, A. P. Semiconductor clusters, nanocrystals, and quantum dots. *Science* **1996**, *271*, 933–937. (b) ALIVISATOS, A. P. Perspectives on the physical chemistry of semiconductor nanocrystals. *J. Phys. Chem.* **1996**, *100*, 13226–13239.
3 BRUS, L. E. Quantum crystallites and nonlinear optics. *Appl. Phys.* **1991**, *A 53*, 465–474.
4 (a) HENGLEIN, A. Small-particle research – physicochemical properties of extremely small colloidal metal and semiconductor particles. *Chem. Rev.* **1989**, *89*, 1861–1873. (b) SCHMID, G. Large clusters and colloids – metals in the embryonic state. *Chem. Rev.* **1992**, *92*, 1709–1727. (c) HAGFELDT, A., GRATZEL, M. Light-induced redox reactions in nanocrystalline systems. *Chem. Rev.* **1995**, *95*, 49–68.
5 WELLER, H. Colloidal semiconductor q-particles – chemistry in the transition region between solid-state and molecules. *Angew. Chem. Int. Ed. Engl.* **1993**, *32*, 41–53.
6 FENDLER, J. H., MELDRUM, F. C. The colloid-chemical approach to nanostructured materials. *Adv. Mater.* **1995**, *7*, 607–632.
7 WANG, Y., HERRON, N. Nanometer-sized semiconductor clusters – materials synthesis, quantum size effects, and photophysical properties. *J. Phys. Chem.* **1991**, *95*, 525–532.
8 GEHR, R. J., BOYD, R. W. Optical properties of nanostructured optical materials. *Chem. Mater.* **1996**, *8*, 1807–1819.
9 MURRAY, C. B., NORRIS, D. J., BAWENDI, M G. Synthesis and characterization of nearly monodisperse cde (E=S, Se, Te) semiconductor nanocrystallites. *J. Am. Chem. Soc.* **1993**, *115*, 8706–8715.

10 KATARI, J. E. B., COLVIN, V. L., ALIVISATOS, A. P. X-ray photoelectron-spectroscopy of cdse nanocrystals with applications to studies of the nanocrystal surface. *J. Phys. Chem.* **1994**, *98*, 4109–4117.

11 HINES, M. A., GUYOT-SIONNEST, P. Synthesis and characterization of strongly luminescing ZnS-Capped CdSe nanocrystals. *J. Phys. Chem.* **1996**, *100*, 468–471.

12 MURRAY, C. B., KAGAN, C. R., BAWENDI, M. G. Self-organization of cdse nanocrystallites into 3-dimensional quantum-dot superlattices. *Science* **1995**, *270*, 1335–1338.

13 ANDRES, R. P., BIELEFELD, J. B., HENDERSON, J. I., JANES, D. B., KOLAGUNTA, V. R., KUBIAK, C. P., MAHONEY, W. J., OSIFCHIN, R. G. Self-assembly of a two-dimensional superlattice of molecularly linked metal clusters. *Science* **1996**, *273*, 1690–1693.

14 (a) HEATH, J. R., WILLIAMS, R. S., SHIANG, J. J., WIND, S. J., CHU, J., D'EMIC, C., CHEN, W., STANIS, C. L., BUCCHIGNANO, J. J. Spatially confined chemistry: Fabrication of Ge quantum dot arrays. *J. Phys. Chem.* **1996**, *100*, 3144–3149. (b) HARFENIST, S. A., WANG, Z. L., ALVAREZ, M. M., VEZMAR, I., WHETTEN, R. L. Highly oriented molecular Ag nanocrystal arrays. *J. Phys. Chem.* **1996**, *100*, 13904–13910.

15 (a) MIRKIN, C. A., LETSINGER, R. L., MUCIC, R. C., STORHOFF, J. J. A DNA-based method for rationally assembling nanoparticles into macroscopic materials. *Nature* **1996**, *382*, 607–609. (b) ALIVISATOS, A. P., JOHNSSON, K. P., PENG, X., WILSON, T. E., LOWETH, C. J., BRUCHEZ, JR., M. P., SCHULTZ, P. G. Organization of 'nanocrystal molecules' using DNA. *Nature* **1996**, *382*, 609–611.

16 COLVIN, V. L., SCHLAMP, M. C., ALIVISATOS, A. P. Light-emitting-diodes made from cadmium selenide nanocrystals and a semiconducting polymer. *Nature* **1994**, *370*, 354–356.

17 DABBOUSI, B. O., BAWENDI, M. G., ONOTSUKA, O., RUBNER, M. F. Electroluminescence from CdSe Quantum-Dot Polymer Composites. *Appl. Phys. Lett.* **1995**, *66*, 1316–1318.

18 KLARREICH, E. Biologists join the dots. *Nature* **2001**, *413*, 450–452.

19 MITCHELL, P. Turning the spotlight on cellular imaging. *Nat. Biotechnol.* **2001**, *19*, 1013–1017.

20 PENG, Z. A., PENG, X. G. Formation of high-quality CdTe, CdSe, and CdS nanocrystals using CdO as precursor. *J. Am. Chem. Soc.* **2001**, *123*, 183–184.

21 GAO, X., CHAN, W. C. W., NIE, S. Quantum-dot nanocrystals for ultrasensitive biological labeling and multiplexed optical encoding. *J. Biomed. Optics* **2002**, *7*, 532–537.

22 BRUCHEZ, M., MORONNE, M., GIN, P., WEISS, S., ALIVISATOS, A. P. Semiconductor nanocrystals as fluorescent biological labels. *Science* **1998**, *281*, 2013–2016.

23 CHAN, W. C. W., NIE, S. M. Quantum dot bioconjugates for ultrasensitive nonisotopic detection. *Science* **1998**, *281*, 2016–2018.

24 HAN, M. Y., GAO, X. H., SU, J. Z., NIE, S. M. Quantum-dot-tagged microbeads for multiplexed optical coding of biomolecules. *Nature Biotechnol.* **2001**, *19*, 631–635.

25 MURRAY, C. B., NORRIS, D. J., BAWENDI, M. G. Synthesis and Characterization of Nearly Monodisperse Cde (E=S, Se, Te) Semiconductor Nanocrystallites. *J. Am. Chem. Soc.* **1993**, *115*, 8706–8715.

26 DABBOUSI, B. O., RODRIGUEZVIEJO, J., MIKULEC, F. V., HEINE, J. R., MATTOUSSI, H., OBER, R., JENSEN, K. F., BAWENDI, M. G. (CdSe)ZnS core–shell quantum dots: Synthesis and characterization of a size series of highly luminescent nanocrystallites. *J. Phys. Chem. B* **1997**, *101*, 9463–9475.

27 HINES, M. A., GUYOT-SIONNEST, P. Synthesis and Characterization of Strongly Luminescing Zns-Capped Cdse Nanocrystals. *J. Phys. Chem.* **1996**, *100*, 468–471.

28 PENG, X. G., SCHLAMP, M. C., KADAVANICH, A. V., ALIVISATOS, A. P. Epitaxial growth of highly luminescent CdSe/CdS core/shell nanocrystals with photostability and electronic accessibility. *J. Am. Chem. Soc.* **1997**, *119*, 7019–7029.

29 QU, L., PENG, Z. A., PENG, X. G. Alternative routes toward high quality CdSe nanocrystals. *Nanoletters* **2001**, *1*, 333–337.

30 Qu, L., Peng, X. G. Control of photoluminescence properties of CdSe nanocrystals in growth. *J. Am. Chem. Soc.* **2002**, *124*, 2049–2055.

31 De Smet, Y., Deriemaeker, L., Parloo, E., Finsy, R. On the determination of Ostwald ripening rates from dynamic light scattering measurements. *Langmuir* **1999**, *15*, 2327–2332.

32 DeSmet, Y., Deriemaeker, L., Finsy, R. A simple computer simulation of Ostwald ripening. *Langmuir* **1997**, *13*, 6884–6888.

33 Chan, W. C. W. Semiconductor Quantum Dots for Biological Detection and Imaging. PhD Thesis, Indiana University, Bloomington, IN, **2001**.

34 Peng, X. G., Manna, L., Yang, W. D., Wickham, J., Scher, E., Kadavanich, A., Alivisatos, A. P. Shape control of CdSe nanocrystals. *Nature* **2000**, *404*, 59–61.

35 Peng, Z. A., Peng, X. G. Mechanisms of the shape evolution of CdSe nanocrystals. *J. Am. Chem. Soc.* **2001**, *123*, 1389–1395.

36 Prieto, J. A., Armeeles, G., Groenin, J., Cales, R. Size and strain effects in the E-1-like optical transitions of InAs/InP self-assembled quantum dot structures. *Appl. Phys. Lett.* **1999**, *74*, 99–101.

37 Micic, O. I., Cheong, H. M., Fu, H., Zunger, A., Sprague, J. R., Mascarenhas, A., Nozik, A. J. Size-dependent spectroscopy of InP quantum dots. *J. Phys. Chem. B* **1997**, *101*, 4904–4912.

38 Schreder, B., Schmidt, T., Ptatschek, V., Winkler, U., Materny, A., Umbach, E., Lerch, M., Muller, G., Kiefer, W., Spanhel, L. CdTe/CdS clusters with "core-shell" structure in colloids and films: The path of formation and thermal breakup. *J. Phys. Chem. B* **2000**, *104*, 1677–1685.

39 Shi, J. Z., Zhu, K., Zheng, Q., Zhang, L., Ye, L., Wu, J., Zuo, J. Ultraviolet (340–390 nm), room temperature, photoluminescence from InAs nanocrystals embedded in SiO_2 matrix. *Appl. Phys. Lett.* **1997**, *70*, 2586–2588.

40 Becerra, L. R., Murray, C. B., Griffin, R. G., Bawendi, M. G. Investigation of the Surface Morphology of Capped Cdse Nanocrystallites by P-31 Nuclear Magnetic Resonance. *J. Chem. Phys.* **1994**, *100*, 3297–3300.

41 Katari, J. E. B., Colvin, V. L., Alivisatos, A. P. X-Ray Photoelectron Spectroscopy of Cdse Nanocrystals with Applications to Studies of the Nanocrystal Surface. *J. Phys. Chem.* **1994**, *98*, 4109–4117.

42 Mitchell, G. P., Mirkin, C. A., Letsinger, R. L. Programmed assembly of DNA functionalized quantum dots. *J. Am. Chem. Soc.* **1999**, *121*, 8122–8123.

43 Willard, D. M., Carillo, L. L., Jung, J., Van Orden, A. CdSe-ZnS Quantum Dots as Resonance Energy Transfer Donors in a Model Protein–Protein Binding Assay. *Nanoletters* **2001**, *1*, 469–474.

44 Mattoussi, H., Mauro, J. M., Goldman, E. R., Anderson, G. P., Sundar, V. C., Mikulec, F. V., Bawendi, M. G. Self-assembly of CdSe-ZnS quantum dot bioconjugates using an engineered recombinant protein. *J. Am. Chem. Soc.* **2000**, *122*, 12142–12150.

45 Peng, X. G., Wickham, J., Alivisatos, A. P. Kinetics of II–VI and III–V colloidal semiconductor nanocrystal growth: ,focusing, of size distributions. *J. Am. Chem. Soc.* **1998**, *120*, 5343–5344.

46 Hines, M. A., Guyot-Sionnest, P. Bright UV-blue luminescent colloidal ZnSe nanocrystals. *J. Phys. Chem. B* **1998**, *102*, 3655–3657.

47 Hu, J. T., Li, L., Yang, W., Manna, L., Wang, L., Alivisatos, A. P. Linearly polarized emission from colloidal semiconductor quantum rods. *Science* **2001**, *292*, 2060–2063.

48 Efros, A. L. Luminescence polarization of CdSe microcrystals. *Phys. Rev. B* **1992**, *46*, 7448–7458.

49 Empedocles, S. A., Neuhauser, R., Bawendi, M. G. Three-dimensional orientation measurements of symmetric single chromophores using polarization microscopy. *Nature* **1999**, *399*, 126–130.

50 Mathies, R., Stryer, L. Single-molecule fluorescence detection, in Applications of Fluorescence in the Biomedical Sciences (Taylor, D. L., Waggoner, A. S., Lanni, F., Murphy, R. F., Birge, R., eds.), *Alan R. Liss*, New York, NY, USA, **1986**, p. 129.

51 Gerion, D., Pinaud, F., Willimas, S.C., Parak, W.J., Zanchet, D., Weiss, S., Alivisatos, A.P. Synthesis and properties of biocompatible water-soluble silica-coated CdSe/ZnS semiconductor quantum dots. *J. Phys. Chem. B* **2001**, *105*, 8861–8871.

52 Henglein, A. Photo-degradation and fluorescence of colloidal-cadmium sulfide in aqueous solution. *Ber. Bunsenges. Phys. Chem.* **1982**, *86*, 301–305.

53 Baral, S., Fojtik, A., Weller, H., Henglein, A. Photochemistry and radiation chemistry of colloidal semiconductors: intermediates of the oxidation of extremely small particles of CdS, ZNS, and Cd_3P_2 and size quantization effects (a pulse radiolysis study). *J. Am. Chem. Soc.* **1986**, *108*, 375–378.

54 Nirmal, M., Dabbousi, B.O., Bawendi, M.G., Macklin, J.J., Brus, L.E. Fluorescence intermittency in single cadmium selenide nanocrystals. *Nature* **1996**, *383*, 802–804.

55 Empedocles, S.A., Bawendi, M.G. Spectroscopy of single CdSe nanocrystallites. *Acc. Chem. Res.* **1999**, *32*, 389–396.

56 Banin, U., Bruchez, M., Alivisatos, A.P., Ha, T., Weiss, S., Chemla, D.S. Evidence for a thermal contribution to emission intermittency in single CdSe/CdS core/shell nanocrystals. *J. Chem. Phys.* **1999**, *110*, 1195–1201.

57 Blanton, S.A., Hines, M.A., Guyot-Sionnest, P. Photoluminescence wandering in single CdSe nanocrystals. *Appl. Phys. Lett.* **1996**, *69*, 3905–3907.

58 Dahan, M., Laurence, T., Schumacher, A., Chemla, D.S., Alivisatos, A.P., Sauer, M., Weiss, S. Fluorescence lifetime study of single qdots. *Biophys. J.* **2000**, *78*, 2270.

59 Dahan M., Laurence T., Pinaud, F., Chemla, D.S., Alivisatos, A.P., Sauer, M., Weiss, S. Time-gated biological imaging by use of colloidal quantum dots. *Opt. Lett.* **2001**, *26*, 825–827.

60 Akerman, M.E., Chan, W.C.W., Laakkonen, P., Bhatia, S.N., Ruoslathi, E. *In vivo* targeting of nanocrystals, **2002**, *99*, 12617–12621.

61 Nicewarner-Pena, S.R., Freeman, R.G., Reiss, B.D., He, L., Pena, D.J., Walton, I.D., Cromer, R., Keating, C.D., Natan, M.J. Submicrometer metallic barcodes. *Science* **2001**, *294*, 137–141.

17
Colloids for Encoding Chemical Libraries: Applications in Biological Screening

Bronwyn J. Battersby, Lisbeth Grøndahl, Gwendolyn A. Lawrie, and Matt Trau

17.1
Introduction

Protein- and gene-based technologies are changing the way in which new drugs are being discovered and developed. New generations of drugs are being developed with the aim of tailoring them to individual patients for disease treatment and also disease prevention. In the not-too-distant future, physicians will be able to remove much of the speculation from diagnosis and drug prescription by sequencing a patient's DNA (Gwynne and Page, 1999). Tests will be developed to guide preventative therapy, signal a need for lifestyle changes or enable early intervention with medicines. Furthermore, by studying the genetic activity of many individuals, the secrets of human diseases and traits will be unlocked and the normal variants that make an individual unique will be identified (Hacia, 1999). To understand gene function, scientists will compare genes from many individuals at different stages of life and under different conditions. The scientific community will eventually determine how the human genome differs from other species and then establish how that makes us distinct on the biochemical level. High-throughput screening (HTS), the process of rapidly testing the biomolecular interactions between large numbers of compounds, is already playing a major role in these developments.

The Human Genome Project (HGP) has caused an explosion in genomics and proteomics research into how genes and proteins affect responses to drugs and how to develop new drugs that will target a specific disease, while decreasing the probability of undesirable side effects. With the completion of the first draft of the human genome sequence (Venter et al., 2001), there has been a massive increase in the amount of new information to process and number of targets to screen. As such, the demand for economical, high-throughput and flexible molecular screening alternatives in genomics, proteomics and drug discovery research has greatly increased. Innovations that are occurring in colloid chemistry are revitalizing the process of HTS by providing unique, robust and faster alternatives to existing technologies. Companies and institutions are beginning to realize the potential of increasing the size and complexity of chemical libraries by producing large colloid-based libraries. In these technologies, there is a requirement to conveniently

Colloids and Colloid Assemblies. Edited by Frank Caruso
Copyright © 2004 Wiley-VCH Verlag GmbH & Co. KGaA, Weinheim
ISBN: 3-527-30660-9

produce large (typically $>10^6$) chemical compound libraries which can be attached to solid support beads (e.g. colloids), where they are rapidly screened for biological activity. Depending on the specific application, chemical libraries can comprise different families of chemical compounds. Genomics applications require a library containing single-stranded DNA molecules (synthetic oligonucleotides or cDNAs) which are all of a different sequence. Proteomics studies engage a library of proteins for exploration of protein diversity, interaction, structure and function. Drug discovery research requires libraries containing a variety of molecular species such as polypeptides and polysaccharides. The ability to perform such analyses for extremely large numbers of probes (e.g. 10^{10}) on an inexpensive and well-defined high-throughput platform is highly desirable and will be the major challenge over the next few years.

17.1.1
Genomics

The entire human genome is composed of approximately 3 billion base pairs of DNA that make up 46 human chromosomes (22 pairs of autosomal chromosomes and 2 sex chromosomes). About 30 000–40 000 protein-coding genes are currently believed to be interspersed on these chromosomes (International Human Genome Sequencing Consortium, 2001). Genomics involves mapping, sequencing and analyzing the function of genomes and may be divided into several closely related categories, including structural genomics, functional genomics and pharmacogenomics. Structural genomics (inextricably linked with proteomics, described below) involves the construction of high-resolution genetic, physical and transcript maps of an organism, and includes the determination of the three-dimensional structure of proteins. Determining the structures of all the key functional sites of any human protein is a major goal of this field (Branca et al., 2001) and this information should make it much easier to develop highly specific drugs, and lead to more effective and safer pharmaceuticals. Functional genomics, on the other hand, encompasses the development and application of experimental approaches to evaluate gene function in cells and is characterized by high-throughput experimental methodologies combined with computational and statistical analysis of the results (i.e. bioinformatics) (Hatzimanikatis, 2000; Hieter and Boguski, 1997). This genomics approach involves expanding the scope of biological investigation from studying single genes or proteins to studying many genes or proteins simultaneously. Finally, the long-term goal of pharmacogenomics is to develop therapies, or even preventative approaches, based on genetic risk factors (Silver, 2001). Drugs specifically targeted to patients with a particular genetic make-up (genotype) will ensure patients have an excellent response to drugs with no side effects (Silver, 2001).

Variations between individual genomes have been found to account for differences in susceptibility to disease and pharmacological response to treatment (Bader, 2001). Change of a single base of a DNA molecule, namely, a single nucleotide polymorphism (SNP), can influence an individual's risk for a particular disease (Bader, 2001). In many cases, combinations of SNPs rather than single mutations might predispose an individual to a disease or condition (Iida et al., 2001). As described in Section 17.2.3, colloid-based gene libraries are beginning to play a major role in identifying these genetic variations, through a technique known as *genotyping*. Diagnostic genotyping, for example, involves linking specific human genetic variation patterns to a multiplicity of clinical conditions in order to improve the ability to diagnose and treat those conditions (Wallin et al., 2002). Human genotyping is also currently being used to investigate disease predisposition and identification of genetic markers for drug efficacy and toxicity (Schur et al., 2001; Wei et al., 1998). Human genotyping research is being conducted on many diseases, some of which are listed in Box 17.1.

Box 17.1 Examples of diseases for which human genotyping tests are being developed

Disease	Review
Alzheimer's disease	Selkoe (2001)
Deep vein thrombosis	de Stefano et al. (2000)
Breast cancer	Friedberg (2001)
Huntington's disease	Gusella and MacDonald (2002)
Colon cancer	Timar et al. (2001)
Sickle cell anemia	Waterfall and Cobb (2001)
Cystic fibrosis	Bobadilla et al. (2002)
Parkinson's disease	Maimone et al. (2001)

Besides SNP genotyping, another important area of genomics is gene expression analysis. One of the key components of future genomic research will be the study of drug- or environment-induced changes in gene expression indicative of disease and/or pharmacological or environmental exposure. Whether a particular cell or tissue type is healthy or diseased can depend on which genes are being expressed and at what levels. Thus by comparative gene expression studies one may be able to determine an altered expression pattern that is indicative of disease or toxic shock. This approach will also allow investigators to determine the effect of therapeutics on gene expression and to discover which genes underlie a given pharmacological or physiological state. Monitoring gene expression quantitatively addresses questions concerning individual gene function, functional pathways and how cellular components work together to regulate and perform cellular processes (Lipshutz et al., 1999). Performing gene expression analysis will enable identification and validation of new molecular targets for drug development, prediction of potential side-effects during preclinical development and toxicology studies, identification of genes involved in conferring drug sensitivity and resistance, and prediction of patients most likely to benefit from trial drugs and those who are likely to suffer an adverse effect (Clarke et al., 2001).

17.1.2
Proteomics

Proteomics, the identification and cataloging of all the proteins expressed at a given time in a given cell, is a rapidly developing field. The proteome varies from cell to cell, and varies with age, environment and disease. Some of the most important events taking place in cells are the interactions between proteins. Multi-protein complexes are known to regulate almost all cellular processes; the absence of particular protein–protein interactions is often the cause of disease in humans (Staglia et al., 2001).

As most of the existing drug targets are proteins, it is essential to understand their function, location and interaction with other proteins in order to design safer and more efficacious drugs. To better understand disease pathways and the manner of action of existing drugs, scientists are focusing on mapping protein–protein interactions and protein complexes. Through investigation of alternative intervention points in these pathways, researchers will be able to find more promising drug targets and will eventually develop lead compounds with higher specificity, efficacy and safety.

Biochemical methods that were initially used to detect protein interactions, are not suitable for large-scale screening (Staglia et al., 2001). Recently, high-throughput techniques have been used to investigate proteins in their native conformation (Cagney et al., 2000; Fields and Song, 1989); however, these techniques have a considerable limitation in that they are unsuitable for the analysis of a variety of proteins (e.g. those associated with membranes). Approximately 30–35% of all proteins in the eukaryotic organism cannot be investigated using currently available techniques (Staglia et al., 2001). Colloid-based libraries are showing promise as new tools for investigating the molecular mechanism responsible for the above-mentioned biological processes, for understanding functional organization in cells and for identifying and validating new drug targets.

17.1.3
Drug Discovery

There are many costly and time-consuming steps involved in bringing a drug to market. To save time and to improve candidate selection for pre-clinical and clinical development, pharmacokinetic characteristics of compounds are examined early in discovery and development (Jonscher, 2001). Detailed knowledge of biochemical responses, such as adsorption, distribution, metabolism, elimination and toxicity (ADMET) allows drug discovery researchers to make rapid and informed decisions about the potential of new drug leads (Viswanadhan et al., 2002).

Minimizing the number of assays (i.e. tests) and maximizing the amount of information obtained helps to avoid bottlenecks in assaying. Although screening of multiple targets simultaneously is an area of HTS that has yet to reach its full potential, the advantages are clear: screening two or more targets at once requires less reagents than running separate screens for each one, and more information can be obtained, at a much lower cost. For example, screening of more than one protein at a time can provide valuable information on the interactions between targets and their selectivity. The use of colloid-based libraries for screening multiple targets simultaneously promises to have a major impact on cost savings and on increasing the quality of therapeutic candidates.

17.2 Current High-throughput Screening Technologies

17.2.1 Screening with Microplates

Screening for novel drug leads is an established field within the pharmaceutical industry. Over the past decade, there has been a tremendous increase in the industry's ability to screen large numbers of compounds against target molecules (Beggs, 2001). This has been made possible through the introduction of high-density microplates (also known as microwell or microtiter plates) containing 96, 384, or 1536 wells, small-volume liquid handling robotics and advanced detection technology. Significant benefits arise from miniaturization of HTS; these include faster turnaround, lower cost and reduced space requirements (Burbaum, 1998).

The preoccupation of the pharmaceutical and biotechnology industries with miniaturizing microplate technologies and assays has been driven by the necessity to keep compounds in libraries separated so they can be identified easily. As a result, most assay methods have been developed for screening compounds in the solution phase rather than attached to a solid support such as a colloid (Harness, 2000). Additionally, most technological development has occurred in the field of miniaturizing microplates: building better robots to handle more plates, engineering better fluid dispensers and building detection units capable of reading the miniaturized plates. Yet, although the achievement is significant compared with the early 1990s, most leading corporations are still only capable of screening approximately 100 000 compounds in a 24-hour period. Emerging methods that have the potential to dramatically increase throughput by managing large colloid-based combinatorial libraries without the restriction of microplates are presented in Section 17.2.3.

17.2.2 Screening with Microarrays

Microarrays are sophisticated tools for applications such as gene expression monitoring of known and unknown genes, genotyping of eukaryotes, micro-organisms and viruses, mutation detection and analysis, antibody-based assays for protein and small molecule analytes (e.g. glycoprotein hormones, steroid hormones, allergen antibodies). Production of microarray devices has become well defined and automated and they are generally created in two formats. In Format I, an array of

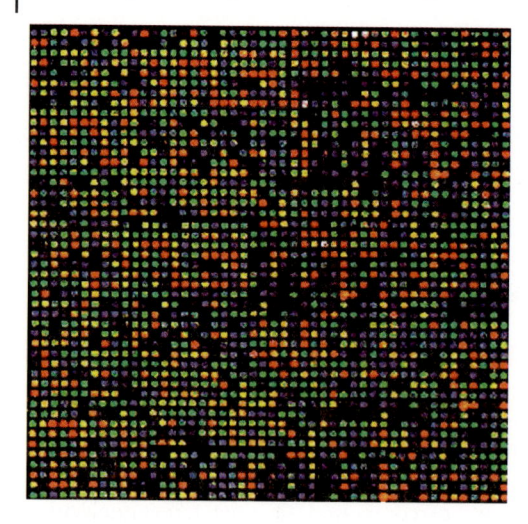

Fig. 17.1 A DNA microarray for gene expression profiling. Microarrays are approximately the size of a postage stamp and may contain 10 000 to 500 000 different oligonucleotide probes arranged in pixels, as shown in this schematic.

oligonucleotide probe molecules (e.g. 20–80 bases in length) is synthesized using photolithographic methods (Fodor et al., 1991). In Format II, probe molecules (e.g. cDNA, 500~5000 bases long) are immobilized to a solid surface, such as a modified glass slide, nylon membrane or silicon wafer, using ink-jet spotting (Schena et al., 1995, 1998).

To perform screening, the microarrays are incubated with target molecules, for example, single-stranded DNA which has been isolated, amplified and labeled with a fluorescent tag. Following completion of the hybridization or binding reaction, the array is inserted into a scanner where hybridization/binding patterns are detected. Probes that are biologically active towards the target are inclined to bind more strongly and produce more intense signals than those which are mismatched (see Fig. 17.1). Through knowledge of the position and sequence of each probe on the array, important structural information about the target can be obtained.

Although microarrays are having a major impact on HTS, they are expensive in terms of capital equipment outlays, sample preparation, analysis times, and array and reagent costs. A major disadvantage of microarrays is that they are currently unable to carry libraries of more than a few hundred thousand compounds because of their two-dimensional geometry and limitations in detection capabilities. Measurements are subject to variability relating to probe hybridization differences and cross-reactivity, differences between elements within microarrays and differences from one array to another (Brenner et al., 2000; Audic and Claverie, 1997; Wittes and Friedman, 1999).

17.2.3
Screening with Colloids

High-throughput screening using colloid-based chemical libraries is emerging as an extremely attractive alternative to microarrays and microplates, for applications such as SNP genotyping and gene expression analysis. Rather than preparing the oligonucleotide or polypeptide probes in pixels on a microarray, the probes are attached to, or synthesized on, colloidal solid support beads 2–300 µm in diameter (see Fig. 17.2). The greatest advantages of colloid-based libraries are that colloids are inexpensive to produce, they can be conveniently stored in small volumes of fluid and they can be screened extremely rapidly using various high-throughput detection technologies (e.g. flow cytometry). The general processes involved in producing and screening colloid-based libraries are shown in Figs. 17.3 and 17.4.

Colloid-based libraries are typically produced using either "combinatorial" or "non-combinatorial" procedures (Fig. 17.3). Production of combinatorial libraries involves progressive synthesis of the library probes on the colloids under nonaqueous conditions. The most powerful combinatorial library synthesis method is the iterative "split and mix" synthesis on insoluble microscopic beads (Fig. 17.3b), which results in each distinct bead containing a single compound. The use of the split-and-mix approach allows enormous numbers of probes to be produced in a very small number of cycles. In a non-combinatorial method (see Fig. 17.3a), however, chemical libraries are generated by sequential attachment of fully synthesized moieties (e.g. cDNAs, small molecules, proteins) onto aliquots of encoded colloids. Different aliquots are mixed together to form a library. This attachment is often performed under aqueous conditions through covalent bonding of the full-length moiety to the colloid.

Screening of a colloid-based library typically involves exposure of the library to one or more fluorescently labeled target molecules (see Fig. 17.4). The target binds or hybridizes to the probes in the library that have biological affinity for the target sequence (Fig. 17.4c). This binding gives rise to a bright fluorescence signal on the colloids under correct illumination, corresponding to the emission wavelength of the target's fluorescent label. The colloids showing the brightest fluores-

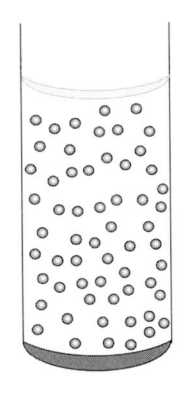

Fig. 17.2 Colloids are ideal as solid support beads for chemical library synthesis and screening. Typical dispersions may contain more than 10^{10} beads.

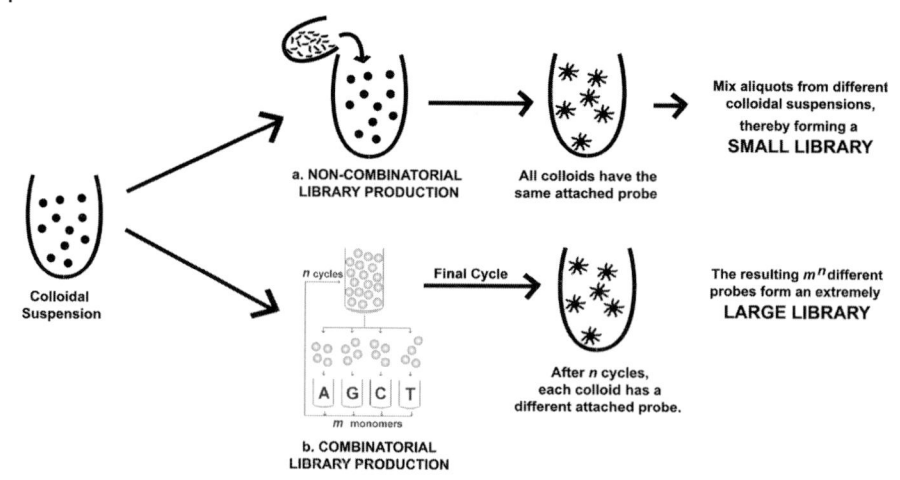

Fig. 17.3 Creation of a chemical library on colloidal support beads. **a** The non-combinatorial method for producing a library involves attachment of identical, fully-synthesized probes to the colloids using physical or covalent bonding. Several different suspensions are mixed together to form a small library. **b** The combinatorial method of library production involves progressive synthesis of probes on functionalized colloids over n cycles of a split-and-mix process. This results in an extremely large chemical library containing m^n compounds.

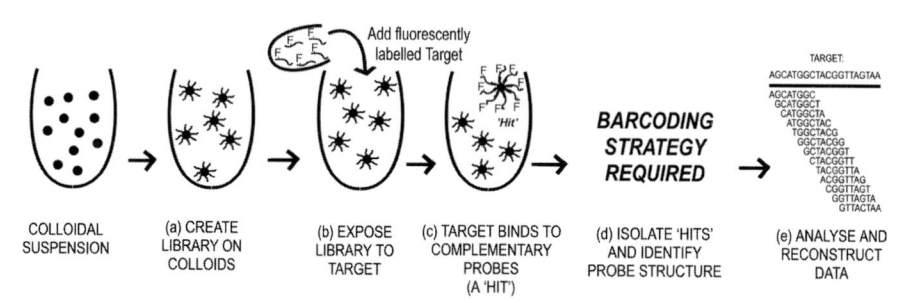

Fig. 17.4 The typical processes involved in producing and screening colloid-based DNA libraries. **a** A library of oligonucleotide probes is created on the colloidal particles using combinatorial or non-combinatorial methods (see Fig. 17.3). **b** Fluorescently labeled target DNA is mixed with the library, and those probes which are complementary will bind with the target. **c** The colloids on which the target is bound will brightly fluoresce the color of the target label, signifying a 'hit'. **d** The 'hits' are distinguished and/or isolated using fluorescence detection instrumentation such as a flow cytometer (see Fig. 17.5) or fluorescence microscope. To determine the structure of a bioactive colloid-based probe (typically present in nanomolar amounts), a barcoding strategy must be in place. **e** Identification of the bioactive probes permits reconstruction of the data, thereby revealing the target sequence.

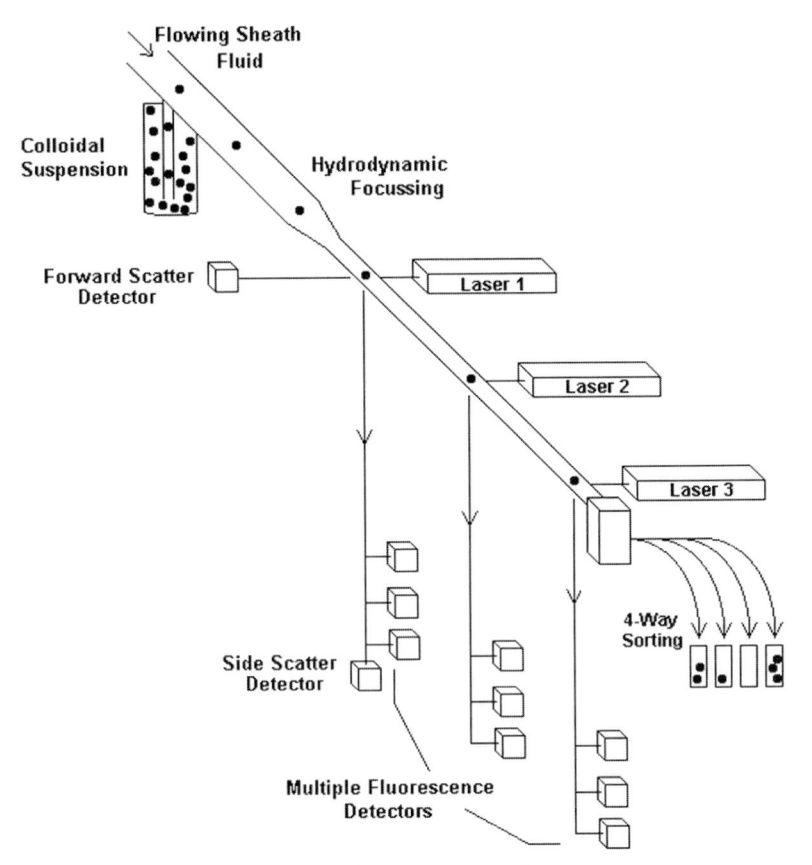

Fig. 17.5 Schematic representation of an 11-parameter flow cytometer. The particles are withdrawn from the colloidal suspension and are hydrodynamically focused in flowing sheath fluid. Precision fluidics align the particles in single file and the particles pass through the various laser beams at a rate of 30 000–50 000 particles s^{-1}. Through the use of an optical array of filters, dichroic mirrors and detectors (FL1 to FL9, side scatter, forward scatter; see Tab. 17.5), the fluorescence and light scattering properties of each particle are measured and recorded.

cence (i.e. the "hits") can be distinguished by a fluorescence detection instrument either in flow, on a stationary platform, or by cleaving the compound off the colloid in a multi-well plate and measuring the fluorescence in solution by, for example, fluorescence polarization (Burbaum and Sigal, 1997), homogeneous time-resolved fluorescence (Burbaum and Sigal, 1997; Grepin and Pernelle, 2000), or fluorescence correlation spectroscopy (Burbaum and Sigal, 1997; Auer et al., 1998).

While the technology for screening combinatorial libraries has advanced for over a decade, screening using non-combinatorial libraries is relatively new. There has been a growing number of reports on the use of colloid-based non-combina-

Tab. 17.1 Encoding/decoding strategies for chemical libraries

	Encoding method	Decoding method	Library size
Probe attached to bead	Polymer beads are non-permanently stained with 2–3 fluorescent dyes	Flow cytometry, optical fibre arrays, digital imaging	100–270,000 probes [1]
Probe synthesized on bead Tag synthesized on bead	Positional encoding. Probes are immobilized in spatially resolved sites on a two-dimensional support	Decoding via position in the array	<10^6 probes [2]
DNA Microarray	Molecular tags are covalently synthesized on support bead, in parallel with combinatorial synthesis of probe	Tag is cleaved from the bead and analysed further using mass spectrometry	<500,000 probes [3]
Probe synthesized on bead	Multi-fluorescent 'reporter' particles are permanently attached to support beads during split-and-mix synthesis (i.e. 'active' encoding)	The color combination bundled into each reporter particle is read in parallel via automated detection instruments	>10^{10} probes [4]
Probe synthesized on bead	The unique 'optical signature' of each multi-fluorescent support bead is tracked by a flow cytometer during the combinatorial synthesis of probe	The optical signature is analyzed by flow cytometry and reaction history of the bead is determined by recalling data stored by the flow cytometer software during probe synthesis	>10^{10} probes [5]

1) Vignali 2000; Walt 2000. 2) Lam et al., 2000. 3) Gwynne and Page 1999; Hacia 1999; Lander 1999. 4) Battersby et al. 2000; Grondahl et al. 2000; Matthews et al. 2001. 5) Trau and Battersby 2001; Battersby et al. 2001, Battersby and Trau 2002; Battersby et al. 2002.

torial libraries for screening, and a variety of techniques are used to detect the bioactivity in these libraries (Spain and Jacobson, 2001; Vignali, 2000; Dunbar and Jacobson, 2000; Oliver et al., 1998; Fulton et al., 1997; Szurdoki et al., 2001; Walt, 2000; Steemers et al., 2000; Brenner, 2000; Han et al., 2001; Nicewarner-Pena et al., 2001; Bruchez et al., 1998). Detection of hits in a stationary system requires a fluorescence microscope (Battersby et al., 2000; Grøndahl et al., 2000; Meldal et al., 1994) or an optical fiber array formed by creating microwells at the tips of optical fiber bundles with the wells filled with probe-labeled colloids (Szurdoki et al., 2001; Walt, 2000; Steemers et al., 2000). The most suitable instrument for detecting hits in a dynamic system is a flow cytometer (see Fig. 17.5) (Trau and Battersby, 2001; Battersby et al., 2001; Needels et al., 1993).

Most of the colloid-based screening technologies currently available have still to reach maturity, and although they show potential of eventually being able to deal with large libraries, they cannot yet compete with microarray devices in terms of number of probes which can be used for each assay (see Tab. 17.1). It is worth remembering that 1 mL of a typical colloidal suspension (containing 3-µm diameter colloids at 20% solids) may contain more than 10^{10} beads. Creating libraries of 10^{10} compounds on colloids is clearly a challenge, but if this can be accomplished, these libraries would greatly surpass microarrays in terms of library numbers, cost-efficiency and value. Indeed, one of the greatest challenges facing researchers using large colloid-based libraries is the ability to expediently identify the chemical structure or sequence of each probe that is found to be "bioactive" (i.e. biologically active, Fig. 17.4d). Because colloid-based probes are randomly located in a suspension (unlike the compounds in microarrays and microplates which are in a fixed, known, position on an array), an encoding system is required to allow the rapid determination of these structures or reconstruction of the target sequences (Fig. 17.4e). Powerful strategies for encoding colloids have been developed for colloid-based HTS over the last few years and these will be discussed in further detail in Section 17.3.

17.3
Encoding Colloids for High-throughput Screening

A relatively new method of encoding colloid-based libraries, optical barcoding, has tremendous potential, and novel methods are being developed by several companies and institutions worldwide (Bruchez et al., 1998; Nicewarner-Pena et al., 2001; Fulton et al., 1997; Battersby et al., 2000; Trau and Battersby, 2001). One of the greatest advantages of using optical methods to barcode colloids is the speed and accuracy with which the structures of biologically active compounds can be determined after library screening. It is perhaps due to these inherent advantages, that all non-combinatorial libraries reported so far have been optically encoded.

17.3.1
Encoding Colloids for Non-combinatorial Libraries

Current methods for optically encoding non-combinatorial libraries include the use of fluorescent dyes, metals and nanocrystals (refer to Box 17.2). The technique of using multiple fluorescence intensities and multiple emission wavelengths (i.e. multiplexed encoding) to encode colloids (3–6 µm in diameter) for small library applications has been employed by a number of groups (Fulton et al., 1997; Steemers et al., 2000). By entrapping various ratios of two fluorescent dyes in the interior of colloidal particles, up to 100 differently encoded colloidal suspensions have been produced (Fulton et al., 1997; Steemers et al., 2000). For each suspension, the polymeric colloids are swollen in a solvent/dye mixture containing a certain ratio of the two dyes. Typical solvents used are dimethylformamide or tetrahydrofuran. Rapid contraction of the colloids occurs upon exposure to an aqueous or alcoholic solution (Steemers et al., 2000), thereby entrapping the fluorescent dyes within the colloids.

A vastly different encoding technique adopts suspensions of colloidal rods which are encoded by complex striping patterns (Nicewarner-Pena et al., 2001; Walton et al., 2002). The patterned rods (15 nm–12 µm in width and 1–50 µm in length) are prepared using sequential electrochemical deposition of metal ions into Al_3O_2 membrane templates with a nominal pore diameter of 200 nm (Nicewarner-Pena et al., 2001). The templates are dissolved under appropriate conditions, leaving patterned metal rods onto which biomolecules can be physically or chemically adsorbed (Walton et al., 2002). Analysis of the differential reflectivity of adjacent stripes in a conventional light microscope permits decoding of the striped patterns. To date, two colloidal rod suspensions with a different code have been used to demonstrate that these rods can be used as supports for biological screening. The encoded rods were successfully used to distinguish between human and rabbit immunoglobulin (IgG) in a "sandwich" type hybridization assay (Nicewarner-Pena et al., 2001).

Another method of optical encoding involves the incorporation of light-emitting nanocrystals (2–15 nm) into 1.2-µm polymeric support beads in controlled ratios (Han et al., 2001; Bruchez et al., 1998; Zhou et al., 2001). Nanocrystals (or quantum dots) are nanometer-sized crystallites of one semiconductor embedded in another semiconductor (e.g. cadmium selenide capped with zinc sulfide). The nanocrystals are often composed of atoms from groups II–VI or III–V elements in the periodic table and are defined as particles with physical dimensions less than the exciton Bohr radius (Chan et al., 2002; Henglein, 1989; Alivisatos, 1996). A unique property of nanocrystals is the change in the electronic structure as a function of particle size. For example, size effects begin to influence the absorption and emission spectra of CdSe particles smaller than ~ 15 nm in diameter. Because of the confinement of electrons and holes in the nanocrystals, the energy level scheme resembles that of an atom, with many discrete energy levels. As the particle size decreases, the separation between energy levels increases and results in a blue shift in the absorption and emission spectra of the nanocrystals. Many sizes of nanocrystals can be excited at a single wavelength, resulting in several emission wavelengths (colors) that can be detected simultaneously (Bruchez et al., 1998). Nie and colleagues recently reported a DNA hybridization experiment which involved four colloidal suspensions encoded with nanocrystals (Han et al., 2001). A different oligonucleotide probe was attached to the colloids in each suspension and these were screened against a fluorescently labeled target DNA using a flow cytometer. Encoding of the colloidal support beads was performed by adding a different ratio of three nanocrystal sizes to four sets of polymer colloids (0.1–5.0 µm) in a propanol (or butanol)/chloroform mixture before transferring to an aqueous solution. Nanocrystals are inherently unstable under aqueous conditions so, to improve their stability, the support beads were coated with a thin polysilane layer which seals in the nanocrystals (Han et al., 2001).

Box 17.2 Summary of optical encoding and decoding methods for colloid-based non-combinatorial libraries

Library	Optical encoding method	Detection instrumentation	References
Non-combinatorial	Colloids stained with fluorescent dyes	Flow cytometry	Fulton et al. (1997)
	Colloids stained with lanthanide complexes	Fiber-optic arrays	Steemers et al. (2000)
	Striped metal rods	Differential reflectivity on optical microscope	Walton et al. (2002); Nicewarner-Pena et al. (2001)
	Nanocrystals entrapped in colloids	Flow cytometry	Han et al. (2001); Bruchez et al. (1998); Zhou et al. (2001)

17.3.2
Encoding Colloids for Combinatorial Libraries

Colloidal solid supports used for combinatorial library synthesis differ markedly from colloids required in non-combinatorial library preparation in that they must withstand harsh solvents and reagents (e.g. acetonitrile, dimethylformamide, tetrahydrofuran) and their encoding system must be impervious to these conditions. The optically-encoded colloids discussed in Section 17.3.1 are stable under aqueous conditions only and will lose their code should they be placed in an organic solvent. Several methods have been devised for encoding solid support beads in combinatorial libraries (Czarnik, 1997). The most common approach is to covalently bind molecular "identifier" tags to the beads in parallel with the compound synthesis (see Tab. 17.1b). These tags may be oligonucleotides (Needels et al., 1993), electrophoretic molecular tags (Ohlmeyer et al., 1993), cleavable dialkylamine tags (Fitch et al., 1999) or trityl mass-tags (Shchepinov et al., 1999). However, additional chemical steps are needed to synthesize the tags on the beads, and artifacts may arise as a result of interfering chemistries between the compound synthesis and the tag synthesis. The requirement for compatible probe and tag synthesis places a considerable restriction on the molecular tagging procedure. Finally, identification of the compound on the bead, through the analysis of the tag, involves added procedures and expensive equipment.

For combinatorial libraries, there is great scope to exploit the advantages of optical encoding. The concept of tagging solid support beads with fluorophores has been present in the literature for several years (Egner et al., 1997; Scott and Balasubramanian, 1997; Campian et al., 1994a,b; Yan et al., 1999; Nanthakumar et al., 2000). In the mid-1990's, Furka and colleagues recognized that fluorophores and chromophores could be covalently synthesized onto aminoalkyl colloids before or during split-and-mix synthesis (Campian et al., 1994a,b). Later, Egner et al. (1997) presented a study in which six dyes were covalently coupled to six different portions of solid support colloids to encode the first reaction step in the synthesis of a small polypeptide library (448 compounds).

Scott and Balasubramanian (1997) used fluorescence spectroscopy to explore the fluorescence properties of support bead suspensions. A variety of commonly-used fluorophores were covalently attached to the beads at high and low loading levels, and implications for on-bead screening were briefly discussed. Yan et al. (1999) took this research a step further by investigating the fluorescence properties of single beads using fluorescence microspectroscopy. A self-quenching effect was found and Yan et al. (1999) called for careful selection of fluorophores and rigorous control of the labeling reaction yield in order to generate labeled beads for combinatorial chemistry.

More recently, Nanthakumar and co-workers (2000) demonstrated oligonucleotide synthesis on fluorescently encoded support beads generated by covalent attachment of linkers and dye and showed the feasibility of using flow-cytometry sorting to identify and separate four bead sets. The sequences synthesized on each bead set were identified by performing hybridization with fluorescently labeled complementary sequences.

Despite the fact that fluorescent encoding of combinatorial libraries has been attempted numerous times over the past few years, the maximum library size which can be encoded is surprisingly small. The common theme behind all of the above-mentioned examples is that the fluorophores are covalently attached to pre-synthesized colloids, where they may be detrimentally exposed to the solvents used for the combinatorial library synthesis. Two alternative encoding strategies which require the use of fluorescent colloids to optically encode large combinatorial libraries have recently been reported (Battersby et al., 2000; Grøndahl et al., 2000; Battersby et al., 2001; Matthews et al., 2001; Trau and Battersby, 2001; Battersby and Trau, 2002; Battersby et al., 2002). These two strategies are described in Sections 17.3.2.1 and 17.3.2.2 and are summarized in Box 17.3. Methods of synthesizing fluorescent colloidal particles, and their detailed characterization, are discussed in Sections 17.4 and 17.5.

Box 17.3 Summary of optical encoding and decoding methods for colloid-based combinatorial libraries

Library	Optical encoding method	Detection instrumentation	References
Combinatorial	Covalent attachment of fluorescent dyes to pre-synthesized colloidal support beads	Fluorescence spectroscopy; fluorescence microspectroscopy; fluorescence microscopy; flow cytometry	Egner et al. (1997); Scott and Balasubramanian (1997); Yan et al. (1999); Nanthakumar et al. (2000)
	Attachment of multi-fluorescent colloidal particles (reporters) onto support beads	Fluorescence microscopy; flow cytometry	Battersby et al. (2000); Grøndahl et al. (2000); Matthews et al. (2001)
	Covalent binding of multiple fluorescent dyes into support beads during their synthesis	Flow cytometer equipped with high-performance sort classifier	Battersby et al. (2001); Trau and Battersby (2001); Battersby and Trau (2002a); Battersby et al. (2002b)

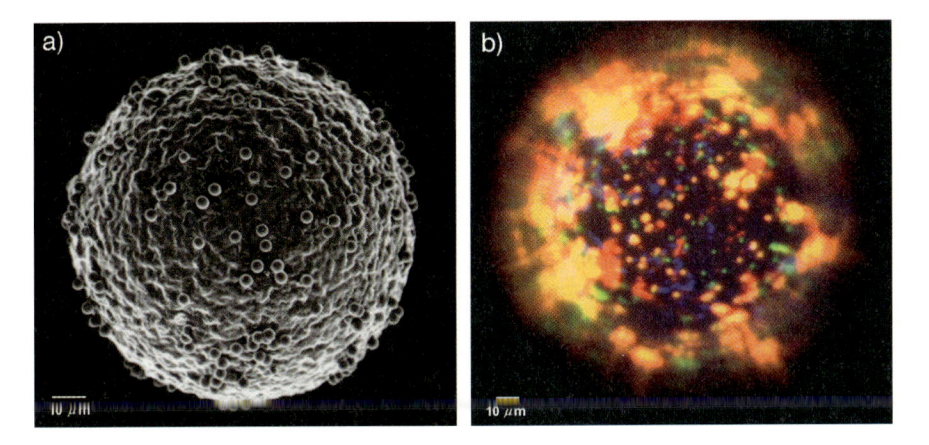

Fig. 17.6 **a** Scanning electron micrograph of a solid support bead (100-μm polystyrene/divinylbenzene particle), encoded with silica reporter particles (2.5-μm diameter) during the split-and-mix process shown in Fig. 17.1. **b** This composite of three fluorescence microscopy images shows a solid support bead with the attached reporters forming a record of the reaction history of the bead (i.e. encoding method I). Determining the dye combination in each type of reporter, using various optical filters in a fluorescence microscope, permits unique identification of the attached compound for libraries of more than 10^{10} compounds.

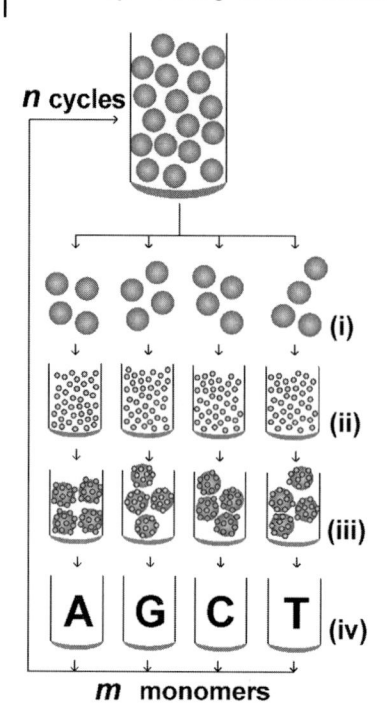

Fig. 17.7 Active optical barcoding of combinatorial libraries. **a** A schematic diagram of a split-and-mix DNA library synthesis on support beads: (i) a large number of colloids is equally partitioned into $m = 4$ vessels; (ii) each portion of beads are mixed with a unique type of reporter that contains a distinct combination of fluorescent dyes; (iii) fluorescent silica reporters become attached to each colloid and (iv) a different monomer (i.e. one of 4 nucleic acids, A, G, C or T) is reacted with each portion, and the beads are recombined to complete the cycle. The split-and-mix process is repeated for a chosen number of cycles, n, resulting in a large DNA library consisting of all combinations of oligonucleotides of length 'n'. The number of oligonucleotides in the library is given by m^n.

17.3.2.1 "Bead-on-Bead" Barcoding of Combinatorial Libraries – Strategy I

The so-called "bead-on-bead" barcoding strategy involves the generation of a *fluorescent barcode* on each support bead through attachment of multi-fluorescent reporter particles (0.5–1.0 µm) during combinatorial synthesis (Battersby et al., 2000; Grøndahl et al., 2000) (Figs. 17.6 and 17.7, Tab. 17.1d). In the split-and-mix process, the beads are first split into several portions and each portion is mixed with a different reporter suspension (Fig. 17.7). The reporter suspensions are distinguishable by the reporters they contain; reporters within a suspension are identical (i.e. same dye combination and size), but each suspension contains a different type of reporter. The support beads in each portion become coated with multiple reporters (>50 reporters per bead) of the same type (Fig. 17.6). Each portion is then reacted with a different monomer (e.g. amino acid, nucleic acid, etc.) and the portions are recombined to complete a cycle (Fig. 17.7). The above processes are repeated for a chosen number of cycles, resulting in an encoded chemical library ideally consisting of all monomer combinations. The various fluorescent barcodes that are generated on the support beads during library synthesis are a record of the reaction history of each bead. Each reporter encodes for a particular monomer in a particular cycle in the split and mix synthesis. Reading the barcode, by imaging the bead through various filters in a fluorescence microscope, permits accurate and inexpensive decoding of the combinatorial sequence generated on the bead during compound synthesis. Thus, the compound attached to the bead is unambiguously identified.

The advantage of this encoding strategy lies in the effective use of very few fluorescent dyes to record an enormous amount of information. The number of different reporter suspensions, R, required to encode every reaction in a split-and-mix process is equal to the number of different monomers used per cycle, m, multiplied by the number of cycles performed, n (see Eq. 1). A chosen dye can be either present or absent in a reporter, so the number of reporters possible from c different dyes is 2^c (see Eq. 1). Since the maximum number of compounds, M, in a split-and-mix synthesis is given by the number of monomers used, raised to the power n (see Eq. 2), a simple combinatoric analysis of this process reveals the general equation (Eq. 3):

$$R = m \times n = 2^c \tag{1}$$

$$M = m^n \tag{2}$$

$$M = \left(\frac{2^c}{n}\right)^n \quad \text{when} \quad 2^c > n \tag{3}$$

where R=number of reporters required, M is the maximum number of compounds uniquely encoded by this method, n=number of cycles in the split-and-mix synthesis and c=total number of colors/dyes used in the encoding procedure.

Because of the nature of Eq. (3), that is, a power raised to a power, the number of compounds which can be encoded by this method becomes explosive with increasing c (i.e. number of fluorescent dyes used). A simple calculation reveals that libraries containing more than 4×10^9 oligonucleotides (4^{16} probes) could be encoded with just $c=6$ fluorescent dyes. To encode 10^{32} compounds (as would be obtained from a 25-cycle split-and-mix process using all twenty naturally occurring amino acids), only $c=9$ colors would be required (Battersby et al., 2001). Of course, this library would be impossible to synthesize given that 10^{32} bead-based compounds would weigh more than the mass of the earth. However, this example serves to demonstrate the incredible information storage capacity of this technique. For any practical use, it is envisaged that no more than seven or eight spectrally distinguishable dyes will ever be required.

As can be appreciated, robust adhesion of the small reporters to the solid support beads is absolutely critical for the viability of this method. This adhesion is achieved through the use of controlled-polyelectrolyte (charged-polymer) multilayer coating of reporters (Battersby et al., 2000; Grøndahl et al., 2000). These coating methods will be described in further detail in Sections 17.6.1 and 17.6.2.

17.3.2.2 Colloidal Barcoding of Combinatorial Libraries – Strategy II

A second colloidal barcoding strategy for combinatorial libraries, illustrated in Fig. 17.8, is very different from the encoding strategies already described in earlier sections (see Tab. 17.1 e). This method involves production of internally encoded, solvent-resistant colloidal support beads with a view to tracking each colloid through the split-and-mix synthesis based on its "optical signature" (Battersby et al., 2001;

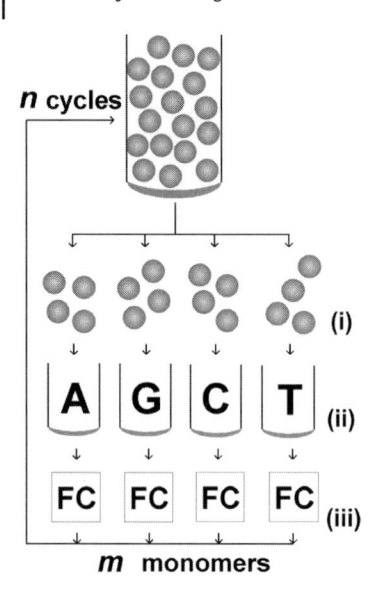

Fig. 17.8 Combinatorial library synthesis on optically unique colloids. In the split-and-mix cycle, a large number of colloids is partitioned into several vessels (i), a different nucleic acid (A, G, C or T) is reacted with each portion (ii), the colloids are passed through the flow cytometer (FC) (iii) and then the particles are recombined to complete the cycle. The process is repeated for a chosen number of cycles, n, resulting in an oligonucleotide library consisting of all m^n combinations. The flow cytometer tracks the synthesis performed on each colloidal particle by detecting its optical signature and storing the reaction history of each particle.

Trau and Battersby, 2001; Battersby et al., 2002). The attributes of each colloid that compose the optical signature, such as fluorescence emission, fluorescence intensity, size and refractive index profile, can be accurately detected at an extremely high rate (up to 50 000 particles s^{-1}) using a high-performance flow cytometer (see Fig. 17.5). Using a flow cytometer equipped with a special high-performance sort classifier, the data from several detectors can be stored as matrices, thereby preserving the association of events (i.e. the optical signature). Before probe synthesis is initiated, the modified flow cytometer can be used to separate those colloids with unique optical signatures from a population of optically diverse colloids. After collection of a chosen number of uniquely encoded colloids, split-and-mix library synthesis is performed, tracking each colloid at each cycle via its optical signature and recording its reaction history. The structure of any probe is then identified by analyzing the optical signature of the colloid on which it resides. This is done automatically, using the data that was stored by the flow-cytometer software during the compound synthesis.

Colloidal suspensions containing millions of beads with diverse optical signatures can be prepared using the particle synthesis techniques described in Section 17.5.4. These techniques involve the synthesis of multiple fluorescent silica shells around core particles. As shown in Fig. 17.9, by examining only three attributes of an optical signature in a flow cytometer, it appears that many particles possess overlapping signatures and that they are impossible to differentiate. However, by incorporating more dyes into the colloids and detecting the optical signature in a larger number of parameters, many colloids may be distinguished. That is, the use of 11 parameters increases the number of particles that are optically distinguishable in the colloid population to more than 99% (Miller and Trau, personal communication).

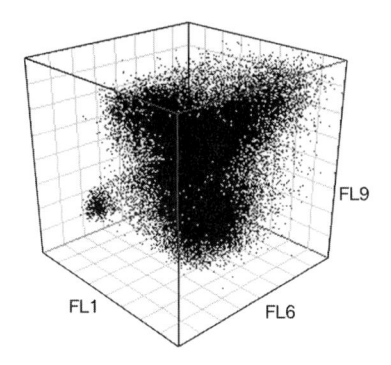

Fig. 17.9 Typical three-dimensional plot of intensity parameters from three detectors (FL1, FL6 and FL9; refer to Tab. 17.5) in a flow cytometer, which demonstrates the optical diversity in a combinatorially synthesized set of shell particles (as described in Section 17.6.3).

17.4
Synthesis of Colloidal Silica

17.4.1
Synthetic and Mechanistic Aspects of Colloidal Silica Particle Formation

Colloidal silica particles have a myriad of commercial uses that depend on the ability to carefully control the size and structure of these particles. A relatively monodisperse population of colloidal silica spheres can be synthesized by the sol–gel method involving the hydrolysis and condensation of silicon alkoxide in mixtures of ethanol and water, using either acid or base catalysis. This method was first developed as a controlled synthesis by Stöber et al. (1968) based on the work of Kolbe (1956) and hence is known as the "Stöber synthesis". Their initial work has been the foundation of multiple studies as efforts to precisely identify the factors and mechanisms of the underlying reactions continue.

The fundamental reaction is shown as the hydrolysis and condensation of tetraethyl orthosilicate (TEOS) below.

Stöber et al. examined the hydrolysis of a series of tetraalkyl silicates (methyl to pentyl) using ammonia as the catalyst in the presence of water and alcohol

Scheme 17.1

Tab. 17.2 Influence of reagents on Stöber synthesis of monodisperse silica particles

	Tetraalkyl silicate	Alcohol	Ammonia	Water	Reaction temperature
Chain length	TMOS: Fastest reactions and smallest particles (0.2 μm). TEOS: Particles cannot be made larger than 1 μm (in ethanol). TPOS: Reacts slowly to give 2 μm particles but wide dispersity (can be narrowed with 1:3 methanol/n-propanol)	CH_3OH: Fastest reaction rates. C_4H_7OH: Slowest reaction rates. C_4H_7OH: Biggest particles but wide size distribution. 1:1 CH_3OH/C_4H_7OH gives uniform large particles. Higher alcohols slow down reaction but cause median size and size distribution to increase simultaneously			
Low			Ammonia influences morphology and makes spherical particles. Silica irregularly shaped if absent		282K: Monodispersity not always attained
High	Maximum size achievable by increasing conc. of TEOS is 800 nm but bimodal particle size distributions common		Increasing conc with other parameters constantly increases particle size which goes through a maxima	Increasing amount of water increases size but exhibits a maxima	Narrow size distributions for TEOS. Particle sizes decrease as temp. increases to 388K

TMOS=tetramethylorthosilicate; TEOS=tetraethylorthosilicate; TPOS=tetrapentylorthosilicate; conc=concentration.

(methanol to n-propanol) and reported that the condensation reaction typically commenced 10 min after the hydrolysis had been initiated. Bogush et al. (1988) expanded the initial studies of Stöber et al. (1968) by publishing more detailed reagent concentrations and reaction environments controlling the particle size of the resultant silica. Tab. 17.2 summarizes the observations of Stöber and Bogush in terms of the variable influences of each participant in the reaction. This method of synthesizing colloidal silica offers great versatility in the control of the ultimate particle size by altering the concentration of water and ammonia in ethanol.

In using the above approach to synthesize colloidal silica particles, it was found that a maximum size of 2 µm (using tetrapentylorthosilicate in 1:3 methanol/n-propanol) could be achieved, but these large particles were not truly monodisperse. If TEOS was used then the maximum particles size was 1 µm (Stöber et al., 1968) and again the size distribution was poor for these larger particles. Bogush et al. (1988) improved the monodispersity of particles produced from TEOS by using a seeded-growth approach whereby small core silica particles are coated with further TEOS shells while retaining the initial monodispersity.

During their extensive studies into the impact of the different factors on size, polydispersity and structure of the final silica particles Bogush et al. (1988) were able to speculate about the mechanism of particle growth. They concluded that the growth mechanism was "self sharpening", i.e. small particles grow faster than large particles, and that this mechanism does not alter significantly during the course of the reaction. They postulated that particle growth occurs primarily through an aggregation mechanism where monodispersity is achieved as a result of size dependence, that is, the probability of aggregation between two particles of the same size decreases as the particles grow.

The mechanism of particle growth has since been the subject of an active debate based on two significantly different models: (a) the "growth-only" model which neglects aggregation (Byers et al., 1987; Matsoukas and Gulari, 1989, 1991) and holds that monodispersity is achieved by a brief "burst" of nucleation followed by diffusion-limited growth; (b) the "aggregation-only" model in which a narrow size distribution is formed through the continuous nucleation and aggregation of the nuclei and subsequent aggregation amongst all the particles. Bogush and Zukoski (1991) and Harris (1992) both arrived at this second model from different perspectives; however, both successfully predicted the size distribution resulting from the Stöber synthesis. Lee et al. (1997) assessed the applicability of both models by monitoring intermediate concentrations in the Stöber synthesis. Their findings were consistent with the aggregation model based on several observations. They found that the number density of detectable particles increased over time and that there was no evidence of a single burst of nucleation. The second hydrolysis of the monomer is the rate-limiting step for nucleation and precipitation of this doubly hydrolyzed monomer is more likely as a nucleation mechanism than auto-accelerating polymerization. They also established that the final particle size is related to both the nucleation rate and the ionic strength, suggesting aggregation of charged colloidal structures.

The actual mechanism of silica particle nucleation and growth has only recently been substantiated satisfactorily (Boukari et al., 1997, 2000; Pontoni et al., 2002) where the technique of small-angle X-ray scattering (SAXS) has been the common tool in these studies. As a result of this research there is overwhelming evidence to support the nucleation and growth mechanism as opposed to the "burst" of nucleation model proposed by Matsoukas and Gulari (1989, 1991). This evidence suggests that the first particles to appear have a diameter between 10 and 20 nm (Boukari et al., 1997, 2000) which subsequently aggregate and grow to form larger particles. These first or "primary" polymeric particles become more dense by intraparticle condensation and subsequently grow through surface addition. In fact, the concept of the formation of "primary" particles had been first introduced in 1992 by Bailey and Mecartney in a cryo-electron microscopy study now recognized as very significant in this field. Boukari et al. (2000) proposed that these small primary particles are formed during an induction period during which the hydrolysis of TEOS dominates and is incomplete due to low concentrations of water and ammonia. This mechanism is supported by NMR data which demonstrates that, preceding condensation, only the unhydrolyzed monomer and the first hydrolysis product are present in solution (van Blaaderen and Kentgens, 1992; Lee et al., 1997).

In summary, the mechanism of the formation and growth of colloidal silica particles has two stages: (1) small primary particles of diameter less than 20 nm formed during an initial hydrolysis reaction aggregate to form larger particles where cross-linking continues to increase the density of the particles during condensation; (2) subsequent hydrolysis and condensation occurs to grow these large particles. This mechanism continues to be supported by the most recent time-resolved SAXS study (Pontoni et al., 2002) which reported the use of high-brilliance synchrotron radiation and high-sensitivity detection to monitor the complete reaction of the Stöber synthesis from nucleation to stable particles.

17.4.2
Determination of Particle Size

In producing colloidal silica particles by controlled synthesis for a particular application, the ability to accurately determine the particle sizes and distributions is vital for quality assurance. The determination of the particle diameter in studies that describe the synthesis of colloidal silica particles is typically based on data gained from either dynamic light scattering or electron microscopy techniques. In comparing the diameters generated by these two methods there is often a discrepancy in the sizes reported from each. In electron microscopy the samples are dried prior to being examined under a vacuum and this inevitably causes shrinkage of the particles. In dynamic light scattering the diameter is a hydrodynamic parameter derived from diffusion constants determined from the Brownian motion of the particles and as such represents an indirect number. Small-angle X-ray scattering studies have provided valuable insight into the validity of the numbers generated by the above techniques but have also allowed the investigation of the internal structure of the silica particles. Megens et al. (1997) found that it was possible to resolve both particle sizes and dispersities producing diameters which were greater than those determined by transmission electron microscopy (TEM) and smaller than those determined by dynamic light scattering. However, they were in excellent agreement with values determined by static light scattering (SLS) which is essentially very similar in methodology to SAXS. Both SLS and SAXS determine the intensity of scattered light using detectors at multiple angles but in the latter the internal structure of the particles can be probed because the wavelength of the light is short compared to the particle size and can be measured at small angles. In their studies of silica particles provided by van Blaaderen and Imhof (University of Utrecht) using SAXS, Megens et al. (1997) were able to demonstrate that there were clear differences in electron densities between the core and the shell of particles grown in a stepwise manner such as the seeded growth approach. Thus SAXS has offered a valuable method of quantifying the diameter, dispersity and particle structure but is generally not readily accessible to most laboratories.

17.4.3
Mono-fluorescent Colloidal Silica Particles

A modified Stöber synthesis (described below) has been used to generate two forms of colloidal silica particle for the two different barcoding strategies described in Sections 17.3.2.1 and 17.3.2.2. For the first strategy, small particles are grown to an optimal size of around 800 nm (Fig. 17.10a and b). The second strategy requires core silica particles (2–5 µm) grown larger through a modification of the procedure based on the seeded growth method. These colloids, termed "shell particles" are illustrated in Fig. 17.10c and d. The formation of mono-fluorescent 800-nm particles (Fig. 17.10a), as well as shell particles containing a single fluorescent dye (Fig. 17.10c), is described in this section. The synthesis and characterization of particles containing multiple fluorescent dyes is more complex (Fig. 17.10b and d) and thus, a complete section (Section 17.5) is devoted to this subject.

The process of incorporating one fluorescent dye in a silica colloid has historical precedence in the work of van Blaaderen and co-workers (1992) and Nyffenegger (1993). The first reports of the incorporation of fluorescent molecules into colloidal silica particles, either during nucleation and growth or as a subsequent seeded-growth process, were developed by van Blaaderen and Vrij (1992). Their approach was to include fluorescein isothiocyanate (FITC, Tab. 17.3) which had been conjugated to an organosilica molecule in different ways within the silica particle to create a variety of particles each distinguished by the location of the dye within the particle. This resulted in particles which possessed a FITC core, an outer FITC shell and a FITC shell subsequently coated by a silica shell. Nyffenegger et al. (1993) extended the work of van Blaaderen by preparing small monodisperse particles (100 nm) that had a high degree of FITC labeling. They used the co-condensation of TMOS and a thiourea derivative to form both core particles and particles onto which a fluorescent shell was formed by seeded growth leading to larger particles. Monodispersity was only achieved through the stepwise addi-

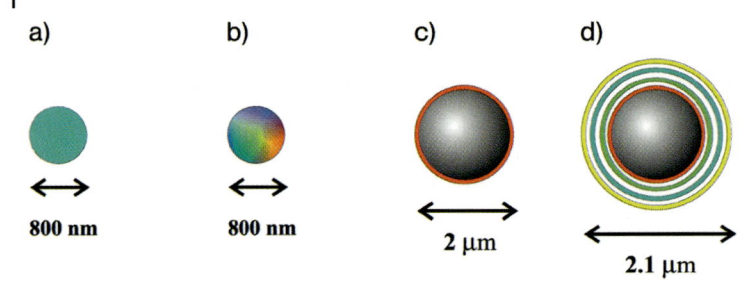

Fig. 17.10 Schematic illustration of the different colloidal silica particles: **a** mono-fluorescent core particle; **b** multi-fluorescent core particle; **c** mono-fluorescent shell particle; **d** multi-fluorescent shell particle.

tion of minute amounts of TMOS. FITC is the fluorophore of choice in all of the studies mentioned above, although the protocol was extended successfully to rhodamine isothiocyanate (XRITC, Tab. 17.3) (Verhaegh and van Blaaderen, 1994) indicating great potential for other dyes.

The first step in producing fluorescently labeled silica involves conjugation of an isothiocyanate or succinimidyl ester derivative of the fluorescent dye to the coupling agent 3-aminopropyl trimethoxysilane (APS) as illustrated in the equation below.

To form fluorescent core particles, the APS-dye conjugate formed by this process is subsequently added to a solution of ethanol/water/ammonia before TEOS is added to initiate particle formation and growth. The molar ratio of APS-dye

Tab. 17.3 Absorption (λ_{abs}) and emission (λ_{em}) maxima for dyes used in the mono-fluorescent and multi-fluorescent particles.[1]

Dye	Free dye		Free dye[2]		Bound dye in shell particles[3]	Bound dye in core particles[2,4]
	λ_{abs} (nm)	λ_{em} (nm)	λ_{abs} (nm)	λ_{em} (nm)	λ_{em} (nm)	λ_{em} (nm)
A350	346	445 (pH7)	350	429	437	420
A430	430	545 (pH7)	430	524	535	514
DPITC	335	536	335	522	–	515
FITC	494	519 (pH9)	494	538	518	515
OG488	495	521	–	–	522	–
NFSE	602 (pH10)	672 (pH10)	602	660	675	652
XRITC	–	–	580	603	597	594
TAMRA	546	576	546	576	581	–
BTRX	–	–	588	616	619	–
B630	632	650	–	–	663	–
DAMCA	–	–	437	472	481	–

1) Solvent = water.
2) Solvent = DMF
3) Single dye layer + single silica layer shell particles, solvent = 25% ethanol in water
4) Prepared using APS-dye conjugate: TEOS ratio of 1:54, 1.7–4.4 µmol dye/g silica.

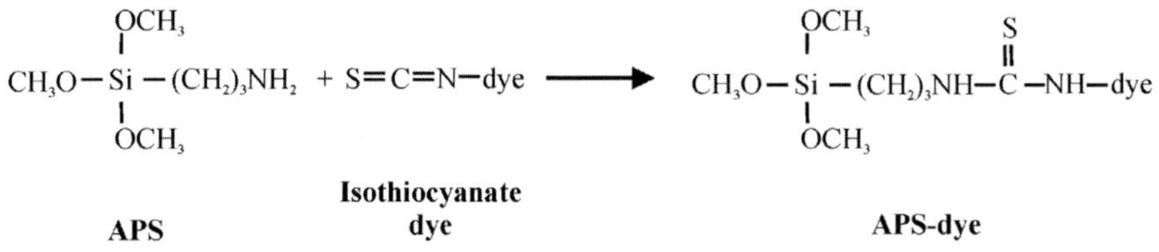

Scheme 17.2

$$CH_3O-\underset{\underset{OCH_3}{|}}{\overset{\overset{OCH_3}{|}}{Si}}-(CH_2)_3NH_2 + S=C=N-dye \longrightarrow CH_3O-\underset{\underset{OCH_3}{|}}{\overset{\overset{OCH_3}{|}}{Si}}-(CH_2)_3NH-\overset{\overset{S}{\|}}{C}-NH-dye$$

APS · Isothiocyanate dye · APS-dye

conjugate to TEOS affects the "brightness" or intensity of fluorescence as well as the particle size (Imhof et al., 1999; Matthews et al., 2001). Imhof et al. (1999) observed a large red-shift of 10 nm in both absorption and emission spectra of the FITC dye upon increasing the dye concentration in the silica particles. Similarly, a red-shift was observed for silica particles incorporating the dye NFSE (Matthews et al., 2001). An increase in size with increased amount of APS-dye conjugate was observed in both studies (Imhof et al., 1999; Matthews et al., 2001). Using APS-FITC conjugate to TEOS molar ratios of 1:215 yields a particle size of 260 nm, whereas a molar ratio of 1:13 yields a particle size of 620 nm (Matthews et al., 2001). This size increase has been attributed to a lowering of the particle surface charge (Imhof et al., 1999).

To form shell particles the fluorescent dye is conjugated to APS and then added to a suspension of core particles in a water/ethanol/ammonia mixture. TEOS is added to generate a silica/dye network on the surface of the core particle (Lawrie et al., 2003; van Blaaderen and Vrij, 1992) as demonstrated in the schematic below.

In both types of silica particles the spectroscopic features of the fluorescent dye is affected by its local environment, i.e. the silica network and the solvent (Section 17.5.1). Synthesizing a silica shell onto the fluorescent shell encases the fluorescent dye molecules in a silica environment, thus somewhat shielding the dye from the solvent.

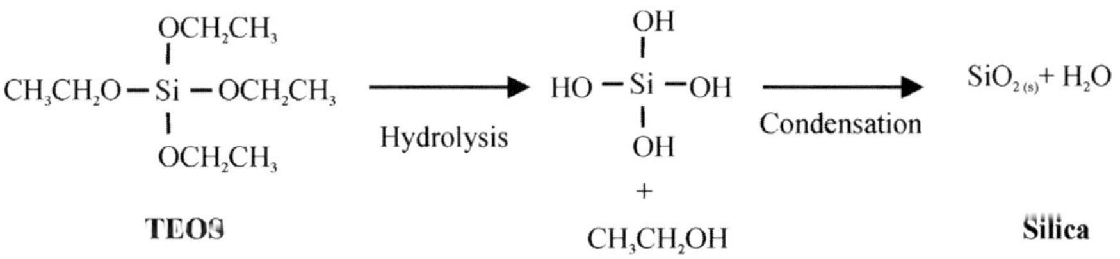

Scheme 17.3

$$CH_3CH_2O-\underset{\underset{OCH_2CH_3}{|}}{\overset{\overset{OCH_2CH_3}{|}}{Si}}-OCH_2CH_3 \xrightarrow{\text{Hydrolysis}} HO-\underset{\underset{OH}{|}}{\overset{\overset{OH}{|}}{Si}}-OH + CH_3CH_2OH \xrightarrow{\text{Condensation}} SiO_{2(s)} + H_2O$$

TEOS · Silica

17.5
Multi-fluorescent Colloidal Silica Particles

For the application of colloidal silica in biological screening (Strategy I and II, Sections 17.3.2.1 and 17.3.2.2), the particles are required to display unique optical properties; this is achieved by encapsulating multiple fluorescent dyes in the silica framework. As described in Section 17.4.3, the synthesis of silica colloids containing one fluorescent dye in its core or in a shell has been studied widely. However, the incorporation of multiple dyes into a particle is a challenging and interesting subject. In this section we describe novel methods for incorporating fluorescent dyes either (1) into the core of a silica particle during its synthesis (Matthews et al., 2001) or (2) into a number of shells on the surface of the core particle, thereby generating optically complex colloidal particles which are multiply fluorescent (Battersby et al., 2002).

17.5.1
Considerations of Multiple Dyes Within Close Proximity

In bringing several different fluorescent dyes together in a single silica particle, either as a mixture for encapsulation during the nucleation and growth process or as concentric shells synthesized onto a core silica particle, a number of fluorescent phenomena are encountered and must be addressed. Fluorescence occurs as a result of emission of light when an electron returns to the ground state orbital from an excited state orbital. The Jablonski diagram in Fig. 17.11 summarizes the processes experienced by the electrons involved in the excitation and emission of fluorescence for a fluorescent dye molecule (fluorophore). This diagram also illustrates the potential process of reabsorption of emitted energy by a second fluorophore. The excited state, S_1, for either fluorophore in Fig. 17.11 is generated by absorption of photons, hv_A, and subsequent relaxation to the ground state, S_0, by emission of photons, hv_F, or fluorescence (a). However, if a second fluorophore (Dye 2) is bound in the path of the emitted photons from the first fluorophore (Dye 1), there is the potential for process (b) to occur where the emitted photons of the correct energy are absorbed in creating the excited state for the second fluorophore (Dye 2). This would result in a loss of observed fluorescence intensity for the first fluorophore (Dye 1).

Two processes other than reabsorption have the potential to interfere with the fluorescence emission of the dyes when they are resident in the silica particles: fluorescence self-quenching and resonance energy transfer (RET). Fluorescence quenching results in a loss of fluorescence intensity and will occur when the con-

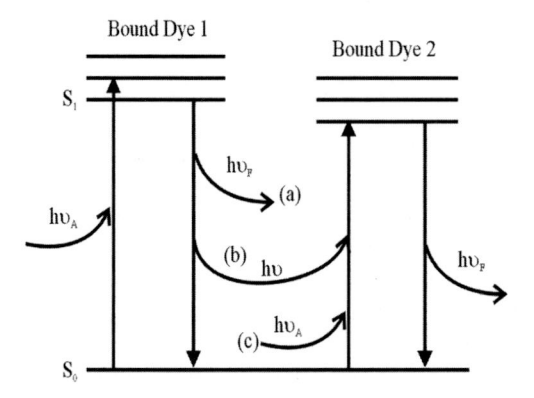

Fig. 17.11 A Jabloñski diagram illustrating possible processes for energy transfer between fluorophores bound in silica colloids: **a** represents emission of photons or fluorescence; **b** represents reabsorption of emitted energy and **c** represents absorption of photons.

centration of dye molecules in the particles reaches a critical value. Most fluorescent dyes are self-quenching if their concentration is high enough in solution and this translates into a similar effect for the bound dye molecules. Indeed, van Blaaderen and Vrij (1992) observed this effect for silica-bound FITC molecules.

Förster energy transfer (Förster, 1948) is a form of RET and is not a simple reabsorption of emitted photons. As for the reabsorption, there must be a spectral overlap between the emission spectrum of the donor and the excitation spectrum of the acceptor. Furthermore, the Förster distance is a critical distance between the donor and acceptor molecule where maximum energy transfer is equal to the decay rate of the donor fluorophore. The distance between the two fluorophores must be typically less than 6 nm (Förster, 1948). Indeed, the efficiency of energy transfer has been found to be 100% at a distance of 1.2 nm and 16% at 4.6 nm (Stryer and Haugland, 1967).

A significant consideration in encoding the silica colloids is the impact of the solvent environment. Where fluorescent silica particles are bound to polymer support particles to encode each step of a synthetic process, they may be suspended in differing solvents including dimethylformamide (DMF), ethanol and aqueous buffers. Any particles which are detected by flow cytometry may be suspended in ethanol or water. Solvents are known to have an impact on the position and sometimes the intensity of an electronic absorption or emission peak (Reichardt, 1988). This is known as solvatochromism and often accompanies a change in polarity of the solvent. This effect can be illustrated by comparing water and DMF as solvents for a suspension of fluorescent particles. DMF is a polar aprotic solvent with a dielectric constant, ε, of 37.4 and refractive index, n, of 1.43 and water is a polar solvent with a dielectric constant, ε, of 80.4 and refractive index, n, of 1.33. The effect of the dielectric constant and refractive index of the solvent on the wavelength of the Stokes' shift can be demonstrated by the Lippert equation (von Lippert, 1957):

$$\bar{v}_A - \bar{v}_F = \frac{2}{hc} \cdot S \cdot \frac{(\mu_E - \mu_G)^2}{a^3} + \text{const.} ; \quad S = \left[\frac{\varepsilon - 1}{2\varepsilon + 1} - \frac{n^2 - 1}{2n^2 + 1} \right]$$

where: $\bar{v}_A - \bar{v}_F =$ Stokes' shift, $\bar{v}_A$ is the wavenumber of the absorbance peak and $\bar{v}_F$ is the wavenumber of the fluorescence peak; ε is the dielectric constant of the solvent, n is the refractive index of the solvent, $(\mu_E - \mu_G)$ is the difference in the dipole moments of the excited and ground states of the fluorophore respectively, a is the radius of the cavity in which the fluorophore resides, h is Planck's constant (6.2656×10^{-27} erg s); and c is the speed of light (2.9979×10^{10} cm s^{-1}).

Assuming all other terms are constant, the value of the solvent term, S, for water will be twice that for DMF resulting in a larger energy difference or Stokes' shift when water is the solvent compared to DMF. This is demonstrated by the blue spectral shift of the dye emission peaks for the DMF systems (Tab. 17.3). The Lippert equation also demonstrates how the Stokes' shift is a function of the interactions between the fluorophore and its immediate environment such as the radius of the cavity where the fluorophore resides, a, and the dipole moments of the ground and excited state fluorophore molecules, μ_G and μ_E.

The selection of the dyes for inclusion in the multiply fluorescent particles (either internal or shell incorporation) is entirely dependent on the potential mode of detection or observation of the optical signature. The fluorophores that are used to internally encode silica particles must be distinguishable through appropriate filters on a fluorescence microscope. Those fluorophores used to synthesize the encoded shell particles are selected according to the available laser excitation source wavelengths (for example 333.6–363.8 nm, 488 nm and 635 nm).

17.5.2
Selection of Dyes for Strategy I Encoding

It is well known that both absorbance and emission bands are broad for organic fluorescent molecules and although new commercial fluorescent dyes have bandwidths at half height of 25 nm, there is still a limit, due to energy transfer reactions, as to how many dyes can be observed independently in the visible and near infrared-infrared (NIR-IR) spectrum. Detection of dyes in multi-fluorescent silica colloids is achieved by fluorescence microscopy and fluorescence spectroscopy. A conventional fluorescence microscope is equipped with optical filters permitting only a narrow range of excitation wavelengths to be activated and, at the same time, permitting only a narrow range of emission wavelengths to be observed. Thus, the filters allow for just a few fluorescent dyes to absorb and emit while other dyes are prevented from doing so. It is therefore not necessary to have unique and separate excitation as well as emission bands for the dyes to be individually detected. For example, the dyes A430 and DPITC (Tab. 17.3) have similar emission maxima (λ_{max}=524 and 522 nm, respectively); however, since they are excited at wavelengths ~100 nm apart (430 and 335 nm, respectively), optical filters can be used to distinguish them (Matthews et al., 2001). Fig. 17.12 shows that under Filter Set 1 (U-MWU, λ_{ex}=330–385 nm, λ_{em}>420 nm) A430 and DPITC silica colloids are very different in color (Fig. 17.12a and c). The A430 colloids are yellow, while the DPITC colloids are blue-green and therefore distinguishable. Under Filter Set 2 (U-MWB, λ_{ex}=450–480 nm, λ_{em}>515 nm), the A430 and DPITC colloids are both gold in color (Fig. 17.12b and d). Thus, the two dyes (A430 and DPITC) are distinguishable by observation through optical filters in a fluorescence microscope.

Observations using fluorescence microscopy are in agreement with fluorescence spectroscopy investigations. Thus, exciting DPITC at 430 nm yields no spectral features in the emission spectrum, and exciting A430 at 340 nm yields only a very minor emission band at 520 nm compared to its emission band when excited at 340 nm (relative intensities at the two excitation wavelengths=1:6).

Energy transfer (either by reabsorption or RET, Section 17.5.1) between dyes is a phenomenon generally to be avoided in this type of multi-fluorescent silica colloid. For example, if one dye has an emission wavelength overlapping the excitation wavelength of another dye, the second dye can consume the emitted energy from the first, thereby making the first dye invisible. However, in situations where the energy transfer reaction is only partial (so that both dyes can still be ob-

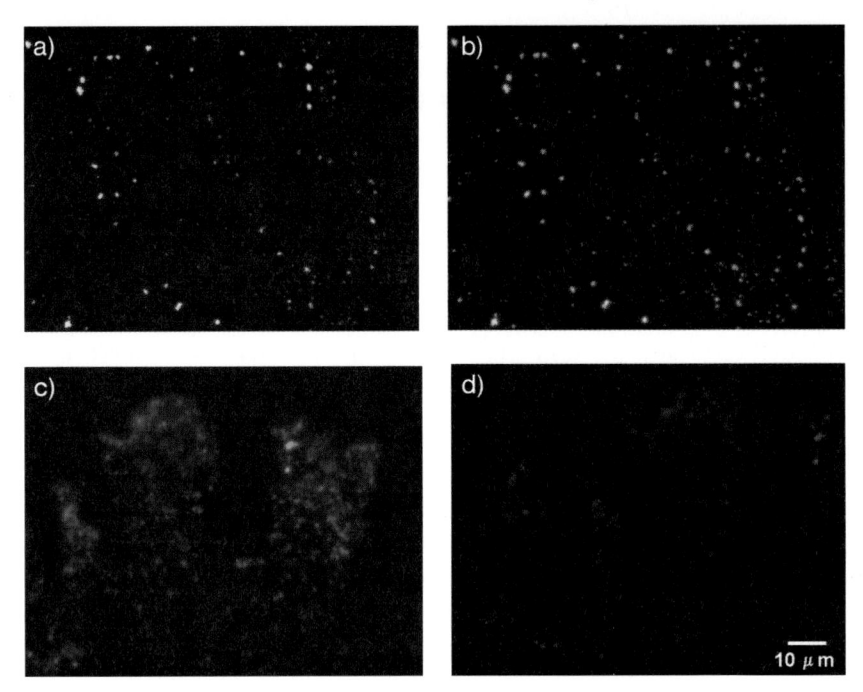

17.5 Multi-fluorescent Colloidal Silica Particles | 535

Fig. 17.12 Individual dyes in reporter particles can be distinguished through multiple optical filters in a fluorescence microscope. This figure shows the case where two dyes (DPITC and A430) emit at a similar wavelength (~524 nm), but are excited at very different wavelengths (335 and 430 nm, respectively). See text for details.

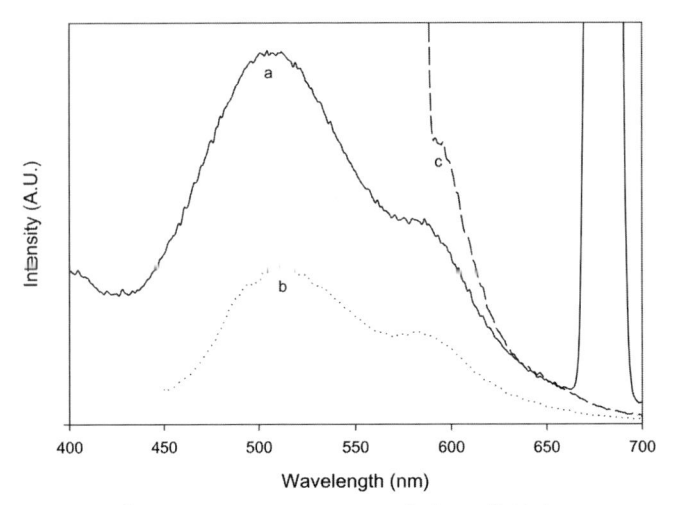

Fig. 17.13 Fluorescence-emission spectra of silica colloids incorporating the dyes A430 (λ_{ex}=430 nm), XRITC (λ_{ex}=580 nm), NFSE (λ_{ex}=602 nm), and DPITC (λ_{ex}=335 nm). Excitation wavelengths: **a** 340 nm; **b** 428 nm; **c** 580 nm.

served), the dyes are compatible. Such a partial energy transfer was observed by Matthews et al. (2001) for multi-fluorescent silica colloids containing the dyes A430, XRITC, NFSE and DPITC. Both A430 (excited at 428 nm) and DPITC (excited at 340 nm) displayed double peaks. Their major peak was observed at the expected wavelength of 523 nm and in addition, a peak was observed at 590 nm (Fig. 17.13). This additional peak is believed to be due to XRITC absorbing some of the energy emitted by A430 and DPITC, as the absorbance band of XRITC (580 nm) somewhat overlaps with the emission band of A430 and DPITC (523 nm). As can be seen in the fluorescence spectrum (Fig. 17.13), the dyes A430 and DPITC still have sufficient intensity at their expected wavelengths in the presence of XRITC. Clearly, the reabsorption of energy from A430 and DPITC does not restrict the use of these dyes with XRITC.

17.5.3
Multi-fluorescent Reporter Colloids for Strategy I Encoding

It has been demonstrated that it is possible to include at least six different fluorescent dyes into colloidal silica particles by combining their APS-conjugates during the silica synthesis (Matthews et al., 2001). The peaks in the fluorescence spectra of each dye are generally red-shifted in wavelength in comparison to the mono-fluorescent particles of the same dye. Upon increasing the amount of APS-dye conjugate in the reaction mixture, further red shifts in the emission bands are observed. This red shift is accompanied by either a minor increase or minor decrease in the fluorescent intensity.

The challenge of producing silica colloids with several fluorescent dyes in the core is for all the dyes to be discernable. The final intensity of a fluorescent dye in a mono-fluorescent colloidal silica particle is dependent on the amount of APS-dye conjugate incorporated into the silica network, on the inherent fluorescence intensity of the dye and on the way the local environment of the silica affects the intensity. In multi-fluorescent silica colloids, however, additional factors affect the final intensities. These factors are the relative rates of APS-dye conjugate hydrolysis, and energy transfer reactions between dyes in the silica colloid (Matthews et al., 2001). It is therefore not possible to use the relative intensities of the dyes in mono-fluorescent silica particles to optimize the intensities in multi-fluorescent particles. Instead, the relative intensities of the dyes in multi-fluorescent particles, which are prepared by using the same amount of APS-dye conjugate of each of the dyes, are used as a guide. Thus, to manipulate the final intensities of the dyes in the silica particles, the relative amounts of APS-dye conjugate is varied in a manner that offsets the relative intensities in the particles prepared by using same amount of all APS-dye conjugates. Fig. 17.14 illustrates this by comparing the fluorescence spectra of two different sets of five-dye particles. Both sets contain the same dyes, but different amounts of APS-dye conjugate were used in their preparation. In the first set (Fig. 17.14a), all APS-dye conjugates were added to the reaction mixture in the same concentration. It is clear that the dyes A350 and DPITC dominate and that the dye NFSE is barely observable. By contrast, in

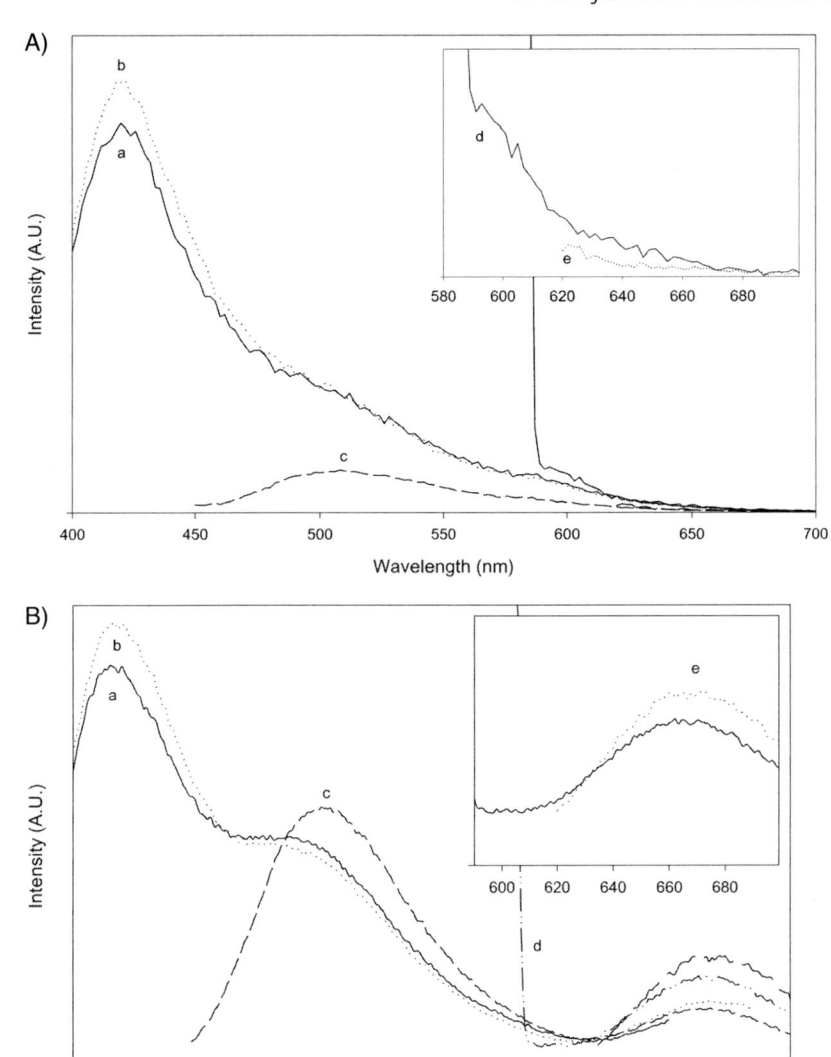

Fig. 17.14 Multi-fluorescent silica particles incorporating the dyes A350, A430, XRITC, NFSE, and DPITC. Excitation wavelength: **a** 340 nm; **b** 353 nm; **c** 428 nm; **d** 580 nm; **e** 603 nm. (A) APS-dye conjugate molar ratios of the dyes used in the synthesis: 1:1:1:1:1. (B) APS-dye conjugate molar ratios of the dyes used in the synthesis: 1:4:4:16:3 (reproduced from *Australian Journal of Chemistry*, Vol. 54, D.C. Matthews, L. Grøndahl, B.J. Battersby and M. Trau, 2001, with permission of CSIRO publishing).

the second set (Fig. 17.14b) a grading of the APS-dye conjugate amounts (see fig. caption) was used in the synthesis. In the spectra of the resulting particles the four dyes A350, A430, DPITC and NFSE are all of comparable intensity; however, XRITC is now barely observable. This disappearance of XRITC is likely to be due to energy transfer between XRITC and NFSE since the emission wavelength of XRITC (603 nm) matches the excitation wavelength of NFSE (602 nm). Clearly, due to the greater amount of NFSE compared to XRITC incorporated in this second set of particles, all the energy from XRITC is transferred to NFSE. With fine-tuning of the APS-dye amounts in these multi-fluorescent particles, it is possible to overcome or minimize this energy transfer phenomenon.

17.5.4
Multi-fluorescent Shell Colloids for Strategy II Encoding

Silica shell particles are required for Strategy II encoding (see Section 17.3.2.2) where a high-performance flow cytometer is used as the primary analytical tool. This instrument can have up to three laser excitation sources (typically, 351 nm, 488 nm and 635 nm) and a combination of fluorescence detectors as well as detectors for the scattering of light. The dyes selected for incorporation into multi-fluorescent shell particles must necessarily have an excitation peak in common with the laser sources. However, multiple detectors covering a range of wavelengths on

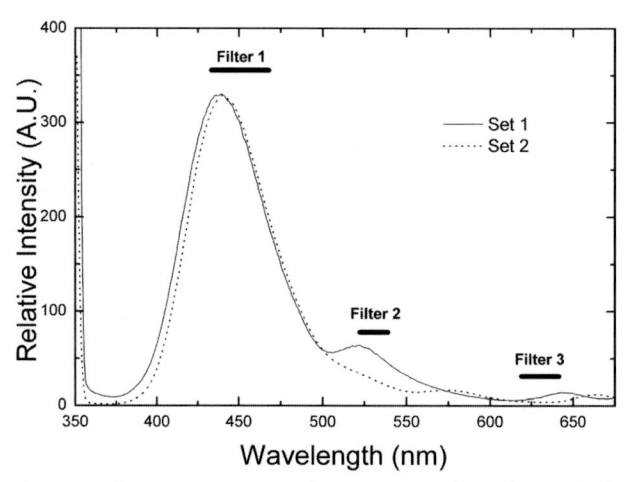

Fig. 17.15 Fluorescence spectra demonstrating where photomultiplier detectors respond to fluorescent emission as a result of the selection of filters. Filter 1 is 450/30 nm, Filter 2 is 530/40 nm and Filter 3 is 630/40 nm. Set 1 shell particles contain the following dyes synthesized combinatorially onto 2.9-µm core silica particles: B630 (λ_{ex}=632 nm)/OG488 (λ_{ex}=495 nm)/A350 (λ_{ex}=346 nm)/DAMCA (λ_{ex}=437 nm). Set 2 shell particles contain the following dyes synthesized combinatorially onto 5.2-µm core silica particles: B630 (λ_{ex}=632 nm)/OG488 (λ_{ex}=495 nm)/A350 (λ_{ex}=346 nm)/TAMRA (λ_{ex}=546 nm).

each laser path allows a range of fluorescent emission wavelengths to be detected. The wavelength-specific filters are interchangeable between the detectors on any laser path. The combination of fluorescence emission wavelengths detected for each excitation source is therefore variable, depending upon the dyes present in the particles. An example of a filter combination is demonstrated in Fig. 17.15 where the spectral ranges of three filters on the 351-nm laser path are superimposed on the fluorescence spectra for two sets of particles excited at 346 nm (Filters 1–3 in Fig. 17.15). The filters indicated in Fig. 17.15 would only permit photons emitted in the wavelength range represented by the bar to pass. The reader should be reminded at this point that fluorescence spectra are a representation of the fluorescence emission of the bulk suspension of particles whereas flow cytometry is sensitive to the fluorescence emission of each individual particle. The two sets of particles in Fig. 17.15 have a different number of photons with different fluorescence intensities detected through Filters 2 and 3.

This versatility in detection provides the opportunity to manipulate the refractive index profiles and optical signatures of the particles. The multi-fluorescent shell beads were developed to limit the energy transfer effects observed when fluorophores are in close proximity. By synthesizing a silica layer between the fluorescent layers, Förster energy transfer processes (Section 17.5.1) can be avoided. However, it has been shown that other energy transfer processes may occur, for example emission from excited fluorophores in a particular silica layer can excite other fluorophores within the same particle. This provides a powerful system in that there are three parameters that have the potential to be altered:

- The number and thickness of shells synthesized on the core particle
- The choice of dye for each of up to six different fluorescent dye layers
- The nature of the spacer layer by using materials other than silica (e.g. TiO_2, Al_2O_3)

In the following sections the two first principles are utilized to produce shell particles displaying various properties.

17.5.5
Emission and Morphology of Multi-fluorescent Shell Particles

The synthesis of multiple fluorescent and silica shells on a core particle has a major impact on the spectroscopic characteristics of dye shells that have been formed early in the synthetic sequence. Fig. 17.16 illustrates the impact of the number of deposited layers on the intensity of the fluorescence emission for a series of particles (Lawrie et al., 2003; Battersby et al., 2002). In this example, fluorophore A350 forms the first dye layer incorporated onto the surface of the colloids and up to 11 dye and silica layers are subsequently deposited on top of this layer. The subsequent dye layers contain the dyes BTRX, FITC, NFSE, TAMRA, DAMCA, respectively, none of which are excited significantly at 346 nm. As shown in Fig. 17.16, there is a significant reduction in the fluorescence emission intensities from the dye A350 as an increasing number of layers are deposited on top. The intensity reduction in the

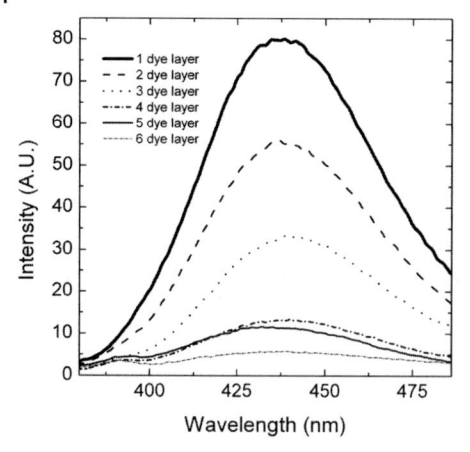

Fig. 17.16 Decrease in fluorescence emission from the innermost dye layer of shell particles, as additional silica layers are synthesized on top. The impact of increasing the number of silica layers deposited on an inner dye shell is visualized by exciting (at 346 nm) the innermost dye A350. Fluorescence emission from the A350 layer clearly decreases as more layers are deposited. These multi-fluorescent shell particles incorporated six dye layers (A350, BTRX, FITC, NFSE, TAMRA, DAM-CA) and six silica spacer layers (reproduced with permission from Battersby et al., 2002, *Chem. Commun.* 14, 1435–1441. Copyright 2002, Royal Society of Chemistry).

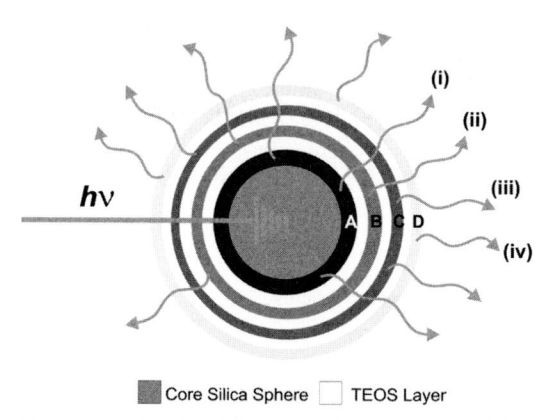

Fig. 17.17 Schematic diagram illustrating the potential influence of shells of variable refractive index on the scattering of light from a multilayered particle (see Section 17.5.5 for detailed description).

emission peaks for each fluorophore is most likely to be caused by the addition of concentric shells each possessing a different refractive index which increases the scattering of the fluorescence emitted from the inner dye layers of the particle.

Fig. 17.17 illustrates the optical properties that influence the scattering of light from a multilayered sphere. Each optically separate layer or region has a refractive

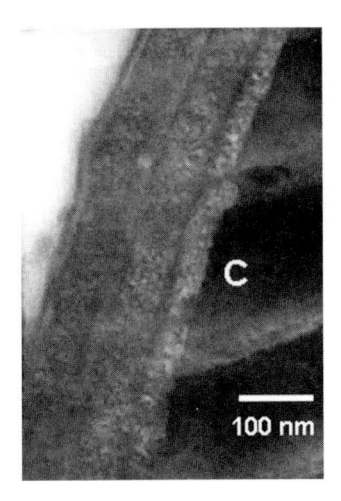

17.5 Multi-fluorescent Colloidal Silica Particles

Fig. 17.18 Transmission electron microscope image of a particle in which seven silica shells have been synthesized around a 2.9-µm silica core bead. The particles were prepared by synthesizing four fluorescently labeled silica shells, with each layer separated by a SiO_2 spacer shell (i.e. seven shells in total). This cross-sectional image clearly shows the shell region (on left of image) and the core particle (labeled 'C'). The contrast variation across the shell layers is caused by variations in silica cross-linking density due to the presence or absence of dye molecules. This complex and random refractive index profile gives rise to a unique scattering signature for each particle. The ultrathin sectioning process used to prepare the sample has revealed that the shell layers are robust and remain intact upon slicing with a diamond knife, while the core silica particles fragment.

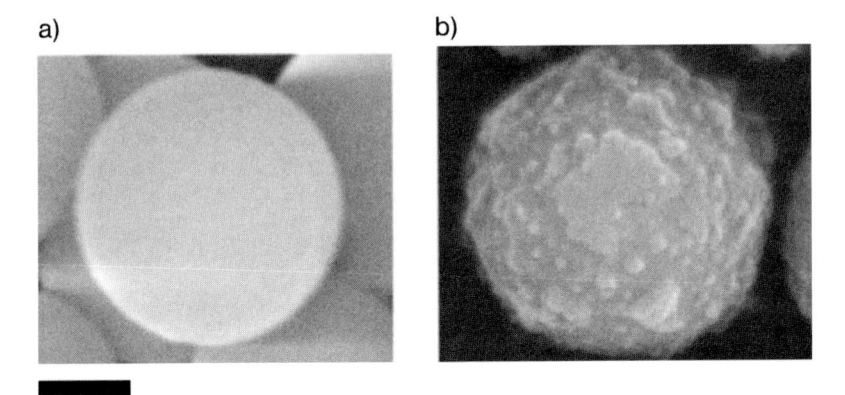

Fig. 17.19 Scanning electron microscopy image of **a** precursor silica particle (diameter 2.9 µm) and **b** multilayered sphere possessing twelve alternating dye and silica layers. Surface roughness is an advantage here as it leads to increased optical complexity of the particles, as measured by light scattering.

index m_i and size parameter, y_i, for a $(r-1)$ layered sphere. The size parameter is described by $y_i = 2\pi r_i/\lambda$, where r_i is the radius of layer i and λ is the wavelength of the radiation incident on the sphere. It is well established that the light scattering from a multilayered sphere is significantly more complex than a simple homogenous sphere (Kokhanovsky, 1999). The scattering patterns of the latter can be described by simple Mie theory or Fraunhofer theory depending on the size of the particle. The scattering from a multilayered sphere (Fig. 17.17) can still be described in terms derived from Mie coefficients and by applying van de Hulst's treatment to obtain scattering coefficients (Kokhanovsky, 1999). However, it is evi-

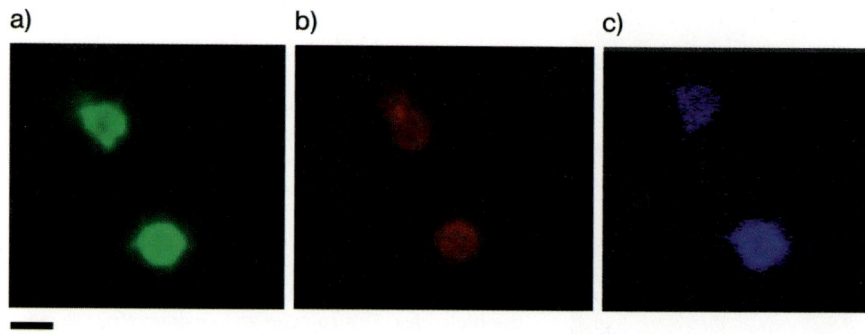

Fig. 17.20 Fluorescence microscopy of 2.9-µm core particles coated with four dye and four silica spacer layers of the fluorescent dye sequence XRITC/FITC/A350/A430. Different fluorophores can be imaged under the various optical filters in the microscope, e.g. **a** U-MWB filter (λ_{ex}=450–480nm, λ_{em}>515 nm) to observe emission from FITC (λ_{ex}=494 nm, λ_{em}=519 nm) and A430 (λ_{ex}=430 nm, λ_{em}=545 nm); **b** U-MWG filter (λ_{ex}=510–550 nm, λ_{em}>590 nm) to observe emission from XRITC (λ_{ex}=580 nm, λ_{em}=603 nm); **c** U-MWU filter (λ_{ex}=330–385 nm, λ_{em}>420 nm) to observe emission from A350 (λ_{ex}=346 nm, λ_{em}=445 nm).

dent that the increasing radii and increase in complexity of the refractive environment experienced by emitted photons traveling paths (i) to (ii) in Fig. 17.17 will increase the scattering of emitted fluorescence photons, thereby decreasing the intensity of fluorescence emission. This phenomenon in fact provides an additional parameter that can be exploited to achieve optical diversity. Indeed, since silica shells that contain dyes have a complex refractive index which is different from that of pure SiO_2 that does not contain dyes, a diverse range of particle optical signatures can be obtained. The individual shells can clearly be observed in Fig. 17.18, a transmission electron micrograph of a cross-section of a core-shell particle (Lawrie et al., 2003). The contrast variation between shell layers in Fig. 17.18 is caused by the difference in silica cross-linking density in the presence or absence of dye molecules. The ability to produce complex refractive index profiles across particles, through the synthesis of silica shells on core particles, permits a wide variety of particle optical signatures.

Examination of the morphology of multi-fluorescent colloids (Fig. 17.19) reveals that the addition of multiple shells has impacted on the smooth surfaces of the precursor core silica spheres (Fig. 17.19a) (Lawrie et al., 2003). As shown by the SEM images, the surface structure of the shells appears to be discontinuous or laminated in places (Fig. 17.19b). Rather than being unfavorable, this surface roughness can be exploited as a method of obtaining optically unique particles for encoding Strategy II. Fluorescence microscopy confirms each fluorescent dye is present (Fig. 17.20). Thus, in the filter U-MWB (Fig. 17.20a) the emission from FITC and A430 can be observed, in filter U-MWG (Fig. 17.20b) the emission from XRITC can be observed, and in filter U-MWU (Fig. 17.20c) the emission from A350 is observed. Transmission electron microscopy demonstrates the variation in density of the core shell compared with the multiple shell coats (Fig. 17.18). This distinction in density accounts for the response to ultrathin sectioning by a diamond knife whereby the lower density core particle often fragments during the process. The shell usually remains intact. Some irregularities in the thickness of the shell are observed but the structure of the shell appears to be continuous. The shell layer thickness typically varies from ~ 15 to 50 nm when dyes are present in the layer, compared with ~ 5–10 nm when dyes are absent (such as in the silica spacer layers between those layers containing dye). Minimal adverse colloidal behavior such as aggregation of the particles occurs and so the sequential addition of multiple coats to create unique individual particles has been successful.

17.5.6
Selection of Dyes for Strategy II Encoding

When several different fluorophores are bound as shells onto a single particle the interaction between dye excitation/emission behaviors becomes very complex (Lawrie et al., 2003). Far from being detrimental, this situation is in fact favorable for our objective of extending the diversity of the fluorescent signatures of individual particles. Increasing the number of dye shells from one to six (i.e. not including the six silica spacer layers synthesized between the dye shells) increases the complexity of the spectra (Fig. 17.21). The emission spectra in Fig. 17.21a and b show two sets of particles that were coated by two different series of six fluorescent dye shells. Two of the dyes common to both batches of particles are fluorophores A350 and FITC. Comparing the spectra of each of these dyes in the two batches illustrates that the spectra exhibit quite differently shaped emission peaks. This indicates that there is an interaction between the different types of fluorophore in adjacent shells within each particle.

The spectra in Fig. 17.22 are the result of excitation at 346 nm yet emission peaks are clearly observed for the fluorophores FITC (517 nm) and TAMRA (578 nm) whose excitation bands are at 494 nm and 546 nm, respectively. These dyes possess very broad excitation peaks and it is probable that they are excited by the energy provided at 346 nm, particularly the dye FITC. However, the process of reabsorption of emitted photons created by fluorophore A350 (excitation maxima at 346 nm, emission maxima at 445 nm) is also possible in this system and must be considered where fluorophores FITC (third dye shell) and TAMRA (fifth dye shell) have subsequently been bound to the particles. The implication is that, with a limited number of dyes, a large diversity in spectral properties (or optical signature) is achieved. How this is utilized in the flow-cytometer application is described in Section 17.6.3.

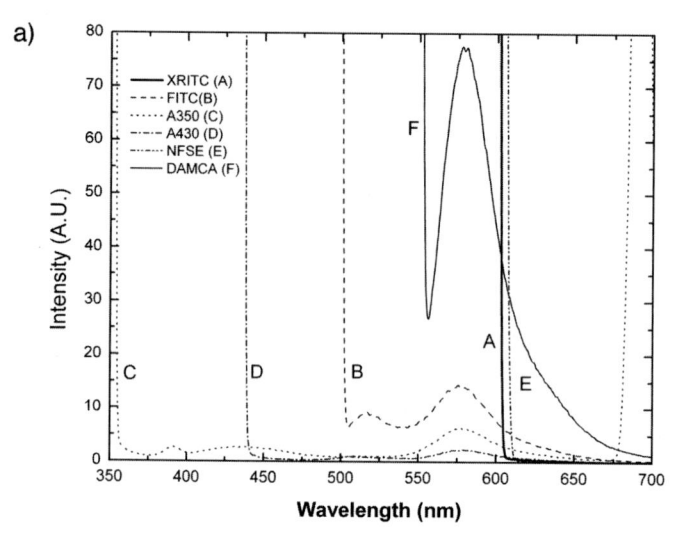

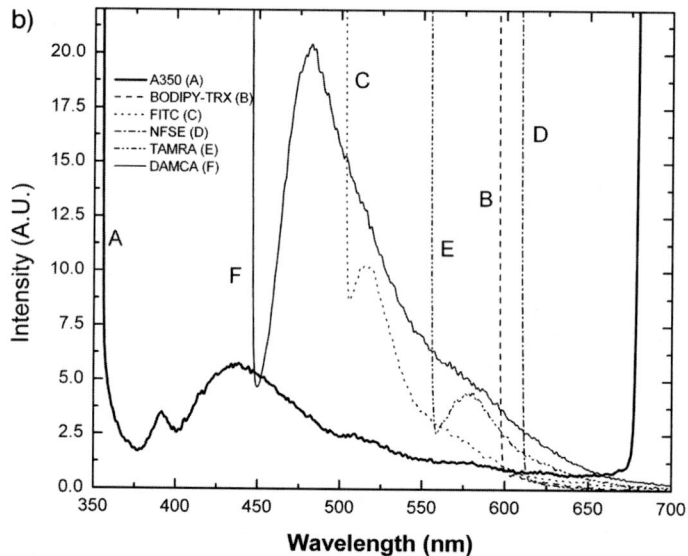

Fig. 17.21 Example of fluorescence emission spectra for 2.9-µm colloids coated with the layer sequence **a** XRITC (λ_{ex}=580 nm)/silica/FITC (λ_{ex}=494 nm)/silica/A350 (λ_{ex}=346 nm)/silica/A430 (λ_{ex}=430 nm)/silica/NFSE (λ_{ex}=602 nm)/silica/TAMRA (λ_{ex}=546 nm)/silica; **b** A350 (λ_{ex}=346 nm)/silica/BTRX (λ_{ex}=588 nm)/silica/FITC (λ_{ex}=494 nm)/silica/NFSE (λ_{ex}=602 nm)/silica/TAMRA (λ_{ex}=546 nm)/silica/DAMCA (λ_{ex}=437 nm)/silica.

17.6
Optimizing Fluorescent Colloids for Use in Combinatorial Libraries

In order for the multi-fluorescent colloids described in Section 17.5 to be used in encoding combinatorial libraries, certain aspects have to be optimized. For encoding Strategy I, which involves attachment of reporter particles to large solid support beads during the synthesis of biomolecular probes (Section 17.3.2.1), key requirements are that the reporter size should be in the order of 0.5–1.0 μm and they should be monodisperse (Battersby et al., 2000; Grøndahl et al., 2000). Furthermore, the reporters must be able to form stable attachments to the polymer surface and, to optimize this, colloidal forces are manipulated through surface modification of the reporters. This optimization is discussed below in Sections 17.6.1 and 17.6.2.

The silica colloids used in encoding Strategy II are the fluorescent shell particles (2–5 μm in diameter) which are used as support beads for biomolecular probe synthesis (Section 17.3.2.2). Their versatility for this application is explained further in Section 17.6.3.

17.6.1
Modification of the Colloidal Silica Surface

The surface of a silica colloid consists of a mixture of silanol groups (Si–OH) and siloxane groups (Si–O–Si). The silanol groups fall into two classes: isolated single silanols (Si–OH) and isolated geminal silanols (Si(OH)$_2$) (Iler, 1955). In aqueous solution the surface is covered with physically adsorbed water, which forms a network of hydrogen bonds with the silanol groups (Zhuravlev, 2000). Since it is the deprotonation of the silanol groups which gives rise to the negative surface charge of the silica colloids, the charge density is dependent on the pH of the solution.

The concept of adsorbing multiple layers of polyelectrolytes or other charged molecules onto a surface was developed by Decher in early 1990 and is termed the layer-by-layer approach.

The amount of positive polyelectrolyte adsorbed on colloidal silica depends on a number of factors such as pH and ionic concentration of the solution (affecting the polyelectrolyte conformation in solution and after adsorption) (Bauer, 1998). In neutral to basic solution, positively charged polyelectrolytes adsorb onto the surface of colloidal silica particles resulting in charge reversal on the surface. The resulting surface charge after saturation depends on the polyelectrolyte used (Tab. 17.4), thus, PDADMAC (Fig. 17.23) coated silica displays a zeta-potential of +88 mV whereas PEI (Fig. 17.23) coated silica yields a zeta-potential of +98 mV. After adsorption onto a silica surface, the polyelectrolyte chains change their conformation and the resulting assembly is stable even upon dilution because of multiple contact points between polyelectrolyte and colloid (Chaplain et al., 1995). A consequence of charge reversal is that it allows oppositely charged molecules to be adsorbed in a subsequent step (Decher, 1997).

Detailed investigations of the structure of polyelectrolyte multilayers have been carried out using surface modified glass slides carrying a constant positive charge

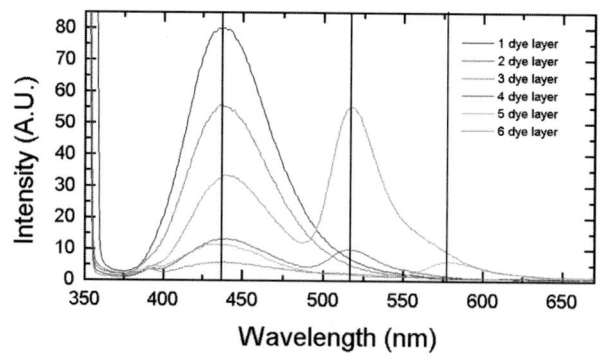

Fig. 17.22 Fluorescence emission spectra (excitation at 346 nm) for 2.9-μm colloid batches coated with increasing numbers of alternating silica and dye layers. Dye layer 1 = A350, dye layer 2 = BTRX, dye layer 3 = FITC, dye layer 4 = NFSE, dye layer 5 = TAMRA, dye layer 6 = DAMCA. Impact of both number of subsequent layers and identity of subsequent dyes is evaluated.

Fig. 17.23 Chemical structures of polyelectrolytes used for coating reporter particles. PAA = poly(acrylic acid), PSSS = poly(sodium 4-styrene-sulfonate), PEI = Poly(ethyleneimine), PDADMAC = poly(diallyldimethylammonium chloride).

(Decher, 1997). It was found that the over-compensation of charge is a property of the polyelectrolyte rather than a property of the underlying surface. Furthermore, polyelectrolyte multilayers were found to have similar surface roughness independent of the surface roughness of the substrate. By using neutron and X-ray reflectometry, it has been clearly demonstrated that an internal layer structure is present in multilayered systems (Decher, 1997). In addition, it was established that a large overlap exists between adjacent layers. This large overlap results in identical concentrations of anionic and cationic groups throughout the polyelectrolyte multilayer film.

In the application of colloidal silica particles as "reporters" in combinatorial synthesis (described in detail in Section 17.6.2), two polyelectrolyte layers are coated onto the colloids in order to increase electrostatic attraction and polymeric bridging between silica colloids and support beads (Battersby et al., 2000; Grøndahl et al., 2000). As discussed above, the surface charge in these multilayered systems

Tab. 17.4 Variation in ζ-potential and the number of silica colloids within a 2500 μm² area on PEGA beads.[1]

Polyelectrolyte multi-layer combination[2]	"Experiment set"	Irradiation solution[3]	Dose (kGy)[4]	ζ-Potential (mV)	Number of silica colloids on support bead[5]	% red colloids on green-labeled carrier beads after 10 days[6]
none		–	–	–41±7	11±6	26±19%
PEI		–	–	+98±12	–	–
PEI	a	H₂O	2.6	+80±10	–	–
PEI	a	H₂O	3.8	+69±11	–	–
PDADMAC		–	–	+88±12	–	–
PDADMAC	a	H₂O	2.6	+70±12	–	–
PDADMAC	a	H₂O	3.8	+61±9	–	–
PEI/PAA		–	–	–42±11	54±23	0.2±0.5%
PEI/PAA	b	H₂O	1.3	–33±12	38±24	–
PEI/PAA	b	H₂O	3.8	–33±10	14±17	–
PEI/PAA	c	0.5% PAA	12.4	–40±10	90±39	–
PDADMAC/PAA		–	–	–54±14	28±21	0.8±0.9%
PDADMAC/PAA	b	H₂O	3.8	–28±8	8±7	–
PDADMAC/PAA	c	0.5% PAA	12.4	–35±8	59±34	–
PEI/PSSS		–	–	–51±16	53±45	–
PEI/PSSS	c	0.5% PSSS	12.4	–48±11	71±24	–
PDADMAC/PSSS		–	–	–64±15	21±16	–
PDADMAC/PSSS	c	0.5% PSSS	12.4	–52±12	24±12	–

1) Data from reference (Grøndahl et al. 2000).
2) PEI = polyethyleneimine, M_n 10000; PAA = poly(acrylic acid), M_n = 250000; PDADMAC = poly(diallyldimethylammonium chloride), M_n = 400000; PSSS = poly(sodium 4-styrene-sulfonate), M_n = 1000000.
3) Polyelectrolyte coated silica colloids were suspended in N_2 saturated solution during gamma irradiation.
4) Gamma irradiation dose, dose rate = 7.6 kGy/hour.
5) 10 mg PEGA resin in 500 μl DMF was added colloidal silica particles to a concentration of 0.026 g/L and mixed for 5 min. Silica colloids within the central focused area of 2500 μm² of a PEGA support bead were counted using Image-Pro Plus software.
6) 10 mg PEGA resin in 500 μl DMF was added colloidal silica particles to a concentration of 0.026 g/L and mixed for 16 hours before washing away excess reporters. The resulting red and green labeled carrier beads were mixed and the number of contaminant silica colloids within the central focused area of 2500 μm² of a PEGA support bead were counted using Image-Pro Plus software.

depends on the polyelectrolyte pair used (Tab. 17.4 and Fig. 17.23). Thus, PSSS yields higher surface charge than PAA in both pairs, and having PDADMAC as underlying positive electrolyte yields the most negative surface charge of the double polyelectrolyte layers.

17.6.2
Adhesion of Reporters to Solid Support Beads

Studies on polymeric solid support beads used in peptide synthesis (PEGA resin that carries positively charged amine groups on its surface) have shown that manipulation of colloidal forces is necessary in order to obtain efficient and stable adhesion (Grøndahl et al., 2000). Coating the silica colloids with double polyelectrolyte layers only increases the surface charge of the colloids in some cases, but always significantly increases the number of colloids that will attach to a support bead (Tab. 17.4). Interestingly, there is no correlation between the surface charge and the number of colloids adhering. Out of four polyelectrolyte double-layer combinations the one with the highest zeta-potential showed the least colloids adhering (PDADMAC/PSSS; Tab. 17.4). This indicates that not only electrostatic but also polymeric bridging flocculation between the silica colloid and the support bead is occurring.

In order to give further evidence for the polymeric flocculation phenomenon, the colloidal particles were modified by gamma irradiation. Gamma irradiation has been used to polymerize low molecular weight polyelectrolyte multilayers (Saremi et al., 1995) and to produce highly cross-linked hydrogels from poly(acrylic acid) (PAA) (Jabbari and Nozari, 2000). In studies on polyelectrolyte-coated colloids for colloidal barcoding two different approaches were investigated:

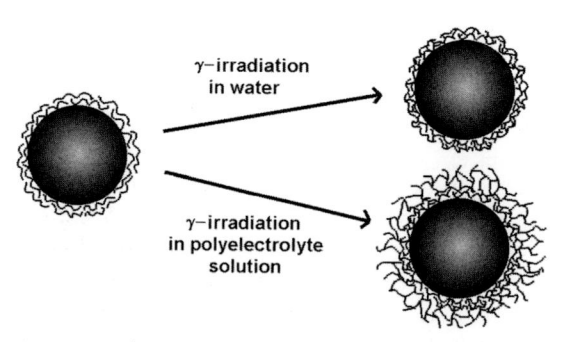

Fig. 17.24 Schematic illustration of how adsorbed polyelectrolyte layers on colloids are affected by γ-irradiation. **a** γ-irradiation of polyelectrolyte-coated colloids in water results in the formation of a cross-linked mesh due to radical-induced chain-breaking and/or cross-linking; **b** γ-irradiation of polyelectrolyte-coated colloids in the presence of excess polyelectrolyte results in radical-induced chain-breaking, cross-linking and polymeric grafting. A mesh is formed, with polymer chains extending into solution.

(1) gamma irradiation in water where radical-induced chain breaking and/or cross-linking of double polyelectrolyte layers can occur (Tab. 17.4b); and (2) gamma irradiation in polyelectrolyte solution where a combination of radical-induced chain breaking, cross-linking and polymeric grafting of polyelectrolyte onto the polyelectrolyte layers can occur (Tab. 17.4c). In all the modified polyelectrolyte coated colloids a decrease in surface charge was observed after gamma irradiation (Tab. 17.4) indicating that scission is occurring. This is also observed for colloids coated with one positive polyelectrolyte layer only (Tab. 17.4a). In experiment set (b) this decrease in surface charge was accompanied by a decrease in the number of colloids attaching to the support beads. In contrast, for experiment set (c) it led to an increase in the number of colloids attaching. The interpretation of these results is that in the first set of experiments (a and b), a lowered flexibility and/or molecular weight of the polyelectrolyte coats by scission and/or cross-linking has occurred (Fig. 17.24a). In contrast, in the second set of experiments (c), extended polyelectrolyte chains have been introduced onto the surface of the colloids thus allowing a higher degree of polymeric bridging flocculation (Fig. 17.24b).

Another way of providing evidence for polymeric flocculation comes from looking at the exchange of silica colloids between carrier beads. Colloids containing a red dye were adhered to one set of carrier beads whereas colloids containing a green dye were adhered to another set of carrier beads. After removing excess colloids the two sets of coded carrier beads were mixed. In such 10-day exchange studies it was clearly shown (Fig. 17.25) that polyelectrolyte-coated colloids out-perform non-coated colloids and this is irrespective of the surface charge (Tab. 17.4).

Clearly, it is possible to manipulate colloidal forces so as to obtain a stable system which can withstand not only peptide synthesis but also an additional large range of organic solvents and reagents (Battersby et al., 2000). An example of an encoded resin bead containing a pentapeptide sequence is given in Fig. 17.26. Here four different reporter beads were attached to the carrier bead during peptide synthesis and decoded using fluorescence microscopy.

17.6.3
Optical Diversity in Eleven Dimensions

As described earlier, preformed silica particles have been used as core particles on which multi-fluorescent silica shells are grown sequentially (Section 17.5). These shell particles are used as optically unique support beads on which biomolecular probes are synthesized. Tracking and sorting multi-fluorescent shell particles on the basis of their optical signature is done through the use of a flow cytometer (Section 17.3.3). This section describes how the optical diversity in these shell particles can be optimized.

Synthesis of fluorescent shell layers in a *direct* manner (i.e. synthesis of shells in a controlled sequence onto the core silica particle) results in a population of particles, all of which contain the same sequence of dyes on the core particle. In principle, this should result in all particles possessing identical optical signatures. As an example, six dye shells in the dye sequence A350/BTRX/OG488/B630/

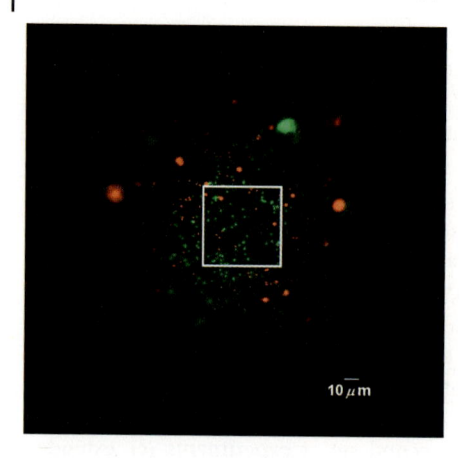

Fig. 17.25 Fluorescence microscope image of cross-contamination on a support bead labeled with uncoated green reporters. On this green-labeled bead, a high number of contaminant red reporters (19) were found among the 60 correct green reporters within the 2500-μm^2 area of interest (i.e. the white square) (reproduced with permission from Grøndahl et al., 2000, *Langmuir* 16, 9709–9715. Copyright 2000, American Chemical Society).

Tab. 17.5 Filter parameters for the fluorescence detectors in the flow cytometer used in the shell silica particle studies.

Detector (Laser excitation λ, nm)	Wavelength (band width) (nm)	Dyes in detector range
FL1 (488)	530 (± 20)	FITC, OG488, A350, A430, DAMCA
FL2 (488)	580 (± 15)	XRITC, DAMCA, TAMRA
FL3 (488)	630 (± 20)	XRITC, TAMRA, BTRX, B630
FL4 (488)	670 (± 15)	NFSE, B650
FL6 (630)	670 (± 20)	NFSE, B630, B650
FL8 (350)	405 (± 20)	A350
FL9 (350)	450 (± 15)	A350, DAMCA

DAMCA/B530 were synthesized onto a batch of 3-µm core silica particles by a *direct* synthesis procedure. Silica spacer shells were also synthesized in between each dye shell (Lawrie et al., 2003). All of the dyes in these particles can be excited, to various extents, by the 488-nm and 351-nm lasers in the flow cytometer. The overall optical signature from a single particle with multi-fluorescent shells is a composite of all individual fluorescence emissions, with more than one dye contributing to the fluorescence detected by each photomultiplier tube (PMT). Each PMT is located behind a carefully selected narrow band-pass filter, an example of the arrangement of these filters and details of the wavelength ranges used for the particles discussed in this section is provided in Tab. 17.5. Thus, the fluorescence emission from the particles possessing the six-dye-shell sequence described above could be detected by at least six of the detectors in Tab. 17.5 (FL1, FL2, FL3, FL6, FL8 and FL9). Flow-cytometry data is often represented as the number of particles (events) of a particular intensity detected by the PMT in the form of a histogram. The fluorescence intensity data for this set of particles from the detectors FL1, FL6 and FL9 are provided in the flow-cytometry histograms in Fig. 17.27a, c and e, respectively. In these histograms, there is a distribution of fluorescence intensi-

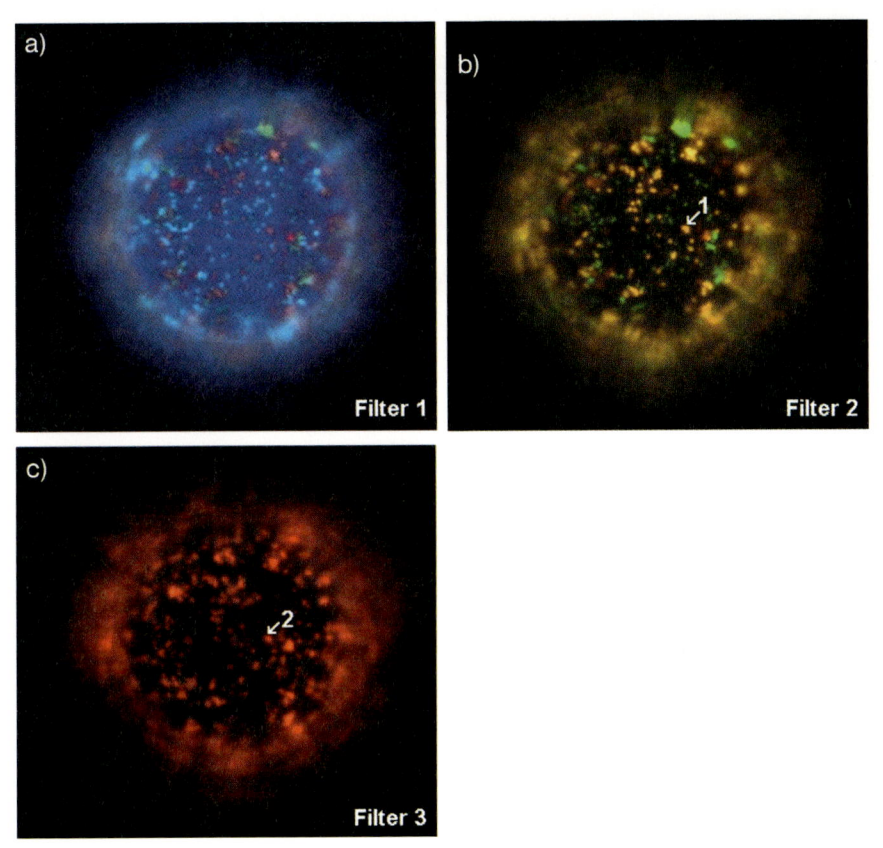

Fig. 17.26 Fluorescence microscope images of carrier bead encoded with four colloidal reporters. The bead is analyzed under three different filters (**a**, **b**, and **c**) to identify the individual dyes present within each colloidal reporter. A yellow dye within a particular reporter is detected under Filter 2 (arrow 1) and under Filter 3 red dye can be detected within the same reporter (arrow 2). Filter sets: **a** U-MWB filter set (λ_{ex}=450–480 nm, λ_{em}>515 nm); **b** U-MWG filter set (λ_{ex}=510–550 nm, λ_{em}>590 nm); **c** U-MWU filter set (λ_{ex}=330–385 nm, λ_{em}>420 nm) (reproduced with permission from Battersby et al., 2000, *J. Am. Chem. Soc.* 122, 2138–2139. Copyright 2000, American Chemical Society)

ties around a peak indicating that, despite each particle possessing the same six-dye sequence, the optical signatures can vary considerably.

In order to create an even greater range of fluorescence signatures within the particles and in sufficient numbers to provide a platform for a vast library of optically unique particles, a novel approach was undertaken. A combinatorial method of synthesizing the dye shells proved to be the method for which this diversity is best achieved in the shell particles. Combinatorial synthesis of multiple dye shells by the split-and-mix approach results in a collection of fluorescent particles each possessing a unique combination of dye shells. There will be particles which con-

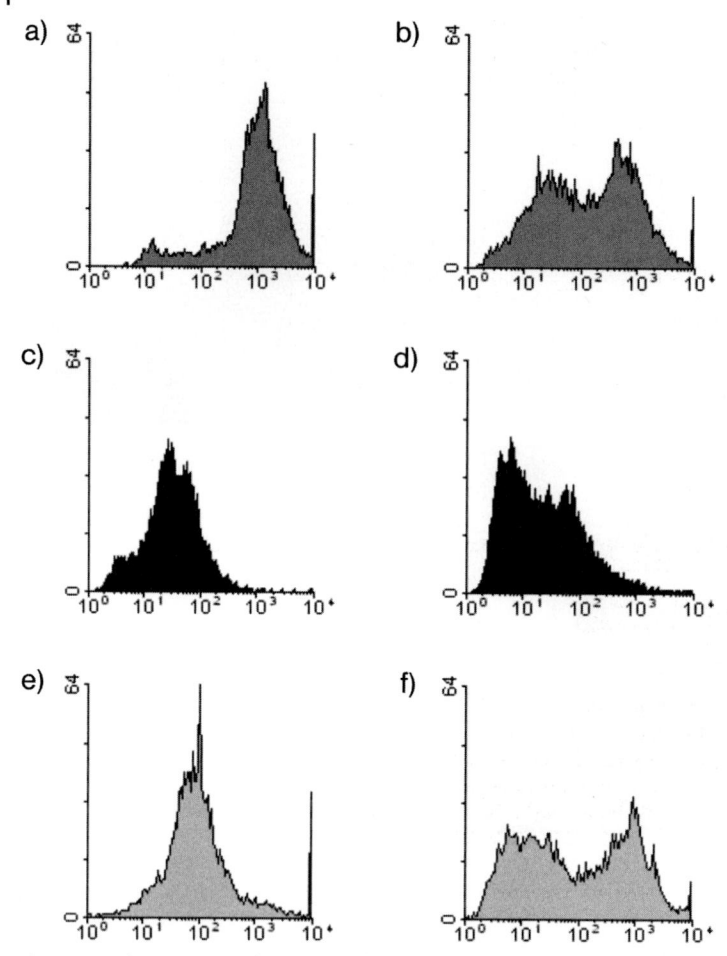

Fig. 17.27 Flow cytometry histograms for particles with six silica spacer and six fluorescent shells (A350/BTRX/OG488/B630/DAMCA/B530) synthesized in a directed sequence (**a, c, e**) and for particles with four silica spacer and four fluorescent shells (A350/DAMCA/OG488/B630) synthesized in a combinatorial sequence (**b, d, f**). Detectors: **a, b** FL1 (530 (±20) nm), **c, d** FL6 (780 (±20) nm), **e, f** FL9 (450 (±15) nm), see Tab. 17.3.

tain all of the dyes but also particles where all the shells contain the same dye. The fluorescence microscopy images in Fig. 17.29 show the range of signatures observed from a set of particles where the shells have been synthesized combinatorially onto core particles over four split-and-mix cycles using the four dyes A350, TAMRA, OG488 and B630. The increase in range of the fluorescent emission intensities achieved by this combinatorial synthesis of the dye shells is demonstrated in the flow-cytometry histograms obtained for this set of particles (Fig. 17.27 b, d and f) from the detectors FL1, FL6 and FL9 (detailed in Tab. 17.5).

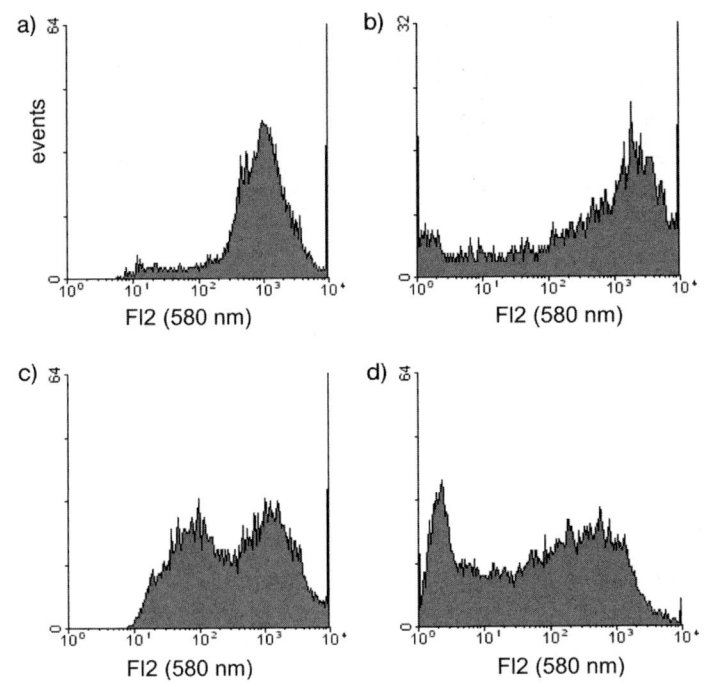

17.6 Optimizing Fluorescent Colloids for Use in Combinatorial Libraries | 553

Fig. 17.28 Influence on optical diversity of multi-fluorescent particles prepared by different methods. The data in these flow cytometry histograms was measured by the FL2 detector (580 nm). **a** Every particle has six fluorescent dye shells in identical sequence (A350/BTRX/OG488/B630/DAMCA/B530; **b** four fluorescent dye shells combinatorially synthesized (A350, OG488, B630, DAMCA); **c** four fluorescent dye shells combinatorially synthesized (A350, OG488, B630, TAMRA); **d** two fluorescent dye shells combinatorially synthesized (OG488, TAMRA).

A broad band of intensities is observed in each histogram reflecting the variety of dye-shell combinations. A comparison between the histograms resulting from this combinatorial set of particles (Fig. 17.27b, d and f) and those for the particles where the shells were synthesized by the direct method using six dyes (Fig. 17.27a, c and e) clearly demonstrates that the combinatorial particles display a greater diversity of optical signatures, achieved with only four dyes.

The power of combinatorial synthesis is further illustrated by using differing numbers of dyes to create the shells. The flow-cytometry histogram obtained for a directed six-layer set of beads (Fig. 17.28a) is compared with those obtained for three different combinatorially synthesized sets of beads using only four (Fig. 17.28b and c) or even only two dyes (Fig. 17.28d). The fluorescence emission detected using the 580-nm filter (FL2) is examined where the dyes TAMRA, B530 and the tail of the OG488 emission peak can all be detected simultaneously. Again the distribution of intensities observed from all the combinatorial sets is

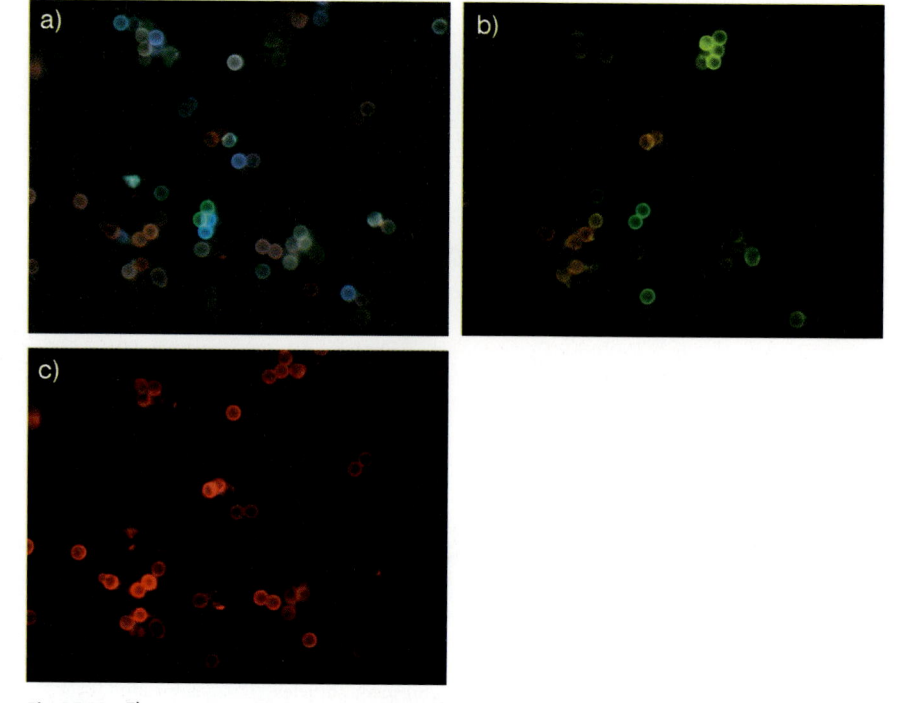

Fig. 17.29 Fluorescence microscope images of particles with four fluorescent shells synthesized in a combinatorial manner using the dyes A350 ($\lambda_{ex}=346$ nm)/OG488 ($\lambda_{ex}=495$ nm)/TAMRA ($\lambda_{ex}=546$ nm)/B630 ($\lambda_{ex}=632$ nm). Different fluorophores are imaged under the various optical filters in the microscope **a** U-MWU filter set ($\lambda_{ex}=330–385$ nm, $\lambda_{em}>420$ nm); **b** U-MWB filter set ($\lambda_{ex}=450–480$ nm, $\lambda_{em}>515$ nm); **c** U-MWG filter set ($\lambda_{ex}=510–550$ nm, $\lambda_{em}>590$ nm).

greater than the six-shell-controlled sequence particles even when there is no TAMRA or B530 (Fig. 17.28b) and only two dyes (Fig. 17.28d).

Clearly, the combinatorial multi-dye shell synthesis procedure is an ideal strategy for producing an enormous amount of optical diversity in solid support beads for chemical library synthesis.

17.7
Conclusions

As described in this chapter, a multitude of opportunities exist for innovative design and application of colloid-based devices in high-throughput biomolecular screening. These devices promise to improve our ability to develop new drugs, to diagnose the cause of disease, to sequence DNA and identify single nucleotide polymorphisms (SNPs), to study gene expression, to study various types of ligand/receptor interactions, and to do so in a manner far more time and cost effectively than has previously been possible.

Exploiting the optical properties of individual colloids for HTS will require the innovative development of new colloidal materials. With the intrinsic advantages that they possess, colloid-based libraries will increasingly challenge microarrays and microplates in the future for a wide variety of HTS applications.

17.8
References

ALIVISATOS, A. P. (1996) Semiconductor clusters, nanocrystals, and quantum dots. *Science* 271:933–937.

AUDIC, S. and CLAVERIE, J. (1997) The significance of digital gene expression profiles. *Genome Res.* 7:986–995.

AUER, M., MOORE, K. J., MEYER-ALMES, F. J., GUENTHER, R., POPE, A. J. and STOECKLI, K. A. (1998) Fluorescence correlation spectroscopy: lead discovery by miniaturized HTS. *Drug Discovery Today* 3(10):457–465.

BADER, J. S. (2001) The relative power of SNPs and haplotype as genetic markers for association tests. *Pharmacogenomics* 2(1):11–24.

BAILEY, J. K. and MECARTNEY, M. L. (1992) Formation of colloidal silica particles from alkoxides. *Colloids Surf* 63:151–161.

BATTERSBY, B. J., BRYANT, D., MEUTERMANS, W., MATTHEWS, D., SMYTHE M. L. and TRAU, M. (2000) Toward larger chemical libraries: encoding with fluorescent colloids in combinatorial chemistry. *J. Am. Chem. Soc.* 122:2138–2139.

BATTERSBY, B. J., LAWRIE, G. A. and TRAU, M. (2001) Optical encoding of colloids for gene screening: alternatives to microarrays. *Drug Discovery Today* 6 *(Supplement)*: S123–S130.

BATTERSBY, B. J. and TRAU, M. (2002) Novel miniaturized systems in high-throughput screening. *Trends in Biotech.* 20:167–173.

BATTERSBY, B. J., LAWRIE, G. A., JOHNSTON, A. P. R. and TRAU, M. (2002) Optical barcoding of colloidal suspensions: Applications in genomics, proteomics and drug discovery. *Chem. Commun.* 14:1435–1441.

BAUER, D., KILLMANN, E., JAEGER, W. (1998) Adsorption of poly(diallyl-dimethyl-ammoniumchloride) (PDADMAC) and of copolymers of DADMAC with N-methyl-N-vinyl-acetamide (NMVA) on colloidal silica. *Progr. Colloid Polym. Sci.* 109:161–169.

BEGGS, M. (2001) HTS – where next? *Drug Discovery World* 2:25–30.

BOBADILLA, J. L., MACEK, M., Jr, FINE, J. P., FARRELL, P. M. (2002) Cystic fibrosis: a worldwide analysis of CFTR mutations-correlation with incidence data and application to screening. *Human Mutation* 19:575–606.

BOGUSH, G. H., TRACY, M. A. and ZUKOSKI, C. F. (1988) Preparation of monodisperse silica particles: Control of size and mass fraction. *J. Non-Cryst. Solids* 104:95–106.

BOGUSH G. H. and ZUKOSKI, C. F. (1991) IV. Studies of the kinetics of the precipitation of uniform silica particles through the hydrolysis and condensation of silicon alkoxides. *J. Colloid Interface Sci.* 142:1–18.

BOUKARI, H., LIN, J. S. and HARRIS, M. T. (1997) Small-angle x-ray scattering study of the formation of colloidal silica particles from alkoxides: Primary particles or not? *J. Colloid Interface Sci.* 194:311–318.

BOUKARI, H., LONG, G. G. and HARRIS, M. T. (2000) Polydispersity during the formation and growth of the Stöber silica particles from small-angle x-ray scattering measurements. *J. Colloid Interface Sci.* 229:129–139.

BRANCA, M. A., HABERMAN, A. B. and LOCKWOOD, D. (eds) (2001) Structural proteomics: High throughput approaches fuel drug discovery and development. Cambridge Healthtech Institute, MA, USA, p. 86.

BRENNER, S., JOHNSON, M., BRIDGHAM, J., GOLDA, G., LLOYD, D. H., JOHNSON, D., LUO, S., MCCURDY, S., FOY, M., EWAN, M., ROTH, R., GEORGE, D., ELETR, S., ALBRECHT, G., VERMAAS, E., WILLIAMS, S. R., MOON, K., BURCHAM, T., PALLAS, M., DUBRIDGE, R. B., KIRCHNER, J., FEARON, K., MAO, J. and CORCORAN, K. (2000) Gene expression analysis by massively parallel signature sequencing

(MPSS) on microbead arrays. *Nature Biotech.* 18(6):630–634.

BRUCHEZ, M. Jr, MORONNE, M., GIN, P., WEISS, S. and ALIVISATOS, A. P. (1998) Semiconductor nanocrystals as fluorescent biological labels. *Science* (Sep 25) 281(5385):2013–2016.

BURBAUM, J. J. (1998) Miniaturization technologies in HTS: how fast, how small, how soon? *Drug Discovery Today* 3:313–322.

BURBAUM, J. J. and SIGAL, N. H. (1997) New technologies for high-throughput screening. *Curr. Opin. Chem. Biol.* 1:72–78.

BYERS, C. H., HARRIS, M. T. and WILLIAMS, D. F. (1987) Controlled microccrystalline growth studies by dynamic laser-light-scattering methods. *Ind. Eng. Chem. Res.* 26:1916–1923.

CAGNEY, G., UETZ, P. and FIELDS, S. (2000) High-throughput screening for protein-protein interactions using two-hybrid assay. *Methods Enzymol.* 328:3–14.

CAMPIAN, E., SEBESTYÉN, F., MAJOR, F. and FURKA, Á. (1994a) Synthesis of support-bound peptides carrying color labels. *Drug Development Res.* 33:98–101.

CAMPIAN, E., SEBESTYEN, F. and FURKA, Á. (1994b) Colored and fluorescent solid supports in Innovation and Perspectives in Solid Phase Synthesis and Complementary Technologies (EPTON, R., ed.), pp 469–472, Mayflower, Kingswinford, UK.

CHAN, W. C. W., MAXWELL, D. J., GAO, X., BAILEY, R. E., HAN, M. and NIE, S. (2002) Luminescent quantum dots for multiplexed biological detection and imaging. *Curr. Opin. Biotechnol.* 13:40–46.

CHAPLAIN, V., JANEX, M. L., LAFUMA, F., GRAILLAT, C., AUDEBERT, R. (1995) Coupling between polymer adsorption and colloidal particle aggregation. *Colloid Polym. Sci.* 273:984–993.

CLARKE, P. A., TE POELE, R., WOOSTER, R. and WORKMAN, P. (2001) Gene expression microarray analysis in cancer biology, pharmacology and drug development: progress and potential. *Biochem. Pharmacol.* 62(10):1311–1336.

CZARNIK, A. W. (1997) Encoding methods for combinatorial chemistry. *Curr. Opin. Chem. Biol.* 1:60–66.

DECHER, G. (1997) Fuzzy nanoassemblies: Toward layered polymeric multicomposites. *Science* 277:1232–1237.

DE STEFANO, V., CASORELLI, I. D. A., ROSSI, E., ZAPPACOSTA, B., LEONE, G. (2000) Interaction between hyperhomocysteinemia and inherited thrombophilic factors in venous thromboembolism. *Seminars in Thrombosis and Hemostasis* 26(3):305–311.

DUNBAR, S. A. and JACOBSON, J. W. (2000) Application of the Luminex LabMAP in rapid screening for mutations in the cystic fibrosis transmembrane conductance regulator gene: a pilot study. *Clinical Chemistry* (Washington, DC) 46(9):1498–1500.

EGNER, B. J., RANA, S., SMITH, H., BOULOC, N., FREY, J. G., BROCKELSBY, W. S. and BRADLEY, M. (1997) Tagging in combinatorial chemistry: the use of coloured and fluorescent beads. *Chem. Commun.* 8:735–736.

FIELDS, S. and SONG, O. (1989) A novel genetic system to detect protein–protein interactions. *Nature* 340:245–246.

FITCH, W. L., BAYER, T. A., CHEN, W., HOLDEN, F., HOLMES, C. P., MACLEAN, D., SHAH, N., SULLIVAN, E., TANG, M. and WAYBOURN, P. (1999) Improved methods for encoding and decoding dialkylamine-encoded combinatorial libraries. *J. Comb. Chem.* 1:188–194.

FODOR, S. P. A., READ, J. L., PIRRUNG, M. C., STRYER, L., LU, A. T. and SOLAS, D. (1991) Light-directed, spatially addressable parallel chemical synthesis. *Science* (Washington, DC 1883:251(4995):767–773.

FÖRSTER, TH., (1948) Intermolecular energy migration and fluorescence. *Ann. Phys. (Leipzig)* 2:55–75.

FRIEDBERG, T. (2001) Cytochrome P450 polymorphisms as risk factors for steroid hormone-related cancers. *Am. J. PharmacoGenomics* 1(2):83–91.

FULTON, R. J., MCDADE, R. L., SMITH, P. L., KIENKER, L. J. and KETTMAN, J. R. Jr (1997) Advanced multiplexed analysis with the FlowMetrix system. *Clinical Chemistry* (Washington, DC) 43(9):1749–1756.

GREPIN, C. and PERNELLE, C. (2000) High-throughput screening: evolution of homogeneous time resolved fluorescence (HTRF) technology for HTS. *Drug Discovery Today* 5:212–214.

GRØNDAHL, L., BATTERSBY, B. J., BRYANT, D. and TRAU, M. (2000) Encoding combinatorial libraries: a novel application of fluorescent silica colloids. *Langmuir* 16:9709–9715.

GUSELLA, J., MACDONALD, M. (2002) Opinion: No post-genetics era in human disease research. *Nature Reviews Genetics* 3(1):72–79.

GWYNNE, P. and PAGE, G. (1999) Microarray analysis: the next revolution in molecular biology. *Science* 285:911–938.

HACIA, J.G. (1999) Resequencing and mutational analysis using oligonucleotide microarrays. *Nat. Genet. Suppl.* 21:42–47.

HAN, M., GAO, X., SU, J.Z. and NIE, S. (2001) Quantum-dot-tagged microbeads for multiplexed optical coding of biomolecules. *Nature Biotechnology* 19(7):631–635.

HARNESS, J. (2000) An automated synthesis programme for drug discovery. *Innovations Pharm. Technol.* 5:37–45.

HARRIS, M.T. (1992) Ultrafine precursor powders by homogeneous precipitation and electrodispersion. PhD dissertation, University of Tennessee.

HATZIMANIKATIS, V. (2000) Bioinformatics and functional genomics: challenges and opportunities. *AIChE Journal* 46(12):2340–2343.

HENGLEIN, A. (1989) Small-particle research: physiochemical properties of extremely small colloidal metal and semiconductor particles. *Chem. Rev.* 89:1861–1873.

HIETER, P. and BOGUSKI, M. (1997) Functional genomics: It's all how you read it. *Science* 278:601–602.

IIDA, A., SAITO, S., SEKINE, A., KITAMURA, Y., KONDO, K., MISHIMA, C., OSAWA, S., HARIGAE, S. and NAKAMURA, Y. (2001) High-density single-nucleotide polymorphism (SNP) map of the 150-kb region corresponding to the human ATP-binding cassette transporter A1 (ABCA1) gene. *J. Human Genetics* 46(9):522–528.

ILER, R.K. (1955) The Colloid Chemistry of Silica and Silicates, Cornell University Press, Ithaca, NY, 324 p.

IMHOF, A., MEGENS, M., ENGELBERTS, J.J., DE LANG, D.T.N., SPRIK, R. and VOS, W.L. (1999) Spectroscopy of Fluorescein (FITC) Dyed Colloidal Silica Spheres. *J. Phys. Chem. B* 103:1408–1415.

International Human Genome Sequencing Consortium (2001) Initial sequencing and analysis of the human genome. *Nature*. 409:860–921.

JABBARI E., and NOZARI, S. (2000) Swelling behavior of acrylic acid hydrogels prepared by γ-radiation crosslinking of polyacrylic acid in aqueous solution. *Eur. Polym. J.* 36:2685–2692.

JONSCHER, K.R. (2001) MS plays a vital role in drug discovery. *Drug Discovery and Development* 4(10):73–76.

KOKHANOVSKY, A.A. (1999) *Optics of Light Scattering Media: Problems and Solutions*. John Wiley & Sons/Praxis Publishing Ltd, Chichester, 217 p.

KOLBE, G. (1956) Das komplexchemische Verhalten der Kieselsäure. Dissertation, Jena.

LAM, K.S., LEBL, M., KRCHNÁK, V. (1997) "One-Bead-One-Compound" Combinatorial Library Method. *Chem. Rev.* 97:411–448.

LANDER, E.S. (1999) Arrays of hope. *Nat. Genet. Suppl.* 21:3–4.

LAWRIE, G.A., BATTERSBY, B.J. and TRAU, M. (2003) Synthesis of Optically Complex Core-Shell Colloidal Suspensions: Pathways to Multiplexed Biological Screening. *Adv. Func. Mater.* 13:887–896.

LEE, K.T., LOOK, J.L., HARRIS, M.T. and MCCORMICK, A.V. (1997) Assessing extreme models of the Stöber synthesis using transients under a range of initial composition. *J. Colloid Interf. Sci.* 194:78–88.

LIPSHUTZ, R.J., FODOR, S.P.A., GINGERAS, T.R. and LOCKHART, D.J. (1999) High density synthetic oligonucleotide arrays. *Nature Genetics* 21(1, Suppl.):20–24.

MAIMONE, D., DOMINICI, R., GRIMALDI, L.M.E. (2001) Pharmacogenomics of neurodegenerative diseases. *Eur. J. Pharmacol.* 413:11–29.

MATSOUKAS, T. and GULARI, E. (1989) Monomer-addition growth with a slow initiation step: a growth model for silica particles from alkoxides. *J. Colloid Interface Sci.* 132:13–21.

MATSOUKAS, T. and GULARI, E. (1991) Self-sharpening distributions revisited polydispersity in growth by monomer addition. *J. Colloid Interface Sci.* 145:557–562.

MATTHEWS, D.C., GRØNDAHL, L., BATTERSBY, B.J. and TRAU, M. (2001) Multi-fluorescent silica colloids for encoding large combinatorial libraries. *Aust. J. Chem.* 54:649–656.

MEGENS, M., VAN KATS, C.M., BÖSECKE, P. and VOS, W.L. (1997) In-situ characterisation of colloidal spheres by synchrotron small-angle x-ray scattering. *Langmuir* 13:6120–6129.

MELDAL, M., SVENDSEN, I., BREDDAM, K. and AUZANNEAU, F. I. (1994) Portion-mixing peptide libraries of quenched fluorogenic substrates for complete subsite mapping of endoprotease specificity. *Proc. Natl. Acad. Sci. USA* 91:3314–3318.

NANTHAKUMAR, A., PON, R. T., MAZUMDER, A., YU, S. and WATSON, A. (2000) Solid-phase oligonucleotide synthesis and flow cytometric analysis with colloids encoded with covalently attached fluorophores. *Bioconj. Chem.* 11:282–288.

NEEDELS, M. C., JONES, D. G., TATE, E. H., HEINKEL, G. L., KOCHERSPERGER, L. M., DOWER, W. J., BARRETT, R. W. and GALLOP, M. A. (1993) Generation and screening of an oligonucleotide-encoded synthetic peptide library. *Proc. Natl. Acad. Sci. USA* 90:10700–10704.

NICEWARNER-PENA, S. R., FREEMAN, R. G., REISS, B. D., HE, L., PENA, D. J., WALTON, I. D., CROMER, R., KEATING, C. D. and NATAN, M. J. (2001) Submicrometer metallic barcodes. *Science* (Oct 5) 294(5540):137–141.

NYFFENEGGER, R., QUELLET, C. and RICKA, J. (1993) Synthesis of fluorescent, monodisperse, colloidal silica particles. *J. Colloid Interface Sci.* 159:150–157.

OHLMEYER, M. H. J., SWANSON, R. N., DILLARD, L. W., READER, J. C., ASOULINE, G., KOBAYASHI, R., WIGLER, M. and STILL, W. C. (1993) Complex synthetic chemical libraries indexed with molecular tags. *Proc. Natl. Acad. Sci. USA* 90:10922–10926.

OLIVER, K. G., KETTMAN, J. R. and FULTON, R. J. (1998) Multiplexed analysis of human cytokines by use of the FlowMetrix system. *Clin. Chem.* (Sept) 44(9):2057–2060.

PONTONI, D., NARAYANAM, T. and RENNIE, A. R. (2002) Time-resolved SAXS study of nucleation and growth of silica colloids. *Langmuir* 18:56–59.

REICHARDT, C. (1988) Solvents and solvent effects in organic chemistry, Chapter 7, 2nd Edn. VCH, Weinheim.

SAREMI, F., MAASSEN, E., TIEKE, B., JORDAN, G. and RAMMENSEE, W. (1995) Self-assembled alternating multilayers built-up from diacetylene bolaamphiphiles and poly(allylamine hydrochloride) – polymerization properties, structure, and morphology. *Langmuir* 11(4):1068–1071.

SCHENA, M., SHALON, D., DAVIS, R. W. and BROWN, P. O. (1995) Quantitative monitoring of gene expression patterns with a complementary DNA microarray. *Science* 270:467–470.

SCHENA, M., HELLER, R. A., THERIAULT, T. P., KONRAD, K., LACHENMEIER, E. and DAVIS, R. W. (1998) Microarrays: biotechnology's discovery platform for functional genomics. *Trends Biotechnol.* 16:301–306.

SCHUR, B. C., BJERKE, J., NUWAYHID, N. and WONG, S. H. (2001) Genotyping of cytochrome P450 2D6*3 and *4 mutations using conventional PCR. *Clin. Chim. Acta* 308(1/2):25–31.

SCOTT, R. H. and BALASUBRAMANIAN, S. (1997) Properties of fluorophores on solid phase resins; implications for screening, encoding and reaction monitoring. *Bioorg. Med. Chem. Lett.* 7:1567–1572.

SELKOE, D. J. (2001) Alzheimer's disease: genes, proteins, and therapy. *Physiol. Rev.* 81(2):741–766.

SHCHEPINOV, M. S., CHALK, R. and SOUTHERN, E. M. (1999) Trityl mass-tags for encoding in combinatorial oligonucleotide synthesis. *Nucleic Acids Symposium Series* 42:107–108.

SILVER, M. (ed.) (2001) Pharmacogenomics: Finding the competitive edge in genetic variation. Summit Report, Cambridge Healthtech Institute, MA, USA, 250 p.

SPAIN, M. and JACOBSON, J. (2001) Gene and protein analysis using a MAPping system. *Am. Genomic/Proteomic Technology* 1(3):36–39.

STAGLIA, I., HOTTIGER, M., AUERBACH, D. and GALEUCHET-SCHENK, B. (2001) Protein–protein interactions as a basis for drug target identification. *Innovations Pharm. Technol.* 1(9):66–69.

STEEMERS, F. J., FERGUSON, J. A. and WALT, D. R. (2000) Screening unlabeled DNA targets with randomly ordered fiber-optic gene arrays. *Nature Biotechnol.* 18(1):91–94.

STÖBER, W., FINK, A. and BOHN, E. (1968) Controlled growth of monodisperse silica spheres in the micron size range. *J. Colloid Interf. Sci.* 26:62–69.

STRYER, L. and HAUGLAND, R. P. (1967) *Proc. Natl. Acad. Sci. USA* 58:719.

SZURDOKI, F., MICHAEL, K. L. and WALT, D. R. (2001) A duplexed colloid-based fluorescent immunoassay. *Anal. Biochem.* 291(2):219–228.

TIMAR, J., CSUKA, O., OROSZ, Z., JENEY, A., KOPPER, L. (2001) Molecular pathology of tumor metastasis: I. Predictive pathology. *Path. Onc. Res.* 7(3):217–230.

TRAU, M. and BATTERSBY, B.J. (2001) Novel colloidal materials for high-throughput screening applications in drug discovery and genomics. *Adv. Mater.* 13:975–979.

VAN BLAADEREN, A. and KENTGENS, A.P.M. (1992) Particle morphology and chemical microstructure of colloidal silica spheres made from alkoxysilanes. *J. Non-Cryst. Solids* 149:161–178.

VAN BLAADEREN, A. and VRIJ, A. (1992) Synthesis and characterisation of colloidal dispersions of fluorescent, monodisperse silica spheres. *Langmuir* 8:2921–2931.

VENTER, J.C., ADAMS, M.D., MYERS, E.W., LI, P.W., MURAL, R.J., SUTTON, G.G., SMITH, H.O., YANDELL, M., EVANS, C.A., HOLT, R.A., GOCAYNE, J.D., AMANATIDES, P., BALLEW, R.M., HUSON, D.H., WORTMAN, J.R., ZHANG, Q., KODIRA, C.D., ZHENG, X.H., CHEN, L., SKUPSKI, M., SUBRAMANIAN, G., THOMAS, P.D., ZHANG, J., MIKLOS, G.L.G, NELSON, C., BRODER, S., CLARK, A.G., NADEAU, J., MCKUSICK, V.A., ZINDER, N., LEVINE, A.J., ROBERTS, R.J., SIMON, M., SLAYMAN, C., HUNKAPILLER, M., BOLANOS, R., DELCHER, A., DEW, I., FASULO, D., FLANIGAN, M., FLOREA, L., HALPERN, A., HANNENHALLI, S., KRAVITZ, S., LEVY, S., MOBARRY, C., REINER, K., REMINGTON, K., ABU-THREIDE, J., BEASLEY, E., BIDDICK, K., BONAZZI, V., BRANDON, R., CARGILL, M., CHANDRAMOULISWARAN, I., CHARLAB, R., CHATURVED, K., DENG, Z., DI FRANCESCO, V., DUNN, P., EILBECK, K., EVANGELISTA, C., GABRIELIAN, A.F., GAN, W., GE, W., GONG, F., GU, Z., GUAN, P., HEIMAN, T.J., HIGGINS, M.E., JI, R., KE, Z., KETCHUM, K.A., LAI, Z., LEI, Y., LI, Z., LI, J., LIANG, Y., LIN, X., LU, F., MERKULOV, G.V., MILSHINA, N., MOORE, H.M., NAIK, A.K., NARAYAN, V.A., NEELAM, B., NUSSKERN, D., RUSCH, D.B., SALZBERG, S., SHAO, W., SHUE, B., SUN, J., WANG, Z.Y., WANG, A., WANG, X., WANG, J., WEI, M., WIDES, R., XIAO, C., YAN, C., YAO, A., YE, J., ZHAN, M., ZHAN, W., ZHANG, H., ZHAO, Q., ZHENG, L., ZHONG, F., ZHONG, W., ZHU, S.C., ZHAO, S., GILBERT, D., BAUMHUETER, S., SPIER, G., CARTER, C., CRAVCHIK, A., WOODAGE, T., ALI, F., AN, H., AWE, A., BALDWIN, D., BADEN, H., BARNSTEAD, M., BARROW, L., BEESON, K., BUSAM, D., CARVER, A., CENTER, A., CHENG, M.L., CURRY, L., DANAHER, S., DAVENPORT, L., DESILETS, R., DIETZ, S., DODSON, K., DOUP, L., FERRIERA, S., GARG, N., HOSTIN, D., HOUCK, J., HOWLAND, T., IBEGWAM, C., JOHNSON, J., KALUSH, F., KLINE, L., KODURU, S., LOVE, A., MANN, F., MAY, D., NELSON, K., PFANNKOCH, C., PRATTS, E., PURI, V., QURESHI, H., REARDON, M., RODRIGUEZ, R., ROGERS, Y., ROMBLAD, D., RUHFEL, B., SCOTT, R., SITTER, C., SMALLWOOD, M., STEWART, E., STRONG, R., SUH, E., THOMAS, R., TINT, N, TSE, S., VECH, C., WANG, G., WETTER, J., WILLIAMS, S., WILLIAMS, M., WINDSOR, S., WINN-DEEN, E., WOLFE, K., ZAVERI, J., ZAVERI, K., ABRIL, J.F., GUIGO, R., CAMPBELL, M.J. SJOLANDER, K.V., KARLAK, B., KEJARIWAL, A., MI, H., LAZAREVA, B., HATTON, T., NARECHANIA, A., DIEMER, K., MURUGANUJAN, A., GUO, N., SATO, S., BAFNA, V., ISTRAIL, S., LIPPERT, R., SCHWARTZ, R., WALENZ, B., YOOSEPH, S., ALLEN, D., BASU, A., BAXENDALE, J., BLOCK, L., CAMINHA, M., CARNES-STINE, J., CAULK, P, CHIANG, Y., COYNE, M., DAHIKE, C., MAYS, A.D., DOMBROSKI, M., DONNELLY, M., ELY, D., ESPARHAM, S., FOSTER, C., GIRE, H., GLANOWSKI, S., GLASSER, K., GLODEK, A., GOROKHOV, M., GRAHAM, K., GROPMAN, B., HARRIS, M., HEIL, J., HENDERSON, S., HOOVER, J., JENNINGS, D., JORDAN, C., JORDAN, J., KASHA, J., KAGAN, L., KRAFT, C., LEVITSKY, A., LEWIS, M., LIU, X., LOPEZ, J., MA, D., MAJOROS, W., MCDANIEL, J., MURPHY, S., NEWMAN, M., NGUYEN, T., NGUYEN, N., NODELL, M., PAN, S., PECK, J., PETERSON, M., ROWE, W., SANDERS, R., SCOTT, J., SIMPSON, M., SMITH, T., SPRAGUE, A., STOCKWELL, T., TURNER, R., VENTER, E., WANG, M., WEN, M., WU, D., XIA, A., ZANDIEH, A., ZHU, X. (2001) The sequence of the human genome. *Science* 291(5507):1304–1351.

VERHAEGH, N.A.M. and VAN BLAADEREN, A. (1994) Dispersions of rhodamine-labeled silica spheres: Synthesis, characterisation and fluorescence confocal scanning laser microscopy. *Langmuir* 10:1427–1438.

Vignali, D.A.A. (2000) Multiplexed particle-based flow cytometric assays. *J. Immunol. Methods* 243:243–255.

Viswanadhan, V.N., Balan, C., Hulme, C., Cheetham, J.C. and Sun, Y. (2002) Knowledge-based approaches in the design and selection of compound libraries for drug discovery. *Curr. Opinion Drug Disc. Devel.* 5(3):400–406.

Von Lippert, E.Z. (1957) Spektroskopische Bestimmung des Dipolmomentes aromatischer Verbindungen im ersten angeregten Singulettzustand. *Electrochem.* 61:962–975.

Wallin, J.M., Holt, C.L., Lazaruk, K.D., Nguyen, T.H. and Walsh, P.S. (2002) Constructing universal multiplex PCR systems for comparative genotyping. *J. Forensic Sci.* 47(1):52–65.

Walt, D.R. (2000) Techview: Molecular biology: Bead-based fiber-optic arrays. *Science (Washington, DC)* 287(5452):451–452.

Walton, I.D., Norton, S.M., Balasingham, A., He, L., Oviso, D.F. Jr, Gupta, D., Raju, P.A., Natan, M.J. and Freeman, R.G. (2002) Particles for multiplexed analysis in solution: Detection and identification of striped metallic particles using optical microscopy. *Anal. Chem.* 74(10):2240–2247.

Waterfall, C.M. and Cobb, B.D. (2001) Single tube genotyping of sickle cell anaemia using PCR-based SNP analysis. *Nucleic Acids Res.* 29(23):e119/1–e119/8.

Wei, X., Elizondo, G., Sapone, A., McLeod, H.L., Raunio, H., Fernandez-Saiguero, P. and Gonzalez, F.J. (1998) Characterization of the human dihydropyrimidine dehydrogenase gene. *Genomics* 51(3):391–400.

Wittes, J. and Friedman, H.P. (1999) Searching for evidence of altered gene expression: a comment on statistical analysis of microarray data. *J. Natl. Cancer Inst.* 91:400–401.

Yan, B., Martin, P.C. and Lee, J. (1999) Single-bead fluorescence microspectroscopy: Detection of self-quenching in fluorescence-labeled resin beads. *J. Comb. Chem.* 1:78–81.

Zhou, H., Roy, S., Schulman, H. and Natan, M.J. (2001) Solution and chip arrays in protein profiling. *Trends Biotechnol.* 19(10, Suppl.): S34–S39.

Zhuravlev, L.T. (2000) The surface chemistry of amorphous silica. Zhuravlev model. *Colloids Surf.* 173:1–38.

18
Polyelectrolyte Microcapsules as Biomimetic Models

Gleb B. Sukhorukov and Helmuth Möhwald

18.1
Introduction

In many biochemistry text books the cell is described as a community (Fig. 18.1). There is a plan for its construction in the nucleus, there are the mitochondria as power plants, there are factories like the endoplasmatic reticulum, there are rails like the tubulin and actin network and motor proteins transporting cargo. Transport also occurs after packing of molecules into lysosomes, especially if larger amounts of materials have to be shipped and protected. Since the cell as a whole functions very well there have been many attempts to make use of this in daily life or in technical systems. These early attempts date back thousands of years,

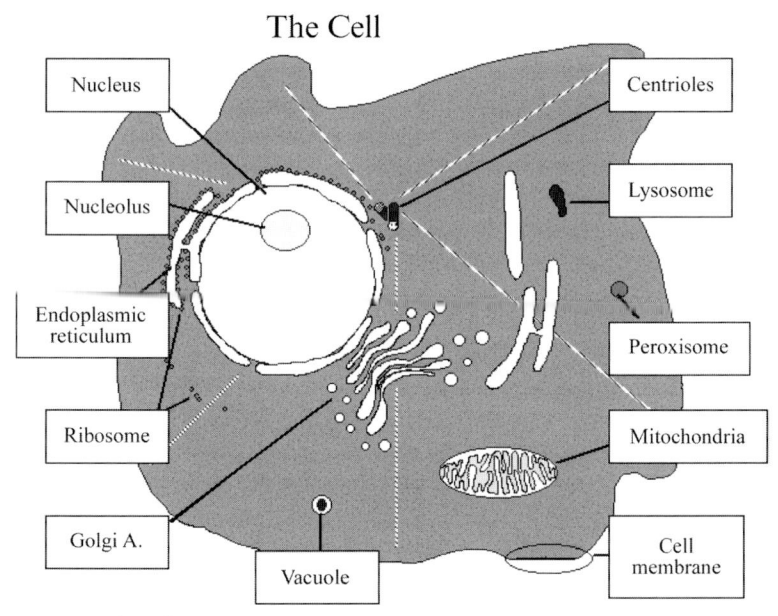

Fig. 18.1 The cell

Colloids and Colloid Assemblies. Edited by Frank Caruso
Copyright © 2004 Wiley-VCH Verlag GmbH & Co. KGaA, Weinheim
ISBN: 3-527-30660-9

e.g. in fermentation, but modern biotechnology has still many opportunities for further progress. Equally, there is also great potential in using only *part* of a biological system in a technical device, e.g. in biosynthesis or biosensing. For all these applications it is obvious that they are feasible and their commercial applicability will be increased with further improvements. However, it is also clear that a qualitative step could be made if one could make use of the complex interplay of functions that a cell exhibits. For this one has to understand more about this interplay, and this is one of the challenges of life sciences in the near future: Understanding the organization and function of matter from the nanometer to the micrometer length scale.

One way to achieve this is to arrange, in a well-defined way, either components of the cellular system or synthetic molecules that are expected to have the same function, study the function and continuously increase the complexity. Although this method may not lead to an artificial cell, it is worthwhile pursuing because of the large gains in knowledge that will follow.

The main features of the cell are that reactions occur in compartments of submicrometer dimensions and/or at interfaces. There has to be transport of information or matter between these compartments, and this occurs across membranes. Thus to build a cell biomimetic system we have to construct micro- and nanocontainers with internal functions and provide them with membranes to study their communication with their environment. In this chapter we will present a specific system which we believe to be very promising, study its properties and deficiencies and comment on its further development. A major part of the chapter will also be devoted to "spin-offs", where this development has already led to systems that may already be used technically or in pharmaceutical applications.

18.2
Construction and Properties of a Biomimetic Model

18.2.1
Polyelectrolyte Micro- and Nanocapsules

The system we propose here makes use of the great progress of recent years in building well-defined organic mono- and multilayers by self-assembly processes. These techniques have been adapted to coat colloidal particles with dimensions from below 100 nm to some micrometers [1]. We thus obtain systems with very high surface/volume ratio which is suitable for many (e.g. chromatographic) applications as well as for surface studies with techniques requiring large sample areas (NMR, flash spectroscopy). If we now use a colloidal core that can later be destroyed with fragments penetrating the organic film we obtain hollow capsules with walls defined with the nanometer precision of the precursor film (Fig. 18.2) [2, 3]. To build up the film we mostly adapt the technique of layer-by-layer adsorption of oppositely charged polyelectrolytes (LbL) because of its robustness and versatility [4]. It can be employed with synthetic and natural polyelectrolytes, pro-

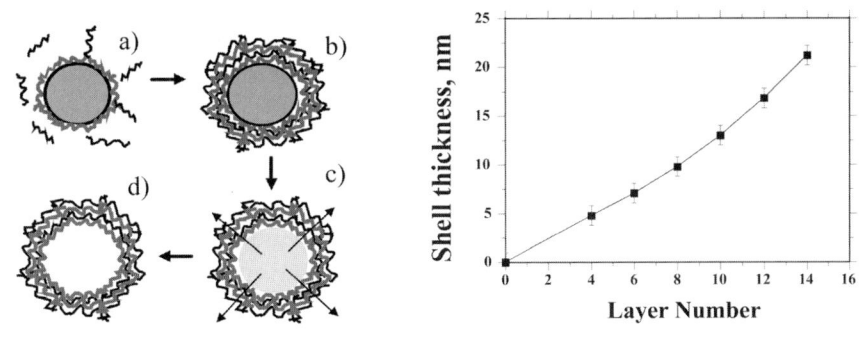

Fig. 18.2 Left: Consecutive adsorption of positively (gray) and negatively (black) charged polyelectrolytes onto negatively charged colloidal particles **a,b**. After dissolution of the colloidal core **c,d** a suspension of hollow polyelectrolyte capsules is obtained. Right: shell thickness as a function of layer number

teins, inorganic particles and multivalent metal ions or dyes. In the latter case we may loosen the electrostatic bond using salt and thus also destroy the film. As cores we initially used weakly cross-linked melamine formamide (MF) particles that are available with high monodispersity. Then, however, special care has to be taken that some residual material, not fully depolymerized on going to low pH, is not allowed to stick to the wall. Other cores have been biological cells that after coating could be destroyed by an oxidizing agent (NaOCl) [5, 6]. This agent, however, also affects polyelectrolyte walls, among others, which provide chemical networks that give the walls elasticity. Meanwhile many other types of cores, including inorganic particles ($CaCO_3$, $CdCO_3$, $MnCO_3$, SiO_2), organic crystals or oil droplets have been developed, all requiring specific removal chemistry. As a variant in the preparation of inorganic hollow porous spheres, SiO_2 has been used as the wall component, and during calcination the SiO_2 particles were sintered to obtain rigid spheres [7, 8]. The LbL deposition process is also promising as it provides the possibility of controlling the chemical composition along the normal to the wall. For some applications, however, it may be too time consuming, and therefore controlled deposition of polyelectrolyte complexes was developed as an alternative [9, 10]. In this case one starts with a solvent in which the colloidal core and the two wall components are soluble and then reduces the solvent power, e.g. by evaporating one solvent component or adding ethanol to water. The precipitate is then deposited on the particle and one obtains a surprisingly homogeneous coating with thickness depending on the amount of material in solution.

Since the capsules are charged they are stabilized by Coulomb forces against precipitation over times of at least a year. More critical are aging processes that can be accelerated by mild heating. With storage time we observe a reduction of permeability which may be ascribed to annealing of defects and misfits of local bonds.

One most important issue is capsule permeability and its control. In most cases the polyelectrolyte wall is impermeable to polymers, proteins and nanoparticles, as revealed by confocal fluorescence microscopy (see below). For low molecular weight dyes and drugs, diffusion coefficients above 10^{-12} cm^2/s have been determined [11]. This is six orders of magnitude lower than the value for water but two orders of magnitude higher than the value previously determined for multilayer films on planar supports. Considering the fact that swelling and shrinking may lead to variations in permeation between these extremes one has to be aware that slight variations in preparation and structure may have a tremendous effect on permeation. On the other hand understanding the underlying mechanisms provides a means to tune permeability over a broad range.

We may conceive many possibilities of tuning or switching permeability, e.g. by osmotic pressure, light, temperature, pH, ionic strength or solvent composition [12–14]. The last three possibilities have been verified at least qualitatively, showing that permeation of macromolecules can be switched reversibly. As an example that we shall refer to later, we consider the encapsulation of an enzyme via a second solvent that does not inactivate the enzyme [14]. In the example of Fig. 18.3 dye-labeled urease is added to a capsule solution in water. Under the preparation conditions the enzyme does not penetrate the shell, and thus the inner volume does not show fluorescence (Fig. 18.3, left). Adding ethanol the shell appears to develop pores through which the enzyme penetrates. This explains the brightness developing in the capsule interior (Fig. 18.3, center). The bright ring obviously results from a large fraction of protein sticking to the wall. Washing the capsules again in water apparently closes them and the enzyme is entrapped as demonstrated again by fluorescence microscopy (Fig. 18.3, right). For this specific case we can show that the process does not harm the function. Also it has been shown that, by incorporating sorbents inside the shell, the encapsulation efficiency can be increased above 95%. Still we have to stress that we do not yet understand the mechanism and have no information on the number and size of pores which would enable the shell to operate as a tunable molecular sieve.

The fact that macromolecules like polyelectrolytes can be encapsulated has also been used to establish a Donnan potential between the inside and the outside.

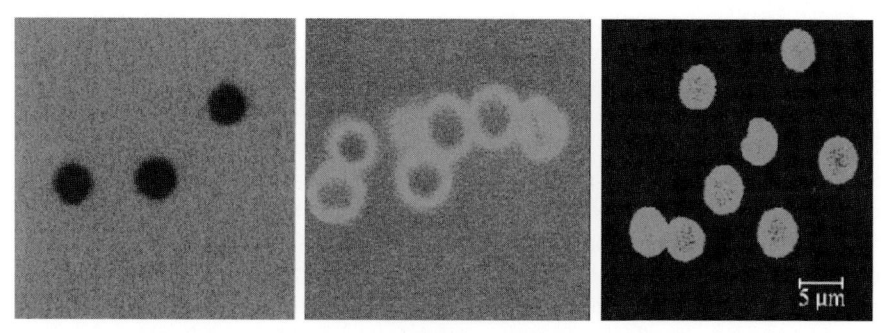

Fig. 18.3 Permeation and encapsulation of urease-FITC into polyion multilayer capsules. Left, in water; center, in a water/ethanol mixture; right, the capsule with encapsulated urease in water

Thus a pH difference up to 4 units could be established enabling specific chemical processes inside [15].

Concluding this section on polyelectrolyte capsules we should also comment on their ionic conductivity, the measurement of which will be detailed below. At a first glance it may seem surprising that it was found to be about 1 S/m, corresponding to about 100 mM NaCl solution. This rather large value can be rationalized if about 10% of the ionic groups of the polyelectrolyte are not counterbalanced by charges on the other polyelectrolyte but by a diffuse ion distribution responsible for the conductivity. This is not unreasonable and is supported by measurements with ionic dyes in planar multilayers that probed the ionic sites within the film. Also the dielectric constant of the shell was found to be rather high ($\varepsilon \approx 60$), suggesting a high water content. This is in agreement with optical studies of local polarity using pyrene probes.

18.2.2
Lipid Bilayer Coupled to a Synthetic Skeleton

A biological cell consists of a polymer network, the interior cytoskeleton, surrounded by a bilayer membrane with integrated or peripheral proteins and peptides. One of the purposes of the network is to provide mechanical stability and at the same time shape flexibility via reversible polymerization and depolymerization processes. A polyelectrolyte shell can be considered similar to a cytoskeleton, with mechanical strength varying in a broad range from that of inorganic spheres to those of composites and loosely connected polyelectrolytes. In addition we may envision forming an actin and tubulin network via entrapment of monomeric proteins and polymerizing them inside via changes in the ionic milieu.

The next task is to assemble the bilayer on the shell and later incorporate functional proteins into it. This can be achieved in two conceptually different ways [16]:

(1) The positively charged shells and lipid vesicles containing anionic phospholipids are incubated for some 10 min, and electrostatic interaction causes a bilayer to form. Lipid multilayers are prevented from forming if the bilayer exhibits sufficient negative charge.
(2) Since the capsules are stable in methanol/water mixtures one may add a lipid-saturated methanol solution from which the lipids assemble as monomers on the capsule surface.

Both processes work, at this stage of development, with similar quality. The first one appears to be extendable to proteoliposomes in a straightforward way, the second one may be preferred for annealing small defects. Both methods work with zwitterionic lipids as well as with ionic ones. In that case, however, the partial formation of multilayers is a critical issue.

A first assessment of the lipid coating can be deduced from the confocal fluorescence micrographs of Fig. 18.4 [16]. By adding a small percentage of fluorescent lipid probes we can identify a homogeneous coating at the micrometer level

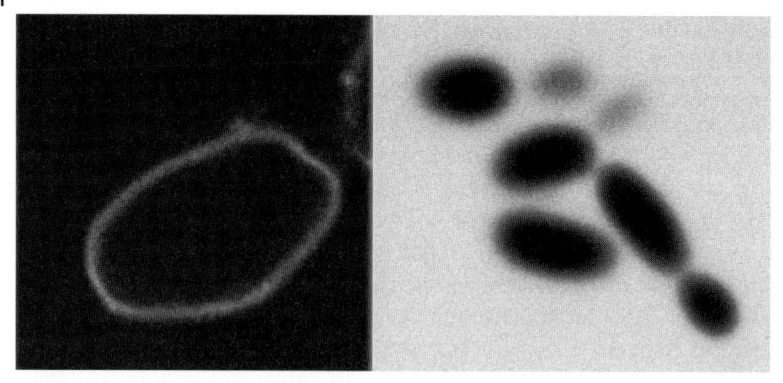

Fig. 18.4 Confocal fluorescent microscopy images of polyelectrolyte capsules covered with labeled lipids (left) and in the presence of 6-CF (right)

(Fig. 18.4, left). Not using a lipid probe but adding a dye probe to the outside we can demonstrate that the dye does not cross the bilayer to enter inside. The dark interior is observed over hours, showing that the bilayer does not exhibit defects through which low molecular weight dyes could penetrate (Fig. 18.4, right). Differential scanning calorimetry demonstrates that the zwitterionic lipid coupling is sufficiently weak not to shift the main phase transition. Ionic lipids, however, cause a drastic reduction of the transition temperature, hence tending to decrease lipid ordering. These findings contrast with results on lateral diffusion obtained for planar bilayers on polyelectrolyte support: For both zwitterionic and anionic lipids we find a drastic reduction of the diffusion coefficient, indicating a strong coupling of at least a fraction of the lipids such that diffusion over micrometer dimensions is impeded. Hence to achieve diffusion as fast as in the natural system we will have to prepare bilayers and polymer supports with a majority of nonionic groups exposed and only some ionic groups embedded to enable interaction.

A most important property of the bilayer membrane is its electrical insulation. It is a prerequisite before inserting pumps and channels to make use of selective ion transport. To assess electrical and dielectrical properties of a particle or a cell the technique of electrorotation spectroscopy has been introduced [17]. With this technique an individual particle is fixed in an electrical potential well under a microscope. An electrical field with frequencies varied from 10^2 to 10^7 Hz is applied and the torque on the particle is measured via its low frequency rotation. The torque is due to the difference in polarizability of the particle and its medium. For example, if the medium is less polarizable than the particle one observes a cofield rotation, in the opposite case an anti-field rotation. The frequencies at which these occur characterize the material and the particle. As an example Fig. 18.5 compares electrorotation spectra for polyelectrolyte capsules in the presence or absence of a lipid coating.

In the absence of a lipid bilayer one observes only a co-field rotation peak, i.e. the capsule is more conducting than the surrounding medium. Increasing the

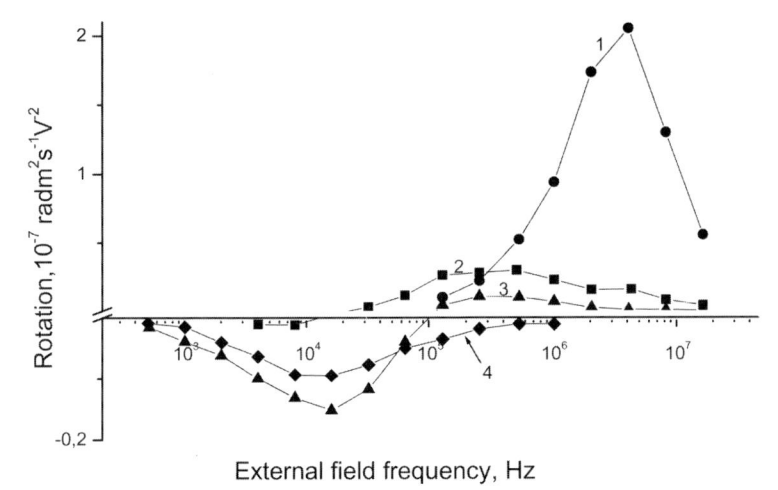

Fig. 18.5 Typical electrorotation spectra of (PAH/PSS)$_5$ PE (1), DPPA-(PSS/PAH)$_5$ PE (2), DPPC/DPPA(95%/5%)-(PAH/PSS)$_5$ PE (3) and DPPC-(PAH/PSS)$_5$ PE (4) capsules. The conductivity of bulk solution was 1 mS/m in the curves (1), (3) and (4) and 0.5 mS/m in curve (2)

ionic strength (the conductivity of the environment) reduces the height of this peak and for an ionic strength of 0.1 N it even reverses sign. From this we conclude, as stated before, that the conductivity of the shell equals 1 S/m. The spectrum can be modeled assuming a spherical shell of uniform thickness, conductivity and dielectric constant, and the parameter easiest to vary is the conductivity, Ge, of the external and internal volume (assumed equal). In the specific example we obtain the (high) value $\varepsilon \approx 60$. To model the capsules with an additional lipid coat we have to assume a second concentric shell with different conductivity, ε, and thickness. The spectrum exhibits an additional anti-field peak at lower frequencies, which is typical for an insulating layer like a bilayer membrane. From measurements with different lipid mixtures and Ge one then derives the following picture of the bilayer membrane [17]:

- The conductivity of the anionic layer of dipalmitoylphosphatidic acid (DPPA) changed between 0.01 mS/m and 0.1 mS/m increasing with ionic strength. For the mixed lipid dipalmitoylphosphatidylcholine (DPPC)/DPPA (95:5) the conductivity is an order of magnitude lower and for the purely zwitterionic lipid DPPC values as low as 10^{-4} mS/m were found. This latter value is lower than that found for a biological membrane. Thus it is conceivable that the system can be used to study inserted ion channels (see below). On the other hand the conductivity is more than five orders of magnitude lower than that of a bilayer membrane, indicating that the bilayer coupled to the polyelectrolyte is rich in defects.

- To estimate the area fraction A of pores within the bilayer we assume they exhibit the same conductivity as the surrounding medium (Ge). This is also supported by the finding that the bilayer conductivity scales with Ge. From the measurement of 10^{-4} mS/m for Ge = 1 mS/m we can then deduce $A = 10^{-4}$. Consequently approaching the insulating properties of a black lipid membrane (BLM) involves annealing of defect areas below the part-per-million range, a most difficult task.

- The dielectric constants of the bilayer ($\varepsilon = 5.15$) are in all cases rather large compared to what one would have expected for the hydrophobic region of the bilayer ($\varepsilon \approx 2.3$). This may be because the model is too simple, and hydrophilic and hydrophobic membrane regions may require individual shells. Also as a consequence of this we derive a high membrane capacitance ($\geq 1\ \mu F/cm^2$) as compared with 0.5 $\mu F/cm^2$ known for the BLM.

As stated above the ionic insulation of the bilayer is far from perfect but it may be already good enough to give rise to biophysical and biological effects. One of the most frequently studied measurements is the ion-specific conduction of valinomycin. This peptide in the membrane presents a channel with a specific conductance for K^+ that is much greater than for other ions. We shall describe an experiment that uses the fact that a potential may exist even in the absence of ion pumps. This is because there is a disequilibrium of nondiffusible macroions (Donnan effect) and that low molecular weight cations and anions have different membrane permeabilities. If we change the membrane conductivity for a specific ion and increase the concentration of the corresponding ion we expect a reduction of the membrane potential. We can measure the membrane potential using the distribution of the membrane-permeating cationic dye TMRE [18]. This dye has been established to be suitable for this as it hardly bleaches or aggregates and has a high fluorescence quantum yield. The fluorescence intensity $F_{i,o}$ is proportional to its concentration inside and outside the capsule, respectively, and thus from measurement of the fluorescence one can determine the potential difference ΔV via the Nernst equation

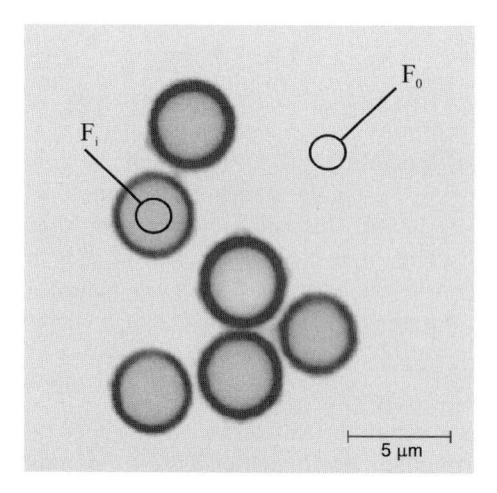

Fig. 18.6 Fluorescence confocal microscopy image of DPPC-HPM in the presence of TMRE: calculation of the membrane potential value

$$\Delta V = -\frac{RT}{F} \ln \frac{F_i}{F_o} \tag{1}$$

where R is the ideal gas constant and F is the Faraday constant.

From a typical fluorescence micrograph such as the one in Fig. 18.6 we can deduce the intensities F_i, F_o with reasonable accuracy and thus obtain potentials for different conditions, such as gramicidin or valinomycin incorporation, different lipid coatings or ion addition. As described in [18] the addition of valinomycin, as expected, does not change the membrane potential, and neither does addition of 100 mM NaCl.

However, addition of 100 mM KCl changes the potential by 5–10 mV, i.e. more than the error limit. Finally, we have been able to reconstitute the specific ion channel and demonstrate that we can observe its activity. Obviously these measurements are merely a proof of principle and will have to be extended to obtain more quantitative information. Also it is clear that this is only a tiny step to mimic a biological cell, the next steps being the incorporation of receptors and specific ligands to study adhesion and ligand-gated ion channels.

18.2.3
Coated Cells

A different approach, not building up a cell from parts but varying the natural system at will, is to coat biological cells. For this there are different objectives and hence the procedures are different.

The first objective was to use a template with a peculiar shape to obtain capsules with this shape [5, 6]. Typical examples are erythrocytes and echinocytes, and the confocal fluorescence micrographs of Fig. 18.7 show that this is possible with the procedure described in Section 2.1. In this case the shape was fixed with glutaraldehyde cross-linking, so the cell was not meant to function. Still it is surprising that, although the cell surface is rough and heterogeneous, nominally four polyelectrolyte bilayer coatings suffice to obtain a continuous elastic shell.

One application of a coated erythrocyte may result from the fact that receptors on the surface, including those for clotting, are covered and thus rendered inactive. However, the coating still allows oxygen exchange and therefore, with haemoglobin inside, the oxygen transport capacity, the most important reason for blood transfusion, may be preserved. This will remain an active area of research under the keyword "artificial blood".

In order to protect cells against immune attack or to guide them to specific sites in the body it may also be interesting to coat living cells and to keep them alive. This has been reported recently with, admittedly very robust, yeast cells [19]. Then it was shown that many of the cellular functions were still retained. Even cell division was possible, in that case with destruction of the coat (Fig. 18.8).

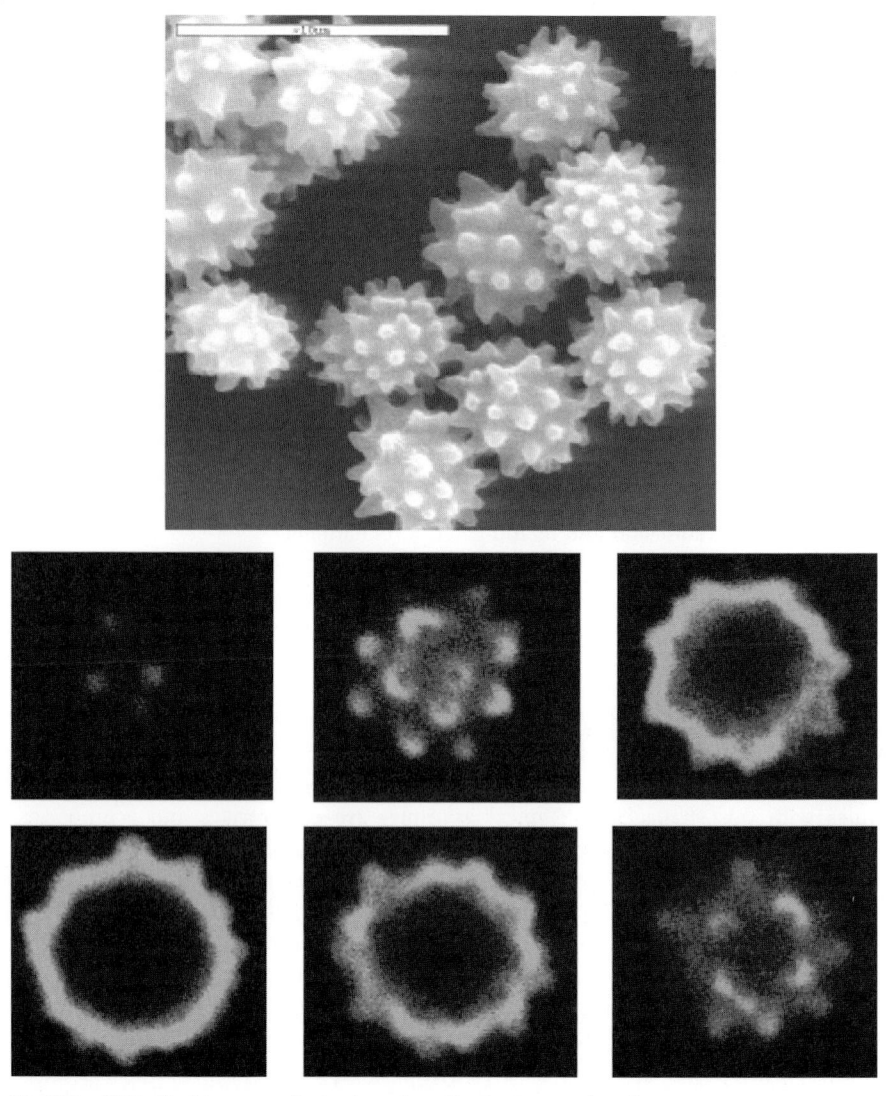

Fig. 18.7 SEM of echinocyte cells (top), and confocal micrographs of 11 layers of PSS-PAH shell templated on a single echinocyte. The outer layer is FITC-PAH. Scans (left–right, top–bottom) run through upper part of the shell on planes separated by 1 μm

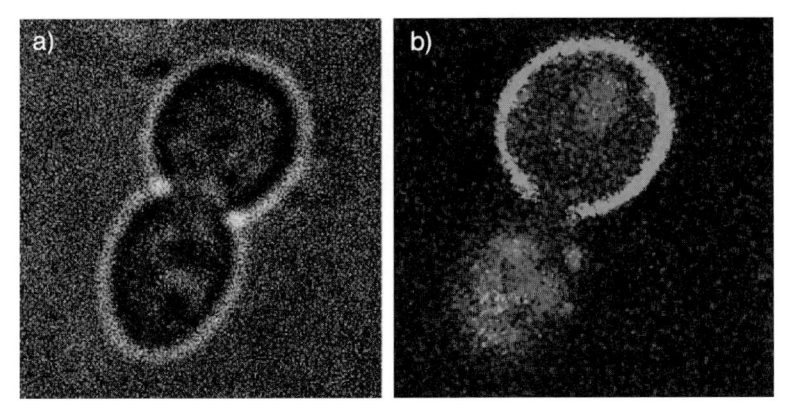

Fig. 18.8 Transmission image **a** of a duplicating cell. This is confirmed by the fluorescence image **b** where the lack of green fluorescence in the mother cell coating reveals where the bud has formed. The DAPI-labelling of the DNA distribution allows the imaging of both mother and uncoated daughter cell

18.3
Biological Processes

18.3.1
Protein (Enzyme) Incorporation into the Capsules

The described polyelectrolyte capsule technology covers a size ranging from 100 nm to tens of microns. Working with a size of about 3–10 µm there is a temptation to model biochemical processes in biological cells. The first related task to step onto "bio-modeling" would be incorporation (encapsulation) of different proteins, which can perform their function while associated with the capsules. There are several possibilities for the insertion of proteins (enzymes) into core-shell (or empty shell) structures fabricated by LbL technology: (1) LbL assembly of proteins as shell constituents; (2) formation of a core out of protein microcrystals or aggregates; (3) impregnation of preformed hollow capsules by proteins by means of controllable opening and closing of pores in polyelectrolyte multilayers. Certainly these three methods differ in several parameters, such as encapsulation efficiency, controlled release, protein concentration and activity, applicability for a wide class of enzymes and possibilities for multienzyme system incorporation. These parameters indicate the further use of protein-containing core-shell structures which might be envisaged for elaboration of drug delivery systems, enzymatic microreactors or for biomimetic modeling. Below, we consider these approaches in more detail.

18.3.1.1 Enzyme/Polymer Composite Shells on Colloidal Particles

The first way is alternating assembling of protein molecules with oppositely charged macromolecules onto the surface of colloidal particles. The LbL approach is suitable to fabricate a protein-polymer multilayer shell assembled on colloidal particles. Using alternate deposition of protein and oppositely charged polymer on PS spheres, multilayer films of various proteins have been formed, including bovine serum albumin (BSA), immunoglobulin G (IgG), glucose oxidase (GOD), horseradish peroxidase (POD), β-glucosidase (β-GLS) and urease [20–22]. The thickness of the protein coating obtained by SPLS varied from several to tens of nanometers. It depends on the kind of proteins and oppositely charged components, number of protein layers and, cannot be precisely measured because Single Particle Light Scattering (SPLS) thickness estimation requires the assumption of a refractive index which might be different for varying protein-polymer compositions. Protein assembling on PS particles has been also proved by TEM studies. In [20] the PS particles coated with a BSA multilayer show both increasing surface roughness and an increase of diameter compared with bare PS spheres. Actually the film thickness determined by this method is consistent with the SPLS data.

Coating particles with enzyme-containing films is attractive mainly because of the inherently high surface area of colloids, which provides the potential for higher enzymatic reaction efficiencies than planar surfaces. The surface area, which becomes the "working" area for enzymatic reactors, can be further increased by alternating assembly of the proteins together with nanoparticles (8–40 nm) and oppositely charged polyelectrolytes on submicron-sized colloids. This has been demonstrated [23] with the example of alternating GOD and silicon oxide or magnetite nanoparticles. In the last case, besides the higher surface area, the particles can be mobilized by an applied magnetic field.

When enzyme coating of colloidal particles is considered, the question of preserving the enzymatic activity often arises. It has been reported [21] that the enzymatic activity (per particle) increased with the number of layers immobilized. But this increase is not always linear because of diffusion-limited penetration of the substrate into the enzyme layers hidden beneath the outer polyelectrolyte layers. Basically, only the outermost enzyme layer is working at 100% capacity. Nevertheless, an outer polyelectrolyte coating that was permeable only for special solutes might make the enzyme more stable in rather aggressive media.

18.3.1.2 LbL Absorption of Polyelectrolytes on Protein Aggregates and DNA Particles

Multilayer assembly can be performed not only on solid particles like silica, PS-latex particles or organic crystals, but also on "soft" particles formed immediately before multilayer build up, such as protein aggregates, crystals and a compact form of DNA. The use of micron-sized protein aggregates or crystals as templates for polyelectrolyte multilayer assembly has been proposed for catalase microcrystals [24], for lactate dehydrogenase [25] and for chymotrypsin aggregates [26]. Polyelectrolyte multilayer coating of these aggregates captures the proteins inside the capsules and in parallel provides a selective barrier for the diffusion of different

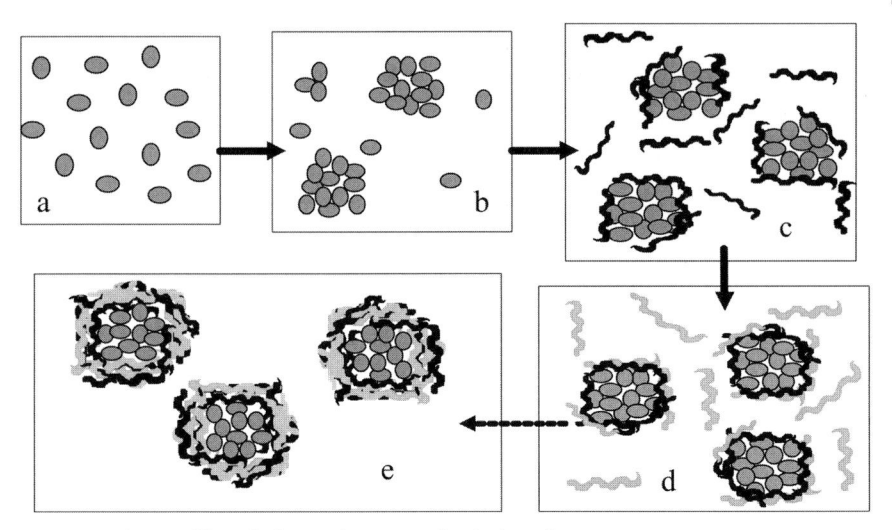

Fig. 18.9 Scheme of layer-by-layer adsorption of polyelectrolytes onto protein aggregates formed in salt solutions

species (substrates, inhibitors) from the exterior. A general scheme of protein aggregate coverage with polyelectrolyte multilayers is presented in Fig. 18.9. The concept of polyelectrolyte multilayer assembly on aggregates is similar to assembly on the surface of colloidal particles [2]. Here, instead of solid colloidal particles, the preformed protein aggregates have been used as templates for polyelectrolyte multilayer assembly.

Chymotrypsin has been chosen as a model protein whose activity can be easily monitored [26]. Chymotrypsin aggregates were prepared by salting out (Fig. 18.9a and b). The precipitation of protein started immediately after mixing of the enzyme and saline solutions and was followed by the increasing turbidity of the solution. After a certain time the turbidity increase stopped. The aggregation process could be reversed, and the aggregates redissolved, by decreasing the NaCl concentration in the suspension. Variations in a-chymotrypsin and sodium chloride concentrations were used permitting optimization of the enzyme precipitation. Because of the positive charge on chymotrypsin at pH 2.3 the negatively charged polyelectrolyte PSS was taken as a first layer to start the multilayer assembly (Fig. 18.9c). PAH was assembled as a polycation constituent (Fig. 18.9d). Finally, these aggregates could be covered with a definite number of layers (Fig. 18.9e).

A typical scanning electron microscopy image of chymotrypsin aggregates uncoated and covered with three (PSS/PAH/PSS) layers is shown in Fig. 18.10. Micron-sized aggregates were found. The morphology of the aggregates was rather amorphous. Each aggregate consisted of numerous smaller protein particles, so-called primary aggregates, in the shape of small spheres with a diameter of the order of 100–300 nm. This indicates a hierarchy in the formation of the protein ag-

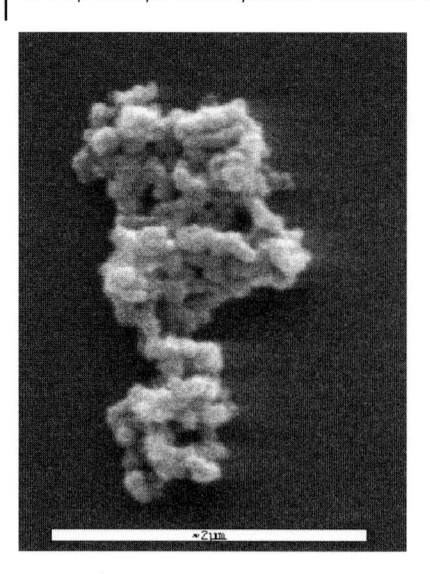

Fig. 18.10 Scanning electron microscopy image of protein aggregates covered by three polyelectrolyte layers (PSS/PAH/PSS). Scale bar is 2 µm

gregates. A characteristic feature of the aggregates is their very large surface. Pores within the aggregates can be observed. Such structures are supposedly formed as a result of secondary aggregation of the primary aggregates being enhanced upon polyelectrolyte deposition. Further sequential polyelectrolyte layer assembly then preserved the structure of the secondary aggregates. Generally, the morphology of aggregates should depend on the preparation conditions, such as the time of aggregate growth, the nature of the polyelectrolyte used and the number of layers, which might influence the composition of aggregates. An enzymatic activity study revealed about 70% activity of encapsulated chemotrypsin. The encapsulated enzyme is protected against a high molecular weight inhibitor, which cannot penetrate the capsule wall and influence enzyme functionality.

Controlled release of encapsulated material is a significant topic for many tasks in biotechnology, pharmacy and medicine. Polyelectrolyte multilayers present a rather unique opportunity to obtain the desired release characteristics. Being very sensitive to a pH change (see above and [12]) the release of encapsulated materials can be controlled by pH variations. When the pH is shifted across the pK value of the polyelectrolytes used the whole multilayer becomes permeable. Protein molecules are released from coated aggregates at a pH higher than 8 as shown for the chymotrypsin release from the capsules composed of PSS/PAH [26]. For such systems the protein release can be stopped by switching the pH back to that at which the polyelectrolytes were assembled (Fig. 18.11). A similar behavior was also shown for catalase release from crystals coated with a PSS/PAH multilayer film at a pH higher than 12 [24].

The method of LbL shell formation on protein aggregates provides an encapsulation efficiency close to 100%. Indeed, one can always find conditions for protein aggregation, for instance at an isoelectric point. At the same time the release of

Protein release, %

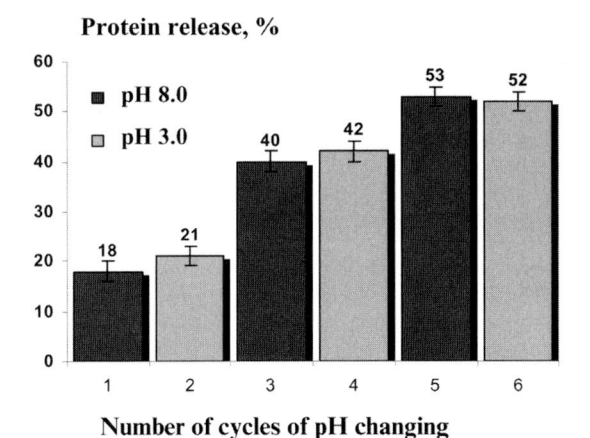

Fig. 18.11 Step-wise chymotrypsin release from 11-layer microcapsules assembled on protein aggregates. Each step shows the percentage of protein released

proteins might be tuned by proper polyelectrolyte composition that makes such systems very suitable as drug-delivery systems with controlled release properties.

18.3.1.3 Enzyme Encapsulation in Preformed Hollow Polyelectrolyte Capsules

The third approach to enzyme encapsulation is based on the concept of macromolecule encapsulation in hollow microcapsules by the opening and closing of pores, and can be applied to the fabrication of enzymatic micro- and nanoreactors. Selective wall permeability allows for substrates and reaction products to diffuse freely through the capsule wall, whereas the encapsulated enzymes are kept in the capsules once the pores have closed again. This approach, with permeability variation for PSS/PAH capsules by pH change, was first applied for encapsulation of dextran and albumin [12] and chymotrypsin [27]. The main problem for such systems is the very high protein adsorption at the PSS/PAH wall. In fact, permeability variation only allows variation of the general contents of enzymes associated with capsules. Only about 10% of chymotrypsin was found by Tiourina et al. [27] in the capsule interior and not adsorbed to the capsule wall. The revealed enzyme activity per protein (chymotrypsin) molecules was about 40–50%. A high molecular weight inhibitor was not able to block chymotrypsin activity in the capsules. The protein content inside the capsule can be essentially increased if some of the core material remains after core dissolution. Gao et al. [28] found that the rest of the MF-core, which forms a complex with PSS molecules, has high affinity for peroxidase, which is sucked into the capsule interior from outside. This mechanism is interesting because it may lead to 100% encapsulation efficiency.

An important task would be to design a capsule wall that is non-adhesive to most proteins. A proper choice of layer constituents for protein encapsulation was proposed in [29]. Capsules made of dextran sulfate or alginate and protamine do not exhibit protein adsorption on the wall of 6.5-µm capsules templated on the MF-core. Formation of a Ca alginate gel inside the capsule provides homogeneous

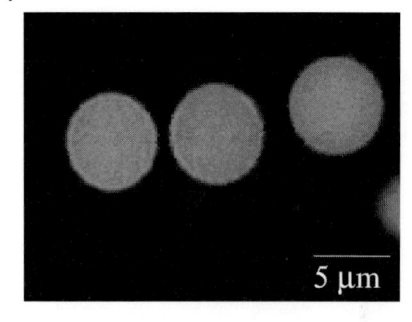

Fig. 18.12 Confocal fluorescence image of alginate/protamine microcapsules loaded with rhodamine-labeled chymotrypsin

distribution of chymotrypsin and peroxidase in the capsule interior (Fig. 18.12) [29]. These microcapsules allow a high loading capacity of about 10^8–10^9 protein molecules per capsule depending on the pH at which the enzyme was embedded. That amount of protein per capsule equals about 10^2 g/L which is higher than for most studies of biochemical reactions. The embedded enzymes (chymotrypsin or peroxidase) retain a high physiological activity of 60–70%. While chymotrypsin is kept for days by the alginate matrix in water or in slightly acidic conditions, transfer of the capsule to 0.05 M Tris buffer (pH=8) results in protein release from the capsules in a matter of several hours. The rate of chymotrypsin release can be regulated by additional adsorption of polyelectrolyte layers onto the microcapsules with encapsulated protein.

Besides pH variation, another approach to reversible polymer segregation in the multilayers is treatment in water/ethanol mixtures. Capsules dried from ethanol suspension show formation of many pores, which apparently also exist in the suspended state as defects in polyelectrolyte multilayer networks. Water/ethanol mixtures were utilized for urease encapsulation in hollow capsules as described above (see Fig. 18.3 and [14]). The figure illustrates confocal images corresponding to each stage of the encapsulation process. In water FITC-labeled urease cannot penetrate the capsule wall and the interior remains dark (left). After addition of ethanol (1:1 water/ethanol ratio) the fluorescence coming from the interior of the capsule is the same as the outside background fluorescence (center). This points to penetration of protein into the capsules and indicates an open state of the capsule wall. Sequential washing in water or ethanol evaporation resulted in defect repair, and the enzyme molecules are enclosed in the interior. After washing, the fluorescence confocal image (right) demonstrates urease distribution in the capsule interior.

The urease activity in the capsule has been measured by a colorimetric assay based on the hydrolysis of urea monitored by the pH-sensitive dye bromocresol purple. The encapsulated urease preserves 13% of its activity compared with free enzyme. As shown in [14] this decrease is due to substrate-diffusion limitations. Nevertheless, the urease activity inside was also stable as compared with free urease: after five days' storage at 7 °C the encapsulated enzyme completely preserved its activity while free urease kept under the same conditions in aqueous solution retained only 45% activity.

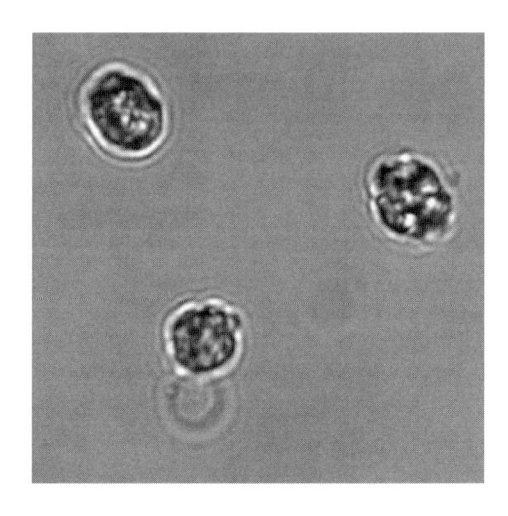

Fig. 18.13 Optical microscopy image of urease-catalyzed CaCO₃ crystal growth inside polyelectrolyte capsules

Micron-sized polyelectrolyte capsules containing encapsulated urease can be a good biomimetic model, because urease catalyses the reaction:

$$CH_4N_2O + 2\,H_2O \quad \rightarrow \quad 2\,NH_3^+ + CO_3^{2-}$$

In the presence of Ca^{2+} ions, calcium carbonate is formed.

$$Ca^{2+} + CO_3^{2-} \quad \rightarrow \quad CaCO_3 \downarrow$$

Carbonate ions generated by urease in the capsule interior immediately react with calcium ions diffusing freely through the capsule wall. As shown by Antipov et al. [30] the calcium carbonate crystals presumably grow in the capsule interior until they occupy the whole capsule lumen (Fig. 18.13). The formation of calcium carbonate particles may also occur out of the capsule owing to diffusion of carbonate but in certain conditions the calcium carbonate particles may be almost completely restricted to inside the capsule. As investigated by electron diffraction, the lattices of calcium carbonate grown in the capsule interior exhibit a vaterite structure. Remarkably there was no growth of calcium carbonate across the capsule wall. Thus, microcapsules incorporating enzymes can be suitable models to mimic inorganic processes in biological cells.

Mimicking the processes in biological cells and cell compartments anticipates involving a number of enzymes, which work together at the same time. The method of controllable opening and closing of pores also provides the possibility to encapsulate several proteins at the same time and, at appropriate concentrations in the capsules, to catalyze sequential enzymatic reactions, for instance glucose oxidation by GOX followed by utilization of hydrogen peroxide by POD. The loaded enzyme, as we have shown, might be released out of the capsule after a certain pH treatment, which could find some application for systems with controlled release.

The use of hollow polyelectrolyte microcapsules as a model artificial cell with size 1–5 µm allows investigation of bioprocesses with protein participation such as enzymatic reactions, conformational changes of protein molecules and protein folding and unfolding in restricted microvolumes imitating biological cells. Indeed, the technology allows incorporation of a wide class of enzymes. Different enzymes being forced together in micron-sized capsules should exhibit other cumulative characteristics of enzyme function, such as kinetic catalytic constant, maximal rate of enzymatic reaction, inactivation constant, inhibition constant and thermodynamic parameters, thermal stability, stability against denaturating agents and extreme values of pH, different from water solutions. Furthermore, the role of the polyelectrolytes that can be considered as a polymer network in biomembranes, i.e. the cytoskeleton, may possibly be revealed in these processes. Talking about encapsulation of enzymes, it should be stressed that selective permeability provides the construction of enzymatic microreactors, where the proteins are placed in the capsule interior. The enzymes are protected by the outer shell against high molecular weight inhibitors and proteolytic agents. While encapsulated enzymes could perform their catalytic reaction if substrates for the reactions are small molecules, the products of one-step or multi-step enzymatic reactions in the capsules could be released or collected in the capsule interior. It is possible to perform an unlimited spectrum of reactions – synthesis, disintegration, polymerization – using different classes of enzymes. Capsules filled with one or a few types of enzymes, when placed as a sensitive layer on transducers, should be of interest for biosensor applications.

18.3.2
Artificial Viruses

DNA delivery in biological cells and sequential gene expression in the cell nucleus are currently hot topics in applied biotechnology. To construct artificial systems providing non-viral gene therapy we have to overcome several obstacles towards simulating natural viruses: compaction of long DNA molecules, formation of sufficiently stable circulating shell with specific receptors on their outer surfaces to be taken up by certain cells, DNA delivery to the nucleus inside the cell and gene expression. The compact form of negatively charged DNA is mainly formed by complexation with molecules bearing several positive charges. Polycations, cationic surfactants and multivalent ions such as spermine, can fold DNA to form particles with sizes between 50 and 100 nm. Remarkably, the compact DNA also has the same size in natural viruses. The following step is the assembly of a "virus envelope" imitation which is the polyelectrolyte shell in our case. Thus, the LbL core-shell structure could work as non-viral vector. Up to now among non-viral systems lipoplexes and polyplexes have been widely investigated, but the transfection efficiency of these systems is not yet adequate to achieve a therapeutic effect in in vivo studies. The main limiting factors are colloidal instability, resulting in a high aggregation tendency under physiological conditions, and attaining the desired size and affinity to certain membrane receptors. The size of coated DNA particles is given by the compact form of DNA and does not change significantly upon polyelectrolyte layer assembly as revealed by dynamic light scattering [31]. Suspensions of DNA nanoparticles with negative or positive charge can be produced that are stable (in physiological solutions) [32]. Sequential gene expression of cells after uptake of such particles has been demonstrated [32, 33]. LbL-coated DNA nanoparticles with an outermost layer bearing a certain receptor, for example a peptide comprising K_{16} [32], give opportunities for the study of a mechanism of receptor-mediated gene transfer. Although the mechanisms of nanoparticle uptake by the cell, DNA delivery in the cell to the nucleus and further gene expression is not yet understood, the LbL method of DNA-nanoparticle design, which provides an effective tool for easier variation of the particle shell composition, size and binding properties, opens an avenue to systematic study. At the same time, this LbL-based artificial virus design could fulfill the demands of gene therapy formulations and can find application in this field.

18.4 Conclusions

In this chapter we wanted to show that polyelectrolyte micro- and nanocapsules may be a promising starting point en route towards an artificial cell. This route is long and will probably not be completed in a single scientist's lifetime, but there is much to be gained on the way. We have pointed to some more-or-less obvious applications; more important, however, is the fact that we learn about the interplay of cellular components at the supramolecular level. This should be considered as a start to learning about the hierarchical structure of biological systems and its importance for their function.

18.5 References

1   E. Donath, G. B. Sukhorukov, F. Caruso, S. A. Davis, H. Möhwald, *Angew. Chem. Int. Ed.* **1998**, *37*, 2202.
2   G. B. Sukhorukov, E. Donath, S. Davies, H. Lichtenfeld, F. Caruso, V. I. Popov, H. Möhwald, *Polym. Adv. Technol.* **1998**, *9*, 759.
3   G. B. Sukhorukov, E. Donath, H. Lichtenfeld, E. Knippel, M. Knippel, A. Budde, H. Möhwald, *Colloids Surf. A* **1998**, *137*, 253.
4   G. Decher, *Science* **1997**, *277*, 1232.
5   B. Neu, A. Voigt, R. Mitlöhner, S. Leporatti, C. Y. Gao, E. Donath, H. Kiesewetter, H. Möhwald, H. J. Meiselman, H. Bäumler, *J. Microencapsul.* **2001**, *18*, 385.
6   E. Donath, S. Moya, B. Neu, G. B. Sukhorukov, R. Georgieva, A. Voigt, H. Bäumler, H. Kiesewetter, H. Möhwald, *Chem. Eur. J.* **2002**, *8*, 5481.
7   F. Caruso, R. A. Caruso, H. Möhwald, *Science* **1998**, *282*, 1111.
8   F. Caruso, *Chem. Eur. J.* **2000**, *6*, 413.
9   A. Voigt, E. Donath, H. Möhwald, *Macromol. Mater. Eng.* **2000**, *282*, 13.
10 V. Dudnik, G. B. Sukhorukov, I. L. Radtchenko, H. Möhwald, *Macromolecules* **2001**, *34* 2329.

11 A. A. Antipov, B. B. Sukhorukov, E. Donath, H. Möhwald, *J. Phys. Chem. B* **2001**, *105*, 2281.
12 G. B. Sukhorukov, A. A. Antipov, A. Voigt, E. Donath, H. Möhwald, *Macromol. Rapid Commun.* **2001**, *22*, 44.
13 G. Ibarz, L. Dahne, E. Donath, H. Möhwald, *Macromol. Rapid Commun.*, **2002**, *23 (8)*, 474.
14 Y. Lvov, A. A. Antipov, A. Mamedov, H. Möhwald, G. B. Sukhorukov, *Nano Lett.* **2001**, *1(3)*, 125.
15 G. B. Sukhorukov, M. Brumen, E. Donath, H. Möhwald, *J. Phys. Chem. B* **1999**, *31*, 6434.
16 S. Moya, E. Donath, G. B. Sukhorukov, M. Auch, H. Bäumler, H. Lichtenfeld, H. Möhwald, *Macromolecules* **2000**, *33*, 4538.
17 R. Georgieva, S. Moya, S. Leporatti, B. Neu, H. Bäumler, C. Reichle, E. Donath, H. Möhwald, *Langmuir* **2000**, *16*, 7075.
18 O. P. Tiourina, I. Radtchenko, G. B. Sukhorukov, H. Möhwald, *J. Membrane Biol.* **2002**, *190*, 9.
19 A. Diaspro, D. Silvano, S. Krol, O. Cavalleri, A. Gliozzi, *Langmuir* **2002**, *18*, 5047.
20 F. Caruso, H. Möhwald, *J. Am. Chem. Soc.* **1999**, *121*, 6039.
21 F. Caruso, H. Fiedler, K. Haage, *Colloids Surf. A* **2000**, *169*, 287.
22 C. Schuler, F. Caruso, *Macromol. Rapid Commun.* **2000**, *21*, 750.
23 M. Fang, P. S. Grant, M. J. McShane, G. B. Sukhorukov, V. O. Golub, Y. M. Lvov, *Langmuir* **2002**, *18*, 6338.
24 F. Caruso, W. Yang, D. Trau, R. Renneberg, *Langmuir* **2000**, *16*, 8932.
25 M. E. Bobreshova, G. B. Sukhorukov, E. A. Saburova, L. I. Elfimova, B. I. Sukhorukov, L. I. Sharabchina, *Biophysics* **1999**, *44*, 813.
26 N. G. Balabushevitch, G. B. Sukhorukov, N. A. Moroz, D. V. Volodkin, N. L. Larionova, E. Donath, H. Möhwald, *Biotechnol. Bioeng.* **2001**, *76*, 207.
27 O. Tiourina, A. A. Antipov, G. B. Sukhorukov, N. I. Larionova, Y. Lvov, H. Möhwald, *Macromol. Biosci.* **2001**, *1*, 209.
28 C. Y. Gao, X. Y. Liu, J. C. Shen, H. Möhwald, *Chem. Commun.* **2002**, *17*, 1928.
29 O. P. Tiourina, G. B. Sukhorukov, *Intern. J. Pharmaceutics* **2002**, *242*, 155.
30 A. Antipov, D. Shchukin, Y. Fedutik, I. Zanaveskina, V. V. Klechkovskaya, G. Sukhorukov, H. Möhwald, *Macromol. Rapid Commun.* **2003**, *24*, 274.
31 V. S. Trubetskoy, A. Loomis, J. E. Hagstrom, V. G. Budker, J. A. Wolff, *Nucleic Acids Res.* **1999**, *27*, 3090.
32 R. Dallüge, A. Haberland, S. Zaitsev, M. Schneider, H. Zastrow, G. Sukhorukov, M. Böttger, *Biochim. Biophys. Acta* **2002**, *1576*, 45.
33 D. Finsinger, J. S. Remy, P. Erbacher, C. Koch, C. Plank, *Gene Therapy* **2000**, *7*, 1183.

Subject Index

a

absorption 219
acid-base interaction 409
acoustic
– cavitation 123
– field 123
acrylamide in a miniemulsion polymerization 194
acrylic acid 194
adaptive chemistry 104
adsorption, colloidal 323–325
– charged multilayer surface 331
– controlled 290, 323
AFM (atomic force microscopy) 256
$Ag@Au$ 223
$Ag@SiO_2$ 232
$Ag@TiO_2$ 232
aggregate/aggregation
– controlled aggregation 40
– diffusion-limited 451
– network, aggregated 40
– primary aggregates 40
AgI 239
agitation, mechanical 177, 188
– high 188
AIDS 503
air-water interface 369–393
– nanoparticle formation by reaction of metal ions at the air-water interface 384–387
– as an organizational medium for nanoparticles 393
aligning rods 447
alkane chalcogenide 372
alkanethiol, gold nanoparticles capped with 376
alkanethiol-alkylamine-capped gold 372
alloy structures 106

alumina 360
aluminium oxide 344
amine
– compulsory use of an aminic coupler 205
– residual groups 197
– – primary 197
– – secondary 197
ammonia catalyzation 238
amphiphilic copolymers
– block copolymers 153
– in miniemulsion 196
anisotropic superstructure in miniemulsions 209
annealing 75
antibody-antigen
– coupling 353
– cross-linking 354
aqueous synthesis 56, 62
area
– effect 123, 125
– fraction 567
artificial
– latexes 3, 5–8
– – preparation 6
– opals 69, 271
– solid 85
assembly/assembled
– chemical methods 370–371
– collective nature 452
– core-shell assembly 481–482
– physical methods 370
– programmed assembly 353
atomic force microscopy (AFM) 256
$Au@Ag$ 222, 226
Au_{55} cluster, phosphine-stabilized 97
$Au@SiO_2$ 230
$Au@SnO_2$ 234

Colloids and Colloid Assemblies. Edited by Frank Caruso
Copyright © 2004 Wiley-VCH Verlag GmbH & Co. KGaA, Weinheim
ISBN: 3-527-30660-9

b

Bacillus subtilis 348
bacteria 348–349
– S-layer proteins from bacteria 352
Bancroft's rule 6, 12
band
– gap 227
– plasmon band 221
barcodes, rods 446
bar-coding technology 502
barium ferrite 372
$BaSO_4$ 384
bead-on-bead 522–523, 536, 548
– adhesion of 548
– reporter particles 522
– split-and-mix process 522
reporter colloids for strategy I encoding 536–538
Bibette emulsion 189
bilayer 565, 568
– conductivity 567
– dielectric constants 568
– ionic insulation 568
– lipid bilayer structures 391
– phospholipid 347
– stacks 382
bioactive glass 487
bioarrays 455
bioassays 501
biocolloids 265
bioelectronics 339
– interfacing, bioelectronic 454
biolabeling, nanocrystals 89–90
biological
– cells 265
– sensors 455–457
biomacromolecules 162–163, 370
biomimetics 342, 344, 360, 364
biomineralization 151, 342
biominerals 343, 350–351, 360–361, 364
biomodeling 571
biomolecules 342–343, 346, 360–361, 364
– structures, biomolecular 343, 360
bioreactors 152
biosensors 455
biostreptavidin 353
biostructures 342
biotemplates 343–353, 364
biotin/streptavidin 355
blanket layer 331
block copolymer
– amphiphilic block copolymers 153
– diblock copolymer 154
– – polypeptide-based 156
– dispersions 3
– triblock copolymer 154
– vesicles 153
blood, artificial 469
Bohr radius 54, 74
bottom-up approach 53
Bragg equation 276, 302
branching units 168
– hyperbranched molecules 169
bubble(s) 122–129
– in acoustic field 123–125
– cavitation bubbles 178
– collapse 125
– dynamics 125
– growth 123–125
– implosion 125
– microbubbles 129
– temperature 125–129
buckyballs 372
building blocks 398
– of superstructures 81–86
bulk plasma frequency 424
butyl acrylate 195

c

$CaCO_3$ 384
cadmium
– alkanoates 390
– sulfide 344, 347
– thioalkanoates 390
calcium carbonate 42, 351, 360
– particles 577
calixarene Cd salt LB films 390
cancer 500, 503
capacitance 429
capillary forces, attractive, crystallization 298–299
capsule permeability 563
carbon
– encapsulation of carbon particles 201
– sonochemical formation of colloids 142
4-carboxythiophenol (4-CTP) 378
catalase
– crystals 271
– enzyme 265
catalysis 113
catalyst 225
cavitation 129
– acoustic 123
– bubbles 179
CCMV (cowpea chlorotic mottle virus) 151
CdS 137, 352, 371, 374–375, 384

- nanocrystals 73
- nanoparticles 347, 374–382, 390
- - 2-aminoethanethiol-capped 375
- - films (see there) 382
- - monolayers, lamellar growth 382
- - size-quantized 377
- - superlattice 381
- particles 137, 374–375
- Q-state 372
- quantum 378
CdS@SiO$_2$ 232
CdSe
- hydrophobized 372
- nanocrystals 65–72, 80, 83–84
- nanoparticles 85
- quantum dots 374, 494
CdSe/CdS core-shell nanocrystals 70–72, 89
CdSe/ZnS nanocrystals 70–72, 89
CdTe 372, 377
- nanocrystals 62, 64, 80, 83, 89–90
Cd$_x$Hg$_{1-x}$Te nanocrystals 62, 64
cells
- biological 265
- biology 500
- biomimetic system 562
- coated 569
- description 561
- division 569–571
- fluidic cells 291
- Grätzel cells 87
- living cells 503
- micelles (see there) 159, 161, 169
- photovoltaic 458–459
- of a polymer network 565
- solar cells 87–88
cetyl trimethylammonium chloride 182
chaining force 438
chalcogenide, alkane 372
charges
- energy, charging 428
- number of 20
- surface charge 20
- - electric charge 121
chemical
- deposition 446–451
- - of shells on metals and semiconductors 222–236
- etching 217
- functionalization rods 447
- interactions 400
- libraries, encoding 516
- reactions 236
- sensors 454–455

chemiresistance effects 454
chemistry
- adaptive 104
- of heterophase polymerization 9
chymotrypsin 573, 575
- aggregates 573
citrate-stabilized particles 406
- gold particles 411
coagulation 39
- controlled 39
- heterocoagulation 40–41
coalescence 24, 218
- by collisions 177
coating 246–247
- cells, coated 569
- formation 247
- inorganic 404–406
- of metal particles with silica 228–232
- particle, coated 221
- polymer coating 403–404, 422
- silica as coating material 404
cobalt particles 420
cohesion 207–208
- forces 208
- miniemulsions of liquid with high cohesion energy 207
collision mechanism 23
colloid/colloidal
- adsorption 323–325
- - charged 323
- - pH effect 324
- biocolloids 265
- core, colloidal 563
- crystals 274–275, 348, 439–444, 466
- - assembly by AC or DC fields 439
- - body-centered cubic 297
- - face-centered cubic 297
- - hexagonal close-packed (hpc) 296, 374
- - of nanocrystals, colloidal 85–86, 274–275
- - 100 planes/100-oriented 299, 301
- - 110 planes 299
- - 111 planes 293
- - random hexagonal close-packed (rhpc) 296–297
- - by self-assembly of colloidal spheres 439
- - templates 467–471
- - - surface-templating 470–471
- - - volume templating 471
- crystallization, template-directed 422
- 2D colloidal crystallization 441
- electrical

– – circuits via colloidal assembly 444–454
– – forces in colloidal suspensions 437–439
– ellipsoidal colloids 310
– encoding colloids (*see there*) 517–525
– for encoding chemical libraries 507–555
– entropically driven colloidal crystallization 105
– gold colloid 357
– monodisperse 284–289
– – silica 289
– – spherical colloids 307
– nonspherical colloids 309
– particles 161, 246–278
– – preparation 246
– – surface modification 247, 249, 258, 327
– polygonal clusters 302, 310
– polyhedral clusters 310
– polymeric complexes, colloidal 3, 43
– polystyrene 288
– preservation of the colloidal entities 209
– reporter colloids for strategy I encoding 536–538
– rod-shaped 310
– self-assembly 286
– silica colloids 346, 525–531
– – sol-gel method 525
– sonochemical synthesis of inorganic and organic colloids (*see there*) 120–145
– spherical 284–291
– – silica spheres 288
– stability/stabilization of colloidal particles 13, 180, 217, 229
– surface-directed colloid patterning 317–339
– synthesis of semiconductor nanoparticles 53, 55–81
– ζ-potential 328
colloid-based libraries 513
– combinatorial 513
– non-combinatorial 513
– screening 513
– split and mix synthesis 513
comminution 6–8
compartments 210
– nanocompartments 176
competitive growth 23–25
complementary match of the patterns 445
composite particles 31–33, 42
– latex particles 32
condensation 6–8
conductivity 567
– electronic/electrical (*see there*) 391, 425, 428

– single electron 428
confinement of two species 209
conformation 157–159
– random-coil 157
– three-dimensional 159
– two-dimensional 159
continuous phase 9, 22, 27–30
– polarity 28
– soluble parameter 27–29
– solvency of 22, 27
controlled or triggered release 153
copolymerization
– of monomers 195
– radial copolymerization in miniemulstions 190–196
– of styrene and butyl acrylate 195
copolymer-protein hybrid materials 156
copolymers 3, 43, 153–154, 156
– block copolymer (*see there*) 3, 153–154, 156
– diblock copolymer 418–419
– double hydrophilic copolymers 43
copper
– fingers 449
– sonochemical formation of colloids, colloidal copper 137
core
– dissolution 236
– particle 165
core-shell
– assembly 481–482
– colloids 291
– gold cores 293
– nanocrystals 77
– nanoparticles, bimetallic 222
– particle 161, 163, 290, 408
– quantum dots 228
– silica 293
– system 63
corona thickness 11
cosonication 201
co-surfactant, methyl tricaprylyl ammonium chloride 29
Coulomb
– blockade 110
– charging 109
– staircase 109, 459
covalent
– binding 411
– coupling strategy 374
coverage of droplets by surfactants 179
cowpea chlorotic mottle virus (CCMV) 151, 344–345

cross-linkers/cross-linking agents   37, 374, 427
cross-section, extinction   219
crystal
- colloidal (*see there*)   274–275, 348, 439–444, 466
- 2D crystals, switchable   443
- growth   152
- nanocrystals (*see there*)   52–57, 86–90
- nucleation, template-directed   422
- photonic   285, 439, 467, 482–485
crystalline
- assemblies   384
- liquid   359
crystallization
- via attractive capillary forces   298–299
- centrifugation   296
- colloids/colloidal
- – entropically driven   105
- – with mixed sizes   308
- 2D colloidal crystallization   441
- during the polymerization within the droplets   193
- long-range   443
- polymer   210
- via repulsive electrostatic interactions   296–298
- sedimentation   295–296
- sonication   293
cubic-close-packed (ccp)   291
cubic-packed monolayer   421
CuS, sonochemical formation of colloids   138
cuttlefish bone   351
cytoskeleton   153

d

decorative paints   112
defocusing size distribution   60–62
degenerative transfer process   192
Deinococcus radiodurans   352
dendrimer   168–171
- starburst   169
dendritic boxes   169
deposition   324–325, 330
- electrodeposition   477
- electrophoretic deposition   440
- layer-by-layer (LbL)   324
- nanoparticles, chemical deposition of shells on metals and semiconductors   222–236
- selective   324–325
- speed   440

- template metal deposition   450
Derjaguin-Landau-Verwey-Overbeek (DLVO) theory   122
diatom   342, 348, 350, 361–363
dielectric
- effective dielectric constant   424
- function   220
dielectrophoretic force   437
diffusion
- degradation   188
- exchange by diffusion process   183
- growth, diffusion-controlled   60
- mechanism   23
- of reactants, constrained lateral   384
diodes, light-emitting   458–459
dispersion   2–6, 8
- block copolymer dispersions   3
- non-polar dispersion media   185
- polymer dispersions   2–5
- – classification   2
- – primary   2
- – – via heterophase polymerizations   8–41
- – secondary   2
- polymerization   28–30
- – stabilizer-free dispersion polymerization   29
displacement current   423
displays   457–458
- LET display   338
- phage display   364
dithiol layer   426
dithiothreitol   496
DLVO (*Derjaguin-Landau-Verwey-Overbeek*) theory   122
DNA   347
- biomacromolecule   370
- hybridization   210, 353, 500
- molecule   358
- nanoparticles   579
- particles   265
- stranded DNA   360
- structure   356
DNA-hybridized particle   419
DNA-mediated assembly   411–412
DNA-processing enzymes   115
DNA-streptavidin   359
3DOM materials (inverse opals)   301, 307, 466, 468–482
- amorphous   472
- application   482
- catalytic   486
- chemical vapor deposition   479
- defects   484

- electrochemical deposition 477–478
- electrode 486–487
- – anodes 487
- – cathode 486–487
- electroless deposition 478
- hybrid organosilicates 468–469
- macroporous oxides 468
- melt-imbibing 480
- mesoporous 472
- metals/metallic 473–475, 478, 485
- – metal oxides 475
- – porous metallic 485
- microporous 472
- nanocrystalline 472
- polymerization 480–481
- – coordination 481
- pyrolysis and spraying techniques 479–480
- refractive index (RI) 483
- salt-precipitation and chemical conversion 473–474
- shrinkage 470, 475, 477
- solvothermal synthesis 476

Donnan
- effect 568
- potential 156

dot pattern 337

double-layer
- capacitance 425–426
- charging 429
- electrical 121
- interactions 217

droplet
- controlled droplet fission and heteroaggregation process 201
- deformation of droplets 178
- disruption of droplets 178
- minidroplet (see there) 183
- miniemulsion droplets 183
- nanodroplet (see there) 177, 182
- nucleation 188
- size 3, 184–185
- – indirect methods for measuring the droplet size 185

Drude model 220, 424

drug
- delivery 487
- discovery 510–511

e

echinocytes 569

EDX (energy-dispersive X-ray) 286, 288

electrical/electrically
- circuits
- – via colloidal assembly 444–454
- – from nanoparticles 451–454
- conductivity 391
- connections
- – self-assembling 453
- – self-repairing 453
- forces in colloidal suspensions 437–439
- functional microstructures 437–459
- networks, three-dimensional 445
- readable sensors 455

electric-field-induced diffusion-limited aggregation (EDLA) 451

electrochemical
- deposition 446–451
- impedance spectroscopy 425
- nano-switch 111–112
- scanning electrochemical microscope (SECM) 109
- sensing 427

electrodeposition 477

electrohydrodynamic attraction 441

electrolyte
- latex particles, electrolyte stability 12
- polyelectrolyte (see there) 81, 156, 159, 234, 250–251, 266–269, 321, 325, 523, 545–548, 564–565

electromagnetic energy, waveguides for 369

electron/electronic
- beam lithography 376
- conductivity 425
- – hopping conductivity 109
- – single electron conductivity 428
- irradiation process 391
- paper 457
- transfer 425, 428
- – resistance 425

electrophoretic
- deposition 440
- force 437
- ink 457

electrorheological fluids 442

electrorotation spectroscopy 566

electrostatic
- assembly 406
- – of nanoparticles 377
- attraction 326
- binding 416
- encapsulation 403
- interactions 318, 377–381, 411
- – repulsive interactions 381
- self-assembly 412
- self-organization 420–422

– stabilization 11, 178
emulsion/emulsification
– *Bibette* emulsion 189
– initiator-emulsifier system (*see* IES) 35
– mechanical emulsification 178
– membrane emulsification 189
– microemulsion 187, 227
– miniemulsion (*see there*) 167, 175–211
– oil-acrylate hybrid emulsions 200
– polymerization 4, 17–22, 138, 165–168, 175, 286–287
– – sonochemical (*see there*) 138–140
– power ultrasound emulsification 178
– surfactant-free 17–22
– polymers 4
– preparation 6
– – monomer emulsion preparation 33
– self-emulsification 7
– stability against *Ostwald* ripening 177
– templating 476–477
encapsulation 40
– efficiency 564
– electrostatic 403
– enzyme 575, 578
– graphite oxide (GO) 405
– host-guest 152
– of liquid 202–204
– of particles 150
– – carbon particles 201
– – magnetite particle 201
– – of nanoparticles 199
– – TiO_2 particles 200
– of pigments by miniemulsification 200
– silica 405
– in situ 403
encoding colloids for high-throughput screening 517–525
– colloidal rods 518
– combinatorial libraries 519–524
– fluorescent dyes 518
– molecular identifier tags 519
– nanocrystals 518
– non-combinatorial 518–519
– optical encoding 520
– quantum dots 518
– self-quenching 520
endocytosis, receptor-mediated 500
endoplasmic reticulum 561
energy-dispersive X-ray (EDX) 286, 288
engulfment 40–41
environmental benign application 175
enzyme/enzymatic
– activity 572

– biomacromolecule 370
– catalase 265
– DNA-processing enzymes 115
– encapsulation 575
– multilayer-coated particles 270–271
enzyme-containing films 572
equilibrium
– concentration of free surfactant 181
– dynamic rate equilibrium 180
– – *Fokker-Planck*-type 180
– fusion-fission rate equilibrium 180
– nanodroplets in 177
– state 186
– thermodynamic 183
erythrocytes 569
etching 77
excitons/excitonic 386
– fluorescence 497

f

fabrication process, sample 322
faradaic impedance 425
Fe_2O_3 372
Fe_3O_4 377
FeCo, sonochemical formation of colloids 143
feed process 14
ferrihydrite 344
ferritin 151, 344, 355
ferroelectric lead zirconium titanate nanoparticles 371
film
– CdS nanoparticle films 382
– – nanoparticle-ODA LB film 382
– enzyme-containing films 572
– gold nanoparticles, thin films of 100
– *Langmuir* films (*see there*) 107–108
– *Langmuir-Blodgett* films 369–393
– macroporous 276–278
– metal particle films 423
– multilayer films 431, 572
– polyelectrolyte film infiltration 417
– polystyrene film 418
– precursor film 562
– soft magnetic films 432
– thermally evaporated films 392
– thin film fabrication 432
monodisperse colloids 284–286
filters, polarizing 113
fission process 186
flocculation 162
– limited 25
Flory-Huggins-Rehner expression 32

flow cytometry/flow cytometer 515, 523–524, 541
– emission intensity 524, 552
– filter parameters 550
– histogram 550
– *Mie* theory 541
– optical signature 523
– refractive index 524
fluids
– biological 162
– cell, fluidic 291
fluorescein 499
– diacetate (FDA) 272
fluorescence/fluorescent
– correlation spectroscopy 501
– emission 494
– excitonic 497
– microscopy image (FMI) 254
– particles 529–539
– – core shell 529
– – *Förster* energy transfer 533, 539
– – *Jablonski* diagram 532
– – multi-fluorescent 531
– – – silica 534
– – reabsorption 534, 536
– – resonance energy transfer (RET) 532, 534
– – self-quenching 532
– – shell 531
– – solvent environment 533
– probes 163
– resonance energy transfer (FRET) 501
focusing size distribution 60–62
Fokker-Planck-type dynamic rate equilibrium 180
folic acid 500
Förster energy transfer 533, 539
free-energy minimization 41
free-radical polymerization 154
fusion process 186

g

GaAs nanocrystals 78
gallery-mode laser, whispering 89
GaP nanocrystals 78
gas sensors 112
gel
– silica gels 232
– sol-gel reactions 267
gene therapy 156
generation 168
genomics 508–509
– functional 508
– gene expression 509
– genotyping 509
– pharmacogenomics 508
– single nucleotide polymorphism (SNP) 508
– structural 508
Gibbs surface excess 134
Gibbs-Thompson equation 57
glass transition temperatures 9–11
glutaraldehyde 375
gold 98–115, 347, 350–354, 357–360, 377, 380, 384
– alkanethiol/alkylamine-capped 372
– clusters 96–98, 413
– colloids 96–98, 356
– core dissolution 236
– core-shell 293
– hydrosols 97, 369
– – citrate-stabilized 97
– monolayer protected clusters (MPCs) of gold and silver 96–115
– nanoparticles 98–115, 160, 263–264, 357–360, 374–376, 404, 410, 452
– – applications 112–114
– – assemblies 454
– – capped with alkanethiols 376
– – electronic properties of 108–112
– – for gene delivery 114
– – ligand mobility 101–104
– – light scattering 114
– – network 374
– – NMR 101
– – polyelectrolyte modification 263
– – reactivity 101–104
– – self-organization 104–108
– – size-exclusion chromatography 99
– – size-selective precipitation 99
– – structure 101–104
– – thin films of 100
– – thiol-stabilization 98
– – water-soluble 113
– nanowires 377
– particles 359, 408–409
– – citrate-stabilized 411
– Se-, Te-capped gold 372
– sonochemical formation of colloids
– – colloidal gold 132
– – gold sols 133
– synthesis of thiolate-stabilized gold nanoparticles 98–100
Gouy-Chapman model 121
graft polymerization 290
graphite oxide (GO) encapsulation 405

Grätzel cells   87
V-grooves   299
growth
– bubble   123–124
– competitive   23–25
– crystal growth   152
– diffusion-controlled   60
– particle   23–24
– seeded   97
– shot   31

h

Hamaker constant   229
HDA (hexadecylamine)   66
heavy metal ions   486
helical secondary structure   157
hemispheres   43
heteroaggregation/heterocoagulation process   40–41, 201
hexadecylamine (HDA)   66
hexagonal
– close-packed structures (hpc/rhpc)   296–297, 374
– packed monolayer   421
HgTe   64
high-throughput screening (HTS) technologies   507, 511–517
– encoding colloids (*see there*)   517–525
– microarrays   511
– microplates   511
– screening with colloids   513–517
HLB values   6
hole-conducting polymers   87
hollow
– capsules   250, 255, 274, 562
– microcapsules   575, 578
– particle   170, 290
– polymer
– – nanocapsules   202
– – particles   161
– – silica   237
homogenization   178–179
– high-pressure   201
– mechanical   178
– process   183
– techniques of miniemulsion preparation and homogenization   178
homogenizer, high-pressure   178
homopolymerization, radial homopolymerization in miniemulsion   190–196
host-guest encapsulation   152
hot
– injection   56, 65, 77

– spot   128
HTS (high-throughput screening)   507
hybrid
– copolymer-protein hybrid materials   156
– nanohybrids   343–347
– nanoparticles by miniemulsions technologies   198–206
– oil-acrylate hybrid emulsions   200
– polymer-particle hybrids   200–202
– polymer-polymer hybrids (*see there*)   199
hydrodynamics   35
– layer thickness, hydrodynamic   11
hydrogel   161
hydrogen
– bonding   331, 409
– – interactions   420
– sensors   450
hydrophilic
– double hydrophilic copolymers   43
– monomers   191
– nanoparticulate   200
– pigments   200
hydrophobe/hydrophobic agent   182–183
– CdSe, hydrophobized   372
– monomers, hydrophobic   158
– nanoparticles (*see there*)   371–377, 383
– surfactant-stabilized hydrophobic silver   371
– ultrahydrophobe (*see there*)   178, 183–185
hyperbranched molecules   169

i

IES (initiator-emulsifier system)   35
– anionic   35
– cationic   35
immiscible oils, thermodynamic considerations of, by *Torza* and *Mason*   202
immunoassays   272, 500
immunosensors   457
in vivo molecular imaging   503
InAs
– nanocrystals   77–78
– quantum dots   498
incorporation   198
initiator(s)
– oil-soluble initiator   191
– as osmotic agents   191
– PEGA-initiators   22
initiator-emulsifier system (*see* IES)   35
injection, hot   56, 65, 77
ink
– e-ink   458
– electrophoretic   457

InP
- nanocrystals 75–77
- quantum dots 498

interaction
- acid-base 409
- additional 325
- chemical 400
- double-layer interaction 217
- electrostatic 318, 377, 379–381, 410
- energy 218
- hydrogen bonding 420
- interparticle interactions 222
- non-covalent interactions 150
- particle-particle interaction 421
- π-π interactions 413
- protein-protein interactions 510
- substrate-particle interaction 421
- *van der Waals* interactions 218

interface/interfacial/interfacing
- air-water interface 369–393
- area, interfacial 185
- bioelectronic 454
- energy 188
- layer 11
- properties 11
- tension values 187

interparticle collisions 144

ions/ionic
- chain transfer agents 21
- chemical manipulation 384
- co-monomers 21
- heavy metal ions 486
- metal ions 384
- strength 19
- transport of inorganic ions within the anisotropic spaces 389

$IrCl_6^{2-}$ 416

IR-emission 88

iron, sonochemical formation of colloids 140

iron-oxide 344–347, 364
- high-melting iron oxides in miniemulsions 209

isotherms, π-A 372

isotope effect, kinetic 127

j

Jablonski diagram 532

joint-nucleation 42

k

Keggin salts (polyoxometallates) 370

Kelvin equation 32–33

kinetics
- of complexation 380
- of deposition 421

l

labile domains 153

lamellar
- c-axis-oriented structure 382
- growth of the CdS nanoparticle monolayers 382

Langmuir
- films 107–108
- – – MPCs 107
- – – phospholipid 107
- monolayers 377–383
- – – complexation of aqueous nanoparticles with 377–383
- – – oppositely charged *Langmuir* monolayers 377

Langmuir-Blodgett (LB) films 369–393
- calixarene Cd salt LB films 390
- chemical methods for generation of nanoparticle *Langmuir-Blodgett* films 383–393
- LB precursor matrixes 389
- nanoparticle
- – – CdS nanoparticle-ODA LB film 382
- – – nanoparticle-LB film composites 390
- of salts and fatty lipids 387–393
- technique 378

lanthanum phosphates 258

Laplace pressure 122, 177, 183

laser
- diffraction patterns 443
- whispering gallery-mode 89

latex particles, polymerics 1–44, 163, 260–262
- artificial latexes (*see there*) 3, 5–8
- armored latexes 204
- close packing latex spheres 17
- cyclic esters 30
- definition 1–5
- electrolyte stability 12
- high solids polymer latexes 17
- historical overview 1–5
- importance 1–5
- microspheres 82
- monodisperse latex particles (*see there*) 17, 22–31
- nonsymmetrical latex particles 42
- polymer-polymer hybrid latex 199
- polyurethane latexes 198
- post-polymerization modification of latexes 39
- preparation 5–41

- spheres 377
- swollen latex particles 32
lauryl methacrylate (LMA) 191
- polymerization of the ultrahydrophobic monomer 191
layer
- bilayers, phospholipid 347
- blanket 331
- dithiol layer 426
- double-layer (*see there*) 121, 217, 425–426, 429
- enzyme multilayer-coated particles 270–271
- hydrodynamic layer thickness 11
- interfacial 11
- metal layers 223
- monolayers (*see there*) 96–115, 290, 370–374
- multilayers (*see there*) 161, 266–271, 321, 325–326, 331, 335, 370–372, 413–414, 427
- of particles 440
- polyelectrolyte multilayers 161, 266–269, 321, 325, 335
- protective 401
- S-layer proteins from bacteria 352
- three-layer micelles 159
layer-by-layer (LbL)
- adsorption 562
- application 248
- assembly 81–83, 398, 572
- coating 403
- deposition 161
- electrostatic assembly 248
- method 247–249
- - layering process 250
- - surface modification 249
- self-assembly of metal nanoparticles on planar substrates 398–433
LB (*see Langmuir-Blodgett*) films 369–393
lead sulfide 347
lepidocrocite particles 345
LET display 338
Lifshitz-Slyozov-Wagner theory 57–59
ligand(s) 164
- exchange reactions 103
- mobility, gold nanoparticles 101–104
ligand-protected particles 412–413
light-emitting
- devices 86–87
- diodes 458–459
lipid 152, 259, 344
- bilayer structures 391
- coating 565

- coupling 566
- nanoparticle formation by reaction of metal ions in *Langmuir-Blodgett* films of salts and lipids 387–393
lipofullerenes 158
lipophobe 185
liposomes 152
liquid
- crystalline 359
- encapsulation 202–204
lithium niobate 268
lithography/lithographic
- electron beam 376
- patterning 472
lumazine synthase 345–346
luminescence 86–88
- efficiency 77, 80
- linear polarized 69, 71
- photoluminescence quantum yield 228
- quantum yields 65, 73–74
- sonoluminescence 126
Lycurgus cup 96
lysosomes 561

m
macromolecules, single 503
macromonomers 154
macroporous
- films 276–278
- materials 285, 465–489
- - 3D ordered 465–489
- monolith, silicalite (zeolite) 276
- oxides 469
- structures 465–469
- TiO_2/SiO_2 structure 277
magnet/magnetic
- electromagnetic (*see there*) 369
- particles 420
- properties 431–432
- quantum magnetic disks (QMDs) 420
- soft magnetic films 432
magnetite 349, 371
- dioxide 351
- inside 344
- particle, encapsulation of 201
- surface charge 349
magnetorheological fluids 442
Mason and *Torza*, immiscible oils 202
mass approximation, effective 54
Maxwell equations 305, 424
Maxwell-Garnett model 408
Mayo-Lewis equation 195
melamine formamide (MF) particles 563

membrane
- emulsification 189
- proteins 156
- shell-forming 152
mercaptoacetic acid 496
mercaptosilane 233
mesh size 159
mesoporous structures 232, 465–467
metal
- chemical deposition of shells on metals and semiconductors 222–236
- cluster 420
- complexes 416
- core-shell bimetallic nanoparticles 222
- depositions, templated 450
- fibers, metallic 452
- heavy metal ions 486
- ions 384
- layers 223
- in miniemulsions 208
- nanocrystal monolayers 374
- nanoparticles 262, 266, 299, 398–433
- – formation by reaction of metal ions at the air-water interface 384–387
- – infiltration of 266
- – layer-by-layer self-assembly of metal nanoparticles on planar substrates 398–433
- – optical properties 423
- – superlattices of 419
- – synthesis 401, 413
- nanoparticle-containing multilayers 423–432
- optical properties of metal nanoparticles 219
- oxides
- – shells 228
- – sonochemical formation of colloids 143
- particles 423
- – particle films 423
- rods 446
- selective metallization 338
metal-insulator transitions 109
metallo-dielectric structures 454
micelles 159, 161, 169
- polymeric 159
- shell cross-linked 161
- three-layer 159
- unimolecular 169
microactivity 88–89
microarrays 503
microbubbles 129
microcapsules 457

microcontact printing 292
microelectrophoresis 253
microemulsion 187, 227, 232
- inverse 187
microfluidics 503
- channels, microfluidic 450
microlenses 285
microporous structures 465–467
microrods 446
microscope
- scanning electrochemical microscope (SECM) 109
- tunneling microscope (STM) 110–112
microscopy/microscopic
- atomic force microscopy (AFM) 256
- cryo-electron microscopy 528
- fluorescence microscopy image (FMI) 254
- scanning electron microscopy (SEM) 290–291, 380
- sensors, microscopic 455
- transmission electron microscopy (TEM) 254, 286–288, 291, 385
microstreaming 129
microwires
- arrays of 447
- current-to-voltage response 452
- electrically functional 452
Mie theory 219
mineralization, biomineralization 151
minidroplet size 184
miniemulsion 167, 175–211
- ad-miniemulsion polymerization 202
- amphiphilic copolymers 196
- anionic polymerization in miniemulsions 198
- anisotropic superstructure in miniemulsions 209
- checklist for the presence of miniemulsions 187–188
- for the convenient synthesis of organic and inorganic nanoparticles 175–211
- critically stabilized 188
- direct 185
- droplets 184
- encapsulation of pigments by miniemulsification 200
- high-melting iron oxides in miniemulsions 209
- of inorganic droplets/synthesis of inorganic nanoparticles 206–209
- inverse miniemulsion 185, 194
- of liquid with high cohesion energy 207

Subject Index

- material synthesis   190–209
- metals in miniemulsions   208
- monomers in miniemulsions, water-soluble   190
- nonionic surfactants for miniemulsions   182–183
- PMMA miniemulsion latex   204
- polyaddition of epoxy-resins in miniemulsions   196
- polymerization in miniemulsion
- – acrylamide in a miniemulsion polymerization   194
- – living free-radical polymerization in miniemulsion   191
- – non-radical organic polymerization reactions in miniemulsions   196–198
- post-stabilized miniemulsions   183
- precursor miniemulsion   179
- principle of miniemulsion synthesis   176
- process of miniemulsification   208
- radial homo- and copolymerization in miniemulsion   190–196
- relation between inter- and intramolecular reactions in miniemulsion   210
- stabilization with inorganic nanoparticles   204
- steady-state miniemulsification   183, 208
- – steady-state dispersed miniemulsion   188
- surfactants   180–183
- – cationic surfactants for miniemulsions   182
- – influence of the surfactant   180–183
- techniques of miniemulsion preparation and homogenization   178
- TiO_2 particles, encapsulation via miniemulsions   200
mitochondria   561
molecular
- diffusion degradation   177
- imaging, in vivo   503
- recognition   343, 353
- relation between inter- and intramolecular reactions in miniemulsion   210
- weight   36
- – low molecular weight material   266
monodisperse
- latex particles   17, 22–31
- – sub-micrometer   25–27
- particles   21, 33
- – polystyrene particles, micron-size   27
- spherical colloids   307

monolayers
- CdS monolayers, lamellar growth of   382
- complexation of aqueous nanoparticles with *Langmuir* monolayers   377–383
- cubic packed   421
- hexagonal packed   421
- *Langmuir* monolayers   377–383
- metal nanocrystal monolayers   374
- nanoparticles   370
- ODA monolayers   379–380
- protected clusters (*see* MPCs)   96–115
- self-assembled   290
- silver (*see there*)   372–373
monomer   60
- concentration   19
- emulsion preparation   33
- hydropholic   191
- hydrophobic   158
- ultrahydrophobic, polymerization of   191
- water-soluble monomers in miniemulsions   190
Monte Carlo simulation   57
Morton-Kaizerman-Altier equation   32
MoS_2
- nanoparticles   372
- sonochemical formation of colloids   142
MPCs (monolayer protected clusters)   96–115
- gold (*see there*)   96–115
- *Langmuir* films   107–108
- multilayered films of   431
- platinum   100
- potential of   429
- silver   96, 100
multicolor quantum   497
multilayers
- assemblies   424
- colloidal adsorption, charged multilayer surface   331
- enzyme multilayer-coated particles   270–271
- films of MPCs   431, 572
- nanoparticles   370–372
- – close-packed nanoparticle multilayers   371
- – metal nanoparticle-containing multilayers   423–432
- PAH/PAA multilayers   414
- polyelectrolyte   161, 266–269, 321, 325–326, 335, 413, 573–574
- sensing   427
multiply-twinned particles (MTPs)   388
murein network   153

n

nanoboxes 166
nanobubbles 235
nanocages 160
nanocapsule
– formation 203
– hollow polymer 202
– polyelectrolyte 159
nanoclusters 63
nanocompartments 176
nanocomposite
– dispersion 206
– polymeric 31
nanocontainers, colloidal 150–171
nanocrystals (see also nanoparticles) 52–57, 86–90, 494
– II–VI 73–74
– III–V 74–78
– biolabeling 89–90
– $Cd_xHg_{1-x}Te$ 62, 64
– CdS 73
– – shape control 73
– CdSe (see there) 65–72, 80, 83–85, 494
– CdSe/CdS core-shell 70–72, 89
– CdSe/ZnS 70–72, 89
– CdTe 62, 64, 73–74, 80, 83, 89–90
– colloidal crystals of nanocrystals 85–86
– core-shell 77
– GaAs 78
– GaP 78
– HgTe 64
– InAs 77–78
– InP 75–77
– metal nanocrystal monolayers 374
– PbSe 74
– semiconductor nanocrystals, application of 86–90
– shape of 68, 73
– solar cells 87–88
– for telecommunication amplifiers 88
– two-dimensional arrays of monodisperse nanocrystals formed by self-assembly 83–84
– ZnSe 74
nanodroplet(s)
– in equilibrium 177
– size 182
nanohybrids 343–347
nanolithographic techniques 420
nanomaterials design, evolving strategies 342–365
nanoparticles 256–261, 343, 347, 369–393, 401

– air-water interface as an organizational medium for nanoparticles 393
– aqueous nanoparticles with Langmuir monolayers 377–383
– assembly 474–476
– CdS particles (see there) 347, 374–382, 390
– chemical nanoparticle assembly 371
– collective properties of ensembles of nanoparticles 369
– conductive 427
– controlled assembly of nanoparticles in solution 369
– core-shell bimetallic nanoparticles 222
– coupling between nanoparticles and oil 206
– electrical circuits from nanoparticles 451–454
– encapsulated 199
– equilibration 380
– ferroelectric lead zirconium titanate nanoparticles 371
– formation by reaction of metal ions
– – at the air-water interface 384–387
– – in Langmuir-Blodgett films of salts and lipids 387–393
– gold (see there) 98–115, 160, 263–264, 358, 374, 403, 410, 452, 454
– hetero-nanoparticle assembly 383
– hybrid nanoparticles by miniemulsions technologies 198–206
– hydrophilic, nanoparticulate 200
– hydrophobic 371–377, 383
– – monodisperse 374
– – organization 371–377
– inorganic 189, 204–206, 370
– – assembled at the air-water interface 370–371
– – miniemulsions of inorganic droplets/synthesis of inorganic nanoparticles 206–209
– – miniemulsion stabilized with inorganic nanoparticles 204–206
– – water-dispersible inorganic nanoparticles 204
– macroporous materials from nanoparticles 475
– metal (see there) 262, 299, 398–433
– miniemulsions for the convenient synthesis of organic and inorganic nanoparticles (see there) 175–211
– modification via chemical reactions 216–241
– metal nanoparticles, optical properties of 219

- monolayers 370
- multilayers (see there) 370–372
- organization 369–393
- – air-water interface 369–393
- – Langmuir-Blodgett films 369–393
- physical assembly of 377
- Prussian blue nanoparticles 185
- self-assembling 370
- semiconductor (see also nanocrystals) 52–91, 227–228, 458
- – applications of semiconductor nanocrystals 86–90
- – chemical deposition of shells on metals and semiconductors 222–236
- – colloidal synthesis 55–81
- – gold nanoparticles (see there) 98–115, 160
- – mixed-phase 64
- – quantum confinement 54–55
- – silicon nanoparticle, sonochemical formation of colloids 142
- – superstructures, building blocks of 81–86
- – thiol-stabilization 98
- silica/silica-coated 206, 229
- silver (see there) 372, 406, 408
- SiO$_2$ nanoparticles 380
- surface modification 370, 372
- synthesis 400
- TiO$_2$ nanoparticles 380
nanoreactors, colloidal 150–171, 176, 196
- bio-nanoreactors 151
nanorods 261
nanosized sheets 404
nanostructures
- inorganic 189
- material, nanostructured 343, 364
nano-switch, electrochemical 111–112
nanotubes 151, 236
- polymer 261
nanowires
- gold 377
- molybdenum 449
- palladium 449
natural rubber 1–5
negative resist 376
nematic liquid crystal 307
Nernst equation 431, 568
net pressure, effective 183
network
- aggregated 40
- murein 153
- polyfunctional network component 197

- two-dimensional polymer networks 158
neurotransmitters 427
nickel
- nanorods 261
- sonochemical formation of colloids 142
non-covalent interactions 150
nonionic surfactants for miniemulsions 182–183
- efficiency 183
nucleation 23, 25
- crystal nucleation, template-directed 422
- droplet 188
- face-selected 384
- joint-nucleation 42
- particle 23
- secondary 189
- surface 239
- thioalkanoate nucleation centers 390
- titania-coated 408
nucleus 561
- nucleic acids 355–360
number of charges 20
Nyquist plots 426

O

octa(3-aminopropyl)silsesquioxane 409–410
octadecyl mercaptane 389
ODA monolayers 379–380
- charge on 380
oil
- coupling between nanoparticles and oil 206
- immisciple (see there) 202
oil-acrylate hybrid emulsions 200
oil-soluble initiator 191, 196
oleic acid-stabilized silver nanoparticles 372
opals
- artificial 89, 274
- color of 285
- inverse (see 3DOM materials) 301, 307, 466, 468–469, 472–482
optical/optically
- encoding 494
- – multiplexed 502
- properties of metal nanoparticles 219, 423
- tunable micropatterns 442
organo-inorganic complex 384
organometallic synthesis 56, 65
organosilicon particles 166
osmotic/osmotically agent
- active agent 177
- influence 188
- initiator as osmotic agent 191

Ostwald
- equation 32
- ripening (OR) 56, 62, 71, 80, 177, 183
- - emulsion stability against *Ostwald* ripening 177
oversaturation 60, 85

p

PAH/PAA multilayers 414
paints, decorative 112
palladium, sonochemical formation of colloids 136, 142
paper, electronic 457
parachute-like structures 158
particle
- $Au@SiO_2$ particles 230
- average particle sizes 35–37
- biofunctional 272
- CdS 137
- citrate-stabilized 406
- coated 221
- cobalt particles 420
- colloidal (*see there*) 161, 246–272
- composite 31–33, 42
- core 165
- core-shell 161, 163, 290, 408
- DNA particle 265
- - DNA-hybridized 419
- encapsulation 150
- - carbon particle 201
- - magnetite particle 201
- - TiO_2 particles 200
- enzyme multilayer-coated particles 270–271
- fluorescent particles 529–531
- formation 15, 168
- functionalized 196–197, 401
- gold (*see there*) 359, 408–409, 411
- growth 23–24
- hollow 170, 290
- interparticle
- - collisions 144
- - interactions 222
- latex (*see there*) 1–44, 163, 260–262
- layer of particles 440
- lepidocrocite particles 345
- ligand-protected particles 412–413
- magnetic particles 420
- melamine formamide (MF) particles 563
- metal 423
- multiply-twinned particles (MTPs) 388
- nucleation 23
- organosilicon 166
- palladium 136
- polyelectrolyte-coated particles 267
- polymers (*see there*) 14
- polymer-particle hybrids 200–202
- preservation
- - of particle character 187
- - of particle identity 187
- - of particle number 187
- reporter particles 522
- seed 14
- shell particle (*see there*) 524
- silica-coated 408
- silver 410
- size 14
- - distribution 60–62, 75
- - - defocusing 60–62
- - - focusing 60–62
- - electron microscopy 528
- - SAXS 528–529
- - static light scattering (SLS) 529
- sonochemical synthesis of colloidal particles 120
- substrate-particle interaction 421
- unsymmetrical 44
- virus particles (*see there*) 151
particle-particle interaction 421
particular identity 176
pattern
- dot pattern 337
- ring pattern 337
- surface-directed colloid patterning 317–339
- templates 348–353
- writing 376
PBGs (*see* photonic band gaps) 285, 303–305, 467, 479, 482
PbS 384
- sonochemical formation of colloids 138
PbSe nanocrystals 74
PDMS stamp 337
PEGA-initiators 22
peptosomes 156
permeability 159
- stimuli-sensitive 167
phage display 364
phosphine-stabilized Au_{55} cluster 97
phospholipid
- bilayers 347
- headgroups 346
- *Langmuir* films 107
- vesicles 344
phosphonic acid 67, 69
photobleaching rates 499

photodegradation 233, 238
photoluminescence quantum yield 228
photonic crystals 88–89
photonic
– band gaps (PBG) 467, 472, 479, 482
– – crystals 303–305
– – materials 467
– – structure 285
– crystals 439, 467, 482–485
– – crystal structure 285
photosensitization 427–428
photostability 499
photovoltaic cells 458–459
phycoerythrin 499
π-A isotherms 372
π-π interactions 413
pickering
– stabilization 205
– stabilizers 205
pigment
– encapsulation by miniemulsification 200
– hydrophilic pigments 200
pigment-polymer interface 200
Pincus brush 13
plants 349–350
plasma
– absorption peak 424
– bulk plasma frequency 424
plasmon
– absorbance, surface 373
– band 221
– – band absorption 425
– coupled plasmon modes 369
platinum 100
– sonochemical formation of colloids, colloidal platinum 136
PMMA miniemulsion latex 204
polyaddition reactions 196–198
– of epoxy-resins in miniemulsions 196
polyamines 362–363
polyelectrolyte 81, 130, 139, 231, 250 251, 266–269, 321, 523, 545–548, 564, 573–574
– capsules 565–566, 571, 577
– charge density 545
– colloidal forces 545
– electrostatic 548
– film infiltration 417
– flocculation phenomenon 548
– γ irradiation 547
– layer-by-layer 545
– multilayers 161, 266–269, 321, 325–326, 335, 413, 545, 573–574
– nanocapsules 159

– particles, polyelectrolyte-coated 267
– shell 565
– strong 415
– surface
– – charge 545
– – modification 545
– weak 413
– ζ-potential 548
polymers
– abbreviations and chemical structure 407
– capsules 203
– cell of a polymer network 565
– coating 403–404, 422
– colloidal complexes, polymeric 3, 43
– copolymers (*see there*) 3, 43, 153–154, 156, 418–419
– crystallization 210
– dispersion (*see there*) 2–6, 8
– emulsion 4
– grafting 403
– hole-conducting polymers 87
– hollow polymer
– – nanocapsules 202
– – particles 161
– latex particles, polymeric (*see there*) 1–44
– micelles, polymeric 159
– nanocomposites, polymeric 31
– nanotube 261
– particles 14
– – size 14, 184
– – size distribution 14
– pigment/polymer interface 200
– rods 446
– single (polymer) molecule applications in materials chemistry 175–211
– stamping 335–338
– surface 336
– two-dimensional polymer networks 158
– water/polymer interface 200
polymerization
– ad-miniemulsion polymerization 202
– anionic, in miniemulsion 198
– copolymerization (*see there*) 190–196
– crystallization during the polymerization within the droplets 193
– dispersion (*see there*) 28–30
– 3DOM materials (inverse opals) 480–481
– emulsion polymerization (*see there*) 4, 17–22, 138, 165–168, 175, 286–287
– graft polymerization 290
– hardware 34–39

– heterophase 3–4, 8–41
– – chemistry 9
– – inverse techniques 195
– – primary polymer dispersions via heterophase polymerizations 8–41
– – techniques 4
– homopolymerization (see there) 190–196
– inverse suspension polymerization 187
– in lipid molecules 152
– miniemulsion polymerization (see there) 194
– non-radical organic polymerization reactions in miniemulsions 196–198
– post-polymerization modification of latexes 39
– progress 190
– radical polymerization (see there) 154, 191–193
– seeded 14, 31–34
– suspension 189
– of ultrahydrophobic monomer 191
– unseeded (ab initio) 14
– vesicular 157
polymer-particle hybrids 200–202
polymer-polymer
– complexation 210
– hybrids 199
– – hybrid latex 199
– stamping 335
polyoxometallate (POM) 486
– *Keggin* salts 370
poly(phenylene vinylene) (PPV) 87
polystyrene
– colloid 288
– film 418
– particles, micron-size 27
– spheres 254, 257, 269
– – metal nanoparticle-coated 267
poly(tetrafluoroethylene) (PTFE) 422
polyurethanes latexes 198
poly(4-vinylpyridine) 418
potassium peroxodisulfate concentration 19
potential of the MPC 429
powder X-ray diffraction (XRD) 63
precipitation 289–290
– controlled 289–290
– size-selective 66, 75, 79, 99
precursor
– miniemulsion 179
– single-precursor 228
pressure
– effective 184
– – net pressure, effective 183

– – real zero effective pressure 186
– homogenizer, high-pressure 178
– *Laplace* 122, 177, 183
– swelling pressure 32
– zero-effective pressure 188
pressure-temperature phase diagrams 374
printing applications, electronic 202
programmed assembly 353
protein 344–345, 353–355, 362
– aggregates 265, 572–573
– biomacromolecule 370
– coating, thickness 572
– copolymer-protein hybrid materials 156
– and DNA 358
– folding 210
– incooperation 571
– membrane proteins 156
– release 574
proteomics 510
– protein-protein interactions 510
Prussian blue nanoparticles 185
pseudo-steady state 179
Pt 352, 372

q
quantum
– confinement, semiconductor nanoparticles 54–55
– – quantum confinement effect 54–55
– dots (QDs) 52, 64, 227–228, 357, 374, 459, 494–503
– – InAs 498
– – InP 498
– – semiconductor quantum dots as multicolor and ultrasensitive biological labels 494–503
– efficiency 68
– luminescence quantum yields 65, 73–74
– magnetic disks (QMDs) 420
– multicolor 497
– size effect 67, 75
– wells 64
– yield 67, 495
quantum-size effect 497

r
radial homo- and copolymerization in miniemulsions 190–196
radical polymerization 154, 191–193
– free-radical 154, 191–192
– – living free-radical polymerization in miniemulsion 191

Subject Index

– – stable free-radical polymerization (SFRP)   192
– living radical   191–192
– primary radicals   131
– reverse atom transfer (ATRP)   193
– secondary radicals   130–131
radiolytic reduction   224
Raman
– scattering, surface-enhanced   299
– spectroscopy, surface-enhanced   485
random-coil conformation   157
Raoult's law   177
rate equilibrium
– dynamic   180
– – *Fokker-Planck*-type dynamic rate equilibrium   180
– fusion-fission rate equilibrium   180
reaction vessels or templates   154
reactor material, influence of   34–36
rectified/rectifying
– diffusion   123
– diodes   448
– rods   447
rhodamine B   154
ring pattern   337
rods
– aligning   447
– barcodes   446
– chemical functionalization   447
– metals   446
– microrods   446
– nanorods   261
– polymers   446
– rectifying   447
– semiconductors   446
rotor-stator dispersers   178
rubber, natural   1–5

s

sacrificial templates   305
salt
– calixarene Cd salt LB films   390
– *Keggin* salts (polyoxometallates)   370
– *Langmuir-Blodgett* films of salts and fatty lipids   387–393
– nanoparticle formation by reaction of metal ions in *Langmuir-Blodgett* films of salts and lipids   387–393
SANS (small angle neutron scattering) measurement/experiments   185, 187
saponification   167
scanning
– electrochemical microscope (SECM)   109
– electron microscopy (SEM)   290–291, 380
– tunneling microscope (STM)   110–112
scattering   219
– gold nanoparticles, light scattering   114
– SANS (small angle neutron scattering) measurement/experiments   185, 187
Schottky and rectifying diodes   448
screening technologies, high-throughput (*see there*)   511–517
Se, Te-capped gold   372
sea urchins   350
SECM (scanning electrochemical microscope)   109
sedimentation, crystallization   295–296
seed
– growth, seeded   97
– particles   14
– – activated   33
– polymerization
– – seeded   14, 31–34
– – unseeded (ab initio)   14
selenides, sonochemical formation of colloids   143
self-assembly/self-assembled
– approach   152
– colloid   286
– connections, electrically self-assembling   453
– cooperative   466
– electrostatic   412
– LbL self-assembly of metal nanoparticles on planar substrates   398–433
– molecular arrays   465
– monolayers   290
– multilayers   562
– nanoparticles   370
– processes   398
– template-assisted self-assembly (TASA)   310
– two-dimensional arrays of monodisperse nanocrystals formed by   83–84
self-emulsification   7
SEM (scanning electron microscopy)   290–291, 380
semi-batch process   14
semiconductors   305–307
– nanoparticles (*see also* nanoparticles)   52–91
– rods   446
– semiconductors on semiconductors   227–228
sensitization   87

sensors
- biological   455–457
- chemical   454–455
- electrically readable sensors   455
- gas   112
- hydrogen   450
- immunosensors   457
- microscopic   455

SERS   226
shear deformation   155
shell
- chemical deposition of shells on metals and semiconductors   222–236
- core-shell (see there)   63, 77, 161, 222, 228
- dye sequences   549
- effect   123, 125
- membrane, shell-forming   152
- metal oxide shells   228
- micelles, shell cross-linked   161
- particles   524, 538–539, 542, 549–551
- – combinatorial synthesis   551
- – flow cytometer   524, 538–539
- – morphology   542
- – optical signature   524, 539, 550
- polyelectrolyte shell   565
- polymeric   235
- porosity   236
- semiconductor shells   233
- silica shell   230

shielding   327
shot
- addition   31
- growth   31

silaffins   362
silane
- coupling agents   230
- mercaptosilane   233

silica/silicia
- catalysts for   362
- charge   348–349
- as coating material   404
- coating of metal particles with   228–232
- colloids   346
- core-shell   293
- deposition   364
- encapsulation   405
- gels   232
- hollow silica   237
- mesoporous   232
- nanoparticles, silicia/silica-coated   206, 229
- shell   230
- sol-gel-derived   351

- sols   229
- spheres, colloid   288
- structures   363
- synthetic routes to   361

silica-coated particles   408
silicalite (zeolite) macroporous monolith   276
silicatein   361–362
silicon sonochemical formation of colloids   142
silver   354, 372–373, 377, 380, 384
- core dissolution   236
- monolayer protected clusters (MPCs) of gold and silver   96, 100
- nanoparticles   372–373, 406, 408
- – domains   373
- – monolayer   373
- – oleic acid-stabilized   372
- particles   410
- sonochemical formation of colloids, colloidal silver   135
- surfactant-stabilized hydrophobic silver   371

single
- electron conductivity   428
- molecule experiments   211
- particle light scattering (SPLS)   252
- (polymer) molecule applications in materials chemistry   175–211

single-precursor   228
SiO_2 377
- $Au@SiO_2$ 230
- $CdS@SiO_2$ 232
- macroporous TiO_2/SiO_2 structure   277
- nanoparticles   380

size
- average particle sizes   35–37
- distribution   14, 58, 60–62, 65–68, 75, 80
- droplet size (see there)   3, 185
- minimum size limit   197
- quantum size effect   67, 75

size-quantification effect   52, 64
size-selective
- fractionation   78–81
- precipitation   66, 75, 79, 99

SnO_2 372
Sogami-Ise potential   296
solar cells   87–88
sol-chemistry   468–473
sol-gel reactions   267
solid, artificial   85
solid-state properties   155
sonochemistry/sonochemical   120, 129–144

- emulsion polymerization 138
- – poly(butyl acrylate) 139
- – poly(methyl methacrylate) 139
- formation of colloids 131–144
- – aqueous media 131–140
- – carbon 142
- – CdS particles 137
- – colloidal copper 137
- – colloidal gold 132
- – colloidal platinum 136
- – colloidal silver 135
- – CuS 138
- – Fe-Co alloys 143
- – *Gibbs* surface excess 134
- – gold sols 133
- – iron 140
- – metal oxides 143
- – MoS_2 142
- – nickel 142
- – non-aqueous media 140–144
- – palladium 136, 142
- – PbS 138
- – selenide 143
- – silicon nanoparticles 142
- synthesis 120–145
- – colloidal particles 120
- – of inorganic and organic colloids 120–145

sonoluminescence 126
sonolysis 126
space fillers 465
spectroscopy
- electrochemical impedance spectroscopy 425
- electrorotation spectroscopy 566
- fluorescence correlation spectroscopy 501
- surface-enhanced *Raman* spectroscopy 485
- UV-vis spectroscopy 385

spin-coating 298
spinodal decomposition 106
SPLS (single particle light scattering) 252–253
sponge
- spicules 361
- starch, gel sponge 351–352
square pyramidal pits 299–300
stability/stabilization/stabilizing agents 55
- colloidal particles, stabilization of 13, 180
- critical borderline between stability and instability 180
- critically stabilized 180
- electrostatic stabilization 11, 178

- emulsion stability against *Ostwald* ripening 177
- gold, citrate-stabilized hydrosols 97
- latex particles, electrolyte stability 12
- phosphine-stabilized Au_{55} cluster 97
- semiconductor thiol stabilization 98
- shelf-life 184
- steric stabilization (*see there*) 11, 183
- synthesis of thiolate-stabilized gold nanoparticles 98–100
- ultracentrifugation stability studies 184

stabilizer-free dispersion polymerization 29
stable free-radical polymerization (SFRP) 192
stamp/stamping
- PDMS 337
- polymer 335–338
- polymer-on-polymer 335
starch sponges 351–352
- gel sponge 352
steric
- barrier 218
- stabilization 11, 178
- – efficiency 182
- – electrosteric stabilization 11
stirrer speed 18–20, 36–37
STM (scanning tunneling microscope) 110–112
- images 375
Stöber
- method 230, 286, 289, 405
- synthesis 525–527, 529
- – aggregation-only 527
- – growth-only 527
- – monodispersity 527
- – mono-fluorescent 529
- – particle size 527
- – polydispersity 527
- – reagents on 526–527
- – seeded growth 527, 529
strain gauges 112
streaming, microstreaming 129
structure/structured
- directors 465
- interfaces 352
styrene acrylate 195
styrene-*n*-butyl acrylate 195
substrate-particle interaction 421
sulfonium surfactants 182
superlattice 86, 98, 104–106
- bimodal 106
- structure 383

superstructures
- ordering of nanocrystals in superstructures 84
- semiconductor nanoparticles as building blocks of superstructures 81–86

surface
- capping 497
- charge 20, 323, 327
- – electric charge 121
- confinement effects 333–334
- coverage 180, 188
- functionalism 400
- loading 400
- modification 217–222, 251–260, 543–546
- – aggregation 543
- – electrostatic attraction 546
- – γ irradiation 548
- – materials for 251–260
- – multilayers 545
- – polyelectrolytes (see there) 545, 564
- – polymeric bridging 546
- – surface charge 545
- – ζ-potential 545
- plasmon absorbance 373
- potentials 122
- sorting 317, 330, 338
- tension 59

surface-directed colloid patterning 317–339

surfactants
- adsorption 329
- cationic surfactants for miniemulsions 182
- co-surfactant (see there) 29
- coverage of droplets by surfactants 179
- emulsion polymerization, surfactant-free 17–22
- equilibrium concentration of free surfactant 181
- miniemulsion, influence of the surfactant 180–183
- molecules 327
- nonionic surfactants for miniemulsions 182
- reactive 21–22
- sulfonium 182
- titrations 183

surfactant-stabilized hydrophobic silver 371
swelling 32–34
- activated swelling method 34
- dynamic swelling method 33
- entropy swelling equation 33
- latex particles, swollen 32
- pressure 32

- promoter 34
- transition 152

synthesis
- aqueous 56, 62
- colloidal 53
- miniemulsions for the convenient synthesis of organic and inorganic nanoparticles (see there) 175–211
- nanoparticles 400–401
- – metal nanoparticles 413
- organometallic 56, 65
- sonochemical synthesis of inorganic and organic colloids (see there) 120–145
- template synthesis 161
- of thiolate-stabilized gold nanoparticles 98–100

t

TASA (template-assisted self-assembly) 310
TBBT (4,4'-thiobisbenzenethiol) 374
telecommunication amplifiers, nanocrystals for 88
TEM (transmission electron microscopy) 254, 286–288, 291, 385
temperature
- bubble 125–129
- glass transition temperatures 9–11
- low-temperature alloying 393
- pressure/temperature phase diagrams 374

templates 154, 161, 299, 465
- pattern templates 348–353
- reaction 154
- sacrificial 305
- synthesis 161

template-assisted self-assembly (TASA) 310
tension
- interface tension values 187
- measurements 179

TEOS (tetraethylorthosilicate) 289
tetraethoxysilane 230
tetragonal crystals, body-centered 442
thermodynamic
- considerations 202
- equilibrium 183

thioalkanoate 390
- nucleation centers 390

4,4'-thiobisbenzenethiol (TBBT) 374
thiols 62–64, 81
TiO_2 (dioxide titan) 371, 377
- $Ag@TiO_2$ 232
- macroporous TiO_2/SiO_2 structure 277

- particles
- – encapsulation via miniemulsions 200
- – nanoparticles 380
titan/titanium 349
- coatings 268
- dioxide (see TiO$_2$) 200, 277, 351, 371, 377
titania-coated particles 408
TMV (tobacco mosaic virus) 151, 347
tobacco mosaic virus (TMV) 151, 347
TOPO-TOP (trioctylphosphine oxide-trioctylphosphine) 66–67, 73, 75, 78
- quantum yield 67
Torza and *Mason*, immiscible oils 202
transferrin 500
transistors, organic 458
transmission electron microscopy (TEM) 254, 286–288, 291, 385
trapped species 177
triggered or controlled release 153
trioctylphosphine oxide-trioctylphosphine (see TOPO-TOP) 66–67, 73, 75, 78
tungstate ions 345
turbidity measurements 179, 208
two-dimensional arrays of monodisperse nanocrystals formed by self-assembly 83–84

u

ultracentrifugation
- experiments 206
- stability studies 184
ultrafiltration 166
ultrahydrophobe/ultrahydrophobic 178
- influence of 183–185
- monomer 191
ultrasonication 7
ultrasound 120
- high-energy ultrasound 201
- power ultrasound emulsification 178
ultrasound-induced effects 129
- chemical 129
- physical 129
ultra-turrax 178
unseeded (ab initio) polymerization 14
urease 564, 577
- encapsulation 564
- permeation 564
UV-vis spectroscopy 385

v

valinomycin, conduction of 568
vesicles 152–153
- block copolymer vesicles 153
- polymerization, vesicular 157
vessels, reaction 154
V-grooves 299
viologen 111
virus 344–345, 347
- artificial viruses 578–579
- cowpea chlorotic mottle viruses 344–345
- particles 151
- – cowpea chlorotic mottle virus (CCMV) 151
- – tobacco mosaic virus (TMV) 151, 347
- single 503
vitreophilic modification 404

w

van der Waals interactions 218
Warburg impedance 425
water, air-water interface 369–393
water-dispersible inorganic nanoparticles 204
water-insoluble materials 198
water-polymer interface 200
water-soluble
- gold nanoparticles 113
- initiator 190
- monomers in miniemulsions 190
whispering gallery-mode laser 89
wood 349

x

X-ray
- diffraction (XRD) 63, 84
- – powder X-ray diffraction 63
- energy-dispersive X-ray (EDX) 286
XRD (X-ray diffraction) 63, 84
- powder X-ray diffraction (XRD) 63

y

Y-Ba-Cu superconductors 372

z

zeolite 276, 349–351
- (silicalite) macroporous monolith 276
zero
- larger than zero 188
- pressure, zero-effective 188
ζ-potential
- colloids 328
- measurement of 206
zirconium titanate nanoparticles, ferroelectric lead 371
ZnSe nanocrystals 74